ISBN 978-0-331-79622-3
PIBN 11050198

1 MONTH OF
FREE
READING

at

www.ForgottenBooks.com

By purchasing this book you are
eligible for one month membership to
ForgottenBooks.com, giving you
unlimited access to our entire
collection of over 1,000,000 titles via
our web site and mobile apps.

To claim your free month visit:
www.forgottenbooks.com/free1050198

English
Français
Deutsche
Italiano
Español
Português

www.forgottenbooks.com

Mythology Photography **Fiction**
Fishing Christianity **Art** Cooking
Essays Buddhism Freemasonry
Medicine **Biology** Music **Ancient**
Egypt Evolution Carpentry Physics
Dance Geology **Mathematics** Fitness
Shakespeare **Folklore** Yoga Marketing
Confidence Immortality Biographies
Poetry **Psychology** Witchcraft
Electronics Chemistry History **Law**
Accounting **Philosophy** Anthropology
Alchemy Drama Quantum Mechanics
Atheism Sexual Health **Ancient History**
Entrepreneurship Languages Sport
Paleontology Needlework Islam
Metaphysics Investment Archaeology
Parenting Statistics Criminology
Motivational

Bulletin 173

DEPARTMENT OF THE INTERIOR

FRANKLIN K. LANE, Secretary

BUREAU OF MINES

VAN. H. MANNING, Director

MANGANESE

USES, PREPARATION, MINING COSTS AND THE PRODUCTION OF FERRO-ALLOYS

BY

C. M. WELD and OTHERS

WASHINGTON
GOVERNMENT PRINTING OFFICE
1920

Bulletin 173

DEPARTMENT OF THE INTERIOR
FRANKLIN K. LANE, Secretary
BUREAU OF MINES
VAN. H. MANNING, Director

MANGANESE

USES, PREPARATION, MINING COSTS
AND THE PRODUCTION OF
FERRO-ALLOYS

BY

C. M. WELD and OTHERS

WASHINGTON
GOVERNMENT PRINTING OFFICE
1920

First edition.   March, 1920.

CONTENTS.

CONTENTS.

PREFACE.

The history of the domestic production of manganese ores and alloys during the war, in common with that of several other materials equally essential for war purposes, is of interest because of its showing how a hitherto latent industry responded quickly to the spur of necessity.

In the past the supplies of manganese ores used in this country have come largely from Russia and India. More recently imports from Brazil began to assume importance, but up to 1914 they were still a relatively minor factor. Also, small amounts came from Cuba and Central America. The domestic output, however, was practically negligible.

The European war soon shut off Russian sources, and receipts from India declined. The domestic output became somewhat larger, but the deficit was principally made up by greatly increased imports from Brazil.

When this country entered the war, the manganese situation became acute. Manganese being essential in the manufacture of steel, the insuring of an adequate supply was imperative. On account of the increasing difficulty of obtaining foreign ores, through uncertain output and shortage of shipping, it was obvious that this country must turn to its own deposits. In July, 1917, the war minerals committee was established, and surveys of the domestic manganese situation were begun. By the end of the year the domestic output had reached hitherto unprecedented figures.

Early in 1918 the need of diverting every available ship to the transportation of troops and military supplies made necessary the reducing to the minimum of all imports of materials that could be produced in this country. The manganese situation was studied most thoroughly by various governmental bodies, in cooperation with the ferro-alloys subcommittee of the American Iron and Steel Institute, and a program was formulated for reducing overseas imports to the minimum figure consistent with the production of an adequate supply of steel. To stimulate domestic production, a higher price schedule was announced by the American Iron and Steel Institute, with the approval of the War Industries Board. At the same time the market for leaner domestic ores was broadened by lowering the standard grade for manganese alloys.

As a result, had the war continued through the year 1918, there would have been produced more than 300,000 tons of domestic ores containing not less than 35 per cent metallic manganese, with larger

1

quantities than ever before of the lower grade ores suitable for the manufacture of spiegeleisen. In addition, alloys of domestic manufacture of the alloys had entirely replaced those formerly imported. In November, 1918, when hostilities ceased, there was nearly a year's supply of ore and alloys on hand. As a result, the buying stopped almost at once and production fell rapidly. Mines were shut down and furnaces were blown out or used for other products.

Various proposed measures for stabilizing the industry were considered, but none seemed to meet the situation. In 1919 the war minerals relief bill was passed, establishing a commission for investigating losses incurred in the production of manganese and certain other minerals and for reimbursement where the investments were due to representations of the Government.

During the war the engineers connected with the war minerals investigations of the Bureau of Mines kept in touch with developments in the industry. Wherever possible advice and help in mining and milling problems were given by correspondence or in person. Metallurgical problems were carefully studied, not only those bearing on the manufacture of the alloys, but those relating to the use of the alloys in the manufacture of steel. The chief end kept in view was to eliminate waste and to widen and popularize the use of the leaner domestic ores.

In order to supplement the field work and make public the results of its investigations, the Bureau of Mines published a series of mimeographed bulletins dealing with those phases of the industry, which were studied in some detail. The object of this report is to present these papers in more permanent form, in the hope that the information may be of present or future value to the industry.

J. E. SPURR,

Executive, War Minerals Investigations.

MANGANESE USES: PREPARATION, MINING COSTS, AND THE PRODUCTION OF FERRO-ALLOYS.

By C. M. WELD and Others.

INTRODUCTION.

During the past two years the Bureau of Mines has issued a series of mimeographed reports giving the results of research work and experiments conducted as part of its war minerals investigations.

In this bulletin the reports on manganese are presented, which range in scope from the beneficiation of the ore to the utilization of the metal. The bulletin is in eleven chapters, each comprising a separate report, arranged in the order given below.

1. General information regarding manganese, by C. M. Weld.

2. Uses of manganese other than in steel making, by W. C. Phalen.

3. Problems involved in the concentration and utilization of domestic low-grade manganese ores, by Edmund Newton.

4. Preparation of manganese ore, by W. R. Crane.

5. Leaching of manganese ores with sulphur dioxide, by C. E. Van Barneveld.

6. The Jones process for concentrating manganese ores; results of laboratory investigations, by Peter Christianson and W. H. Hunter.

7. Cost of producing ferromanganese ores, by C. M. Weld and W. R. Crane.

8. Production of manganese alloys in the blast furnace, by P. H. Royster.

9. National importance of allocating low-ash coke to the manganese-alloy furnaces, by P. H. Royster.

10. Electric smelting of domestic manganese ores, by H. W. Gillett and C. E. Williams.

11. Use of manganese alloys in open-hearth steel practice, by Samuel L. Hoyt.

As each chapter was originally prepared as a unit, there is necessarily some duplication and overlapping. With certain exceptions, however, each paper is reproduced in practically its original form.

In chapter 1, covering certain general phases of the industry, such as uses, specifications, prices, and statistics, the subject matter has been somewhat modified in view of changes in conditions.

3

About 95 per cent of the manganese consumed in the United States goes into the manufacture of steel. The remaining 5 per cent is used in a number of minor industries, the chief of which is the manufacture of dry cells. These minor uses are described in chapter 2.

No discussion of the geology of manganese ores is presented, but a short bibliography on the subject is appended. With few exceptions the manganese deposits of the United States are irregular, pockety, and uncertain. Largely for this reason, mining methods are crude and hardly warrant descriptions.

The concentration of manganese ores, however, is an important problem and involves some details not common to other ores. Chapter 3 discusses the concentration and utilization of low-grade ores, notes the relationship of concentration to metallurgical practices, outlines concentrating processes, and treats of commercial problems.

In general, as regards concentration methods, manganese ores may be divided into two groups, as follows: (1) Ores permitting mechanical separation of the manganese minerals and the gangue, and (2) ores in which the manganese minerals and the gangues are so intimately associated that separation requires some hydrometallurgical or pyrometallurgical process. Heretofore, attention has been confined almost wholly to mechanical separation, chiefly by gravity. In fact, so far as known, all production of concentrates has been by wet gravity methods, although during the war some companies investigated the commercial possibilities of magnetic separation and reported favorable results, the construction of one magnetic concentration mill being started. The usual wet gravity methods are described in chapter 4.

During the war, the Bureau of Mines undertook investigations of methods applicable to ores of group 2. A hydrometallurgical process involving leaching with sulphur dioxide was studied at the mining experiment station of the bureau at Tucson, Ariz. The results of this work, presented in chapter 5, indicate that the process is metallurgically feasible but the cost makes it unattractive.

The results of an investigation of a pyrometallurgical process known as the Jones process are presented in chapter 6. The process, which is still in the experimental stage, was found to be metallurgically sound. The manganese product is not a concentrate but an alloy, made directly from a lean unconcentrated ore. The customary method is to manufacture alloys from high-grade ores or from concentrates of leaner ores.

A discussion of the costs of ferro-grade manganese ores is presented in chapter 7. This discussion is in general terms only, but should be of use in so far as it relates to competitive conditions.

The results of a study of blast-furnace practice on ferromanganese and on spiegeleisen are presented in chapter 8. On account of the difficulty of obtaining complete and reliable records, the work is to be regarded as largely preliminary. Some of the conclusions tentatively advanced may be erroneous because of the nature of the data on which they are based; in the main, however, they are believed to be correct and are put forward in the hope that they may arouse criticism and thereby stimulate discussion and research. The Bureau of Mines proposes to continue this investigation.

The importance of allocating good grades of coke to manganese-alloy furnaces is discussed in chapter 9. The need for such allocations has passed but it is felt that the paper has more than an historic interest. Conservation and economy would both be served if the principles laid down were more effectively observed.

The chapters already cited deal with the manufacture of manganese alloys in the blast furnace with coke fuel. During the war there was considerable development in the electric smelting of such alloys. The Bureau of Mines was actively interested in the applicability of electric smelting to the leaner, more siliceous domestic ores, and conducted experiments at the Ithaca (Cornell University) field station. The results are presented in chapter 10. The general conclusions are that lean ores and manganiferous slags probably can not be smelted at a profit in the electric furnace except in times of high prices.

The purpose of the investigation described in chapter 11, on the use of manganese alloys in open-hearth steel practice, was chiefly to point the way to conservation of resources. The results show that conservation could best be attained by developing the use of those alloys, such as spiegeleisen and silicomanganese, which could be produced from lean domestic ores, thus conserving the high-grade domestic ores and reducing the need for high-grade foreign ores.

In addition to those members of the staff whose names appear on the several papers, acknowledgment is due to F. H. Probert, G. D. Louderback, Theodore Simons, W. S. Palmer, W. R. Eaton, F. B. Foley, C. F. Julihn, and all others who contributed to the success of the work by their active and whole-hearted cooperation.

CHAPTER 1.—GENERAL INFORMATION REGARDING MANGANESE.

By C. M. Weld.

USES OF MANGANESE.

Approximately 95 per cent of the manganese consumed in this country is used in making steel not only to deoxidize and recarburize the molten metal, thereby making possible the production of cleaner and sounder ingots containing the desired amount of carbon, but also to impart certain qualities to the finished product. Small proportions of manganese, 0.5 to 0.8 per cent, make the steel easier to work and stronger in service. At the same time slight proportions of impurities remaining in the steel are taken into combination and rendered less harmful. Relatively small quantities of so-called "manganese" steel are made which contain 11 to 14 per cent or more of manganese and possess special qualities of hardness and strength.

To be used in manufacturing steel, the ore must first be smelted to an alloy, in which the manganese is combined in varying proportions with iron, carbon, and silicon. Formerly the two standard alloys were ferromanganese, containing 78 to 82 per cent metallic manganese, and spiegeleisen, containing 18 to 22 per cent metallic manganese. During the war the composition of these two alloys underwent considerable modification, the standard manganese content being lowered in order to make practicable the use of leaner domestic ores. At the same time two other alloys, silico-manganese and silico-spiegel, assumed more or less prominence. The approximate range of composition of these four alloys is as follows:

TABLE 1.—*Range of composition of manganese alloys.*[1]

Alloy.	Mn.	Fe.	Si.	C.
	Per cent.	*Per cent.*	*Per cent.*	*Per cent.*
Ferromanganese	50–80	40–8	0.5–1.5	5–7
Spiegeleisen	10–35	85–60	About 1.0	4–5
Silico-manganese	55–70	20–5	About 25.0	.35
Silico-spiegel	20–50	67–43	4–10	1.5–3.5

[1] Data largely from Newton, Edmund, manganiferous iron ores of the Cayuga district, Minnesota: Univ. of Minnesota Bull. 5, 1918, 126 pp.

The phosphorus must be so low that adding the alloy will not cause the phosphorus in the steel to exceed the specified limit, which is usually about 0.05 per cent. Consequently the phosphorus limit in the alloy will vary more or less directly with its manganese and carbon content.

Practice in utilizing the alloy varies widely. The alloy may be added to the molten metal either molten, red-hot, or cold, with or without other additions such as ferrosilicon or coal; it may be added in the open hearth, the converter, or the ladle. The average consumption of metallic manganese per ton of steel is approximately 15 pounds.

Minor uses of manganese ore are as follows: As an oxidizing agent in dry-cell batteries, for decoloration of glass, as a drier in certain paints and varnishes, in the preparation of oxygen on a small scale, and in the manufacture of certain disinfectants. It is also used to color glass, pottery, brick, and tile, and in calico printing and dyeing. Of these minor uses the first two named are by far the most important.

SPECIFICATIONS.

Ores suitable for the manufacture of alloys are generally termed "metallurgical" or "furnace" ores. They may have a wide range in composition, according to the particular alloy to be made from them.

Formerly, when the standard grade of ferromanganese contained 78 to 82 per cent metallic manganese, ores of ferro grade were required to contain not less than 40 per cent metallic manganese, dried at 212° F. Prices were based on not more than 8 per cent silica and not more than 0.2 per cent phosphorus, and ores containing more than 12 per cent silica or more than 0.225 per cent phosphorus were subject to rejection. When, in May, 1918, the standard content of ferromanganese was lowered to 70 per cent, the specified minimum manganese content of ferro-grade ores was correspondingly lowered to 35 per cent metallic manganese. At the same time the rejection limits for silica and phosphorus were considerably extended, but much more severe penalties were imposed for silica exceeding 8 per cent. The rejection of ores with more than 0.25 per cent phosphorus was left to the option of the buyer, who at the same time was urged to use the ore, whatever the phosphorus content, if he could use it to advantage. The effect of silica and phosphorus, and their relation to the alloy to be produced are discussed in chapter 3. The purpose of relaxing specifications was to widen the market for the leaner and more siliceous domestic ores. It is probable that former specifications will be revived with return of normal conditions.

Once it was the custom to pay for the iron in manganese ores of ferro grade, but this practice was discontinued several years ago. The permissible high limit for iron is governed by the nature of the alloy to be produced, as practically all the iron goes into the alloy and degrades it by crowding out manganese. In general, the allowable ratio of iron to manganese in the ore would be approximately 1 to 10 for 80 per cent ferro, or 1 to 5 for 70 per cent ferro.

Ores suitable for the manufacture of spiegeleisen may contain a much larger proportion of iron to manganese and are usually termed manganiferous iron ores. No exact specifications for ores of this class have been customary. In general, the silica and phosphorus requirements have been about the same as for ferro-grade ores, the balance consisting of manganese plus iron in varying ratio, plus gangue materials, such as alumina, lime, and magnesia. Ores containing 15 to 40 per cent metallic manganese were formerly classed as manganiferous, and these limits will probably be adopted again in the future. The war-time classification included ores containing 10 to 35 per cent manganese.

Certain manganiferous ores from the weathered parts of silver-lead deposits contain enough silver to warrant their shipment to lead smelters for use in fluxing. Hence these ores are not available to the steel industry.

When the alloy to be produced is silico-manganese or silico-spiegel a much higher content of silica is acceptable than with ores to be used in making ferromanganese or spiegeleisen, but the former alloys can advantageously be made only in the electric furnace. Ores with 30 to 40 per cent of manganese and 20 to 25 per cent of silica can be used to advantage in making silico-manganese. The ratio of silica to manganese may be still higher if the usual slag-making constituents are relatively absent. It is also probable that at least a part of the phosphorus in the ore may be volatilized in the electric furnace, thus raising the permissible limit of this element. Roughly, the same holds true for silico-spiegel, with the substitution of iron for a part of the manganese.

A large class of manganiferous ores comprises those that are essentially iron ores containing small proportions of manganese, generally about 5 per cent but occasionally as much as 10 per cent. These ores are not available for making manganese alloys, but enter into the manufacture of manganiferous pig iron, which in turn contributes its manganese to the steel made from it.

The ores suitable for oxidizers and chemical use are commonly called "battery," "chemical," or "dioxide" ores. As their function is to act as an agent for carrying oxygen, it is essential that they contain a large proportion of manganese dioxide, either as pyrolusite or other minerals that readily liberate oxygen. The following table shows the proportions of MnO_2 and available oxygen in the common manganese minerals.[a]

TABLE 2.—*Manganese dioxide and available oxygen in the common manganese minerals.*

Mineral.	MnO_2.	Available O.
	Per cent.	*Per cent.*
Pyrolusite	100.00	18.39
Psilomelane	42.46 to 77.33	7.81 to 13.06
Manganite	49.44	9.09
Braunite	43.11	7.93
Hausmannite	37.99	6.99
Rhodochrosite	None.	None.

a Fermor, L. L., Manganese deposit of India: Geol. Survey of India, Memoirs, vol· 37, p. 508.

The two most important industrial uses of manganese ore as an oxidizer are in the manufacture of dry cells and of glass. Specifications generally call for 80 to 90 per cent MnO_2, but it is understood that ores with as little as 70 per cent MnO_2 were accepted during the war. Formerly it was customary to require that the iron content should not exceed 1 per cent, but here also specifications were greatly relaxed, particularly as it has been shown that the presence of several per cent of iron does not greatly affect the efficiency of the battery. Copper, nickel, and cobalt, on the other hand, are probably harmful when present in excess of a few tenths of 1 per cent, though there is some difference of opinion as to this. The point is discussed in the chapter following.

Leaner argillaceous and siliceous ores, with less than 40 per cent metallic manganese, are used for coloring pottery, tiles, and brick. The amounts used annually for this purpose and for paints, dyeing and printing calicoes, and other purposes are unimportant.

PRICES.

Prices paid for imported metallurgical ores have always been subject to individual contracts, based on the chemical and physical characteristics of the particular ores. The terms of these contracts have, of course, never been made public, but their general trend has no doubt been reflected in the price schedules for high-grade domestic ores issued from time to time by the Carnegie Steel Co. The latter are summarized in the following table:

TABLE 3.—*Carnegie Steel Company's price schedules for domestic ores.*[a]

Year.	Prices in cents per unit for percentages of manganese ranging—				Cents per unit of iron.	Silica standard.	Phosphorus standard.
	From 40 to 43.	From 43 to 46.	From 46 to 49.	49 and more.			
						Per cent.	Per cent.
1892	27	28–29	29–30	31	10	8	0.10
1897	25	26–27	27	28	6	8	.10
1903	22	23–24	24	25	5	8	.10
1906	27	28–29	29	30	6	8	.10
1910	23	24–25	25	26	5	8	.20
1914			24	25	26		8	.20
1915	36	40	43	45		8	.20
1916–17	46	50	53	55		8	.20
	From 38 to 43.	From 42 to 46.	From 46 to 50.	50 and more.			
1918	90	100	110	120		8	.20

a Data largely from Mineral Resources U. S., various years, U. S. Geological Survey.

These prices are based on long tons for material delivered at the furnace; material dried at 212° F. silica penalty, 15 cents per ton for each unit over 8 per cent; phosphorus penalty, up to 1910, 1 cent per unit of manganese for each 0.02 per cent phosphorus over 0.10 per cent; thereafter, 2 cents per unit of manganese for each 0.02 per cent phosphorus over 0.20 per cent.

In May, 1918, a new price schedule was adopted by the Ferro Alloys Committee of the American Iron and Steel Institute and was approved by the War Industries Board. It became effective on May 28. It is given below in full as announced, including penalties and terms of payment, although it is no longer in force and its present value is chiefly historical.

PRICE SCHEDULE OF MAY 28, 1918.

Schedule of domestic metallurgical manganese ore prices per unit of metallic manganese per ton of 2,240 pounds for manganese ore produced and shipped from all points in the United States west of South Chicago, Ill. This schedule does not include chemical ores as used for dry batteries, etc.

Following prices are on the basis of delivery f. o. b. cars South Chicago, and are on the basis of all-rail shipments. When shipped to other destination than Chicago, the freight rate per gross ton from shipping point to South Chicago, Ill., is to be deducted to give the price f. o. b. shipping point.

Prices for ore dried at 212° F.

Content of metallic manganese, per cent.	Price per unit	Content of metallic manganese, per cent.	Price per unit.
35 to 35.99	$0.86	45 to 45.99	$1.12
36 to 36.99	.90	46 to 46.99	1.14
37 to 37.99	.94	47 to 47.99	1.16
38 to 38.99	.98	48 to 48.99	1.18
39 to 39.99	1.00	49 to 49.99	1.20
40 to 40.99	1.02	50 to 50.99	1.22
41 to 41.99	1.04	51 to 51.99	1.24
42 to 42.99	1.06	52 to 52.99	1.26
43 to 43.99	1.08	53 to 53.99	1.28
44 to 44.99	1.10	54 and over	1.30

For manganese ore produced in the United States and shipped from points in the United States east of South Chicago, 15 cents per unit of metallic manganese per ton shall be added to above unit prices.

Above prices are based on ore containing: Not more than 8.00 per cent silica; not more than 0.25 per cent phosphorus, and subject to:

SILICA PREMIUMS AND PENALTIES.

For each 1 per cent of silica under 8 per cent down to and including 5 per cent, premium at rate of 50 cents per ton. Below 5 per cent silica, premium at rate of $1 per ton for each 1 per cent.

For each 1 per cent in excess of 8 per cent and up to and including 15 per cent silica there shall be a penalty of 50 cents per ton; 15 per cent and up to and including 20 per cent silica, there shall be a penalty of 75 cents per ton.

For ore containing in excess of 20 per cent silica, a limited tonnage can be used; but for each 1 per cent of silica in excess of 20 per cent, and up to and including 25 per cent silica, there shall be a penalty of $1 per ton.

Ore containing over 25 per cent silica subject to acceptance or refusal at buyer's option, but if accepted shall be paid for at the above schedule with the penalty of $1 per ton for each extra unit of silica.

All premiums and penalties figured to fractions.

PHOSPHORUS PENALTY.

For each 0.01 per cent in excess of 0.25 per cent phosphorus there shall be a penalty against unit price paid for manganese of one-half cent per unit figured to fractions.

In view of existing conditions, and for the purpose of stimulating production of domestic manganese ores, there will be no penalty for phosphorus so long as the ore shipped can be used to advantage by the buyer. The buyer reserves the right to penalize excess phosphorus as above by giving 60 days' notice to the shipper.

The above prices to be net to the producer; any expenses, such as salary or commission to buyer's agent, to be paid to the buyer.

Settlements to be based on analysis of ore sample dried at 212° F. The percentage of moisture in ore samples as taken to be deducted from the weight.

PAYMENTS.

Eighty per cent of the estimated value of ore (less moisture and freight from shipping point) based on actual railroad scale weights to be payable against railroad bill of lading with attached certificates of sampling and analysis of an approved independent sampling chemist, balance on receipt of ore by buyer.

Actual railroad scale weights to govern in final settlement.

Cost of sampling and analysis to be equally divided between buyer and seller.

It will be observed that the value per long ton of natural (undried) manganese ore f. o. b. mine was to be calculated from this schedule in the following manner; (1) Multiply the percentage of manganese in material dried at 212° F. by the corresponding unit price as given in the table, adding 15 cents to this price if the ore was produced east of Chicago; (2) add premiums or deduct penalties, if any, for silica and phosphorus; (3) convert to wet or natural basis; (4) deduct freight per long ton to Chicago.

The following freight rates from points in the several Western manganese-producing States to Chicago were still in force in December, 1919. These were for carload lots with a minimum weight of 60,000 pounds.

TABLE 4.—*Freight rates on manganese ore from Western States to Chicago.* a

State.	Rate per ton of 2000 pounds.	Rate per ton of 2240 pounds.
Oregon	$11.00	$12.32
Washington	11.00	12.32
California	11.00	12.32
Montana	8.00	8.96
Arizona	9.00	10.08
Colorado	7.00	7.84
Nevada	10.00	11.20
Utah	9.00	10.08
New Mexico	7.00	7.84

a Personal communication from H. H. Porter of U. S. Shipping Board.

The ore prices given above applied only to ferro-grade ores. Prices for manganiferous ores have usually been subject to individual contract. In 1918 the better class of Leadville ores sold for about 30 cents per unit, with a premium of 6 cents per unit for iron and a moderate penalty for silica.

Comparison of the prices for ferro-grade ore with those for ferro-manganese is of interest. Eighty per cent ferromanganese was worth $40 to $60 per ton in prewar times. The price from May to November, 1918, was $250 for 70 per cent material. In the tables below are shown (1) the price of ferro per ton; (2) the price per unit of metallic manganese in ferro; (3) the price per unit of 46 per cent manganese ore; (4) the unit price of metallic manganese in ferro divided by the unit price of metallic manganese in the ore, to show the ratio between the two. Since November, 1918, the market has been in a demoralized condition, quotations falling continuously and rapidly.

TABLE 5.—*Comparison of prices of alloy and ore.*[a]

Year.	Price of 80 per cent ferro per ton.	Price per unit of manganese.	Price per unit of domestic ore containing 46 per cent manganese, delivered at furnace.	Column 3 divided by column 4.
(1)	(2)	(3)	(4)	(5)
1910	$40.49	$0.51	$0.25	2.04
1911	37.25	.44	.25	1.76
1912	50.40	.63	.25	2.52
1913	57.87	.72	.25	2.88
1914	55.80	.70	.25	2.80
1915	92.21	1.15	.40	2.88
1916	164.12	2.05	.50	4.10
1917	300.17	3.86	b.75	5.15
1918	285.00	3.56	b1.23	2.93

a Data largely from Metal Statistics, 1918, published by the American Metal Market and Daily Iron and Steel Report.
b Estimated average.

As the last column in the Table 5 indicates, although the price of ore relative to alloy fell far below during 1916 and 1917, the price adjustments of 1918 brought this ratio back again nearly to the figures for 1913–1915; a fact that appears to controvert the opinion prevalent among producers that the proportionate prices of 1918 strongly favored the furnaces. In fact, the ratio shown for 1918 was possibly less favorable to the furnaces than that for 1913–1915, owing to the great increase of conversion cost through high costs of labor, fuel, and supplies.

The prices of "battery" ore also rose enormously during the war. Such ores were formerly worth $20 to $35 per ton. The best grades at last accounts were being sold for $80 to $110 per ton delivered.

STATISTICS.

In the following tables of imports, production, etc., the number of tons of contained metallic manganese are shown as well as the number of tons of ore and alloys, for in no other way can a true comparison be made. The metallic manganese in the ores is also multiplied by a factor representing the conversion loss from ore to alloy to give the "recoverable" manganese in a form suitable for use in steel manufacture. The recovery in making ferromanganese is assumed to be 73 per cent, and in making spiegel 65 per cent. Where the exact analysis of the ores and alloys is not known, the figures for tons of contained metallic manganese are estimated on the basis of the best data available.

TABLE 6.—*Imports of manganese ores, 1903–1918.*

(Long tons.)

Year.	Russia.	India.	Brazil.	Other countries.[a]	Total ores.	Total manganese.	Recoverable manganese, 73 per cent.
1903	5,576	35,900	76,910	....	146,056	55,350	40,450
1904	11,959	10,200	66,875	....	108,579	41,200	30,100
1905	24,650	101,030	114,670	....	257,033	114,340	83,500
1906	13,805	154,180	30,260	....	221,260	97,600	71,350
1907	1,000	95,800	52,922	....	209,021	71,950	52,500
1908	250	143,813	17,150	....	178,203	79,700	58,150
1909	14,486	145,140	35,600	....	212,765	95,800	70,000
1910	33,120	140,965	55,750	....	242,348	110,250	80,500
1911	19,108	106,580	41,600	....	176,832	81,590	59,000
1912	83,334	128,645	81,580	....	300,661	142,700	104,200
1913	124,337	141,587	70,200	8,976	345,090	168,142	123,000
1914	52,681	103,583	113,924	13,106	283,294	134,640	98,300
1915	....	36,450	268,786	8,749	313,985	142,680	104,000
1916	....	51,960	471,837	52,524	576,321	259,317	189,300
1917	....	48,975	512,517	68,450	629,942	282,500	206,400
1918	....	29,275	345,877	116,151	491,303	204,137	149,200

[a] Chiefly Cuba and Central America. In 1918 the sources were as follows (figures are long tons): Cuba, 82,974; Costa Rica, 9,680; Panama, 5,607; Mexico, 5,251; United Kingdom (probably reexported Indian), 4,362; Chile, 2,908; Argentine, 849; China, 3,000; Japan, 709; miscellaneous small lots, 721.

TABLE 7.—*Domestic production of manganese ores.[a]*

(Tons of 2,240 pounds.)

Item.	1912	1913	1914	1915	1916	1917	1918
Ferro grade:[b]							
Total ores	1,664	4,048	2,365	9,709	26,997	114,216	294,497
Total manganese	765	1,800	1,090	4,470	11,880	48,000	118,000
Recoverable manganese	560	1,360	795	3,260	8,670	35,000	80,000
Spiegel grade:[c]							
Total ores[d]	107,569	138,089	158,582	207,511	364,486	732,618	831,000
Total manganese	20,994	20,642	25,170	32,093	58,318	116,880	125,000
Recoverable manganese	13,650	13,420	16,360	20,850	37,900	76,000	81,200
Total recoverable manganese, both grades	14,210	14,780	17,155	24,110	46,570	111,000	167,200

[a] Data in part from Hewett, D. F., Our mineral supplies; manganese: U. S. Geol. Survey Bull. 666–C, 1917, p. 4, and in part from Mineral Resources U. S., U. S. Geol. Survey, for several years.
[b] Ferro grade, 1913–1917, 40+ per cent manganese; in 1918, grade was dropped to 35+ per cent manganese.
[c] Spiegel grade, 1912–1917, 15 to 40 per cent manganese; in 1918, grade was dropped to 10 to 35 per cent manganese.
[d] Includes manganiferous zinc residuum, but excludes manganiferous ores.

TABLE 8.—*Imports and production of manganese alloys, 1905-1918.*[a]

[Tons of 2,240 pounds.]

Year.	Ferromanganese.				Spiegeleisen.				Total manganese, both alloys.	Steel produced.
	Imports.	Production.	Total alloy.	Total manganese.	Imports.	Production.	Total alloy.	Total manganese.		
1905......	52,841	62,186	115,027	92,020	55,457	227,797	283,254	56,650	138,670	20,023,947
1906......	84,359	55,520	139,879	112,000	103,268	244,980	348,248	69,650	18,1650	23,398,136
1907......	87,400	55,918	143,318	114,800	48,994	283,430	332,424	66,500	181,300	23,362,594
1908......	44,624	40,642	85,266	68,150	4,579	111,376	115,955	23,200	91,350	14,023,247
1909......	88,934	82,209	171,143	137,000	16,921	142,831	159,752	31,950	168,950	23,955,021
1910......	114,228	71,376	185,604	148,300	25,383	153,055	178,438	35,700	184,000	26,094,919
1911......	80,263	74,602	154,865	123,900	20,970	104,013	124,983	25,000	148,900	23,676,106
1912......	99,137	125,378	224,515	179,700	1,015	102,561	103,576	20,700	200,400	31,251,303
1913......	128,070	119,495	247,565	198,050	77	110,338	110,415	22,000	220,050	31,300,874
1914......	82,997	106,083	189,080	151,250	2,870	79,935	82,805	16,560	167,810	23,513,030
1915......	55,263	149,521	204,784	163,800	200	97,885	98,085	19,600	183,400	32,151,036
1916......	90,928	221,532	312,460	249,970	....	194,002	194,002	38,800	288,770	42,773,680
1917......	45,381	260,125	305,506	244,400	....	192,985	192,985	38,600	283,000	45,060,607
1918......	26,200	333,027	357,227	274,000	....	283,853	283,853	51,000	325,000	42,212,000

[a] Data compiled from various sources, largely from the statistical bulletins of the American Iron and Steel Institute.

The figures in these tables tell briefly the history of the manganese industry for a number of years past. Thus, Table 6 shows how imports from Russia ceased early in the war and imports from India declined, whereas imports from Brazil increased enormously up to 1918, when restrictions on account of the shortage of ships caused a decrease. During 1918, however, imports from Cuba and from Central American countries practically doubled the 1917 figure.

The domestic industry shows the first effects of the war in 1916. From that time on, domestic production, both of ores and of ferromanganese, increased rapidly.

Table 8 shows that dividing the total pounds of metallic manganese in all alloys produced and imported during the 14 years, 1905 to 1918, by the total tons of steel produced during the same period gives a quotient of 15.5 pounds of manganese per ton of steel. As there were heavy stocks on hand at the end of 1918, it is probable that the true average figure for metallic manganese consumed per ton of steel is more nearly 15 pounds.

No comprehensive figures showing the consumption of "chemical" ore are available. The requirements of the United States are understood to be 25,000 to 40,000 tons annually. In the past these ores were largely imported from Russia. Recently they have come in part from domestic sources and in part from Japan, Brazil, and other countries.

CHAPTER 2.—USES OF MANGANESE OTHER THAN IN STEEL MAKING.

By W. C. PHALEN.

INTRODUCTION.

This report is divided into two parts—use of manganese dioxide ore; use of manganese bronze.

USES OF MANGANESE DIOXIDE ORE.

GENERAL REMARKS.

Estimates of the quantity of manganese ore used in the industries, other than metallurgical, range from 25,000 to 50,000 tons per annum. The latter figure is probably more nearly correct. Most of the ore used is the highest grade of pyrolusite (MnO_2) obtainable and commands a much higher price than metallurgical ore. The manganese dioxide ore is generally termed chemical manganese ore.

Manganese ore, as an oxidizing agent, is used in the manufacture of dry cells, as a decolorizer in certain kinds of glass, and as a drier in oils, paints, and varnishes. The ore is also used directly or indirectly in the manufacture of various manganese chemicals. Many of the textbooks describe its reaction with hydrochloric acid as a source of chlorine. It is no longer used commercially in making chlorine gas, as the latter is an abundant by-product in the manufacture of caustic soda and potash. Some chlorine may be made by this method when the object is the production, not of chlorine, but of manganese chloride. The ore has been mentioned as a soil stimulant, but its value in the fertilizer industry has not been established, and so far as known it is not a constituent of any commercial fertilizers. Only those uses of commercial importance are mentioned here.

THE DRY CELL.

GENERAL DESCRIPTION.

To show the importance of manganese dioxide in the dry cell, the general make-up of the cell, the chemistry of the processes involved, and the constituents entering into the reactions are described herewith.

The modern dry cell, commonly but erroneously called the dry battery, is a portable modification of the "Disque" Leclanche cell invented by Leclanche about 1868. That consisted of a porous cup

15

containing a carbon electrode and a mixture of rather coarsely ground retort coke and pyrolusite. The cup was sealed, two vents being left for the escape of gases, and was placed, with a zinc rod, in a glass jar containing a solution of ammonium chloride. Because of the simplicity and ease of operation of this cell, many attempts were made to fix the electrolyte in various media, such as sawdust, gelatin, asbestos, and silicic acid, so as to make the cell portable.

In 1888 Gassner brought out the first really successful dry cell. The positive pole consisted of a cylindrical mass of ground pyrolusite and coke packed in a canvas bag around a carbon electrode. This was placed in a zinc container which also served as the negative pole, and a paste of plaster of Paris, zinc chloride, and ammonium chloride was poured under and around it. As soon as the paste had set the cell was sealed with a rosin or pitch composition.

The Gassner cell, though a great improvement over various other dry cells, did not meet with much favor because of its high internal resistance and low voltage. The highest current that could be obtained was about 6 amperes, and its voltage was about 1.3. Hence its use was limited to service requiring only small current drains.

DEVELOPMENT.

The advent of the gasoline engine greatly stimulated the development and production of the present type of dry cell. Ignition systems demanded a cheap portable cell able to recuperate after comparatively heavy current drains and by 1897 the manufacture of such cells had attained considerable volume. The present normal yearly requirements of dry-cell production include about 25,000 tons of high-grade manganese dioxide ore, an equal amount of carbon (petroleum coke and graphite) and 8,000 to 10,000 tons of sheet zinc, besides corresponding quantities of zinc chloride, ammonium chloride, paper, carbon electrodes, pitch, and sundry other substances.

METHOD OF MANUFACTURE.

In the modern American dry cell of the usual type, the negative pole is a cylindrical zinc can which also serves as a container. Several sizes of cells are on the market, the great majority, however, being about $2\frac{1}{2}$ inches in diameter by 6 inches high (the so-called No. 6 cell). The inner surface of the zinc is lined with a special grade of absorbent paper, which acts as a reservoir for the electrolyte and as a diaphragm between the zinc and the positive pole. The depolarizing mass, called the "mix," is tightly tamped into the can, around a centrally placed carbon electrode, to about 1 inch from the top. This "mix" is composed of ground carbon (usually calcined petroleum coke and graphite) manganese dioxide ore, and the electrolyte. The mix with the carbon electrode, constitutes the positive pole. After the mix has been tamped, the paper is turned down over the mix; sand or sawdust is poured in, to a depth of about one-half inch, and the cell is sealed with a hot pitch composition. The object of the layer of sand is to provide an expansion chamber for the electrolyte and the excess gas, and also to provide a dry bed for the hot pitch.

CONSTITUENTS USED.

The sheets used for the cans are substantially pure zinc, prime western spelter being generally used, and are usually about 0.016 to 0.019 inch thick in the 2½ by 6 inch cells. Comparatively little of the zinc is used up during the service life of a cell. The mechanical strength to withstand filling and the fact that the can corrodes unevenly, in spots, patches, or streaks, must be taken into account when considering the necessary thickness. One of the problems of the dry cell manufacturer is to construct the cell so as to make this corrosion uniform.

Formerly, blotting paper was largely used for lining the cans, but at present most manufacturers use a special grade of pulp board. This is of ground wood pulp and sulphite pulp, and is 0.03 to 0.04 inch thick. The pulp board should be porous enough to allow the electrolyte to diffuse readily through it, but still retain the smallest particles of carbon and manganese; and it should be capable of absorbing several times its weight of water. Furthermore, it should obviously be free from metallic particles.

FUNCTIONS OF DIFFERENT CONSTITUENTS.

The depolarizing mass or "mix" is the vital part of the cell. To render the most efficient service its different components must be properly proportioned.

The ammonium chloride should be as pure as the usual chemically pure article. It should be substantially free from alkalies, sulphates, inert material, and heavy metals. When the cell is discharging, by virtue of the carbon, manganese dioxide, and zinc couple the ammonium chloride is split up into hydrogen, ammonia, and chlorine, which attacks the zinc can, forming zinc chloride. The reaction is: $Zn + 2NH_4Cl = ZnCl_2 + 2NH_3 + 2H$. The liberated gases cause polarization.

Artificial graphite is generally used, several grades of it being made for dry cells. The graphite does not enter into the chemistry of the cell, merely serving, with the coke, to render the mix more conductive. The coke used is calcined petroleum coke—that is, residues remaining in petroleum stills. When raw it is practically a nonconductor. However, on being calcined at a high temperature it becomes denser and as good a conductor as the better grades of amorphous natural graphite.

The relative fineness of the graphite, coke, and manganese ore and their distribution should be so uniform as to make the mix approach a solid porous mass. The current should flow along radial lines from the carbon electrode to every point on the surface of the mix adjacent to the paper lining. The ideal condition is to have each particle of manganese coated with enough carbon to render it a good conductor but still porous enough to permit efficient depolarization, while the voids should be filled with the porous coke.

The zinc chloride should be as free as possible from heavy metals. Its function is to depolarize the ammonia, which it does by the formation of double salts of zinc and ammonium chloride. No exact information is available as to the reactions involved, but it appears that a slightly soluble double chloride of zinc and ammonium is formed as the end product. During the earlier part of the service life of a cell, inefficient depolarization of the ammonia is probably often the cause of failure under heavy drains. A cell that has been short-circuited or subjected to a heavy drain smells strongly ammoniacal when opened. Another cause of dry-cell failure is the formation of a highly resistant, nonporous crust between the paper lining and the zinc can, probably through the formation of the double salt mentioned or the formation of zinc hydrate.

ROLE OF MANGANESE.

The manganese serves to depolarize the hydrogen. The reaction involved is usually given as follows: $2H + 2MnO_2 = Mn_2O_3 + H_2O$. The manganese reacts almost instantaneously in depolarizing the hydrogen, very likely while the latter is still nascent.

CHARACTER OF MANGANESE ORE USED.[a]

CHEMICAL REQUIREMENTS.

Several factors determine the suitability of manganese ore for dry cells. The ore should have a high available oxygen content present in the form of pyrolusite (MnO_2), should have a minimum amount of iron, and should be free from copper, nickel, cobalt, arsenic, and other metals electronegative to zinc. Copper is particularly harmful. If these impurities are present in the electrolyte or insoluble compounds they do no harm other than as inert or poor conducting materials. If soluble, however, their solutions diffuse to the zinc can, where they are deposited, forming an electrocouple which causes local and useless corrosion of the zinc and consequent deterioration of the cell. When the cell is in service, the deleterious action of these impurities is greatly hastened.

---

a Storey, O. W., Determination of manganese dioxide in pyrolusite: C. F. Burgess Laboratories, Madison, Wis.

PHYSICAL REQUIREMENTS.

The physical properties of manganese ores influence their suitability for use in the dry cell. An ore should be somewhat porous to perform its function efficiently. A somewhat hard but porous ore is likely to give better results than a hard, dense ore, even though the latter is higher in available oxygen. In a dense ore the depolarizing reaction takes place only on the surface, whereas in a porous ore it can occur throughout the mass. Better service life is obtained from a dry cell containing rather coarsely ground ore, as this can hold more electrolyte than finely ground ore. Other factors of importance are the porosity of the ore and the fact that more contact resistance exists between particles of fine ore than those of coarse ore. Careful grading of the ore greatly influences the performance of a cell. The ore, therefore, should not be of an earthy nature like wad, as this mineral does not lend itself to efficient milling and grading.

Before the war the manganese ore used in dry cells was Caucasian pyrolusite. Common specifications called for material containing 80 to 85 per cent MnO_2 and less than 1 per cent iron. No particular attention was given to other ingredients, at least by most buyers, because of the purity and uniformity of the Caucasian ore. There are considerations other than the content of MnO_2 which determine the usefulness of manganese for depolarizer purposes, such as the screen analysis, hardness, density, and other physical qualities. Various manufacturers employ different specifications as to the screen analysis, a common specification being that the run of material shall pass through a 10-mesh or a 20-mesh screen. Some manufacturers specify the removal of the fine particles.

During the war manganese from many other sources was used in making dry cells because of the scarcity of the Caucasian ore. Most of these ores run lower in MnO_2 and higher in iron and have larger percentages of impurities, some of which are decidedly harmful. During the war users accepted material that ran 70 to 80 per cent MnO_2 and as high as 3 to 4 per cent iron. Users have found by experiment and manipulation how to get results with domestic ores and foreign ores other than Caucasian, closely approximating the results obtained with Caucasian.

An important source of manganese dioxide during the war was the ore from old dry cells, which was rejuvenated by processes generally kept secret by the firms employing them. Such processes doubtless contributed to the conservation of our high-grade domestic manganese ore.

In the manufacture of dry cells two classes may be recognized—the standard or so-called No. 6 cell, which is used for ignition, telephone, signal, and other similar purposes; and the small size, or flash-light type, which is used for portable lighting. The quantity of pyrolusite ores used in the standard No. 6 cell is far larger than that used in other sizes. In the manufacture of flash lights, which is growing rapidly, the higher grades of materials are required, such as 80 to 85 per cent ore that has been purified and also various grades of chemically prepared manganese dioxide and hydrates of manganese.

MANGANESE ORES IN THE CERAMIC INDUSTRIES.

GLASS MAKING.

Practically all the raw materials used in glass contain some iron, usually in the form of ferric oxide. The iron, when present even in small quantity, imparts to the glass a pale green color that increases rapidly in intensity as the iron content increases. If a colorless glass is desired, this green color must be removed by some decolorizer. Manganese, selenium, cobalt, and nickel are the most common decolorizers in use, and of these manganese has been most widely employed because it permits easy control of the color. In using selenium and nickel, the quantity must be carefully controlled, but these latter substances are desirable, especially in window and plate glass, because glass decolorized with manganese often changes to a pink color on exposure to the light. A decreasing quantity of manganese is being used by makers of tank glass, and its place is being taken by selenium.

The quantity of manganese used varies considerably, depending on the character of the glass, the method of its manufacture, the iron content of the raw materials, and the character of the manganese ore used. Each manufacturer has his own ideas on this subject. The quantity used is figured in terms of pounds of manganese dioxide per 1,000 pounds of sand, which constitutes 50 to 75 per cent by weight of the entire batch. The temperature employed in the glass-making process helps to determine the quantity of manganese dioxide used, for because of volatilization the higher the temperature, the more manganese is necessary. The maximum limit is 10 to 15 pounds of manganese dioxide per 1,000 pounds of sand, and the minimum may be 2 to $2\frac{1}{2}$ pounds.

CHEMISTRY OF USE OF MANGANESE IN GLASS-MAKING PROCESS.

Compounds of manganese, when other coloring ingredients are absent, produce pink, purple, and violet hues according to the chemical nature of the glass. Manganese dioxide neutralizes the green color caused by iron compounds. Used in excess, it imparts an amethyst tint, and when used in considerable excess, the color is so dark as to appear black.

The neutralization of the iron tint by manganese dioxide is explained by some chemists on a physical, and by others on a purely chemical basis. The green tint is due to the presence of ferrous silicate. Some chemists think that this green compound is oxidized to the ferric silicate, which has an almost imperceptible pale straw-yellow color. According to this view, the oxidizing agent used must not completely decompose at high temperatures, and manganese dioxide seems to be the most available compound fulfilling this condition. At red heat, the dioxide loses one-third of its oxygen, leaving the tetraoxide (Mn_3O_4) which, at still higher temperatures, is an oxidizing agent.

Red lead and other oxidizing agents have not this decolorizing power. Hence some chemists have thought that the result is not due to oxidation, or chemical reaction, but is purely physical. It is possible, however, that other compounds may lose their oxygen at too low temperatures to be effective as oxidizing agents.

SPECIFICATIONS FOR MANGANESE ORE USED IN MAKING GLASS.

Before the war, the ordinary specifications for manganese ore used in glass making were 85 to 90 per cent manganese dioxide and less than 1 per cent metallic iron. Outside of these two ingredients, each manufacturer has his own requirements. Special glasses may require ore carrying more than 90 per cent manganese dioxide and less than 0.5 per cent iron. The higher the manganese content and the lower the iron, the better the ore is for glass making. In general, the grades of ore are similar to those used in making dry cells. Obviously, siliceous pyrolusite is not objectionable but carbonaceous pyrolusite is.

Manganese ore for glass making is sold in powdered, granulated, or lump form. There are objections to the lump form because of the time required to melt it into the batch. Powdered ore is used principally when the batch is melted in pots; lump, or granular ore is used when melting is done in tanks.

Before the war, high-grade pyrolusite for glass making and other chemical purposes was imported from Russia, Saxony, Japan, Nova Scotia, and other foreign countries. As the war progressed, such ore became scarce and, as a consequence, specifications were relaxed and low-grade ores were purchased. During the war some excellent domestic ore was developed, which found a ready market.

OTHER CERAMIC USES.

Another use for manganese ore in glass making has developed in the last few years, namely, for producing black glass used for ornamental purposes. About 3 per cent of ore is added to the batch in making this opaque glass.

Pyrolusite is added to the constituents of glazes and enamels to produce purple tints. Black enamels are those containing manganese. Manganese oxide is also used in brick making.

USE OF MANGANESE SALTS IN DRIERS.

DEFINITION.

Driers are substances, generally metallic oxides or their compounds, that are added to linseed or other drying oils at high or low temperatures to make them capable of readily absorbing oxygen from the air, or of drying by its action. Some chemists consider the action to be catalytic, the manganese compound acting as a catalyser or carrier of oxygen. The principal manganese compounds used as driers are: Manganese sesquioxide (Mn_2O_3), pyrolusite (MnO_2), also known in the trade as dioxide, binoxide or peroxide, manganese hydrate, sulphate, borate, resinate, linoleate, oxalate, and possibly other salts. Certain of the corresponding double salts of manganese and lead are often used.

Some persons claim that pyrolusite is now little used in driers because the manufactured hydrate, on account of its purity, gives better results. This claim does not agree with statements made by dealers in the trade. Of the various substances named above, each acts in a way peculiar to itself. These driers are added only in small quantities, usually less than 0.5 per cent.

MANGANESE DIOXIDE.

Manganese dioxide, extensively used as a drier, is marketed in two forms, the natural and the artificial. The natural mineral, pyrolusite, is simply ground to a powder with water and then dried. The mineral is essentially a peroxide, a class of substances containing more oxygen than is required to satisfy the valence of the metal present. This extra oxygen is loosely combined and readily enters into combination with oxidizable bodies. This feature in the composition of manganese compounds makes them useful in oil boiling, because the oxygen combines with the oil, oxidizing it, while some of the manganese dissolves and forms a compound with the linoleic acid of the oil. In consequence of this action manganese compounds are powerful driers. The quantity of manganese dioxide added in the process of boiling is small, not more than a quarter of a pound to a hundred weight of oil to get the best results. The use of the black dioxide, however, tends to make the oil dark.

MANGANESE SULPHATE.

The methods of preparing this compound and its uses are described on page 24. Rather less than one-half a pound is added to each hundred weight of oil or paint. Oil boilers use it largely as a dryer of pale boiled oils.

MANGANESE BORATE.

Manganese borate is perhaps the least objectionable of all the manganese salts used as drying agents, although the black oxides are more used. To make the borate, 1 part of manganese sulphate is dissolved in 10 parts of distilled water and a little of this solution is added to some soda solution to determine whether any iron is present. If pure, the manganese sulphate solution is added to the hot borax solution as long as a precipitate forms. The precipitate is filtered, washed with hot water, and dried.

MANGANESE RESINATE.

Resinates, formed by the combination of resin or rosin with certain metallic oxides, are much used in making varnish and paint and with the exception of the linoleates are most readily soluble in linseed oil. Manganese resinate is made as follows: Soda ash is dissolved in water and the solution is boiled with steam, the proper proportion of light coarsely powdered rosin is added, and then more soda ash. The clear solution is run into a clear manganese chloride, the resinate separating as a white flocculent precipitate, which is filtered, washed, and dried.

MANGANESE LINOLEATE.

Manganese linoleate is made by pouring a solution of a soap, made by boiling linseed oil and caustic soda into a solution of manganese chloride or manganese sulphate. The dark-brown plaster-like mass is liable to oxidation. When exposed to air, the surface becomes covered with a hard, rather insoluble, protective skin. The material should, therefore, be kept in tightly closed vessels. It acts both as a bleaching agent and a drier for linseed oil. One pound, mixed first with 5 pounds of linseed oil and the whole poured into 10 gallons of linseed oil at 250° F. gives a good drying oil.

The double resinates and linoleates of manganese and lead are also in use.

MANGANESE OXALATE.

Manganese oxalate is prepared by precipitating manganese salts with oxalate of soda or potash or by treating manganese hydrate or carbonate with oxalic acid. One of the advantages of the oxalate is said to be that it decomposes during oil boiling. The manganese dissolves in the oil in combination with linoleic acid, and the oxalic acid is decomposed with evolution of carbonic acid. One-fourth to one-half a pound may be used per hundred weight of oil.

USE OF MANGANESE IN MISCELLANEOUS CHEMICALS.

MANGANESE CHLORIDE.

Manganese chloride ($MnCl_2$) is prepared by the action of hydrochloric acid on the dioxide, chlorine being a by-product of the reaction. It is a rose red, deliquescent salt used in dyeing cotton cloth a manganese brown or bronze. The fabric to be dyed is soaked in the salt solution and passed through a caustic alkali, whereby manganese hydroxide is precipitated in the fabric and on subsequent oxidation turns brown. The material thus treated may be used also for subsequent dyeing by anilin black.

MANGANESE SULPHATE.

Manganese sulphate ($MnSO_4$) may be prepared on a large scale from the black dioxide by heating to redness with ferrous sulphate and subsequently extracting with water. The salt forms pink crystals which are readily soluble in water and are used in calico printing and in porcelain painting. It is also used as a drier for pale oils or for conversion into the oxalate or borate which are used for the same purpose.

MANGANESE PERSULPHATE.

Manganese persulphate, $Mn(SO_4)_2$ is prepared by the electrolytic oxidation of manganous sulphate ($MnSO_4$) and forms a black substance that can be obtained in solution only in the presence of sulphuric acid. It is used as an oxidizing agent in the manufacture of organic products.

POTASSIUM PERMANGANATE.

Potassium permanganate ($KMnO_4$) is prepared industrially by mixing a solution of caustic potash (KOH), specific gravity 1.44, with powdered manganese dioxide and an oxidizing agent, such as potassium chlorate. The mixture is boiled and evaporated, and the residue is fused in crucibles and heated until it has a pasty consistency. The potassium manganate (K_2MnO_4) thus obtained is dissolved by boiling with much water while a current of chlorine, carbon dioxide, or ozone is passed through the liquid. Potassium permanganate separates in crystalline form from concentrated solutions even in the presence of the caustic potash formed during the reaction, and is separated from the dissolved substances in a hydroextractor.

The permanganate is used for preserving wood; it is also used for bleaching textile fibers, by immersing them for a time in an aqueous solution of it and then dissolving the manganese dioxide with sodium disulphite. The permanganate is an energetic disinfecting and oxidizing agent and is used for purifying various gases.

USE OF MANGANESE IN MANGANESE BRONZE.

GENERAL REMARKS.

During the last twenty years manganese bronze has been widely used. The requirements of marine construction, of mining machinery, and wherever corrosion has presented a serious problem, have created a demand for a nonferrous metal to replace steel. Probably the most popular of such substitutes has been manganese bronze.

Manganese bronze made its first appearance about 1876. Its name is somewhat misleading for the alloy contains only a small percentage of manganese. Indeed, it is simply a brass to which have been added by proper methods of alloying, small quantities of aluminum, iron, or manganese, for the purpose of strengthening the alloy and making it denser and closer grained than the average yellow-brass casting.

FUNCTION OF MANGANESE IN MANGANESE BRONZE.

Manganese bronze should not contain much manganese, in fact not more than 0.05 per cent in high-grade bronze. Consequently the consumption of manganese in such alloys is small. The object of the manganese is not so much to act as an ingredient of the alloy, as to serve as a carrier of the iron necessary to insure the required strength and elastic limit. The manganese serves one purpose only, to introduce the iron, for without the manganese the iron would not alloy with the copper. Usually the manganese is added in the form of ferromanganese. If added in large quantities it hardens the alloy, but not nearly as much as iron; such an addition also lowers the elastic limit. Aluminum imparts a good sand-casting quality to the bronze.

Two grades of manganese bronze are now in common use. One is used for rolling into sheets, or drawing into wire or tubes, and for forging. This grade contains no aluminum, and has slightly less zinc than the other and can not be cast in sand. The second alloy is used for sand casting and is the one employed in making propellers and other common appliances.

The method of making manganese bronze—that is, the materials used, the methods of combining them, and the process for casting—is discussed by Sperry [a] and will not be discussed here.

McKinney [b] describes a process of manufacturing manganese bronze wherein is used, instead of virgin metals and raw materials of the highest purity, by-products and scrap. The methods described are, therefore, timely.

The composition of manganese bronze is as follows:

Composition of manganese bronze.

Constituent.	Per cent.
Copper	57.00–59.00
Zinc	38.00–40.00
Iron, manganese, aluminum, and tin	.25– 1.00
Lead	.10– .50

Evidently there is no particular need of using high-grade materials, provided the finished product is properly refined. Among the low-grade materials suggested for use in making manganese bronze are

---

a Sperry, E. F., Manganese bronze and its manufacture: Brass World, vol. 1, December, 1905, pp. 399–406.
b McKinney, P. E., Manganese bronze: Am. Inst. Min. Eng. Bull. 146, February, 1919, pp. 421–425.

the following: Skimmings from the foundry, especially skimmings and dross ordinarily recovered from brass rolling mills or cartridge-case plants; zinc dross recovered from galvanizing plants; aluminum turnings that are generally unrecoverable without serious loss and deterioration of product through oxidation, and other by-products and scrap metals that ordinarily are not usable in foundry practice as remelting scrap.

High-grade manganese bronze can not be made from the above raw materials on a small scale and thus manufacture in crucible furnaces is excluded. A reverberatory furnace or other equipment with which a bath of considerable proportion may be employed is necessary. McKinney [a] discusses a typical charge, the materials being melted in the presence of charcoal with salt as a flux.

USES OF MANGANESE BRONZE.

The most important use of manganese bronze is in propeller blades. A strong, tough alloy is necessary which will resist the action of sea water. The blades are made thin to save weight.

---

[a] McKinney, P. E., Manganese Bronze: Am. Inst. Min. Eng. Bull. 146, February, 1919, pp. 421-425.

CHAPTER 3.—PROBLEMS INVOLVED IN THE CONCENTRATION AND UTILIZATION OF DOMESTIC LOW-GRADE MANGANESE ORE.

By EDMUND NEWTON.

INTRODUCTORY STATEMENT.

In the past, the steel industry of the United States has depended almost wholly on imports for its supplies of manganese. Many of the important domestic sources yield ores that in their natural condition contain less manganese than the foreign ores the steel industry has been accustomed to use. To make these domestic ores available, therefore, they must be concentrated or practice in the steel industry must be modified.

Roughly, 25,000 tons of high-grade manganese ores is used annually for dry batteries, for chemical purposes, and in other minor ways, and approximately 750,000 tons is required for making steel.

By present practice every ton of steel takes an average of about 15 pounds of metallic manganese, which generally is added to the steel in the form of an alloy. The standard alloys are 80 per cent ferromanganese and 20 per cent spiegeleisen. During the year 1917, 286,000 tons of ferromanganese and 193,291 tons of spiegeleisen were made in this country, the former largely from imported ores; and 45,381 tons of ferromanganese was imported. The metallic manganese represented by these alloys was 304,000 tons, being roughly the product of 800,000 tons of high-grade ore and 345,000 tons of low-grade ore.

There is an abundance of low-grade ore in this country suitable for the manufacture of spiegel, but higher grade ore is necessary to make ferromanganese. For this reason the concentration of domestic ore presents a field for constructive and practical research.

MANGANESE DEPOSITS IN THE UNITED STATES.

Before the war manganese ore was mined in relatively small quantities in the Appalachian region, which includes parts of Virginia, Tennessee, and Georgia, and in Arkansas, but in consequence of higher ore prices because of the rise of ocean freight rates, manganese mining has been undertaken in Montana, California, Arizona, New Mexico, Nevada, Utah, and Minnesota as well as in the Appalachian region.

Data now available indicate that in this country deposits of high-grade manganese ores are usually small, but some deposits of ore lower in manganese and higher in iron are of considerable size.   In the aggregate, the total quantity of manganese-bearing material is relatively large, but the difficulty of mining small deposits of the better grades of material and the seeming undesirability of low-manganese alloys in the steel industry, make the outlook for large production of manganese in this country uncertain.

As regards the geologic origin of the majority of the manganese and manganiferous iron ores in this country, Harder[a] states that they are largely the result of secondary concentration. Most of the ores of the Eastern United States, Arkansas, the Lake Superior region, Leadville and other silver districts, and of western California are of this type. The rhodonite and rhodochrosite in the unoxidized parts of the silver veins at Butte, however, are primary concentrations derived from igneous intrusion.   The ores of northern Arkansas are largely reconcentrations from low-grade secondary deposits, derived by decomposition of crystalline rocks, and the California ores are concentrations within chert lenses of material originally present in a disseminated form.

Manganese-bearing materials of the United States may be roughly classified as follows:

1. Manganese ore proper.
2. Manganiferous iron ore.
3. Miscellaneous material:
    (a) Manganiferous silver and lead ore.
    (b) Zinc residuum from manganiferous zinc ore.

Manganese ore, as now defined by the trade, is material that contains more than 35 per cent manganese and is suitable for the manufacture of 70 per cent ferromanganese.   Manganiferous iron ore contains less manganese and more iron.   In general, the iron predominates, but there is no hard and fast line of demarcation between manganese ore and manganiferous iron ore.   Manganese and iron are so closely associated in nature that all gradations from low-manganese, high-iron ore, to high-manganese, low-iron ore may be found in various deposits or in the same deposit.

Manganiferous silver ore is similar to manganiferous iron ore; it carries enough silver to make it valuable for that metal.   Commercial considerations alone control the balance between the manganese or the silver value.

Zinc residuum is a by-product of the smelting of zinc ores from Franklin Furnace, N. J., which contain considerable manganese. After the zinc is removed the remaining product, called residuum,

─────────────

a Harder, E. C., Maganese deposits of the United States: U. S. Geol. Survey Bull. 427, 1910, p. 4.

has nearly the same composition as natural manganiferous iron ore, and for years it has been smelted to spiegeleisen.

In general it may be said that the domestic manganese ores consist of an aggregate of minerals of manganese and iron with various impurities, such as silica, alumina, lime, and magnesia and accessory constituents, such as phosphorus, sulphur, silver, lead, and zinc. The determining characteristics of the manganese minerals and their relation to impurities are discussed in subsequent pages.

In the steel industry manganese is chiefly used in the form of alloys. A less important use is for increasing the manganese content of pig iron to give particular grades, as foundry irons, or to assist in metallurgical operations of certain steel-making processes, as in the basic open-hearth process. The alloys of manganese generally used in this country are ferromanganese, formerly containing 80 per cent, but now 70 per cent, metallic manganese, and spiegeleisen, having 15 to 20 per cent metallic manganese. The rest of these alloys consists principally of iron with small quantities of carbon, silicon, and phosphorus.

During the last few years ferromanganese has been gaining popularity among steel manufacturers, with spiegeleisen declining proportionately. Until recently approximately nine-tenths of the metallic manganese used in the steel industry was in the form of the standard 80 per cent alloy. Ferromanganese, or "ferro," as it is usually called, is easier to use in steel making than alloys containing smaller quantities of manganese, as the required quantity of that metal is contained in a smaller bulk.

The difficulty of obtaining ores suitable for the production of "ferro" has led to a consideration of the possibility of using what may be called intermediate alloys with manganese contents varying between 20 and 80 per cent. In the electric furnace certain alloys can be made with a relatively large content of silicon, in addition to the manganese and iron. The extent to which such alloys may satisfactorily be used in steel manufacture is not alone a technical or economic problem, but is largely controlled by the human element and the unwillingness of steel masters to deviate from long-established practice.

Phosphorus is undesirable in finished steel. In the manufacture of manganese alloys all the phosphorus contained in the ore will be recoverable in the alloy and will enter the steel when the alloy is used. It is permissible, however, for an alloy high in manganese to contain more phosphorus than one low in manganese, for less of the former alloy is needed to carry a given quantity of manganese. It is interesting to note that when 17 pounds of 80 per cent ferromanganese is added to a ton of steel the alloy may contain 1.33 per cent phosphorus and yet increase the phosphorus content of the steel by only 0.01 per cent. To add the same amount of manganese in the form of spiegel, the alloy may contain 0.33 per cent phosphorus and increase the phosphorus content of the steel 0.01 per cent.

For many years manganese alloys have been made chiefly in the blast furnace, although recently certain plants have produced them in the electric furnace. Blast-furnace practice on manganese alloys is generally similar to ordinary pig-iron practice, but there are high metal losses, principally in the slag, by volatilization and as flue dust. The amount of manganese contained in a given weight of slag may be partly controlled by furnace manipulation, but it is evident that the total amount of manganese lost in this manner is directly proportional to the "slag volume."

The results of increased slag volume and greater loss of manganese are cumulative and rather serious. More ore must be used per ton of alloy produced. This additional ore carries more slag-forming constituents. More coke is required to melt it, which in turn tends to produce more slag, and increased slag volume cuts down the daily output of alloy from the furnace. The greater loss of manganese decreases the ratio of manganese to iron in the alloy, so that unless proper allowance is made the alloy will be below the standard grade and therefore subject to a penalty by the purchaser. Not only will the alloy sell for less money, but the decreased daily output of the blast furnace will lessen the total profits.

A manufacturer of manganese alloys endeavors to protect himself against these decreased profits and adjusts his schedules and penalties for ore purchase with that end in view. Although he endeavors to equalize the effects of poor ores, the alloy producer would prefer to buy better ores at correspondingly higher prices.

CONCENTRATION OF DOMESTIC LOW-GRADE MANGANESE ORES.

The comprehensive term "concentration" as here used is intended to cover the improvement of low-grade ore by any suitable means preliminary to smelting. The requirements of metallurgical practice control the classification of manganese ore as low grade and high grade. Thus the term low grade may refer to a low content of manganese with respect to iron or to large quantities of nonmetallic impurities. The detrimental effects in metallurgical practice and the resulting penalties are the incentive for attempts to improve the ore or raise its grade before smelting.

FACTORS CONTROLLING THE POSSIBILITIES OF CONCENTRATION.

In order to interpret properly the possibility of concentrating at a profit any type of manganese-bearing material, many technical and economic factors must be considered. For a particular property, district, or class of material it is necessary to obtain data on these factors:

1. Character and size of the deposit,
2. Conditions affecting mining and marketing,
3. Character of ore material as affecting the possible improvement of grade,
4. Metallurgical value of crude ore and possible concentrate,
5. Commercial considerations.

SIZE AND CHARACTER OF DEPOSIT.

Obviously a large deposit of low-grade manganese ore would warrant considerable experimental work in order to determine satisfactory methods of treatment. Conversely, if a particular type of ore occurred in only one deposit of only a few thousand tons, the value to the industry of the product from such a deposit would be relatively small, even if concentration were feasible. Therefore, no undue proportion of time should be devoted to a concentrating problem that although of considerable technical and individual interest, gives little promise of furnishing any considerable part of the industry's needs.

The mineralized mass must be of such size and character as to justify the expenditure of money in its development and beneficiation and to promise return interest on the investment proportional to the risk taken. This factor of cost is of vital importance, and it is feared that under the stimulus of production incident to national needs during the war sound business principles have at times been temporarily forgotten.

CONDITIONS AFFECTING MINING AND MARKETING.

Not only must the quantity and character of material available be considered, but the natural factors controlling the mining methods, the transportation facilities, and the marketing of products must be carefully studied.

The manganese deposits of the United States, although widely scattered and comparatively small, may nevertheless be mined by relatively simple and therefore cheap mining methods. The mines are for the most part shallow, and extensive development and elaborate equipment are not necessary. Intricate problems of ventilation and drainage do not have to be solved, and if all operations are competently directed common mine labor will generally suffice. Small deposits mean short life, the ore is mined out rapidly and more or less crude mining practice prevails. The cost of mining will, however, be more or less governed by the necessity of selective mining, which in turn is determined by the variability of the ore, the feasibility of economic concentration, transportation facilities, and distance from a consuming center.

All these factors must be coordinated and their combined influence carefully studied before intensive production from individual properties begins. Concentration by eliminating waste may yield a product desired by the steel industry, but its cost may be prohibitive. Discarding waste may enable the producer to offset excessive freight rates, but geographic isolation will invariably handicap an enterprise.

Foreign ores will always find a market in the United States because they come from larger and more uniform deposits, are mined with cheaper labor, and the ocean freights are lower than rail.

CHARACTERISTICS OF ORE AFFECTING BENEFICIATION.

CHARACTER OF MANGANESE MINERALS.

Although many minerals contain manganese, only a few are commercially important. Usually it is rather difficult accurately to identify the manganese minerals in domestic oxidized ore. Several minerals may be intimately associated, and one may have been formed by alteration of another. The hardness of the individual minerals varies widely. Pyrolusite is soft and may be readily pulverized between the fingers. Hence difficulty might be expected in attempts to recover this mineral by the common processes of wet concentration. The other minerals are harder, but usually brittle. Although the character of the individual minerals is important, the association of the several manganese minerals with various gangue materials often has a more important bearing on methods of concentration.

IMPURITIES ASSOCIATED WITH MANGANESE MINERALS.

Wherever manganese ore is mined on a commercial scale, the product of the mine always contains impurities. Some of these are obvious on inspection; others may require chemical analysis for their determination. The impurities associated with manganese minerals may be classified as (1) those derived from associated rocks, or rocks partly replaced by manganese-bearing solutions, (2) those associated with the manganese in solution, and deposited simultaneously, and · (3) those chemically combined with manganese in the mineral.

From the viewpoint of the metallurgist, all are impurities, and must be removed either before or by metallurgical treatment.

For convenience, the common impurities in manganese ores may be classified according to certain general physical and chemical principles, as follows:

1. Metallic. Iron, lead, zinc, silver, and in some ores, nickel, copper, and tungsten.

2. Gangue. "Basic" lime, magnesia, baryta, "acid" silica, and alumina.

3. Volatile. Water (atmospheric moisture and molecular water), carbon dioxide, and organic matter.

4. Miscellaneous. Phosphorus and sulphur.

The chemical behavior of these impurities affects metallurgical operations, and the physical form in which they occur controls the possibility of removal previous to smelting, and the choice of methods for accomplishing such removal.

The proportion of manganese to the useless or harmful constituents of an ore determines the value and desirability of the ore.

The presence of appreciable quantities of any impurity means that more ore must be mined and smelted in order to produce a given weight of manganese alloy. Some impurities are more detrimental to alloy manufacture than others.

Metallic impurities, of which iron is the most common, are usually reduced in smelting and are retained by the alloy. The quantity present naturally affects the character of the alloy produced, which in turn controls its desirability for use in steel manufacture. Other metallic impurities occur usually in such small amounts that they are not detrimental to the resulting alloy. Zinc is an exception. This metal is largely volatilized in smelting, and if it is present in appreciable quantities its fume condenses as oxide in the hot-blast stoves and may hinder furnace operation. Unless the furnace-top gases are washed, the stoves must frequently be cleaned, with consequent loss of time. When the price of zinc is high, the zinc oxide recovered from the stoves yields a substantial sum.

Silver, from the standpoint of the steel manufacturer, is neither detrimental nor advantageous to manganese alloys. The silver content of a manganese alloy has no value; consequently no credit is allowed the miner for silver contained in an ore to be used for manganese-alloy manufacture. In some ores the silver content is such that the ore has greater value for the lead smelter. The manganese then acts as a flux and the silver may be recovered by purification of the lead.

The gangue impurities classed as basic and acid may also be called "slag-forming impurities." In smelting, these impurities must be fluxed to form slag. Slag is usually considered a waste product, but it has important metallurgical functions, and just enough slag must be present for performing these functions properly and economically. An excess of slag must be avoided. In manganese-alloy manufacture the slag contains more or less manganese, the quantity of manganese thereby lost being dependent on the basicity of the slag, the temperature, and the slag volume. The first two factors control the quantity of manganese in a given weight of slag, and it is obvious that the greater quantity of slag will result in a greater loss of manganese. A large slag volume will rapidly decrease the daily alloy output of a blast furnace. The overhead charges must be distributed over a smaller daily tonnage of alloy and the profit from the sales for a given unit of time will decrease. It follows that more than the quantity of slag required to provide for the metallurgical functions is highly undesirable.

Silica is usually the predominant slag-forming constituent in domestic manganese ore. Some silica is reduced to the metallic state in smelting and is recovered in the alloy as silicon, but the larger part must be fluxed with lime, magnesia, or other bases to form slag. Manganese-alloy slags should be basic; hence a larger quantity of slag will be produced from an ore with acid gangue than in normal iron blast-furnace practice. Alumina is a slag-forming constituent and although usually classed with silica, it acts somewhat differently in the blast furnace. Brazilian ores are notably high in alumina, but most domestic ores contain relatively small quantities.

Lime, magnesia, and baryta in an ore are also slag-forming constituents, but they combine with the silica and alumina present and thereby reduce the quantities of those bases necessary in the form of limestone or dolomite for the furnace charge. Baryta is not common as a gangue mineral. It is not as strong a base as either lime or magnesia. While these constituents offset the metallurgical effects of silica or alumina, as regards evaluating an ore, they represent weight, and if the ore must be transported a considerable distance to the point where it is smelted, it is doubtful whether·their value as bases would equal the additional freight charge. Limestone can generally be obtained at low cost close to the smelter.

Volatile impurities are removed from the top of a blast furnace largely by the surplus heat. It is desirable, however, in order to reduce the loss of manganese to keep the top of a manganese-alloy blast furnace cool. Volatile compounds are not particularly detrimental to smelting. When carbonate ores are being treated the case is somewhat different. Some metallurgists claim that in treating rhodochrosite ores the ratio of carbon monoxide to dioxide in the furnace gases is disturbed, which has a detrimental effect on the reduction of the oxides in the upper part of the furnace. It has also been suggested that the carbon dioxide driven off combines with carbon of the coke, forming carbon monoxide in the upper part of the furnace, and thus increases the coke consumption. Definite data are not available on these points.

In practice, all the phosphorus in the ore mixture and that contained in the coke and limestone is recovered in the resulting alloy. The permissible quantity of phosphorus in an alloy (the proportion that does not produce detrimental effects when the alloy is added to steel) has not been definitely determined. The higher the manganese content of an alloy, the larger the proportion of phosphorus that may safely be contained. Ordinarily steel makers desire as large a margin of safety as possible, and therefore have specified that the phosphorus in an ore shall not exceed a certain percentage.

Sulphur is usually present in small quantities in oxidized manganese ores, but in the primary rhodochrosite ores of Butte and other parts of the West there may be considerable quantities of sulphides of iron and zinc. Sulphur is not a serious factor, as the conditions of blast-furnace operation when making manganese alloys are such that sulphur combines with manganese or lime and is readily retained by the slag, only traces entering the alloy.

Knowing the effect of impurities in manganese ores on blast-furnace practice, the methods of eliminating them may now be considered. Ore dressing deals with the problems of separating deleterious or useless materials from the more valuable minerals, thereby raising the grade and reducing the quantity of the concentrated product. To accomplish this, it is essential that the physical and chemical characteristics of the ore be determined. These factors are conditioned largely by the type of deposit from which the ore is mined. As types of ore, entirely disregarding genesis, we recognize:

1. Rhodochrosite and rhodonite; carbonate and silicate ores, deposited in fissure veins or replacing original rocks.

2. Nodular ores, accretions of manganese oxide in soft plastic clays.

3. Manganese oxides deposited in small fissures or fracture planes, as braccia fillings, or as more or less impure beds.

4. Manganese oxides occurring as infiltrations, deposited in minute pore spaces, as partial replacements, or otherwise intimately mixed with the rock or gangue.

In the first class of deposits, the principal gangue impurity is silica, although sulphides of silver, lead, zinc, and iron are often found in appreciable quantities. The silica occurs both as quartz and chemically combined in rhodonite ores. In the carbonate ore, the carbon dioxide may be removed by calcination, thus effecting a concentration of manganese, reducing the bulk, and lowering the freight rate per unit of manganese contained in the original ore, but rhodochrosite decrepitates strongly when heated to a temperature where the oxide is formed, tending to produce an excessive amount of fines which is undesirable in practice. The breaking up of the particles by calcination will isolate some of the free silica which on account of the larger sized particles might be screened out. The sulphide minerals may occur in such quantity that it is desirable to remove them by gravity methods of separation.

In the second class of deposits, the nodules are of variable size and usually high in manganese. They do not appear to be contaminated internally with the inclosing material. The clays are soft, whereas the nodules are generally hard. This type is common in the Appalachian region of the United States. The clay may be separated from the manganese nodules by log washers, followed where necessary, and where the size of the deposits warrants the installation, by picking belts, crushers, screens, and jigs.

In deposits of the third class, the manganese minerals, although closely associated with the inclosing rock, are generally not contaminated by it and may be relatively pure. The method of treatment varies with the size of the manganese particles and the hardness of the rock, but does not differ essentially from the treatment of the second class. If there is little or no clay, the log washer will be omitted, and crushing, screening, jigging, and possibly tabling will make up the concentrating process.

If the manganese mineral is largely pyrolusite, and therefore friable and soft, crushing may produce an excessive proportion of fines difficult to recover by gravity or water methods of separation. If, however, the manganese mineral is hard, dense, and massive, and the inclosing rock is more friable, the problem is simpler. When the specific gravity of the minerals and the gangue approximate each other, wet concentration is difficult unless the particles differ decidedly in size.

Obviously the association of gangue materials with the desired mineral in ores of the fourth class is so intimate that the finest crushing imaginable will not permit separation by mechanical means. To this type the siliceous manganese ores of the Western States may be assigned. Ore-dressing tests have conclusively shown that where the silica is chemically combined with the manganese or where colloidal silica envelops the manganiferous particles, any wet process or gravity concentration will not give the desired results.

CONCENTRATION PROCESSES.

This paper does not describe in any detail actual ore-dressing or concentration practice. It is axiomatic that small deposits or mines of questionable life, as determined by tonnage and markets, do not warrant elaborate plants or the adoption of intricate beneficiation processes. A general classification of methods applicable to the manganese industry is given below. The processes mentioned are all preliminary to the greater and final concentration of the desirable elements in the blast furnace.

I. Simple methods of concentration.
    (a) Selective mining.
    (b) Hand picking.
    (c) Jigging.
    (d) Screening.
    (e) Log washing.
    (f) Water classification.
    (g) Roughing-table treatment.
    (h) Slime-table or vanner treatment.
    (i) Pneumatic separation.
    (j) Combination of two or more of the above methods.

II. Complex methods.
   (a) Magnetic separation.
      (1) Without preliminary thermal treatment.
      (2) With preliminary thermal treatment
   (b) Electrostatic separation.
   (c) Hydrometallurgical processes.
      (1) Leaching with various acids. Precipitation by chemical substances.
      (2) Leaching with various acids. Precipitation by electrolysis.
      (3) Leaching with various acids. Evaporation of solution and heat treatment in rotary kiln.
   (d) Preliminary thermal processes.
      (1) Drying, to remove hygroscopic moisture.
      (2) Calcining, to remove carbon dioxide or combined water.
      (3) Agglomerating fine concentrates, to make them desirable for blast-furnace use.
      (4) Volatilizing manganese at high temperatures in the presence of constituents that form readily volatile compounds.
      (5) Direct reduction of oxides by carbon, under temperature control.
   (e) Miscellaneous processes.
      (1) Flotation.
      (2) Use of heavy solutions.

There are many standard machines for the concentration of ores, but it is unwise to think that a certain machine will accomplish the necessary result on manganese-bearing materials. As the character of the manganese materials varies greatly in different districts, it is more logical to determine in detail physical, as well as the chemical, characteristics of the material. When such preliminary study has shown the nature of the impurity, and its relation to the manganese mineral, it is easier to outline a suitable method of treatment. The flow sheet, however, must be determined by experiment.

COMMERCIAL CONSIDERATIONS REGARDING BENEFICIATION.

If the technical possibilities of beneficiating any particular ore are favorable, it is then necessary to ascertain whether such an operation on a commercial scale would yield a reasonable profit. The cost of the plant and its installation must be justified either by the available ore in the deposit, or by the length of time during which the profit could be made. The amortization of capital and the interest on the investment must be included in the estimation of cost.

The effect of concentrating an ore is not always clearly appreciated. Concentration implies that an improvement of metallic content is made by the intentional elimination of impurities, but there is always a loss of the valuable mineral itself. When the grade of the product is increased, its weight is decreased. In other words, 2 to 25 tons of crude manganese-bearing material may be required to produce 1 ton of high-grade concentrate. The income results from the sale of the smaller quantity of concentrate, but chargeable against this will be the cost of mining some tons of crude ore, the cost of treating the ore, the freight to market, and the special overhead charges. Concentration, however, may be necessary to make the material marketable at all.

ESTIMATED COST DATA.

In order to illustrate the above points a little more clearly, the essential economic factors have been applied to three ores from California.

It is estimated that the three ores were obtained from deposits of such size that it would be possible to mine 75 tons of crude ore a day. In the table below the figures from actual concentrating tests have been combined with estimates of the cost of mining and treatment. The mining-cost figures are those reported by one of the three mines for the season of 1917.

A rough estimate of the cost of a small concentrating plant capable of treating 75 tons of crude ore a day is $20,000. It is assumed that this amount must be charged off in a year of 300 operating days. Instead of the amortization rate per ton being actually computed, the $20,000 and the interest charge of 7 per cent are distributed over the annual production of concentrates. The interest charge at 7 per cent is also distributed over the total tonnage of concentrates.

Royalty is considered to remain at $1.50 a ton of crude ore, even if the material is concentrated. Based on concentrates, it therefore increases proportionally to the ratio of concentration.

The value of crude ore and concentrates has been computed on the basis of the schedule approved by the War Industries Board in May, 1919. These prices no longer hold, and new computations should be made on the basis of existing prices. The methods to be followed are, however, indicated by the examples given. A freight rate of $11.20 per ton has been used from California points to Chicago on a long ton of 2,240 pounds.

All figures in these estimates are based on dry analyses and on a long ton of 2,240 pounds. Because of the length of haul, it is assumed that moisture would be negligible in both crude ores and concentrates. The figures follow in table 9.

TABLE 9.—*Estimated cost data applied to three California ores.*

MANGANESE ORE 1009.

	Fe	Mn	SiO$_2$	P
Analysis of crude ore, per cent..................	2.55	40.01	28.91	0.047

Concentration results:

	Crude ore.	Concentrates.	Tailings.
Per cent by weight................................	100.00	50.55	49.45
Per cent manganese................................	38.59	51.10	25.81
Per cent manganese by weight.......................	100.00	66.93	33.07
Ratio of concentration.............................	1.987	1.00	....

	Fe	Mn	SiO$_2$	P
Analysis of concentrate, per cent..............	1.48	51.55	13.14	0.053

Daily crude ore capacity, tons.. 75
Daily concentrate production, tons... 38

Estimated cost of mining crude ore, per ton:

Mining...	$3.00
Tramming and loading..	1.25
Royalty...	1.50
Unloading...	.25
Trucking..	2.50
Overhead..	1.75
Total...	10.25

Value of crude ore, per ton:

Manganese value, @ $1.02..............................	40.01 × $1.02 = $40.81
Silica penalty...	7 × 0.50 = 3.50
	5 × 0.75 = 3.75
	8.91 × 1.00 = 8.91
Total..	16.16

Net f. o. b. Chicago.. $40.81 − 16.16 = 24.65
Net value, railroad shipping point, $11.20 freight............ 24.65 − 11.20 = 13.45
Net profit on sale of crude ore............................ 13.45 − 10.25 = 3.20

Estimated cost of concentrating plant................ $20,000
Estimated amortization period....................... 1 year—300 days
Charge per ton concentrate.......................... $20,000 ÷ (300 × 38) = $1.75
Interest on investment at 7 per cent per ton of concentrate.. (0.07 × $20,000) ÷ 11,400 = 0.12
Estimated cost of concentration per ton.............. $0.75 per ton of concentrates

Total estimated cost of concentrate on cars:

Mining...........................@ $3.00 per ton of crude	1.987 × $3.00 = $5.96
Tramming.......................@ .60 per ton of crude	1.987 × .60 = 1.19
Concentrating...................@ .75 per ton of concentrates........	1.000 × .75 = .75
Amortization..	1.75
Interest..	.12
Royalty at $1.50 per ton of crude.........................	1.987 = 1.50 × 2.98
Unloading at .25 per ton of concentrate...................	1.000 = .25 × .25
Trucking at 2.50 per ton of concentrate...................	1.000 × 2.50 = 2.50
Overhead at 1.75 per ton crude...........................	1.987 × 1.75 = 3.48
Total..	18.98

137338°—20——4

TABLE 9.—*Estimated cost data applied to three California ores*—Continued.

MANGANESE ORE 1009—Continued.

Value of concentrate per ton:
Manganese value, at $1.24............................. 51. 55× $1. 24=$63. 92
Silica penalty....................................... 5. 14× . 50= 2. 57
Net value, f. o. b. Chicago................................$63. 92— $2. 57= 61. 35
Net value, railroad shipping point, $11.20 freight........... 61. 35— 11. 20= 50. 15
Net profit on concentrate per ton......................... 50. 15— 18. 98= 31. 17

Daily profit on crude ore................................. 75×$3. 20= $240. 00
Daily profit on concentrate............................... 38×31. 17=1, 184. 46
Balance in favor of concentrating......................... = 944. 46

Annual profit on crude ore, 300 days...................... 300×$240. 00=$72, 000
Annual profit on concentrate, 300 days.................... 300× 944. 46×283, 338
Balance in favor of concentration......................... =211, 338

Annual production of crude ore, tons....................... 22, 500
Annual production of concentrate, tons..................... 11, 400

MANGANESE ORE 1011.

	Fe	Mn	SiO₂	P
Analysis of crude ore, per cent....................	1. 27	32. 01	39. 55	0. 029

Concentration results:

	Crude ore.	Concentrate.	Tailings.
Per cent by weight...............................	100. 00	46. 47	53. 53
Per cent of manganese...........................	32. 19	45. 05	21. 02
Per cent of manganese by weight..................	100. 00	65. 05	34. 95
Ratio of concentration...........................	2. 152	1. 000	

	Fe	Mn	SiO₂	P
Analysis of concentrate, per cent..................	1. 50	45. 05	18. 64	....

Daily crude ore capacity, tons... 75
Daily concentrate production, tons.. 35

Estimated cost of mining crude ore, per ton:
Mining.. $3. 00
Tramming and loading.. 1. 25
Royalty... 1. 50
Unloading... .25
Trucking.. 2. 50
Overhead.. 1. 75

Total... 10. 25

Value of crude ore—Too siliceous.

Estimated cost of concentrating plant.............. $20, 000
Estimated amortization period..................... 1 year—300 days
Charge per ton of concentrate...................... $20, 000÷(300×35)=$1. 90
Interest on investment at 7 per cent per ton concentrate... (0. 07 ×$20, 000)÷10, 500=. 12
Estimated cost of concentration per ton............ $0. 75 per ton of concentrate.

TABLE 9.—*Estimated cost data applied to three California ores—*Continued.

MANGANESE ORE 1011—Continued.

Total estimated cost of concentrate on cars:

Mining, at $3 per ton of crude................................ 2. 152×$3. 00=$6. 46
   Tramming, at $0.60 per ton of crude...................... 2. 152× . 60= 1. 29
   Concentrating, at $0.75 per ton of concentrate............ 1. 000× . 75= . 75
   Amortization.. 1. 90
   Interest.. . 13
   Royalty, at $1. 50 per ton of crude....................... 2. 152× 1. 50= 3. 23
   Unloading, at $0.25 per ton of concentrate................ 1. 000× . 25= . 25
   Trucking, at $2.50 per ton of concentrate................ 1. 000× 2. 50= 2. 50
   Overhead, at $1.90 per ton of crude...................... 2. 152× 1. 90= 4. 09

     Total... 20. 60

Value of concentrate per ton:

   Manganese value, at $1.12............................. 45. 05×$1. 12=$50. 46
   Silica penalty.. 7. 0 × 0. 50= 3. 50
Net value... 3. 64× 0. 75= 2. 73

     Total... 6. 23
Net value, f. o. b. Chicago................................. $50. 46 —$6. 23= 44. 23
Net value, railroad shipping point, $11.20 freight........... 44. 23 —11. 20= 33. 03
Net profit on concentrate per ton.......................... 33. 03 —20. 60= 12. 43

Daily profit on crude ore..................................
Daily profit on concentrate................................. 35×$12. 43=$435. 05
Balance in favor...

Annual profit on crude ore, 300 days......................
Annual profit on concentrate—300 days................. 300×$4. 35. 05=$130, 515
Balance in favor...

Annual production of crude ore, tons...................... 22, 500
Annual production of concentrate, tons.................... 10, 500

MANGANESE ORE 1020.

	Fe	Mn	SiO₂	P
Analysis of crude ore, per cent................	1. 50	39. 95	25. 10	0. 075

Concentration results:

	Crude ore.	Concentrates.	Tailings.
Per cent by weight....:	100. 00	33. 14	66. 86
Per cent of manganese............................	38. 77	51. 21	32. 60
Per cent of manganese by weight..................	100. 00	43. 78	56. 22
Ratio of concentration¹...........................	3. 018	1. 00..........	

	Fe	Mn	SiO₂	P
Analysis of concentrate, per cent.............	1. 23	51. 55	11. 65	0. 074

Daily crude ore capacity, tons... 75
Daily concentrate production, tons... 25

TABLE 9.—*Estimated cost data applied to three California ores*—Continued.

MANGANESE ORE 1020—Continued.

Estimated cost of mining crude ore, per ton:

Mining..	$3.00
Tramming and loading...	1.25
Royalty..	1.50
Unloading...	.25
Trucking..	2.50
Overhead...	1.75
Total..	10.25

Value of crude ore, per ton:

Manganese value, at $1.00..............................	$39.95×$1.00=	$39.95
Silica penalty..	7× .50=	3.50
	5× .75=	3.75
	5.1 × 1.00=	5.10
Total...		12.35
Net f. o. b. Chicago..................................	$39.95−$12.35=	27.60
Net value, railroad shipping point, $11.20 freight.......	27.60− 11.20=	16.40
Net profit on sale of crude ore..........................	16.40− 10.25=	6.15

Estimated cost of concentrating plant...................................	$20,000
Estimated amortization period..................................	1 year—300 days
Charge per ton concentrate.......................	$20,000÷(300×25)=$2.67
Interest on investment at 7 per cent per ton of concentrate...	(0.07×20,000)÷7,500= 0.19
Estimated cost of concentration per ton..................	$0.75 per ton concentrates

Total estimated cost of concentrate on cars:

Mining, at $3 per ton of crude.......................	3.018×$3.00=	$9.05
Tramming, at 60 cents per ton of crude....................	3.018× .60=	1.82
Concentrating, at 75 cents per ton of concentrates..........	1.000× .75=	.75
Amortization..		2.67
Interest..		.19
Royalty, at $1.50 per ton of crude.......................	3.018×$1.50=	$4.53
Unloading, at 25 cents per ton of concentrate.............	1.000× .25=	.25
Trucking, at $2.50 per ton of concentrate.................	1.000× 2.50=	2.50
Overhead, at $1.75 per ton of crude.....................	3.018× 2.67=	8.06
Total..		29.82

Value of concentrate per ton:

Manganese value, $1.24...............................	51.55× $1.24=	$63.92
Silica penalty..	3.65× .50=	1.83
Net value, f. o. b. Chicago.............................	$63.92−	1.83= 62.09
Net value, railroad shipping point, $11.20 freight...........	62.09−	11.20= 50.89
Net profit on concentrate per ton......................	50.89−	29.82= 21.07

Daily profit on crude ore.............................	75× $6.15=	$461.25
Daily profit on concentrate...........................	25× 21.07=	526.75
Balance in favor.....................................	Concentrating=	65.50

TABLE 9.—*Estimated cost data applied to three California ores*—Continued.

MANGANESE ORE 1020—Continued.

Annual profit on crude ore, 300 days.....................	300 × $461.25 = $138,375
Annual profit on concentrate, 300 days...................	300 × 526.75 = 158,025
Balance in favor...	Concentrating = 19,750

Annual production of crude ore, tons.....................................	22,500
Annual production of concentrate, tons..................................	7,500

DISCUSSION OF ESTIMATED COST DATA.

ORE 1009.

The manganese content of this crude ore is within the range of the schedule. The silica content is a few points above, but it has been assumed that a small tonnage of such ore might be accepted. The value of the crude ore has, therefore, been computed.

By a concentration of approximately 2 to 1 a high-grade product can be obtained. Shipment of the crude ore shows a small profit per ton. On concentrate the profit per ton is comparatively large, and, although it weighs much less, the daily profit is shown to be much larger.

ORE 1011.

The manganese content of ore 1011 is below the range of the present schedule and the silica content is very high. It is, therefore, considered that this ore could not be sold as mined, but would require concentration. The concentrate is not very high in manganese, being 5.50 per cent lower than that obtained from ore 1009, although the ratio of concentration is nearly the same. The computation shows a daily profit by concentrating considerably less than for ore 1009.

ORE 1020.

The manganese content of ore 1020 is within the range of the present schedule and the silica content is only 0.1 per cent higher than the rejection limit. Therefore, the value of the crude ore has been computed. The manganese content of the concentrate produced is almost the same as for ore 1009, but the ratio of concentration is much larger. Hence, the profit per ton is less than for ore 1009, although more than for ore 1011. The daily profit is considerably decreased by the smaller output. Consequently, while there is a decided balance in favor of concentrating ore 1009, the balance for ore 1020 is comparatively rather small.

FINENESS OF CONCENTRATES.

The concentration tests of the three ores embodied crushing to pass 10 mesh, classifying, and treating on roughing tables. Ore buyers have said that they were not interested in such fine material.

Indeed, it would not be suitable as a major part of a manganese blast-furnace mixture, but up to 5, or possibly 10 per cent, it should not be excessively harmful.   There would be a larger loss in flue dust, and if material of better structure were obtainable it would be greatly preferred by blast-furnace operators.   Because of the larger losses attending the use of this fine material, if it were purchased at all, it would probably be penalized.

Possibly such fine concentrates could be agglomerated by briquetting, sintering, or nodulizing.   But in practice, the application of any such process to a particular mine is rather doubtful.   In the first place, a small daily output would not suffice to keep such a plant in continuous and effective operation.   A sintering plant might be erected for the joint use of several mines, or might be erected independently and operated on a custom basis.   The first cost of the equipment and the operating costs that would probably prevail in California being high, it is doubtful whether such plants would be profitable, unless there were some assurance of high prices prevailing for a considerable period of time.

At present nodulizing plants at iron blast furnaces in Chicago and elsewhere are running on flue dust.   In an emergency they might be turned over to manganese concentrates, and the operating company reimbursed in some manner, possibly by a joint assessment of miner and alloy manufacturer.   It is presumed that the sinter would be superior to natural manganese ores as mined and shipped direct, and that a certain part of the cost might be charged off against improved furnace practice.

The time required for the erection of a plant may be considerable, and private capital might not be attracted to an enterprise of this character unless there were some assurance that a given margin of profit might be possible for a period long enough to permit retirement of the original investment for the plant and the gaining of a satisfactory return.

CHAPTER 4.—PREPARATION OF MANGANESE ORE.

By W. R. CRANE.

INTRODUCTORY STATEMENT.

The great demand for manganese during the war caused many persons to engage in mining and preparing manganese ores for market. As a result much money was spent on plants that were not adapted to the work. This condition was recognized by the Bureau of Mines early in 1918, and an effort was made to assist the operators of manganese properties by sending engineers into the field to advise as to the proper methods of mining and treating the ores.

A preliminary study of the operations showed that the methods in general use were wasteful; consequently, if practice could be improved it was estimated that the production of the various districts could be increased, by saving a large proportion of the mineral formerly lost—thousands of tons a year. With this object in mind, a standard washing plant was designed, which was based on the best and most successful practice in manganese properties in the country.

The methods employed in cleaning and concentrating manganese ores and the scale of operations depend largely upon the size and character of the deposits worked. The ores obtained from the different forms of deposits are so varied in character that were it economically possible to treat all of them, a wide variation in practice would be necessary. The irregular deposits, consisting largely of manganese and rock fragments inclosed in clay, are probably more largely treated than the regular forms as they produce the bulk of the ores treated. A relatively large amount of ore also comes from the regular or blanket formations, which consist of replacement deposits in the limestone bed rock.

The blanket deposits may be hard ore lying upon or within the limestone bed rock, or may be soft or wad ores. The deposits of wad, some of which are of large size, are usually of low grade and present special problems in washing and concentration; as yet no serious attempt has been made toward their economic treatment.

Washing and concentration methods are discussed herein in the following order: (1) Description of the practice in the various districts and (2) description of a standard plant the design of which is based upon the best practice in the treatment of manganese ore in this country.

45

The principal difficulty in treating the ore, or wash dirt, from the irregular deposits is the uncertain supply. Next in importance is the character of the manganese mineral and the proportion of ore to waste, such as clay, sand, and iron. The kind of ore best adapted to washing and concentration is the nodular or pebble-like forms of small size; next is the granular form; both are usually of convenient size for log washing and subsequent jig work. The kidney, dornick, and massive ores require reduction before treatment by log washers and on the picking belt, but owing to the size and purity of many of the masses, it is often possible to make a high recovery prior to treatment in logs and jigs.

On account of the uncertain and variable conditions mentioned, the tendency in treating manganese is to reduce the equipment to a minimum, both with regard to kind and number of parts employed. This is frequently done irrespective of the desirability of such limitation. Imperfect cleaning and excessive losses usually result from the curtailment of equipment. However, inadequate equiment of plants is not always due to the wish to limit the first cost, but often to lack of definite knowledge regarding proper methods of treating manganese ores, and the following of local practice, however poor it may be. It is, therefore, important to present details regarding a standard plant that incorporates the most useful and desirable features of the best practice in this country.

METHODS EMPLOYED.

The preparation of manganese ores may be divided into two separate and distinct parts—namely, dry mining, and washing or concentration.

DRY MINING.

Dry mining does not refer to mining, except indirectly, but more to the cleaning of the ore, or wash dirt, by screening it as mined. Not all wash dirt is adapted to dry mining, as some clays are wet and sticky, or occur in large masses, rendering their separation from the manganese mineral by screening practically impossible. Dirt that can be best treated by dry mining is dry and granular, and can be easily broken and separated.

The dry-mining method is unsatisfactory at best and is scarcely ever employed in operations of any size except as a temporary expedient prior to the erection of a washing plant. In small-scale operations where the deposits are small, or sufficient funds to properly equip the mine are lacking, such a method may be permissible. Also, deposits distant from a water supply may require dry mining if they are to be worked at all. In any event the method is wasteful, the loss of fine ore frequently amounting to 25 to 35 per cent of the recoverable mineral in the bank.

On the other hand, the high-grade manganese occurring in solution cavities can be mined and cleaned by hand with or without screening. In such deposits the bulk of the ore is of suitable size and is easily cleaned; however, even under the most favorable conditions, large losses result.

The proportion of manganese in the wash dirt is the controlling factor in dry mining; dirt having less than 10 per cent mineral can not be successfully separated by screening unless the bulk of the mineral is of large size.

WASHING AND CONCENTRATION.

The principal considerations affecting the washing and concentrating of manganese ores are: Quantity of wash dirt available, percentage of manganese in the dirt, and water supply. No plant should be erected until the available supply of ore has been determined within reasonably close limits; failure to test a property properly before erecting a plant shows lack of good business judgment. Ore may occasionally be mined and hand picked under exceptionally favorable conditions of occurrence even when the proportion of ore to clay is 1 to 50, or 2 per cent, but mining must be done on a large scale by machinery and the bulk of the ore must be in fairly large masses. Log washers and jigs can not readily treat materials that are leaner than 1 to 35; the ratio of ore to dirt should preferably be not less than 1 to 10, or 10 per cent, which is a fair average for wash dirt treated.

The design of a plant for the proper treatment of any manganese ore must be based upon the following points: Character of the manganese mineral and its physical properties and occurrence; the associated minerals, and the inclosing rock formations; and the practice previously followed, if any. Of only slightly less importance are the local conditions, such as lay of the ground, water supply, and timber for construction and other purposes.

The character of manganese ore varies widely in different parts of the same district, and even in openings on one property. The mineral may be dense and hard, or porous and hard, or porous and soft. The specific gravity also varies considerably, ranging from 3.5 to nearly 5; the average for the bulk of the mineral treated is between 3.5 and 4. The small rounded or nodular forms and the small granular pieces can be treated to advantage, but fine material and large masses require special treatment.

The principal minerals associated with manganese, as indicated in a previous chapter, are quartz and iron, mostly limonite; barite is of rare occurrence, although occasionally found. As the specific gravity of iron is similar to that of manganese it is practically impossible to effect separation mechanically; consequently it is common practice to hand pick the ore coming from the finishing apparatus when a product low in iron is desired. As a rule, however, no attempt is made to separate the small proportion of iron present in ores from the ordinary manganese deposit.

The specific gravity of quartz is about 2.65, but that occurring in connection with manganese deposits, particularly the sands resulting from the decay of impure limestone and other formations, is often rough and porous and has a specific gravity of 1.8 to 2. It is evident, then, that the average specific gravity of manganese that must be treated is about twice that of quartz, the principal impurity that must be separated. This difference is ample to insure clean separation.

PRACTICE IN CLEANING MANGANESE ORE.

The work of cleaning manganese ores as practiced in the various districts can hardly be considered standard except in general outline—that is, the use of grizzlies, log washers, screens, and picking belts, with a growing tendency toward employing jigs. Rarely are plants constructed fully equipped with the apparatus that have proven useful and satisfactory for treating manganese ores; few plants have a full installation of essential equipment, while a number are operating with the irreducible minimum of equipment, namely, a log washer and a screen, or simply hand jigs.

Practice in the treatment or cleaning of manganese ore varies from fair to bad with a wide range between, but as a whole is poor. Plants are modeled after others already operating, with the assumption, apparently, that because a plant is in operation it must be doing good work, which is often far from being true. However, the essential principles of a prevailing practice are likely to be sound and particularly adapted to the special conditions and needs of the district and are, therefore, worthy of careful consideration and adoption in part at least.

PRACTICE AT A VIRGINIA PLANT.

The Crimora mine is the largest manganese producer in Virginia and until recently has had the reputation of being the largest in the United States. The flow sheet given below shows the method employed in treating a high-grade but rather lean ore.

Flow sheet of Crimora mill, Virginia.

Wash dirt to *1*.

1. Grizzly, 4 by 8 feet, 80-pound rails, spaced 7 inches, to *2*.

2. Revolving screen, 60 inches in diameter by 12 feet long, 3½-inch perforations; oversize—clay to *3*, manganese to *4*; undersize to *5*.

3. Waste bank.

4. Blake crusher, 1½ inches, to *6*.

5. Log washer, double, 25 feet long; slope 1 inch to 1 foot; discharge to *6*; overflow to *7*.

6. Revolving screen, 9 feet 8 inches long by 48 inches in diameter, ¼-inch perforations; oversize by elevator to *8;* undersize to *9*
7. Settling pond.
8. Revolving screen, 6 feet long by 40 inches in diameter, 1-inch and 1¼-inch perforations; oversize by elevator to *10;* undersize—from 1-inch screen to *11,* from 1¼-inch screen to *12.*
9. Revolving screen, 9 feet 8 inches long by 48 inches in diameter. ⅜-inch perforations; oversize by elevator to *13;* undersize by elevator to *14.*
10. Picking belt, waste, 24 inches wide by 22 feet long; ore to *15;* waste to *16.*
11. Bull jig No. 1, two cells, 36 by 36 inches; gate discharges to *15;* overflow to *10.*
12. Bull jig No. 2, two cells, 36 by 36 inches; gate discharges to *15;* overflow to *10.*
13. Harz jig, three cells, 30 by 40 inches; gate discharges to *17,* hutches to *17,* overflow to 7.
14. Woodbury jig, two cells; gate discharges to *17;* hutches by elevator to *17;* overflow to 7.
15. Picking belt, 30 inches by 22 feet long; ore to *17,* waste to *10.*
16. Waste bin by cars to *3.*
17. Ore bin, finished ore.

PRACTICE AT A PLANT IN GEORGIA.

The flow sheet of the Aubrey plant of the Georgia Iron & Coal Co., given below, represents the best practice and most extensive operation in the Cartersville district, Georgia. The capacity of the plant is 50 tons of finished ore per day of 10 hours. There are three large open cuts that are connected with railroad tracks, so that strippings as well as wash dirt can be handled with equal facility. It is possible that certain parts of the plant might have been arranged to better advantage, yet taken as a whole the design and arrangement are fair.

Flow sheet of the Georgia Iron & Coal Co. plant at Aubrey, Ga.

Wash dirt to *1.*
1. Grizzly (four sets), five railway rails, four spaces of 1 inches each, length 7 feet 3 inches; oversize—rock to *2;* manganese broken by sledge to *3;* undersize to *3.*
2. Rock dump.
3. Log washers (four double logs, 30 feet long); discharge to 4, overflow to *5.*
4. Screen (perforated metal, openings 1 by 2 inches); oversize to *6;* undersize to 7.
5. Mud or settling pond.
6. Picking belts; rock to *2;* manganese and attached rock to *8.*
7. Belt conveyors to *9.*
8. Crusher reducing to one-half inch to *9.*
9. Ore bins by elevator or lift to *10.*
10. Revolving screens for jig feed, Nos. 3, 4, and 5; for jig No. 3, $\frac{1}{16}$-inch, ¼-inch, and ½-inch perforations; for jig No. 4, $\frac{1}{16}$-inch, ¼-inch, and ⅜-inch perforations; for jig No. 5, ¼-inch, ⅜-inch, and ½-inch perforations; products from these screens to *11.*
11. Jigs Nos. 3, 4, and 5 (four single cells); gate discharges to *12;* hutches to *13;* overflow to *14.*
12. Concentrates bin to *17.*
13. Flat screen, ½-inch perforations, for jig No. 1; oversize to cell No. 1 of *15;* undersize to cell No. 2 of *15.*

14. Revolving screen, for jig No. 2, $\frac{1}{4}$-inch, $\frac{3}{16}$-inch, and $\frac{1}{8}$-inch; oversize to cell No.
1, undersize to cells Nos. 2, 3, and 4, No. 2 jig.

15. Jig No. 1 (two cells); gate discharges to *12;* hutches to *12;* overflow to *5.*

16. Jig No. 2 (four cells); gate discharges to *12;* hutches to *12;* overflow to *5.*

17. Hand-picking table, where ore is sorted by attendants, who pick out iron and
other impurities. The iron is sold as manganiferous iron ore.

PRACTICE IN THE BATESVILLE DISTRICT, ARKANSAS.

There were 11 washers built in the Batesville district, Arkansas,
during 1917 and 1918. None of these did satisfactory work, largely
because the builders and operators did not appreciate the necessity
of providing screens for properly sizing the material treated on the
jigs.

The practice in this district is simple and may be outlined as follows:

Flow sheet used in Batesville district, Arkansas.

Wash dirt to *1.*

1. Grizzly, size 5 by 6 feet, 10 to 14 railroad rails spaced 4 inches; oversize—waste
rock thrown on dump, lump ore reduced to size that will pass through grizzly
to *2;* undersize to *2.*

2. Double log washer, 28 feet long, slope 1 inch to 1 foot; discharge to *3;* overflow to *4.*

3. Revolving screen, 6 feet long by 30 inches in diameter, with $\frac{1}{4}$-inch perforations;
oversize to *5;* undersize to *6.*

4. Waste bank.

5. Picking belt, 22 feet long by 24 inches wide; ore to *7;* waste to *4.*

6. Harz jig (four cells); size of cells 22 inches by 42 inches; gate discharges to *7;* hutches
to *8.*

7. Finished ore bin.

8. Harz jig (two cells), size 28 inches by 36 inches; gate discharges to *7,* hutches to *7,*
tails to *4.*

Rarely are more than two jigs employed in the plants for washing
manganese ores, and in many plants only one jig is used. Revolving
screens or trommels are commonly used, but flat screens are occasionally employed. In either case it is the exception rather than the
rule to find more than one size of opening, which ranges from $\frac{3}{4}$ inch
to $1\frac{1}{4}$ inches. The wide range in size fed to the jigs from such screens
renders good separation next to impossible. Few of the jig products
are clean, and they require extensive hand picking to complete the
preparation. This hand sorting could be obviated by simply rerunning the hutch products of the rougher jigs; fairly clean products
could then be made on the second or cleaner jigs. When only one
jig is employed, few or none of the jig products is suitable for market.
Fairly close sizing would greatly improve the work done by the jigs
and reduce the expense of preparation by eliminating hand picking.

Probably a dozen hand jigs have been in use in the district at
different times. As with the power jigs, incomplete separation of
waste was the almost universal rule, little or no attention being
paid to sizing of the feed.

PRACTICE AT A MILL AT PHILIPSBURG, MONT.

The mill of the Philipsburg Mining Co. is one of the largest in the West, having a capacity of 75 to 100 tons per day. The ores treated are the oxides mined at Philipsburg, Mont. The flow sheet of the mill follows:

Flow sheet of Philipsburg Mining Co. at Philipsburg, Mont.

Wash dirt (ore) to *1*.
1. Grizzly, 2-inch spaces; oversize to *2*, undersize to *3*.
2. Gyratory crusher, No. 4, 1¼ inches, to *3*.
3. Ore bin to *4*.
4. Revolving screen, 1½-mm. mesh; oversize to *5*, undersize by elevator No. 1 to *6*.
5. Picking belt; clean ore to *7*, rejected by elevator No. 1 to *6*.
6. Revolving screen, two sections—first, ⅜-inch mesh, second, ¼-inch mesh; oversize of first to second, undersize of first to *8*, oversize of second to *6*, undersize of second to *9*.
7. Clean-ore bin.
8. Revolving screen, 4-mm. mesh; oversize to *10;* undersize to *11*.
9. Bull jig; concentrates to *12;* overflow to *13*.
10. Middling jig; concentrates to *12;* overflow to *13*.
11. Revolving screen, 1½-mm. mesh; oversize to *14;* undersize to *15*.
12. Dorr classifier; spigot to *16;* overflow to *17*.
13. Rolls, ¼-inch space, by elevator No. 2 to *18*.
14. Fine jig; concentrates to *12*, overflow to *13*.
15. Dewatering tank; spigot to *19*, overflow to *17*.
16. Frue vanner; concentrates by No. 3 elevator to *20*, overflow to *17*.
17. Settling pond.
18. Revolving screen; 2½-mm. mesh, oversize to two coarse jigs, undersize to *11*.
19. Distributor to *21*.
20. Drier to *22*.
21. Wilfley tables; concentrates to *16*, tailings to *23*.
22. Concentrates bin.
23. Waste bank.

DESCRIPTION OF A STANDARD WASHING PLANT.

The cleaning of manganese ores by washing and jigging can readily be accomplished and that, too, with slight change in the present practice. The flow sheet of a plant properly arranged to do satisfactory work is given below.

Flow sheet of standard washing plant.

Wash dirt to *1*.
1. Grizzly, 2-inch to 4-inch spaces between bars; oversize rock to *2*; manganese by sledge to *3*; undersize to *3*.
2. Rock dump.
3. Log washer, double log, 20 to 30 feet long; discharge to *4;* overflow to *5*.
4. Revolving screen, cylindrical or conical, perforations or mesh ½-inch and $\frac{1}{16}$-inch; oversize, everything above ½-inch, to *6;* undersize, ½-inch to $\frac{1}{16}$-inch, to *7;* $\frac{1}{16}$-inch and smaller, to *8*.
5. Mud or settling pond.
6. Picking belt; rock to *2*; manganese to *9*.

7. Rougher jig, four cells; gate discharges to *9*; first hutch to *9*; second, third, and fourth hutches to *10*.

8. Sand jig, three cells; gate discharges to *9*; hutches to *9* or *10*.

9. Finished ore bin.

10. Cleaner jig, three cells; gate discharges to *9*; hutches to *9* or *11*.

11. Shaking table; finished product to *9*; tailing to *5*.

A standard washing plant having the above flow sheet may be called a single washer or unit, and has a capacity of 40 to 50 tons of finished ore per 10 hours. An increase in capacity of such a plant, aside from a small increase gained through crowding, would mean doubling, trebling, or quadrupling of most of the equipment listed, with corresponding increase in the volume of wash dirt treated.

By single washer is not meant a single log, but a single washer with double logs. Single logs do good work, yet in a plant as outlined, a double log should be employed.

The sizes of the various apparatus of a standard plant are indicated in the flow sheet; the sizes of screens and adjustments of the various apparatus for proper operation are given below.

LOGS.

The number of revolutions of logs varies from 12 to 15 per minute. The amount of water required varies widely with the nature of the bank dirt, but is usually 50 to 75 gallons per minute. The capacity likewise varies with the bank dirt; average dirt should yield 40 to 50 tons product per 10 hours. Roughly, 25 horsepower is required to drive logs.

SCREENS.

The number of revolutions is 15 to 20 per minute, being approximately the same for cylindrical and conical screens 36 to 48 inches in diameter. The power required for driving varies somewhat with feed and size of material handled, but is about one horsepower. The capacity of screens of the sizes given ranges from 45 to 55 tons per 10 hours for the smaller, and 50 to 75 tons for the larger. For the ordinary material treated the screen should be not less than 60 nor more than 72 inches long. If a larger number of sizes is desired, it might be better to shorten the length of the separate screening surface than to increase unduly the length of the whole screen.

PICKING BELTS.

The sizes of picking belts commonly employed, which have proved satisfactory for the usual range of work done, are: 30 feet long by 18 inches wide to 50 feet long by 30 inches wide. The speed of the picking belt ranges from 50 to 60 linear feet per minute. The capacity is limited by the character of the work; if the material sorted is largely ore and the waste rock is fairly coarse, the capacity is large, similarly where the percentage of ore is small and the individual pieces are large; but where the ore and waste are about equal in amount and there is a wide range in sizes, the capacity may be small. Occasionally two picking belts are used, one for coarse, the other for the finer materials, but such practice is neither necessary nor desirable in a standard plant.

JIGS.

The following data show the proper adjustment and operation of jigs:

Data regarding size and operation of jig equipment.

(Dimensions in inches.)

	Width.	Length.
Size of sieve compartments	20 to 24	30 to 38
Size of plunger compartments	18 to 22	30 to 38

Size of sieve openings:	1st cell.	2nd cell.	3rd cell.	4th cell.
Rougher jig	¼	¼	$\frac{3}{16}$	$\frac{3}{16}$
Sand jig	$\frac{3}{16}$	$\frac{3}{16}$	⅛	⅛
Cleaner jig	$\frac{3}{16}$	$\frac{3}{16}$	⅛	⅛
Length of stroke of jigs:				
Rougher jig	1½	1½	1¼	1¼
Sand jig	1	1	⅞	⅞
Cleaner jig	1	1	⅞	⅞
Number of strokes per minute:				
Rougher jig	150	150	150	150
Sand jig	200	200	200	200
Cleaner jig	200	200	200	200

Power for 4-cell jig, 2.5 to 3.0 hp.
Power for 3-cell jig, 2.0 to 2.5 hp.
Hydraulic water pipes: Main pipe 2 inches; branch pipes 1 inch.
Amount of water used by a jig, 350 to 400 gallons per minute.
Height of tailboard or dams, 3½ to 4½ inches.
Slope of bed (sieve) should not exceed 1 inch to length of bed; with heavy mineral, bed should be level. Slope should be against, not with movement of mineral.
Drop between tailboards or dams, 1 inch.
Height of gate discharge above bed, 2 to 2½ inches. This must be varied to suit conditions and character of material treated.

Jigs to have full or large capacity must be fed regularly; therefore jigs receiving the feed directly or indirectly from logs usually do unsatisfactory work for the reason that irregular feeding does not permit the maintenance of a bed of uniform depth. Further, the feeding of materials which vary from sandy to clayey in character also renders it difficult to keep a full bed; clayey ores tend to clog the jig beds and are preferably cleaned by trommels and jets of water.

With a suitably arranged plant, practically all materials treated may be handled by gravity, but in places elevators are required. Elevators, although adding to the equipment and the expense of operation, permit a better arrangement, in that more of the apparatus may be on the same level, and therefore more accessible.

Elevators are particularly useful for raising the hutch products from the rougher and sand jigs to the cleaner jigs. Should it be found desirable to employ a shaking table in reworking the hutch products from the cleaner jigs, elevators should also be employed in handling them.

TABLES.

It is doubtful whether tables can be used to advantage in the preparation of the usual run of manganese ore, as outlined and discussed. However, there is a well-defined field for their application in treating certain ores, such as the low-grade blanket deposits and breccia ores.

DRY CONCENTRATION.

Largely through the lack of an adequate supply of water a number of attempts have been made to clean manganese ores by dry methods. Probably the most extensive and elaborate equipment is that of the Southern Hill Manganese Co. on the Southern Hill property, in the Batesville district, Arkansas.

It is proposed to separate the ore from the clay and rock fragments by passing the material as excavated through a revolving drum, where it will be thoroughly dried and the clay pulverized. On being discharged from the drum the fine material will be removed by a revolving screen, which is a continuation of the drum. The coarse manganese and waste will then pass to a picking belt, where final separation is to be made. The smaller sizes of ore are to be treated on hand jigs. This plant has never been operated on a commercial scale; consequently no definite statement can be made regarding its practicability.

The water supply for manganese washing and concentrating plants is important, as many of the workable properties are often at considerable distance from an available supply. Pipe lines 3 to 4 inches in diameter are sufficiently large for a standard plant. The conservation of water by passing it through settling or impounding ponds usually solves the problem of water supply where the quantity of fresh water is limited. Once a plant has been put in operation the addition of 25 per cent of the total consumption not only provides for the loss due to wastage, but furnishes the required amount of fresh water for those steps of the process requiring clean water.

GENERAL SUMMARY OF CONDITIONS AFFECTING CONCENTRATION.

On account of the irregularity of manganese deposits and the uncertainty of an adequate supply of ore, great care should be exercised in connection with the various operations, particularly with respect to prospecting and mining. Similarly, preparatory to the erection of a washing or concentrating plant, the factors that have to do with the success of the work must be carefully considered.

The principal considerations affecting the cleaning of ores are:

1. The character and grade of ore.
2. The recoverable percentage of mineral.
3. Relative value of crude to cleaned ore.
4. Basis upon which royalty is paid, whether crude ore or the cleaned product.

Aside from the clay and other materials more or less intimately mixed with manganese considerable silica is associated with the ores. The silica may be "free" or "attached." The free silica can be readily removed by washing, but the attached silica, being embedded in the ore or attached to it can be separated, if at all, with difficulty.

High-grade ores, particularly when occurring in large masses, and soft ore, as pyrolusite, should receive the minimum preparation consistent with proper cleaning. Low-grade ores usually require much more careful treatment than the high-grade ores, and the work and expense of concentration depend largely upon the impurities present. Free silica is not difficult to separate from the manganese. Soft ore, or wad, although of high grade, is difficult to clean without great loss from fines, particularly when much fine sand is mixed with the clays and ore.

The recoverable percentage of mineral in the wash dirt depends largely upon the character of the mineral. Certain clays are readily broken and separated from the manganese, whereas others become pasty when washed, adhering tenaciously to the particles of mineral. As a rule, the larger the pieces of ore and the higher the grade the more readily is separation from the waste effected, owing probably to the smoother surfaces. Nodular ore of small and fairly uniform size is readily washed and jigged, but fragments from large masses and rough particles resulting from decay of limestone and possible partial solution of manganese are very difficult to clean satisfactorily.

The relative value of crude ore as compared with that of cleaned ore may be the deciding factor in determining whether a concentrating plant should be erected. During the past year considerable quantity of low-grade ores was shipped at a low price, simply because there was a market for it. The question is whether such ores could not have been raised in grade by concentration, so as to have brought a price that would have warranted the erection of a suitable plant. However, uncertainty as to extent of the deposits and the length of time the prevailing schedule of prices would be maintained did not foster experiments of this sort.

137338°—20——5

The grade of ore upon which royalties are assessed has been the cause of considerable trouble in different districts, but in most of the districts during the past year royalties were paid on all ores coming within the schedule unless otherwise specified. In the future high-grade ores will alone be subject to royalty charges.

In future careful mining in well-proved deposits of high-grade ore will be necessary in the various manganese districts. The ores mined will, in turn, require either close hand-picking or concentration in well-designed plants in order to produce a high-grade ore, low in silica and phosphorus. With a dependable output of such ore, it should be possible to continue operation in the face of foreign competition wherever freight rates to consuming furnaces are reasonably favorable.

CHAPTER 5.—LEACHING OF MANGANESE ORES WITH SULPHUR DIOXIDE.

By C. E. van Barneveld.

INTRODUCTORY STATEMENT.

Early in 1918 the Bureau of Mines, through its mining experiment station at Tucson, Ariz., began an investigation of possible methods of recovering manganese from the numerous deposits of low-grade, siliceous, secondary manganese ores that are scattered throughout Arizona, New Mexico, and Utah. The aggregate quantity of such ore is large, but as a rule these ores are not amenable to ordinary methods of concentration, and the proportion of ore that can be brought to shipping grade by sorting and screening is usually small. Much of this ore is found in localities tributary to the copper-smelting districts where waste sulphurous roaster gases are available. The possibility of using sulphur dioxide as a leaching agent for manganese had long been recognized, and it was thought that a process developed at the Tucson station for leaching certain copper ores with hot sulphurous fumes might be successfully applied to this type of manganiferous material. In May, 1918, some leaching tests were made at this station on manganese ores from the Patagonia district in Arizona. The results were sufficiently encouraging to lead to an extensive program of research into the possibility of developing an SO$_2$-leaching process which would produce a high-grade sinter containing over 60 per cent metallic manganese in the form of oxides, free from silica and phosphorus. The investigation had not been completed when the economic situation as regards manganese was entirely changed by the armistice and further experimentation was abandoned. However, sufficient work had been done to warrant certain conclusions. As these may have a scientific interest and a bearing upon other chemical and metallurgical operations, the following brief report is presented.

The work was undertaken with due recognition of certain facts:

1. That prompt action was essential.
2. That the process would probably have to be applied at smelting centers where sulphurous fumes were available as waste products, and that the cost of transportation of ores to the smelter would be a limiting factor in its application.

57

3. That a plant of 250-ton capacity would be the desirable unit and that in all probability two such units would be the maximum installation for any one smelter.

4. That from 4 to 6 months would be required for construction and that the cost of the plant would have to be written off over an operating period of 12 or even 6 months.

5. That the cooperation of the copper-smelting companies in Arizona and Utah and of the large consumers of manganese would be necessary in order to arrange for prompt plant construction; also that it might be possible and necessary to use public funds.

The investigation was carried through the laboratory stage on a sufficiently large scale to warrant as a wartime measure the construction on the results obtained of a commercial unit. On October 1 negotiations were under way for the erection of a 250-ton unit at one of the copper smelters, in cooperation with a large producer of lead-zinc-silver-iron-manganese ore in which only the fluxing value of the manganese was being realized. While this negotiation was in progress the rapid improvement in the military situation so changed the outlook that the undertaking was abandoned before final plans were drawn.

RESULTS OF TESTS.

The tests made included ores from the Clifton, Bisbee, and Patagonia districts in Arizona; also representative ores from New Mexico, Utah, Nevada, and California. The results of the tests are presented in Table 10 following:

TABLE 10.—*Results of tests of manganese ores by sulphur dioxide leaching with continuous-drum process.*

Source of ore.	Mn in heads.	Weight of ore taken for test.	Size of ore (mesh).	Pulp ratio.	Temperature of solution.	Rate of feed per hour.	Mn in tails.	Percentage of Mn recovered.	Remarks.
	Per cent.	*Pounds.*			*° C.*	*Pounds.*	*Per cent.*		
Parker, Ariz.	11.9	500	20	2-1	50	200	1.25	89.5	Highly siliceous ore containing 58 per cent SiO₂, 4 per cent Fe, 5 per cent CaO.
Bisbee, Ariz.	15.4	200	20	2-1	51	100	4.2	73	High lime ore (33 per cent CaO).
Do.	16.6	200	20	2-1	53	100	6.3	58	High iron ore (32 per cent Fe).
Do.	15.2	200	20	2-1	50	100	2.56	83	General ore.
Tombstone, Ariz.	7.42	100	20	4-1	47	100	1.85	75	Highly siliceous ore (39 per cent SiO₂).
Do.	16.6	102	20	4-1	47	50	.99	93.9	Silver-copper ore.
Do.	20.1	101	20	4-1	45	50	.85	95	Do.
McCarty, Delta, Utah	30.4	100	20	4-1	50	50	3.4	90	Contained 39 per cent SiO₂, 7 per cent Fe, 16 per cent CaO.
Hardshell, Patagonia	26.3	250	30	2-1	50	200	4.2	84	Silver ore (10 ounces Ag per ton).
Prince consolidated ore, Pioche, Nev.	11.5	100	20	2-1	50	100	.6	94.8	Lead-silver-gold ore containing 35 per cent Fe, 4 per cent CaO, 11 per cent SiO₂.
Wad ore, California.	33.		20	4-1	50		2.3	93	Sample by C. E. van Barneveld.
Montgomery, Nev.	13.3	100	20	4-1	46		5.5	59	High lime ore (31 per cent CaO), containing considerable manganese as carbonate and silicate.

OBSERVATIONS ON RESULTS OF TESTS.

The following observations, based entirely on work done with dilute hot sulphurous gases (under 6 per cent SO_2) used countercurrent to the flow of the pulp, are presented:

THE ORE.

The general run of western ores may be described as impure psilomelane resulting from alteration of silicate or carbonate. The impurities may include numerous substances such as lime, silica (which may or may not be combined), and iron in the form of limonite or siderite.

Such an ore is readily leached with SO_2 and need rarely be crushed finer than 20 mesh to insure a good extraction.

THE PULP.

The pulp should be broken into a fine spray to insure contact between the mineral particles and the reagent. A leaching drum was designed at this station for the lixiviation of nonsulphide copper ores which gives the necessary agitation and contact.

A patent (U. S. patent No. 243015) was applied for and was granted on November 8, 1918.

Briefly, this apparatus comprises a drum set horizontally, to rotate slowly. The interior is divided into compartments by a series of transverse partitions, each compartment having longitudinal baffles and peripheral lifters. The transverse partitions are perforated so that the pulp may pass through the successive compartments to the end or discharge compartment. The pulp is fed into the first compartment and drops to the bottom of the compartment; as the drum rotates the peripheral lifts raise the pulp until it is spilled onto the horizontal staggered baffles below in such manner that the pulp is splashed and distributed over the surfaces of the bars and against the transverse partitions in fine descending drops and particles, thereby insuring intimate and prolonged contact between the ore particles and the countercurrent sulphurous and oxidizing gases. With a constant feed a flow is established within the drum, so that the pulp is gradually passed through the successive compartments into the last or discharging compartment, the lifters of which raise the pulp and drop it into a discharge or exit pipe whereby it is conducted through a trap opening into a discharge launder.

The pulp enters one end of the drum cold, and the SO_2-charged air or gas enters the opposite end hot. The countercurrent flow developed results in the pulp becoming progressively warmer and the gas becoming correspondingly cooler, until the pulp at the discharge end is heated to any desired temperature. When SO_2 is introduced as a hot dilute gas containing less than 6 per cent SO_2 by volume (the balance of the gas being largely air, which has lost part of its oxygen and which may contain the various impurities commonly found in roaster gases) and is projected into a fine spray of hot pulp or solution, absorption of SO_2 is practically negligible and the hot pulp or solution will contain practically no free SO_2 and will discharge from the drum in a practically neutral condition.

SOLUBILITY OF THE MANGANESE MINERALS.

The various higher manganese oxides, especially pyrolusite, psilomelane, and wad, were found to be readily soluble in the hot sulphurous acid formed by introducing hot air or furnace gas containing 2 to 6 per cent SO_2 (by volume) countercurrent to a pulp flow having a consistency of 2 of water to 1 of ore crushed to 20 mesh. Manganite and braunite are not commercially soluble.

Silicates of manganese are insoluble in sulphurous acid at atmospheric pressure and only slightly soluble under pressure.

Carbonates of manganese may, from an operating standpoint, be considered insoluble. The reagent slowly attacks and decomposes the carbonate, but some of the dissolved manganese at once reprecipitates as a sulphite; this precipitate coats the remaining undissolved carbonate and effectually prevents further dissolution.

IRON.

The effect of SO_2 on iron minerals differs markedly from its effect on manganese minerals: Iron in the oxide form, such as magnetite, hematite, and limonite, is practically insoluble in SO_2. Metallic iron and iron in the form of carbonate (siderite) is readily soluble. This iron will be in solution as a ferrous sulphite or sulphate; it will, however, precipitate as a basic ferric compound in the presence of either $CaCO_3$ and oxygen or $ZnCO_3$ and oxygen; it will also be similarly precipitated from solution so long as there is undissolved MnO_2 in the ore charge.

PHOSPHORUS.

In some forms phosphorous dissolves readily in SO_2. No large scale tests in the drum were made on high phosphorus ores. Laboratory tests, however, indicate that by the use of tandem drums phosphorus may be eliminated in the following manner: The phosphorus will not be dissolved in the first drum (the discharge from which still contains much undissolved manganese and other bases). Whatever dissolved phosphorus is discharged from the second drum will return to the ball mill. The addition of large quantities of fresh ore will at once neutralize these solutions, and the phosphorus will thereupon precipitate. If sufficient soluble iron and phosphorus are present to cause the building up of these elements by resolution in the drum after precipitation in the ball mill, then the filtrate from the second drum may be passed over any available carbonates, such as $CaCO_3$, prior to being returned to the ball mill.   The iron and phosphorus in solution would then be precipitated and eliminated from the circuit.

LIME.

In the presence of an appreciable amount of MnO_2 the concentration of free SO_2 is insufficient to decompose $CaCO_3$.   As dissolution of MnO_2 progresses, some $CaCO_3$ will be dissolved and precipitated as $CaSO_4$ (insoluble).   The catalytic action of the iron and manganese oxides will convert a small proportion of the entering SO_2 to SO_3; the resulting sulphuric acid will convert $CaCO_3$ to $CaSO_4$.   Some of the $CaCO_3$ particles become coated with gypsum and are not further acted upon by the acids in the drum.

ZINC AND COPPER.

Zinc and copper in nonsulphide form are readily dissolved by SO_2, but sulphides of these metals are not attacked.

TREATMENT OF THE PREGNANT SOLUTION.

The discharge from the drum will consist of a pulp containing in solution as sulphates the manganese, zinc, copper, and perhaps some iron; as solids the gangue, the sulphides, the other insoluble minerals, and any dissolved lime reprecipitated as $CaSO_4$.

The pregnant solution may be separated from the solids by filtration and washing in pressure filters.

Any copper present in the filtrate may be recovered by precipitation on iron.

The manganese-zinc content is removed together, either by crystallization, in large-scale permanent installations, where the lowest operating costs are to be sought, or by evaporation in pans with artificial heat in smaller or in temporary installations, where minimum installation expense would regulate the choice.

The resulting sulphate is roasted for two hours, more or less, in a rotary type of clinker furnace at a temperature ranging from 825° to 1,000° C.

The product from the roaster is a hard, compact clinker which in the zinc-free ores will run from 60 to 64 per cent manganese.   Any remaining manganese sulphate may be removed by water leaching; this leaching does not seem to cause the cake to crumble.   The waste heat from this operation is used to heat the cast-iron evaporation pans.   The SO_2 gas may be used over again as a solvent if desired.

The manganese sulphate commences to decompose around 700° C. and should break up completely at 830° C.   It was found that in order to effect complete decomposition of the sulphate with reasonable quickness (say in 30 minutes) the roast should be preformed at considerably higher temperature—around 950° C. Plenty of oxygen is necessary. In a reducing atmosphere some manganese sulphide will be formed; once formed, this can only be broken up by first reoxidizing to the sulphate.

CONDITIONS ESSENTIAL IN COMMERCIAL PRACTICE.

Three essential conditions must be complied with in commercial operations.

(a) In order that dissolution may be effected rapidly a reasonable strength of gas is required—not less than 2 per cent and up to 6 per cent of SO_2—the balance of the gas being heated furnace products.

(b) Heat aids dissolution. It is desirable to have an admission temperature of gas which will give the discharging solutions a temperature of 40° C. minimum to 60° C. maximum.

(c) The time of contact is regulated by the rate of flow of pulp through the drum. The efficiency of extraction is directly proportionate to the pulp surface exposed to the gas. This in turn depends upon proper agitation and breaking up of the pulp into a very fine spray which must be so directed countercurrent to the gas that no particle of mineral can escape contact with the reagent. This can be satisfactorily accomplished in 15 to 30 minutes in the drum described.

LEACHING IN TWO STAGES.

Manganese sulphate solutions are near the point of saturation when they contain 11 per cent manganese (specific gravity 1.5). As the solution approaches the point of saturation the rate of dissolution decreases markedly. On the other hand, maximum concentration of solutions minimizes the expense of subsequent evaporation to dryness, which is the next step in the process. Therefore, leaching should be performed in two stages in two tandem drums as indicated on the flow sheet shown in figure 1.

Wet crushing in ball mills is advised in circuit with drag classifiers, the filtrate from the second drum discharge being used to pulp the ore.

The discharge from the first drum will be a practically neutral pulp made up of partly leached ore, which will still contain considerable undissolved manganese, and a solution containing $MnSO_4$ near the point of saturation, and also $ZnSO_4$ and $CuSO_4$ if nonsulphide zinc and copper are present in the ore. This solution will not contain iron. This pulp is then filtered on a pressure type of filter and the various products are disposed of as follows:

(a) The filtrate is treated for recovery of valuable metals as previously described.

(b) The filter cake and wash water are repulped in a pulping tank, together with the wash water from the second drum filtration, to form the feed for the second drum, where the remaining SO₂-soluble minerals, including perhaps some iron, are dissolved.

The discharge from the second drum is filtered and the various products are disposed of as follows:

(a) The residue or filter cake goes to waste, or to a briquetting plant in case the lead-silver or copper-silver or other values warrant further treatment such as smelting.

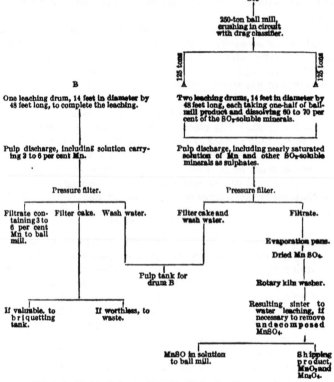

FIGURE 1.—Flow sheet of proposed plant for SO₂ leaching of manganese ores.

(b) The filtrate, which may contain up to 5 or 6 per cent manganese, goes to the ball mill to pulp the original ore for the first drum.

(c) The washwater from filtration goes to the pulping tank in which the filtercake from the first filtration is pulped prior to being fed into the second drum.

The crushing, leaching, filtration, and washing are thus carried on in a circuit. Any iron dissolved in the second drum is returned to the ball mill and is there immediately precipitated.

FILTERS.

A pulp made of 20-mesh ore and a saturated, slimy $MnSO_4$ solution presents an interesting filtration problem. Tests on a semi-commercial scale proved that the vacuum type of filter could be used only by the addition of filter cells and by dilution of the $MnSO_4$ solution considerably below the saturation point. In a series of tests made under the direction of A. W. Hudson, of the Phelps Dodge Corporation, on manganiferous silver ore from Tombstone, Ariz., vacuum filters proved inadequate. The addition of filter cells partly nullified the concentration of silver values obtained by extraction of the manganese. The dilution of the pregnant liquid appreciably below the point of saturation increased the cost of subsequent evaporation of $MnSO_4$.

A test made on a pressure type of filter gave the following results: The quantity of pulp passed through the filter contained 3,200 pounds of solid matter and yielded 4,000 pounds of cake containing 20 per cent water. The filter yielded 6.6 pounds of filter cake (20 per cent water) per square foot per hour. On this basis a plant to treat 250 tons of dry ore would yield 312.5 tons of filter cake (20 per cent water) and would require $(312.5 \times 2,000) \div (6.6 \times 24) = 3,946$ square feet of filtering surface.

CORROSION.

Corrosion of metal in contact with the solution was found by experiment to be due entirely to the action of ferrous sulphate and liberated sulphuric acid. Leaching in two stages presents a simple solution of this difficulty. As sulphur dioxide has a marked affinity for manganese, conditions in the first drum may be so regulated as to avoid the dissolution of any iron, and the resulting pregnant liquid will not corrode the evaporation pans. The solution from the second drum may be neutralized and the dissolved iron may be precipitated over carbonates prior to the return of this solution into the crushing end of the circuit, or this may be done in the ball mill.

SULPHUR DIOXIDE.

Sulphur dioxide for leaching purposes may be obtained (a) by burning flower sulphur in sulphur burners; (b) by roasting high-grade iron pyrites containing sufficient copper or silver to give the calcines a market value; (c) as a waste product from roasting furnaces at a copper smelting plant.

Approximately three-fourths of a unit of sulphur is required on an average for each unit of manganese, iron, zinc, and lime dissolved. Most of the sulphur dioxide consumed in leaching is recovered in the final roasting of the sulphate and can be reused. Sulphur in combination with lime is lost. In view of the plant requirement, the relatively small size of the average western manganese deposit, and the uncertainty of the manganese market over a period of years, it would seem that with sulphur dioxide reduction of straight man-.ganese ores must be confined to favorably situated copper smelting plants where sulphur dioxide of the proper strength is available as a waste product.

COMPLEX ORES.

The foregoing observations relate largely to ores in which manganese is the principal constituent. Some study was also given to the more complex manganiferous ores, such as silver ores, lead-silver ores, copper-silver ores, and zinc ores, containing sufficient manganese and occurring in sufficiently large quantities to warrant their consideration as a possible emergency source of manganese. Much laboratory work was done in an attempt to determine the relative solubility in sulphur dioxide of the various minerals that may be present in a complex ore, in the hope that some selective action might be discovered which would effect a separation between manganese and zinc. However, it was impossible to reproduce or to maintain in the drum the conditions that in the laboratory might produce certain preferential action on oxides over carbonates. Under operating conditions a large excess of acid is always present in some part of the drum; and the result is that the various SO_2 soluble manganese, zinc, and copper minerals are attacked practically simultaneously. If anything, zinc minerals (with the exception of crystalline zincite) will be dissolved fastest; the higher manganese oxides are next in the scale and the copper minerals last.

As the three minerals soluble in SO_2, copper, zinc, and manganese, are recovered together as sulphates in the filtrate, their separate recovery from sulphate solution was investigated. Copper offers no difficulty; it may be recovered by passing the solution over iron. In order to retain the manganese in solution during the precipitation of the copper, the specific gravity of the pregnant solutions in treating copper-bearing ores, must be lessened. The copper in solution will be replaced with iron which will, of course, result in a slightly lower manganese content in the sinter. Some preliminary work was done on the separation of zinc and manganese, which may be briefly summarized as follows:

(a) A straight oxidizing roast around 1,050° C. converts the sulphates to oxides. If these oxides are mixed with powdered coal and are then heated in a retort to 1,500° C., the zinc volatilizes with a recovery of 80 per cent of condensed metallic zinc.

(b) A 60 per cent zinc recovery as oxide is possible by subjecting the sulphates to a straight oxidizing roast until converted to oxides and then adding coal or other reducing agent and continuing the roast in the presence of air at 1,350° C.

(c) A chloridizing roast will cause the zinc to volatilize as chloride. If sodium chloride is used the soda fuses with the mass, thus preventing complete chloridization. If calcium chloride is used there is no fusion. The lime is converted to sulphate (insoluble in SO_2), part of the manganese becomes oxidized and part remains as sulphate. The latter is water-soluble and the oxidized manganese can be extracted by redissolving it with SO_2, an operation which calls for too much manipulation and expense to be commercially practicable except under very favorable conditions.

(d) Electrolysis does not seem practicable. The zinc goes to the cathode and the manganese to the anode. The H_2SO_4 set free at both poles attacks the zinc. Apart from this, there are the troubles engendered by foul solutions.

This investigation is incomplete and should receive further attention.

COSTS.

GENERAL ESTIMATE.

The cost of a 250-ton unit erected at the Salt Lake smelters or at the Douglas, Arizona smelter, would in normal times be roughly $100,000. A 500-ton unit should not cost to exceed $150,000. Under present (1918) conditions a 250-ton unit would cost perhaps as much as $200,000. The operating cost would be $12 to $14 per ton of shipping product based on an 80 per cent recovery of manganese from an ore which has a manganese content of 20 per cent. This includes crushing, leaching, and filtering 4 tons of ore; drying the filtrate and roasting the dried manganese sulphate. The resulting shipping product would be a sinter, free from silica, alumina, and phosphorus, which in the ores free from nonsulphide zinc or copper would contain 60 to 64 per cent manganese as oxides.

The cost of SO_2 treatment per ton of ore at a smelter where waste SO_2 gases are available should fall well within the estimated cost of $3.50 indicated in the following table.

Estimate cost of treating one ton of manganese ore.

Crushing in gyratory and ball mill	$0.40
Disposal of tailings	.10
Operation of drums and filtration plants	.15
Evaporation of $MnSO_4$ and roasting (heat, labor, and supplies)	[a] 1.50
Incidental handling	.25
Power	.30
Water	.10
General overhead	.20
Charge for SO_2	.50
	3.50

[a] This item is high.

CALCULATED ESTIMATES FOR TYPICAL ORES.

The following estimates cover typical examples in Utah and Arizona. A railroad freight rate of $12 per long ton on shipment of sinter to eastern points has been assumed.

EXAMPLE 1.

Assume that a large siliceous ore body is located 10 miles from the main line of a railroad. The manganese content varies from 15 to 40 per cent, and the silica, 30 to 20 per cent, the balance being alumina, lime, and a little iron, with no zinc or copper. Average manganese content of the run-of-mine ore is 20 per cent, and all the manganese is soluble in SO_2.

Four tons of ore will (on an 80 per cent recovery basis) produce 1 short ton of high-grade sinter containing 60 per cent manganese, no silica, alumina, or phosphorus. The price, according to May, 1918, schedule is $1.30 per unit plus $6.50 silica premium = $84.50 per long ton, f. o. b. Chicago, or $84.50 − $12 freight = $72.50 per long ton f. o. b. smelters. The costs will be as follows:

Market value of one short ton shipping product, f. o. b. smelters, based on above schedule.	$64.74
Cost of mining 1 ton of ore	$2.00
Haulage to railroad at 30 cents per ton-mile	3.00
Loading, sampling, switching and general handling	1.00
Railroad freight to smelter	2.50
SO_2 treatment charge	3.50
SO_2 plant extinguishment	2.00
Cost producing and treating 1 ton of ore	14.00
Cost of producing 1 short ton of shipping product	56.00
Net profit on 4 tons of ore	8.74
Net profit per ton of ore mined	2.18

It is assumed that the market will hold for 12 months after completion of the SO_2 plant, or that 12 months' production has been contracted for, and that the entire cost of the plant (at war-time cost of $200,000) must be charged off against 100,000 tons of ore. The plant-extinguishment charge will therefore, be $2 per ton, as given in the table.

EXAMPLE 2.

Assuming that the shipping grade of the ore in example 1 could readily be raised to 25 per cent manganese at an additional mining cost of 50 cents per ton and to 30 per cent manganese at an additional cost of $1 per ton, the returns would be, respectively, as follows:

Four tons of 25 per cent ore will yield 1.19 long tons of 60 per cent manganese sinter, having a market value of [($84.50 − $12 freight)] × 1.19 = $86.30. The profit under these conditions would be [($86.30 − ($56 + 2)] ÷ 4 = $7.08 per ton of sorted ore at the mine.

Four tons of 30 per cent ore will yield 1.4 long tons of 60 per cent manganese sinter, having a market value of ($84.50−$12 freight) ×1.4=$101.50 f. o. b. smelter. The profit would be [($101.50− ($56+4)]÷4=$10.37 per ton of sorted ore at the mine.

Evidently both these sorted ores would stand a larger haulage charge.

EXAMPLE 3.

Assume that a manganiferous lead-silver ore has a lime-silica gangue with sufficient iron and lime to balance the silica after removal of the manganese. The manganese content is 20 per cent. This has a fluxing value of $1.40. The removal of the manganese effects a shrinkage of smelter tonnage, which is in turn offset by a briquetting charge on the lixiviation tailings. The lead-silver contents are assumed to have a net value of $4 per ton of original ore if the manganese is removed and $6.40 per ton of original ore if smelted without removal of manganese.

Assuming the same general conditions and prices as obtained in example 1, it is evident that this would be a highly profitable wartime operation, as the $4 smelter return per ton of original ore would be added to the $2.18 profit on the manganese content per ton of ore mined, making a total profit of $6.18 per ton of ore mined. An otherwise unprofitable operation would thus be made highly profitable and an important source of manganese would be developed.

EXAMPLE 4.

Assuming that under normal economic conditions there would be a ready eastern market for a 60 per cent manganese sinter at 75 cents per unit, the value of the manganese f. o. b. smelter would shrink to about $28 and the maximum return per ton of original ore would be $28+$16=$44.

Under normal conditions there would be an appreciable scaling down of operating costs and this operation would show a small profit as follows:

Market value at smelter of the manganese and other metals in four tons of ore	$44.00
Mining per ton of ore	$1.50
Haulage to railroad at 20 cents per ton-mile	2.00
Loading, sampling, and general handling	1.00
Railroad freight to smelter	2.50
SO_2 treatment charge	3.00
SO_2 plant extinguishment and interest on investment (based on 4 years)	.50
Cost of producing and treating 1 ton of ore	10.50
Cost of mining and treating 4 tons of ore necessary to produce 1 short ton 60 per cent manganese shipping product	42.00
Profit on 4 tons of ore at mine	2.00
Profit on 1 ton of ore at mine	.50

A slight increase in either the manganese content or in the lead-silver content on the one hand or a decrease in transportation costs on the other hand would show a satisfactory balance.

POSSIBLE FUTURE APPLICATION OF SO₂ METHOD.

The war-need incentive to further investigation of this question by the Bureau of Mines no longer exists. It is probable, however, that there are certain properties where the foregoing information might be profitably applied, as these figures offer some encouragement when considered in connection with fairly large ore bodies which, in addition to the manganese, may contain other valuable minerals, such as (a) 1 to 2 per cent of copper (whether sulphide or nonsulphide); (b) sufficient lead or silver, or combined lead and silver to leave a residue that would pay to smelt; (c) sufficient zinc to warrant separating the manganese and the zinc at the initial treatment plant or to warrant experimentation along the lines of electric smelting and subsequent recovery of the zinc from the slag.

CHAPTER 6.—THE JONES PROCESS FOR CONCENTRATING MANGANESE ORES; RESULTS OF LABORATORY INVESTIGATIONS.[a]

By Peter Christianson and W. H. Hunter.

INTRODUCTORY STATEMENT.

During the summer of 1918 a "direct-reduction" process for treating manganese ores, known as the Jones process from the name of its inventor, John T. Jones, was investigated at the laboratory of the Minneapolis experiment station of the Bureau of Mines. This work was undertaken because the process offered possibilities for concentrating the manganese in the manganiferous iron ores of the Cuyana Range and elsewhere.

The importance of this work in relation to war needs has passed, but a record of the investigation may be of general interest to the industry.

The aim of the Jones process is to effect by metallurgical means a separation and consequent concentration of the manganese in ores in which the manganese is so intimately associated with the iron that ordinary methods of gravity or magnetic separation are impracticable.

The process consists of two stages. In the first, called the low-temperature reduction stage, two products are made, namely, (a) metallic iron, suitable for direct use in the manufacture of steel, and (b) slag. Most of the manganese is concentrated in the slag. Separation of these two products, as outlined by the inventor, is accomplished by grinding the sinter and passing it over a magnetic separation apparatus, producing thereby a high-grade magnetic iron concentrate, and a nonmetallic manganiferous sinter. This method was used in the preliminary tests. In the final test, however, separation was accomplished by pouring the material under treatment in a liquid condition and recovering the iron metal in the form of a button. This last method of separation, developed at the Minneapolis station of the bureau, has obvious advantages, and should the process become a commercial success, would doubtless be the method used.

The second stage, which may be called the high-temperature stage, involves smelting the manganiferous slag or sinter derived from the first stage, to produce a manganese alloy.

---

a Prepared in cooperation with the school of mines of the University of Minnesota.

LOW-TEMPERATURE REDUCTION.

PRELIMINARY TESTS.

OUTLINE.

The ore used in preliminary tests was typical of those locally known as disseminated Cuyuna ores, containing approximately 24 per cent iron, 18 per cent manganese, and 26 per cent silica. The following temperatures were tried: 1050°, 1150°, 1250°, 1300°, and 1350° C.

The time was varied from one-half hour to three hours.

The reducing agent used was Elkhorn coal, containing 6.6 per cent ash and 40.8 per cent volatile matter; the amount was varied from 10 to 50 per cent of the ore.

APPARATUS USED.

The apparatus used · in the tests were as follows:

Dixon graphite crucibles, size No. 2 for 200 grams of ore, and others in proportion to size of charge.

Gas muffle furnaces for temperatures of 1050° and 1150° C.; oil-fired muffles for 1250° C.; electric carbon resistor furnace for 1300° and 1350° C.

Pyrometers: Hoskins base-metal couples and Hoskins meters for temperatures up to and including 1250° C.

Platinum-and-platinum-rhodium couples and Leeds & Northrup potentiometer for temperatures above 1250° C.

Hand-crushing apparatus, screens, hand magnets, and panning apparatus for separating magnetic from nonmagnetic material.

The temperatures indicated are those of the muffle outside the crucibles. The temperatures of the charges within the crucibles were probably nearly the same, except as regards those charges heated only 30 minutes. Temperatures of heats above 1250° C. were taken within the charge. Readings were generally taken every. five minutes.

PROCEDURE.

CHARGING AND HEATING.

When muffle furnaces were used the ore and the reducing agent, crushed to pass 10 mesh, were mixed and charged into cold crucibles; these were placed in the muffle heated to nearly the desired temperature. When crucible furnaces were used the charge was put into crucibles heated to nearly the desired temperature. By using the muffle furnaces a number of crucibles could simultaneously be subjected to practically the same temperatures, thus keeping this variable constant; and a variation in time could be obtained by removing the crucibles at the end of different periods of time.

DISCHARGING.

At the end of the reduction period the contents of the crucibles were discharged into cast-iron molds, so as to effect a rapid cooling.

SEPARATION.

After cooling, the content of each crucible was generally separated into the following:

1. Metal +30 mesh } magnetic part.
2. Metal −30 mesh }

3. Slag } nonmagnetic part.
4. Carbon. }

This separation was effected by a series of crushings, screenings, and magnetic separations, frequently supplemented by panning. Each of these parts, if panned, was weighed and sampled for analysis.

ANALYSES.

The determinations generally made were: Per cent soluble iron; per cent total iron; per cent soluble manganese; per cent total manganese; per cent insoluble.

CALCULATIONS.

The results obtained from the analyses, together with the weights of the various parts, were combined by calculations to obtain the metal content and percentages of iron and manganese recovered in each of total magnetic and nonmagnetic parts.

The following formulas were used in calculations:

Formulas used in calculating recoveries.

Percentage Extraction.

For magnetic part.

Grams Fe.

Grams of magnetic part+30×per cent Fe content=...................... ....
Grams of magnetic part−30×per cent Fe content=...................... ....
Sum=total grams Fe in magnetic part=............................ ....

$$\frac{\text{Grams Fe in magnetic part}\times100}{\text{Grams Fe in charge}}=\text{per cent extraction of Fe.}$$

For nonmagnetic part.

Grams Fe.

Grams of slag×per cent of Fe content=............................... ....
Grams of carbon×per cent of Fe content=............................ ....
Sum=total grams Fe in nonmagnetic part=.......................... ....

$$\frac{\text{Grams Fe in nonmagnetic part}}{\text{Grams Fe in charge}}\times100=\text{per cent extraction Fe.}$$

Recovery Per 100 Units of Ore.

Per cent Fe or Mn in ore×per cent extraction=Recovery per 100 units of ore.

Per Cent Soluble Iron in Concentrates.

Grams soluble Fe.

Per cent soluble Fe×grams of magnetic part+30=...................... ....
Per cent soluble Fe×grams magnetic part−30=.......................... ....
Sum=Total grams soluble Fe in magnetic part=...................... ....

$$\frac{\text{Sum}\times100}{\text{Grams of magnetic part}}=\text{per cent soluble Fe in magnetic part.}$$

Per Cent Soluble Iron in Soluble Part of Magnetic Part.	Grams soluble matter.
Per cent soluble matter $\times$ magnetic $+30=$...............................	....
Per cent soluble matter $\times$ magnetic $-30=$...............................	....
Sum $=$ Total grams soluble matter in magnetic part $=$................	....

Total grams soluble Fe in magnetic part $\times 100=$ per cent soluble Fe in soluble part of magnetic part.

$$\frac{\text{Total grams soluble Fe in magnetic part}}{\text{Grams soluble matter in magnetic part}} \times 100 = \text{per cent soluble Fe in soluble part of magnetic part.}$$

Results obtained in the tests are given in Table 11.

TABLE 11.—*Results of preliminary low-temperature reduction tests with the Jones process.*

RESULTS WITH VARYING TEMPERATURES.

(a) DATA.

Test No..	16-D.	17-D.	35-D.	31.	32.
Temperature, °C................................	1,050	1,150	1,250	1,300	1,350
Ore, grams.....................................	363	363	160	363	363
Coal, grams....................................	91	91	40	91	91
Time, minutes..................................	120	120	120	120	120

(b) EXTRACTION.

Temperature °C.	Product.	Weight, grams.	Iron.		Manganese.	
			Total, per cent.	Recovery, per cent.	Total, per cent.	Recovery, per cent.
	Ore...............................	363.0	24.36	....	18.12	......
1,050	Magnetic part	233.5	32.0	84.7	19.8	70.5
1,050	Nonmagnetic part	84.6	5.7	5.5	19.3	24.8
1,150	Magnetic part	131.4	58.5	86.9	15.0	30.0
1,150	Nonmagnetic part	189.4	8.18	17.5	20.9	60.2
a 1,250	Magnetic part	40.5	90.6	93.9	3.2	4.5
a 1,250	Nonmagnetic part	85.3	2.0	4.4	28.9	84.8
1,300	Magnetic part	94.8	87.2	93.5	7.0	16.0
1,300	Nonmagnetic part	209.0	2.4	5.7	26.2	83.0
1,350	Magnetic part	96.6	80.4	91.0	9.4	14.3
1,350	Nonmagnetic part	202.2	4.4	10.0	23.7	72.8

a In test 35-D, 160 grams of ore was used.

(c) RECOVERY.

Test No.	Variable temperature, °C.	Iron.			Manganese.		
		Grams in 100 grams of ore.	Recovery.		Grams in 100 grams of ore.	Recovery.	
			Per cent.	Grams.		Per cent.	Grams.
16-D......................	1,050	24.36	84.7	20.5	18.12	70.5	12.80
17-D......................	1,150	24.36	86.9	21.1	18.12	30.0	5.44
35-D......................	1,250	24.36	93.9	22.7	18.12	4.5	0.80
31.........................	1,300	24.36	93.5	22.6	18.12	5.7	1.03
32.........................	1,350	24.36	91.0	22.0	18.12	14.3	2.00

RESULTS WITH VARIATION OF TIME, AT 1,150° C.

(a) DATA.

Test No...............................	17-A.	17-B.	17-C.	17-D.	17-E.
Time, minutes..........................	30	60	90	120	150
Ore, grams.............................	363	363	363	363	363
Coal, grams............................	91	91	91	91	91
Temperature, °C........................	1,150	1,150	1,150	1,150	1,150

RESULTS WITH VARIATION ON TIME, AT 1,150° C.—Continued.

(b) EXTRACTION.

Minutes.	Product.	Weight, grams.	Iron. Total, per cent.	Iron. Recovery, per cent.	Manganese. Total, per cent.	Manganese. Recovery, per cent.
	Ore....	363.0	24.36	.....	18.12	.....
30	Magnetic part	156.0	38.7	68.4	15.8	87.5
30	Nonmagnetic part	163.0	10.0	18.6	18.8	34.3
60	Magnetic part	204.4	38.4	89.0	17.7	55.5
60	Nonmagnetic part	125.5	7.1	10.1	21.3	40.6
90	Magnetic part	166.0	44.8	84.3	18.7	47.0
90	Nonmagnetic part	154.2	8.4	14.6	22.7	53.3
120	Magnetic part	131.4	58.5	86.9	15.0	30.9
120	Nonmagnetic part	189.4	8.18	17.5	20.9	60.2
150	Magnetic part	122.7	57.0	79.3	13.6	25.2
150	Nonmagnetic part	178.5	8.1	16.4	22.1	50.9

(c) RECOVERY.

Test No.	Variable, minutes.	Iron. Grams per 100 grams of ore.	Iron. Recovery. Per cent.	Iron. Recovery. Grams.	Manganese. Grams per 100 grams of ore.	Manganese. Recovery. Per cent.	Manganese. Recovery. Grams.
17-A	30	24.36	68.4	16.6	18.12	37.5	6.8
17-B	60	24.36	89.0	21.7	18.12	55.5	10.1
17-C	90	24.36	84.3	20.5	18.12	47.0	8.5
17-D	120	24.36	86.9	21.2	18.12	30.0	5.4
17-E	150	24.36	79.3	19.1	18.12	25.2	4.6

(d) SOLUBLE IRON IN MAGNETIC PART.

Minutes.	Product.	Weight, grams.	Total, grams.	Percentage of soluble iron.	Grams of soluble iron.	Total grams of soluble iron.	Percentage of soluble iron.
30	Magnetic part {+30 / −30}	.....	.....	.....	.....	.....	.....
60	Magnetic part {+30 / −30}	131.5 / 72.9	204.4	46.0 / 22.9	60.5 / 16.7	77.2	37.7
90	Magnetic part {+30 / −30}	135.1 / 30.9	166.0	47.8 / 25.4	64.7 / 7.9	72.6	43.6
120	Magnetic part {+30 / −30}	84.15 / 47.2	131.4	62.8 / 33.0	52.8 / 15.6	68.4	52.0
150	Skull / Powder	50.9 / 71.8	122.7	44.0 / 61.6	22.4 / 44.2	66.6	54.4

(e) SOLUBLE IRON IN SOLUBLE PART OF MAGNETIC PART.

Minutes.	Product.	Weight, grams.	Total, grams.	Percentage of soluble matter.	Grams of soluble matter.	Total grams of soluble matter.	Total grams of soluble iron.	Percentage of soluble iron in soluble matter.
30	Magnetic part {+30 / −30}	.....	.....	.....	.....	.....	.....	.....
60	Magnetic part {+30 / −30}	131.5 / 72.9	204.4	75.5 / 62.2	98.8 / 45.6	144.4	77.2	53.4
90	Magnetic part {+30 / −30}	135.1 / 30.9	166.0	73.0 / 59.3	98.8 / 18.6	117.4	72.6	61.5
120	Magnetic part {+30 / −30}	84.15 / 47.2	131.4	81.7 / 64.2	68.8 / 30.3	99.1	68.4	69.7
150	Skull / Powder	50.9 / 71.8	122.7	67.2 / 79.2	34.2 / 56.8	91.0	66.6	73.4

RESULTS WITH VARIATION OF TIME, AT 1,250° C.

(a) DATA.

Test No.	35-A.	35-B.	35-C.	35-D.	35-E.
Time, minutes.	30	60	90	120	150
Ore, grams.	160	160	160	160	160
Coal, grams.	40	40	40	40	40
Temperature, °C.	1250	1250	1250	1250	1250

(b) EXTRACTION.

Minutes.	Product.	Weight, grams.	Iron.		Manganese.	
			Total per cent.	Recovery per cent.	Total per cent.	Recovery per cent.
	Ore.	160.0	24.36	....	18.12	....
30	Magnetic part	67.4	38.26	66.2	15.7	38.6
30	Nonmagnetic part	78.7	10.8	21.8	20.5	55.5
60	Magnetic part	44.0	79.2	89.2	4.8	7.3
60	Nonmagnetic part	92.0	3.3	7.7	27.4	86.9
90	Magnetic part	39.7	88.6	90.4	4.3	5.9
90	Nonmagnetic part	93.8	4.9	11.8	26.3	85.2
120	Magnetic part	40.5	90.6	93.9	3.2	4.5
120	Nonmagnetic part	85.3	2.0	4.4	28.9	84.8
150	Magnetic part	42.7	91.0	98.5	4.0	5.9
150	Nonmagnetic part	81.7	1.6	3.3	29.0	81.7

(c) RECOVERY.

Variable, minutes.	Iron.			Manganese.		
	Grams. per 100 grams of ore.	Recovery.		Grams. per 100 grams of ore.	Recovery.	
		Per cent.	Grams.		Per cent.	Grams.
30	24.36	66.2	16.1	18.12	38.6	7.1
60	24.36	89.2	21.7	18.12	7.3	1.3
90	24.36	90.4	22.0	18.12	5.9	1.1
120	24.36	93.9	22.8	18.12	4.5	0.8
150	24.36	99.5	24.2	18.12	5.9	1.1

(d) SOLUBLE IRON IN MAGNETIC PART.

Minutes.	Products.	Weight, grams.	Total, grams.	Percentage of soluble iron.	Grams of soluble iron.	Total grams of soluble iron.	Percentage of soluble iron.
30	Magnetic part {+30 / −30}	....	....	....	....	....	....
60	Magnetic part {+30 / −30}	34.3 / 9.7	44.0	81.0 / 71.8	27.8 / 7.0	34.8	79.0
90	Magnetic part {+30 / −30}	33.7 / 6.0	39.7	90.8 / 77.0	30.5 / 4.0	35.2	88.6
120	Magnetic part {+30 / −30}	36.4 / 4.1	40.5	92.3 / 71.96	33.6 / 3.0	36.6	90.5
150	Skull / Powder	39.1 / 3.6	47.7	91.5 / 71.75	35.8 / 2.6	38.4	9.00

Results with Variation of Time, at 1,250° C.—Continued.

(c) SOLUBLE IRON IN SOLUBLE PART OF MAGNETIC PART.

Minutes.	Product.	Weight, grams.	Total, grams.	Percentage of soluble matter.	Grams of soluble matter.	Total grams of soluble matter.	Total grams of soluble iron.	Percentage of soluble iron in soluble matter.
30	Magnetic part {+30 / −30}	....	....	....	....	....	....	....
60	Magnetic part {+30 / −30}	34.3 / 9.7	44.0	90.0 / 88.0	20.9 / 8.5	39.4	34.8	88.5
90	Magnetic part {+30 / −30}	33.7 / 6.0	39.7	98.2 / 90.0	33.1 / 5.4	38.5	35.2	91.5
120	Magnetic part {+30 / −30}	36.4 / 4.1	40.5	100.0 / 87.8	36.4 / 3.6	40.0	36.6	91.2
150	Magnetic part {+30 / −30}	39.1 / 3.6	42.7	99.6 / 87.7	39.0 / 3.2	42.2	38.4	91.0

Results With Variation of Amount of Reducing Agents.

(a) DATA.

Test No.	34-A.	34-B.	34-C.	34-D.
Coal, grams	20	40	60	80
Ore, grams	200	200	200	200
Temperature, °C	1,250	1,250	1,250	1,250
Magnetic part, grams	64.6	53.5	50.1	52.7
Nonmagnetic part, grams	94.2	109.7	121.6	127.6

(b) EXTRACTION.

Coal, per cent of ore.	Product.	Weight, grams.	Iron. Total, per cent.	Iron. Recovery, per cent.	Manganese. Total, per cent.	Manganese. Recovery, per cent.
	Ore	200.0	24.36	....	18.12	....
10	Magnetic part	64.6	68.3	90.5	10.7	19.0
10	Nonmagnetic part	94.2	6.1	11.7	27.4	71.3
20	Magnetic part	53.5	90.0	98.8	4.7	13.0
20	Nonmagnetic part	109.7	3.6	7.4	28.3	85.8
30	Magnetic part	50.1	89.9	92.5	4.8	6.5
30	Nonmagnetic part	121.6	4.1	10.2	26.0	87.2
40	Magnetic part	52.7	70.5	76.2	13.5	19.6
40	Nonmagnetic part	127.6	10.7	27.9	21.1	74.3

(c) RECOVERY.

Coal, percentage (by weight) of ore.	Iron. Pounds per 100 pounds of ore.	Iron. Recovery. Per cent.	Iron. Recovery. Pounds.	Manganese. Pounds per 100 pounds of ore.	Manganese. Recovery. Per cent.	Manganese. Recovery. Pounds.
10	24.36	90.5	22.0	18.12	19.0	3.5
20	24.36	98.8	24.0	18.12	7.2	1.3
30	24.36	92.5	22.5	18.12	6.5	1.2
40	24.36	76.2	18.5	18.12	19.6	3.6

The foregoing results have been plotted in figures 2 to 9.

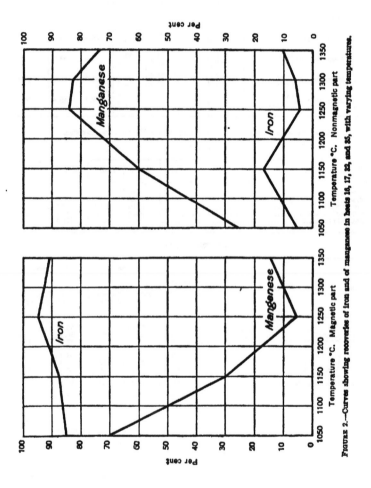

FIGURE 2.—Curves showing recoveries of iron and of manganese in heats 14, 17, 22, and 25, with varying temperatures.

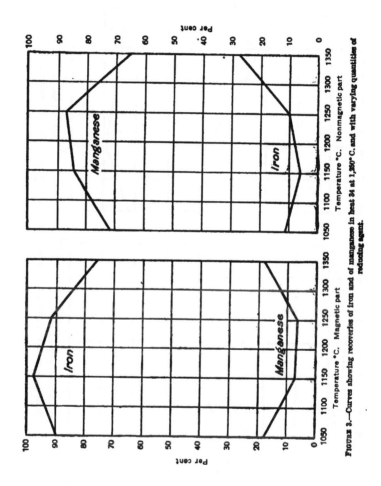

FIGURE 3.—Curves showing recoveries of iron and of manganese in heat 34 at 1,360° C. and with varying quantities of reducing agent.

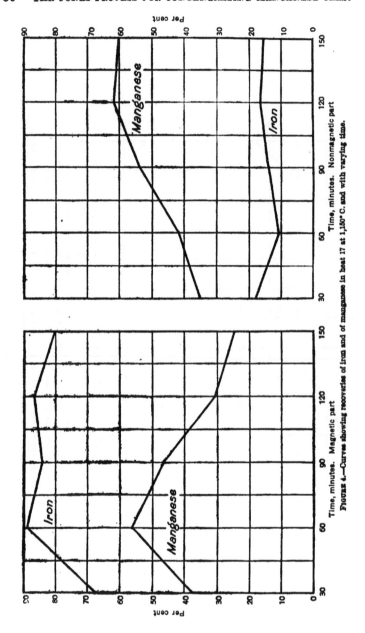

FIGURE 4.—Curves showing recoveries of iron and of manganese in heat 17 at 1,150° C. and with varying time.

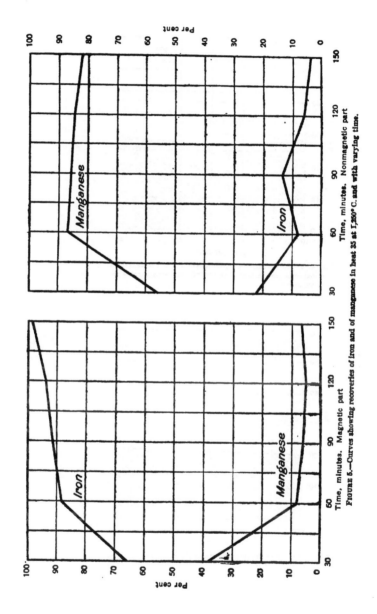

FIGURE 5.—Curves showing recoveries of iron and of manganese in heat 25 at 1,250° C. and with varying time.

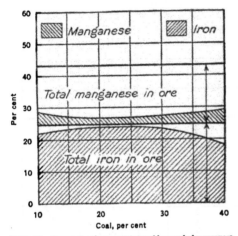

FIGURE 6.—Curves showing percentages of iron and of manganese in magnetic part in heat 34 at 1,250° C. and with varying reducing agent.

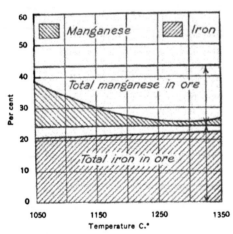

FIGURE 7.—Curves showing percentages of iron and of manganese in magnetic part in heats 16, 17, 31, 32, and 35 with varying temperatures.

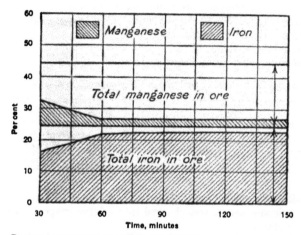

FIGURE 8.—Curves showing percentages of iron and of manganese in magnetic part in heat 35 at 1,250° C.

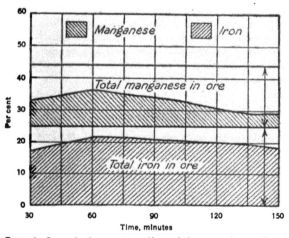

FIGURE 9.—Curves showing percentages of iron and of manganese in magnetic part in heat 17 at 1,150 °C.

CONCLUSIONS.

Figures 2 to 9 indicate that the following are the requisite conditions for the low-temperature reduction of the iron:

(a) Time, preferably 30 to 40 minutes, although satisfactory results were obtained up to 2 hours.

(b) Temperature, 1,250° C.

(c) Reducing agent, 20 per cent coal.

Analyses of ores used in final tests with low-temperature reduction.

Ore.	Fe.	Mn.	SiO₂.	P.
	Per cent.	*Per cent.*	*Per cent.*	*Per cent.*
West End...	24.36	18.12	26.50	0.097
Ferro...	28.55	24.55	17.60	.091
1019 E..	35.15	13.68	9.46	.179
1016 Cts..	40.30	13.08	10.06	.107
381 Cts...	35.40	18.25	21.36	.659

Two kilograms of each ore, crushed to 10 mesh and mixed with 20 per cent reducing agent, was charged into a large graphite crucible heated to redness.

The temperature was raised to 1,250° C. and kept as closely as possible between 1,250° and 1,275° C. until reaction was complete.

The time was varied to some extent according to the appearance of the charge, with a maximum time of two and one-half hours, except in heat 43.

The reducing agent used was Elkhorn coal like that used in the preliminary tests, except that in heat 41 coke was used.

The furnace used was a Case oil-fired melting furnace No. 40.

The method of handling the furnace products was practically the same as in the preliminary tests. However, the entire contents of the crucible were poured in a liquid condition, and most of the metal produced was recovered in the form of a large button.

TABULATED RESULTS.

Detailed results of the tests are presented in Table 12 following:

TABLE 12.—*Results of final tests with low-temperature reduction.*

Test No.		36	37	38	39	40	41	42	43	44	45	46
Charge.												
Weight:												
Ore	grams	2,000	2,000	2,000	2,000	2,000	2,000	2,000	2,000	2,000	2,000	2,000
Coal	do.	300	400	400	500	400	300	400	400	400	400	400
Lime	do.											100
Ore analysis:												
Fe	per cent.	24.98	23.55	(b)	35.15	(c)	(c)	40.30	(d)	(d)	35.40	(d)
Mn	do.	18.12	24.55		13.66			13.08			10.25	
SiO_2	do.	20.00	17.00		9.46			10.08			21.36	
P	do.	0.097	0.081		0.179			0.107			0.059	
Metal.												
Weight, by weight of ore	grams	468	691	571	927	697	683.00	781	818	708	691	695
		24.15	34.55	28.55	34.35	34.85	34.18	39.05	40.90	35.15	34.55	34.75
Analysis:												
Fe	per cent.	92.8	88.90	95.30	96.00	95.90	92.90	98.30	94.80	94.80	94.70	95.00
Mn	do.	2.64	2.90	0.70	1.25	0.56	0.45	0.28	0.25	0.53	0.46	0.35
P	do.	0.44	0.25	0.25		0.45	0.50	0.50	0.25	0.46	0.184	0.184
S	do.	0.04	0.008	0.085	0.043	0.057	0.071	0.062	0.097	0.064	0.096	0.079
Recovery:												
Fe	per cent.	87.8	91.8	95.2	98.3	95.0	90.3	98.8	98.0	94.5	98.0	93.6
Mn	do.	3.34	4.7	0.8	3.33	1.4	1.1	2.6	2.6	1.4	1.6	1.2
P	do.	96.0	80.2	80.2		81.6	82.7	82.0	98.5	85.0	100.0	100.0
Slag.												
Weight, by weight of ore	grams	1,127	1,066	1,107	851	816	980	785	819	905	965	967
Percentage, by weight of ore	per cent.	56.9	53.3	55.35	42.55	40.8	46.5	39.3	40.9	45.3	47.7	48.4
Analysis:												
Fe	per cent.	4.9	3.9	2.9	2.2	4.4	6.8	6.4	3.6	4.6	4.6	4.2
Mn	do.	27.8	40.0	42.7	28.2	29.55	20.0	20.8	28.7	27.75	20.6	18.75
Recovery:												
Fe	do.	11.4	7.3	5.6	2.6	5.1	9.0	6.2	2.6	5.8	6.2	5.7
Mn	do.	85.5	86.0	96.0	90.7	88.2	88.3	92.4	98.8	93.5	95.7	58.4

ᵃSame as for No. 37,   ᵇSame as for No. 36,   ᶜSame as for No. 37.   ᵈSame as for No. 42,   ᵉSame as for No. 42.   ᶠSame as for No. 44.

DISCUSSION.

The foregoing tabulation is self-explanatory. The results were obtained practically under the conditions indicated under preliminary tests. However, the appearance of the charge, and in particular the condition of the slag and metal, were also used to determine the end of the reaction.

PRODUCTS.

On averaging all the results 93.5 per cent of the iron and 2.0 per cent of the manganese contained in the ore were reduced and entered the metal, while 4.3 per cent of the iron and 90.3 per cent of the manganese remained in the slag. The composition of the metal obtained in all the tests varied as follows:

Iron, per cent, 89.9 to 96.3; average, 94.2.

Manganese,[a] per cent, 0.28 to 3.9; average, 1.0.

Phosphorus, per cent, 0.1842 to 0.50, or 88.0 per cent of that contained in the ore.

Sulphur, per cent, 0.003 to 0.097, or about 10 per cent of that contained in the charge.

ORES.

All the ores tested were of the intimately disseminated Cuyuna manganiferous type and represent a wide range of iron and manganese content. All were crushed to pass a 10-mesh screen, except that used in tests 45 and 46, which was crushed to 8-mesh. It is probable that if coarser material were used the time necessary to complete the reaction would increase. The uniform results obtained in testing such a variety of ores would tend to show that this low-temperature reduction is applicable to a much greater variety of ores.

TIME.

A considerable range of time at which the charge was kept at a temperature between 1,250° and 1,300° C. was used. From the data thus obtained it is apparent that, under the conditions given, the best results are obtained when the charge is kept at this temperature for 30 to 40 minutes.

REDUCING AGENT.

Elkhorn coal crushed to 10 mesh was used in all tests, except No. 41, in which foundry coke, crushed to the same mesh, was used. The Elkhorn coal analyzed 6.66 per cent ash, 40.98 per cent volatile matter, and 0.645 per cent sulphur. The coke analyzed 13.9 per cent ash, 3.33 per cent volatile matter, and 1 per cent sulphur.

No difference was noticed in the reaction when coke was used, except that the slag fused at a lower temperature. This was probably

---

[a] Discarding the manganese content of metal obtained in tests 36 and 37, the reactions of which were not completed, the average manganese content of the metal would be reduced to 0.60 per cent.

due to the larger percentage of ash in the coke. The use of coke also tends to increase the sulphur in the metal. The proportion required did not seem to vary with the character of the ore.

LIME.

The addition of lime in test 46 did not materially change the results. Its use as a flux would probably be necessary to obtain a high recovery of iron from ores containing small amounts of manganese and large percentages of silica. The lime used was of the commercial air-slaked type.

FINAL TESTS.

OUTLINE.

In the final tests with low-temperature reduction, ores of the following analyses were used:

HIGH-TEMPERATURE REDUCTION TESTS.

GENERAL OUTLINE.

The slag obtained by the process of low-temperature reduction was crushed to pass a 30-mesh screen, mixed with 25 per cent coke crushed to pass a 10-mesh screen, and charged into a carbon crucible which was heated in an ordinary electric carbon resistor furnace.

CHARGE.

An analysis of the slag used for the charge was: Iron, 4.9 per cent; manganese, 28.68 per cent; SiO_2, 31.1 per cent; Al_2O_3, not determined. An analysis of the coke used as a reducing agent is given above. Slag, coke, and lime were mixed before putting the charge into the crucible. The only variable used in these tests was lime. This was air-slaked material similar to that used in the low-temperature final tests.

PYROMETER.

The temperature was measured by means of a Scimatco pyrometer. This instrument, though not very satisfactory, was the best available. It was first calibrated against a Hoskins thermocouple, having a temperature capacity up to 1,100° C. With great care in observing by keeping the comparison lamp at a constant amperage fairly concordant results were obtained. The temperature measured was that of the surface of the charge only, but when a charge is in constant agitation, owing to the carbon monoxide produced by the reaction, it is assumed that the surface temperature serves as a fair indication of the temperature of the crucible contents.

CRUCIBLES.

A number of standard crucibles were tried, as follows: Ordinary assay crucible, graphite crucible, and alundum crucible. None of these could withstand the conditions. The crucibles finally used were cut on a lathe from 4-inch carbon electrodes. With care they would hold 200 grams of crushed slag mixed with the requisite coke and lime. These crucibles suffered deterioration only from oxidation at the upper edge and could be used until they became too shallow to hold the charge.

TEMPERATURE.

The temperature at which a satisfactory reaction took place was found to be about 1,450° C. This satisfactory reaction may be designated as uniform and rapid. Above 1,450° C. the reaction was so rapid that it caused violent boiling, and below this temperature the reaction was so slow and inactive that no appreciable metal was reduced. Hence, the condition of the reaction was used as a criterion for temperature regulation.

TABULATED RESULTS.

The detailed results of the high-temperature reduction tests are presented in the tabulation following:

TABLE 13.—*Results of high-temperature reduction tests.*

Test No.	54	55	56	57	58
Charge.					
Weight:					
Slag............................grams..	200	200	200	200	200
Coke..............................do....	50	50	50	50	50
Lime..............................do....	40	30	20	30	00
Analysis:					
Fe............................per cent..	4.9	(a)	(a)	(a)	(a)
Mn...............................do....	28.68	(a)	(a)	(a)	(a)
SiO₂..............................do....	31.10	(a)	(a)	(a)	(a)
Alloy produced.					
Weight............................grams..	66.0	66.3	66.7	76.4	81.5
Percentage of slag charged............	33.0	33.15	33.35	37.7	40.75
Analysis:					
Fe............................per cent..	16.8	15.40	16.40	12.5	12.0
Mn...............................do....	66.9	65.00	6.42	62.4	58.4
Si...............................do....	10.7	13.77	13.56	23.6	25.0
Recovery, per cent of content of slag charged:					
Fe............................	b 112.0	b 102.6	b 106.0	b 97.0	b 100.0
Mn............................	77.2	75.4	74.8	83.2	84.0
Si............................	24.1	31.4	31.0	62.0	70.5
Slag produced.					
Weight............................grams..	134.0	108.7	112.3	83.3	82.9
Percentage of slag charged............	67.0	54.4	56.15	36.9	41.45
Analysis:					
Fe............................per cent..	1.26	1.94	2.15	1.37	1.30
Mn...............................do....	6.15	4.25	7.00	4.70	5.44
SiO₂..............................do....	36.25	34.85	37.80	38.10	30.60

a See test 54 for analysis.
b Recoveries are calculated on content of slag charged. Discrepancy shown by getting more than 100 per cent recovery is due to fact that coke contains iron as part of the ash.

DISCUSSION.

GENERAL.

Coke was used as a reducing agent because it was thought to be the best commercial source of carbon suitable for this high-temperature reduction. It was mixed well with the rest of the charge and seemed to remain more or less disseminated through the charge after the slag portion of the charge had melted. The amount used in the tests does not indicate the proportions required on a commercial scale where no other source of carbon would be present, because with the use of carbon crucibles there was always an excess of carbon present from this source.

Lime was added to the charge to render the resulting slag sufficiently liquid to pour. The amount added was varied for the purpose of controlling the reduction of silicon. However, the effect of lime in this respect can not be considered very definite, although the tendency is to decrease the silicon in the alloy as the lime is increased in the charge.

The data obtained indicate that the reduction of silica to metallic silicon increases as the temperature is increased and is a function both of the temperature and of the lime added. Possibly, if the lime were incorporated in the slag portion of the charge, its effect would be more pronounced.

ALLOY.

The metal produced is essentially a silico-ferro-manganese, consisting of about 15 per cent iron, 60 to 67 per cent manganese, and 10 to 25 per cent silicon. The percentage of silicon increases with the temperature of the reaction; and as the silicon increases the percentage of manganese in the alloy will correspondingly decrease.

RECOVERY.

The excess of iron recovered, together with some iron remaining in the resulting slag, is explained by sources of iron being present in the charge, which were not taken into account in making the calculations. The recovery is based on the slag portion of the charge only. Under the given conditions, the manganese remaining in the resulting slag was not less than about 5 per cent. It is, therefore, evident that as the slag volume is decreased the recovery of manganese is increased. This is shown in experiments Nos. 57 and 58. The recovery of manganese, as given in the tabulated results, varied from 75 to 84 per cent, and averaged 79 per cent.

SLAG.

As the iron and manganese is reduced from the charge, the fusibility of the slag decreases and its viscosity increases as a result of the increase in unbalanced silica. The presence of lime in the charge is advantageous, increasing the fusibility of the slag, and decreasing its viscosity, so that it may be handled in a liquid condition.

SUMMARY.

The essential results contained in the foregoing report may be summarized in semi-tabular form as follows:

Summary of results of experiments with the Jones process.

LOW-TEMPERATURE REDUCTION, 1,250° TO 1,300° C.

Charge:

Reducing agent, 20 per cent of a good grade of coal.

Ore:

Fe,	per cent	24.0 to 40.0
Mn,	per cent	10.0 to 24.0
SiO_2,	per cent	10.0 to 26.0
Al_2O_3,	per cent	Not determined.

Products:

	Approximate composition, per cent.	Recovery, per cent.
Liquid metal:		
Fe,	94.00	94.00
Mn,	1.00	2.0
Si,	.10 (maximum)	.....
P,	.50 (maximum)	88.0 (average)
Liquid slag:		
Fe,	5.00	6.0
Mn,	29.00	90.0
SiO_2,	31.00	.....
Loss......Manganese		8.0

HIGH-TEMPERATURE REDUCTION, 1,450° C.

Charge:

Liquid slag from low-temperature reduction.

Coke, 25 per cent of slag.

Products:

	Approximate composition, per cent.	Recovery, per cent.
Liquid metal:		
Fe....15		100.0
Mn....55 to 65		80.0
Si.....10 to 25		.....
Liquid slag:		
Fe.... 1.0 to 2.0		
Mn.... 4.0 to 6.0		
SiO_2...30.0 to 40.0		

CONCLUSIONS.

The foregoing results seem to indicate that by the Jones process concentration of manganese in the finely disseminated manganiferous iron ores is metallurgically possible.   A discussion of the commercial possibilities of the process will not be attempted.

The metal produced in the low-temperature reduction is suitable for conversion into steel by any basic steel process.   It could not be used as a foundry metal. Whether the low-temperature reduction could be modified to yield a foundry metal would require further experiments.   By a suitable regulation of temperature and by the use of iron ore in the place of manganiferous ores, it is possible that a foundry metal could be produced.

The high-temperature reduction produces an alloy that may be termed a silico-ferromanganese.   Although an alloy of this nature would be new to the steel industry, it could probably be used to advantage.   About 72 per cent of the manganese in the ore appears in the alloy.

CHAPTER 7.—COST OF PRODUCING FERRO-GRADE MANGANESE ORES.

By C. M. WELD and W. R. CRANE.

INTRODUCTORY STATEMENT.

In collecting and disseminating information on methods of mining and preparing domestic manganese ores, careful consideration was given by the bureau to the important matter of costs. Unfortunately few operators kept systematic cost sheets, hence the data collected are incomplete and unsatisfactory, particularly for an industry having so little uniformity. Domestic manganese deposits, especially deposits of ferro-grade ores (35 + per cent metallic manganese), vary widely, not only in the character of the ore, but also in the conditions of occurrence. Costs must therefore be expected to vary correspondingly, even when cost-keeping methods are uniform. But when the methods are as variable as the costs, computation becomes still more difficult.

It has, nevertheless, been possible by carefully compiling the material at hand to arrive at certain estimated average figures which are believed to be approximately correct, and to draw therefrom certain general conclusions. It must be clearly understood, however, that these average figures may be entirely misleading if applied to some one particular deposit. The conditions are so variable that it is necessary to consider each deposit strictly on its individual merits when considering its commercial possibilities. The average figures, however, have distinct value as guides to the probable competition which will have to be faced.

In fact these average figures have been compiled with this latter point especially in view. For this reason the more immediate foreign competitors, Brazil and Cuba, are considered as well as the domestic sources.

The subject may be approached in two ways—by determining so far as possible the actual cost, and by determining what may be called the "proper" cost. It is thoroughly appreciated that at many properties, owing to a multitude of reasons, costs have been unduly and unnecessarily high. However, as regards the costs of producing manganese ores during the war, it would serve no useful purpose to estimate what the cost ought to have been. War costs have become a

92

matter of history, but the actual figures as bearing on probable future costs are of vital importance. When it comes to estimating the future, the opposite angle of approach may be taken. Domestic ores will soon once more have to face active and unrestricted foreign competition, and "proper" costs will have to be attained in order to keep the industry alive. Furthermore, if the question of protection be raised, only "proper" costs should be considered. It is appropriate, therefore, to discuss future possibilities from the second point of view, namely, as regards what the costs should be.

Another vital question, in every discussion of this sort, is what constitutes cost. Strange as it may seem, the important factors of amortization of plant and depletion of ore reserves are more often overlooked than not. This has no doubt been especially true of a business like that of the manganese industry during the war, chiefly characterized by the hurried and ofttimes ill-considered undertakings of many small operators with little financial strength. Nevertheless, owing to the uncertain life of the business, the writing off of the investment should have been the prime consideration.

In most instances the cost data collected evidently do not include allowances for amortization and depletion. It is now obviously impossible to say what these items should have amounted to; therefore, where they do seem to have been included, they will be omitted from the following discussion, for uniformity. This fact, however, should be kept in mind, if any comparison of the average estimated war costs with the prices then prevailing is attempted. The apparent profit which such a comparison shows probably failed, in a large number of cases, to offset the investment before the business collapsed upon the signing of the armistice. In fact, were not this the case, there would be little ground to-day for any claims under the War Minerals Relief act.

The same considerations must be kept in mind as regards the estimated average future costs. These are not intended to include appropriate charges for writing off investment.

COSTS OF DOMESTIC MANGANESE ORES.

The domestic manganese ores of ferro grade fall into two distinct classes, which vary widely from one another in their cost of production. These are (1) the carbonate ores, and (2) the oxide ores. A third class, namely, the silicate (rhodonite) ores, should perhaps be mentioned, but such ores in the natural state are too siliceous, and concentration too costly, to make their use feasible even with a war-time market. Consequently the small amounts produced were incidental to the production of carbonate and oxide ores, and they need not be considered separately.

CARBONATE ORES.

Carbonate ores were mined in quantity in only one locality, namely, Butte, Mont., but the output for 1918 was more than 20 per cent of the total output of manganese ores mined in the United States. The great bulk of the ore required no concentrating and little if any hand sorting. Mining was of the underground type, but the deposits were persistent and the mining operations were efficient. Furthermore the haul to the railroad was short, and accomplished by motor trucks over well-paved roads. It is understood that the cost on board railroad cars was not more than $5 per long ton—probably less. As the product contained 36 to 37 per cent metallic manganese, the unit cost was about 14 cents.

A small amount of siliceous carbonate ore was produced in the Butte district and concentrated at a neighboring customs mill, where the milling cost is said to have been $4.50 per ton of product.

OXIDE ORES.

The oxide ores occur mixed with clay, chert, and other gangue materials and invariably require careful hand cleaning or mechanical washing or concentrating. The deposits are generally erratic. Production costs have therefore necessarily been much higher than with the carbonate ores. In a few places carbonate ores were encountered as the weathered oxide ores were followed downward, but these carbonate deposits were too small and generally too impure, to lessen the otherwise high costs. Occasionally silicate (rhodonite) ores were encountered with depth, when mining as a rule had to be abandoned.

The oxide ores are widely scattered over the United States. They represented nearly 80 per cent of the 1918 production. In estimating the average cost of producing them, they will be treated as a group, although individual costs as between States, districts, and even mines in the same district, frequently varied within wide limits.

These variations were principally due (a) to inherent differences in the deposits as regards their size and richness, and (b) to their situation as affecting the cost of hauling to the railroad. Costs may best be discussed therefore under two main heads, namely (a) cost at the mine, and (b) cost of transportation to the railroad.

COST AT THE MINE.

Ordinarily it would be found convenient to subdivide the cost at the mine into mining cost and treatment or preparation cost. A large part of the manganese ore produced, however, was hand cleaned by picking or screening, and the available cost data do not differentiate between mining and hand cleaning. Consequently, three kinds of costs must be considered under mining and preparation, namely,. mining, milling, and mining plus hand cleaning.

Figures for these costs as given below include labor, supplies, and the usual overhead charges, such as royalty. It will be of interest to examine these items separately before taking them up as lumped together under mining and treatment.

LABOR.

The labor supply, never abundant in many of the manganese dis-- tricts, was seriously affected by conditions resulting from the war, not the least of which was the draft. Table 14 following shows data. regarding wages recently paid.

TABLE 14.—*Wages paid in several of the manganese-producing States.*

(Dollars per day.)

	Ari-zonia.	Arkan-sas.	Geor-gia.	Mon-tana.	Neva-da.	Ten-nessee.	Utah.	Vir-ginia.	Wash-ington.
Carpenter			3.25	6.00		3.00			4.00
Common labor	4.75	3.00	2.75	5.50	5.00	2.50	4.00	2.00	
Underground mines			3.75	5.50					5.00
Open-cut mines		3.00	2.75						4.50
Steam-shovel operator			5.75	5.00					
Steam-shovel helper			2.75	4.00					
Dinky driver				3.75					
Fireman			2.75	2.75					
Mill men			3.75	3.00	5.50		3.00		4.50
Hand picking				2.50					
Mine foreman			6.00		7.50		3.50		
Mill foreman			4.00		8.30				
Blacksmith			3.25		6.00		5.00		5.00

Sufficient data are not at hand for satisfactory comparison between prewar and war times, but a few figures in this direction are available. For example, in 1916 the wage paid common labor in the Cartersville district, Georgia, was $1.25 per 10-hour day; a year later it had risen to $1.75, while in 1918 it was $2.50 per 8-hour day. In certain localities a ·bonus was paid for a maximum number of days' work per week, which raised the wage to $3 per 10-hour day. A similar condition existed in the Batesville district, Arkansas, and undoubtedly such conditions were universal through the manganese fields of the country. A wage of $2.50 for eight hours is probably a fair average.

SUPPLIES.

No detailed information as to cost of supplies per ton of product is available. The item was far higher, possibly double, the prewar cost.

ROYALTY.

A royalty commonly demanded during the war was 10 per cent of the gross value of the product, that is to say, of the selling price. The royalty thereby varied automatically with the grade of the ore.

Occasionally an arbitrary sliding scale was used, of which the following is an example:

Example of arbitrary sliding scale of royalty for manganese ore.

Manganese content of ore, per cent.	Royalty per ton.
20 or less	$0.50
20 to 30	1.00
30 to 40	1.50
40 and over	2.00

This scale provided for the higher grades of manganiferous ores as well as for ferro-grade ores. At some places, where the ore ran fairly uniformly as to grade, a single fixed rate of royalty was paid, as for instance 50 cents per ton at the Cason mine, Batesville, Ark.; or $1.50 per ton in the Erickson district, Utah.

The average manganese content of all ferro-grade oxide ores mined in the United States was approximately 41 per cent, and the average royalty paid for such ores may be taken to be $1.50 per long ton. Before the war, the current royalty for rather better grade ore was $1.00.

MINING AND TREATMENT.

The mining cost per ton of product naturally varied widely, not only with the cost of handling the bank dirt, or crude ore, but also with the yield of concentrates. As a rule, however, cost and yield roughly offset each other. Thus, a lean bank dirt had to be mined more cheaply than a rich one. Yields varied from 2 or less up to as much as 10 tons of bank dirt per ton of concentrates; and mining costs, where the ore was mined as it came and was sent to a mill for treatment, were found to range from $5 to $10 per ton of product, with the average not far from $7.50.

Milling costs likewise varied with the grade of the bank dirt, as also with the degree of refinement to which the process was carried. Concentrating plants varied from simple log washers, with or without picking belts, to elaborate mills with screens, jigs, and even occasionally tables. Water was not infrequently a costly item. Costs per ton of product varied from $5 to $8.50, the average being approximately $6.50.

At many small operations the bank dirt was simply picked or screened by hand. A profit in such cases was only possible when the crude material was fairly rich, but frequently cheaper costs were attained than at the larger and more elaborate operations, chiefly because overhead charges were reduced to a minimum, and the "mine" was readily moved from point to point, following the richer and cheaper ore. The average cost per ton of product was $12 to $13; individual costs were frequently less.

Considering the respective proportions of ore hand cleaned and milled, the average cost of the total product is estimated to have been $13.50. The figures that contribute to this average include a certain amount of the overhead charge in many cases, but probably $1.50 should be added to cover royalty, sampling, and similar overhead charges, which in other cases have not already been included. Thus it is concluded that the total average cost at the mine of the oxide ores was approximately $15.

The following individual examples are given as illustrating some of the details underlying the above conclusion.

Mr. F. G. Moses of the bureau's staff reported mining costs per ton of product in the Erickson district of Utah as given below. Costs at the mine were comparatively low at that point.

Costs per ton of product in the Erickson district, Utah.

Mining	$5.00
Overhead	2.00
Sampling	1.00
Royalty	1.50
Total	9.50

The average cost of mining per ton of product in the Cartersville district, Georgia, was about as follows:

Average costs per ton of product in the Cartersville district, Georgia.

	Underground.	Open cut.
Mining	$3.00	$2.50
Timbering	1.50	......
Royalty	1.50	1.50
Handling and miscellaneous	2.25	.50
Total	$8.25	$4.50

The bulk of the ore was obtained from open-cut workings and the average cost per ton of product is assumed to be $5.50. This is by far the most important manganese producing district in the State of Georgia.

The following rates paid for contract mining in the Batesville (Arkansas) and Cartersville (Georgia) districts are interesting.

Rates for contract mining per ton in Batesville and Cartersville districts.

Ore.	Batesville.[a]	Cartersville.[b]
Low grade	$8	$17
Medium grade	$12 to $15	....
High grade	$25 to $40	...:

A high price was paid in Batesville for high-grade ores, yet the amount of ore that could be mined was much less than with low-grade ores, so that the wage earned was about the same.

---

[a] The wash dirt mined to be cleaned at company's expense.   [b] For all lump and cleaned ore.

The following figures are given as approximate milling costs per ton of concentrates at a large operation in Virginia:

> 1917: 729 tons were cleaned at a cost of $10.77 per ton.
> 1918: 945 tons were cleaned at a cost of $6.53 per ton.
> 1918: Estimated cost for remodeled mill, $6.50 per ton.

The total estimated cost per ton of product on board railroad cars, including all charges, was given for the latter part of 1918 as $15. This is only one of a number of relatively important operations in Virginia.

Theodore Simons reported the following average costs for July, 1918, in the Philipsburg district, Montana. These costs are per ton of product on board railroad cars and include all charges for labor, timbering, hand cleaning or concentrating, road construction, development, repairs, and overhead:

Average costs in Philipsburg district.

Mining Company A	$15.14
Mining Company B	21.00
Mining Company C	16.00
Mining Company D	17.33
Mining Company E	17.00
Mining Company F	18.00

The arithmetical average of the above is $17.41 per ton of the product. As company A was by far the largest single producer its lower cost would bear down this figure in a weighted average to approximately $16.50, or roughly $15, excluding the cost of hauling and loading into railroad cars.

A representative milling cost at Philipsburg was reported as being $7.14 per ton of product. Custom mill work on Butte and Philipsburg oxide ores is said to have cost $2.50 per ton of crude ore. As the ratio of crude to concentrates was approximately 2.6 to 1, this latter figure would become about $6.50 per ton of concentrates.

COST OF TRANSPORTATION TO RAILROAD.

The location of the mines worked for manganese during the war with reference to the nearest railroad shipping point varied as widely as the other factors entering into cost. Not infrequently the distance hauled differed greatly even within the same district. For instance, in the Batesville district, Arkansas, hauls were commonly 2 to 3 miles, but hauls of 5 to 10 miles were not uncommon. The ore from a mine in Tennessee was hauled 14 miles to the railroad, and in the Buena Vista Valley, Nev., the distance hauled was 20 miles.

Transportation was by wagons and motor trucks. The average load was 1 ton. This average figure frequently applied even when 2-ton trucks were used, on account of poor roads. A few 5-ton trucks were used, but the tonnage handled in such units was small.

Table 15 following gives cost data and other data relating to a number of typical and important camps. The average cost of hauling for all oxide ores was roughly $3.50 per ton. To this should be added an average charge of 35 cents per ton for handling the ore at the two terminals of the haul.

TABLE 15.—*Cost of handling and hauling ore in manganese-producing States.*

State.	Average distance hauled, miles.	Load hauled, pounds.	Method employed.	Cost per ton.			
				Loading cars.	Hauling to railroad.	Loading wagons.	Weighing ore.
Arkansas	3½	3,300	Wagons and motor trucks	$0.25	$2.60		$0.10
Georgia	3.5	2,800	Wagons		1.10		
Montana	a 1½	3,000	Wagons and motor trucks	b .29	1.45		
Nevada	10	4,000	do		5.95		
New Mexico	18	2,800	Wagons	.45	3.50		
Tennessee	2½	3,000	do		2.50	$0.35	
Utah	13	2,800	do		2.50		
Virginia	2½	4,000	do	1.00	6.50	.50	
				.50	1.50		

a Philipsburg district only.   b Cost of loading both wagons and cars.

SUMMARY OF COST OF OXIDE ORES ON BOARD CARS.

Adding the cost per ton of product at the mine, $15, to the cost for transportation to railroad plus handling, $3.85, the total estimated average cost on board railroad cars for the oxide ores is seen to be $18.85, or, say, $19 per long ton. With 41 per cent average metallic manganese content the unit cost is 46.4 cents.

COST OF RAILROAD TRANSPORTATION TO MARKET.

Railroad freight rates advanced materially during the war. Certain reductions were later made in favor of western shippers of manganese ores, but were still high. Some typical rates are given below.

Freight rates between Southern points and Chicago.

From—	Rate per ton of 2,240 pounds.
Northeastern and northwestern Georgia	$6.90
Eastern Tennessee and western North Carolina	5.90
Virginia	6.30
Abbeville and Greenville, S. C., and Lincolnton, Ga.	8.40
Batesville, Ark.	4.60

Freight rates between western points and Chicago, Philadelphia, and Birmingham.

State (any point).	Destination.	Rate per short ton.	Destination.	Rate per short ton.
Colorado	Chicago, Ill	$7.00	Philadelphia, Pa	$8.50
Oregon	do	11.00	Birmingham, Ala	12.50
Washington	do	11.00	do	12.50
California	do	11.00	do	12.50
Montana	do	8.00	do	10.50
Arizona	do	9.00	do	9.00
Nevada	do	10.00	do	11.00
Utah	do	9.00	do	10.00
New Mexico	do	7.00	do	7.00

The Government schedule of prices on which ores were sold during the war fixed the price per unit delivered at Chicago. The producer calculated the value of his ore at Chicago and then deducted the freight rate from point of production to Chicago to get the selling price of his ore f. o. b. railroad cars at shipping point. The buyer paid the actual freight from shipping point to point of consumption.

Whether ores are sold in the future f. o. b. cars at the shipping point or delivered at the furnace makes no difference to the producer. It is at the furnace that he must compete with other producers, whether domestic or foreign, and he must therefore compute the freight as part of his cost of production.

A small amount of ferro was made in western electric furnaces during the war, but the great bulk of it was made east of the Mississippi, at Chicago and in Alabama, Virginia, Pennsylvania, and New York. The production of Pennsylvania furnaces largely predominated, and this will unquestionably be even more true in the future, as production outside of Pennsylvania was principally stimulated by war prices.

It has already been pointed out that costs at the mine and costs of transportation to the railroad varied widely from point to point. These variations were not due to the section of the country in which the mines were situated, however, but rather to the nature of the deposits themselves. Two distinct groups of deposits have been differentiated, namely, the carbonates and the oxides. Within these groups it mattered little whether the mine was situated, for example, in Virginia or in California; the cost might be high or low in either place. Therefore it was found desirable to average all costs for each group and so determine an estimated average cost for producing ore of that group, irrespective of its locality.

It seems necessary to consider the matter of railroad freights from another angle. If all oxide ores, for instance, had been produced in Virginia, their average cost delivered at Pennsylvania furnaces would have been less than if a part had been shipped from points west of the Mississippi. Therefore, the freight for each State or district to the principal center should be weighted in an average according to the amount of ore shipped from that State or district. It has been necessary to make certain assumptions, as exact freight rates are in many cases not at hand, but the resulting figures without doubt approximate the truth.

Proceeding by this method and assuming that the principal consuming center is Pennsylvania, it is found that the average freight rate for the oxide ores during the latter part of 1918 was approximately $10.75 per long ton and for carbonate ores approximately $11.25, including war tax.

SUMMARY OF COST OF DOMESTIC ORES.

The foregoing estimates are summarized as follows:

Summarized data on estimated average cost of domestic ores.

	Carbonate ore.	Oxide ore.	Average.
Estimated tons produced in 1918	63,000	231,500	294,500
Estimated cost on railroad cars	$5.00	$19.00	$16.00
Estimated railway freight	$11.35	$10.70	$10.85
Total estimated cost delivered	$16.35	$29.70	$26.80
Manganese, per cent	36.5	41.0	40.0
Estimated cost per unit, cents	45.0	70.7	65.0

Of course, the more favored districts could produce oxide ores more cheaply than the average figures given. Virginia, for instance, probably delivered ore to the furnaces at a cost of about 55 cents per unit. Cartersville (Ga.) and Batesville (Ark.) ores should have reached Birmingham, Chicago, or Pennsylvania furnaces for 55 to 60 cents per unit. Nearly 85 per cent of the oxide ores, however, came from the far western States and were subject to high railway freights. The carbonate ores paid a high railway freight but other costs were so low that these were probably the cheapest ores produced either on a ton or unit basis.

DISCUSSION OF FUTURE POSSIBILITIES.

The average costs computed in the foregoing pages are for war times and more particularly for the last six months before the signing of the armistice.

In estimating future costs we have little to guide us in the way of prewar costs. Then only small amounts of domestic ores were mined, frequently in a small way by farmers when not busy with farm work. The production was in the southeastern States, where freight rates to consuming centers were not excessive. Probably only small profits were made, but these were satisfactory under the circumstances. Costs were $4 to $8 per ton on board railroad cars, but these costs are not fair guides to future possibilities.

Aside from railway freights, the wage scale enters more largely into the cost of production than any other single item; and it is probable that reduction in cost will come chiefly from reduction in wages. Better practice should account for some reduction, and cost of supplies may perhaps be properly expected to drop about in proportion to reduction in wages.

The consideration of future wages is a far-reaching problem, involving infinitely more important national industries than manganese mining, and an infinite number of correlated problems. The subject is ventured upon only because it seems desirable at this time to attempt some reasonable forecast of the probable future cost of manganese.

There is admittedly much difference of opinion in this respect, but it may fairly be assumed that wages must inevitably drop. The cost of labor is now approximately double the prewar figure in many of the more important manganese-producing districts. An instance has already been given where it has more than doubled. Its advance has frequently outsped the increase in cost of living, which is said in general not to have increased more than approximately 50 per cent in most manganese mining camps. The economic pressure of competition must either depress these wages or else the industry must be abandoned. If it be assumed that wages will drop, to fall in line with the increased cost of living, and that the latter will not decrease, then the drop in wages will be equivalent to 25 per cent of the war wage.

In this connection it may be further assumed that the cost of supplies will decrease in like proportion; whereas better practices will prevail and more efficient labor will be available than under war conditions. The net result is an estimated future reduction in cost of manganese ore on board railroad cars of 25 per cent.

It does not seem likely that railway freights will be materially lowered in the near future, except perhaps to the extent of the removal of the war tax.

On the basis of the foregoing premises, the summarized costs for war times given on page 101 become as follows:

Revised estimated average cost data on domestic manganese ores.

	Carbonate ore.	Oxide ore.	Average.
Estimated proportionate tonnage, per cent...........................	21	79	100
Estimated cost on railroad cars..	$3.75	14.25	11.72
Estimated railway freight...	$11.00	10.40	10.53
Total estimated cost, delivered.......................................	$14.75	24.65	22.25
Manganese, per cent...	36.5	41.0	40.0
Estimated cost per unit, cents..	40.4	60.0	55.6

Here again, in certain more favored districts, as regards freight rates, the oxide ores can undoubtedly be delivered to the furnaces for less than 60 cents per unit. For the Southern States, generally, rates of 50 cents per unit or less should be attained, whereas for Western ores the cost would probably exceed 60 cents. The Butte carbonate ores are the one exception in the far west, owing to exceptionally cheap mining and hauling costs.

COSTS OF FOREIGN ORES.

The principal source of foreign ore during the war was Brazil. The next most important single source was Cuba. Smaller amounts came from other points in the West Indies, Central and South America, and the Orient, chiefly India.

In prewar times, Russia and India were more important sources than Brazil. It seems probable, however, that Russia will not for a few years become a large factor once more, and that the Indian output may also be slow in reaching this country; though English ferromanganese made from Indian ores is likely to appear before long in quantities sufficient to affect the market.

Our more immediate concern is with Brazilian and Cuban ores. Some information regarding costs in these countries is reviewed in the following pages.

CUBA.

Manganese mining in Cuba is generally by open-cut methods. In only one instance has there been anything approaching true underground work. Preparation for market is generally by hand cleaning, though log washers are used at the most important mine in the island. One concentrator, involving screens and jigs, was built and operated during the war.

The price of labor rose rapidly during 1917–18, being at the last as much as $1.80 to $2 per 9-hour day.

Mining and treatment costs naturally varied from point to point, ranging between $5 and $8 per ton of product.

Much of the ore was hauled from mine to railroad by mules or oxen, the cost varying, with the distance, from $2 to $10 per ton. A small amount was even packed on mules at a cost of $10 and upward.

Freight rates to shipping port ranged from less than $1 a ton up to $2.50 and more, according to the length of haul. Charges for loading into ship were sometimes as high as $1.50. Lighters were used in Santiago harbor.

Royalties were advanced materially during the war period. Before the war, $1 per ton was usual, and some leases were made for 80 cents. The figures for 1918 ranged from $1.50 to $2.50. In one instance a sliding scale was used, based on the delivered value of the ore. This scale started at $1,75 for ore worth $35 per ton, and advanced 25 cents for every $5 additional value for the ore. This meant approximately $2 for 40 per cent ore.

The foregoing figures are assembled for convenient comparison in the following table:

Manganese ore costs in Cuba.

Mining and treatment	$5.00 to $8.00
Hauling to railroad	2.00 to 12.00
Cuban railroad freight	1.00 to 2,50
Port costs and charges	0.75 to 1.50
Royalty	1.50 to 2.50

The two largest producers in Cuba were situated, as regards the bulk of their output, directly on the railroad or on tram lines leading to it. This factor must be considered with others in estimating the average cost. The total cost of ore in the ship is taken as $14 per ton, including the above items and some further miscellaneous and general charges. It is, of course, understood that no amortization of investment is included in this figure.

The actual costs at a small operation whose daily output was 12 to 20 tons of hand-cleaned shipping products were as follows:

Detailed costs for one Cuban operation.

Mining and cleaning	$6.50
Hauling to railroad	2.05
Surface rights	.80
Miscellaneous and general	2.00
Railroad freight to Cuban port	2.02
Handling into ship	1.50
Total per ton of product	14.87

The operators owned the ore deposit and therefore paid no royalty. They were heavily penalized, however, by the owners of the surface.

The 1916 costs at another operation, situated on the railroad and at which the ore was washed in log washers, were as follows:

Detailed cost at another Cuban operation, 1916.

Mining and washing	$5.50
Miscellaneous and general	1.00
Surface rights and royalty	1.10
Railroad freight and switching	1.79
Handling into ship	.72
Total	10.11

The costs of this latter operation were probably $3 to $5 higher during 1918.

BRAZIL.

Below is a statement of costs of mining manganese ore at four operations in Brazil.

Cost data from four operations in Brazil.

Item.	Operation.			
	A.	B.	C.	D.
Mining, f. o. b. cars	$5.00	$4.00	$5.00	$5.00
Export tax	1.85	1.85	1.90	1.85
Railway freight rate to Rio de Janeiro	1.50	1.50	a 5.35	a 5.40
Dockage at Rio de Janeiro	.65	.65		
Lighterage at Rio de Janeiro			2.50	2.50
Transfer from narrow to broad gage railroad			.25	.25
Haulage to station				2.50
	9.00	8.00	15.00	17.50

a Companies C and D are handicapped by increased freight rates. The other two companies are still operating under old contracts.

In addition to the above figures a States tax of $3.20 was imposed during the war. Further, royalty charges up to $1.75 per ton were not uncommon. The total range of costs for Brazilian ore into ship would therefore be $12 to $22.50 per ton. As the low cost is for one of the largest and most important mines, it is probably that a weighted average would be approximately $15.

OCEAN FREIGHT RATES.

Freight rates on manganese ores from Brazil, Cuba, and India to the United States and England are given below:

Freight rates on manganese ores by vessel to United States and England.

Pre-war, India to the United States...................................		$4.32
Do.	India to England..	3.84
Do.	Brazil to the United States or England.....................	2.88
Do.	Cuba to the United States................................	1.50 to 2.50
1916,	Brazil to the United States..............................	5.50 to 6.50
1918,	Brazil to the United States..............................	15.00
1918,	Cuba to the United States...............................	9.50

The rates of $9.50 and $15 per long ton of manganese ore may be taken as representative, respectively, for Brazil and Cuba during 1918. These figures are just about five times the pre-war rates. It is understood that they include insurance.

The average rail freight during the war from Atlantic seaboard to consumer was probably about $3 per long ton.

SUMMARY OF COST OF FOREIGN ORES.

Assembling the foregoing figures, we have a total estimated average cost for ore delivered at the furnace in 1918 of $23.50 for Cuba and $30 for Brazil. On a unit basis this would be 62 cents for Cuba (for ore containing 38 per cent manganese) and 67 cents for Brazil (for ore containing 45 per cent manganese).

DISCUSSION OF FUTURE POSSIBILITIES.

Before the war such ore as was shipped from Cuba was of a higher grade than that mined during the war, and was sold at the Atlantic seaboard for $10 to $12 per ton, or for, say, $8 to $10 per ton into ship at a Cuban port. This was very close to the average cost at that time. It is not likely that the cost of ore from Cuba will drop in the future much below $11 to $12, but it is certain that ocean freights will drop materially, not improbably to the pre-war figure.

On this assumption we may expect to see Cuban ores reach the United States Atlantic seaboard in the not remote future at a cost of about $13.50, or, say, $16.50 a ton delivered at the furnace if the United States rail freight is not reduced. The quality of the ore, in order to be saleable, will have to be better than it was during 1918. If it contains an average of, say, 41 units of manganese, the estimated future cost delivered to the furnace becomes 40 cents per unit.

The Brazilian problem is somewhat more complicated, as it is uncertain what will be done by the Brazilian Federal and State

Governments with regard to the present high taxes. Other Brazilian costs will probably not be much reduced. If we assume that the taxes, which in 1918 amounted to practically $5 per ton, are reduced two-thirds, the estimated average cost of $15 becomes $11.67, or say $11.50 to include some possible slight reduction in other directions.

It is interesting to compare this estimate with the selling price of Brazilian ores at Rio de Janeiro in former years, which were as follows:

Selling prices of Brazilian ores at Rio de Janeiro, 1914–1918.

Year.	Selling price per ton, ries.	Approximate equivalent.
1914	22, $000	$7.50
1915	29, $000	10.00
1916	55, $000	13.50
1917	93, $000	23.25
1918	117, $000	29.25

The real difference in the future will come in the matter of ocean freights. If these drop to nearly their former level, it should be possible to deliver Brazilian ores to the United States consumer for approximately $18.50 per ton. This figure includes $13.50 for cost into ship at Rio de Janeiro, $3 ocean freight, $1, say, for insurance, and $3 for United States rail freight to the furnace.

Brazilian ores reaching this country were formerly of higher grade than those imported during the war. If the grade now improves slightly, say to 46 per cent manganese, the estimated average future unit cost becomes 40 cents.

Thus it may fairly be expected that both Brazilian and Cuban ores can before long be delivered to the United States consumer for about 40 cents per unit. At an equal price, Brazilian ore will always have the preference owing to the higher content of metallic manganese.

COMPARISON OF DOMESTIC AND FOREIGN COSTS.

The estimated average costs for manganese ores, per long ton delivered to the consumer, are compared in the following table:

Estimated average costs of manganese ores delivered to the consumer during 1918 and in the future.

Kind of ore.	Estimated average cost during 1918.			Estimated average cost in the future.		
	Cost per long ton.	Average manganese content, per cent.	Cost per unit, cents.	Cost per long ton.	Average manganese content, per cent.	Cost per unit, cents.
Domestic:						
Carbonates	$16.35	36.5	45.0	$14.75	36.5	40.4
Oxides	29.70	41.0	70.7	24.65	41.0	60.0
Foreign:						
Cuba	23.50	38.0	62.0	16.50	41.0	40.0
Brazil	30.00	45.0	67.0	18.50	46.0	40.0

This comparison indicates that the oxide ores could not compete with foreign ores even when protected by high ocean freights. Had foreign ores moved freely and been sold at strictly competitive prices, the domestic manganese industry could not have developed except perhaps in certain more favored districts in the Southern States, and in the Butte district (carbonate ores). The urgent need for diverting shipping into other channels, with the consequent restrictions and embargoes on imports, together with the fixing of high prices in order to stimulate production to replace such imports, created highly artificial conditions which practically removed the element of competition. In short, the ores had to be had "at any price."

The comparison further shows that when ocean freight rates return to more nearly their normal level, even the most favored oxide ores, as well as the carbonate ores, will have hard going in a competitive field.

Looking into the more remote future, it is not impossible that the costs of Cuban and Brazilian ores will have to be reduced still further in order to compete with Indian and Russian ores, thereby placing domestic ores at a still greater disadvantage unless domestic costs are correspondingly decreased.

It must furthermore be remembered that at the same price per unit the higher grade ore will always have the preference. This fact is reflected in the invariable sliding scale of prices, the higher price per unit being paid for the higher grade ore. In smelting manganese ore, not only does it cost money to melt and remove the impurities as slag, but the loss of manganese is more or less directly proportionate to the volume of the slag. Domestic ores, being generally of lower grade than foreign ores, command a lower price.

Other factors which militate against domestic ores in competition with foreign ores are that they are variable in grade while foreign ores are far more uniform, and their production is irregular while the foreign supplies are generally dependable.

CHAPTER 8.—PRODUCTION OF MANGANESE ALLOYS IN THE BLAST FURNACE.

By P. H. ROYSTER. [1]

INTRODUCTORY STATEMENT.

This chapter gives data collected by the Bureau of Mines on the operation of blast-furnace plants in the United States producing manganese alloys. This information should be of value to the furnace operator; certainly, without necessarily agreeing with any generalizations and conclusions introduced, he can determine from the data assembled in Table 17 what has been done by others with raw materials and furnace conditions approximating his own, provided comparable combinations can be found in the results presented.

The operation of a blast furnace even for making pig iron is highly empirical. Past furnace records and previous experience are nearly as necessary to successful operation as good coke and a hot blast. As in the past, the production of manganese alloys in this country has been comparatively small, and most of the men in charge of manganese furnaces have had relatively little experience on such ores. Hence, the information obtainable from the few records extant is presented herein in considerable detail, to compensate in part for the paucity of data that can be consulted by furnace operators.

The attempt has been made to have this investigation cover the widest possible range of materials and operating conditions. Data from 18 furnaces will be found in the tables following. The information was obtained in the summer of 1918, and includes every furnace in blast that was making manganese alloys, with the exception of the furnaces of the Carnegie Steel Co., and of the Lavino Co., which were unwilling to cooperate in the investigation. The companies which by their generous cooperation have made the report possible are the American Manganese Co., Bethlehem Steel Co., B. & B. Trading Co., Buffalo Union Furnace Co., Colorado Fuel and Iron Co., John B. Guernsey & Co., Donner Steel Co., E. E. Marshall, Miami Metals Co., New Jersey Zinc Co., Seaboard Steel & Manganese Corporation, Southeastern Iron Corporation, and Wharton Steel Co.

---

[1] Assistant physicist of the Bureau of Mines, with headquarters at the mining experiment station at Minneapolis, Minn. The investigation reported in this chapter was made in cooperation with the School of Mines of the University of Minnesota.

FURNACES INVESTIGATED.

The 18 furnaces investigated have been assigned letters and are designated throughout this report by these letters. The physical dimensions of 14 of these furnaces are given in Table 16 following.

TABLE 16.—*Dimensions of 14 of the 18 blast furnaces investigated.*

[All dimensions in feet.]

Furnace.	Distance from stock line to tuyères.	Bosh diameter.	Hearth diameter.	Stock-line diameter.	Distance from tuyères to hearth bottom.	Bosh height.
A	62.1	13.4	8.8	(?)	5.9	8.8
B	55.0	16.5	9.5	(?)	(?)	11.2
C	66.2	14.3	9.5	10.0	5.0	9.3
D	55.8	14.8	9.5	10.0	8.3	17.7
E	68.5	16.0	10.0	10.0	6.5	(?)
F	57.8	15.2	10.5	12.5	7.6	12.5
G	63.0	18.0	11.5	13.5	8.3	12.1
H	53.6	17.0	12.5	13.0	5.7	12.8
I	55.3	18.5	12.5	12.5	7.6	12.0
J	67.4	18.3	13.0	12.5	15.9	12.5
K	63.0	18.6	13.2	14.5	6.0	14.0
O	(?)	17.7	11.0	13.0	(?)	11.9
Q	65.5	17.5	12.5	12.9	6.3	12.5
R	66.0	19.0	13.0	13.0	6.0	12.5

COLLECTION OF DATA.

CHARACTER.

On every visit of the Bureau of Mines field party to the different furnaces the companies have opened for inspection all the furnace records, charge sheets, and chemical analyses in their possession. Periods of ten days' continuous operation were selected during which the various factors of blast temperature, silicon in the alloy, manganese in the slag, etc., were relatively constant, and all the figures applicable to such a period were copied and averaged. The average figures for such 10-day periods constituted the data on a single "experiment" or "run." The guiding principle in the selection of the experimental periods was to cover the extreme range of furnace conditions; for example, operation with high and low grade ores, and with both high and low blast temperature.

COMPLETENESS.

The information obtainable without undue trouble from a blast furnace includes the chemical analyses, the physical condition, and the weights of the ores, stone, coke, coal, scrap, and wind charged and of slag, metal, dust, gas and scrap produced, with the temperature observable in the hearth, the temperature of the blast entering the furnace, and the temperature of the metal, slag, and gas coming out of the furnace. Even with the aid of such data, it is seriously to be doubted whether anything like a complete description, of what takes place in a furnace can be given. In the present investigation, the data obtained fell short of what must be considered essential to a complete understanding of the smelting problem. Information regarding gas analyses, moisture in the blast, the amount and the analysis of the dust blown from the furnace was not obtainable in any case. Analyses of coke, by far the most important single analysis in the whole group, were rather fragmentary. Furthermore, the custom of limiting the slag analysis to a determination of silica, alumina, and manganese is widespread. Naturally no measurements of metal, slag, or tuyère temperatures had been made by the companies; nor had any gas analyses been taken, save in one case. The bureau's field party made a limited number of such temperature measurements at six furnaces and determined also the CO and CO_2 contents of the gas at five furnaces. On the whole, however, the information was far from complete.

ACCURACY.

Aside from purely analytical errors, the applicability of any given analysis is questionable for two reasons: (a) The sample analyzed may not have been representative of the pile or car from which it was taken; (b) the furnace records failed to give any connecting link between car and pile analysis and the material charged on any given day. Coke was charged by volume and not by weight, making difficult an estimation of the weight of carbon charged. It is obvious that greater accuracy would be obtained from experimental periods taken over a whole month or, better, six months. It must be remembered, however, that differences in operating conditions were being sought and long-period averages flatten out most of the variations from which it was hoped to obtain information.

In a number of runs, it was possible to obtain samples of slag, metal, stone, and ore, which were reanalyzed by the bureau's chemists. No detailed comparisons are given here but the combined effect of all the differences in the analyses introduces in the manganese balance an average uncertainty of about 15 per cent. Even if, therefore, the blast furnace were a precise and sensitive heat engine with a number of definite relations existing between its operating quantities, it is not to be hoped that these relations could be determined from the kind and quantity of data collected here.

TABULATED OPERATING DATA.

Data on ferromanganese production were obtained from 11 furnaces, and are presented in Table 17 in the form of 40 10-day periods. Similiar data from 10 furnaces making speigeleisen are presented in Table 18, in the form of 30 10-day periods. The data in these tables are classified under 23 items as applied to each of the 70 periods.

Items 1, 2, 2a, and 3 give figures on the ore, coke, coal, and stone charged, the units being in pounds per ton of 2, 240 pounds of alloy made. Items 4, 5, and 6 give a partial analysis of the ore mixture. Item 7 shows the tons of alloy made in 24 hours (not including scrap made). Item 8 gives the slag weight as calculated by the bureau and is a rather uncertain quantity. Two methods of calculating the weight of the slag, both of which should give the same answer, are as follows:

b=weight of stone,

c=per cent base in slag;

assuming there is no appreciable base in the coke or ore, then

$$\text{Slag weight}=\frac{ab}{c} \tag{1}$$

Method 2.

If d=per cent silica in ore

e=weight of ore

f=per cent silica in stone

g=weight of stone

h=per cent silica in coke

i=weight of coke

j=0 .478 × per cent silicon in metal

k=per cent silica in slag, then

$$\text{Slag weight}=\frac{de+fg+hi-j}{k} \tag{2}$$

On account of the various errors and omissions in the weights and analyses involved, it is unusual to find any close agreement between the results of the two equations; in some computations the slag weight may be given as 2,500 pounds by equation 1 and as 3,500 pounds by equation 2. Each value under item 8 is the average of the two answers resulting from equations 1 and 2.

Metal analyses for Mn and Si are given by items 9 and 10. The largest disagreement in checking manganese in the metal was 10 per cent, and in checking Si, 1.1 per cent. Values for the temperature of the blast and of the top gas appear in items 11 and 12. For a given furnace, these have some significance. A comparison of figures for different furnaces should be made cautiously, however. At some furnaces, the pyrometer will indicate a blast temperature of 1650° F. with a black blowpipe; at others the pyrometer will show only 1350° F. with a red-hot blowpipe.

Still less importance should be attached to the wind blown per minute as given by engine displacement in item 13. In period 1 on furnace A 12 pounds of carbon per minute was charged and 12,870 cubic feet of air per minute indicated. This is 105 cubic feet of air per pound of carbon. Even with no carbon blown out as dust, none taken up by the metal, and none absorbed in the stack, this means a tuyère combustion producing 35 per cent CO_2, which is impossible.

Items 14 to 18 show the slag analysis. In a number of calculations the lime and magnesia had to be computed together by difference. Item 19 gives the fixed carbon in the coke and coal charged per ton of metal. Items 20 to 23 give the manganese balance. Item 23 shows the percentage of the manganese not accounted for by that in the metal and in the slag. This figure includes practically every error in weights, analyses, and computations in the preceding 22 items and is undoubtedly the least satisfactory figure in the table:

TABLE 17.—*General data from operation of ferromanganese blast furnaces.*

Period No.		1	2	3	4	5	6	7	8	9	10	11	12	13
Furnace		A	A	A	A	A	A	A	A	A	B	B	B	B
1 Ore	pounds per ton	5,981	5,308	5,870	6,000	5,090	5,475	5,720	6,166	6,990	7,560	6,212	7,545	6,025
2 Coke	do	4,658	5,475	6,290	7,133	5,016	5,628	5,565	7,352	7,836	8,070	7,776	8,435	7,534
3 Stone	do	1,950	1,780	1,826	2,320	1,632	1,805	1,860	2,172	2,112	2,700	2,150	2,475	2,365
Ore analysis:														
4 Mn	per cent	41.18	39.11	39.43	42.01	42.25	39.80	38.45	38.06	37.14	35.75	36.70	36.00	38.00
5 Fe	do	4.54	6.65	7.42	3.33	6.07	11.28	14.95	4.60	3.69	6.35	4.41	6.14	3.43
6 SiO$_2$	do	7.78	5.64	6.33	8.45	6.73	4.34	2.88	6.14	6.65	11.06	17.06	14.97	15.28
7 Ferromanganese	tons per day	45.7	37.2	28.0	33.3	35.7	40.9	41.5	29.5	31.2	83.8	85.0	83.8	34.7
8 Slag	pounds per ton	2,671	2,445	2,605	3,035	2,248	2,180	2,070	3,335	3,460	4,630	4,110	4,460	3,670
9 Mn in alloy	per cent	78.2	70.6	71.3	80.90	73.5	66.4	68.3	78.3	86.35	22.34	77.18	72.88	77.7
10 Si in alloy	do	.76	1.47	.92	1.24	1.24	.73	.85	.70	1.86	.62	.58	.74	1.43
11 Blast temperature	°F	1,308	1,268	1,196	1,232	1,432	1,279	1,303	1,175	1,150	786	770	896	782
12 Top temperature	°F	935	1,025	1,049	1,054	998	1,147	1,103	1,014	980				
13 Air per minute	cubic feet	12,870	12,820			12,510								
Slag analysis:														
14 CaO	per cent	27.2	24.1	23.9	30.1	26.5	30.1	33.1	38.2	50.2	39.5	44.6	39.1	44.8
15 MgO	do	9.7	8.9	8.6	10.3	9.6	10.9	11.5	11.8	11.7	11.1	12.5	11.0	14.1
16 Al$_2$O$_3$	do	13.0	16.2	12.2	12.2	13.5	17.0	15.4	14.1	20.2	29.9	28.7	30.5	29.8
17 SiO$_2$	do	27.8	26.2	26.3	29.8	28.7	27.9	26.3	25.5	6.3	13.6	9.5	12.3	9.1
18 Mn	do	14.6	13.6	18.9	11.7	14.9	9.4	9.4	6.9					
19 Carbon	pounds	3,810	4,440	5,360	5,830	4,060	4,560	4,480	5,992	6,380	7,090	6,810	6,700	6,600
20 Mn charged	do	2,462	2,675	2,316	2,530	2,149	2,180	2,200	2,260	2,600	2,700	2,290	2,715	2,292
21 Mn to alloy	per cent	71.2	76.4	69.0	72.0	76.3	71.2	71.0	77.8	66.8	60.8	75.8	60.1	76.0
22 Mn to slag	do	15.8	18.4	22.0	14.0	15.5	9.5	9.4	10.1	7.0	22.5	17.1	21.9	14.6
23 Mn lost in stack	do	12.9	6.2	8.1	14.0	8.0	19.3	19.6	12.0	23.7	17.5	7.1	18.0	9.4

Period No. Furnace	14 B	15 F	16 F	17 F	18 G	19 G	20 I	21 I	22 J	23 J	24 J	25 J	26 K
1 Ore..........pounds per ton	8,168	7,098	4,518	4,881	7,525	6,060	8,159	5,980	8,560	5,040	4,900	4,990	4,040
2 Coke..........do	9,510	7,539	6,092	6,431	5,460	4,863	7,283	4,659	6,940	500	4,900	4,955	6,276
2a Coal..........do					546	459				432	415	419	
3 Stone..........do	3,820	3,166	2,091	2,084	2,941	2,675	3,160	2,166	2,516	1,216	128	1,255	1,985
Ore analysis:													
4 Mn..........per cent	34.00	42.91	43.68	45.76	32.53	32.33	39.26	41.64	34.83	44.75	45.20	43.75	50.65
5 Fe..........do	7.75	4.09	6.71	5.90	5.60	6.60	2.40	5.54	4.00	4.08	3.80	4.18	7.37
6 SiO2..........do	17.84	11.30	6.87	8.44	12.62	12.63	15.54	7.43	14.92	2.50	2.53	2.86	6.02
7 Ferromanganese..........tons per day	35.0	32.0	44.9	44.7	56.5	60.7	60.0	82.7	60.8	94.1	70.0	97.9	46.9
8 Slag..........pounds per ton	5,740	4,100	2,470	2,610	4,600	3,765	5,040(?)	2,556	6,290	2,050	2,020	1,985	2,190
9 Mn in alloy..........per cent	74.5	78.3	79.1	78.8	78.2	67.5	71.4	77.1	69.1	79.2	79.7	79.3	75.9
10 Si in alloy..........do	.58	.58	.83	1.12	1.53	1.61	1.92	.56	2.05	0.75	0.63	0.61	2.43
11 Blast temperature..........°F	880	726	993	1,027	1,140	1,183	963	1,277	1,108	1,156	1,257	1,172	900(?)
12 Top temperature..........°F		757	328	143	730	780	1,016	934	849	495		852	
13 Air per minute..........cubic feet		14,380	22,500	22,500	17,500	16,900	17,100	16,500	20,000	15,900	20,000	19,500	
Slag analysis:													
14 CaO..........per cent	35.5	28.4	35.0	33.3	25.2	28.2	26.6	31.4	24.2	26.0	27.3	27.9	45.4
15 MgO..........do	10.7	6.5	9.9	11.2	10.7	12.8	14.1	12.6	11.1	8.9	7.1	7.1	1.7
16 Al2O3..........do	31.3	11.1	10.3	11.5	11.6	11.9	15.2	11.3	11.4	24.2	28.5	24.5	14.7
17 SiO2..........do	15.6	29.1	27.5	27.8	29.8	30.9	27.7	26.0	30.4	21.7	21.3	20.9	27.3
18 Mn..........do		19.8	13.2	11.3	13.3	11.6	11.4	12.0	15.1	8.1	13.3	12.1	5.4
19 Carbon..........pounds	8,340	6,220	5,030	5,310	5,140	4,563	5,990	3,825	5,550	4,395	4,340	4,434	5,420
20 Mn charged..........do	2,760	3,043	2,100	2,218	2,450	1,999	3,200	3,435	2,985	2,256	2,228	2,168	2,048
21 Mn to alloy..........per cent	60.6	56.1	84.4	80.2	71.5	77.0	82.7	71.0	81.2	78.7	80.3	81.6	83.0
22 Mn to slag..........do	32.4	25.6	15.5	13.3	24.9	21.8	18.0	12.6	17.2	7.4	12.1	11.1	6.8
23 Mn lost in stack..........do	7.0	18.3	.1	6.5	3.6	1.2	39.3	16.4		13.9	7.6	7.3	11.2

TABLE 17.—General data from operation of ferromanganese blast furnaces—Continued.

	27	28	29	30	31	32	33	34	35	36	37	38	39	40	Average
Period No. / Furnace	K	K	K	L	L	L	L	L	L	P	P	C	M	M	
1. Ore..........pounds per ton	4,480	4,650	5,215	6,770	6,565	5,312	6,222	5,465	5,200	7,000	4,117	5,245	6,598	6,598	5,992
2. Coke..........do	3,802	6,388	6,329	7,438	8,459	6,500	6,421	4,855	5,164	5,350	5,905	5,704	7,118	7,100	6,320
2a. Coal..........do													998	987	6,200
3. Stone..........do	2,266	2,184	2,386	2,965	3,100	2,582	2,528	2,200	2,155	3,591	1,686	2,773	3,042	3,510	2,349
Ore analysis:															
4. Mn..........per cent	51.13	50.11	46.94	39.97	39.68	35.45	35.39	32.38	33.53	38.70	45.50	40.70	45.11	46.36	40.33
5. Fe..........do	7.10	7.30	10.84	3.51	3.69	9.54	4.50	8.02	7.81	4.25	6.53	7.45	3.66	4.15	6.93
6. SiO$_2$..........do	6.35	5.57	6.24	7.53	7.40	8.91	6.28	10.11	8.63	15.99	4.85	7.01	6.59	4.08	8.60
7. Ferromanganese....tons per day	85.8	49.5	77.5	48.1	43.0	64.7	50.5	77.2	77.1	61.3	78.0	35.0	42.0	38.8	51.7
8. Slag..........pounds per ton	2,580	2,315	3,285	3,126	3,305	2,896	2,727	2,452	2,256	4,468	1,880	2,700	3,900	3,480	3,196
9. Mn in alloy..........per cent	74.8	75.9	62.5	80.7	81.1	69.8	78.4	70.4	71.8	78.1	80.9	74.9	75.3	74.5	74.9
10. Si in alloy..........do	0.29	1.82	0.26	1.29	1.47	0.75	2.72	1.46	2.22			2.19	1.20	1.30	1.15
11. Blast temperature..........°F	970	960(?)	998	1,065	1,043	1,099	1,204	1,457	1,207	1,450(?)	1,450(?)	1,387	1,242	1,192	1,135
12. Top temperature..........°F	965		845	1,162	1,152	1,162	1,127	1,123	1,140	750(?)	750(?)	52	905	910	690
13. Air per minute..........cubic feet	17,500		19,100	17,800	17,700	18,800	17,800	18,800	18,800						
Slag analysis:															
14. CaO..........per cent	37.1	46.6	33.3	40.4	47.7	42.0	43.3	41.9	44.3	41.0	45.6	29.8	38.8	43.3	41.75
15. MgO..........do	1.1	1.5	1.4	6.8	5.7	5.6	4.3	6.3	4.9	14.9	18.8	18.2	14.4	11.1	14.0
16. Al$_2$O$_3$..........do	11.4	13.8	11.1	11.7	12.8	11.8	14.5	10.5	13.2	35.4	20.3	10.7	25.7	28.7	28.1
17. SiO$_2$..........do	28.8	27.6	20.5	31.1	27.4	29.9	27.3	31.6	31.0	5.0	9.9	32.2	15.2	11.8	10.6
18. Mn..........do	14.3	8.8	19.6	8.8	6.9	7.2	9.8	4.5	4.4	9.9		7.3			
19. Carbon..........pounds	4,690	5,450	5,280	6,740	7,655	5,951	5,650	4,432	4,568	5,050	4,700	5,280	6,500	6,471	5,323
20. Mn charged..........do	2,288	2,330	2,442	2,710	2,608	1,888	2,200	1,770	1,745	2,718	2,060	2,138	2,970	2,970	2,362
21. Mn to alloy..........per cent	73.1	73.0	57.2	66.9	69.7	82.8	79.7	89.0	92.1	64.5	88.0	78.7	57.1	54.5	72.0
22. Mn to slag..........do	16.2	5.7	26.3	10.3	5.9	10.1	11.1	8.1	8.7	6.4	4.9	12.3	22.9	13.4	14.7
23. Mn lost in stock..........do	10.7	21.3	16.5	22.8	24.4	7.1	9.2	2.9	-0.8	29.1	7.1			32.1	12.8

TABLE 18.—General data from operation of spiegeleisen blast furnaces.

Period No.	41	42	43	44	45	46	47	48	49	50	51	52	53	54	55
Furnace	C	C	C	C	D	D	D	D	D	E	E	E	E	E	H
1. Ore......pounds per ton	5,192	5,795	5,488	5,365	4,510	4,505	4,516	4,420	4,282	5,290	6,560	6,945	6,550	6,900	4,525
2. Coke......do	3,327	3,831	2,923	3,150	3,110	3,180	3,060	3,185	3,200	3,600	4,560	4,640	4,518	4,100	2,970
2a. Coal......do	749	926	725	693											
3. Stone......do	1,568	2,052	1,655	2,328	1,436	1,691	1,721	1,684	1,530	2,981	3,290	3,090	3,310	3,380	1,331
Ore analysis:															
4. Mn......per cent	12.72	12.77	12.77	12.77	14.62	13.62	14.13	13.84	14.06	13.35	12.55	11.27	12.29	12.91	14.85
5. Fe......do	36.72	36.90	37.65	37.65	40.20	39.90	37.83	39.33	41.46	31.87	30.84	34.71	32.27	32.18	40.68
6. SiO₂......do	17.46	16.65	17.41	17.41	9.78	10.20	10.22	10.55	9.09	19.80	20.81	21.13	22.30	20.30	9.62
7. Spiegeleisen......tons per day	44.7	30.2	46.0	51.0	88.3	93.0	95.8	94.7	74.9	83.0	73.0	64.0	66.0	69.0	113.3
8. Slag......pounds per ton	2,614	2,990	2,330	2,930	2,035	2,050	2,155	2,135	1,940	3,765	4,615	4,400	4,755	4,430	1,870
9. Mn in alloy......per cent	14.9	16.50	16.2	17.1	19.9	19.2	18.9	17.6	19.4	16.7	14.7	15.7	14.6	16.7	20.0
10. Si in alloy......do	3.1	6.9	2.1	1.2	2.3	2.6	1.0	1.3	1.3	0.7	0.8	2.4	0.8	1.1	0.9
11. Blast temperature......°F	1,234	1,304	1,344	1,211	1,178	1,166	1,132	1,194	1,100	983	914	1,088	914	1,005	1,206
12. Top temperature......°F	613	749	540	538	700	600	500	400	393			693			
13. Air per minute......cubic feet	8,660	8,740	8,660	8,800					15,660						20,600
Slag analysis:															
14. CaO......per cent	26.0	24.3	27.8	31.6	26.1	26.2	25.5	25.8	25.4	42.0	39.3	39.5	37.1	42.0	25.1
15. MgO......do	10.1	11.1	12.6	11.6	15.2	15.4	15.0	13.2	14.4	9.4	10.9	13.2	10.9	9.9	14.1
16. Al₂O₃......do	10.3	13.3	9.8	7.8	17.6	16.9	15.2	15.1	16.2	36.6	35.8	37.3	38.1	36.9	14.7
17. SiO₂......do	28.3	37.2	38.3	37.2	31.1	31.0	30.4	30.3	30.3	8.40	9.88	6.73	9.88	7.73	30.7
18. Mn......do	10.79	10.07	10.48	8.91	6.39	6.31	9.95	10.66	9.35						9.68
19. Carbon......pounds					2,930	2,880	2,780	2,990	2,910	3,000	3,330	3,570	3,760	34.20	2,710
20. Mn charged......do	660	740	695	681	660	614	638	612	601	793	823	790	802	879	672
21. Mn to alloy......per cent	49.0	50.6	53.4	55.1	67.5	70.0	66.1	64.3	72.2	47.2	40.0	46.4	40.7	42.1	66.6
22. Mn to slag......do	42.6	40.7	35.0	38.7	19.7	21.1	33.5	37.2	30.2	39.8	55.4	38.0	58.0	39.0	26.9
23. Mn lost in stack......do	8.4	8.7	11.6	6.2	12.8	8.9	0.4	-1.5	-2.4	13.0	4.6	15.6	1.3	18.9	6.5

Period No.		56	57	58	59	60	61	62	63	64	65	66	67	68	69	70	Average
Furnace.		H	H	J	J	M	M	M	M	P	B	Q	Q	R	R	R	
1. Ore	pounds per ton	5,100	4,300	5,710	5,000	5,025	4,025	5,145	5,065	4,481	5,522	6,080	5,852	5,495	5,818	5,855	5,380
2. Coke	do.	2,980	2,870	4,900	4,210	3,506	3,258	3,110	3,987	3,290	4,220	5,300	5,147	5,853	7,147	5,048	3,960
2a. Coal	do.			290	260			575	580	340						187	187
3. Stone	do.	1,423	1,280	2,155	1,980	2,104	2,026	1,882	1,703	1,460	1,880	3,491	3,921	3,728	3,946	3,460	2,358
Ore analysis:																	
4. Mn	per cent	14.80	15.88	9.02	11.86	18.8	13.3	11.1	15.96	12.76	11.06	11.10	9.83	10.25	9.85	9.08	12.76
5. Fe	do.	37.67	40.92	31.60	31.84	31.8	32.3	35.1	32.14		20.42	32.43	32.87	32.35	31.24	31.24	35.10
6. SiO₂	do.	10.79	9.19	17.70	15.37	107.0	12.9	9.6	79.6	10.98	60.0	16.60	14.59	19.97	17.41	16.48	14.84
7. Spiegeleisen	tons per day		134.0	91.7	85.5		88.0	81.3	69.0	163.5	3,980	117.3	127.0	71.4	50.6	60.3	58.3
8. Slag	pounds per ton	2,150	1,572	3,450	2,830	2,800	2,500	2,100	2,460	1,810	17.9	4,535	4,000	4,585	4,000	4,735	3,109
9. Mn in alloy	per cent	18.7	20.2	16.7	19.8	17.0	69.2	18.0	16.1	19.3	17.8	17.8	17.0	4.5	2.6	16.1	17.48
10. Si in alloy	do.	1.0	2.3	2.8	1.3	0.7	0.5	10.8	91.6			0.8	0.9	2.9	2.5	1.4	1.77
11. Top temperature	°F	1,283	1,196	880	1,068	1,008	1,316	1,172	943			980	755	708	600	600	1,080
12. Blast temperature	°F	1,500			490	505	750	804	942			511	522	533	668	488	672
13. Air per minute	cubic feet	19,300	20,500	18,700	19,500							30,000	30,300	28,800	25,000	21,300	
Slag analysis:																	
14. CaO	per cent	23.9	27.1	30.2	27.7	40.2	49.7	53.5	39.8	43.7	46.4	49.0	50.8	44.0	44.7	46.1	42.18
15. MgO	do.	14.0	16.2	11.1	9.0	13.9	12.2	12.2	14.8	18.4	11.1	9.4	0.8	0.7	11.0	11.8	12.41
16. Al₂O₃	do.	15.7	15.8	0.5	8.3	24.8	30.9	31.2	30.2	30.5	24.5	35.1	24.1	26.5	25.4	26.3	23.48
17. SiO₂	do.	10.24	20.9	30.8	33.9	7.37	4.36	11.71	10.56	6.02	4.93	3.83	4.195	4.790	4.14	3.66	7.56
18. Mn	do.		6.35	6.50	7.37					2,870	3,680	4,370	576	563	5,920	4,110	3,444
19. Carbon	pounds	2,720	2,620	4,398	3,780	2,800	2,796	2,864	3,785	564	564	571	576	563	574	576	671
20. Mn charged	do.	667	667	516	593	942	615	570	908							496	
21. Mn to alloy	per cent	66.7	67.4	72.5	75.0	40.6	69.8	70.7	41.1	76.7	68.0	69.8	66.0	63.4	65.2	78.6	60.22
22. Mn to slag	do.	29.9	15.0	18.6	31.0	20.2	17.7	39.3	28.8	19.3	33.5	32.4	51.8	51.8	33.2	33.6	32.75
23. Mn lost in stack	do.	13.4	17.6	8.9	-6.0	89.2	12.5	-10.0	30.1	4.0	-1.5	-2.2	12.8	-16.2	3.1	-10.2	6.72

DISCUSSION OF DATA ON FERROMANGANESE PRACTICE.

GENERAL DESCRIPTION OF FERROMANGANESE PRODUCTION.

No specific information on the production of ferromanganese in the blast furnace is available in technical literature. Certain principles are generally understood to be true; these probably being: (a) The blast temperature should be high, (b) the slag should be basic, (c) the hearth temperature should be high.

HIGH BLAST TEMPERATURE.

Table 17 shows that temperatures as low as 735° F. have been used. The temperature for all 40 runs was only 1,135° F. In 1876, ferromanganese was made at the Diamond furnace, Cartersville, Ga.[a] A cold blast driven by a water wheel through a single 3-inch tuyère was employed. In making a ton of 55 per cent ferromanganese, 6,050 pounds of siliceous native ore, with 35 per cent Mn, was used, indicating a 58 per cent recovery. Comparing this with period 22, furnace J, Table 12, where a manganese recovery of only 55 per cent was made with 1,100° F. blast temperature and a 35 per cent ore, it will be seen that whatever economy results from the hot blast, it has not been found at all necessary to actual and profitable operation.

BASIC SLAG.

The average ferromanganese slag represented in Table 17 has the composition 41.7 per cent base, 14 per cent alumina, 28.1 per cent silica, and 10.6 per cent manganese. The average pig-iron slag[b] runs 48.7 per cent base, 13.1 alumina, and 35.3 silica. If a "more basic slag" means that the percentage of base is greater, then the ferro slag is 7 per cent less basic than the pig-iron slag. This method of comparison, however, is hardly proper. The function of the ferro furnace is to reduce the MnO from the slag and leave a practically irreducible calcium and magnesium aluminosilicate. This reduction is never complete. What actually exists may be considered to be reducible MnO as a solute dissolved in the quarternary mixture of irreducible oxides (CaO, MnO, Al_2O_3, SiO_2) as a solvent. It is the composition of the solvent that will affect the extent of the MnO reduction reaction. Compared on the basis of base + alumina + silica = 100 per cent, the two slags analyze as follows:

Analysis of pig-iron slag and of ferromanganese slag.

	Pig iron.	Ferro-manganese.
Bases..per cent..	50.3	49.8
Alumina..do....	13.4	16.7
Silica...do....	35.3	33.5

[a] Ward, W. P., Manufacture of ferromanganese in the blast furnace: Trans. Am. Inst. Min. Eng., vol. 5, February, 1876, p. 611.
[b] Feild, A. L., and Royster, P. H., Slag viscosity tables for blast-furnace work: Tech. Paper 187, Bureau of Mines, 1918, p. 17.

The percentage of bases is identical for the two slags. The only difference is that the ferromanganese slag carries a little more alumina and a little less SiO_2. This difference, moreover, is due to the character of the materials used and is not the result of any metallurgical design.

HIGH HEARTH TEMPERATURE.

At furnaces A, B, F, I, J, and K, temperatures were taken by means of several Leeds & Northrup optical pyrometers, with telescope sighted (1) through the tuyères both with and without a tuyère glass interposed; (2) on the surface of the slag at flush; and (3) on the surface of the metal at cast. As was to be expected, the temperatures observed opposite the tuyères varied over a rather wide range. To indicate the normal temperature variation in practice, the following comparison was made: Two hundred individual readings taken at seven furnaces, chosen to give a minimum, averaged 1,438° C. These measurements at each furnace were read by at least two observers, with either two or three separate pyrometer sets, through two or more tuyères, and at five well separated periods of time scattered through two to five days. The same sort of an average selected to give a maximum read 1,653° C. The difference between the maximum and minimum here can not well be attributed to any unusual furnace operation or to errors of observations or to instrumental errors. In the face of these figures, it must be accepted that as furnaces are run at present, a day-to-day variation of 225° C. in hearth temperature is not abnormal.

Laboratory tests strongly indicate that 1,265° C. is the lowest temperature at which MnO is reducible by either carbon or carbon monoxide. The lowest practical hearth temperature therefore may safely be set at 1,350° C. Superpose a variation of 225° C. on this minimum and the limits read 1,350° and 1,575° C. The maximum tuyère readings at the hottest furnace observed averaged 1,750° C. The minimum readings at the coldest furnace observed averaged 1,375° C. Where a temperature phenomenon such as this covers almost the whole range of practicable metallurgical temperatures, there is no room left on the temperature scale which the ferro furnace can monopolize as exclusively its own. Average temperatures taken on these seven furnaces were approximately as follows, the tuyère readings being corrected for the absorption by the tuyère sight glass when present and the temperature of the metal and slag corrected for emissivity as given by Burgess: [a]

	°C.
Temperature opposite tuyères	1,550
Temperature of metal at cast	1,386
Temperature of slag at flush	1,426

---

[a] Burgess, G. K.: Temperature measurements in Bessemer and open-hearth practice: Bureau of Standards Technologic Paper 91, 1917, pp. 8-9.

RÉSUMÉ OF CONDITIONS IN FERROMANGANESE FURNACE.

It has been shown that in changing a furnace over from iron to ferromanganese it is neither necessary nor usual to raise the blast temperature, to increase the basicity of the slag, or to run with a hotter hearth.

True, there are certain resultant changes in the practice, the most important one being that the coke requirements are tripled. The daily tonnage is reduced to about one-third; the top temperature rises to about 1,000° to 1,500° F.; the carbon monoxide content of the top gas is increased about 50 per cent; and the blast pressure drops to 3 to 8 pounds per square inch. The operation of the furnace becomes easier; stock descends without hanging, slips become infrequent, fewer tuyères burn out, sulphur is taken care of completely as MnS in the slag, and further, there are no silicon requirements to be met. All furnace difficulties practically resolve themselves into one problem—to get into the metal the maximum percentage of the manganese charged.

THEORY AS TO FUEL REQUIREMENTS FOR PRODUCING FERRO-MANGANESE.

A number of heat balances with values in B. t. u. have been published on the iron furnace, but it seems hardly justifiable to construct a heat balance for the ferro furnace for comparison, for the following reasons: At room temperature, the heat of combustion of $C + O_2$ to CO_2 is either 4,250 or 4,430 B. t. u. per pound of C; at 1,500° C. it is either 4,550 or 4,950 B. t. u., depending on the authority selected. There is a wide lack of agreement among the few existing determinations of the heat of combustion, and greater lack of agreement among the experimental values for the specific heat of carbon, of oxygen, and of CO. At 600° C. the specific heat of carbon may be 0.31 or 0.44. At metallurgically interesting temperatures (between 1,400° and 1,800° C.), the discrepancy in figures for the total heat of carbon is about 30 per cent.

The specific heat of CO and of N_2 at 1,600° C. may be either 0.266 or 0.298. For the iron furnace the heat imparted to the gaseous products of reaction per pound of metal is either 972 or 1,280 B. t. u. The rise in temperature of gases above that of the burning coke is, therefore, either 696° or 1,070° F., 54 per cent discrepancy. The continued reprinting of some arbitrarily selected figure for the necessary thermal constants is both practically and scientifically objectionable and tends almost invariably to erroneous conclusions.

It is possible, however, to give an idea of the reason for the difference between the fuel requirements of the iron and those of the ferromanganese furnace and yet avoid excessive accumulation of meaningless B. t. u. figures. The comparison following is of interest.

Comparative operating data for pig iron furnace and for ferromanganese furnace.

	Pig Iron.	Ferromanganese.
Carbon charged, pounds	1,727	5,524
Stone charged, pounds	984	2,349
Slag, pounds per ton	1,160	3.196
Gas analysis:		
CO_2 (by weight), per cent	21.43	10.44
CO, per cent	23.13	31.00
N, per cent	55.02	58.36
Blast temperature, °C	555	613
Top temperature, °C	204	363

The figures for the iron furnace are the averages for eight of the largest blast furnace plants in the country. The figures for the ferro furnace are the averages from Table 17. The calculations of the weight of the blast and of the gas and the determination of the amount of carbon absorbed by CO_2 and that burned at the tuyères for the ferro furnace are given in Table 19 following.

TABLE 19.—*Selected data on operation of ferromanganese blast furnaces.*

	Pounds.
Carbon charged per ton of alloy	5,523
Carbon combined with metal (6.5 per cent C)	145
Carbon blown out of stack as dust (9.6 per cent assumed)	530
Carbon available as fuel	4,848
Carbon from CO_2 in stone	263
Total carbon gasified	5,111
Carbon in top gas by weight, per cent	0.1615
Top gas weight (per ton of alloy: 5,111÷0.1615)	31,647
Weight of N_2 in top gas: (0.5836×31,647)	18,469
Weight of air blast: (18,469÷0.7672)	24,073
Weight of O_2 in blast: (24,073×0.2310)	5,561
Weight of carbon burned at tuyères: (5,561÷1.333)	4,172
Carbon absorbed in stack (4,848−4,172)	576

The two most important thermal reactions in the blast furnace are:

(1) $C + O = CO + 4,450$ (or 4,950) B. t. u. per pound of C (at 1,600° C.).

(2) $CO_2 + C = 2CO + 5,410$ (or 5,900) B. t. u. per pound of C (at 850° to 1,300° C.).

In reaction 1 both the carbon and the oxygen are supposed to be at the temperature of the solid stock in the combustion zone. As already shown this temperature is probably 1,550° C. for the ferro furnace and 1,590° C. for the iron furnace. Subtracting from the total heat of the reaction sufficient heat to bring the blast up to the temperature of the solid stock in the combustion zone gives a net heat to the gaseous product of reaction of 972 (or 1,280) B. t. u. per pound of metal made for the iron furnace and of 3,690 (or 4,800) B. t. u. for the ferro furnace.

In the ferro furnace, either in the hearth, at the tuyères, or just above them all the manganese in the metal was reduced from MnO. All of the CO_2 resulting from this reduction is itself reduced by solid carbon to CO; there is involved in this reaction 0.218 pound of carbon per pound of metal, resulting in a loss of 1,180 (or 1,290) B. t. u., leaving 2,500 to 3,500 B. t. u. to be used in heating the ascending gas above the temperature of the solid stock. The ferro-furnace gas weighs 12.6 pounds per pound of metal and the iron-furnace gas 4.2 pounds. Hence, in the ferro furnace the amount of heat available for raising the temperature of the gases is two to three times greater than in the iron furnace, and the weight of gas to be heated is three times as great. The rise in temperature, therefore, is either the same or less in the ferro furnace than in the iron furnace.

It is unwise to go here into any detailed consideration of why it is necessary to heat the ascending gas to this rather definite but somewhat unknown elevation above the temperature of the descending stock. It may be pointed out, however, (a) that some 250 tons of solid stock must be heated from 70° F. to above 2,800° F. in 24 hours; (b) that the greater part of this stock, being coke and mineral oxides, has a low thermal conductivity; (c) that the ascending gas stream does not have a chance to flow around every separate piece of coke, stone, and ore; and (d) that the transfer of heat from gas to solid is, under the most favorable conditions, a slow process. In order to heat the stock in the required time, therefore, the gases must at every point in the furnace be appreciably hotter than the stock; and this means not several degrees hotter but several hundred degrees hotter. Thus, the 4,165 pounds of carbon burned at the tuyères in the ferro furnace gives the gas stream no greater temperature elevation above the solid stock than does the 1,386 pounds burned in the iron furnace.

COMPOSITION OF FURNACE GAS.

At five furnaces samples of downcomer gas were taken and analyses made for the CO and CO_2 contents. The results obtained were somewhat variable but indicated a distinct tendency toward what might be called a normal composition.

As regards oxygen the evidence was that the true furnace gas contains no oxygen. Well taken samples showed 0.0 to 0.2 per cent. It was felt, however, that this oxygen could be safely attributed to leakage of air into the sample. Where the conditions of sampling permit the chance of contamination by the atmosphere, the oxygen content might be as high as 1.0 per cent.

The CO_2 content varied between the extreme limits of 11.6 and 3.5 per cent, with averages for the five furnaces of 6.7, 7.6, 7.1, 5.9, and 6.0 per cent. The CO content varied from 25.0 to 38.0 per cent, with averages for the five furnaces of 31.8, 34.2, 30.9, 30.0, and 31.0. per cent.   The average analysis for the five furnaces was CO_2, 6.67 per cent; CO, 31.7 per cent.

Moissan[a] records an analysis of gas from a furnace making 60 per cent ferromanganese as CO_2, 5.5 per cent; CO, 30.0 per cent. This run, made in 1876, was one of the earliest attempts at the production of manganese in blast furnaces.

The analyses of the gas from nine of the largest iron blast furnace plants in the country show the following CO and CO_2 percentages:

Percentages of CO and of CO_2 in gas from nine iron blast furnace plants.

	CO₂	CO.
Furnace 1	12.6	26.0
Furnace 2	13.2	25.4
Furnace 3	13.0	24.3
Furnace 4	16.3	22.8
Furnace 5	14.3	24.3
Furnace 6	14.9	23.5
Furnace 7	15.6	23.2
Furnace 8	14.6	24.9
Furnace 9	13.3	25.4

The average complete analysis of the nine samples is:

Average analysis of gas from nine iron blast furnaces.

	Per cent by volume.	Per cent by weight.
CO₂	14.34	21.43
CO	24.32	23.13
CH₄	.36	.20
H₂	3.16	.22
N₂	57.82	55.02

Assuming from lack of better information that the H_2 and CH_4 content of the gases of the ferro furnace is the same as that of the iron furnace, and estimating the N_2 by difference, the analysis by volume and by weight of ferro gas is:

Assumed average analysis of gas from nine ferro furnaces.

	Per cent by volume.	Per cent by weight.
CO₂	6.67	10.44
CO	31.17	31.00
CH₄	.36	.17
H₂	31.6	.21
N₂	58.64	58.18

"DIRECT" AND "INDIRECT" REDUCTION.

In Table 19, the weight of the furnace gas is calculated to be 31,647 pounds per ton of metal, which means, from the above analyses, 3,300 pounds of CO_2.   The CO_2 from 2,341 pounds of the

[a] Moissan, H. H., Le manganese et ses composes, Encl. Chimique, 1886, p. 26.

stone is 968 pounds, from the reduction of 190 pounds of Fe_2O_3, 156 pounds, and from the reduction of 3,770 pounds of MnO_2 to MnO by CO, 1,885 pounds; total, 3,009 pounds. This indicates a 9.56 per cent CO_2 content in the gas as against 10.44 per cent given on page 121. There is left 291 pounds of CO_2, which may be due to reduction of MnO by CO in the lower part of the furnace, to carbon deposition from CO in the upper part of the furnace, or—more likely—to experimental error either in the analyses, or in their application to the average practice given in Table 17. The evidence points closely to a 100 per cent reduction of MnO "directly" by carbon.

It is well to point out that "direct" reduction by carbon probably means nothing of the kind. As in most important metallurgical problems, no experiment yet made settles this question. It can be assumed that the reaction:

$$MnO + CO = Mn + CO_2 \qquad (A)$$

is the reaction by which the final reduction of MnO takes place at a temperature somewhere between 1,265° and, say, 1,400° C. At this temperature the following reaction takes place, almost to completion, and with great rapidity:

$$CO_2 + C = 2CO \qquad (B)$$

Thus the result of the two reactions going on simultaneously, although not necessarily in the same part of the furnace, gives the same products as the single reaction:

$$MnO + C = Mn + CO \qquad (C)$$

As A and B together are not distinguishable from C by any calculation based upon gas analysis, one must be content to realize that either of these two reactions may be the one that takes place.

BURDENING AND DRIVING THE FURNACE.

It is all very well to apply to the furnace records accepted metallurgical theory, but practically the important thing is to have at hand something definite in the way of a guide to burdening and driving the furnace. The following discussion, therefore, is based exclusively on the figures in Table 17 and is free from any theoretical considerations whatever. It is an attempt to establish from the whole mass of data certain relationships between the 23 items of each run. There are only three factors that the furnace operators can change at will. They are (1) ratio of coke and ore; (2) ratio of stone and ore; and (3) quantity of wind blown per minute. It is assumed here that the blast temperature is kept as high as stove conditions permit. The number of rounds charged and the tons made per day are decided by the furnace itself. With the above three factors determined, there is nothing left for the operator to do but watch the furnace.

FUEL REQUIREMENTS.

Aside from the relatively constant amount of fuel needed to preheat, reduce, and melt the metal, the amount of carbon fuel required per ton of alloy should depend upon the temperature and humidity of the blast, the weight of the slag, and the temperature maintained in the combustion zone and hearth. That is, there should exist a relationship between five variables. In an attempt to determine this relationship from the data taken, the question of humidity must be dismissed from consideration, not necessarily because it is an unimportant factor but because no information is available concerning it. In the matter of hearth temperatures, no direct measurements are at hand; the pyrometer observations made by the bureau's investigators covered too short a period to prove of material assistance at this stage of the investigation. In a general sense the silicon in the alloy may be regarded as a thermometer, but for a given hearth temperature the amount of silica reduced to silicon undoubtedly depends on a number of other factors, including probably the composition of the slag and the speed of operation. The best indication of the temperature of the hearth is given by the amount of unreduced MnO in the slag.

The following procedure was adopted: The results of the 40 runs were arranged in order of decreasing slag weight and placed in four groups. The averages by groups appear in the first part of Table 20. The results were also arranged in the order of decreasing blast temperature, being again placed in four groups and the results averaged by groups. The averages are given in the second part of Table 20 following.

TABLE 20.—*Results of furnace runs arranged to show variation of carbon fuel with weight of slag and with blast temperature.*

VARIATION WITH WEIGHT OF SLAG.

Slag per ton.	Carbon per ton.	Blast temperature.	Carbon reduced to blast temperature of 1,130° F.
Pounds.	*Pounds.*	*° F.*	*Pounds.*
4,728	6,410	990	6,224
3,334	6,150	1,080	6,063
2,604	4,830	1,210	4,936
2,130	4,550	1,220	4,670

VARIATION WITH BLAST TEMPERATURE.

Blast temperature.	Carbon per ton of alloy.	Slag per ton of alloy.	Carbon reduced to 3,250 pounds of slag.
° F.	*Pounds.*	*Pounds.*	*Pounds.*
1,358	4,670	2,700	5,160
1,210	5,150	3,000	5,380
1,077	5,680	3,400	5,530
862	6,400	3,900	5,820

By the method of successive approximation it was found that 1 pound of extra slag requires 0.88 pound of carbon, and 1° F. extra blast temperature saves 1.33 pounds of carbon. Written in the form of an equation:

$$Ka = 4170 + 0.88 \ V - 1.33 \ T \qquad (1)$$

Where

> Ka = pounds of carbon per ton of alloy used by the average furnace,
> V = pounds of slag per ton of alloy,
> T = blast temperature in degrees F.

This equation shows, for a given slag weight, for a given blast temperature, and for average humidity, the quantity of carbon required to heat the materials in the hearth to the average hearth temperature. As has been shown, this temperature is probably about 1,550° C., but it is better not to place too much confidence in this figure.

It is of course possible to charge either more or less carbon than the amount given by equation 1. The average hearth temperature used by any given furnace operator means merely that temperature found by experience, or assumed from lack of experience, to be the most desirable to maintain. If the operator decides to use a higher temperature in practice, the furnace can be given "excess carbon," a phrase taken to mean the carbon actually charged minus the amount indicated by equation 1. When excess carbon is used, the percentage of manganese in the slag will be lower, provided all the other conditions remain constant. On the other hand, the carbon charged may be considerably lower than that given by equation 1, with a resultant hearth temperature lower than the average.

PERCENTAGE OF MANGANESE IN SLAG.

Determination of the percentage of manganese that goes into the slag is approached in the same manner as the problem of fuel requirements, with the exception that there must not be any "left over" variable such as excess carbon over requirements. The operator can at his own free will charge so little fuel that the hearth freezes, or he can charge so much fuel that he bankrupts the owners. The percentage of manganese in the slag, however, is the result of a number of metallurgical phenomena and is definitely determined by the actual furnace facts.

The data in Table 17 show that the percentage of manganese in the slag depended upon the following factors:

(1) The ratio of bases to silica in the slag;

(2) The hearth temperature as determined by the excess or deficiency of fuel above or below that indicated in equation 1;

(3) The rate of driving.

RATE OF DRIVING.

The expression found most nearly to represent the speed of operation was the pounds of gross slag produced per minute per square foot of hearth area. Gross slag here means the weight of slag-forming materials, manganese being a slag-forming material; that is, the weight of slag that would be flushed from the furnace if none of the manganese was recovered in the metal. There is an implication here—probably warranted—that the MnO_2 charged is reduced to MnO in the upper stack, but that the MnO entering the bosh is fused with the silica, alumina, and bases, and that it is reduced to metallic manganese only after it has descended molten into the hearth. The calculation of the gross slag is simple. To take the figures from the average of the 40 runs represented in Table 17, as an example, the slag weight is 3,196 pounds. The metallic manganese in the metal is 1,680 pounds. This metallic manganese was reduced from 2,170 pounds of MnO. The gross slag was therefore 3,196 plus 2,170, or 5,366 pounds per ton of metal.

It is well to notice that the more usual expression for speed of operation in blast-furnace parlance is the pounds of carbon burned per minute per square foot of hearth area. Without commenting upon this point of view as it applies to the iron furnace, its applicability to the ferromanganese furnace is extremely doubtful. The only factor limiting the speed of operation, assuming that the speed is not so great that the stock is blown out the top of the furnace, is the ability of the hearth to reduce manganese from the slag. As long as the hearth temperature is above 1,265° C., all the manganese in the slag should be reduced, provided the slag remains for a sufficient length of time in the hearth under the reducing influence of a CO atmosphere and in contact with solid carbon. Therefore, no principle of common sense is violated in taking as the criterion of speed the figure for the pounds of slag passing into the hearth from the bosh per minute per square foot of hearth area.

FIGURE 10.—Curve showing relation of percentage of manganese in slag to carbon difference (carbon charged minus carbon required for average hearth temperature conditions; see Table 17).

EQUATION FOR PERCENTAGE OF MANGANESE IN SLAG.

The results of the 40 runs represented in Table 18 were arranged in order of decreasing "excess carbon" as determined by equation 1. The results were then divided into four groups and the averages by groups recorded as in Table 21. The results are also plotted in

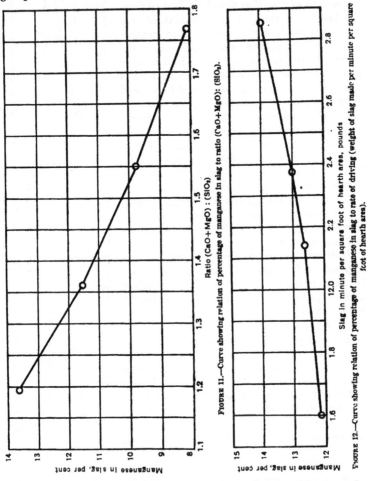

FIGURE 11.—Curve showing relation of percentage of manganese in slag to ratio (CaO+MgO): (SiO₂).

FIGURE 12.—Curve showing relation of percentage of manganese in slag to rate of driving (weight of slag made per minute per square foot of hearth area).

figure 10. The results of the same runs were rearranged in the order of decreasing basicity (ratio of bases to silica), grouped, and averaged, and the averages recorded as in Table 22. The results are also plotted in figure 11. In the same way Table 23 gives the averages for groups taken in order of decreasing rates of driving. The results are also plotted in figure 12. The three tables follow.

TABLE 21.—*Results of furnace runs arranged to show variation of percentage of manganese in slag with excess or deficiency of carbon fuel as required for average hearth-temperature conditions.*

[Results are plotted in figure 10.]

Group.	Runs included.	Furnaces included.	Carbon excess or deficiency (pounds.)	Basicity ratio.	Rate of driving.	Percentage of manganese in slag.	
						Actual.	Reduced to standard basicity[a] and rate of driving.[b]
1	4, 26, 30, 31, 32, 33, 37, 38, 40.	A, C, K, L, M, P..	+754	1.09	1.84	8.2	10.3
2	3, 7, 8, 12, 13, 14, 17, 25, 29, 39.	A, B, F, K, M....	+354	1.48	1.92	11.0	11.0
3	2, 5, 6, 10, 11, 19, 23, 24, 25...	A, B, G, J.......	− 56	1.45	2.13	12.1	11.4
4	1, 15, 18, 20, 22, 27, 34.......	A, F, G, I, J, K, L.	−765	1.37	2.36	13.3	11.8

a Standard basicity (ratio of bases to silica), 1.50.
b Rate of driving (pounds of gross slag produced per minute per square foot of hearth area), 2.1.

TABLE 22.—*Results of furnace runs arranged to show variation of percentage of manganese in slag with basicity of slag (ratio of $CaO + MgO$ to SiO_2).*

[Results are plotted in figure 11.]

Group.	Runs included.	Furnaces included.	Basicity ratio.	Carbon excess or deficiency (pounds).	Rate of driving.	Percentage of manganese in slag.	
						Actual.	Reduced to standard hearth temperature and rate of driving.
9	6, 8, 23, 24, 25, 26, 28, 31, 33..	A, J, K, L......	1.77	+411	1.79	7.4	8.2
10	7, 11, 17, 30, 32, 34, 35, 39, 40.	A, B, F, L, M....	1.55	+187	2.06	9.2	9.9
11	1, 4, 10, 12, 13, 20, 27, 38......	A, B, C, I, K....	1.36	− 10	2.32	11.8	11.5
12	2, 5, 8, 14, 15, 19, 22.........	A, B, F, G, J....	1.19	−110	2.42	14.2	13.6

TABLE 23.—*Results of furnace runs arranged to show variation of percentage of manganese in slag with rate of driving (pounds of gross slag per minute per square foot of hearth area).*

[Results are plotted in figure 12.]

Group.	Runs included.	Furnaces included.	Basicity ratio.	Rate of driving.	Carbon excess or deficiency (pounds).	Percentage of manganese in slag.	
						Actual.	Reduced to standard basicity and temperature.
5	1, 4, 6, 7, 14, 15, 18, 22, 39...	A, B, F, J, M....	1.38	2.86	− 60	13.0	13.9
6	2, 5, 8, 19, 20, 21, 25.........	A, G, I, J.......	1.70	2.38	−312	12.0	12.9
7	3, 10, 11, 12, 19, 23, 27, 40...	A, B, G, J, K, M..	1.51	2.15	− 62	12.6	12.5
8	13, 15, 17, 24, 26, 28, 38......	B, C, F, J, K....	1.64	1.60	+240	10.3	12.2

In deriving an equation for the percentage of manganese in the slag four quantities are involved. If only figures could be selected wherein two of the quantities varied while the other two quantities remained constant, the derivation of the equation would be direct and simple. Unfortunately this is not possible. Thus, in Table 22 there is a distinctly apparent relation between "carbon excess or deficiency," basicity, and rate. As the carbon decreases the basicity decreases and the rate of driving increases. This relationship is due to the human management of the furnace. The increase of basicity is governed by the stone charged; the carbon excess or deficiency by the coke charged; and the rate of driving by the speed of the engines. For any run the stone can be decreased, the coke increased, and the engines slowed down by practically any arbitrary amount, and the furnace will still continue to make ferromanganese, though it may not make a profit. The fact, therefore, that this apparent relationship exists between excess carbon, basicity, and speed, shows only that the furnace operator followed, perhaps unknowingly, a definite rule in burdening and driving the furnaces.

Although in Tables 21, 22, and 23 all the four quantities concerned vary, it is found by successive approximation that—

(a) 100 pounds of excess carbon lowers the manganese in the slag 0.11 per cent;

(b) An increase in the rate of driving of 1 pound of slag per minute per square foot of hearth area increases the manganese in the slag 1.34 per cent;

(c) An increase in the ratio of base to silica of 1 unit decreases the manganese in the slag 9.16 per cent.

In Table 21 values for the actual percentages of manganese in the slag are given in the seventh column. In the last column the values have been corrected for each of the four groups for the amount by which the basicity ratio differs from 1.50, using the correction indicated in (c); the values are further corrected for the amount by which the rate of driving differs from 2.1, using the correction indicated in (b). In the same way for Tables 22 and 23 the percentages of manganese in the slag are given in the last column corrected for variations of the two quantities, in the fifth and sixth columns, in order to show the relationship that would be found between the variable shown in the fourth column and the manganese in the slag, were the other two variables held constant. These relationships are plotted in figures 9, 10, 11, and 12. Combining these three relations in a single equation gives the result—

$$M = 21.4 - 9.16B + 1.34R - 0.0011(K - K_a)(2)$$

where
M = percentage of manganese in the slag;
B = ration of base to silica;
R = rate of driving (pounds of gross slag per minute per square foot of hearth area);
K_a = carbon required by equation 1;
K = actual carbon charged.

Combining equation *2* with equation *1* gives—

$$M = 21.4 + 1.34R - 9.16B - 0.0011K - K_a + 0.00097V - 0.00145T \qquad (3)$$

Equation *3* is the general solution for *M*, the percentage of manganese in slag, *M* being a function of five variables. The assumption has been made in deriving this equation that the function was linear with respect to all five variables. This certainly is not true. The effect of speed on the reduction of MnO will be, itself, a function of the temperature. Moreover, a large number of variables undoubtedly affecting *M* have been omitted, for example, the physical character of the stock, the temperature of the top gas, the amount of carbon absorbed by CO_2, and the humidity of the blast. It can be hoped only that equation *3* includes the more important variables.

It will be noticed that in joining the points in figures 10, 11, and 12, a curved rather than a straight line is drawn in each figure. This was done with the hope of giving a hint as to the nature of the curve. In each figure a straight line can be drawn which will not miss the experimental point so much as 0.5 per cent. In view of the limitations of the data 0.5 per cent is much inside the limit of experimental error.

BURDENING THE FURNACE.

Observance of the relations in equations *1*, *2*, and *3* may be of material assistance in the burdening and driving of the furnace. The first step in solving this problem is the calculation by equation *1* of the ratio of coke and stone to ore with an assumed basicity ratio. Then by looking at figures 10, 11, and 12, the furnace operator can know at a glance the change in the manganese content of his slag to be expected from adding to, or trimming down, his fuel, or from changing his basicity ratio. No general principles can be given for the solution of this problem as the most important considerations governing the burdening problem are the cost of the materials used, the daily pay roll, and the furnace overhead. Every change in the burdening and driving factors cuts both ways. Slow driving reduces slag loss and runs up labor and overhead charges. Increasing the proportion of lime in the slag within reasonable limits results in lowering the slag loss, but increases coke consumption, and cuts down tonnage, resulting in lower ore cost, higher fuel cost, and increased labor and overhead charges. Using excess coke results in a slight decrease in the percentage of manganese in the slag, but in general increases the slag weight more rapidly, so that the slag loss, being the product of the slag weight and the percentage of manganese in the slag, is increased. About the only general criticism that can be made of the burdening and driving data is that there is on the whole a tendency to use too much fuel and to drive too slowly. The saving of ore resulting from slow driving, while appreciable, causes a decided decrease in profits of the furnace and, although in line with conservation of materials, will not appeal to the operator who is to pay the labor and overhead charges.

It may be stated that the most effective methods of reducing the slag loss of manganese are: (1) The use of low-ash coke and of low-silica stone and (2) carrying as basic a slag as is practicable. In the matter of selecting a low-ash coke it was not possible, perhaps, in the past year for a number of operators to do more than take whatever they could get.

It was universally realized at the furnace that the presence of a large percentage of silica in the coke was playing havoc with the practice; but it is believed that, had the office fully appreciated the loss in profits occasioned by excess silica a considerable improvement in coke could have been effected. Actual purchases of stone, it is certain, have not been made with the care the problem merited. Calcites and dolomites have been used with silica content as high as 7 per cent. Naturally the office did not purchase this stone on the market without a considerable preferential price, but, as is shown later, stone with this silica content should not be used even if obtained free of charge f. o. b. the furnace.

STACK LOSS.

The sum of the manganese in the slag and that in the metal does not equal that charged into the furnace. The percentage of manganese in the charge not accounted for by the manganese carried from the hearth is for convenience called "stack loss." The value of this quantity in Table 17 varies from 0.8 to 32.1 per cent. Its average is 12.8 per cent. As the average loss of manganese in the slag is only 14.7 per cent, the stack loss is as important as the slag loss. It is a quantity, however, that has not thus far proved itself susceptible to analysis or definite explanation. It may be due to: (1) Volatilization of metallic manganese from the hearth, (2) volatilization of manganese oxide in the hearth or stack, (3) manganese oxide carried off mechanically as ore fines by the blast, and (4) a fictitious result due to error in chemical analyses and weights.

To a certain extent in any individual run the apparent stack loss is surely a combined result of these four causes. But for the 40 runs taken as a whole it is difficult to understand how errors in weights and analyses would give a *loss* in 39 cases, instead of giving for one-half of the runs a "*gain* of manganese in the stack." Unless it can be shown that the analyses for manganese in the ore are systematically high or those for manganese in the slag and metal are systematically low, it must be concluded that the errors are sufficiently eliminated in the average of a number of runs.

There is a possibility that the loss as ore fines carried out mechanically by the blast may account for the whole loss.

The operating data from nine iron furnace plants indicate that 6.5 per cent of the iron charged is not accounted for by the iron in the slag and metal. This figure does not inspire much confidence and is given here only for the sake of comparison. As to the loss by volatilization, it is probably not important practically to distinguish between volatilization of metallic manganese and volatilization of manganese oxide. The answer to this problem will appear doubtless, when the relation is found between stack loss and some other operating quantity. If the stack loss is a volatilization phenomenon, it should increase with a rise in hearth temperature. It should also decrease with any increase in the speed of operation, as the longer a given pound of slag or metal remains in the hearth the greater the volatilization. The 40 runs in Table 17 were arranged in the order of decreasing stack loss and divided into four groups. The average for each of the four groups appears in Table 24 following:

TABLE 24.—*Results of furnace runs arranged to show manganese lost in stack as related to other factors.*

Group	1.	2.	3.	4.
Mn lost from stack, per cent	24.4	15.7	9.1	4.0
Top temperature, °F.	904	851	995	904
Mn in slag, per cent	8.6	13.5	11.1	10.6
Si in metal, per cent	1.43	.96	1.15	1.26
Ratio of base to silica	1.61	1.47	1.56	1.49
Rate of driving a	2.17	2.32	1.95	2.23
Carbon excess or deficiency, pounds	+253	−86	+82	−121

a Pounds of gross per minute per square foot of hearth area.

The results presented in the table show a manganese loss for each group of runs, but no systematic variation of stack loss with any other probably related quantity. High hearth temperatures as indicated either by high-silicon metal, excess carbon, basic slag, or low percentage of manganese in the slag appear not to affect the amount of stack loss. There is only one practical conclusion to be formed—the furnace man must admit that the stack loss is an important physical constant and should operate his furnace on that assumption, or he must conclude that the data here presented are insufficient to permit a solution of the problem and obtain for himself more complete and exact furnace records.

SOME TYPICAL CALCULATIONS.

To illustrate the use of calculations suggested in this report, to indicate how many factors are involved in the simplest detail of furnace practice, and to serve as an introduction to the operating problem, a brief outline of the factors involved in the selection of coke and stone follows.

For the sake of comparison a computation of the approximate furnace costs for the average run represented in Table 17, apart from investment charges, may be of interest.

The average charge per ton of metal is 2.67 tons of ore, 3.16 tons of coke, and 1.03 tons of stone. At the arbitrary prices of $40, $10, and $3 per ton, f. o. b. furnace, the materials cost $149.49 per ton of metal made. The following daily costs are assumed: Six hundred dollars for labor, and $175 for the aggregate of superintendence, chemical laboratory, and reserve for relining, repairs, liability insurance, clerical work, demurrage, supplies, etc. Thus there is a total fixed charge of $775 against a daily production of 51.7 tons of ferromanganese, or a charge of $14.75 per ton. Hence the total cost per ton is $164.24. The selling price on 74.9 per cent ferromanganese is taken to be $267.59 ($250 per ton for 70 per cent alloy plus or minus $3.50 per unit of manganese), so that the profit per ton is $103.35, or $5,343 per day.

COKE ASH.

The following analyses of coke are taken from furnace records and although the coke represented is not typical of that used in average practice, it is not abnormal, as millions of tons of such coke have been used annually at certain iron furnaces for years.

Analyses of coke used in certain iron furnaces.

Sample No.	SiO_2.	Al_2O_3.	Fixed carbon.	Volatile matter.	Sulphur.	Moisture.
	Per cent.	Per cent.	Per cent.	Per cent.	Per cent.	Per cent.
1	2.78	2.47	90.95	0.79	0.41	1.35
2	1.80	2.70	91.12	.47	.44	1.52
3	3.06	2.43	91.06	1.18	.80	1.40

Let it be assumed that coke 1 is available for a ferro campaign and that stone of grade "A," as described in the section following ("Selection of Stone"), is also available. Let the ore have the average analysis given in Table 17. Then it is possible to construct a charge sheet for these materials to give approximately the maximum daily profits, subject to the limitation that none of the operating quantities shall lie appreciably beyond what has been tried out as shown by Table 17. The charge will be then as follows:

Analysis of charge.

Material.	Weight, pounds.	SiO_2.		Al_2O_3.		Bases.		Mn.		Fe.		C.	
		Per cent.	Lbs.	Per cent.	Lbs.	Per cent.	Lbs.	Per cent.	Lbs.	Per cent.	Lbs.	Per cent.	Lbs.
Ore	5,310	8.60	457	2.50	133	......	......	40.33	2,240	5.93	325	......	......
Coke	4,000	2.74	109	2.47	98	......	......	......	......	......	......	91.95	3,620
Stone	2,102	.42	10	.13	3	53.0	1,115	......	......	......	......	......	......
Total	......	......	576	......	234	......	1,115	......	2,240	......	325	......	3,620

The manganese balance will be as follows:

Manganese balance.

	Pounds.
Charged	2,240
Stack loss (12.66 per cent)	[a] 284
Slag loss (8.43 per cent)	[b] 189
To metal (79 per cent)	1,767
Accounted for	2,240

The weight of the slag will be 2,167 pounds, the analysis of the slag being as follows:

Analysis of slag.

Constituent.	Per cent.	Pounds.
SiO$_2$	26.05	[a] 561
Al$_2$O$_3$	10.85	234
Bases	51.75	1,115
MnO	11.30	241
S		17
Total		2,167

[a] 151 pounds to silicon in metal.

The carbon as required by equation *1* would be:

$$K = 4,170 + 0.88 \times 2,167 - 1.35 \times 1,250 = 4,405$$

The "deficiency" of carbon charged below that required by equation *1* is therefore 785 (4,405 − 3,620) pounds, only 30 pounds less than was used in the runs in group 1, Table 21. According to equation *2*, this lack of fuel causes the manganese in the slag to increase only 0.87 per cent. Referring to figure 9 it will be seen that the actual curved line falls somewhat below this assumed straight line.

The ratio of bases to silica is 51.75 + 26.05, or 1.99. This basicity is 1.99 − 1.48, or 0.51 higher than that of the average run, and lowers the percentage of manganese in the slag by 4.67 per cent, according to equation *1*. Figure 12 shows that the curved line lies above the straight line. To be conservative, a decrease of only 3.75 per cent is taken, the smallest value figure 12 will justify.

The gross slag is 2,167 + 1.29 × 1,777, or 4,485 pounds. For comparison, the hearth diameter of the furnace is taken to be 10.78 feet, which gives a hearth area of 91.3 square feet, the average hearth area in Table 16 if each furnace be weighted in the average according to the number of runs represented in the table. A driving rate of 2.61 pounds of gross slag per minute per square foot of hearth area is taken. This gives the output as 77.5 tons per day. The rate 2.61 is 0.54 greater than the average rate in Table 17, and, according to equation *2*, causes a rise in the percentage of manganese in the slag

---

[a] Average. [b] As shown below.

of only 0.71 per cent. Within this part of the curve in figure 11 it is difficult to take a greater rise than 0.75 per cent; which figure will be used.

From the above, then, the percentage of manganese in the slag is—

	Per cent.
Average in Table 17	10. 60
Increase due to low fuel	. 87
Increase due to fast driving	. 75
Decrease due to high basicity	3. 75
Manganese in the slag	8. 57

Slag loss (slag weight times percentage of manganese in slag: 2,167×8.75). 189 pounds.
Slag loss in percentage of manganese charged.......................... 8.43 per cent.

Summarizing:

Charge:

Ore, 5,310 pounds, 2.48 tons, at $40	$99. 38
Coke, 4,000 pounds, 2 tons, at $10	20. 00
Stone, 2,102 pounds, 0.94 ton, at $3	2. 82
Cost of materials	122. 20

Daily labor and fixed charges, $775.

Charge against 1 ton of metal, at 77.5 tons per day	10. 00
Total furnace cost of 1 ton of metal	132. 20

Selling price, $250 for 70 per cent ferro, with $3.50 per unit bonus and penalty.

Selling price of 1 ton of ferro containing 79 per cent Mn	281. 50
Furnace profit per ton	149. 30
Furnace profit per day, at 77.5 tons per day	11,570. 76
Furnace profit per day of furnaces represented in Table 2	5,323. 00
Gain in profits due to improved materials and practice	6,247. 76

These figures are disturbing, yet it is difficult to see just where they are wrong. The matter is important. If the 11 furnaces represented in Table 17 had operated on this average practice the possible saving would have been $2,061,760.80 per year per furnace, or, for the 11 furnaces, $22,679,368.80. The practical problem would have been to obtain the high-grade coke and stone involved.

An interesting point in the situation is that the iron furnace using the coke best adapted for the production of ferromanganese, namely, that designated as sample 2 in the table on page 112, used stone with 3.36 per cent SiO_2 (to increase the slag volume), and yet made metal of the following composition: Si, 1.52 per cent; S, 0.049 per cent; P, 0.173 per cent.

With the phosphorus content indicated, the metal is neither foundry nor Bessemer grade. As basic pig, 1.52 per cent silicon is objectionably high, unless a better sulphur content than 0.049 per cent can be attained. The answer is that the coke used was lower in SiO_2 than was really desirable. A good coke consumption and a fair tonnage can be shown with low-silica coke and stone, but in basic practice, where the logical function of the blast furnace is to keep down the sulphur and the silicon, the slag volume should not be so low that it can not take care of the sulphur.

Undoubtedly the operator of the basic furnace above mentioned realized this. Enough commercial pressure would have made him release this coke. If the ferro furnace had paid $20 per ton for low-silica coke, the daily profit would still have been $10,000 per furnace, an improvement of $4,523 per day, or roughly 100 per cent. If the desired quality of coke could have been procured for $20 per ton, and if the other assumptions are correct, the increased profit for any one furnace would have been $1,800,000 per annum.

SELECTION OF STONE.

As a typical problem relating to the selection of stone for use in furnaces producing ferromanganese, two grades of stone may be assumed to be under consideration. One may be designated grade A, containing 53 per cent basic constituents and 0.42 per cent silica; the other may be designated grade B, containing 49.5 per cent basic constituents and 7 per cent silica. Stone of grade A has been used in great quantities at one iron furnace plant, and hence is not of purely hypothetical composition. Stone of grade B has been used in producing ferromanganese.

If all the values, except for the grades of stone, be taken as for the average run represented in Table 17, a comparison can be made as follows:

Comparative results obtained with two grades of stone.

Constituent of charge.	Stone A.		Stone B.	
	Quantity per ton of alloy.	Cost.	Quantity per ton of alloy.	Cost.
	Tons.		*Tons.*	
Ore	2.64	$105.56	2.72	$108.80
Coke	3.12	31.15	3.54	35.45
Stone	1.05	3.15	1.70	5.09
Total		139.86		149.34
Manganese in metal, per cent	75.6		72.15	
Slag per ton of alloy, pounds	2,890		3,839	

The foregoing data are based on the assumption that the slag has a constant basicity ratio of 1.48 and that an increase of 1 pound in the rate of slag formation requires an increase in carbon of 0.88 pound.

The gross slag produced per ton of metal when stone A is used is 5,150 pounds, and when stone B is used 5,970 pounds. If the rate of driving be assumed to be constant, the tonnages will be 51.7 tons with stone A and 44.65 tons with stone B. The daily fixed charges will be roughly the same for both. There will, therefore, be a charge of $14.90 against a ton of metal if stone A be used and a charge of $17.35 if stone B be used.

The total costs are $154.61 and $166.66, respectively, per ton of alloy. The selling prices are $269.60 and $257.53; the profit per ton, therefore, when stone A is used is $114.99, and when stone B is used, $90.87. The profits per day are $5,950 and $4,060. Thus, the daily loss in furnace profits when stone B is used is $1,890. This loss must be charged to the 75 tons of B stone used per day. Thus, even if this grade of stone were obtained free of charge, the furnace loss from its use would be $22.20 per ton. Therefore, if stone A could be purchased at some premium less than $22.20 per ton, a furnace having no choice other than buying A or B would profit by the purchase of A. Such computations as these could be widely extended showing the actual premiums and penalties for coke and ore as based on the cost to the furnace, but the above example will suffice here to substantiate the following conclusions:

(1) The blast-furnace superintendent should know at all times the cost and the analyses of all materials purchased, and should also be familiar with all purchasable materials even including $40 coke and $20 stone.

(2) The decision as to the purchase of raw material should be left to the furnace superintendent, who should be in complete charge of the operation of the furnace.

(3) Far greater saving in furnace profits can be attained from judicious selection of raw materials than from any changes in operating conditions.

When the market price of the alloy falls, as it doubtless will, and the costs of stone, coke, ore, and labor, readjust themselves, it is only by a continued revision of some such calculations as the above, covering the coke and stone, that the furnace operator can hope to make the best showing.

SUMMARY OF OBSERVATIONS ON FERROMANGANESE DATA.

Operating data were collected from the furnace records of 11 blast furnaces making ferromanganese. These data are presented in Table 17, together with a maganese balance showing the distribution of the manganese charged between metal, slag, and top gas. The table includes 40 "experimental periods" of 10 days each and covers the extreme range of operating conditions for each furnace.

Several thousand pyrometer measurements were made of the temperature in the furnaces opposite the tuyères; of the slag at flush and cast; and of the metal at cast. It is shown that, in general, the ferromanganese furnace operates with a colder hearth and produces colder metal and slag than does the iron furnace.

The results of gas analyses made by the bureau show that practically all of the MnO is reduced "directly" by carbon. This reduction probably takes place in the hearth.

The metallurgical questions involved in the smelting of manganese are briefly discussed. It is shown that although more than three times as much carbon is burned at the tuyères in the ferro furnace as in the iron furnace, the gaseous products from the combustion zone are probably at a lower temperature.

The two losses of manganese in the furnace—the slag loss and the "volatilization" loss—are discussed. It is shown that the percentage of manganese in the slag is decreased by raising the blast temperature, by increasing the basicity of the slag, and by charging more coke; and that it is increased by fast driving, and by carrying a greater slag "volume." A general equation is given for the percentage of manganese in the slag as a function of these five quantities, by means of which the slag loss can be calculated in advance from the charge-sheet data.

The "volatilization" loss is shown not to be affected by the basicity of the slag, the silicon in the metal, the carbon charged as fuel, the rate of driving, or the temperature of the top gas. The results, however, are not convincing when examined carefully, and additional and more exact figures are needed. It is shown that by making five changes in practice the 11 furnaces examined could have raised their recovery to 79 per cent, using the same ore, could have increased their tonnage 59 per cent, and could have increased their annual net profits by more than twenty million dollars. The changes are as follows:

(a) Using a better grade of coke existing in sufficient quantities and probably purchasable.

(b) Using a better grade of stone.

(c) Using less fuel.

(d) Running with a more basic slag.

(e) Driving faster.

Certain furnaces are shown to have lost $10 to $20 a ton by using high-silica stone.

It is concluded that there is greater room for improvement in the selection of coke and stone than in furnace practice.

DISCUSSION OF DATA ON SPIEGELEISEN PRACTICE.

GENERAL DESCRIPTION OF SPIEGELEISEN PRODUCTION.

The simultaneous smelting of iron and of manganese ores—that is, the production of spiegeleisen—is metallurgically an uneconomical process. The reason can readily be appreciated. The iron furnace and the manganese furnace present problems that, though somewhat similar, are at many points fundamentally different. The most economical operation for making iron is not the most economical for

making manganese. It is desirable, therefore, that the two processes
be kept separate. The only logical reason, of course, for making spiegeleisen at all is to utilize manganiferous iron ores. Spiegeleisen has
not, however, always been made in this way. The manganese content of the metal made is frequently reduced by the addition of iron
ore or of iron scrap to the charge. During the year 1918, pressure
from several sources was brought to bear upon the manganese producers to lower the manganese content of their alloys. The prevailing standard for ferromanganese was lowered from 80 to 72 per cent,
and for spiegeleisen from 20 to 16 per cent. To the extent that this
was done with the use of lower grade ores, it served a real purpose by
creating a market for otherwise unsalable ores. It should be remembered, however, that such a process was wasteful of coke and manganese, and caused decreased furnace tonnage. The rather anala-
gous method of "sweetening" a spiegel ore mixture with high-grade
manganese ore is equally objectionable from the standpoint of a
metallurgist.

The average spiegel analysis given in Table 18 (p. 116) shows that
in order to produce 1 ton of manganese, 5.78 tons of spiegel was required. This alloy contained about 4.4 tons of metallic iron, the
equivalent of 4.64 tons of pig iron. The carbon fuel required to
make 5.78 tons of spiegel is 19,900 pounds (5.78 × 3444). Crediting
this metal with the amount of carbon required for making 4.64 tons
of pig iron, namely, 1,727 pounds of carbon multiplied by 4.64 tons
of pig iron, or 8,000 pounds total, it is seen that the ton of metallic
manganese in the form of spiegel requires 11,900 pounds of carbon.

It is clear that in 1.33 tons of 74.9 per cent alloy there is 1 ton of
metallic manganese and 0.23 ton of metallic iron, the equivalent of
0.25 ton of pig iron. The fuel consumption was 7,100 pounds of
carbon (1.33 × 5323) according to Table 17. Giving credit for 420
pounds (1727 × 0.25) for the iron made, 1 ton of manganese in the form
of ferro requires 6,680 pounds of carbon. Thus, it appears that a
ton of manganese in the form of spiegel requires 78 per cent more
carbon fuel than a ton of manganese in the form of ferro.

This result is by no means due wholly, or even largely, to the
mere combination of the two smelting processes. Most of the excess
fuel requirement is due to the character of the ferro ores and of the
spiegel ores used. The manganese and silica contents of the two
ores are:

	Spiegel.	Ferro.
Manganese, per cent	12. 75	40. 33
Silica, per cent	14. 84	8. 60
Silica per pound of manganese, pounds	1. 16	. 214

The average spiegel ores were therefore 5½ times as siliceous as the
ferro ores.

SLAG COMPOSITION.

Table 18 shows that the average spiegel slag is less basic than either the iron slag or the ferro slag. The following figures will permit a comparison of these three average slags calculated on the basis of base + alumina + silica = 100 per cent.

Comparison of three average slags.

	Base.	Alumina.	Silica.	Ratio (Base to SiO₂).
	Per cent.	Per cent.	Per cent.	
Pig iron	50.3	13.4	35.3	1.42
Spiegel	47.9	14.1	38.0	1.26
Ferro	49.8	16.7	33.5	1.49

Table 18 demonstrates that basicity ratios higher than 1.26 have been successfully used in spiegel practice. In fact, the eight runs showing the highest basicity ratio, namely, Nos. 46, 57, 58, 61, 64, 66, and 67, including furnaces B, D, H, J, M, P, and G, show an average ratio of 1.42—exactly that of the slag in the average iron practice quoted above. The average recovery of manganese by the metal for these eight runs was 70 per cent.

It is of interest to compare these figures with those for the eight runs (42, 43, 50, 51, 52, 53, 54, and 60) which had the lowest basicity ratios, the average being 1.08. The average recovery for these runs was 47.3.

HEARTH TEMPERATURE.

At furnaces C, D, E, H, M, and Q, optical pyrometer observations were made of slag, metal, and tuyère temperatures. The same day-to-day variations in temperature were noted. The averages of all temperatures from the six furnaces are given below and the figures from the ferro-furnace observations are repeated for comparison:

Average temperatures in speigel and in ferro furnaces.

	Spiegel.	Ferro.
Temperature opposite tuyères, ° C	1597	1550
Temperature of metal at cast, ° C	1392	1386
Temperature of slag, ° C	1427	1426

Seemingly, both the metal and the slag flow from a spiegel furnace at the same temperature as from the ferro furnace.

COMPOSITION OF FURNACE GAS.

Analyses of spiegel furnace gas were made by the bureau's field party at only four furnaces. The composition of the gas as indicated by the average of all the observations was as follows:

Average composition of speigel furnace gas.

	By volume.	By weight.
CO_2 ....per cent..	7.08	11.05
CO ....do....	30.68	30.40
N_2 ....do....	58.72	58.08

These figures are not at all satisfactory. The CO content is nearly as high as in ferromanganese-furnace gas. It is difficult to charge this result to ordinary analytical errors. Leakage of air into the sample would lower the CO content; also, the estimation of CO by cuprous chloride absorption when improperly carried out gives a result too low rather than too high. The average analyses for the four furnaces gave 4.62, 5.62, 8.95, and 9.24 per cent for CO_2 content and 24.42, 29.25, 34.20, and 34.85 per cent for the CO content. The extreme limits for CO_2 were 3.9 and 11.1 per cent; and for CO, 23.3 and 35.1 per cent. This range of variation is the same as that encountered in ferromanganese-furnace gas. This fact points strongly to the conclusion that in both of these alloy furnaces, the chemical reactions in the lower part of the furnace are continuously changing. It might almost be said to prove that for a period of time the bosh and hearth reactions are nearly those of the iron furnace, and that following this period as a result of the settling down of the stock, a large amount of manganese oxide is thrown into the hearth, resulting in a rapid reduction of MnO, with a consequent increase in the ratio of CO to CO_2. Therefore, unless more or less continuous samples of furnace gas are taken over a rather long period, the actual analysis might easily misrepresent the true average composition of the furnace gas.

CARBON BURNED AT THE TUYÈRES.

From what has been said in regard to the gas analyses it will be anticipated that a calculation of the amount of carbon burned at the tuyères and of the carbon absorbed by CO_2 (or "the carbon used in direct reduction") will show figures that will have little significance. Table 25 following shows a calculation of the weight of gas, the weight of blast, and the distribution of carbon between the stack and the tuyères. So long as the amount of carbon blown out as flue dust must be guessed at; and until systematic gas analyses are taken covering appreciable periods of time, no satisfactory figures for these quantities can be given. However, the figures 676 for the ferro furnace, 347 for the spiegel furnace, and 169 for the iron furnace undoubtedly indicate, in a general qualitative way, the pounds of carbon per ton of metal absorbed by CO_2.

TABLE 25.—*Selected data on operation of spiegeleisen blast furnaces.* .

Carbon charged per ton of alloy..........................pounds.. 3,444
Carbon combined with metal (5.0 per cent C).............do.... 112
Carbon blown out of stack as dust (9.6 per cent assumed)..do.... 330
Carbon available as fuel................................do.... 3,002
Carbon from CO_2 in stone..............................do.... 260
Total carbon gasified...................................do.... 3,262
Gas analysis (by weight):
   CO_2...per cent.. 11.05
   CO..do.... 30.40
   N_2...do.... 58.08
Carbon in top gas.......................................do.... 16.07
Weight of top gas per ton of alloy......................pounds.. 20,300
Weight of N_2 in top gas................................do.... 11,810
Weight of air blast.....................................do.... 15,280
Weight of O_2 in blast..................................do.... 3,530
Weight of carbon burned at tuyères......................do.... 2,655
Carbon absorbed by CO_2 or involved in direct reduction..do.... 347

FUEL REQUIREMENTS.

No relation can be determined between carbon fuel, blast temperature, and slag weight in the spiegel furnace by the method employed for ferromanganese. Table 18 (p. 116) shows that most of the furnaces operating with high-blast temperatures were working on ore mixtures that gave low slag weights, and the furnaces smelting siliceous ores were running with low-blast temperatures. In attempting to determine K_a (the carbon fuel used for average hearth temperatures) as a function of T (the blast temperature) and of V (the slag weight), taking these last as independent variables, it is difficult to distinguish between the effect of changes in blast temperature and of changes in slag weight, as T and V do not vary independently. The method employed below is not so satisfactory as regards accuracy as that used with ferromanganese, but it is simpler and more direct.

The runs in Table 18 were divided into three groups, selected to conform with the following classification:

(1) High-blast temperature and small slag weight.
(2) High-blast temperature and large slag weight.
(3) Low-blast temperature and large slag weight.

The average values of K_a, T, and V for these three groups are given in Table 26 following:

TABLE 26.—*Average values for carbon fuel, blast temperature, and slag weight in the furnace runs represented in Table 18.*

Group.	Runs included.	Furnaces included.	Carbon per ton.	Blast temperature.	Slag weight.
			Lbs.	*°F.*	*Lbs.*
1	43, 46, 48, 49, 55, 57, 62, 64................	C, D, H, M, P.....	2,821	1,221	1,979
2	41, 42, 44, 52, 54, 58, 59, 60, 61, 70...........	C, E, J, M, R.....	3,525	1,142	3,349
3	50, 51, 53, 65, 66, 67, 68, 69...............	E, B, Q, R........	4,175	825	4,392

On the assumption that the carbon fuel required for average hearth temperature conditions can be written in the form—

$$K_a = A + BT + CV$$

where A, B, and C are constants, the figures given in Table 26 are sufficient to determine the value of the three constants. Substituting in the above equation three times gives—

$$2,821 = A + 1,221B + 1,979C$$
$$3,525 = A + 1,142B + 3,349C$$
$$4,175 = A + 825B + 4,392C$$

From which it follows that—

$$K_a = 2,365 - 0.44T + 0.49V \qquad (4)$$

As previously stated, K is the quantity of carbon per ton of metal used for any particular run. K_a is the quantity that would be required if the same thermal conditions were maintained in the furnace as in the average furnace. Confusion has often arisen in discussions of blast-furnace fuel economy from identifying the carbon "consumed in" the furnace with the carbon "required by" the furnace. It is true in iron furnace practice that for a given product there is a minimum carbon consumption below which the furnace will not work, but there is practically no upper limit. In making manganese alloys, however, it may be said that there is neither a maximum nor a minimum limit. The actual amount of carbon, K, charged into any furnace is merely the amount the operator believes necessary to maintain a given ratio of coke to ore. In each run, therefore, K, the actual carbon will differ from K_a, the carbon that would on the average be charged in a furnace working with that particular blast temperature and slag weight. $K - K_a$ is a measure of the thermal condition of the hearth. It should be true that when $K - K_a$ is positive the furnace is running hotter than when $K - K_a$ is negative, and this difference in available fuel should influence the amount of MnO left unreduced in the slag.

PERCENTAGE OF MANGANESE IN THE SLAG.

It has been shown for the ferromanganese furnace that M, the percentage of manganese in the slag, is a function of three quantities, namely: B, the ratio of base to silica; $K - K_a$, the excess carbon; and R, the rate of slag formation per square foot of hearth area. The relation between the quantities was found to be—

$$M = 21.4 - 9.16B + 1.34R - 0.11(K - K_a)$$

In the same way, as is shown below, for the spiegel furnace.

$$M = 14.3 - 5.31B + 0.0R - 0.09(K - K_a) \qquad (5)$$

This equation is determined as follows: The runs in Table 18, arranged in order of decreasing basicity, are divided into two groups of 15 runs each, group 1 representing runs having the highest basicity and group 2 those having the lowest. The runs are rearranged in order of decreasing $K-K_a$, and two further groups—group 3, representing runs with excess carbon, and group 4, representing runs with deficiency of carbon—are selected. The average values of M B, and $K-K_a$ for these four groups appear in Table 27.

TABLE 27.—*Results of furnace runs arranged to show variation of manganese in slag according to basicity of slag and according to excess or deficiency of carbon fuel.*

Group.	Basicity.	Excess carbon.	Manganese in slag.
		Pounds.	*Per cent.*
1	1.36	123	6.69
2	1.14	137	8.09
3	1.20	292	6.89
4	1.21	331	8.01

From these figures, assuming that the value of M can be expressed in the form given in equation 5, a simple calculation shows: (a) That one unit increase in basicity lowers the manganese in the slag 5.31 per cent; (b) That 100 pounds of carbon lowers the manganese in the slag 0.09 per cent.

To show that the rate of driving does not affect the manganese content of the slag, the runs can be again rearranged in order of decreasing R (rate of slag formation per square foot of hearth area), giving two further groups—group 5 representing runs with fast driving, and group 6, representing runs with slow driving. The average values of M, B, $K-K_a$, and R for groups 5 and 6 are given in Table 28.

TABLE 28.—*Results of furnace runs arranged to show variation of manganese in slag according to rate of driving.*

	Group 5.	Group 6.
Basicity ratio	1.24	1.23
Excess carbon, pounds	256	208
Rate of driving, expressed in pounds of slag made per minute per square foot of hearth area	2.86	1.48
Mn in slag, per cent	7.81	7.47

Although the rate of driving differs by 1.38 pounds of slag per minute per square foot of hearth area, the difference of 0.34 per cent manganese in the slag is exactly accounted for by the difference in the values of B and of $K-K_a$. This might well have been anticipated. The curve in figure 10, (p. 127), showing the relation of the percentage of manganese in the slag to the rate of driving for the ferromanganese furnace, is convex downward. When two or more pounds of slag per square foot of hearth area is being produced, the time factor is of importance. The lower end of this curve, however, shows a distinct flattening with a driving rate as low as 1.6 pounds of slag per minute. It is a peculiar coincidence that the average rate of slag formation in both the spiegel practice and in the ferro practice is just 2.1. As regards merely the reduction of MnO, there seems to be no necessity for running the spiegel furnace as slow as the ferro furnace. There is so much more MnO per pound of gross ferro slag to be reduced that the ferro slag should probably enter the hearth only at one-third the rate that is found economical in spiegel practice.

Seemingly, two conclusions may be drawn from these facts:

(1) That overdriving results in a loss of manganese, while underdriving does not result in any corresponding saving of manganese.

(2) That the spiegel furnaces investigated were driven much slower than is shown necessary by any facts presented in this paper.

It is, of course, possible that through other causes furnace difficulties might arise with fast driving, although no evidence of such an effect is at hand.

STACK LOSS.

The "stack loss" is a factor of less importance in the spiegel furnace than in the ferro furnace; the average value is only 6.72 per cent, whereas for ferromanganese it was shown to be 12.8 per cent. The individual values vary from $+39.2$ to $=19.2$, a range of 58.4 per cent. The individual values for ferro stack loss varied over a range of 32.9 per cent.

The stack loss shown in Table 18 does not seem to be connected with any of the other operating quantities. The 30 runs were divided in half, giving two groups of 15 runs each; one group included runs with the larger stack losses and the other included runs with the smaller stack losses. The average values of the figures that may have some possible connection are given in Table 29 following:

TABLE 29.—*Results of furnace runs arranged to show comparative effects of large and of small stack losses.*

Item.	Furnaces with large stack loss.	Furnaces with small stack loss.
Stack loss, per cent.............................	15. 5	−2. 22
Silicon in the metal, per cent...................	1. 92	1. 57
Basicity..	1. 25	1. 27
Blast temperature, °F..........................	1,130	1,018
Top temperature, °F............................	633	525
Slag per ton, pounds...........................	2,930	3,320
Manganese in slag, per cent....................	7. 42	7. 69
Carbon burned, pounds.........................	3,260	3,665
Recovery, per cent.........	55. 9	64. 5
Excess carbon, pounds..........................	−35	110

The difference between the stack losses in the two groups is 17.7 per cent. None of the other factors, however, show enough variation to be responsible for such a change in the stack loss. In the first group the silicon content in the metal is higher, as is also the top temperature, while the manganese in the slag is lower. These are changes in the right direction to account for a stack loss, although they can not explain it quantitatively. The fact that the basicity is identical in the two groups is important, as it has been thought by some that "liming up" the slag tends to increase the stack loss.

SUMMARY OF OBSERVATIONS ON SPIEGELEISEN PRACTICE.

The average recovery of manganese in the production of spiegeleisen was found to be 60.22 per cent, the loss of 39.78 per cent being distributed, one-sixth to the stack and five-sixths to the slag.

Although the individual values for stack loss were very irregular, it appears that they were not affected by the basicity of the slag, the temperature of the top gas, nor the carbon fuel charged. There seems to be no way to reduce this anomalous quantity. On the other hand, the evidence does show that the stack loss will not be increased by increasing the basicity of the slag.

The slag loss is decreased by increasing the basicity of the slag. The average basicity of the spiegel slags investigated was $CaO + MgO \div SiO_2$ $= 1.26$; it seems that this basicity ratio is uneconomically low. The percentage of manganese in the slag is shown to be independent of the rate of slag formation; it is also shown that the spiegel furnaces are operated much more slowly than the manganese oxide reduction demands.

To make a ton of metallic manganese in the form of spiegel requires 78 per cent more carbon fuel than is required to make a ton of metallic manganese in the form of ferro. This greater fuel consumption is due largely to the enormous amount of silica carried in the spiegel ore with the manganese, namely, 1.16 tons of silica to 1 ton of manganese. In this respect the ferro investigation and the spiegel investigation do not overlap. The most siliceous ferro ores used (containing 0.5 ton of SiO_2 per ton of Mn) have a SiO_2 : Mn ratio just equal to that of the least siliceous spiegel ores used. The successful operation of spiegel furnaces on very siliceous ores, however, indicates that the silica specifications on high grade manganese ores have been much too stringent.

Practically all of the conclusions reached in regard to the production of ferromanganese are applicable to spiegel production. In making speigeleisen, the use of better coke and stone has some advantages, but not sufficient to permit such bonuses for low-silica stone and coke as would be permissible in ferro practice; because the silica in the spiegel slag comes largely from the ore gangue, and because manganese in the form of spiegel ore is less expensive than manganese in the form of ferro ore.

Prices for spiegel ores have not been fixed, as for ferro ores, but are subject to individual contract, so a discussion of profits in spiegel operation will not be attempted.

CHAPTER 9.—NATIONAL IMPORTANCE OF ALLOCATING LOW-ASH COKE TO MANGANESE ALLOY FURNACES.

By P. H. ROYSTER.

INTRODUCTORY STATEMENT.

During 1918 an investigation of manganese alloy furnaces with particular reference to the character and quality of the coke being supplied to the furnaces was conducted by the Bureau of Mines. Twelve typical blast furnaces producing ferromanganese and spiegeleisen were examined. These 12 furnaces produced approximately 40 per cent of the total output of manganese alloys in the United States and offer a fair representation of the industry as a whole.

It developed that the poor quality of the coke in general use was responsible for serious waste, both of manganese and of coke. This fact was of vital interest not only to the Bureau of Mines, but also to the Fuel Administration.

FURNACE DATA.

The 12 furnaces studied furnished the following statistics:

Data regarding 12 blast furnaces producing ferromanganese and spiegeleisen.

Item.	Furnaces making ferromanganese.	Furnaces making spiegeleisen.
Number of furnaces investigated	6	6
Average monthly output of alloys, July and August, tons	9,966	11,736
Average coke per ton of alloy, tons	3.25	2.12
Total coke consumed, tons	32,387	24,880
Average manganese in alloys, tons	73.0	17.5
Metallic manganese in alloys, tons	7,266	2,100
Conversion loss, per cent	29	38
Metallic manganese in ores, tons	10,266	3,380
Average manganese in ores, per cent	40.6	12.8
Total ore consumed, tons	25,276	26,502

IMPORTANCE OF ALLOCATING LOW-ASH COKE TO MANGANESE FURNACES.

A total of about 57,000 tons of coke per month was being consumed by the 12 furnaces under review. This coke was found to contain an average of nearly 14.5 per cent ash, or over 7 per cent silica.

The principal losses of manganese in the manufacture of alloy are in the slag, these losses being roughly proportional to the slag volume. The controlling slag-forming element is silica. Silica in coke is more harmful than silica in ore because it not only increases the slag volume but also diminishes the efficiency of the coke, making more coke necessary, which in turn introduces more silica, thereby again increasing the slag volume and the coke requirement. In fact,

148

if the coke contains as much as 20 per cent ash, the vicious circle may be complete, all further additions of coke serving no purpose whatever. It is obvious that the more nearly this extreme is reached, the more wasteful the process becomes, both as to manganese and as to coke. Conversely, with better coke, less coke is needed and more manganese is saved.

Comparative observations made and recorded demonstrate if the 12 furnaces under review had been supplied with coke containing 8 per cent ash instead of 14.5 per cent, the same monthly tonnage of alloys could have been produced with about 40,000 tons of coke instead of the 57,000 tons actually used. At the same time, the conversion loss of metallic manganese would have been reduced from 29 per cent to 14 per cent for ferro, and from 38 per cent to 29 per cent for spiegel. The net result of this last would have been a saving of 1,826 tons of metallic manganese, equivalent to 4,560 tons of 40 per cent ore in the case of ferro; and a saving of 420 tons of metallic manganese, equivalent to 3,230 tons of 13 per cent ore in the case of spiegel.

Extending these figures, which represent actual conditions as to about 40 per cent of the country's (1918) output of manganese alloys, to cover the entire estimated requirement of such alloys for the year 1919, gives the following results:

1. Estimated requirement for 1919, 280,000 tons metallic manganese of which 20 per cent shall be in the form of spiegel.

Tons of coke.

2. 56,000÷17 per cent=330,000 tons spiegel×2.12=......................	699,600
224,000÷72 per cent=312,000 tons ferro×3.25=......................	1,014,000
Total requirement of 14.5 per cent ash coke....................	1,713,600
Saving through use of good quality coke, 31 per cent..................	532,000
Total requirement of 8 per cent ash coke.......................	1,181,600

	Poor coke.	Good coke.	Saving.
3. Spiegel conversion loss, per cent..................	38	29	....
Tons Mn needed for spiegel......................	90,000	79,000	11,000
Tons 13 per cent ore needed for spiegel...........	690,000	606,000	84,000
Ferro conversion loss, per cent.....................	29	14	....
Tons Mn needed for ferro.........................	315,000	262,000	53,000
Tons 40 per cent ore needed for ferro..............	786,000	655,000	131,000

In short, by allocating low-ash coke during 1919 to the manganese-alloy furnaces, there may be saved over half a million tons of coke; about 131,000 tons of manganese ore of ferro grade; about 84,000 tons of ore of spiegel grade; and a total of about 64,000 tons of metallic manganese.

The effect of such allocation on the production of pig iron must, however, be considered. The pig-iron industry would be deprived of roughly 1,200,000 tons of good coke and receive instead about 1,700,000 tons of relatively poor coke.

In the manufacture of pig iron, poor coke does not cause any increase in loss of metallic iron. The substitution of 14.5 per cent ash coke for 8 per cent ash coke would, however, cause about 10 per cent decrease in production and about 20 per cent increase in the use of coke. The decrease in production is the direct result of having to use an increased amount of coke.

The net effect would be a decrease in the production of pig iron of about 120,000 tons, and an increase in coke consumption of about 240,000 tons. One hundred and twenty thousand tons is approximately one-third of 1 per cent of the total pig-iron output of the country.

However, as the pig-iron industry would receive 500,000 tons more coke than it would be asked to release there would be enough coke left, after supplying the above additional requirement of 240,000 tons, to make up the deficiency in pig iron and still leave a margin of 120,000 tons.

Therefore the net coke supply would be actually increased, while maintaining the pig output at its present level; but about 130,000 tons additional furnace capacity would have to be provided.

The greater part of this, namely, at least 100,000 tons,[a] would be furnished by the manganese-alloy industry, because its furnace requirements would be reduced by this amount, while producing the same tonnage of alloys, on account of having high-grade coke. The small balance must be made up from some other source. As bearing upon this point, it is interesting to note the following statistics:

Data on pig-iron blast furnaces in the United States.

	Number.
In blast, Aug. 31, 1918	371
In blast, Sept. 30, 1918	364
Out of blast, Sept. 30, 1918	74
Total furnaces, Sept. 30, 1918	438
Average yearly capacity per furnace (approximate), tons	100,000

CONCLUSION.

Without doubt, a large proportion of the furnaces idle (Oct. 1, 1918) are small and poorly equipped. But, in view of the immense national importance of allocating low-ash coke to the manganese-alloy furnaces, it would seem justifiable to exert pressure on the pig-iron industry to make up the resulting slight loss of furnace capacity from among the furnaces now idle.

The figures given herein are based on a grade of coke which is to-day admittedly scarce; on the other hand, the total requirements of the manganese-alloy furnaces are only a small proportion of the total coke output of the country. The best coke available, whatever its grade may be, should be allocated to these furnaces; and if the average ash proves to be somwhat more than 8 per cent, the savings made, while proportionately less than shown above, will still be of the utmost importance.

In making these figures, commercial aspects have been constantly kept in mind. The figures are not theoretical; they represent what it is believed can actually be accomplished.

---

[a] Furnaces capable of making 68,000 tons of spiegel per year could be released for pig production. The estimate of 100,000 ton : pig from these furnaces is low. A 100-ton pig furnace will not make 68 tons of spiegel.

By H. W. Gillett and C. E. Williams.

INTRODUCTORY STATEMENT.

The utilization of domestic manganese ores has been an important war-time problem, and its bearing on national self-sufficiency makes it worthy of attention even in times of peace. This paper deals only with the utilization of low-grade domestic ores by electric smelting.

As a rule, blast furnaces can not, in normal times, economically use ores averaging much below 40 per cent Mn and much above 12 per cent SiO_2, although during the war these limits were exceeded. The United States contains large tonnages of ores lower in Mn and higher in SiO_2, which may be amenable to electric smelting. An electric-furnace plant can be constructed more rapidly and can be economically operated in smaller units than a blast furnace; also it can be situated wherever a suitable supply of ore and a suitable water-power development are found near each other. An ore deposit inadequate to supply a modern blast furnace would last a rather large electric furnace for a considerable time. Hence, the smaller electric unit can be so located that the transportation charges are less than if the ore were shipped to a more distant blast furnace.

The writers wish to acknowledge the valuable assistance of Messrs. E. L. Mack and F. C. Ryan, of the Bureau of Mines, and of Lieuts. L. S. Deitz, jr., and J. G. Thompson, Ordnance Department, and Pvt. E. O. Denzler, of the Chemical Warfare Service, who aided in the experimental and analytical work. Grateful acknowledgment is also made to the Chemical Department of Cornell University for the use of its electric-furnace equipment and of its analytical laboratory.

PRESENT DEVELOPMENT OF ELECTRIC SMELTING OF DOMESTIC ORES.

Electric smelting of local ores has been considerably developed. The Southern Manganese Corporation, Anniston, Ala., has eight furnaces, some of them of 3,000 kilowatt size, operating on local ore.[a] The Anaconda Copper Co.[b] will soon be operating five 3,000-

---

[a] Swann, T., (news item): Met. and Chem. Eng., vol. 18, 1918, p. 512.
[b] Anonymous, Electric furnace for manganese ores: West. Elec., vol 73, 1918, p. 218; Chem. Abs., vol 12, 1918, p. 1856.

kilowatt furnaces at Great Falls, Mont., on uncalcined rhodochrosite ore from Butte, Mont., and will probably also use some ore from Philipsburg, Mont. Washed Philipsburg ore is also smelted by the Bilrowe Alloys Co., Tacoma, Wash., in six 350 to 400-kilowatt furnaces, and this firm has also smelted ore from the Olympic Mountains in Washington.

The Noble Electric Steel Co., Heroult, Calif., smelts California ores in two 1,300 to 1,500-kilowatt furnaces. The Pacific Electro Metals Co. makes ferromanganese and silicomanganese in a 3,000-kilowatt furnace at Bay Point, Calif., from California ores.

The Iron Mountain Alloy Co., Utah Junction, Calif., runs a 1,200-kilowatt and an 1,800-kilowatt furnace on Colorado, Utah, and Nevada ores.

The Western Reduction Co., Portland, Oreg., is just starting the production of ferromanganese in a 700-kilowatt furnace.

ELECTRIC SMELTING PRACTICE ON FERROMANGANESE.

FURNACES USED.

Practice in making ferromanganese differs slightly at the various plants, but in general is uniform. Keeney[a] has recently described the Iron Mountain Alloy Co.'s practice. Furnace voltages (electrode to charge) run from 55 volts in the 350-kilowatt size to 85 in the 3,000-kilowatt size. The furnaces are usually three-phase, three-electrode, open-top furnaces, the exceptions being those of the Bilrowe Alloys Co., which are single-phase, and those of the Noble Electric Steel Co., which are of a three-phase, four-electrode type. The general type of furnace used is the rectangular form described by Lyon, Keeney, and Cullen.[b] The Noble furnace is described by Vom Baur.[c]

The furnace lining is usually carbon or magnesite, rarely water-cooled fire brick. The furnaces are charged continuously and tapped every two hours. Slag and metal are tapped together into a settler. A few plants use round electrodes jointed for continuous feed, which should reduce the electrode consumption. However, in most plants the feed is not continuous, and the butts are scrapped. Seemingly, a Scott-connected, two-phase circuit from a three-phase circuit with the conducting bottom of the furnace as the neutral electrode, and two upper electrodes, might reduce the electrode consumption by presenting only two instead of three electrodes to oxida-

a Keeney, R. M., The manufacture of ferro-alloys in the electric furnace. Bull. Am. Inst. Min. Eng., vol. 140, August, 1918, pp. 1321-1373.

b Lyon, D. A., Keeney, R. M., and Cullen, J. F. The electric furnace in metallurgical work: Bull. 77, Bureau of Mines, 1914, p. 113.

c Rodenhauser, W., Schoenawa, J., and Vom Baur, C. H., Electric furnaces in the iron and steel industry, 1917 ed., p. 365.

tion, while still allowing the construction of a furnace of high kilowatt capacity without the upper electrodes being required to carry too much current.

RAW MATERIALS EMPLOYED.

The ores of mixtures used at all these plants vary between 39.5 to 43.5 per cent manganese and 15 to 25 per cent silica with low phosphorus, averaging about 40 per cent manganese, 20 per cent silica, 0.20 per cent phosphorus, and less than 4 per cent ferro. The Great Falls, Mont., plant will be an exception, as it will operate largely on Butte rhodochrosite ore which is extremely low in silica and phosphorus.

The plants all flux the ore with limestone, except the plant at Heroult, Calif., which uses a little fluorspar. Also they produce a slag whose average composition is 10 to 12 per cent Mn., 35 to 40 per cent CaO+MgO, about 30 per cent SiO_2, and such small amounts of Al_2O_3 as are carried by the ores. The latter are all low in alumina with the exception of occasional lots smelted at Anniston, Ala.

AVERAGE RESULTS OBTAINED.

The power consumption and the electrode consumption in the various furnaces, brought to the basis of a 3,000-kilowatt furnace, average 5,000 kilowatt-hours and 170 pounds of carbon electrodes per long ton of 80 per cent ferromanganese.

As the ores are low in iron, a little iron ore, or better and more generally used, some steel scrap, is added to supply enough iron to bring the alloy to 80 per cent manganese. The alloys contain 0.20 to 0.40 per cent phosphorus, the former figure being aimed at, but the latter being taken by the trade while the ferro supply is low. The California ores are low in phosphorus; consequently the California alloys meet the lower figure. The silicon content runs under 2 per cent, though it sometimes goes up to 5 per cent in the Anniston product when ores higher than usual in silica are employed.

The Anniston, Anaconda, and Iron Mountain plants use anthracite coal as reducer; the Bay Point and Tacoma plants use coke; and the Heroult plant uses about half coke and half charcoal.

The recovery of manganese averages 75 per cent, being nearer 80 per cent on ores of 15 per cent silica and nearer 70 on those of 25 per cent silica content. From a sixth to a third of the conversion loss is due to volatilization and dusting, and slag losses account for the rest. In the blast furnace operating on low silica ores these losses are exactly reversed, slag accounting for only a sixth to a third and the rest being due to volatilization and dusting.[a]

---

[a] Newton, Edmund, Manganiferous iron ores of the Cuyuna District, Minn., Bull. 5, Minn School of Mines Sta., University of Minn., 1918, p. 74.

EFFECT OF HIGH SILICA IN ORES.

It is seen that the electric smelting plants use ores higher in silica than are standard for blast-furnace practice, but that 25 per cent silica is the present limit in ores used for making ferromanganese.

The trouble that silica causes is reflected in the manganese price scale of the War Industries Board,[a] by which an ore with 40 per cent manganese and 8 per cent silica is worth $40.80 per ton and one with 35 per cent manganese and 35 per cent silica only, $7.45 per ton. The scale does not run below 35 per cent manganese, but if it did an ore with 27 per cent manganese and 35 per cent silica would be worth nothing, the penalty for silica entirely wiping out the value of the manganese even though calculated at the same price per unit as given in the scale for a 35 per cent manganese content.

CONCENTRATION OF ORES.

The Philipsburg and the Butte, Mont., districts contain vast tonnages of manganese ores containing around 25 per cent manganese and 35 per cent silica, to say nothing of the rhodonite ores of Colorado and of the tailings from concentrating ores by washing or screening.[b]

Some concentration of domestic ores is done,[c] but in general concentration, is not very successful[d] and a slight improvement in the method of handling low-grade ores might make it more desirable to smelt the run-of-mine ore than to attempt concentration. It is stated[e] that some ores could be concentrated after fine crushing; also that leaching and electrolytic deposition of the manganese oxide is possible. The product would, however, be fine powders and would be extremely difficult to smelt in the usual fashion, keeping the shaft full of charge, unless first briquetted or sintered. It is not certain that briquetting or sintering could be successfully applied.

REDUCTION OF SILICON.

Even though the usual basic slag be employed when smelting ores high in silica, silicon is reduced and enters the alloy when the slag volume is high. Newton[f] states that this is true in the blast furnace. Swann[g] also indicates it for electric-furnace operation, stating that slags from electric smelting containing 12 to 14 per cent Mn, 40 per cent CaO, and 30 per cent SiO_2 can be smelted to an alloy con-

a Chapter 1 of this bulletin.

b See Pardee, J. T., Manganese at Butte, Mont.: U. S. Geol. Survey Bull. 690, 1918, pp. 111-130; Harder, E. C., Manganiferous iron ores; U. S. Geol. Survey Bull. 666-ee, 1917, 13 pp.

c See Swann, T., place cited; Anonymous, Manganese concentrator at Philipsburg, Mont.: Met. and Chem. Eng., vol. 18, 1918, p. 625.

d See Pardee, J. T., work cited, p. 11; Harder, E. C., work cited, p. 13; Newton, E., work cited, p. 57.

e News item, Manganese in California: Chem. and Met. Eng., vol. 19, 1918, p. 702.

f Newton, E., work cited, p. 72.

g Swann, T., place cited.

taining 60 per cent Mn and 19 per cent Si. Had the original smelting been carried on far enough to reduce more of the manganese, considerable silicon would have been reduced also. A high temperature and a large amount of reducing agent are needed to drive the manganese out of the slag, and these conditions favor the reduction of silicon.

As in smelting high-silica ores some silicon will in any event go into the alloy; it appears that it will probably be best to make an alloy with 15 to 20 per cent silicon. In fact, utilization of ores high in silica appears to demand such a silicon content.

PRODUCTION OF SILICOMANGANESE.

Very little silicomanganese has been used in the United States, although it is widely used abroad, being made in large amounts in Sweden, France, and Switzerland. Harden [a] gives the Swedish production of ferromanganese as 2,000 tons per year and that of silicomanganese as 3,500 to 4,000 tons. He says the silicomanganese is made from a mixture of manganese ore and quartz. Lyon, Keeney, and Cullen [b] state that it is also made from rhodonite ore. Swedish practice is said [c] to utilize Bessemer and blast-furnace slags for the manufacture of silicomanganese.

While the demand for silicomanganese is still low in the United States, one firm, the Pacific Electro Metals Co., is making it. The ore used contains 40 per cent Mn, 23 per cent SiO, very little phosphorus, and practically no CaO, MgO, or Al_2O_3. Quartz is added to the ore to form a mixture of about 30 per cent Mn and 35 to 40 per cent SiO_2. A low-grade ore of this composition would be acceptable. Steel scrap is also added to the charge, and the alloy produced runs 50 to 55 per cent Mn. 20 to 25 per cent Si, 0.6 per cent C, 0.2 per cent P, and the remainder iron. The power consumption per long ton of alloy produced is said to be 5,800 kw.-hours, and the coke consumption 1,400 to 2,000 pounds in a 3,000-kw. furnace. The furnace is run without flux, low-ash coke being desirable, and the whole charge is reduced to metal, just as in ferrosilicon smelting. This would be possible only with a practically gangue-free ore.

The absence of slag avoids the usual loss of manganese in the slag incident to making ferro-alloy—the manganese recovery in making silicomanganese being 95 per cent. That is, with an ore of this grade, 20 per cent more manganese is recovered in alloy than when ferro-alloy is made—or 4 tons of ore made into silicomanganese gives as much usable manganese as 5 tons made into ferromanganese.

---

[a] Harden, J., Utilization of manganese ores in Sweden: Met. and Chem. Eng., vol. 17, 1917, p. 701.

[b] Lyon, D. A., Keeney, R. M., and Cullen, J. F., work cited.

[c] News item, Manganese from steel slags: Min. and Sci. Press, vol. 116, Apr. 6, 1918, p. 486.

As regards conservation, silicomanganese, rather than ferro, should therefore be made from ores low in lime and alumina.

The addition of iron is probably due to the buyer's specification. It is illogical to add iron, use power to melt it, and then have to ship it to the steel works where all it does is to chill the steel bath when the alloy is added. By leaving out the iron an alloy containing 68 per cent Mn, 2 per cent Fe, 29.5 per cent Si, 0.3 per cent C, and 0.23 per cent P could be produced and should be preferable to the alloy now made.

FUTURE OF HIGH-SILICA ORES DEPENDENT ON USE OF SILICO-MANGANESE.

Any utilization of low-grade ores high in silica postulates that the steel industry can and will use silicomanganese.

Only a few brief calculations are required to prove that there is enough steel produced for castings, forgings, etc., to which both ferromanganese and ferrosilicon are added, to absorb a very large production of silicomanganese. Common sense and foreign practice, as well as actual large-scale experiments in this country, indicate not only that this alloy can be used, but that its use is distinctly desirable. [a]

It should be noted that the high-silicon alloys contain very little carbon, so that where recarburization is now done by ferromanganese or spiegel, other means of recarburization, such as the use of charcoal, coal, or coke, must be resorted to when silicomanganese is substituted for ferro or spiegel. The low carbon content would, on the other hand, often be a distinct advantage.

CARBON CONTENT OF MANGANESE ALLOYS.

The relation of the silicon to the carbon content in alloys containing 60 to 80 per cent manganese is shown by the curve in figure 13. Silicospiegels of lower manganese content will give a curve lying below this one, as iron takes up less carbon than manganese. The smelting temperature, as well as the practice with regard to carbon, affects the carbon content; but for alloys made by electric smelting at normal smelting temperature and with an excess of carbon, the curve is closely accurate. As regards alloys with less than 5 per cent silicon, the carbon content may vary about 0.75 per cent from the values shown by the curve. For alloys with 10 per cent silicon the deviation will be less than 0.5 per cent carbon; for those with 15 per cent silicon and higher, the deviation will be only 0.1 to 0.2 per cent carbon.

---

a See Swann, T., The development of the ferromanganese industry in the United States since 1914: Chem. and Met. Eng., vol. 19, 1918, p. 672.

REASONS FOR EXPERIMENTAL WORK.

It is natural that commercial electric-furnace plants should procure the highest grade raw material available. For this reason little is known about the actual working possibilities of low-grade manganiferous ores. If any experiments have been made on such ores, the records have not been published. Their behavior in the electric furnace can, to a certain extent, be predicted by a metallurgist skilled in smelting high-grade ores, but beyond a certain point, such predictions become little more than speculation.

As the United States has immense known deposits of low-grade ores, many of which are not amenable to ordinary methods of gravity concentration, the study of the smelting of such materials

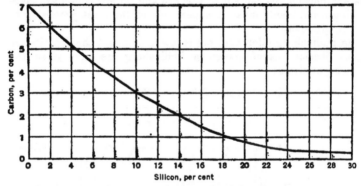

FIGURE 13.—Curve showing relation of carbon to silicon in ferro-silicomanganese alloys.

in the electric furnace, as herein described, was undertaken by the Bureau of Mines. It was felt that if information could thereby be secured that would make any portion of these immense ore reserves commercially available, a real contribution would be made to the country's resources.

CLASSES OF LOW-GRADE ORES.

In the utilization of low-grade domestic ores, one would naturally start with the best of such ores—that is, those highest in manganese or lowest in silica, leaving the leaner ores until experience had been gained on the better ores. Attention should, then, be paid (1) to ores high in silica, low in manganese, and low in phosphorus, like the low-grade Butte ores; (2) to similar ores, but high in ratio of phosphorus to manganese, such as the Philipsburg ores; and (3) to those high in iron, such as the Cuyuna ores.

Exact analyses of the low-grade Butte ores are not available, but from Pardee's statements[a] we may assume that large tonnages

---

[a] Pardee, J. T., Manganese at Butte, Mont.: U. S. Geol. Survey Bull. 690, 1918, pp. 123-125.

of ore containing 25 per cent Mn, 40 per cent SiO, 3.5 per cent Fe, approximately 5 per cent Al$_2$O$_3$, less than 0.05 per cent P, 13 per cent loss on ignition, and only a trace of CaO, should be available. Obviously, such ore would be eminently fitted for the manufacture of silicomanganese, smelting without a slag, provided the alumina did not form a stiff slag of sufficient volume to give trouble. In attempting complete reduction of this ore some aluminium would probably be reduced into the alloy; but as this aluminium would act first in the deoxidizing work of the alloy in steel, being thus eliminated and at the same time protecting the manganese, its presence would probably not be detrimental. In fact silicomanganese-aluminum of high aluminium content is used abroad as a deoxidizer. The low phosphorus content of this ore and its low content of slag-forming constituents make it appear one of the most promising of the low-grade ore.

The Philipsburg (Montana) ore is low in iron and the deposits are large. Special lots of low-grade Philipsburg ore have given the following analyses:

Results of analyses of low-grade Philipsburg (Montana), manganese ores.

Lot No.	Mn.	Fe.	SiO.	CaO-MgO-BaO.	Al$_2$O$_3$.	P.	Loss on ignition.
	Per cent.	*Per cent.*	*Per cent.*	*Per cent.*	*Per cent.*	*Per cent.*	*Per cent.*
1..........................	29.5	2.5	25.0	9.6	1.5	0.105	18.7
2a..........................	26.6	2.8	35.9	6.7	5.8	.24	13.3
2b b..........................	24.1	1.6	37.7	7.7	1.6	.37	12.7
2n c..........................	28.7	2.4	28.5	4.1	3.4	.14	14.6
4c..........................	26.4	2.5	36.4	7.5	1.9	.21	11.8

a Contained both lump and screenings.   b Lump picked off grizzly.   c Screenings, ½-inch screen.

The lump and screenings are obtained in the concentration (by hand picking and screening only) of the run-of-mine ore. The screenings are thought to constitute about 20 per cent of the ore as mined. This type of ore is too basic to allow running without a flux. The ratio of phosphorus to manganese content is high. There would then be two problems, namely, how to flux the ore and how to obtain an alloy low in phosphorus.

EXPERIMENTAL WORK.

To solve these and similar problems, advantage was taken of the electric-furnace equipment of the Bureau of Mines field office at Cornell University, Ithaca, N. Y. Small-scale experiments of this sort are admittedly inconclusive. They go a long way, however, toward replacing speculations and proving theories, and may safely be used as guides to large tests on a commercial scale.

The ores used in the tests were obtained from Philipsburg. Their analyses are given in the foregoing table.

DESCRIPTION OF FURNACE USED.

A single-phase furnace with conducting bottom, with a carbofrax (carborundum) lined shaft 14 inches in diameter and 18 inches deep, surmounted by a fire-brick shaft 12 inches deep, was built. The upper electrode was a square carbon 4 by 4 inches. The carborundum lining worked well, and required few repairs. What little patching was required was readily done with a cement of fine carborundum and water glass. Inasmuch as many of the slags were acid, a magnesite lining would not have been satisfactory. As the runs were intermittent a carbon lining would have been liable to oxidation while the furnace was cooling between runs. The furnace uses about 65 kw., or 1,450 amperes at 50 volts, with an average power factor of 90.

PRELIMINARY TESTS.

Some continuous runs of about 14 hours were made with screenings that passed a ½-inch screen, keeping the shaft full of ore and tapping at intervals. The ore contained much fine dust, and was too fine to bridge well; hence in most runs the furnace was run discontinuously, one charge being smelted and tapped before a succeeding charge was added. The tap hole was closed with a charcoal plug and a wad of fire clay rammed over the plug. In tapping, the fire clay was dug out, when the charcoal could be readily barred out. Starting with a cold furnace, it took about 8 hours to smelt three batches of ore each of 100 pounds. To obtain a slag low in manganese, and a high recovery of manganese in the alloy, it was necessary to run the furnace hot, using not less than 60 to 65 kw. A fairly fluid slag and a considerable excess of reducer were also required. The excess of reducer does not usually come out of the furnace with the slag, but accumulates from heat to heat in the furnace bottom, so that in commercial operation after the first few heats considerably less reducer would be needed than was required in the experimental work. The excess of coke or charcoal holds some slag and metal in its pores so that the "dross" scraped out of the furnace after a series of runs is always higher in manganese than the slag. In commercial operation the "dross" would remain in the furnace and the metal lost in operation on an experimental scale would be recovered.

The manganese content of the slag on the first heat in a cold furnace is always higher than on subsequent heats, even though the slags may be tapped at equal temperatures.

STANDARDIZATION OF FURNACE.

For these reasons the manganese recovery to be expected in the commercial operation would be higher than that obtained in small-scale experiments. Power consumption and electrode consumption in a small furnace, when the furnace is started cold and only a short run is made will be much higher than in commercial practice. A rough factor for calculating power consumption in large-scale operations can be obtained by running a high-grade ore in the small furnace and comparing the power used with that required in commercial operation on such an ore.

Such a standardization was made with a high-grade ore—Philipsburg concentrates—analyzing 47.7 per cent Mn, 1.9 per cent Fe, 8.7 per cent SiO_2, 2.1 per cent CaO, 1.1 per cent Al_2O_3, 0.08 per cent P, 14.2 per cent loss on ignition. The usual basic slag was made. The results follow:

Run 52.—Three charges, each consisting of 100 pounds of ore, 3 pounds of steel scrap, 14 pounds of coke (metallurgical coke, ¼-inch mesh; analyzing 4.75 per cent H_2O, 10.3 per cent ash, of which 1.4 per cent was Fe, 4.3 per cent SiO_2, 2.5 per cent Al_2O_3, 0.8 per cent CaO), 14 pounds of charcoal (crushed to about ½ inch), and 20 pounds of $CaCO_3$ (marble, crushed to ½ inch). Time of run, 8 hours 10 minutes; power used, 440 kw.-hours (180 kw.-hours, first heat; 140, second heat; 120, third heat). Obtained 158.2 pounds of metal, containing 82.0 per cent Mn, 2.0 per cent Si, 0.12 per cent P; 84.3 pounds of slag, containing 8.5 per cent Mn; and 11 pounds of dross containing 34.5 per cent Mn. Mn distribution: In metal 90.5 per cent, in slag 4.2 per cent, in dross 2.7 per cent, volatilization and dusting loss 2.6 per cent. Recovery of P in metal, 79 per cent.

As, in commercial practice, not over 4,000 kw.-hours per long ton of alloy would be required on such an ore in a 3,000-kw. furnace, or 2.2 kw.-hours per pound of metallic manganese in the alloy, and as the last two taps in the small furnace used 3.0 kw.-hours per pound of manganese in the alloy, it appears that 75 per cent of the power used per pound of alloy in the experimental furnace on the last two taps of a run on 300 pounds should give the approximate power required commercially. A previous run (No. 50) on this high-grade ore, using the same flux, and adding 5 pounds of mill scale, to supply iron, 10 pounds of burnt lime, and 33 pounds of charcoal per 100 pounds of ore, gave an alloy of 77 per cent Mn, 2 per cent Si, 0.16 per cent P, with a manganese recovery of 79 per cent. Some of the heats were not given enough power, and some of the slags were high in manganese. It was calculated that under correct operating conditions 88 per cent of the manganese would have been recovered. This calculation was checked by the results of run 52. The results of these runs indicated that a mixture of half coke and half charcoal was as good a reducer as all charcoal.

FLUXING TESTS WITH HIGH-SILICA ORE.

A series of fluxing tests was made with a high-silica ore, namely, the Philipsburg screenings, lot 4 (see table of analyses), containing 26.4 per cent Mn, 2.5 per cent Fe, 36.4 per cent SiO_2, 7.5 per cent CaO + MgO, 1.9 per cent Al_2O_3, and 0.21 per cent P, with 11.8 per cent loss on ignition. In all the runs 10 pounds of coke and 10 pounds of charcoal were used per 100 pounds of ore, except in run 73 in which 15 pounds of each were used. No steel scrap was added except 0.5 pound per 100 pounds of ore in run 70. The results of the runs are summarized as follows:

Results of fluxing tests with high-silica ore.

CHARGES AND POWER CONSUMPTION.

Run No.	CaO per 100 pounds of ore.	CaF₂ per 100 pounds of ore.	CaO per pound of silica, including CaO equivalent to bases in ore.	CaF₂ per pound of silica.	Ore used.	Time of run.	Power consumed.
	Pounds.	Pounds.	Pounds.	Pounds.	Pounds.	H. min.	k.w.-hours.
70	72	....	1.30	....	300	8 20	522
69	28	....	.60	....	300	7 20	450
67	19	7	.36	0.19	300	7 25	453
68		14.5	.20	0.40	300	7 20	452
73	None till end, then 3.5.	10	None till end, then 0.30.	0.28	200	7 8	469

PRODUCTS AND POWER CONSUMPTION.

Run No.	Alloy made per 100 pounds.	Analysis of alloys.			Slag made per 100 pounds.	Mn in slag.	P in slag.	Kw.-hours per pound of alloy, including all heats.	Kw.-hours per pound of Mn, including all heats.
		Mn.	Si.	P.					
	Pounds.	Per cent.	Per cent.	Per cent.	Pounds.	Per cent.	Per cent.		
70	23.8	71.0	13.0	0.60	93	6.0	0.04	7.3	10.2
69	18.4	71.4	11.6	.73	61	11.5	.06	8.2	11.4
67	21.9	73.0	13.0	.68	65	14.0	.04	6.9	9.5
68	26.3	71.0	15.5	.60	59	12.0	.04	5.75	8.1
73	26.7	72.7	19.8	.57	17	23.1	.04	8.8	12.0

DISTRIBUTION OF ELEMENTS IN PERCENTAGES OF TOTAL.

Run No.	In metal.			In slag.		In dross.		Loss.	
	Mn.	Si.	P.	Mn.	P.	Mn.	P.	Mn.	P.
70	63	8½	68	21	18	....	....	16	13
69	50	6	64	26	18	7	5½	17	12½
67	60	8	70	34	3½	(a)		3	13½
68	71	11	75	27	11½	(a)		2	13½
73	73	14½	72½	21	5	6	10½	0	12

a Amount of dross small; included in weight of slag and in sample of slag for analysis.

More power was used in run 70 than in the others because of the large amounts of CaCO₃ charged. The slag was very stiff in run 69 and an arc had to be used to tap. More power might have given a slightly better recovery.

In run 73 more reducer was used than in the other runs, and no flux was added till the end. Two batches of 100 pounds of ore each were added before tapping. The metal tapped, but the slag was too stiff to run; the fluxes were than added, and 45 kw.-hours were used (included in the 469 total in the table) to get the slag out. It then tapped well.

With the same power input for runs 69 and 67 as for run 70, run 67 would have given better results than run 70, but the results of run 69 would not have been as satisfactory as for run 70, the slag being too stiff. The acid slags containing fluorspar, however, tapped well.

Considering the cost of power and fluxes, it appears that the acid slags of runs 67 and 68 were preferable to the basic slag of run 70. It is noteworthy that the basic slag did not prevent the reduction of silicon in run 70. The recovery of manganese was higher in run 73, operating without flux until the furnace got clogged with slag and then fluxing out the slag, but the power consumption was too high. The flux of run 68—all CaF_2, except for the CaO in the ore itself—gave the best results metallurgically. The slag of run 68 was very fluid, and some of the CaF_2 could probably have been replaced by CaO, using a flux intermediate between those of runs 68 and 73. Slags made with fluorspar, are, for an equal Ca content, much more fluid than those made with lime.

ADVANTAGES OF ACID SLAG.

The percentage of manganese is somewhat lower in a strongly basic slag than in an acid slag, but the weight of slag formed is much greater. Hence, the weight of manganese lost in the slag is greater with the basic slag, when the SiO_2 content of the ore is much above 25 per cent. When the ore is fluxed to a basic slag the cost of handling and disposing of the excess slag formed must be considered. The cost of the lime flux for the basic slag would, of course, be lower than that of the fluorspar required to produce an acid slag fluid enough to tap; although in a large furnace, where the tap hole would not freeze as readily as in the small one when tapping a sticky slag, the amount of fluorspar could doubtless be materially reduced. The weight of manganese lost in the slag is greater with the basic slag, when the SiO_2 content of the ore is much above 25 per cent. When the ore is fluxed to a basic slag the cost of handling and disposing of the excess slag formed must be considered. The cost of the lime flux for the basic slag would, of course, be lower than that of the fluorspar required to produce an acid slag fluid enough to tap; although in a large furnace, where the tap hole would not freeze as readily as in the small one when tapping a sticky slag, the amount of fluorspar could doubtless be materially reduced.

It will be noted that in the best run (No. 68), the ratio of phosphorus to manganese in the alloy was 1:118, whereas in the ore the ratio was 1:125. Thus it appears that phosphorus and manganese were recovered in about the same proportion.

TEST OF REDUCERS.

On account of the variation in commercial practice as to the reducer used, anthracite, coke, or a mixture of coke and charcoal being used by different firms, a series of tests was made with reducers.

Some preliminary runs in a small furnace taking only about 20 pounds of ore per charge were made. The Philipsburg screenings used contained 29.5 per cent Mn, 2.5 per cent Fe, 25.0 per cent SiO_2, 9.5 per cent CaO + MgO, 1.5 per cent Al_2O_3, and 0.105 per cent P, and showed 18.7 per cent loss on ignition. The following results were obtained, the same power input being used in each run:

Results of tests of reducers in small electric furnace.

Charge:		Run 20.	Run 21.	Run 22.
Ore	pounds	20	20	20
Coke	do	6	.....	.....
Anthracite	do	.....	5	.....
Charcoal	do	.....	.....	5
Calcium fluoride (CaF₂)	do	2½	2½	2½
Analysis of alloy made:				
Manganese (Mn)	per cent	74.0	77.3	75.3
Silicon (Si)	do	13.0	4.6	12.5
Phosphorus (P)	do	0.24	0.32	0.20
Recovery in alloy:				
Manganese	do	66	43	66
Phosphorus	do	57	52	40

Anthracite gave the lowest recovery. The recovery with coke was the same as that with charcoal.

Runs 48 and 49 were then made in the larger furnace. The slag used was calculated for a ratio of CaO to SiO_2 of 0.4:1, which in the later fluxing tests was shown to be too stiff a slag for good recovery. Seventy-five pounds of ore was used per charge and the furnace was run 12 to 13 hours. The data on the average ore mixture used, the composition of the charge, and the results follow:

Data on reducer tests with large electric furnace.

Ore mixture:		Run 48.	Run 49.
Mn	per cent	28.7	29.0
Fe	do	2.4	2.2
SiO_2	do	28.5	26.5
CaO	do	4.1	4.2
Al_2O_3	do	3.3	2.7
P	do	.14	.16
Loss on ignition	do	14.5	16.5

Charge (weight of constituent to each 100 pounds of ore):

		Run 48.	Run 49.
CaCo$_2$	pounds..	12	12
Charcoal	do....	12.3	None.
Coke	do....	None.	13.5
Quantity of ore charged	do....	900	825
Time of run	hours..	12.25	13
Power used	kw.-hours..	714	710
Weight of alloy made	pounds..	182.0	146.1
Composition of alloy:			
Mn	per cent..	74.0	76.0
Si	do....	10.0	8.0
P	do....	.34	.38
Weight of slag made	pounds..	496	462
Manganese in slag	per cent..	16.4	18.3
Recovery in alloy:			
Mn	do....	52	46
P	do....	49	42

As regards the two tests, although the ore used in run 49 was slightly higher in manganese and lower in silica, and although more power was used per unit weight of ore, charcoal gave better results than coke. However, neither result was satisfactory, the slag being too acid and holding up too much manganese. A slag of the same acidity made with fluorspar would have shown better results. The amount of reducer in both tests was too small for good results.

COMMENTS ON USE OF RHODOCHROSITE.

As the lot of ore used in run 48 was exhausted before run 49 was finished, the last few charges in run 49 were made with a mixture of uncalcined rhodochrosite and Philipsburg lump (lot 2b). On account of the high loss (31 per cent) on ignition of the rhodochrosite, it gave trouble, even when mixed with the other ore. The large volume of CO_2 given off by the raw rhodochrosite tended toward excessive electrode consumption, and probably also toward wasting of the reducer in the charge. If ore so high in volatile matter falls into the melt before it has been thoroughly calcined in the shaft, the furnace will "blow," or form a "volcano."

For these reasons, and because it should be cheaper to calcine by fuel heat than to expel CO_2 by electric heat in the furnace, it will probably be found desirable to calcine rhodochrosite ore before charging it into the electric furnace. The ore decrepitates somewhat on calcination, and the product may have to be screened, the coarse material being then smelted continuously and the fines fed to a separate furnace as fast as the melt will take the ore, but without keeping the shaft piled full, in order to keep down dust losses. Rhodochrosite, may, in fact, be too high grade an ore for electric smelting without being mixed with ores containing more slag-forming constituents.

The analysis of the rhodochosite used, after complete calcination, was 53.5 per cent Mn, 2.9 per cent Fe, 8.8 per cent SiO_2, 1.65 per cent

CaO + MgO, 1.4 per cent Al_2O_3, and 0.04 per cent P. Fluxed to the ordinary basic slag and with suitable addition of steel scrap, this would give only about 1,050 pounds of slag per long ton of 80 per cent ferromanganese produced. Compared to the 40 per cent Mn 20 per cent SiO_2 ore in common use, which, when similarly slagged, gives about 4,000 pounds of slag per long ton of alloy, the rhodochrosite may cause too short a column of slag, and hence, at the low voltage of a ferromanganese furnace, may not present sufficient electrical resistance to develop enough heat.

In the ordinary three-phase, nonconducting-bottom furnace, the current may flow from one electrode to the other two by two paths— (1) through slag entirely, or (2) from the electrode straight down through the slag to the alloy beneath and then through the alloy to directly beneath the other electrodes, and up through the slag again to the other electrodes. With too short a column of slag, path 1 presents the least resistance and will result in the generation of heat in the slag in three spots only. With a deeper slag column, the resistance of the path from one electrode to another through the slag only is less than that from electrode to electrode by way of slag, alloy, and slag. If the power input is too low, the large mass of metal below the slag may get cool enough to make tapping difficult, and the slag itself may not get hot enough for complete reduction of manganese, resulting in high manganese loss in the slag. If, in order to develop enough heat, the furnace is run as an arc furnace instead of as a resistance furnace, local overheating will occur and manganese will be lost by volatilization. In this connection, the comments of Bardwell [a] on smelting rhodochrosite are of interest.

TESTS OF EFFECT OF SIZE OF REDUCER CONSTITUENTS.

Runs 50 and 52, previously mentioned, had shown that the mixture of half coke with half charcoal worked as well as all charcoal. When all one-fourth-inch coke was used as reducer, some of the excess of fine coke tended to emulsify with the slag, coming out with it and making it harder to tap. When half coke and half charcoal was used this was not the case. It was thought that larger coke might give better results, so a series of runs was made to test this assumption and to compare the results from coke and from anthracite with the results of runs in which a mixture of half coke with half charcoal was used. The results of the tests are presented in the table following. The ore used was Philipsburg screenings, lot 4, containing 26.4 per cent Mn, 2.5 per cent Fe, 36.4 per cent SiO_2, 7.5 per cent CaO + MgO, 1.9 per cent Al_2O_3, and 0.21 per cent P, with a 11.9 per cent loss on ignition. It was crushed to pass a one-half-inch screen. As a flux,

[a] Bardwell, E. S., Discussion; Bull. 43, Am. Inst. Mining Eng., Nov., 1918, p. 1651.

only 14.5 pounds of CaF was used per 100 pounds of ore. Three heats were made on each run; that is, 300 pounds of ore was used in each run.

Results of tests of reducers to determine effect of size of constituents.

Run No.	Coke a (½-inch).	Charcoal (½-inch).	Coke b (1 to ¼ inch).	Coke b (less than ¼-inch).	Anthracite c (¼ to ¼ inch).	Anthracite d (less than ¼-inch).	Recovery of Mn.	Recovery of P.
	Pounds.	*Pounds.*	*Pounds.*	*Pounds.*	*Pounds.*	*Pounds.*	*Per cent.*	*Per cent.*
68	10	10	....	....	....	....	71	75
71	....	....	21	....	....	....	54	68
74	....	....	....	21	....	....	35	....
72	....	....	....	....	21	....	46	65
75	....	....	....	....	....	21	32	53

Run No.	Time.	Power used, kw.-hours.	Alloy made.	Composition of alloy.			Slag made.	Mn in slag.	P.	Dross.		
				Mn.	Si.	P.				Weight.	Mn content.	P content.
	Hr.min.		*Lbs.*	*P. ct.*	*P. ct.*	*P. ct.*	*Lbs.*	*P. ct.*	*P. ct.*	*Lbs.*	*P. ct.*	*P. ct.*
68	7 20	452	78.8	71.0	15.5	0.60	177	12.0	0.04	5	(e)	(e)
71	7 15	450	60.8	71.0	16.5	.71	156	15.4	.06	21	14.9	0.13
74	6 45	450	57.6	73.5	13.2	(f)	183	15.0	(f)	29	21.0	(f)
72	7 0	450	50.7	72.0	15.0	.81	200	15.6	.08	29	13.9	.31
75	6 55	450	g 33.6	73.6	11.5	1.00	205	13.1	.03	66	20.7	.14

a Metallurgical coke, 4.75 per cent H₂O, 10.3 per cent ash.
b Gas coke, 13.8 per cent ash.
c 4.5 per cent H₂O, 8.5 per cent ash.
d 3.5 per cent H₂O, 13 per cent ash.
e Included in slag.
f Not determined.
g The fine coal lying on the slag at the end of a tap ran out with the slag, leaving less excess reducer in the furnace on the last two heats than in the other runs. In run 75, 20 to 24 pounds of dross was obtained on each tap. In the other runs the dross was all taken out after the last tap only.

These results indicate that anthracite is not as good a reducer as coke, perhaps because anthracite is less porous, giving less area of contact with the melt. It would be interesting to know whether the difference in the performance of coke and anthracite would be as marked in a large furnace as it was in the experimental one. The experimental work indicates that, at least with a highly siliceous ore, the mixture of coke and charcoal used by the Noble Electric Steel Co. should be highly desirable where charcoal is available.

SLAG-SMELTING TESTS.

In the summer of 1917 Prof. F. F. McIntosh, consulting chemist of the Bureau of Mines, and assistants, made some tests at the Carnegie Institute of Technology, Pittsburgh, Pa., on making manganese alloys from slags. A 50-kw. furnace was used. The slags were not fluxed, and as the iron and manganese were reduced out the slags became stiff, so that the furnace had to be torn down after each tap. The results, therefore, could hardly be applied directly to commercial practice. Table 30 gives a summary of some of the results.

TABLE 30.—*Results of slag smelting tests in electric furnace.*

Run No.	Slag or slag mixture.	Kind of slag.	Analysis of slag or mixture.					Reducer.		Time.	Kw-hour used.	Alloy made.	Composition of alloy made.				Recovery of Mn.
			Mn.	Fe.	SiO₂.	CaO.	Al₂O₃.	Coke.	Charcoal.				Mn.	Fe.	Sl.	P.	
	Lbs.		*P. ct.*	*P. ct.*	*P. ct.*	*P. ct.*	*P. ct.*	*Lbs.*	*Lbs.*	*Hr. min.*		*Lbs.*	*P. ct.*	*P. ct.*	*P. ct.*	*P. ct.*	*P. ct.*
G	240	Bessemer ladle	8.5	13.5	57	2.5	7	44	......	7 45	280	59.5	23	50	26	......	55
H	1,000	Bessemer ladle plus basic open-hearth.	11	21	43	12	10	178	......	22 0	900	108.5	14.5	71	12	0.50	12
J	536	Bessemer ladle plus manganese steel plus spiegel cupola slags.	24	7	42	6	11	30	116	14 45	650	126.5	52.5	26	19	.11	53
K	548	....do.	22	8.5	45	6	12	......	126	13 15	660	142.0	53	28	21	.14	63
L	280	Manganese steel plus spiegel cupola slag.	24	6.5	40	a 3	12	122	......	6 15	320	51	51.5	27	17	.06	36

a 19 pounds of CaO also added to the charge.

Runs K and L indicated that charcoal was a better reducer than coke. Evidently such manganese-bearing slags need to be fluxed, so as to make a fluid slag, as they contain 11 to 22 per cent $Al_2O_3 + CaO$, but those of about 25 per cent Mn content could be used to make silicomanganese. The basic open-hearth slag used in run H was rather high in phosphorus. The other slags were low in phosphorus, and hence made alloys very low in that element.

EFFECT OF PHOSPHORUS CONTENT OF ORES.

In the experimental runs low-grade ores high in phosphorus gave an alloy rather high in that constituent. If smelted with the shaft kept full of ore, the alloy might be still higher in phosphorus. A considerable loss of phosphorus by volatilization took place, and the slags were always rather low in phosphorus, no matter what the original content in the ore was, or whether the slag was acid or basic. The percentage recoveries of phosphorus and of manganese in the alloy were usually within a few per cent of each other.

Lyon, Keeney, and Cullen [a] state that in the electric smelting of ferromanganese all the phosphorus in the charge goes into the metal, and that in making ferrosilicon most of it does, owing to the strongly reducing conditions.

Newton [b] makes a similar statement in regard to blast-furnace ferromanganese smelting. Continuous smelting with the shaft kept full of ore will tend to condense any volatilized phosphorus, so that the losses found in the experimental smelting by batches may not occur in normal practice. Doubtless the lower the ore is in phosphorus the less will escape. The remarks cited are, of course, based on normal practice with low phosphorus ores. However, W. W. Clark, former manager of the Noble Electric Steel Co., states [c] that phosphorus is volatilized in the Noble ferromanganese furnace.

Lonergan [d] cites a test run at the Iron Mountain Alloy Co. works on ore having a very low phosphorus content (less than 0.04 per cent P) in which 52.6 per cent of the phosphorus contained in the ore, coal, and limestone was recovered in the ferromanganese, 14.75 per cent in the slag, and 6.06 per cent in "dirty metal and slag," leaving 31.4 per cent phosphorus lost by volatilization and in dust. Lonergan concluded that the greater part of this loss is purely mechanical, the phosphorus in the dust being retained in its original form. He assumes that any volatilized phosphorus would go off as phosphine.

The writers see nothing to prevent elemental phosphorus being volatilized, as it is in electric-furnace production of phosphorus, and burning to the oxide as soon as it reaches the air.

[a] Lyon, D. A.; Keeney, R. M.; and Cullen, J. F.: The electric furnace in metallurgical work: Bull. 77, Bureau of Mines, 1917, pp. 144, 166.

[b] Newton, E., Manganiferous iron ores of the Arizona District, Minn.: Bull. 5, Minn. School of Mines Exp. Station, Univ. of Minn., 1918, p. 75.

[c] Personal communication.

[d] Lonergan, J., Eliminating phosphorus and sulphur in electric ferromanganese furnaces: Met. and Chem. Eng., vol. 20, 1919, p. 245.

If 31.4 per cent of the manganese also was lost in dust, to say nothing of volatilization, then Lonergan's assumption that phosphorus was lost only mechanically would be justified; but according to Keeney,[a] the average loss by volatilization and dusting in the Iron Mountain Alloy Co.'s practice is 7.3 per cent. It would therefore appear that even on ores so low in phosphorus an appreciable amount of the phosphorus is volatilized.

TESTS WITH HIGH-PHOSPHORUS ORES.

Runs 60 and 61 were made with Philipsburg lump ore, lot 3b, containing 24.1 per cent Mn, 1.6 per cent Fe, 37.7 per cent SiO_2, 7.7 per cent $CaO+MgO$, 1.6 per cent Al_2O_3, and 0.37 per cent P, with 12.8 per cent loss on ignition. The procedure and results were as follows:

Results of run 60—discontinuous smelting, each charge being tapped before adding the next.

[Furnace cold at start.]

	Charge a.	Charge b.	Total for run.
Make-up of charge:			
Ore...pounds..	100	100	....
$CaCO_3$ (0.32 pound $CaO+O$ per pound of SiO_2)..........do....	8	8	....
CaF_2 (18 pounds per pound of SiO_2)..................:....do....	7	7	....
Coke...do....	7.5	8.5	....
Charcoal...do....	7.5	8.5	....
Steel scrap..do....	0.5	None.	....
Time of run..	3 h. 25 m.	2 h. 25 m.	5 h. 50 m.
Power used..kw.-hours..	149	130	279
Metal tapped..pounds..	6.0	18.7	24.7
Slag tapped [a]..do....	65.0	70.5	135.5

Analysis of metal:	Per cent.	Distribution of metal:	Per cent.
Mn.............................	66.0	Mn.............................	34.0
Si.............................	13.5	P..............................	27.5
P..............................	0.71	Distribution of slag:	
Analysis of slag:			
Mn.............................	15.0	Mn.............................	42.5
P..............................	0.7	P..............................	14.5

[a] Some slag stuck in furnace.

Results of run 61—continuous smelting, shaft being kept full.

[Total charge in pounds: Ore, 400; $CaCO_3$ (0.35 pound CaO per pound of SiO_2), 40; CaF_2 (1.18 pound per pound of SiO_2) 28; coke, 38.5; charcoal, 38.5.]

Tap No.	After.	Kw.-hours used.	Metal tapped.	Analysis of metal.			Slag tapped.	Analysis of slag.	
				Mn.	Si.	P.		Mn.	P.
	Hr. min.		Pounds.	Per ct.	Per ct.	Per ct.	Pounds.	Per ct.	Per ct.
1.....................	2 30	142	13.5	66.5	11.6	0.80	92.8	17.1	0.07
2.....................	1 50	141	30.1	70.5	16.5	.95	84.0	13.2	.09
3.....................	2 55	141	31.1	70.0	18.3	.93	30.0	11.8	.12
4.....................	40	28	.8				[a]42.5	10.5	.15
Total................	7 55	452	75.8	....	....	....	249.3	....	....
Average.............				69.0	16.0	.93		14.0	.095

Distribution.	Mn.	P.
	Per cent.	Per cent.
Metal...	54.5	57.5
Slag..	35.5	21.5
Loss..	10.5	21

[a] Includes dross.

[a] Keeney, R. M., The manufacture of ferroalloys: Bull. Am. Inst. Min. Eng., August, 1918, p. 1233.

These results, compared with those from run 68, show that the power supply was insufficient for good recovery of the manganese, 452 kw.-hours being used on 400 pounds of ore in run 61 and 452 on 300 pounds of ore in run 68. The reducer was probably not in too small excess, and a slag higher in CaF₂ would have given better results. Under proper conditions this ore should give the 71 per cent recovery of run 68, when the increased reduction of manganese and silicon should bring the phosphorus down to about 0.75 per cent.

TESTS ON VOLATILIZATION OF PHOSPHORUS.

As the volatilization of phosphorus is probably due to to its reduction from the calcium phosphate in the slag to elemental phosphorus, which will then either volatilize or be absorbed by the metal, it would appear that if only enough reducer were used to reduce Fe and Mn to FeO and MnO, and P_2O_5 to P, and the charge were held molten for a time before more reducer were added to cause a metal fall, the elimination of phosphorus would be favored. With this consideration in mind, run 62 was made.

In this run each of the three batches charged consisted of 100 pounds of lot 2b ore, 10 pounds of $CaCO_3$, 7 pounds of CaF_2, and 2 pounds each of coke and charcoal. After the batch was melted it was kèpt molten for 20 minutes, then 8 pounds each of charcoal and coke were added. The results were as follows:

Results of run 62.

Tap.	Time.	Kw.-hours used.	Quantity of metal obtained.	Analysis of metal.			Quantity of slag obtained.	Analysis of slag.		Quantity of dross obtained.	Analysis of dross.	
				Mn.	Si.	P.		Mn.	P.		Mn.	P.
	Hr. min.		*Lbs.*	*P. ct.*	*P. ct.*	*P. ct.*	*Lbs.*	*P. ct.*	*P. ct.*	*Lbs.*	*P. ct.*	*P. ct.*
a	3 5	182	13.5			1.29	67	13.3	0.06			
b	2 35	131	19.8			.97	78	11.8	.06			
c	2 20	130	15.0			1.16	70	11.7	.10	27	17.1	0.16
Total	8 0	443	48.3				215					
Average				71.0	14.0	1.12		12.6	.07			

Distribution:	Mn.	P.
Metal .... per cent..	42	49
Slag and dross .... do....	43	31
Loss .... do....	15	20

As there was a considerable excess of reducer left after each tap, which would cause too early a metal fall and collect phosphorus, another run (No. 63) was made, omitting the reducer from the charge till all had been melted 20 minutes (except on the first charge), letting the excess reducer remain in the furnace after each tap, and then adding 10 pounds of coke and 10 pounds of charcoal. Five pounds of NaCl was added to each charge. Otherwise the charges were as in run 62. The results obtained were as follows:

Results of run 63.

Tap.	Time.	Kw.-hours used.	Quantity of metal obtained.	Analysis of metal.			Quantity of slag. obtained.	Analysis of slag.	
				Mn.	Si.	P.		Mn.	P.
	Hr.min.	*P. ct.*	*Lbs.*	*P. ct.*	*P. ct.*	*P. ct.*	*Lbs.*	*P. ct.*	*P. ct.*
a...............	3 15	182	13.0	....	....	1.42	72	12.0	0.02
b...............	3 30	136	20.6	....	....	.75	81	11.9	.03
c...............	2 15	131	21.2	....	....	.50	93	11.0	a.03
Total.......	9 0	449	54.8	....	....	......	246	......	......
Average.......			....	71.6	14.5	.93		11.5	.03

a Slag included in dross.

Distribution:

	Mn.	P.
Metal.....................................per cent..	54.5	46
Slag......................................do....	39.5	6
Loss......................................do....	6	48

On the assumption that a more acid slag might help, another run (No. 64) was made, the total charge being practically the same as in runs 62 and 63, but divided as follows:

Charge for run 64, in pounds.

	Batch a.	Batch b.
Added at start:		
Ore..	100	100
Charcoal...	4	None.
$CaCO_3$..	3	5
CaF_2...	2	1
Added after heating 20 minutes:		
Charcoal...	6	10
Coke..	10	10
$CaCO_3$..	7	8.5
CaF_2...	5	6

Results of run 64.

Tap.	Time.	Kw.-hours used.	Quantity of metal obtained.	Analysis of metal.			Quantity of slag obtained.	Analysis of slag.		Dross.
				Mn.	Si.	P.		Mn.	P.	
	Hr.min.		*Pounds.*	*Per ct.*	*Per ct.*	*Per ct.*	*Pounds.*	*Per ct.*	*Per ct.*	*Lbs.*
a.................	4 30	210	9.3	68.0	12.8	1.45	65	18.0	0.11	
b.................	2 35	148	10.9	71.8	11.8	1.57	63	18.0	.09	17
Total...........	7 5	358	20.2	......	......	......	128	......	......	......
Average...........			....	70.0	12.5	1.52		18.0	.10	......

Distribution:

	Mn.	P.
Metal......................................per cent..	29.5	42
Slag......................................do....	48	17.5
Loss......................................do....	22	40.5

As the results of this run were not promising, another run (No. 59) was made in which the excess reducer was carefully scraped out after each tap, so that only the four pounds of reducer charged with the ore, as in run 62, would be present.

In run 59, the charge at the start consisted of 100 pounds of ore (lot 2b), 8 pounds of $CaCO_3$, 5 pounds of CaF_2, and 4 pounds of charcoal. After all was melted for 20 minutes, 6 pounds of charcoal and 6 pounds of coke were added; then the furnace was tapped. This procedure was repeated four times. The results follow:

Results of run 59.

Tap.	Time.	Kw.-hours used.	Quantity of metal recovered.	Analysis of metal.			Quantity of slag and dross recovered.	Analysis of slag and dross.	
				Mn.	Si.	P.		Mn.	P.
	Hr. min.		Pounds.	Per ct.	Per ct.	Per ct.	Pounds.	Per ct.	Per ct.
a	4 0	200	12.5	66.0	11.9	0.70	83.3	17.0	0.10
b	2 50	130	10.2	66.0			80.5	17.1	.06
c	2 45	116	9.6	66.2	12.3	.64	85.7	17.1	.06
Total	9 35	446					249.50		
Average			32.3	66.0	12.0	.68		17.0	.09

Distribution:

		Mn.	P.
Metal	per cent..	29.5	19.5
Slag	do....	58.5	20
Loss	do....	12	60.5

These results, owing to the poor slag and the lack of sufficient excess reducer, are far from satisfactory as regards recovery of manganese, but they are somewhat promising as regards volatilization of phosphorus.

The low recoveries of manganese in this series were doubtless largely due to the ore not being properly fluxed. This defect was not realized till the later runs on fluxing were made. However, it is evident that one requirement for high recovery of manganese, namely, an excess of reducer, and one for high volatilization of phosphorus, namely, no excess reducer beyond that necessary to form MnO, FeO and P (that is, no metal fall to collect phosphorus), are incompatible. It is hardly practicable to clean the furnace from all excess carbon after each tap.

DEPHOSPHORIZATION AND SMELTING IN TWO STAGES.

Therefore, the only procedure metallurgically possible seems to be to carry out the dephosphorization in one furnace, running with just enough reducer to form MnO, FeO, and P, and holding the charge molten for a while to allow the phosphorus to escape; then to tap the hot slag into a second furnace, add excess reducer, and smelt the previously dephosphorized slag.

As it would be difficult to determine the exact point where there is complete reduction to MnO, FeO, and P, and as the electrode will also supply a little carbon, it would probably be more feasible to provide the dephosphorizing furnace with metal and slag taps, the latter being at a higher level. The slag can thereby be tapped into the second furnace, and a very little high-phosphorus metal can be taken either at each slag tapping or at longer intervals. This scheme would eliminate phosphorus not only by volatilization but also, in part, by concentration into the first metal to fall.

That this method is metallurgically possible is shown by the results of runs 56 and 57, presented in the following tabulations:

Charges used in run 56.

[Figures in pounds.]

Charge.	Ore.a	Char-coal.	Coke.	CaCO₃.	CaF₂.	Steel scrap.	Mill scale.
a	100	9.5	9.5	15	6	....	....
b	100	2.5	2.5	8	5	10	....
c	100	2.5	2.5	8	5	....	....
d	100	5.5	5.5	6	5	....	15
e	75	4	4	5	4	....	11

a From lot 2b.

NOTES.—
a. Run to reduce considerable Mn.
b. Run to reduce very little Mn, steel scrap added to collect P.
c. Run to reduce very little Mn.
d. Run to reduce very little Mn, and leave Fe in slag from mill scale.
e. Run to reduce very little Mn, but all the mill scale.

Charges used in run 57.

Charge.	Ore.	Char-coal.	CaCO₃.	NaCl.
a	50	1.9	6.6	4.3
b	50	1.9	6.6	4.3

NOTES.—
a. Run to reduce very little Mn.
b. Run to reduce a good deal of Mn.
It was hoped that the addition of NaCl would form phosphorus chlorides or oxychlorides that might be more readily volatile.

Results of runs 56 and 57.

Run.	Time.	Kw.— hours used.	Quantity of alloy made.	Analysis of alloy.		Slag.	Analysis of slag.	
				Mn.	P.		Mn.	P.
	Hr. min.		Pounds.	Per ct.	Per ct.	Pounds.	Per ct.	Per ct.
56a	3 15	136	13.2	66.5	0.96	74.8	15.8	0.05
56b	1 25	73	12.8	41.1	1.02	99.5	23.0	.12
56c	1 40	60	2.0	69.6	1.41	87.6	23.0	.11
56d	1 20	59	2.6	18.8	1.23	86.6	21.6	.10
56e	2 15a	105	11.1	41.8	1.32	59.0	21.6	.05
57a	1 10b	41	.4	32.3	1.07	49.0	24.0	.09
57b	2 25	88	9.0	50.1	.39	41.0	16.6	.11

a Furnace cold at start.   b Furnace hot at start.

Results of runs 56 and 57—Continued.

DISTRIBUTION OF MN. AND P.

Run.	Mn.			P.		
	Alloy.	Slag.	Loss.	Alloy.	Slag.	Loss.
	Per ct.	*Per ct.*	*Per ct.*	*Per ct.*	*Per ct.*	*Per ct.*
56a	37	49	14	34.5	11	53.5
56b	17	91	a8	34.5	32.5	33
56c	6	83	11	7.5	26	52.5
					b 41	
56d	2	78	20	9	23	53
					b 38	
56e	26	74	0	40	7.5	52.5
57a	1	97.5	1.5	3	24.5	72.5
57b	33	56	11	19.5	24.5	56

a Gain.   b Including dross.

The results of runs 56b, 56c, and 57a indicate that by proper care 85 to 90 per cent of the manganese could be recovered in the slag, with only about 25 to 30 per cent of the phosphorus, the slag having about 23 per cent manganese and 0.10 per cent phosphorus where the original ore had 24 and 0.37 per cent, respectively, or, on the basis of the calcined ore (deducting loss on ignition), 27.5 and 0.42 per cent.

Attempts to smelt the dephosphorized slags without the proper flux and with too little reducer gave low recoveries of manganese. After tests had indicated the flux and reducer needed, similar tests were again made. Not enough ore from lot 2b was left for a run, so ore from lot 4 was added. The charges were as follows:

In run 76 (dephosphorization), 57 pounds of lot 4 ore, averaging 25.4 per cent Mn, 2.1 per cent Fe, 37.2 per cent SiO_2, and 7.6 per cent $CaO+MgO$; 43 pounds of lot 2b ore, averaging 1.7 per cent Al_2O_3, 0.29 per cent P, and 12.1 per cent loss on ignition; 4 pounds of charcoal; and 7½ pounds of CaF_2, were charged in two batches. The results were as follows:

Results of run 76 (dephosphorization).

	Charge 1.	Charge 2.
Time	2 h. 10 m.	1 h. 35 m.
Kw.-hours used	125	100
Alloy made pounds	2.9	4.5
Manganese in alloy per cent	54.	64.1
Phosphorus in alloy do	2.5	2.12
Slag made pounds	96.7	82
Manganese in slag per cent	23.4	20.3
Phosphorus in slag do	.1	.09

Distribution of manganese:
Metal per cent	9	
Slag do	79	
Loss do	12	

Distribution of phosphorus:
Metal do	29.5	
Slag do	29.5	
Loss do	41	

The second tap was run a little too long and too much metal reduced.

The slags were then smelted. In order to determine what the results would be if both slags and metal were run into the smelting furnace from the dephosphorizing furnace, the metals as well as the slags were recharged.

In run 77 (smelting) the first charge consisted of 96.7 pounds of slag, 2.9 pounds of metal from run 76 (1), 17 pounds of charcoal (charcoal only was used, the coke supply being exhausted), and 7½ pounds CaF$_2$. The second charge consisted of 82 pounds of slag, 4.5 pounds of metal from run 76 (2), 17 pounds of charcoal, and 7½ pounds CaF$_2$. The results were as follows:

Results of run 77 (smelting).

	Charge 1.	Charge 2.	Total (or average.)
Time...................................	2 h. 35 m.	1 h. 55m.	4 h. 30 m.
Kw.-hours used........................	175	130	305
Metal recovered.............pounds..	23.7	21.5	45.2
Manganese in metal...........per cent..	75.2	72.5	74
Silicon in metal...:.............do...........		....	17.2
Phosphorus in metal.............do....	.53	.53	.53
Slag recovered................pounds..	50	66	116
Manganese in slag.............per cent..	9.8	8.8	9.2
Phosphorus in slag.............do....	.06	.07	.65

Based on the manganese and phosphorus in the slag and metal charged in run 77, the distribution was as follows:

Percentage distribution of Mn and P based on analysis of slag and metal charged.

	Mn.	P.
Metal..	75	70.5
Slag..	24	20.5
Loss..	1	9

Based on the analysis of original ore charged in run 76, the distribution when both slag and metal of that run were recharged was as follows:

Percentage distribution of manganese and phosphorus, based on original ore.

	Mn.	P.
Final metal..	66	41.5
Final slag..	21	12
Loss, dephosphorizing run.............................	12	41
Loss, smelting run.....................................	1	5.5

The fact that the manganese loss was 12 per cent in the first (dephosphorizing) run, as compared with 1 per cent in the second (smelting) run, although the temperature was much higher in the second, indicates that the loss in smelting is more by dusting than by volatilization. The true volatilization loss appears to be small as long as the furnace is run as a resistance furnace, or with only a small submerged arc. A large, open, high-voltage arc would doubtless volatilize manganese.

To find what the results would have been had the metal obtained in run 76 not been recharged in run 77, we may subtract the 7.4 pounds of alloy, with its content of 4.4 pounds of manganese and 0.17 pound of phosphorus, from the 45.2 pounds of alloy obtained in run 77, with its 33.4 pounds of manganese and 0.24 pound of phosphorus. This would leave 37.8 pounds of alloy, containing 29 pounds of manganese, 7.75 pounds of silicon, and 0.07 pound of phosphorus, and analyzing about 77 per cent of manganese, 20 per cent silicon, and 0.18 per cent phosphorus, most of the iron being taken out in the first alloy.

The distribution, based on the content in the original ore, would then have been:

Percentage distribution of Mn and P when the alloy is not resmelted.

	Mn.	P.
Final metal	57	12
Final slag	21	12
Loss, dephosphorizing run	12	41
Loss, smelting run	1	5.5
Discarded in first metal	9	29.5

If the second charge in heat 76 had not been run so long, so that only 2 or 3 pounds of metal instead of 4½ pounds had been collected, the recovery, based on the original ore, would have been about 60 per cent.

However, if the high phosphorus alloy charged in run 77 had been left out, more of the phosphorus left in the dephosphorized slag might have gone into the metal, as the phosphorus may be expected to divide itself between the two liquid layers somewhat according to the concentration of phosphorus in each, the recovery of the 0.075 pound of phosphorus left in the two slags might have been practically complete. This would have given, in 37.8 pounds of alloy, 0.20 per cent phosphorus.

Two more slags were made from the same charges used in run 76, but the reduction was carried further, more metal being reduced than in run 76. The slags contained an average of 20 per cent manganese and 0.10 per cent phosphorus. These were smelted without the addition of the metal obtained in the dephosphorizing run and gave an alloy of about 72 per cent manganese, 27 per cent silicon, and 0.17 per cent phosphorus, with a recovery of 71 per cent manganese, based on the manganese content in the slags charged.

As in runs 56 and 57, the slags from an ore containing 0.37 per cent phosphorus were brought down to 0.10 per cent phosphorus in the dephosphorizing run, and as in run 76 an ore containing 0.29 per cent phosphorus came down also to 0.10 per cent phosphorus, it is probable that the slag of ores containing even more than 0.37 per cent phosphorus could also be brought down to about the same phosphorus content without much greater loss of manganese.

In order to get an idea of the power that would be needed in smelting the hot dephosphorized slag, the power consumption was noted in run 77 from the time the slag was thoroughly melted—that is, in about the condition it would be when tapped from the dephosphorizing furnace till the end of the heat. On the first heat, starting with a cold furnace, 95 kw.-hours was used and on the second 85 kw.-hours was used. On a third heat 70 kw.-hours would probably suffice.

The dephosphorization heats in run 76 were doubtless continued longer than necessary, as in heats c and d, run 56, and only 60 kw.-hours was needed.

CONCLUSIONS AS TO TWO-STAGE PROCESS.

It would appear that with a 3,000-kw. installation, using a 1,350-kw. furnace for dephosphorization and a 1,650-kw. furnace for smelting, 12,500 pounds of ore (containing 24 per cent Mn, 1.5 per cent Fe, 38 per cent SiO_2, 7.5 per cent CaO +MgO, 0.4 per cent P, and 1.5 per cent Al_2O_3, with 13 per cent loss on ignition), mixed with 250 pounds of coke, 250 pounds of charcoal, and 940 pounds of fluorspar (CaF_2), and given 5,500 kw.-hours, would produce 250 pounds of high phosphorus alloy, with about 50 per cent Mn and 2.5 per cent P, containing about half of the Fe in the ore and about 11,000 pounds of slag, with 23 per cent Mn and 0.10 per cent P. The latter tapped hot into the second furnace and mixed with 950 pounds of coke, 950 pounds of charcoal, and 860 pounds of fluorspar, and smelted with 7,000 kw.-hours, should produce 2,350 pounds (equivalent to a long ton of 80 per cent alloy) of an alloy containing 76 per cent Mn, 4.5 per cent Fe, 18.3 per cent Si, 1.0 per cent C, 0.20 per cent P, with a recovery of 60 per cent of the metallic manganese in the original ore.

The ore smelted direct without dephosphorization would give an alloy containing about 73 per cent Mn, 8 per cent Fe, 17 per cent Si, 1.25 per cent C, and 0.75 per cent P, with a recovery of 70 per cent of the manganese. A 3,000 k. v. a. installation including two furnaces as suggested above would produce daily 4.5 long tons of the dephosphorized alloy, containing 7,700 pounds of Mn, 1,850 pounds of Si, 456 pounds of Fe, 100 pounds of C, and 20 pounds of P, while a single 3,000 k. v. a. furnace would produce daily 5.5 long tons of the high-phosphorus alloy, containing 9,000 pounds of Mn, 2,100 pounds of Si, 950 pounds of Fe, 185 pounds of C, and 92 pounds of P.

Whether the lower phosphorus content resulting from the two furnace treatments would compensate for the lower manganese recovery and the lower output depends on whether the high-phosphorus alloy can be utilized.

If 0.9 per cent of standard 80 per cent ferromanganese—that is to say, 0.72 per cent ferromanganese—is added to the average steel bath, a content of 0.4 per cent phosphorus in the ferromanganese (the ratio of P to Mn being 1 to 200) would raise the phosphorus content of the steel by 0.0036 per cent. Ferro alloy of this ratio has been accepted by some steel makers during the manganese shortage.

The alloy containing 73 per cent manganese, 17 per cent silicon, and 0.75 per cent phosphorus would raise the phosphorus content of the steel by 0.0074 per cent, while the alloy containing 76 per cent manganese, 18.3 per cent silicon, and 0.2 per cent phosphorus would raise it only 0.0019 per cent. Either alloy would raise the silicon content of the steel 17 points, or a little less, as some silicon will be lost by oxidation.

An indefinite quantity of manganese alloys high in phosphorus can doubtless be absorbed by the steel industry. Cromlish,[a] in fact, suggested making a spiegel with 13 to 25 per cent manganese and 2.2 to 3.4 per cent phosphorus from "flush" and "tapping" cinders and utilizing this in the manufacture of the steel sheet, where a phosphorus content higher than normal is required to keep the sheets from sticking together in pack-rolling.

During the preliminary runs, and in some runs not reported herein, with ores high in phosphorus, when the proper flux or reducer was not used, a number of alloys running from 65 to 75 per cent Mn, 3 to 17 per cent Si, 1.3 to 5.3 per cent C, and 1.6 to 0.7 per cent P were produced. An assortment of these alloys has been sent to the Minneapolis station of the Bureau of Mines, where it is planned to study the possibility of eliminating the phosphorus.

According to Lang,[b] by melting ferromanganese high in phosphorus under manganese oxide (MnO_2) at 1,200° C., the phosphorus may be eliminated. A couple of tests were made at Ithaca by melting the ferro alloys high in phosphorus under high-grade manganese-oxide ores. No dephosphorization of the alloy, but instead a dephosphorization of the ore occurred.

TESTS WITH HIGH-SILICA ORES.

An important class of low-grade manganese ores is that in which manganese and iron occur in about equal amounts, with or without a high proportion of silica or phosphorus, or both, to manganese.

---

a Cromlish, A. L., U. S. Patent 1261907, Apr. 9, 1918.
b Lang, G., German Patent 252166, class 18 b, group 2, Oct. 14, 1912.

Many Cuyuna ores, such as the products of the Ferro and Mille Lacs mines, some of the Appalachian ores, and some ores from Leadville, Colo., fall into this class.[a]

A small lot of such an ore was at hand from the Sultana mine, Cuyuna district. The lump ore analyzed 29.0 per cent Mn, 24.2 per cent Fe, 4.5 per cent SiO_2, 1.5 per cent CaO +MgO, 3.7 per cent Al_2O_3, 0.20 per cent P, and showed 11.6 per cent loss on ignition.

Such an ore, smelted direct, would give about a 50 per cent manganese alloy. The ratio of phosphorus to manganese is high. It seems desirable to produce manganese alloys either of the composition of spiegel, which will bear cupola melting, for adding hot without excessive loss of manganese; or of that of 70 to 80 per cent ferromanganese for adding cold. Alloys of intermediate composition, say 40 to 60 per cent manganese, where the balance is iron instead of silicon, will chill the steel bath badly if added cold on account of the large amounts of inert iron carried. In order to add such alloys hot, electric melting of the alloys is necessary, as they would lose too much manganese in cupola melting. Electric melting, even of 80 per cent ferromanganese, is probably desirable, but few American steel plants are doing so as yet, and the necessity of installing electric furnaces in order to utilize the 50 per cent manganese alloys would be an obstacle to the adaption of electric melting. Where silicomanganese containing 60 per cent manganese and 20 per cent silicon is used instead of both ferromanganese and ferrosilicon, the inert iron carried by the silicomanganese is less than that in an equivalent mixture of 80 per cent ferromanganese and 50 per cent ferrosilicon.

APPLICABILITY OF TWO-STAGE PROCESS TO CUYUNA ORES.

As so many Cuyuna ores carry an undesirably high proportion of phosphorus to manganese, it appears that if electric smelting is used, a differential, two-stage, two-furnace process is worth consideration. The bulk of the iron, a little manganese, and much of the phosphorus may be reduced in the first furnace to a high-phosphorus, manganiferous pig iron, or, in an ore sufficiently low in phosphorus, to a spiegel, leaving in both cases a slag high in manganese, low in iron, and low in phosphorus to be tapped hot into the second furnace and there smelted to ferromanganese.

RESULTS OF TESTS.

Tests with the two-stage process were made with Cuyuna ores. In run 54, three heats were made, each charge including 100 pounds of Sultana lump ore, $3\frac{1}{4}$ pounds limestone (the furnace bottom had been patched with dolomite before this run, and some of the

---

a Harder, E. C., Manganiferous iron ores: Bull. 690, U. S. Geol. Survey, 1918, pp. 5, 8, 11.

dolomite went into the slags), and 12 pounds of charcoal (except in heat c, where 10 pounds was used).   The results were as follows:

Results of run 54.

Heat.	Time.	Kw.-hours used.	Alloy.	Analysis of alloy.					Slag.	Analysis of slag.		
				Mn.	Fe.	Si.	C.	P.		Mn.	Fe.	P.
	H. min.		*Lbs.*	*P. ct.*	*P. ct.*	*P. ct.*	*P. ct.*	*P. ct.*	*Lbs.*	*P. ct.*	*P. ct.*	*P. ct.*
a.....................	2 15	100	30.2	24.5	68.1	0.07	2.35	0.17	60.0	32.0	2.7	0.04
b.....................	1 20	58	20.3	11.4	82.5	.23	1.7	.19	41.0	44.8	5.9	.06
c.....................	1 20	50	20.6	10.5	84.0	.02	2.25	.19	51.0	46.9	4.4	.03
da....................	0 15	10	13.0	17.1	75.3	.40	2.10	.21	｛21.0 ᵇ15.5	36.1 26.1	3.0 13.8	Tr. .08

DISTRIBUTION OF Mn Fe, AND P.

	Mn.	Fe.	P.
	Per cent.	*Per cent.*	*Per cent.*
In metals................	13.5	75	22
In slags.................	75.5	12	13
Volatilisation and dust loss...........	11.0	13	65

a A little CaF₂ was added and a longer run was made to get the slightly stiff slag to tap more cleanly from the furnace.

b Mixed slag and metal shot taken from furnace bottom when cold.

The reduction in heat a was carried too far, reducing too much Mn.   With experience it should be possible to leave about 80 per cent of the Mn, 15 per cent of the Fe, and 20 per cent of the P in the slag; take out 10 per cent of the Mn, 75 per cent of the Fe, and 20 per cent of the P in the metal; and lose 10 per cent Mn, 10 per cent Fe, 60 per cent P by volatilization and dusting.

Two thousand pounds of ore plus 200 pounds of reducer (half charcoal, half coke) plus 200 pounds of limestone, smelted with 1,000 kw.-hours or less, would then produce about 425 pounds of alloy (low-grade spiegel) containing 13 per cent Mn, 89.5 per cent Fe, 2 per cent C, and 0. 20 per cent P; and 1,100 pounds of slag containing 42 per cent Mn, 6.5 per cent Fe, 8 per cent SiO₂, and 0.07 per cent P, that is the equivalent of a high-grade ore.

This slag tapped hot into the second furnace, with the addition of 50 pounds of limestone and 330 pounds of reducer (half coke, half charcoal), and smelted with 700 kw.-hours, should give, assuming an 80 per cent recovery of the Mn in the slag, 460 pounds of alloy, analyzing 78.5 per cent Mn, 14.4 per cent Fe, 1.0 per cent Si, 6.0 per cent C, and 0.15 per cent P, and about 550 pounds of waste slag.

The recovery of the Mn in the final alloy would be 64 per cent of that in the original ore.   A 3,000-kw. installation would then use a 1,750-kw. furnace for the first operation and one of 1,250-kw. for the second.

In run 55, 190 pounds of slag chosen from the slags made in run 54 and in another similar run were mixed, the composition of the mixed slags being 41.0 per cent Mn, 15 per cent SiO_2, 3.4 per cent Fe, and 0.06 per cent P.

Results of run 55.

Heat.	Charge.		Charcoal.	Coke.	Mill scale.	Steel scrap.
	Mixed slags.	CaCO₃				
	Pounds.	Pounds.	Pounds.	Pounds.	Pounds.	Pounds.
a	100	10	18	....	2.5	....
b	90	9	8	8	....	1.5

Heat.	Time.	Kw.-hours used.	Alloy.	Analysis of alloy.			Slag.	Analysis of slag.	
				Mn.	Si.	P.		Mn.	P.
	H. min.		Lbs.	Per ct.	Per ct.	Per ct.	Lbs.	Per ct.	Per ct.
a	2 50	176	43.5	74.2	3.2	0.09	48.2	11.8	(a)
b	2 10	106	41.0	74.5	5.7	.10	45.0	5.4	(a)
Total	5 0	282	84.5	......	......	......	93.2	......	......
Average	......	......	......	74.3	4.4	.10	......	8.7	......

DISTRIBUTION.

	Mn.	P.
	Per cent.	Per cent.
Metal	82½	75
Slag	6½	....
Loss	11	....

a Not determined.

The results of this run checked the assumptions made as to recovery and power comsumption in smelting the slag.

If the ore were high in phosphorus, the first furnace would be so operated as to throw as little manganese into the alloy as possible, and produce a pig iron with a manganese content of about 10 per cent and a phosphorus content according to that in the ore. Such pig iron would doubtless find use when mixed with other pig of suitable composition and should have some value. If the ore is low enough in phosphorus, a little more reducer would be used in the first furnace so as to cause enough manganese to be thrown down to form a standard spiegel. If too little iron in proportion to the manganese was left in the slag for the second smelting, steel scrap or ore low in manganese but high in iron would be added, so that the products from a low phosphorus ore could be speigeleisen and ferromanganese.

COMPILATION OF DATA.

The probable performance of several types of domestic manganese ore under electric smelting, compiled from the data available on present practice, and from the results of experimental work, is given in Table 31 following.

The electrode consumption per ton of product has been calculated on the basis of kw.-hours required per ton of product, taking as standard the normal commercial consumption of 175 pounds of electrode per ton of 80 per cent ferromanganese produced from an ore containing 40 per cent manganese and 20 per cent silica, and using 5,500 kw.-hours. In other words, the electrode consumption per 1,000 kw.-hours used is 31 pounds. The actual consumption of electrodes, not including stub ends in the experimental runs, varied from 6 to 10 pounds per 450 kw.-hours used, which is of the same general order of magnitude as the figures assumed.

In Table 32 an attempt has been made to calculate costs of operation for the eight ores represented in Table 31, on the assumption that the ores are priced as follows: Ores 1 and 2 in accordance with the War Industries Board's schedule, ores 3 and 4 at $15 per ton, ore 5 at $7.50 per ton, ores 6 and 7 at $7 per ton, and ore 8 at $8 per ton. Limestone is assumed to cost 0.2 cent a pound, fluorspar 3 cents a pound, coke 0.5 cent a pound, charcoal 1¼ cents a pound, steel scrap 2 cents a pound, electrodes 15 cents a pound, and power 0.5 cent per kw.-hour used. Selling prices are calculated at $285.80 per long ton of 80 per cent ferromanganese (equivalent to $250 per long ton of 70 per cent ferromanganese), and $150 per long ton of 50 per cent ferrosilicon; and the manganese and silicon contents in a silicomanganese alloy are calculated as having the same value as in the separate ferro alloys. The alloy of ore 5, with 0.60 per cent phosphorus, is figured as worth $260 per equivalent ton of 80 per cent ferro, while that of ore 7, with 0.75 per cent phosphorus, is figured as worth $235 per equivalent ton. All these figures are assumed, and would vary widely according to the situation of the plant. They are included merely to give a rough idea of the costs. Peace-time figures would greatly alter the costs and profits.

Lyon, Keeney, and Cullen[a] give the following prices per metric ton in Germany, f. o. b. Louisberg, January 1, 1913:

50 per cent ferrosilicon... $77.55
80 per cent ferromanganese... 68.00
Silicomanganese (68 to 75 per cent Mn, 20 to 25 per cent Si)......... 106.50

It is seen that the price paid for contained manganese and silicon in the silicomanganese is higher than that paid for the same amount of manganese and silicon in ferro alloys.

---

[a] Lyon, D. A., Keeney, R. M., and Cullen, J. F., work cited, p. 140.

The greater daily output per furnace when using the higher grade ores makes their use more profitable to the electric-smelting plant. On the assumptions made, the practice of smelting ores that can be run without a slag to silicomanganese shows the largest profit as well as the highest recovery of manganese from the ore.

The figures for ores 1 to 4 have been calculated on the basis of commercial practice, and for ores 5 to 8 on the basis of the experimental work in which expensive fluorspar and charcoal was used. Experience in large-scale operation would probably allow the substitution of some limestone for fluorspar and some coke for charcoal. No experiments to test this possibility were made, as the work was aimed to show the metallurgical rather than economic possibilities, and as such tests on a laboratory scale would have to be checked by large-scale tests before accurate information would be obtained.

According to Willcox,[a] Germany made silicomanganese from blast-furnace slags, and ferro or silico from low-grade ores during her manganese shortage.

The Taylor-Wharton Steel Co., High Bridge, N. J., which makes manganese steel, had a slag from manganese steel running 35 to 42 per cent silica and 31 to 43 per cent manganese.[b] The company had some of this slag electrically smelted into an alloy of 47 per cent manganese and 20 per cent silicon, which was used successfully in the manufacture of steel. The cost of smelting was too high to make the use of the alloy economically desirable at the peace-time prices of ferro then prevailing.

Tables 31 and 32 follow on next pages.

CONCLUSION AS TO ELECTRIC SMELTING.

The conclusion to be drawn from this investigation is that although the electric smelting of manganiferous slags and low-grade domestic ores is unlikely to be profitable at times of normal costs and prices, such smelting is metallurgically possible, and could be done profitably in times of high prices.

---

[a] Willcox, F. H., The significance of manganese in American steel metallurgy: Trans. Am. Inst. Min. Eng. Vol. 56, February, 1917, pp. 412-420.
[b] Personal communication.

TABLE 30.—*Data showing probable results of electric smelting of domestic manganese ores.*

[Based on results of experiments and on data available from present practice.]

Ore No.	Kind of ore.	Results of electric smelting of domestic manganese ore.								Analysis of alloy made.					Analysis of slag made.		
		Analysis of ore.						Loss on igni-tion.	Recovery of Mn in ore as alloy.								
		Mn.	Fe.	SiO_2.	CaO+MgO.	Al_2O_3.	P.			Mn.	Fe.	Si.	C.	P.	Mn.	CaO.	SiO_2.
		P. ct.	P. ct.	P. ct.	P. ct.	P. ct.	P. ct.	P. ct.	P. ct.	P. ct.	P. ct.	P. ct.	P. ct.	P. ct.	P. ct.	P. ct.	P. ct.
1	Mixed Colorado and Utah	40.0	1.0	13.0	5.0	5.0	0.15	13.0	75	80.0	12.8	1.0	6.0	0.20	12	35	30
2	Washed Phillipsburg	42.0	1.3	20.0	6.5	4.0	.10	13.0	72	80.0	11.8	2.0	6.0	.18			30
3	Mixed California ore plus quartz	30.0	1.0	35.0	.5	.5	.11	12.0	96	53.0	23.8	23.5	.6	.18			
4	Same as No. 3 without steel scrap	30.0	1.0	35.0	.5	.5	.11	12.0	96	68.0	2.0	29.5	.3	.23			
5	Phillipsburg screenings	28.4	2.5	38.4	6.5	2.0	.21	14.5	70	71.0	11.4	15.5	1.5	.60	11	25	55
6	Phillipsburg lump—one operation	24.0	1.6	38.0	7.5	1.6	.37	12.8	74	73.0	1.7	17.0	1.5	.75	12	26	55
7	Phillipsburg lump—two operations	24.0	1.6	38.0	7.5	1.6	.37	12.8	44 / 60	50.0 / 76.0	41.0 / 4.5	.5 / 18.8	6.0 / 1.0	2.50 / .20	23	25	55
8	Sultana lump—two operations	29.0	24.2	4.5	1.5	.3	.15	12.3	10 / 64	13.0 / 78.5	84.5 / 14.4	.3 / 1.0	2.0 / 6.0	.20 / .15	42 / 10	40	30

Ore No.	Kind of ore.	Charges required to produce 1 long ton of 80 per cent ferro or its equivalent in Mn (1,792 pounds).						Slag made per 1,792 pounds of Mn.	K W H used per 1,792 pounds of Mn	K.W.H. per long ton of the alloy made.	Carbon electrodes used.		Output per day per 3,000 k. v. a. connected load; 90 per cent power factor, 90 per cent load factor assumed.		Silicon output per day, calculated to equivalent 50 per cent 81 ferro-silicon.	Long tons of high-P Mn pig iron.
		Ore or mixture.	CaO_1.	CaF_2.	Coke.	Charcoal.	Steel scrap.				Per 1,792 pounds of Mn made.	Per long ton of alloy made.	Alloy.	Equivalent 80 per cent ferro.		
		Lbs.	Lbs.	Lbs.	Lbs.	Lbs.	Lbs.	Lbs.			Lbs.	Lbs.	Long tons.	Long tons.	Long tons.	
1	Mixed Colorado and Utah	6,000	1,100		1,800		220	2,600	5,000	5,000	155	155	11.7	11.7		
2	Washed Phillipsburg	6,000	2,100		1,900		190	4,000	5,500	5,500	175	175	10.6	10.6	6.8	
3	Mixed California ore plus quartz	6,300			2,400		700		8,750	5,800	270	180	10.0	6.8	6.8	
4	Same as No. 3 without steel scrap	6,300		1,450	1,200	1,200		6,000	8,700	7,400	215	185	7.9	6.7	6.7	
5	Phillipsburg screenings	10,250		1,500	1,200	1,200		6,200	10,000	9,000	310	280	6.5	5.8	2.0	
6	Phillipsburg lump—one operation	10,700		940	260	250		a11,000	11,500	10,500	355	325	5.5	5.0	1.9	
7	Phillipsburg lump—two operations	{12,500 / a11,000		880	960	960		6,000	{45,500 / 47,000		390	365	4.5	4.2	1.65	7.1
8	Sultana lump—two operations	{10,000 / b5,500	1,000 / 250	1,900	1,200 / 500 / 825	1,200 / 500 / 825		a5,600 / 2,750	12,500 / 45,000 / b3,500	11,800	265	260	6.9	6.8		
			1,250		1,325	1,325			8,500	8,350						

c250 pounds of waste metal.   b Slag A.   a In 1,850-k. w. furnace.   d In 1,650-k. w. furnace.   e In 1,750-k. w. furnace.   f In 1,250-k. w. furnace.

TABLE 31.—Data showing calculated costs of electric smelting of domestic manganese ores.

Ore No. (See Table 31)	1	2	3	4	5	6	7	8
Kind of ore	Colorado and Utah	Washed Phillipsburg	California ore and quartz	Same as 3 without steel scrap	Phillipsburg screenings	Phillipsburg lump—one operation	Phillipsburg lump—two operations	Sultana lump—two operations
Cost of ore per ton	$38.30	$36.25	$15.00		$7.50	$7.00	$7.00	$8.00
Cost per long ton of 80 per cent ferro or equivalent Mn (1,792 pounds):								
Ore	102.50	97.50	42.00	42.00	34.30	34.40	39.00	35.80
CaCO₃	2.20	4.20						
CaF₂					43.50	45.00	54.00	
Coke	9.00	9.50	12.00	12.00	6.00	6.00	6.00	6.65
Charcoal					18.00	18.00	18.00	19.90
Steel scrap	4.60	3.80	14.00					42.50
Power	25.00	27.50	43.75	43.50	50.00	57.50	62.50	40.00
Electrodes	23.15	26.20	40.50	40.50	46.50	54.50	58.50	
Total	166.45	168.70	151.25	138.00	198.30	215.30	238.00	147.35
Value of product:								
On basis of long tons of 80 per cent ferromanganese at $25	(1) 285.00	(1) 285.00	(1) 285.00	(1) 285.00	(1) $260.00	(1) c235.00	(1) 285.00	(1) 285.00
On basis of long tons of 50 per cent ferrosilicon at $130					(0.35) 52.50	(0.38) 57.00	(0.38) 52.00	
On basis of long tons of pig iron with 10 per cent Mn at $30			(0.88) 132.00	(0.88) 132.00			1.05	31.50
Calculated value of alloy per 1,792 pounds of Mn contained	285.00	285.00	417.00	417.00	312.50	292.00	337.00	316.50
Difference—value of product less cost of raw material	119.55	116.30	265.75	279.00	114.20	76.70	99.00	169.15
Difference times equivalent tons of ferromanganese made per day	(11.7) 1,400.00	(10.6) 1,235.00	(6.6) 1,750.00	(6.7) 1,870.00	(5.8) 605.00	(5.0) 383.00	(4.3) 425.00	(6.5) 1,150.00
Assumed cost of superintendence, labor, overhead, interest, depreciation, etc., per day, per 3,000 k. v. a. of plant capacity	400.00	400.00	400.00	400.00	400.00	400.00	400.00	400.00
Calculated profit per day per 3,000 k. v. a.	1,000.00	835.00	1,350.00	1,470.00	205.00	(loss) 17.00	25.00	750.00

a Figures in parentheses represent tons.
b Assumed penalty for 0.00 per cent P, $25 per ton of product equivalent to 80 per cent ferro manganese.
c Assumed penalty for 0.75 per cent P, $50 per ton of product equivalent to 80 per cent ferro manganese.

CHAPTER 11.—USE OF MANGANESE ALLOYS IN OPEN-HEARTH STEEL PRACTICE.

By Samuel L. Hoyt.

INTRODUCTORY STATEMENT.

This report presents the results of an extensive study of the use of manganese alloys in open-hearth steel practice in the United States. The magnitude of the work and the number of men who aided in one way or another make difficult acknowledgment to all who have contributed to its progress. However, some expression should be made of the hearty cooperation accorded to the members of the bureau by the various manufacturing interests. With such a sympathetic attitude, the investigation, depending so largely, as it did, upon cooperation between the steel plants and the investigators, was early assured of every possibility of ultimate success. The war conditions that rendered this investigation necessary have passed, but it is hoped that the results may prove of permanent value to the open-hearth steel industry.

PURPOSE AND SCOPE OF INVESTIGATION.

The purpose of making this investigation was to determine the extent to which domestic or low-grade manganese alloys could properly be substituted in open-hearth steel practice for high-grade alloys without materially impairing the steel production either as to quality or quantity. Moreover, it was held that such an extensive investigation of this important step in the manufacture of steel would undoubtedly yield valuable results to the steel industry as well as contribute, in no small way, toward directing future investigations in the same field.

It was recognized at the start that data bearing on the projected study should be available at individual plants, and that a compilation and digest of such results would be the logical method of approaching the proposed investigation. This consideration somewhat controlled the selection of the steel plants at which to conduct the detailed investigations.

The evolving of a definite experimental program from the statement of the general problem, considering the time element and the many and varied factors involved, was not reached without due consideration of competent metallurgical advice. After making

187

a preliminary survey, it seemed important to determine (a) the
conditions in open-hearth practice that would lead to a conservation
of manganese, both during the working of the heat and in making
the final additions; (b) the most satisfactory metallurgical conditions
for the use of manganese in the form of low-grade or special alloys;
and (c) the effect on the finished steel, both as to quality and "condition," of the various methods and processes studied. With these
points in mind, the selection of the steel plants was made so that
research work bearing on one or more of these points could be
conducted.

It was decided to determine slag and metal compositions during
the refining of the heat; also, the temperature was to be noted each
time a sample was taken, in order to determine, if possible, the
temperature effect. The "recovery" of manganese was to be
determined from the residual and final manganese contents and the
weight of the metal. To this end a sample of the finished steel was
taken during teeming. By taking three such samples, one at the
beginning, one toward the middle, and one at the end of teeming,
tests for uniformity were possible. This practice was generally
observed throughout the investigation. The data obtained were
also supplemented by the plant records covering given heats as
well as by personal observation during refining, pouring, and teeming.

When planning steps that should be taken to determine the quality
and "condition" of the steel, it was found that no definite and well-
proved method was available. True, the open-hearth melter
knows whether his heat is in proper condition, but what was needed
was a quantitative estimate of "condition." Without attempting
to discuss the physical chemistry of a heat of molten steel, it may
be said that the condition of the heat must depend, aside from the
temperature, upon the presence in the steel in those substances that
affect the "condition." Of these there are two kinds: (1) Substances that promote "openness," or the gases, which again may be
classified as (a) gases that are products of chemical reactions, being,
in so far as we know, CO and possibly CO_2, and (b) gases that are
absorbed from the furnace gases, such as H, N, CO, and CO_2; and
(2) substances that promote "soundness," such as the reducing and
solidifying agents, C, Mn, Si, and Al.

In general it is held that Mn, Si, and Al inhibit the chemical
reactions producing CO by reducing (or partly reducing) FeO, the
principal constituent that produces the reactions. In this statement only the metal bath is considered and the FeO, and not
Fe_2O_3, is assumed to be in solution in the steel. According to this
idea, reducing action on a slag containing Fe_2O_3 would produce
FeO, part of which would enter the steel to react later with C, Mn,
and other reducing agents present. The reduction of FeO, then,

is the principal means of "settling" the liquid steel, and it is for this reason that Mn is added in the final steps. It is also held that Si and Al produce solidity in the finished steel, aside from reducing FeO and CO, either by keeping the gases H, N, etc., in solid solution, or by preventing the dissociation of the compounds of those elements and iron.

The obvious procedure to get a quantitative estimate of the "condition" of the steel, considering both the behavior of the molten metal and the character of the ingot, would be to determine the amounts of the constituents in each of these two groups and to weigh one set against the other. Even this procedure would not, at present, lead to results that could be interpreted with entire confidence, even though there were no uncertainties in the analytical methods, because we do not know the quantitative effect of each constituent, either by itself or when associated with other constituents in varying amounts. In view of this lack of fundamental data, it was decided to make the analyses and use the results in a qualitative way, at least, to compare the different practices investigated.

THE FUNCTIONS OF MANGANESE.

During such a critical period as that now passed, the question might be raised as to the possibility of eliminating manganese from steel making. This point was duly considered but it was at once held that the use of manganese is not merely an expedient, for which some substitute might readily be had, but is rather one of the basic requirements of successful practice in working steel. It is quite true that in many instances the actual amount of manganese used in a heat of steel is greater than purely metallurgical considerations demand, and any excess could well be considered as so much wasted.

It may be well to review briefly the important functions of manganese as they bear directly on both of the points mentioned above. The first function of manganese, broadly considered, is to refine and "settle" the molten bath of steel. The aim here is to put the metal in a proper condition for pouring, and to produce ingots (or castings) of the desired quality and texture. Manganese is not the most efficient element that can be used for this purpose, calculated from the heat of combustion of the element to its oxide, but is without doubt the most satisfactory because of the excellent condition (freedom from objectionable foreign inclusions) in which it leaves the bath.

The proportion of manganese theoretically required for this operation might possibly be calculated from the amount of oxygen converted from the active form, FeO, to the inactive form, MnO. Assuming an oxygen content of 0.075 per cent in the unsettled steel and of 0.015 per cent in the finished steel (oxygen by the Ledebur method), the amount of manganese used in this way would be 0.2

per cent. The writer is informed by the Bureau of Standards that such a calculation is premature, owing to lack of knowledge on the subject of "deoxidation" and the faultiness of the Ledebur determinations. However it would seem to the writer, from the work done at the Bureau of Standards,[a] that the amount of oxygen determined is the amount present as FeO (active form), subject possibly to an error due to partial reduction of CO during the determination. At any rate, the above is advanced as at least the first approximation of the amount of manganese required simply for destroying the ferrous oxide present in the bath. The amount of manganese required naturally would vary with the condition of the bath and, in order to insure efficient "deoxidation," would be somewhat in excess of the calculated amount. A well-made heat of steel would probably not require more than 0.35 per cent Mn.

Manganese is also desirable in steel to improve the rolling properties, in which capacity it appears to serve a dual purpose. First of all, manganese deoxidizes and refines the molten steel in such a way as to give ingots of the desired texture without robbing the steel of its hot-working properties. Thus, ingots may be rolled into finished shape, without the formation of excessive fissuring or surface defects. Other reducing agents, such as aluminum and silicon, are prone to leave the metal in poor condition for rolling and forging. They eliminate one cause of hot shortness—iron oxide—but fail to convert the sulphur into a harmless form, as manganese does, and leave behind their highly refractory oxides, both of which tend to produce poor rolling qualities. Secondly, manganese, by retarding the rate of coalescence or grain growth, renders steel less sensitive to the effects of the high temperatures used in rolling and is supposed to promote plasticity, at least in ordinary steels, at rolling temperatures. Silicon and aluminum, on the other hand, increase, rather than decrease, the grain size of steel. The proportion of manganese required in this capacity probably does not exceed 0.35 per cent in well-made steel.

Finally, manganese is desired in the finished steel to produce certain physical or mechanical properties or to make the steel more amenable to subsequent heat treatment.

The foregoing discussion indicates that manganese is an important factor in the steel industry. Of course, material such as "American ingot iron" can be successfully rolled, even though no manganese be added, but requires greater time and care.

It is of interest to note that manganese, coming in the periodic system between iron and the strengthening elements on one side and the hardening elements on the other, has the dual function of strength-

---

a Cain, J. R., and Pettijohn, Earl. A critical study of the Ledebur method for determining oxygen in iron and steel: Bureau of Standards Technologic Paper 118, 1919, 33 pp.

ening and hardening steel, which is not possessed by any other element.

Manganese conservation would best be obtained by eliminating the manganese specification except where the amount of manganese present in the finished steel has some definite bearing on the properties of heat treatment of the steel. In other words, whenever only casting and rolling-mill practice (plant problems) are involved, the steel man should be allowed to exercise his own judgment as to the amount of manganese that should be used to give the most satisfactory and economical practice, and the finish and quality of the product should be controlled by adequate inspection. On the basis of Ellicott's figures,[a] by reducing the manganese requirements by 0.2 per cent in making plates and shapes and other low-carbon steel (estimated production, 21,350,000 tons), 54,000 tons of 80 per cent ferromanganese could be saved.

RECOMMENDATIONS FOR THE UTILIZATION OF DOMESTIC ALLOYS.

The investigation here reported indicated that three practices for utilizing domestic alloys in open-hearth steel practice seem to commend themselves above the others. These are as follows (but not in the order of their importance): (1) The use of a "molten spiegel mixture" for deoxidation and recarburization; (2) the practice of melting and refining the steel bath so as to insure a comparatively high residual manganese content, say, 0.3 per cent Mn; (3) the use of manganese alloys containing silicon. In selecting plants for investigating these practices two points were kept in mind. The plant should have either "ordinary" practice, for the sake of comparison, or else one of the three just mentioned, and the product or kind of steel made should be representative of the larger tonnages, such as shell steel, plates, or sections.

"MOLTEN SPIEGEL MIXTURE" PRACTICE.

The practice has been adopted at a few plants of combining in one operation both recarburization and deoxidation by using a mixture of pig iron and spiegel that has been premelted in a cupola. This "molten spiegel mixture" contains 5 to 11 per cent metallic manganese, 4 per cent carbon, and the desired amount of silicon, and is added to the ladle during the tapping in such a way as to cause a thorough and uniform mixture of the two streams.

The principal advantages of interest here, not considering questions of plant and operating economy, are as follows: 1. A low-grade or domestic alloy can be used in the preparation of the "mixture."

---

[a] Ellicott, C. R. Manganese conservation in steel making: Iron Age, vol. 101, June 6, 1918, pp. 1484-1485.

2. The deoxidation is accomplished by means of a dilute solution, with a consequent increase (on theoretical grounds) of the efficiency of the deoxidizer. This point will receive further consideration. 3. The deoxidizer is added in the molten state, insuring certain attendant advantages, which will also be considered at greater length. 4. A special advantage, if a large steel output is desired, is that the amount of the recarburizer is comparatively large and the capacity of the plant is materially (and economically) increased thereby. There is some question as to the propriety of including this advantage as peculiar to this particular practice. The use of pig iron as a recarburizer may be accomplished in other ways with the same economy and increase in plant capacity. 5. Another advantage would seem to the writer to be as follows: As compared with the results in the usual practice of adding carbon and manganese, there should be less likelihood of missing a heat.

This practice, at least at the plant visited, and it is understood to be the same elsewhere, is limited to the manufacture of the high-carbon steels or those running 0.30 per cent or more of carbon. To make steels with 0.20 per cent carbon would require working the bath until the carbon content was about 0.10 per cent, and the molten mixture added would have to contain about 20 per cent manganese (spiegel). The amount of the addition would be reduced from 13,000 pounds to about 4,000 pounds, which would mean that some of the advantages just enumerated would be lessened, and, with the increased loss of manganese the practice would probably not be commercially feasible. However, when the other alternative—the use of ferromanganese, either solid or liquid—is considered, the practice of premelting spiegel in the cupola seems commendable, on grounds to be considered later. In the event of undue shortage of high-grade ferromanganese the practice would doubtless offer a ready solution of the problem of using domestic alloys in making steel for shapes, plates, etc. Against the increased cost of production, as compared with the cost of cold ferromanganese practice, there would be the greater uniformity of product and more uniform practice as an offset.

HIGH RESIDUAL MANGANESE PRACTICE.

At certain plants the practice of preferential oxidation and elimination of carbon and phosphorus has been developed, the residual manganese being kept at a comparatively high value, say, 0.25 to 0.30 per cent, as compared with 0.10 per cent manganese for a final carbon content of 0.10 per cent. This is accomplished, broadly speaking, (a) by rapidly removing the phosphorus and retaining it as stable calcium phosphate during the earlier and colder period of melting; (b) by maintaining a high finishing temperature and working the charge with a high manganese content so that the slag contains about 8 per cent manganese; and (c) by increasing the lime content of the slag to about 47 per cent as a minimum.

This process possesses undoubted advantages, but they are such that they are probably best appreciated by plants in which the process has been developed and where it is now in operation on a sound commercial basis. First of all it may be stated that the practice, correctly applied, leads to the production of high-grade and uniform steel, which in itself means increased rolling-mill output, fewer rejections, and a more ready market. This is largely due to the fact that the steel is made—where it should be made—in the furnace.

A second advantage derived from the high MnO and CaO contents of the slag, is that the final additions of manganese can be added in the furnace, with a recovery comparing favorably with that of ladle additions. A third advantage is that the same pig iron used for the charge, and containing appreciably more manganese than ordinary basic iron does, can be used to recarburize and partly deoxidize the bath. The rest of the manganese is added as ferromanganese. At a steel plant which operates in conjunction with a blast-furnace plant a harmonious and economical cycle of plant operations is made possible. At the same time the open-hearth slag can be resmelted in the blast furnace for the recovery of the iron and manganese and the utilization of the lime.

This practice is largely depéndent upon the amount of phosphorus in the slag, for obviously it would not be worth while to recover the manganese at the expense of unduly increasing the phosphorus content of the pig iron. In this country we are fortunately situated in this respect, as there is still a large amount of ore rather low in phosphorus available. No definite figure can be given at this time as to the maximum allowable phosphorus content of the pig iron, but it is the opinion of at least one steel man who uses this process that a content of 0.6 per cent would not be excessive. Under the conditions prevailing in 1918, this practice had the additional advantages that the high-manganese pig iron could be procured by smelting domestic manganiferous iron ore and that the manganese alloy added to the furnace at the end of the heat could as well be spiegel as ferromanganese, assuming that the finished steel contains more than about 0.10 per cent carbon. There would also be certain disadvantages, particularly that the carbon content of the bath would have to be worked to a lower figure than in present practice. On account of the high cost of spiegel and the greater time required, it is doubtful whether the steel plants would substitute spiegel for ferromanganese. Another interesting point, as regards the utilizing of domestic manganiferous iron ore, is that low-silica ore could be added to the slag as a source of manganese oxide.

The high manganese content of the charge is generally obtained by using a "high-manganese" pig iron (2 to 3 per cent manganese), but may also be obtained by adding manganese ore to the slag or manganese alloys to the bath or by a combination of these methods. This point would be determined by plant economy, but it seems doubtful whether the practice would be worth while unless a high-manganese pig iron were available. The writer is informed by one blast-furnace superintendent that running the manganese up to 2 per cent does not materially affect the production, so that lower pig-iron production would not be a disadvantage in this practice. The loss of manganese by oxidation and transferrence to the slag is considerable. This loss may be kept at a minimum by increasing the basicity of the slag in CaO and FeO, which, combined with the MnO, which also acts as a base, exert the desired effect upon the manganese of the bath.

As the working of the charge progresses its temperature rises until finally with the high CaO, and particularly the high MnO content of the slag, the carbon is eliminated more rapidly than the manganese, with the result already stated, namely, the manganese can be held to about 0.3 per cent at the end of the heat. Present data indicate, unfortunately, that no material decrease in the amount of manganese required and no material increase in the recovery of manganese in the additions may be expected, so that the advantages are derived not from a decreased consumption, but from the form in which it can be added.

Data for one such heat showed that a total of 3,728 pounds of manganese was used in one form or another to produce 1,272 pounds of manganese in the finished steel—that is, 3.54 pounds was used to produce 1 pound in the steel. The manganese added in the recarburizer and as ferromanganese amounted to 1,068 pounds, of which, assuming the manganese loss to come from these two sources, 838 pounds was recovered in the finished steel, a recovery of 78.4 per cent. In this heat the ferromanganese was added to the furnace. Another more or less comparable heat selected at random, but more representative of "standard" practice, used 2,190 pounds of manganese to produce 1,200 pounds, or 1.82 pounds (as compared to 3.54 pounds) to produce 1 pound of manganese in the finished steel.

STANDARD OPEN-HEARTH PRACTICE COMPARED WITH CERTAIN OTHER PRACTICES.

Data regarding the results obtained in standard open-hearth steel practice as compared with the results when molten spiegel is used or when there is a large percentage of residual manganese will be presented in a later report after more complete analytical results have been received.

USE OF MANGANESE-SILICON ALLOYS.

The high silica content of most of our domestic manganese and manganiferous iron ores made it advisable to consider the possible use of manganese-silicon alloys in steel making in both acid and basic practice. For the purposes of the present discussion these alloys will be divided roughly into two classes—high-grade silicomanganese containing about 50 per cent manganese and 25 per cent silicon and low-grade silicospiegel with about 15 to 20 per cent silicon and 30 to 35 per cent manganese with 50 per cent iron. The manganese-silicon ratio of the first alloy is about 2 and of the second alloy $2\frac{1}{4}$ to $1\frac{1}{2}$. Each of these alloys would be made from the silicious manganese ores of California and Montana, and the low-grade alloys from the silicious manganiferous iron ores of Minnesota.

While there is nothing new about the practice of using manganese-silicon alloys in steel making, it may be well to review some of the points connected therewith.

It is understood that silicomanganese has been used fairly extensively in Europe, and in this country it was used at certain plants as standard practice until the supply was cut off by the war. Silicon is always an efficient reducing or settling agent when used in the customary small amounts, but it may or may not be desirable in the finished steel. On this account the possibility of using manganese-silicon alloys depends upon the amount of silicon that can be tolerated in the finished steel in the ingot form. In certain grades of steel, particularly in steel that must be welded, silicon should be low or practically absent. In steel for sheets and plates, which must give a good finished surface, the most efficient rolling-mill practice requires that the silicon be kept tolerably low; but it is believed that 0.10 to 0.15 per cent could be used, provided the manganese content were not too high. In forging steel, high-carbon steels, and castings, where the aim is to produce sound steel, more silicon can be used, or between 0.20 and 0.35 per cent. Of these three fields the latter is the one in which manganese-silicon alloys will find their first application. In the second field it seems quite probable that conditions (to be discussed later) will many times permit their use; but from the nature of things manganese-silicon alloys can not be used to make steels of the first group—those that must be welded.

ACID PRACTICE.

It is with considerable hesitation that the discussion of manganese-silicon alloys in open-hearth practice is approached, particularly as the controversial character of many of the points is so clearly recognized. Consequently, it may be well, at the outset, to state briefly the manner in which the writer first became interested in the possibilities of their use. A number of years ago the writer was conducting a series of experiments on the occurrence and identification of foreign inclusions in acid open-hearth steel, principally ordnance steel.   In this work ferromanganese, ferrosilicon, and a mixture of ferromanganese and ferrosilicon were added to a steel sample taken shortly after "oreing"—that is, to "wild" steel—in an attempt to produce an excess of the constituent, or constituents, supposed to form as a result of the addition.

It seemed fairly clear as a result of this work that the use of silicon was apt to be dangerous, not on account of any harmful effect of the residual metallic silicon but because it produced a constituent (assumed to be SiO_2 or at least a highly refractory silicate) that was likely to remain in the ingot and produce hot shortness. Hence the idea was suggested that a manganese-silicon alloy might, and probably would, form a manganese silicate containing some ferrous oxide (a true slag) which would be fluid and would more readily coalesce into larger particles than SiO_2 would, and therefore free itself more readily from the steel. By using such an alloy it would then be possible to take full advantage of the use of silicon as a deoxidizer without suffering the usual attendant disadvantages of its use. None of the manganese-silicon alloy was available at the time so a parallel experiment could not be conducted.

As binary alloys are known to be generally more active, or powerful, than the weighted sum of the two constituents would indicate, it was also assumed that, aside from the possibility of obtaining a better separation of the insoluble products of the deoxidation process, the alloy of manganese and silicon would prove to be a more powerful reducing agent than ferromanganese and ferrosilicon used separately. On reflection, the thought occurs that manganese and silicon, reacting separately with FeO, would produce the oxides MnO and FeO or a silicate of iron. Manganese and silicon reacting as an alloy with FeO would produce a silicate of manganese, which may or may not form a double silicate with FeO. In either case we would expect to find the advantage in favor of the manganese-silicon alloy.

The relative weights of the silicomanganese and of the mixture of the ferromanganese and ferrosilicon will be considered at another place.

Another point of great technical importance is the percentage recovery of manganese when added as silicomanganese and as ferromanganese along with ferrosilicon. It should be stated that a 100 per cent recovery, based on the present theory of "deoxidation," is hardly possible, nor is it desirable. Such recovery would mean retention of the deoxidation products, to be determined later as metallic manganese and silicon. A method of addition that would lead to satisfactory deoxidation and yet would eliminate the loss due to admixture with the slag, volatilization, etc., and could be accomplished with the minimum amount of manganese, would be very desirable because it would lead to both conservation of manganese and uniformity of composition of the steel. Conservation of manganese would be given by the actual percentage recovery, and uniformity of composition would be assumed by the constancy of the percentage recovery.

Fortunately the writer was able to examine records of heats made with silicomanganese covering a period of several years, from which some fairly satisfactory conclusions may be drawn bearing on these points. During this time when there were periods when the silicomanganese alloy was not available and a mixture of ferromanganese and ferrosilicon had to be substituted. Thus, direct comparison of these two methods of deoxidation was afforded. Certain results taken from the heat records, and believed to be typical, are given in Table 33 following. Obviously, figures showing the variation in heat composition and the average manganese recovery of several years' practice can not be given in this table. The records themselves clearly show greater uniformity for the silicomanganese heats.

TABLE 33.—*Comparative results obtained with silicomanganese and with a mixture of ferromanganese and ferrosilicon.*[a]

Heat.	FeSi.[1]	FeMn.	SiMn.	Total charge.	C.	Mn.	Si.	Mn. added.	Mn. recovered.	
	Lb.	Lb.	Lb.	Lb.	Per ct.	Per ct.	Per ct.	Lb.	Lb.	Per ct.
A	160	300	....	31,160	0.26	0.56	0.294	240	174	72.5
B	215	400	....	40,765	.21	.57	.306	320	232	72.5
C	160	310	....	30,620	.32	.63	.312	248	193	77.8
D	160	300	....	30,910	.27	.70	.318	240	216	90.0
E	215	400	....	40,865	.24	.72	.312	320	294	91.8
F	....	40	470	41,010	.22	.58	.308	281	238	84.7
G	....	35	350	30,685	.26	.60	.302	214	184	86.0
H	....	40	470	40,960	.21	.64	.310	281	261	93.0
I	....	40	470	30,510	.21	.66	.303	281	267	95.0
J	....	40	420	36,760	.24	.68	.310	255	250	98.0

[a]Approximate compositions: Silicomanganese, Mn 53 per cent, Si 20 per cent; ferromanganese, Mn 80 per cent; ferrosilicon, Si 50 per cent. Residual manganese was neglected in calculating recoveries.

It can hardly be claimed that these figures, or the three years' records which they represent with reasonable accuracy, furnish a truly scientific basis for comparison of the two alternate practices, but they do show rather convincingly that the same results (Mn and Si contents of the finished steel), by using silicomanganese can be obtained with consistently smaller amounts of both manganese and silicon, as compared with the combination of ferromanganese and ferrosilicon. In addition there is the advantage of more uniform practice, which in itself would warrant smaller additions. The weights of the additions favor the silicomanganese; thus in heats A, C, and D, 460 pounds was added, as compared with 385 pounds for heat G, and in heats B and E, 615 pounds was added, as compared with 510 pounds in heats F, H, and I. The low carbon content of the silicomanganese may or may not be a material advantage, but is in favor of the single-alloy addition because the carbon need not be worked as low and there seems to be less danger of missing the desired carbon content.

ELECTRIC-FURNACE PRACTICE.

No information is available to the writer bearing on the use of manganese-silicon alloys in electric-furnace practice, but we may at least consider such practice on the basis of the known behavior of such alloys. Considering acid casting practice first, there seems to be no reasonable doubt that either silicomanganese or silicospiegel could be at once substituted for ferromanganese and ferrosilicon. Inasmuch as the usual aim is to make high-grade castings, the manganese-silicon alloys would appear to have the distinct advantage of making sounder and cleaner steel. Silicospiegel, aside from possessing the theoretical advantage of being diluted with iron, could be more readily prepared with the correct manganese-silicon ratio so as to eliminate the use of an additional alloy. In this practice the advantage of greater dilution need not carry with it the disadvantage of increased weight on account of the higher temperature of the electric furnace. The uncertainty as to the relative behavior of the manganese-silicon alloys as compared with that of the ferro-alloys, and the relative efficiencies of low-grade and of high-grade alloys, as well as the importance of this step in the manufacture of steel, suggest the advisability of conducting a definite research to settle such points. It would seem that there is no better place for such a research than in this particular industry.

In basic electric-furnace practice the manganese-silicon alloys, on the same grounds, could likewise be utilized, particularly as the attempt is always to produce sound and clean ingots. However, in this practice, ferrosilicon is used as a reducing agent along with coke, and hence the operator would probably not see any advantage in changing his practice in favor of the manganese-silicon alloys.

BASIC OPEN-HEARTH PRACTICE.

The amount of information available on the use of silicomanganese in basic open-hearth practice is meager, but it can be said that silicomanganese can be used, probably with as satisfactory results as with ferromanganese and ferrosilicon. Through the cooperation of one steel plant the writer was able to follow two shell-steel heats made with silicomanganese which was added to the ladle. The results of the second of these heats are given here to show what was done. To 11,100 pounds of molten pig iron in the ladle was tapped 122,340 pounds (estimated) of steel analyzing 0.09 per cent C,

0.12 per cent P, 0.15 per cent Mn, 0.033 per cent S, and 0.02 per cent Si. At the same time 1,000 pounds of silicomanganese (50 per cent Mn), 300 pounds of 70 per cent ferromanganese, 12 pounds of aluminum, and 50 pounds of coal were added to the ladle. The heat was in excellent condition and the ingots had smooth even tops and displayed no superficial action or evolution of gas. The final analysis was 0.40 per cent C, 0.028 per cent P, 0.58 per cent Mn, 0.041 per cent S, and 0.21 per cent Si. The recovery of manganese, assuming the entire loss to come from the alloy added, was 77.5 per cent; the recovery of silicon was 65.1 per cent. Only 24 pounds of carbon, or 5 per cent of the total was lost. In the first heat, which was thought to be more highly oxidized, the recovery of silicon was only 58 per cent while the recovery of manganese was only slightly less. These results indicate that silicon protects manganese in "oxidized" heats.

137338°—20——14

SELECTED BIBLIOGRAPHY ON MANGANESE DEPOSITS.

GENERAL.

BEYSCHLAG, F., KRUSCH, P., and VOGT, J. H. L. Die Lagerstatten der nutzbaren Mineralien und Gesteine. 3 vols., 1910, pp. 850–869, 1099–1115.

DEMORET, L. Le principaux gisements des minerais de manganese du monde. Ann. des Mines de Belgique, t. 10, 1905, pp. 809–901.

FACH, E. Le mineral de manganese. Paris. 1914, 44 pp.

UNITED STATES.

HARDER, E. C. Manganese deposits of the United States. U. S. Geol. Survey Bull. 427, 1910, pp. 298.

PANAMA.

CHIBIAS, E. J. Manganese deposits of the Department of Panama, Republic of Colombia. Trans. Am. Inst. Min. Eng., vol. 27, 1897, p. 63.

WILLIAMS, E. G. The manganese industry of the Department of Panama, Republic of Colombia. Trans. Am. Inst. Min. Eng., vol. 33, 1902, pp. 197–234.

SOUTH AMERICA.

BRAZIL.

BRANNER, J. C. The manganese deposits of Bahia and Minas Geraes, Brazil. Trans. Am. Inst. Min. Eng., vol. 29, 1899, p. 756.

SCOTT, H. K. The manganese ores of Brazil. Jour. Iron and Steel Inst., vol. 1, 1900, p. 179.

DERBY, O. A. On the manganese deposits of the Queluz (Lafayette) district, Minas Geraes. Am. Jour. Sci., 4th ser., vol. 25, 1901, p. 18.

CHILE.

HARDER, E. C. Manganese ores of Russia, India, Brazil, and Chile. Trans. Am. Inst. Min. Eng., vol. 56, 1916, pp. 31–68.

EUROPE.

AUSTRIA.

NASKE, T. [Manganese ore in Austria]. Stahl und Eisen, Jahrg. 28, 1907, pp. 543–547.

SCOTT, H. K. Manganese ores of the Bukowina. Jour. Iron and Steel Inst., vol, 94, 1916, pp. 288–355.

GERMANY.

KERN, J. Zur Frage der Mangan-versorgung Deutschlands. Berg. Mitteilungen, Jahrg. 4, 1913, pp. 49–59.

SCHEFFER, ·— [The importance of manganese and manganiferous ores in German industry]. Gluckauf, Jahrg. 49, 1913, pp. 2056, 2111, 2151.

RUSSIA.

Drake, Frank. The manganese industry of the Caucasus. Trans. Am. Inst. Min. Eng., vol. 28, 1898, p. 191.

SPAIN.

Hoyer, M. Contributions to the knowledge of the manganese deposits of the Province of Huelva. Ztschr. prakt. Geol., vol. 19, 1911, pp. 407–438.

AFRICA.

WEST AFRICA.

Ford, S. H. Manganese in West Africa. Mining Mag., vol. 17, 1917, p. 271.

ASIA.

INDIA.

Fermor, L. L. The manganese deposits of India. India Geol. Survey Mem., vol. 37, pts. I, II, III, IV, 1909,

Carter, H. A. Manganese mining in British India. Min. and Sci. Press, vol. 103, 1911, pp. 834–835.

India Geological Survey Record, Quinquennial review of the mineral production of India. Vol. 46, 1915, pp. 135–139.

JAPAN.

Snodgrass, J. H. Manganese mining in Japan. Mining World, vol. 30, 1909, p. 790.

PUBLICATIONS ON METALLURGY.

A limited supply of the following publications of the Bureau of Mines has been printed and is available for free distribution until the edition is exhausted. Requests for all publications can not be granted, and to insure equitable distribution applicants are requested to limit their selection to publications that may be of especial interest to them. Requests for publications should be addressed to the Director, Bureau of Mines.

The Bureau of Mines issues a list showing all its publications available for free distribution as well as those obtainable only from the Superintendent of Documents, Government Printing Office, on payment of the price of printing. Interested persons should apply to the Director, Bureau of Mines, for a copy of the latest list.

PUBLICATIONS AVAILABLE FOR FREE DISTRIBUTION.

BULLETIN 67. Electric furnaces for making iron and steel, by D. A. Lyon and R. M. Keeney. 1914. 142 pp., 36 figs.

BULLETIN 73. Brass furnace practice in the United States, by H. W. Gillett. 1914. 298 pp., 2 pls., 23 figs.

BULLETIN 77. The electric furnace in metallurgical work, by D. A. Lyon, R. M. Keeney and J. R. Cullen. 1914. 216 pp., 56 figs.

BULLETIN 84. Metallurgical smoke, by C. H. Fulton. 1915. 94 pp., 6 pls., 15 figs.

BULLETIN 97. Sampling and analysis of flue gases, by Henry Kreisinger and F. K. Ovitz. 1915. 68 pp., 1 pl., 37 figs.

BULLETIN 100. Manufacture and uses of alloy steels, by H. D. Hibbard. 1915, 78 pp.

BULLETIN 110. Concentration experiments on the siliceous red hematites of the Birmingham district, by J. T. Singewald, jr. 1917. 91 pp., 1 pl., 47 figs.

BULLETIN 133. The wet thiogen process of recovering sulphur from sulphur dioxide in smelter gases, a critical study, by A. E. Wells. 1917. 66 pp., 2 pls., 3 figs.

BULLETIN 140. Occupational hazards and accident prevention at blast-furnace plants; based on records of accidents in blast furnaces in Pennsylvania in 1915, by F. H. Willcox. 1917. 155 pp., 16 pls.

BULLETIN 154. Lead and zinc mining and milling, by C. A. Wright. 1918. 134 pp., 17 pls., 13 figs.

BULLETIN 157. Innovations in the metallurgy of lead, by O. C. Ralston. 1918. 167 pp., 13 figs.

BULLETIN 168. Recovery of zinc from low-grade and complex ores, by D. A. Lyon and O. C. Ralston. 1919. 145 pp., 23 figs.

TECHNICAL PAPER 8. Methods of analyzing coal and coke, by F. M. Stanton and A. C. Fieldner. 1913. 42 pp., 12 figs.

TECHNICAL PAPER 50. Metallurgical coke, by A. W. Belden. 1913. 48 pp., 1 pl., 23 figs.

TECHNICAL PAPER 76. Notes on the sampling and analysis of coal, by A. C. Fieldner. 1914. 59 pp., 6 figs.

TECHNICAL PAPER 86. Ore-sampling conditions in the West, by T. R. Woodbridge. 1916. 96 pp., 5 pls., 17 figs.

TECHNICAL PAPER 96. Fume and other losses in condensing quicksilver from furnace gases, by L. H. Duschak and C. N. Schuette. 1918. 29 pp., 4 figs.

TECHNICAL PAPER 102. Health conservation at steel mills, by J. A. Watkins. 1916. 36 pp.

TECHNICAL PAPER 106. Asphyxiation from blast-furnace gas, by F. H. Willcox. 1916. 79 pp., 8 pls., 11 figs.

TECHNICAL PAPER 135. Bibliography of recent literature on flotation of ores, January to June, 1916, compiled by D. A. Lyon, O. C. Ralston, F. B. Laney, and R. S. Lewis. 1917. 20 pp.

TECHNICAL PAPER 136. Safe practice at blast furnaces; a manual for foremen and men, by F. H. Willcox. 1916. 73 pp., 1 pl., 43 figs.

TECHNICAL PAPER 143. The ores of copper, lead, gold, and silver, by C. H. Fulton. 1916. 41 pp.

TECHNICAL PAPER 149. Answers to questions on the flotation of ores, by O. C. Ralston. 1917. 30 pp.

TECHNICAL PAPER 152. The inflammability of aluminum dust, by Alan Leighton. 1918. 15 pp.

TECHNICAL PAPER 153. Occurrence and mitigation of injurious dusts in steel works, by J. A. Watkins. 1917. 20 pp., 4 pls.

TECHNICAL PAPER 156. Carbon monoxide poisoning in the steel industry, by J. A. Watkins. 1917. 19 pp., 1 fig.

TECHNICAL PAPER 157. A method for measuring the viscosity of blast-furnace slag at high temperatures, by A. L. Feild. 1916. 29 pp., 1 pl., 7 figs.

TECHNICAL PAPER 176. Bibliography of recent literature on flotation of ores, July 1 to December 31, 1916, by D. A. Lyon, O. C. Ralston, F. B. Laney, and R. S. Lewis. 1917. 27 pp.

TECHNICAL PAPER 177. Preparation of ferro-uranium, by H. W. Gillett and E. L. Mack. 1917. 46 pp., 2 figs.

TECHNICAL PAPER 182. Flotation of chalcopyrite in chalcopyrite-pyrrhotite ores of southern Oregon, by Will H. Coghill. 1918. 13 pp., 1 fig.

TECHNICAL PAPER 187. Slag viscosity tables in blast-furnace work, by A. L. Feild and P. H. Royster. 1918. 38 pp., 1 fig.

TECHNICAL PAPER 198. Sulphur dioxide method for determining copper minerals in partly oxidized ores, by C. E. van Barneveld and Edmund Leaver. 1918. 12 pp., 1 fig.

TECHNICAL PAPER 200. Colloids and flotation, by F. G. Moses. 1918. 24 pp.

TECHNICAL PAPER 201. Accidents at metallurgical works in the United States during the calendar year 1916, compiled by A. H. Fay. 1918. 18 pp.

TECHNICAL PAPER 211. Approximate quantitative microscopy of pulverized ores, including the use of the camera lucida, by W. H. Coghill and J. P. Bonardi. 1919. 17 pp. 3 pls.

TECHNICAL PAPER 225. The vapor pressure of lead chloride, by E. D. Eastman and L. H. Duschak. 1919. 16 pp., 2 pls., 2 figs.

TECHNICAL PAPER 227. The determination of mercury, by C. M. Bouton and L. H. Duschak. 1920. 39 pp., 2 pls., 1 figs.

TECHNICAL PAPER 230. The determination of molybdenum, by J. P. Bonardi and W. L. Barrett. 1920. — pp., — pls., — fig.

TECHNICAL PAPER 236. The abatement of corrosion in central heating systems, by F. N. Speller. 1919. 12 pp., 2 figs.

PUBLICATIONS THAT MAY BE OBTAINED ONLY THROUGH THE SUPERINTENDENT OF DOCUMENTS.

BULLETIN 3. The coke industry of the United States as related to the foundry, by Richard Moldenke. 1910. 32 pp. 5 cents.

BULLETIN 7. Essential factors in the formation of producer gas, by J. K. Clement, L. H. Adams, and O. N. Haskins. 1911. 58 pp., 1 pl., 16 figs. 10 cents.

BULLETIN 12. Apparatus and methods for the sampling and analysis of furnace gases, by J. O. W. Frazer and E. H. Hoffman. 1911. 22 pp., 6 figs. 5 cents.

BULLETIN 47. Notes on mineral wastes, by O. L. Parsons. 1912. 44 pp. 5 cents.

BULLETIN 54. Foundry-cupola gases and temperatures, by A. W. Belden. 1913. 29 pp., 3 pls., 16 figs. 10 cents.

BULLETIN 63. Sampling coal deliveries and types of Government specificaitons for the purchase of coal, by G. S. Pope. 1913. 68 pp., 4 pls., 3 figs. 10 cents.

BULLETIN 64. The titaniferous iron ores in the United States; their composition and economic value, by J. T. Singewald, jr. 1913. 145 pp., 16 pls., 3 figs. 25 cents.

BULLETIN 70. A preliminary report on uranium, radium, and vanadium, by R. B. Moore and K. L. Kithil. 1914. 114 pp., 4 pls., 2 figs. 15 cents.

BULLETIN 81. The smelting of copper ores in the electric furnace, by D. A. Lyon and R. M. Keeney. 1915. 80 pp., 6 figs. 10 cents.

BULLETIN 85. Analyses of mine and car samples of coal collected in the fiscal years 1911 to 1913, by A. C. Fieldner, H. I. Smith, A. H. Fay, and Samuel Sanford. 1914. 444 pp., 2 figs. 45 cents.

BULLETIN 98. Report of the Selby Smelter Commission, by J. A. Holmes, E. C. Franklin, and R. A. Gould, with reports by associates on the commissioners' staff. 1915. 525 pp., 41 pls., 14 figs. $1.25.

BULLETIN 108. Melting aluminum chips, by H. W. Gillett and G. M. James. 1916. 88 pp. 10 cents.

BULLETIN 122. The principles and practice of sampling metallic metallurgical materials, with special reference to the sampling of copper bullion, by Edward Keller. 1916. 102 pp., 13 pls., 31 figs. 20 cents.

BULLETIN 130. Blast-furnace breakouts, explosions, and slips, and methods of prevention, by F. H. Willcox. 1917. 280 pp., 2 pls., 37 figs. 30 cents.

BULLETIN 150. Electro-deposition of gold and silver from cyanide solutions, by S. B. Christy. 1919. 171 pp., $\frac{3}{8}$ pls., 41 figs. 25 cents.

BULLETIN 171. Tests of rocking electric brass furnace, by H. W. Gillett and A. E. Rhodes. 1918. 131 pp., 4 pls., 1 fig. 25 cents.

TECHNICAL PAPER 31. Apparatus for the exact analysis of flue gas, by G. A. Burrell and F. M. Seibert. 1913. 12 pp., 1 fig. 5 cents.

TECHNICAL PAPER 41. The mining and treatment of lead and zinc ores in the Joplin District, Mo., a preliminary report by O. A. Wright. 1913. 43 pp., 5 figs. 5 cents.

TECHNICAL PAPER 54. Errors in gas analysis due to the assumption that the molecular volumes of all gases are alike, by G. A. Burrell and F. M. Seibert. 1913. 16 pp., 1 fig. 5 cents.

TECHNICAL PAPER 60. The approximate melting points of some commercial copper alloys, by H. W. Gillett and A. B. Norton. 1913. 10 pp., 1 fig. 5 cents.

TECHNICAL PAPER 81. The vapor pressure of arsenic trioxide, by H. V. Welch and L. H. Duschak. 1915. 22 pp., 2 figs. 5 cents.

TECHNICAL PAPER 83. The buying and selling of ores and metallurgical products, by C. H. Fulton. 1915. 43 pp. 5 cents.

TECHNICAL PAPER 90. Metallurgical treatment of the low-grade and complex ores of Utah, a preliminary report, by D. A. Lyon, R. H. Bradford, S. S. Arentz, C. C. Ralston, and C. L. Larson. 1915. 40 pp. 5 cents.

TECHNICAL PAPER 95. Mining and milling of lead and zinc ores in the Wisconsin district, Wisconsin, by C. A. Wright. 1915. 39 pp., 2 pls., 5 figs. 5 cents.

TECHNICAL PAPER 189. Temperature-viscosity relations in the Ternary system $CaO-Al_2O_3-SiO_2$, by A. L. Feild and P. H. Royster. 1918. 36 pp., 1 pl., 16 figs. 5 cents.

INDEX.

Bulletin 174   Law Serial 17

DEPARTMENT OF THE INTERIOR
FRANKLIN K. LANE, Secretary
BUREAU OF MINES
VAN. H. MANNING, Director

ABSTRACTS OF CURRENT DECISIONS

ON

MINES AND MINING

REPORTED FROM MAY
TO   SEPTEMBER, 1918

BY

J. W. THOMPSON

WASHINGTON
GOVERNMENT PRINTING OFFICE
1919

GENERAL SUBJECTS TREATED.

CONTENTS.

CONTENTS.

TABLE OF CASES DIGESTED.

ABSTRACTS OF CURRENT DECISIONS ON MINES AND MINING.

MAY TO SEPTEMBER, 1918.

By J. W. THOMPSON.

MINERALS AND MINERAL LANDS.

MINERALS.

NATURE AS PERSONAL PROPERTY.

Ore after having been mined and removed from the earth is personal property.

Kelvin Lumber & Supply Co. v. Copper State Min. Co. (Texas Civil Appeals), 203 Southwestern 68, p. 69.

NATURE AS LAND—CONVEYANCE.

Minerals are land so long as they are undisturbed and must be conveyed with the same formalities as other lands are conveyed.

Kennedy v. Hicks (Kentucky), 203 Southwestern 318, p. 320.

CONVEYANCE OF MINERALS—SEVERANCE.

The owner of both the minerals and the land may convey the minerals, in which case the corpus passes and thereby a severance is effected, the vendor remaining the owner of that part of the land which does not consist of minerals and the vendee owning the land which consists of minerals.

Kennedy v. Hicks (Kentucky), 203 Southwestern 318, p. 320.

CONVEYANCE—EFFECT AS CONSUMATED SALE.

The owner of land may convey the minerals upon the condition that the vendee extract them by a specified time or in a stipulated mode or that title shall pass only when certain royalties be paid, and in such instance there is no present consummated sale.

Kennedy v. Hicks (Kentucky), 203 Southwestern 318, p. 320.

SALE OF ORE—SPECIFIC PERFORMANCE OF CONTRACT.

A mining company for a period of 12 years had been selling its ore to a certain smelting company for the purpose of obtaining a continuous and steady market for its ore and for the purpose on the part of the smelting company of obtaining a continuous and steady supply of ore for smelting; an agreement was entered into by which the mining company agreed to sell and the smelting company agreed to purcshase at ruling prices all the ore produced by the mining company of certain stated grades, and to this end the mining company agreed to actively operate its mines. It is no longer the rule that a suit will not lie for the specific performance of a contract relating to personal property; but the question for determination in such a case is whether the complainant has a plain, speedy and adequate and complete remedy at law, and if he has no such remedy, specific performance will be granted. In an action by a smelting company for the specific performance of a contract of sale it appeared that the ore produced by the mining company had a peculiar value for smelting purposes and was not readily obtainable elsewhere, and it would be practically impossible to measure the value for determining the loss the complainant would sustain if the smelting company were not able to obtain the ore under the contract. It is comparatively simple for a court of equity to enforce the agreement as it pertains to the normal product of the mine and there is no plausible reason why a court may not require a performance on the part of the mining company where the mining company obligated itself to operate all its mines. There is no delicate or complex condition involved in the production or delivery of the ore or in obtaining the stipulated prices agreed upon, and there is nothing apparent that will necessitate supervisory control in requiring performance on the part of the mining company. A decree for specific performance would require no supervision by the court either for the operation of the mines or for ascertaining and fixing the prices the mining company is to receive for its ore.

American Smelting & Refining Co. v. Bunker Hill & Sullivan Min. & Concentrating Co., 248 Federal 172, p. 182.

BREACH OF CONTRACT FOR SALE OF ORE—PRELIMINARY INJUNCTION.

A mining company agreed to sell all of its ore, and a smelting company agreed to purchase all of the ore of the mining company of a certain designated grade at ruling prices. The evident purpose of the agreement was to afford the mining company a regular and steady market for its ores and to supply the smelting company with a ready and steady product for smelting. In a suit by the smelting company to enforce specific performance of the contract and to enjoin the mining company from breaching the contract and ceasing to sell to the smelting company the product of its mines, a court in an application for a preliminary injunction will not pass upon the sufficiency of the bill except to inquire whether the questions presented are grave and weighty and difficult of solution and such as to call for a maintenance of the status quo during the pendency of the litigation.

American Smelting & Refining Co. v. Bunker Hill & Sullivan Min. & Concentrating Co., 248 Federal 172, p. 182.

AGREEMENT TO PURCHASE ORE—INJUNCTIVE RELIEF.

A mining company on March 20, 1905, entered into an agreement with a smelting company whereby it agreed to sell and the latter to purchase all the lead-silver ores, slimes, and concentrates mined from the property of the mining company at agreed prices, with certain stipulated deductions. The contract provided that the ore should be of a lead assay of between 30 per cent and 75 per cent, and that the average product should be of approximately the same yearly analysis and lead assay as the shipments made from the mine during any year of the 12 years immediately preceding the date of the agreement, and this standard should not be varied unless the smelting company gave its written consent, and in case there was a variation from the standard the smelting company was to be given an option to purchase at the best going market rates and terms. The purpose of the agreement on the part of the smelting company was to secure as nearly as practical a uniform product of an average of about 53 per cent lead assay, and it was the intention of the parties as shown by the agreement that the mining company should sell and the smelting company should purchase the entire product, and it was the purpose of the mining company to have a continuing and suitable market for its normal and usual output and that the smelting company should have a dependable source of supply of a reasonably uniform commodity for its smelting operations. Under such a contract and where it appears that the mining company has constructed a smelter of its own for the purpose and with the intention of smelting its own ores and repudiated the contract to sell its ore to the smelting company, the latter is entitled to a preliminary injunction to maintain the status quo when it is shown that there is a probable right and a probable danger that such right will be defeated without the immediate interposition of a court of equity. Such injunctive relief will be granted whenever the question of law or fact to be ultimately determined is grave and difficult and injury to the moving party will be immediate, certain, and great if denied, while loss and inconvenience to the opposing party will be comparatively small if granted.

American Smelting & Refining Co. v. Bunker Hill & Sullivan Min. & Concentrating Co., 248 Federal 172, pp. 175-182.

CONVERSION OF ORE.

The rule that it constitutes conversion to receive property from one wrongfully in possession of it and thereafter exercise dominion or control over it contrary to the wish of the person rightfully entitled to it applies to ores or minerals as well as to other property. But the rule does not apply to a common carrier, and the carrying of goods or ores by a carrier from terminus to terminus upon the requirement of a person unlawfully in possession of them is not conversion.

Dickson v. Southern Pacific Co. (Nevada), 172 Pacific 368, p. 369.

CONVERSION OF ORE BY CARRIER—NOTICE OF OWNERSHIP.

The owner of certain ores notified a railroad company that a particularly named person would offer ore for transportation and forbade the transportation of the ore if offered by the person named. The person named did offer the ore for transportation and the railroad company required of him an affidavit of ownership, and upon such affidavit being made the railroad company received and shipped the ore. The fact that the railroad company required the affidavit of ownership was sufficient to show that it had notice of the claim of the alleged owner of the ore and was sufficient to render the railroad company liable on a charge of converting the ore.

Dickson v. Southern Pacific Co. (Nevada), 172 Pacific 368, p. 369.

CONVERSION—RECOVERY BY OWNER.

Minerals after they are taken from the mine and from the real estate wherein they are found become personal property, and where there has been a wrongful conversion the owner may sue in any State where the minerals are found and recover either the minerals themselves or the value thereof.

Kelvin Lumber & Supply Co. v. Copper State Min. Co. (Texas Civil Appeals), 203 Southwestern 68, p. 69.

CONVERSION OF ORE—MEASURE OF DAMAGES.

In an action by the owner of ore against a railroad company for damages for an alleged conversion of ore, the measure of damages is the value of the ore at the time of the conversion with legal interest from the date of the conversion to the date of rendering judgment.

Dickson v. Southern Pacific Co. (Nevada), 172 Pacific 368, p. 369.

The measure of damages for a conversion of ore or minerals is the market value of the ore at the time and place of conversion.

Barlettsville Zinz Co. v. Compania Minera Ygnacio Rodriguez Ramos (Texas Civil Appeals), 202 Southwestern 1048.
See Kelvin Lumber & Supply Co. v. Copper State Min. Co. (Texas Civil Appeals), 208 Southwestern 68.

TRESPASS AND CONVERSION—JURISDICTION AND VENUE.

The owner of a valid mining location can not maintain an action in the court of a foreign State to recover the value of minerals from the purchaser of the same from a person who had in good faith made an attempted relocation of a part of the ground included in the original location, and who was in the adverse possession of that part of the mining claim under his relocation; and as such relocator mined and removed the minerals from his relocated claim, such an action necessarily involves the title to the realty and it is the policy of the law that such an issue should be tried in the local court where the property is situated.

Kelvin Lumber & Supply Co. v. Copper State Min. Co. (Texas Civil Appeals), 208 Southwestern 68, p. 70.

MILITARY SEIZURE.

A mining company sued for damages for the conversion of ore, alleging that the ore was taken from the complainant in the Republic of Mexico by unknown parties and subsequently unlawfully appropriated by the defendant company. The defendant pleaded that the ore was seized and confiscated by Francisco Villa as a military necessity, was sold to parties in Mexico and subsequently purchased by the defendant in good faith. An issue was formed on the defendant's plea and the jury found as a fact that a civil war existed in Mexico, but that the ore was not taken by any force or government in possession and control of the territory, or by any agent of such force or government acting by authority of the government, and that the proceeds derived from the ore were not intended for the benefit of the faction dominated by Villa. This was a question of fact entirely within the province of the jury and its finding is binding on the court.

Barlettsville Zinc Co. v. Compania Minera Ygnacio Rodriguez Ramos (Texas Civil Appeals), 202 Southwestern 1048.

EXTRACTION OF MINERAL FROM ORE—JUDICIAL NOTICE.

Court can not take judicial notice of what percentage of mineral can be extracted from a particular class of ore, but this is a matter of proof in each particular case.

Dickson v. Southern Pacific Co. (Nevada), 172 Pacific 368, p. 370.

PROCESS OF TREATING ORES—JUDICIAL NOTICE.

Courts will take judicial knowledge of the fact that processes of crushing, amalgamating, and cyaniding ores will not effect an extraction of 100 per cent of the metallic content. What will be a reasonable percentage of extraction will depend largely upon the process used and the character of the ore.

Dickson v. Southern Pacific Co. (Nevada), 172 Pacific 368, p. 370.

MINERAL LANDS.

MINERAL CHARACTER OF PUBLIC LAND—SELECTION BY RAILROAD COMPANY—KNOWLEDGE OF AGENT.

A railroad company selecting nonmineral lands under its grant was required to attach to its selection list an affidavit by its agent to the effect that the lands mentioned had been examined by the agents and employees of the company as to their mineral and agricultural character and that the lands selected were not mineral lands. In an action to set aside a patent to lands so selected on the ground that the land was in fact mineral in character, the agent in making the substantive representations as to the character of the land is chargeable with knowledge of all the conditions which a careful examination if made would have disclosed, including everything that was known to the company's geologist and other agents or employees.

Southern Pacific Co. v. United States, 249 Federal 785.

SALE AND CONVEYANCE.

CONVEYANCE OF MINERALS IN PLACE.

A written instrument called a lease provided that the first parties in consideration of the stipulations and covenants contained lease, demise, and let unto the party of the second part for the sole and only purpose of quarrying, drilling, and digging for minerals and oils, the exclusive right to a certain tract of land described for a term of 99 years. The consideration to be paid was $200 cash and $300 within six months after commencing quarrying, mining, or drilling for minerals or oil. The land described was in Hardin County, Ky. The grantee lived in Jefferson County and subsequently assigned or conveyed his interest in the premises to a stone company that began quarrying stone on the premises. On failure to make the deferred payment, the grantor or lessor brought suit in Hardin County to enforce a vendor's lien against the land making the original grantee and his assignee parties defendant. The original grantee, a resident of a different county, denied the jurisdiction of the court over his person on the ground that the action was transitory and he must be sued in the county of his residence. The correctness of this proposition depends on whether the instrument was merely a lease of the land with the privilege of removing the minerals during a certain period, or was in reality a sale of the land. When a contract employs the language and observes the formality required for conveyances of real estate and the consideration is paid or made payable before severance the contract is a sale of real property; but when a consideration is payable after the severance the contract may be a sale or an agreement to sell realty or personalty depending upon the language employed. The language of this instrument is sufficient to convey title to real estate. A part of the consideration was paid in cash and the balance payable upon the happening of a contingency other than the removal of the minerals, and there are no provisions for further payments of any kind either as rent or royalty. The instrument conveyed to the original grantee the minerals as land and as the action was to subject real estate to a lien for a part of the unpaid purchase money it was not transitory but local to Hardin County, where the land was located, and the original grantee was a necessary party and properly in court.

Kennedy v. Hicks (Kentucky), 203 Southwestern 318, p. 320.

LIEN FOR PURCHASE MONEY—PLACE OF ENFORCEMENT.

Where the conveyance of minerals vests the title to such minerals in the vendee and retains a lien for unpaid purchase money due and payable before the severance of the minerals from the real estate, then an action to enforce the vendor's lien by a sale of the minerals while a part of the land is local to the county where the land is located under the statute of Kentucky.

Kennedy v. Hicks (Kentucky), 203 Southwestern 318, p. 320.

FRAUD—PRESUMPTION FROM RELATIONSHIP OF PARTIES.

Dealings in the purchase and conveyance of mineral lands between persons standing in the relationship of brother-in-law and sister and affecting injuriously the rights of others are not presumably fraudulent, but such relationship when fraud is charged calls upon a court for a careful and close scrutiny of the transactions and conduct of and evidence offered by the parties, and when the transactions involved the payment of large sums of money the testimony of such parties when uncorroborated by receipts, memoranda, or other documentary evidence must be clear, positive, definite, and consistent with other evidence offered and free from contradiction.

North American Coal & Coke Co. v. O'Neal (West Virginia), 95 Southeastern 822, p. 826.

SURFACE AND MINERALS—OWNERSHIP AND SEVERANCE.

METHOD OF SEVERANCE.

A severance of the surface and the minerals or mineral interest may be by conveyance of the mines or minerals only; or by a conveyance of the land with a reservation or exception as to the mines or minerals.

De Moss v. Sample, 143 Louisiana ———, 78 Southern 482, p. 486.

SALE OR RESERVATION OF MINERALS—NATURE OF ESTATE.

A sale or reservation of mineral rights does not constitute a sale or reservation of so many tons of coal, iron ore, barrels of oil or feet of gas; but a grant may convey or reserve the exclusive right to explore the property for any of the minerals designated in the act of conveyance or reservation. To say that such a contract is void for lack of certainty as to the things sold or reserved, or contrary to any principle of public policy of the State, would relegate the most valuable property in the State to the category of property de hors commerce.

De Moss v. Sample, 143 Louisiana ———, 78 Southern 482, p. 484.

CONVEYANCE OF SURFACE AND RESERVATION OF MINERALS.

The owner of land may sell the surface rights and except from the sale the mineral below the surface and reserve to himself the right to mine such minerals, whether the minerals be in place, as coal, or whether they be migratory, like oil and gas.

De Moss v. Sample, 143 Louisiana ———, 78 Southern 482, p. 483.

EXCEPTION OR RESERVATION.

A reservation or exception of the minerals in a tract of land conveyed is a separation of the estate in the minerals from the estate in the surface, and it makes no difference whether the word used is "excepted" or "reserved."

De Moss v. Sample, 143 Louisiana ———, 78 Southern 482, p. 486.

POSSESSION OF SURFACE.

Where the ownership of minerals has been severed from the ownership of the surface, the possession of the latter does not necessarily carry with it possession of the minerals.

Kentucky Block Cannel Coal Co. v. Sewell, 249 Federal 840, p. 847.

ADVERSE POSSESSION OF MINERALS—STATUTE OF LIMITATIONS.

The statute of limitations does not begin to run against the title to minerals until the adverse possession thereof is taken by another; and the bar of the statute would not be complete unless and until the adverse possession was continued for the full statutory period.

Kentucky Block Cannel Coal Co. v. Sewell, 249 Federal 840, p. 847.

ADVERSE POSSESSION OF COAL—EFFECT ON OWNERSHIP OF OIL AND GAS.

The fact that the owner of the surface of land took possession and mined coal for such length of time as would give title to the coal by adverse possession would not of itself give the surface owner title to oil and gas deposits not then known to exist as against a prior grantee of all mineral rights in the land.

Kentucky Block Cannel Coal Co. v. Sewell, 249 Federal 840, p. 847.

COAL AND COAL LANDS.

COAL LOCATION—CANCELLATION OF PATENT—BONA FIDE PURCHASER.

The United States brought suit to cancel on the ground of fraud a patent issued for coal land. The defendant pleaded that he was an innocent purchaser and purchased the land from a trustee for value and without any knowledge of fraud on the part of the entryman. Under such a plea the burden of proof is on the defendant to sustain the allegations of his plea and to show that he was an innocent purchaser for value. A purchaser who pays his money with knowledge of facts which if investigated would lead to the knowledge of the fraudulent entry is not an innocent purchaser.

United States v. Kirk, 248 Federal 30, p. 35.
United States v. Routt County Coal Co., 248 Federal 485.

FRAUD—CANCELLATION OF PATENT—DATE OF RECEIVER'S RECEIPT.

In an action by the United States to cancel a patent to certain coal lands on the ground of fraud the fact that the receiver's receipt bears a date subsequent to the patent and to the deed by the patentee conveying the land to another, cannot be taken as conclusive on the subject of constructive notice to subsequent purchasers and such a circumstance would not be sufficient to put a purchaser upon inquiry within any sound rule of law.

United States v. Routt County Coal Co., 248 Federal 485, p. 487.

SALE AND CONVEYANCE—FRAUDULENT REPRESENTATIONS—RESCISSION OF CONTRACT.

The Georgia Iron & Coal Company sold to the Georgia Steel Company a tract of 52,000 acres of land containing coal and iron, one furnace, a railroad and its equipment, and other property used in cperating the mine. The purchasing company took possession and continued to develop the property for five years, making partial payments of the purchase money without complaint or objection. Later, believing that false representations were made by the agents of the seller as to the quantity of iron and coal deposits, a suit was brought to rescind the contract on the ground of fraud. In order for a defrauded party to make a proper plea to a court of equity to rescind a contract on the ground of fraud or misrepresentation there must not only be a material misrepresentation intended to mislead and having that effect, but action must be taken as soon as facts are discovered that give notice of the fraud. The purchaser continued after it had an opportunity to know as much as any one knew to take the chances. A person who speaks tardily to a court of equity must speak convincingly.

Peeples v. Georgia Iron & Coal Co., 248 Federal 886, p. 891.

SALE—SECRET COMMISSION TO OFFICER OF PURCHASING CORPORATION—FRAUD AND RESCISSION.

It is a rule of law that when a person by the payment of secret commission, corrupts an officer or employee of the purchasing person or corporation the sale may be set aside at the suit of the latter. But the payment by a selling company of a commission to an officer of the purchasing corporation carries with it only one right in equity and that is to rescind the contract. The buyer has the option to avoid the sale, but he must act promptly after ascertaining the fact and so long as he is in a position where he can restore the original status the purchaser may invoke the remedy of rescission. But this right to rescission does not carry with it any right of readjustment of price. The payment of corrupt commissions is a character of fraud that has no relation to price; and equity will not relieve by rescission a contract after a delay of five years with knowledge of the facts.

Peeples v. Georgia Iron & Coal Co., 248 Federal 886, p. 892.

CONTRACT FOR SALE OF COAL—CONSIDERATION—COMPROMISE OF DOUBTFUL RIGHTS.

The settlement of controverted matters arising out of a contract by a coal-mining company for the sale of all the coal mined by it was a sufficient consideration for a new and different contract, and in an action involving the rights of the parties growing out of the new relation it is not admissible to go back of the compromise or settlement in order to determine which party was right in its contention. When a compromise of a doubtful right is fairly made between the parties to a contract it is binding and can not be affected by any subsequent investigation or result. The rule applies whether the compromise was of a doubtful question of law or fact.

Producers' Coal Co. v. Mifflin Coal Min. Co. (West Virginia), 95 Southeastern 948, p. 951.

OIL AND OIL LANDS.

OIL AND GAS AS MINERALS.

Oil and gas in place are minerals.

De Moss v. Sample, 143 Louisiana ————, 78 Southern 482, p. 485.

NATURE AND CHARACTERISTICS OF OIL—REDUCING TO POSSESSION.

Oil and gas are substances of a peculiar character and decisions of ordinary cases of mining for coal and other minerals that have a fixed situs can not be applied to contracts concerning oil and gas without some qualifications. Oil and gas belong to the owner of the land and are a part of it so long as they are on it or in it or subject to the owner's control; but when they escape and go into other land or come under another's control, the title of the former owner is gone. Where an adjoining owner drills a well on his own land and taps a deposit of oil and gas extending under his neighbor's land so that it comes into his well it then becomes his property.

De Moss v. Sample, 143 Louisiana ————, 78 Southern 482, p. 485.

LANDS VALUABLE FOR OIL—PROOF OF VALUE.

In an action to determine the value of oil lands or of an oil lease of a tract of land on which there was a flowing well, it is proper to show the value of an oil lease on an adjoining tract of the same size with a flowing well where the conditions as to the two leases, improvements, qualities, etc., are alike; and in proving the value of the lease of the adjoining tract it was proper to prove what it could have been sold for under existing conditions.

Peden Iron & Steel Co. v. Jenkins (Texas Civil Appeals), 203 Southwestern 180, p. 183.

GRANT OF RIGHT TO PROSPECT FOR OIL AND GAS.

The right to go upon land for the purpose of prospecting and taking therefrom oil and gas is a proper subject of sale and may be granted or reserved by deed or other conveyance. The title to the oil and gas becomes perfect when discovered and reduced to actual possession.

Ramey v. Stephney (Oklahoma), 173 Pacific 73.

GRANT OF OIL AND GAS—RIGHTS GRANTED.

Oil and gas while in the earth are not the subject of ownership distinct from the soil and the grant of oil and gas is a grant, not of the oil or gas that is in the ground, but of such part as the grantee may find, and such a grant passes nothing but the right to explore for the oil and gas under the terms of the contract.

De Moss v. Sample, 143 Louisiana ——, 78 Southern 482, p. 483.

OWNERSHIP OF SURFACE—PRESUMPTION AS TO OWNERSHIP OF OIL.

The owners of several tracts of land entered as "surface" for the purpose of taxation continued to own the undivided interest in the oil and gas and presumably the value of such interest was included in the valuation of the land entered as "surface;" and a general allegation that such oil and gas interests or estates were not subsequently taxed will not overcome the presumption that such undivided interest continued charged to the owners of the estates entered as surface.

State v. Guffey (West Virginia), 95 Southeastern 1048, p. 1049.

SALE—SEVERANCE OF SURFACE AND MINERALS.

When the owner of land conveys to another the oil and gas which has not been developed, the oil and gas become property distinct from the surface and are under two ownerships; the oil and gas are a real and corporeal property separate from the surface or soil itself. Oil is a part of the land and owned by the owner of the land, and the owner can sever such ownership and except or reserve the oil to himself.

De Moss v. Sample, 143 Louisiana ——, 78 Southern 482, p. 485.

CONVEYANCE OF OIL AND GAS—CONSTRUCTION OF DEED—RIGHT TO PROSPECT.

A quit claim deed by a landowner recited that it "does hereby grant, bargain, sell, convey, and quit claim" to a named grantee "all of its right, title, and interest in the oil and gas for 21 years from March 29, 1912, in and under," the described land. The deed further stipulated: "This is intended to convey only nine-tenths of the oil and gas for 21 years from March 29, 1912, which nine-tenths was reserved by the grantor in a deed by the grantor" in a certain prior deed described. The recital as to the conveyance of only nine-tenths of the oil and gas does not limit the deed to a simple conveyance of nine-tenths of the oil and gas in and under the land, but the deed conveyed to the grantee all the rights of the grantor reserved in its warranty deed to nine-tenths of the oil and gas, including the right to enter upon the land, prospect and drill and take therefrom the oil and gas.

Ramey v. Stephney (Oklahoma), 173 Pacific 72.

CONVEYANCE—INCONSISTENT CLAUSES—CONSTRUCTION—INTENTION OF GRANTOR.

A quit claim deed by a landowner to a purchaser purported to convey all the grantor's right, title, and interest in the oil and gas in the land. A later clause in the deed recited that it intended to convey only nine-tenths of the oil and gas. The rule of construction is that a grant must be construed to effect the plain intention of the grantor and if that intention is plain it controls regardless of inconsistent clauses which are to be reconciled by the intention deduced from the entire instrument. The provision that the deed intended to convey only nine-tenths of the oil and gas is not in conflict with the granting clause nor does it defeat its effect, where the deed otherwise showed that the grantor had owned only nine-tenths of the oil and gas.

Ramey v. Stephney (Oklahoma), 173 Pacific 73.

INJUNCTION AND RECEIVER.

Drilling operations on lands withdrawn from occupation and development by presidential order may be enjoined against an oil locator who has caused to be assigned to him a large number of distinct oil locations, where the injunction against further drilling will not harm the locator and operator, and where there is nothing to indicate that he intended to continue drilling until controversies pending in the Land Office were settled.

United States v. Honolulu Consolidated Oil Co., 249 Federal 167, p. 168.

CONTRACT FOR DRILLING—CONSTRUCTION AND CONSIDERATION—"PROPOSED WELL."

A contract between a well driller and an oil company for the drilling of certain oil wells provided that a further sum of $1,000 shall be paid to the driller "when and after a well proposed to be drilled * * * is drilled in and proves a commercial paying oil or gas well." The word "proposed" must, in the construction of the contract, be defined as a fixed intention on the part of the promisor, fully known to the promisee at the time the contract was entered into that the well was contracted to be drilled upon the land referred to in the contract.

Gem Oil Co. v. Callendar (Oklahoma), 173 Pacific 820, p. 822.

TENANTS IN COMMON—FRAUD OF ONE TENANT—LIABILITY TO OTHERS.

The owner of oil and gas lands conveyed the same to a third person and reserved a full one-sixteenth part of all the oil and gas in and under the land conveyed. The grantee covenanted to lease the land for the production of oil and to keep the same leased while the surrounding premises were leased for that purpose. In consideration of this agreement the grantee was to receive all delayed rentals. The grantee of the land leased the same for five years, the consideration being one-eighth of the oil produced and $200 per year for each gas well. The lessee agreed to drill a well within 30 days or pay a rental of $3.50 quarterly in advance for each three months until the well should be completed. The lease just before its expiration, and after the discovery of gas in paying quantities in the immediate vicinity, was renewed by the original grantee and extended for five years in consideration of $75 in cash and the further payment of a like sum for each three months thereafter until a well was drilled. Wells were drilled in close proximity to the land on other lands owned by the original grantee, the lessor. The gas sand was such as to make probable drainage of the gas from the lands in controversy. The renewal of the lease on the part of the original grantee under the circumstances was a fraud upon the rights of the original grantor of the tract, and entitled him either to have the lease canceled or to ratify the agreement and recover one-third of the consideration paid to the lessor in lieu of drilling.

McMillan v. Connor (West Virginia), 95 Southeastern 642, p. 643.

INSURANCE—LOSS OF OIL—OPTION TO REPLACE.

An insurance policy covered oil and the tanks in which the oil was contained and gave the insurer, in case of loss, an option to repair, rebuild, or replace the property lost or damaged with other like kind and quality within a reasonable time. On the destruction of oil and tanks by fire the insurer can not elect to replace the oil without the tanks in lieu of paying the damages in money. If the insurer proposed to replace the oil it was bound also to replace the tanks as the insurance covered both. The option to replace is an option to replace the entire loss and it is not permissible to replace in part and pay in part. If the policy had covered the oil only the insured may have taken the chance of having at hand suitable tanks or some other container in which to receive the oil if the insurer proposed to replace the oil, but this rule will not be applied where the policy covered the tanks as well as the oil.

Globe & Rutgers v. Prairie Oil & Gas Co., 248 Federal 452, p. 457,

INSURANCE—LOSS OF OIL—DEDUCTION FOR SEDIMENT.

In an action on an insurance policy for the loss of oil stored in tanks the proof of loss stated the contents of the tanks destroyed at a certain number of barrels less a certain stated number of barrels deducted for sediment and less a certain number of barrels saved, showing the net balance of barrels of oil destroyed. The insurer can not claim a certain stated per cent deduction for sediment, amounting to more than that allowed by the insured where the per cent of deduction insisted upon by the insurer is that customary for sediment accumulated in the various tanks through which the oil passes before reaching the final storage tank and where it appears that on reaching the final storage tank having passed over a distance of from 400 to 500 miles the oil is 100 per cent pure, or at most contains about a quarter of 1 per cent of sediment. But in the absence of proof showing a different amount of sediment from that allowed in the proof of loss there was no question of fact on the particular issue.

Globe & Rutger v. Prairie Oil & Gas Co., 248 Federal 452, p. 458.

LOSS OF OIL BY FIRE—MEASURE OF DAMAGES.

An insurance policy covering oil in tanks provided that the company should not be liable beyond the actual cash value of the property at the time of loss and the loss shall be ascertained according to such actual cash value with proper deductions for depreciation. On the loss of the oil insured the actual cash value was to be the measure of damages, but it could not exceed what it would cost the insured to replace it. The cash value of an article is the amount of cash for which it will exchange in fact; and cash value is the market value for which an article will sell for in cash on the market. Where a State had a State corporation commission which fixed the price of oil and no one had a legal right to sell oil in the State for less than the price so established, this is sufficient to establish the cash value of the oil, especially in the absence of countervailing evidence.

Globe & Rutger v. Prairie Oil & Gas Co., 248 Federal 452, p. 457.

NATURAL GAS.

BOND FOR FRANCHISE FOR NATURAL GAS LINE—DAMAGES OR PENALTY.

The fiscal court sold a franchise for a natural-gas pipe line to a public-service corporation, it being the highest bidder. The public-service corporation was required to give a bond in the sum of $1,000, conditioned that it should within six months from the date of the purchase of the franchise comply with the terms of the franchise and furnish gas to the consumers in a certain designated city. The sum stated was held to be liquidated damages in view of the fact that it would be practically impossible to estimate the damages which would accrue from a breach of the bond and that the sum fixed was not unreasonable or oppressive.

Fiscal Court, etc., v. Kentucky Public Service Co. (Kentucky), 204 South-western 77, p. 79.

RECEIVER—REGULATION OF RATES.

The courts of Kansas have no jurisdiction to appoint receivers for the purpose of regulating the rates of natural gas companies, and neither courts nor receivers of such corporations have jurisdiction to change legal rates without the consent of the public utilities commission; but when the legal rates charged by the receiver of a natural-gas company have been enjoined the receiver may put into effect rates to be charged until the public utilities commission establishes a new rate.

State v. Independence Gas Co. (Kansas), 172 Pacific 713, p. 714.

PIPE LINES.

IMPROPER CONSTRUCTION ON HIGHWAY—LIABILITY—PROOF.

A pipe line company laying its pipe line in a highway and leaving the pipe a distance of 12 inches above the ground thereby created a nuisance. The building of such a structure was not only negligent but it was a wrong, which rendered the pipe-line company liable without proof of negligence. But the fact that the court, in an action by a person injured by reason of the construction of the pipe line, in its instructions placed the burden upon the plaintiff of proving negligence did not prejudice the defendant.

Carlson v. Mid-Continent Development Co. (Kansas), 173 Pacific 910, p. 911.

LIABILITY OF OWNER DURING RECEIVERSHIP.

A gas company laid its pipe line in a highway more than 12 inches above the ground and while plaintiff was driving along the highway his horses' feet were caught under the pipe throwing the horses down, upsetting the buggy and injuring the plaintiff. At the time of the accident and injury the pipe line and the property of the pipe line company were in the hands of a receiver, but at the time the plaintiff began his action for damages against the company the property had been restored to the possession of the company. Under such circumstances the company so negligently constructing the line is liable, although the pipe line was in the control of the receiver at the time the plaintiff was injured.

Carlson v. Mid-Continent Development Co. (Kansas), 173 Pacific 910.

RAILROAD GRANTS.

LIEU LANDS—MINERAL CHARACTER.

Conditions which are such as to suggest the probability that lieu lands selected under a railroad grant contain some oil at some depth, but where there is nothing to point persuasively to its quality, extent or value or that the conditions were such as to suggest the possibility of oil in paying quantities and to induce venturesome persons to prospect the field, are not sufficient to justify the cancellation of a patent after seven years' standing, although there was a subsequent discovery of oil in paying quantities.

Southern Pacific Co. v. United States, 249 Federal 785, p. 804.

SELECTION OF NONMINERAL LANDS—KNOWLEDGE OF AGENT.

In the selection of lieu lands under the Southern Pacific Railroad grant, the railroad company was required to attach to its selection list an affidavit by its land agent to the effect that the lands listed had been examined and that they are not mineral lands. An agent making such an affidavit is chargeable with knowledge of all the conditions that a careful examination of the land would have disclosed, and a court must assume that the agent had all such valuable information, including that which was known to the company's geologist and other agents and employees.

Southern Pacific Co. v. United States, 249 Federal 785, p. 804.

MINING TERMS.

ACCIDENT.

An accident is any unlooked mishap or untoward event not expected or designed.

Indian Creek Coal & Min. Co. v. Calvert (Indiana Appeals), 119 Northeastern 519, p. 521.

ALL COAL.

The term "all coal" used in a lease is the equivalent to all minable coal.

Fisher v. Maple Block Coal Co. (Iowa), 168 Northwestern 110, p. 111.

GIANT.

A giant is the nozzle of a pipe used to convey water for hydraulic mining and is used for the purpose of distributing or properly applying and increasing the force of the water.

Roseburg National Bank v. Camp (Oregon), 178 Pacific 312, p. 316.

OIL TERRITORY.

Oil territory does not necessarily imply a real issue of fact as the phrase has no fixed or well recognized meaning and may well be used in one sense and understood in another. But it may mean territory where the observable geological conditions are such as to justify expenditures in prospecting by those who are able to take the chance.

Southern Pacific Co. v. United States, 249 Federal 785, p. 786.

PERSONAL INJURY.

A personal injury under a workmen's compensation act or an employers' liability act, has reference to the consequences or disability that results therefrom.

Indian Creek Coal & Min. Co. v. Calvert (Indiana Appeals), 119 Northeastern 519, p. 525.

SLOW DRAG.

A "slow drag" is a trip or train of cars run by a mine operator for the purpose of carrying men in and out of the mine.

Sloss-Sheffield Steel & Iron Co. v. Crosby (Alabama), 78 Southern 896.

20

SLUDGE.

Sludge is a murky colored sediment flowing from the operations of a lead and zinc mining plant.

Dickensheet v. Chateau Min. Co. (Missouri Appeals), 202 Southwestern 624, p. 625.

TRIP.

A "trip" is a number of mine cars attached together and drawn by a mule or other motive power.

Maize v. Big Creek Coal Co. (Missouri Appeals), 208 Southwestern 683, p. 684.

MINING CORPORATIONS.

Promoters of a mining corporation stand in a fiduciary relation to the corporation and to the subscribers for stock as well as to those it is expected will afterwards purchase stock from the corporation. The promoters owe to such persons the utmost good faith and if they undertake to sell their own property to the corporation they are bound to disclose the whole truth respecting it. If they fail to do this, or if they receive secret profits out of the transaction, either in cash or by way of an allotment of stock, if there are other stockholders, the corporation may elect to avoid the purchase or it may hold the promoters accountable for the secret profits or may require a return of the stock if unsold and if sold an accounting for the profits of its sale.

North American Coal & Coke Co. v. O'Neal (West Virginia), 95 Southeastern 822, p. 825.

PROMOTERS AND DIRECTORS—ACCOUNTING FOR SECRET PROFITS.

Promoters or directors of a corporation and those colluding with them who in breach of their trust and in fraud of the corporation take to themselves secret profits are liable to account to the corporation therefor.

North American Coal & Coke Co. v. O'Neal (West Virginia), 95 Southeastern 822, p. 825.

FRAUD OF PROMOTERS—LIABILITY—PARTIES.

Certain promoters of a corporation who were afterwards selected as directors, officers and agents of the corporation voluntarily surrendered to the corporation all the profits of a conspiracy entered into by them and other promoters to defraud the corporation. Subsequently a bill was filed by the corporation against the remaining promoters, parties to the fraud to clear up the title to property purchased by the corporation being the subject matter of the fraud and conspiracy and for an accounting by them for money and property fraudulently obtained, the promoters and officers who voluntarily surrendered their part of all the profits of the conspiracy are not necessary parties to the action or relief.

North American Coal & Coke Co. v. O'Neal (West Virginia), 95 Southeastern 822, p. 823.

22

LIABILITY OF DEFRAUDING PROMOTERS.

One of several promoters of a mining company fraudulently conspired with other promoters to make certain secret profits in the purchase of a tract of coal land for the corporation and as a part of the scheme fraudulently and secretly agreed with the other promoters to allow and pay them out of the price he was to receive from a syndicate of which he was also a member and by and through which the coal lands were to be fraudulently transferred to the corporation at a price which would net to the syndicate a great profit, may in a suit by the corporation, after receiving a conveyance of the land be held to account to it for the sum so paid or agreed to be paid by him to other promoters in addition to the amount received by such promoter in carrying out the fraudulent scheme.

North American Coal & Coke Co. v. O'Neal (West Virginia), 95 Southeastern 822, p. 824.

CAPITAL STOCK—PAYMENT IN PROPERTY.

When a statute or a charter authorizes the capital stock of a corporation to be paid in property, and the stockholders honestly and in good faith put in property instead of money in payment of their subscriptions, third persons have no grounds for complaint. Under this rule the owners of an oil lease may form a corporation and transfer to it as its capital stock the lease, at a fair and reasonable value, in good faith estimated by the owners—the stockholders and subsequent creditors of the corporation can not attack the validity of the incorporation except for fraud in overvaluation of the property.

Peden Iron & Steel Co. v. Jenkins (Texas Civil Appeals), 203 Southwestern 180, p. 184.

PROPERTY RECEIVED IN PAYMENT OF CAPITAL STOCK—PROOF OF VALUE.

The statute of Texas provides that when the stockholders of a corporation shall furnish satisfactory evidence to the Secretary of State that the full amount of the capital stock has been subscribed and 50 per cent paid in cash or its equivalent in property or labor done, the product of which shall be to the company the actual value at which it was taken or property received, that officer shall, on the payment of the fee and tax, issue a certificate of incorporation. Under this statute, owners of an oil lease may form themselves into a corporation and capitalize the corporation on the value of their lease; and it is sufficient if the incorporators on their oaths state the cash value of the property received, giving its description and location; and when the Secretary of State accepts the value placed upon the property and issues the charter, the valuation of the property is

prima facie correct and can only be called in question on the ground
of fraud.

. Peden Iron & Steel Co. v. Jenkins (Texas Civil Appeals), 208 Southwestern
180, p. 187.

OIL LEASE AS BASIS OF CAPITAL—PROOF OF VALUE OF STOCK.

The joint owners of an oil lease on a ten-acre tract of land with an
oil well that had been flowing for two months, organized a corporation, valued the lease at $120,000, and issued stock to the owners and
subscribers for their respective interest in the lease. 'In an action by
creditors of a corporation to hold the stockholders personally liable
for debts it was proper for a stockholder to testify how he and his associates arrived at the value they placed on the lease, and that one
method was to show the revenue it was then producing. Proof that
other and older wells on adjoining tracts were still producing apparently the same amount of oil as when first brought in, and the owners
of the well in controversy had the right to presume that their well
would continue to flow oil together with other wells to be sunk on the
land and to value it accordingly. They had the right to show that
the well was producing from 2,000 to 2,500 barrels per day, and that
they were receiving 45 cents per barrel for the oil, and that on such
basis they could safely place the valuation at $120,000.

Peden Iron & Steel Co. v. Jenkins (Texas Civil Appeals), 208 Southwestern
180, p. 185.

PROPERTY AS BASIS OF CAPITAL—LIABILITY OF STOCKHOLDERS.

Where a statute or the charter of a corporation authorizes the payment of capital stock in property, the incorporators are not responsible for an honest error in judgment or a mistake in placing valuation on property appropriated or used as capital by a mining company; and where it appears that the persons forming the corporation
honestly believed the property to be worth the amount specified in
the articles and that their mistake was one of judgment only, their
action can not be considered fraudulent either in fact or in law. The
law imposes no penalty of this kind upon the stockholders or trustees
of a company for a mistake or erroneous judgment in the honest and
faithful discharge of their duties.

Peden Iron & Steel Co. v. Jenkins (Texas Civil Appeals), 208 Southwestern
180, p. 187.

SALE OF TREASURY STOCK—OPTION PURCHASE—SPECIFIC PERFORMANCE.

Treasury stock of a mining corporation was sold at 10 cents per
share to be taken and paid for in such sums and amounts as might
be required in the prosecution of development work. The purchaser

was given the right to "throw up" the option on reasonable notice, and the stock was to be issued as each payment was made at the specified rate per share. This was not a contract for an outright unconditional sale of stock, but it was a conditional sale to be consummated in part or in whole as the need for raising funds for the prosecution of the development of the mine might arise; neither was it an option open to acceptance for an unlimited time. It was not intended that the contract should have any binding force upon either the mining company or the purchaser after the mine became a profitable working mine and the need for the money for development work had ceased. After such a condition arose and money for development work was no longer needed the mining corporation could not insist upon specific performance of the contract.

Woldson v. Richmond Min., Smelting, etc., Co. (Washington), 172 Pacific 1162.

INCREASE OF CAPITAL STOCK—RIGHT OF STOCKHOLDERS TO PURCHASE.

When the capital stock of a mining corporation is increased by the issue of new stock, each holder of the original stock has a right to subscribe for and to demand from the corporation such a proportion of the new stock as the number of shares already owned by him bears to the whole number of shares before the increase. The rule applies to such part of the capital stock as is issued after the corporation has commenced business.

Thurmond v. Paragon Collieries Co. (West Virginia), 95 Southeastern 816, p. 819.

RIGHT TO SELL UNISSUED STOCK.

A private corporation issued and sold part only of its authorized capital stock and engaged in business for some time and desired to increase its working capital. It was then within the discretion of the directors of the corporation to determine the manner in which the money should be raised and they may lawfully do so by selling a part or all of its unissued stock and they may sell it at such price per share as previously determined by a majority vote of the stockholders.

Thurmond v. Paragon Collieries Co. (West Virginia), 95 Southeastern 816, p. 819.

PURCHASE OF STOCK—SPECIFIC PERFORMANCE—RIGHTS OF PURCHASER.

Ordinarily complete and satisfactory remedy for a failure to deliver certificates of stock to a purchaser may be had in damages at law but there are exceptions where the value of the stock is not readily to be ascertained or where the stock can not be had in the market except at great inconvenience and expense.

Mutual Oil Co. v. Hills, 248 Federal 257, p. 260.

SPECIFIC PERFORMANCE OF CONTRACT FOR SALE OF STOCK.

Where the stock of a corporation has a recognized market value, courts will ordinarily leave a purchaser to his action at law for damages for breach of an agreement to sell; but where the stock of a mining corporation has no recognized market value, is not purchasable in the market, or has a value which is not settled but is contingent upon the further workings of the corporation, equity may decree specific performance of a contract to sell the stock of the corporation.

Mutual Oil Co. v. Hills, 248 Federal 257, p. 260.

AGREEMENT TO EMPLOY OFFICER—SALE OF STOCK—SPECIFIC PERFORMANCE.

An oil company entered into an agreement with the president of a competing company by which he was to end his relation with the competitor and to take charge of the business of the oil company as district manager. As a necessary part of that agreement he entered into an agreement to purchase a certain number of the shares of the stock of the oil company at an agreed price to be paid for in a certain stated manner. Pursuant to the agreement the party severed his connection with the competing oil company, became manager of the oil company in one of its districts and continued in that relation for three years, when he was discharged. In an action by such manager for specific performance to compel the transfer of the shares of stock purchased it appeared that the value of the stock had risen from the face value to $200 per share; that the business of the company had and continued to increase; that large dividends would be earned; that there was but a limited amount of stock issued and that no stock was for sale upon the market; that it was impossible to determine the actual damage the complainant would sustain if the stock was not delivered to him and that the oil company had sufficient stock to transfer the number of shares agreed upon. Under such a state of facts a court if equity is authorized to decree specific performance of the agreement and compel the issue of the stock.

Mutual Oil Co. v. Hills, 248 Federal 257, p. 260.

PRINCIPAL PLACE OF BUSINESS—MEANING.

The Western Oil Refining Company, a corporation organized in the State of Indiana, owned two stationary oil tanks located at Glasgow, Ky., on which it paid an ad valorem tax; that it had selling agents in Glasgow who used the tanks as storage places for oil and from which they sold many thousands of gallons of oil annually by the use of oil wagons owned by the corporation and for which it procured a license to operate; that frequently the oil was shipped from the home office in Indiana billed to itself at Glasgow and on arrival there was received by its selling agents and put into the tanks from which it subsequently distributed the oil to the customers. These facts are sufficient to show that the company was doing business in the State and had one of its principal places of business in the State within the meaning of the Kentucky statute requiring it to paint or print its corporate name in a conspicuous place upon its premises.

Western Oil Refining Co. v. Commonwealth (Kentucky), 202 Southwestern 636, p. 637.

DUTY TO PUBLISH CORPORATE NAME.

The statute of Kentucky requires every corporation doing business in the State to have painted or printed the corporate name in a conspicuous place and in letters sufficiently large to be easily read and in like manner to have printed or painted the word "incorporated." The fact that a mining corporation engaged in selling oil had the word "incorporated" painted or printed as required by the statute constitutes only a partial compliance with the statute and is not sufficient to exonerate the corporation from the penalty imposed by the statute, but the corporation must go further and paint or print its corporate name as well as the word "incorporated."

Western Oil Refining Co. v. Commonwealth (Kentucky), 202 Southwestern 636, p. 637.

LIABILITY OF STOCKHOLDERS FOR CORPORATE DEBTS.

Creditors of an oil corporation sought by an action to hold the subscribers to the capital stock liable for the corporate debts on the insolvency of the corporation, on the ground that the capital stock had been paid in property at a fictitious price and an overvaluation. A recovery was not permitted under the statute of Texas where the proof showed that an oil lease on a ten acre tract of land was received by the incorporators, the owners of the lease, at an estimated value of $120,000; that there was at the time of the incorporation on the tract a flowing well producing from 2,000 to 2,500 barrels of oil per day; that they were receiving a revenue of from $1,000 to $1,200 per day; that other wells on adjoining tracts, similar in conditions and surroundings, had shown an equal or greater and an undiminished production for a period of two months and more; that offers of $200,000 had been made for similar leases in the same field; that undeveloped tracts had sold for $8,000 per acre, and that experienced oil operators had placed a value of $200,000 on the particular lease.

Peden Iron & Steel Co. v. Jenkins (Texas Civil Appeals), 203 Southwestern 180, p. 184.

FAILURE TO PUBLISH CORPORATE NAME—DOING BUSINESS.

An indictment against a mining corporation under the Kentucky statute for failing to post in a conspicuous place its corporate name and the fact that it was incorporated is sufficient where it charges that an oil company had one of its principal places of business at a certain named city in the state and was "then and there engaged in the business of selling oil," as this is sufficient to show that the company was engaged in "doing business" in the state named.

Western Oil Refining Co. v. Commonwealth (Kentucky), 202 Southwestern 636, p. 637.

DISCRETION OF DIRECTORS—CONTROL OF COURTS.

Courts of equity will not interfere by injunction as a general rule to control the discretion of the directors of a private corporation in the management of its internal affairs.

Thurmond v. Paragon Collieries Co. (West Virginia), 95 Southeastern 816, p. 819.

DIRECTOR INTERESTED—RIGHT TO VOTE.

The reason for denying to a director of a corporation the right to vote on a matter in which he is otherwise interested than as a stockholder in the corporation is because of the fiduciary trust relation he bears toward it.

Thurmond v. Paragon Collieries Co. (West Virginia), 95 Southeastern 816, p. 817.

CONTRACT—INTEREST OF DIRECTOR—RATIFICATION BY STOCKHOLDERS.

A contract made by the directors of a corporation is not affected by the fact that one or more of the directors had an interest in such contract other than as stockholders in the corporation where the contract was either authorized or was subsequently ratified by a majority vote of the stockholders in a meeting regularly assembled.

Thurmond v. Paragon Collieries Co. (West Virginia), 95 Southeastern 816, p. 818.

STOCKHOLDERS INTERESTED—RESTRICTION ON RIGHT TO VOTE.

The only restriction upon the right of a stockholder to vote in a stockholders' meeting where he is personnally interested seems to be that the matter must not be illegal or ultra vires and the action of the majority of the stockholders must not be so antagonistic to the corporation as a whole as to indicate that their interests are wholly outside of the interest of the corporation and destructive of the interest of the minority stockholders.

Thurmond v. Paragon Collieries Co. (West Virginia), 95 Southeastern 816, p. 818.

STOCKHOLDER INTERESTED—RIGHT TO VOTE.

A stockholder in a corporation is not denied his right to vote on any matter properly coming before a stockholders' meeting on account of any private interest he may have which is detrimental to the corporation.

Thurmond v. Paragon Collieries Co. (West Virginia), 95 Southeastern 816, p. 817.

CONTRACT BETWEEN STOCKHOLDERS—EFFECT ON CORPORATION.

A contract between certain stockholders of a mining corporation to which the mining corporation was not a party, can have no binding effect on the corporation, and no action against the corporation can be based on such contract.

Northrop-Bell Oil & Gas Co., In re (Oklahoma), 171 Pacific 1116.

ACTION AGAINST—SERVICE ON OFFICER—SUFFICIENCY OF RETURN.

Under the laws of Oklahoma service can be made upon the corporation's treasurer only when the president, chairman of the board of directors, or other chief officers can not be found in the jurisdiction, and this fact must be stated in the officer's return of service. Where a return shows service on the treasurer of a corporation, but fails to show the absence of other officers of the corporation, the court may direct the officer serving the writ to amend his returns according to the true state of facts.

Stevirmac Oil & Gas Co. v. Dittman, 38 Supreme Court Rep. 116.

POWER TO EXECUTE INDEMNITY CONTRACT.

A mining company as lessee of a mine may execute a valid contract of indemnity by which it agrees to indemnify and save harmless the lessor of certain mining machinery by reason of injuries to employees, although any such injuries might be caused by the negligence of the lessor.

Harden v. Southern Surety Co. (Missouri Appeals), 204 Southwestern 34, p. 35.

CORPORATION SELLING PRODUCTS OF MINES—RESTRAINT OF TRADE.

A contract of agency by which a corporation undertakes to sell the coal produced by several mining companies is not void as being in restraint of trade where the selling corporation is not owned or controlled by the mining companies and where it sells the coal as its own property.

Thurmond v. Paragon Collieries Co. (West Virginia), 95 Southeastern 816, p. 818.

SUPERINTENDENT OF MINE—AGENT—AUTHORITY.

A corporation acts only by and through its officers and agents. The general manager of a coal mining company has authority to bind it by such contracts as are reasonably incident to the management of the business. In an action to recover damages for the breach of a contract for the sale of the output of a mining company's coal mine, it is not incumbent upon the plaintiff to prove that the general manager of the coal company was expressly authorized to make the contract sued on in order to hold the mining company liable. His apparent authority as general manager of the business of the coal company to make the contract was sufficient ground to create a liability.

Producers' Coal Co. v. Mifflin Coal Min. Co. (West Virginia), 95 Southeastern 948, p. 950.

ENFORCING LIEN AGAINST STOCK—JUDGMENT BINDING ON COOWNERS.

In an action to enforce a lien against stock of a mining corporation owned jointly by several parties a judgment enforcing the lien is binding upon all the parties, though some of them were not within the jurisdiction of the court, and were not served with process, but had actual knowledge of the pendency of the suit.

Montana Coal & Oil Co. v. Haskins (Oregon), 172 Pacific 119, p. 120.

TRUST FUNDS FOR INJURED MINERS—MINING CORPORATION AS TRUSTEE.

In the mining regions of Michigan it was the custom for the miners to pay, or for a mining corporation to retain from the pay of every miner, a small sum per month, to be used as a fund to be paid to injured miners. In such cases the fund so accumulated was a trust fund and the mining corporation a trustee.

Walters v. Pittsburgh, etc. Iron Co. (Michigan), 167 Northwestern 834, p. 836.

EMPLOYEE'S BENEFIT FUND—TERMINATION OF PURPOSE—OWNER OF SURPLUS.

A mining corporation received monthly from its miners certain sums and itself contributed an equal sum by which a fund was created to be paid to injured miners. The practice was maintained for many years and a large trust fund accumulated, as an aggregate of the overpayments made by the miners. The fund was deposited with and held by the corporation upon the express trust to disburse so much thereof as was necessary from time to time in the payment of injury or death benefits. When the mining company accepted the Michigan workmen's compensation law the purposes of the trust were discharged and the trust extinguished, and thereupon in the absence of an agreement a resulting trust arose in favor of all who had at any time contributed to the fund. If a trust for a specific purpose fails by the failure of the purpose the property reverts to the donor.

Walters v. Pittsburgh, etc. Iron Co. (Michigan), 167 Northwestern 834, p. 836.

CONTRIBUTIONS TO MINERS' RELIEF FUND—DISTRIBUTION OF SURPLUS.

The employees and miners of a mining company contributed monthly from their wages a fund for the payment of benefits to injured miners, the company paying sums equal to those paid by the miners. The payments were continued over a long period of time and many contributing miners terminated their employment and stopped their contributions, while other miners were during the period employed and made like contributions. A large surplus fund had accumulated at the time of the enactment of the Michigan workmen's compensation act and the contributions ceased on the acceptance by the mining corporation of that act. The surplus remaining in the possession of the corporation was a trust fund and must be distributed to all contributors in proportion to the amount contributed by each. The difficulty of locating some of the contributors does not affect this equitable rule of distribution. The rule that remaining members of a voluntary association on its dissolution own the funds does not apply to the facts as they appear in this case.

Walters v. Pittsburgh, etc. Iron Co. (Michigan), 167 Northwestern 834, p. 837.

DISSOLUTION—LAW GOVERNING.

A mining company organized under the laws of Nevada carrying on its mining operations in Alaska is governed as to the time and manner of its dissolution by the statute of Nevada.

Matson v. Connecott Min. Co. (Washington), 171 Pacific 1040, p. 1044.

DISSOLUTION—EXISTENCE CONTINUED.

A mining company organized under the laws of Nevada, doing business in the State of Washington, was sued in the latter State for damages for injuries to an employee on the ground of its alleged negligence. The complaint was signed and verified on June 1, 1915, but the corporation was properly dissolved June 10, 1915, before service or an appearance by it to the action. The statute of Nevada (Laws 1903, chap. 88) provides that corporations shall be continued for one year from the expiration of the charter or dissolution, and until all

82478°—19——4

litigation to which the corporation is a party, if begun within the year, is ended. The statute of Nevada in effect continued the corporation's existence for the purpose of commencing actions against it, and the corporation had a legal existence as a corporate entity for the purpose of the action in Washington at the time of its commencement; and where the action had been commenced within one year following the disincorporation of a company for other purposes, the action could proceed to final judgment upon the merits in favor of or against the corporate entity.

Matson v. Connecott Min. Co. (Washington), 171 Pacific 1040, p. 1042.

FORFEITURE OF CHARTER—POWER TO CONFESS JUDGMENT.

Under the statutes of Utah (Laws 1909, p. 228; Laws 1913, p. 14), on the forfeiture of the charter of a corporation, or where its franchise has expired by limitation, the corporation may continue for the purpose of winding up its affairs, and for this purpose may sell or otherwise dispose of its property, may make contracts, and sue and be sued. These statutes contemplate that the board of directors of a mining company shall wind up its affairs on the forfeiture of its charter; and under this authority the board of directors may confess a judgment on an indebtedness justly owing by the corporation and which it was unable to pay.

Henriod v. East Tintic Development Co. (Utah), 173 Pacific 134, p. 136.

MINING PARTNERSHIPS.

WHAT CONSTITUTES.

Mining partnerships are indulged between coowners only when they actually engage in working the mining property. Before actual operations began and after actual operations cease the parties are simply cotenants unless the ordinary partnership has in fact been formed.

Huston v. Cox (Kansas), 172 Pacific 992, p. 993.

FORMATION—CONTRIBUTION BY PARTNERS.

A mere agreement to form a partnership does not itself create a partnership, nor does the advancement by any one partner of his agreed share of the capital; but the entire agreement and all the attending circumstances must be taken into consideration in determining whether a partnership was actually formed. The fact that two or three persons proposing to form a partnership may have paid their proportions of the capital provided for would not estabish a partnership composed of the three persons named, with each contributing property of considerable value, when there is no proof that the third party either directly or through one of the others paid his part of the capital.

Costello v. Gleason (Arizona), 172 Pacific 730, p. 734.

CONTRACTS—REFORMATION OF INSTRUMENT.

A mining contract or lease executed in the names of the individual members of the firm may be reformed and enforced in the name of the firm where the contractee knew that he was dealing with the firm as such.

Last Chance Min. Co. v. Tuckahoe Min. Co. (Missouri Appeals), 202 Southwestern 287.

CONVEYANCE OF LAND TO FIRM—RIGHTS OF PARTNERS.

A conveyance of mineral lands to a firm in its firm name vests the title to the land in the partners as tenants in common; and in the absence of partnership debts the partners have a right to personally divide the lands between them as real estate.

Kentucky Block Cannel Coal Co. v. Sewell, 249 Federal 840, p. 848.

TRUST RELATION—PROOF.

A consummated partnership which would have the effect of establishing in one partner a trust relation regarding certain mining claims should be proved by evidence as clear and satisfactory as would be required to show a trust relation direct without the intervention of a partnership.

Costello v. Gleason (Arizona), 172 Pacific 730, p. 734.

FIDUCIARY RELATION—BREACH OF DUTY.

The requirement of the utmost good faith forbids that a partner benefit his private interest by deceiving his copartner by misrepresentations or concealments of the confidential relation. A mining partner who promised, on a sufficient consideration, to perform the assessment work on a mining claim owned by the partners can not be permitted, after failing to perform the assessment work, to procure the ground to be relocated and the new location conveyed to him, thereby depriving the other partner of his interest in the claim.

Kittilsby v. Vevelstadt (Washington), 173 Pacific 744, p. 745.

FRAUD OF PARTNER—PROCURING RELOCATION—LIABILITY.

One of two partners, owners of a mining claim, agreed to have the assessment work done for a certain year. He failed to perform or to have performed the assessment work, but secretly requested a third person to make a relocation and to locate a claim covering practically the same ground. This partner, in conjunction with another third person, located three other claims adjoining such relocated claim. Subsequently all the claims were sold, the partner received $10,000 as his share for all the claims. The original partner, on discovering the failure of the partner to perform the assessment work as agreed and the relocation of the ground and the sale of the claims, brought suit for relief and was awarded the sum of $5,000, one-half of the entire amount received by the defrauding partner for all the claims.

Kittilsby v. Vevelstadt (Washington), 173 Pacific 744, p. 745.

RIGHT TO APPOINT RECEIVER.

Whether the parties owning an oil and gas lease and intending to operate the same are partners or whether they are simply cotenants of an incorporeal hereditament, is no reason why a court of equity should not appoint a receiver to hold the lease, protect the property, and perform other functions where the parties can not agree with respect to their rights, or as to the management of the property or a disposition of it.

Huston v. Cox (Kansas), 172 Pacific 992, p. 993.

ABADONMENT AND DISSOLUTION.

One partner in a mining partnership enterprise accepted the proceeds of certain notes which had been the foundation upon which his interest in the partnership rested and made no arrangements for paying his proportion of the partnership capital and furnished no security therefor. These acts conclusively show an abandonment by him of his interest in the partnership property regardless of his intention. A partner can not withdraw his entire contribution because of the fear of losing both his interest and his money and afterward recover in an equity action the very interest which the money withdrawn was supposed to pay for.

Costello v. Gleason (Arizona), 172 Pacific 730, p. 733.

PRINCIPAL AND AGENT.

An agent is held to the utmost good faith in his dealing with his principal. If he acts adversely to his employer in any part of the transaction or omits to disclose any interest which would naturally influence his employer's conduct in dealing with the subject of employment, it is such a fraud upon his employer as forfeits any right to compensation for his services. This rule was applied in an action by a mining engineer to recover from a third person a part of the compensation paid such third person by a mining and exploration company for services in the purchase of certain mining claims during a time when the mining engineer was an agent of the mining and exploration company under an agreement to devote his entire time to the mining company and not to become interested in or connected in any way with any business similar to that which the mining company carried on.

Beatty v. Guggenheim Exploration Co. (New York), 119 Northeastern 575, p. 576.

GENERAL MANAGER OF COAL COMPANY—AUTHORITY AS AGENT.

The general manager of a coal mining corporation is a general agent and has implied authority to bind the company by such contracts and modifications thereof as are reasonably necessary in conducting the business of mining and selling coal. Where such an agent has such implied power to bind his principal it is error to submit to a jury the question of his authority.

Producers' Coal Co. v. Mifflin Coal Min. Co. (West Virginia), 95 Southeastern 948, p. 950.

PROOF OF AGENCY—MINE FOREMAN ACTING AS AGENT.

The mere fact that a mine foreman assumed to act as the agent of the mine operator is not alone sufficient to show such agency. To establish such agency without proof of express authority by the acts of the mine foreman alone, those acts must have been so open, apparent and notorious that it is evident that they must have been known to the mine operator.

Strother v. United States Coal & Coke Co. (West Virginia), 95 Southeastern 806, p. 807.

GENERAL MANAGER OF COAL MINING COMPANY—AUTHORITY—PROOF.

In an action against a coal mining company for damages for the breach of a contract executed by the general manager of the coal company providing for the sale of the entire output of its mines, the contractee, as plaintiff, is not required to prove the authority of the general manager to make the contract in order to hold the coal company liable, as his apparent authority as general manager of the coal company's business to make the contract is sufficient ground to create a liability.

Producers' Coal Co. v. Mifflin Coal Min. Co. (West Virginia), 95 Southeastern 948, p. 950.

DUTY OF AGENT—DISOBEDIENCE—DISCHARGE.

An agent owes to his principal disinterested and unswerving loyalty and the law does not permit him to place himself secretly in a position in which his own interest is antagonistic to that of his principal, or whereby he may be tempted by self-interest to disregard the interest of his principal. This carries with it the duty of one accepting employment to disclose to his employer, if not already known to him, any and all adverse interests.

Burch v. Conklin (Missouri Appeals), 204 Southwestern 47, p. 49.
See Consumers' Lignite Co. v. James (Texas Civil Appeals), 204 Southwestern 719.

EMPLOYMENT OF SUPERINTENDENT—DISCHARGE—LIABILITY.

A mine owner, by written agreement, dated June 1, 1916, employed a superintendent for a year at a salary of $3,600 to superintend four or five mines owned and operated by him. The employee resigned his position as deputy State mine inspector and accepted the employment and was, on September 1, 1916, discharged by the employer. In an action for damages for the wrongful discharge, an answer to the effect that the plaintiff advised the defendant in regard to the purchase of a certain mine and gave incorrect information because he was trying to induce the defendant to purchase the mine in order that he, the plaintiff, might earn a commission from the vendor, will not be sustained by proof that prior to the employment the plaintiff had induced the defendant to purchase mines and mining property and out of which purchases the plaintiff had received commissions.

Burch v. Conklin (Missouri Appeals), 204 Southwestern 47, p. 49.

VIOLATION OF CONTRACT BY AGENT.

A mining engineer by a written agreement became the agent for a specified time at a stipulated salary of the Guggenheim Exploration Company. The agreement gave the company the sole benefit of the engineer's unbiased judgment and bound him not to become interested in or connected in any way with any business similar to that which the company carried on without the written consent of the company. It was a breach of this contract for the mining engineer as such agent to enter into a contract with a third person by which the mining engineer was to have a share in certain compensation which such third person was to receive from the sale of mining property to the mining and exploration company. The agreement of the agent with such third person was a violation of the agent's contract with the mining and exploration company and a breach of his confidential relation to that company.

Beatty v. Guggenheim Exploration Co. (New York), 119 Northeastern 575, p. 576.

RESTRAINT OF TRADE—CONTRACT OF AGENCY FOR SELLING OUTPUT OF MINE.

A contract of agency by which a coal selling corporation is to act as the sole agent in selling all the coal produced by a number of coal mining companies is not void as being in restraint of trade, where the selling corporation was not owned or controlled by the mining companies and sold the output of the mines of each company as its own individual product and at such prices as it would command in the markets.

Thurmond v. Paragon Collieries Co. (West Virginia), 95 Southeastern 816, p. 818.

LAND DEPARTMENT.

MINING ENTRY—CONTINUING AUTHORITY.

The Land Department has authority at any time before a patent is issued to inquire whether an original mineral entry was in conformity with the act of Congress.

Kirk v. Olson, 38 Supreme Court Rep. 114, p. 115.

AUTHORITY TO DETERMINE CHARACTER OF LAND.

The Land Department on an application for mineral entry and patent is authorized to inquire into the mineral character of the land described in the application. Where on a preliminary hearing the officers of the Land Department determine that the land is mineral in character and permits a mineral claimant to make entry accordingly, the finding is not conclusive, but the department may make further investigation and if the land is found to be nonmineral in character may so classify the same and cancel the entry.

Kirk v. Olson, 38 Supreme Court Rep. 114, p. 115.

MINERAL CHARACTER—DETERMINATION—JURISDICTION OF LAND OFFICE AND COURTS.

The Land Department has jurisdiction to determine the mineral or the nonmineral character of ground located as a mining claim and held as such by the performance of the annual assessment work, although the locator has made no application for patent. Where the officers of the Land Department have undertaken to inquire into the mineral character of such a mining claim a court of equity will not entertain a bill to enjoin the officers of that department from making such inquiry.

Lane v. Cameron, 45 Appeal Cases (D. C.) 404, p. 410.

MINERAL CHARACTER—JURISDICTION TO DETERMINE—CONCLUSIVENESS OF FINDING.

The question on application for patent whether the land described was valuable for placer mining was one of fact to be determined by the officers of the Land Department, but a finding respecting the mineral character of the ground is not in itself final or conclusive, but essentially interlocutory. It was only a step in a proceeding looking to the ultimate disposal of the title and until the issue of a patent was open to reconsideration and reversal.

Kirk v. Olson, 38 Supreme Court Rep. 114, p. 115.

PUBLIC LANDS.

Where Congress has provided for the disposition of a portion of the public lands of a particular character and authorized the officers of the Land Department to issue a patent upon the ascertainment of certain facts that the department had jurisdiction to inquire into and determine as to the existence of such fact, its determination is conclusive as against a collateral attack. Cases arise before the Land Department in the disposition of the public lands where it is difficult on the part of its officers to ascertain with accuracy whether the lands to be disposed of are mineral in character or are agricultural lands. The rule is that they will be considered mineral or agricultural as they are more valuable in the one class or the other. But the determination of the fact by the officers of the Land Department that they are one or the other will be conclusive.

Lane v. Cameron, 45 Appeal Cases (D. C.) 404, p. 409.

40

MINING CLAIMS.

NATURE AND GENERAL FEATURES.

MINING CLAIM AS PROPERTY.

A valid mining claim is property which may be bought and sold and which passes by descent.

Lane v. Cameron, 45 Appeal Cases (D. C.) 404, p. 409.

OWNERSHIP OF MINERALS.

The minerals in a mining claim, or that have been wrongfully taken from a mining claim, belong to the person who made the proper location of the claim and followed the same by the performance of the required annual assessment work; and the person in possession of such a claim on the public lands may sue another for the wrongful taking or conversion of minerals from his claim.

Kelvin Lumber & Supply Co. v. Copper State Min. Co. (Texas Civil Appeals), 208 Southwestern 68, p. 70.

MINERAL CHARACTER OF LAND.

CHARACTER OF LAND—DETERMINATION—JURISDICTION OF LAND DEPARTMENT.

When the question of the mineral character of the land in a mining claim is an issue it is one for the Land Department.

Lane v. Cameron, 45 Appeal Cases (D. C.) 404, p. 409.

SUFFICIENCY OF PROOF.

The finding of minerals within the surface lines of a location to the extent that a person of ordinary prudence would be justified in the further expenditure of his labor and means with the reasonable prospect of success in developing a valuable mine is sufficient to meet the requirements of the statute.

Batt v. Steadman (California Appeals), 178 Pacific 99, p. 102.

KNOWLEDGE AND BELIEF OF MINERAL CHARACTER.

The requirement of knowledge of the mineral character of land as required by the statute is not satisfied by knowledge of conditions which are plainly such to engender the belief that the land is valuable for its minerals but to be mineral in fact the land must contain mineral deposits of such quality and in such quantity as would render their extraction profitable and justify expenditures to that end.

Southern Pacific Co. v. United States, 249 Federal 785, p. 798.

LAND VALUABLE FOR PLACER MINING.

Placer mining claims are sufficiently established where a miner in good faith located the claims on unoccupied public lands and had been in possession of one claim for almost six years and of the other for nearly four years and the annual assessment work had been done by him on both claims and gold had been found in both.

Batt v. Steadman (California Appeals), 173 Pacific 99, p. 102.

LOCATION OF CLAIMS.

METHOD OF MAKING.

A mineral claimant merely stakes out his location, files a notice of his location in the office of the clerk of the county wherein the land is situated and proves each year that he has done a certain amount of work on the claim. By the filing of the notice of his claim he acquires what is known as a mining location, and is not required to file any papers in the Land Office unless and until he applies for patent.

Lane v. Cameron, 45 Appeal Cases (D. C.) 404, p. 408.

POSSESSORY RIGHTS—VALIDITY OF LOCATION.

The legal right to possession of a mining claim can only come from a valid location.

Batt v. Steadman (California Appeals), 173 Pacific 99, p. 106.

VALIDITY OF SECOND LOCATION.

Where there has been a valid location of a mining claim followed by the performance of the annual assessment work by the original locator or by his assigns or a purchaser from him, there can not be another location of any part of the ground originally located, and possession under such second void location confers no rights upon the person making the same.

Kelvin Lumber & Supply Co. v. Copper State Min. Co. (Texas Civil Appeals), 208 Southwestern 68, p. 70.

LOCATION NOTICE.

POSTING AT POINT OF DISCOVERY—SUFFICIENT COMPLIANCE.

In an action to quiet title to a mining claim it is a sufficient compliance with the California statute which requires the posting of a notice at the point of discovery, where the locator testified that he put the notice "on the lode and surrounded it by a monument of rock right where the quartz shows—where the lode shows."

Batt v. Steadman (California Appeals), 173 Pacific 99, p. 100.

POSTING AT POINT OF DISCOVERY—INSUFFICIENT COMPLIANCE.

The Code of California (Civil Code, sec. 1426) requires that the location notice of a mining claim shall be posted at the point of discovery. A posting of such a notice within 75 feet of the point of discovery is not a sufficient compliance with the statute and does not constitute a valid location.

Batt v. Steadman (Caliofrnia Appeals), 173 Pacific 99.

DISCOVERY.

CONDITIONS SUGGESTING PROBABILITY.

Conditions such as only suggest the probability that lands contain some oil at some depth, but without anything to point persuasively to its quality, extent, or value, can not be a discovery within the meaning of the statute.

Southern Pacific Co. v. United States, 249 Federal 785, p. 804.

RECITALS IN LOCATION NOTICE—PRESUMPTION.

Proof that the locator and his successors in interest have been in the possession of a mining claim and have held and worked the claim for a period equal to the time prescribed by the statute of limitations for mining claims of the state where the same may be situated is sufficient to establish a right to a patent, without proof of an actual discovery where the recorded notice of location expressly recites the fact of a discovery as such a recital "creates a presumption of discovery of mineral and of a valid location."

Ralph v. Cole, 249 Federal 81, p. 93.

MARKING BOUNDARIES.

SUFFICIENCY—IDENTIFICATION OF CLAIM.

The marking of the boundaries of a mining claim on the ground, though not exactly what and where they should be, will answer the statute, if they are sufficient when taken in connection with an amended location certificate, to identify the land claimed with reasonable certainty.

Batt v. Steadman (California Appeals), 173 Pacific 99, p. 101.

SURFACE LINES.

SIDE LINES—FORM AND DIRECTION.

The side lines of a mining location are not required to be either parallel or straight. They may have angles and elbows and be converging and diverging so long as their general course is along the vein and the statutory restriction on the width of the claim is respected.

Jim Butler Tonopah Min. Co. v. West End Consol. Min. Co., 38 Supreme Court Rep. 574, p. 575.

END LINES—WHAT CONSTITUTES—PARALLELISM.

The end lines of a mining location in the sense of the statute are those which are laid across the vein to show how much of it in point of length is appropriated and claimed by the locator. The end lines of a mining location must be both parallel and straight.

Jim Butler Tonopah Min. Co. v. West End Consol. Min. Co., 38 Supreme Court Rep. 574, p. 575.

SURVEY—CHANGE OF LINES.

Proceedngs of miners in the location of mining claims are regarded with indulgence and lines are not required to be laid with severe accuracy. A claim is not rendered invalid because the surveyor, on the final survey, was required to draw in some of the lines as they were marked on the ground in order to bring the boundaries of the claim within the limits prescribed by law. It is a rule that for the purpose of ôbtaining parallelism or casting off excess the surface lines may be drawn in.

Batt v. Steadman (California Appeals), 173 Pacific 99, p. 101.

VEIN OR LODE.

OWNERSHIP OF VEINS.

Every vein whose apex is within the vertical limits of the surface lines of a location passes to the locator by virtue of his location. He is not limited to those veins only which extend from one end line to another or from one side line to another or from one line of any kind to another but he is entitled to every vein whose top or apex lies within his surface lines and he is entitled to such veins throughout their entire depth although they may so far depart from a perpendicular in their course downward as to extend outside the vertical side lines of his location.

Jim Butler Tonopah Min. Co. v. West End Consol. Min. Co., 38 Supreme Court Rep. 574, p. 575.

FISSURE VEIN—DESCRIPTION AND APEX.

A fissure vein was shown to have two dipping limbs whose course downward was substantial, regular, and practically free from undulation. For 750 feet out of its total length of 1,150 feet within a mining claim each limb was practically separate with a distinct summit or terminal edge. For the remaining 400 feet of the claim the two limbs were shown to be united and from the point of union the mineralized quartz or rock continued upward for from 20 or 30 to more than 100 feet. This is sufficient to answer all the calls of a summit or terminal edge. Under these circumstances a court would not be warranted in saying as a matter of law that the vein has no top or apex within the claim in the sense of the statute.

Jim Butler Tonopah Min. Co. v. West End Consol. Min. Co., 38 Supreme Court Rep. 574, p. 577.

APEX OF VEIN.

LOCATION ON APEX—TITLE.

The law assumes that a lode has a top or apex and provides for the acquisition of title by a location upon this apex. A court should not adjudge the absence of a top or apex in the case of a fissure vein where it appears that each limb is practically a separate vein with a distinct summit or terminal edge or where the two limbs unite and from the point of union the mineralized quartz or rock continues upward for from 20 to more than 100 feet.

Jim Butler Tonopah Min. Co. v. West End Consol. Min. Co., 38 Supreme Court Rep. 574, p. 577.

EXTRALATERAL RIGHTS.

ORIGIN—WHAT CONSTITUTES.

What in mining cases is termed the extralateral right is the creation of the mining laws of Congress and to determine what it is these laws must be looked to, rather than to some system of law to which it is a stranger. Congress having plenary power over the disposal of the mineral bearing public lands, could say to what extent, if any, the right to pursue a vein on its downward course into the earth should pass to and be reserved for those to whom it granted possessory or other titles in such lands.

Jim Butler Tonopah Min. Co. v. West End Consol. Min. Co., 38 Supreme Court Rep. 574, p. 575.

RIGHT TO PURSUE VEIN BEYOND SIDE LINES.

The locator of a mining claim by section 2322, (United States Revised Statutes), is given a right to pursue any vein, whose apex is within his surface lines, on its dip outside the vertical side lines, but he may not in such pursuit go beyond the vertical end lines of his location.

Jim Butler Tonopah Min. Co. v. West End Consol. Min. Co., 38 Supreme Court Rep. 574, p. 576.

LIMITATIONS ON RIGHT.

The extralateral right created by section 2322 (United States Revised Statutes 1878) is subject to three limitations. One conditions it on the presence of the top or apex inside the boundaries of the surface location. Another restricts it to the dip or course downward and so excludes the strike or onward course along the top or apex of the vein; and the third confines it to such outside parts as lie between the end lines extended outwardly in their own direction and extended vertically downward. Otherwise the right is without limitation or exception and broadly includes "all veins, lodes and ledges throughout their entire depth"—one as much as another and all whether they depart through one side line or the other.

Jim Butler Tonopah Min. Co. v. West End Consol. Min. Co., 38 Supreme Court Rep. 574, p. 575.

VEIN PASSING THROUGH BOTH SIDE LINES.

Where a single vein having its apex within the surface lines of a location in its descent separates into two limbs which depart through the opposite side lines, the owner of such surface location is entitled to follow each limb of the vein beyond the vertical side lines of his location under the statute creating extralateral rights.

Jim Butler Tonopah Min. Co. v. West End Consol. Min. Co., 38 Supreme Court Rep. 575, p. 575.

DIRECTION OF DEPARTING VEIN.

When the other elements of the extralateral right are present it may be exercised by the locator beyond either or both side lines, depending on the direction which the departing vein or veins take in their downward course.

Jim Butler Tonopah Min. Co. v. West End Consol. Min. Co., 38 Supreme Court Rep. 574, p. 576.

ASSESSMENT WORK.

PRIMA FACIE PROOF OF PERFORMANCE.

Section 2324 (United States Revised Statutes) provides that on the failure of a coowner to perform his share of the assessment work his interest on proper notice may be forfeited. The statute of California (Civil Code, sec. 14260) provides that the notice to the coowner of his failure to perform his part of the assessment work may be recorded within 90 days in the office of the recorder of the proper county and that such notice shall be prima facie evidence of the coowner's delinquency. On the failure to file the notice in the recorder's office, or on failure to file it within the time required the interest of the delinquent coowner may be forfeited on actual proof of his failure to contribute to the assessment work. The filing of the notice makes it prima facie evidence only and the failure to file deprives the coowner performing the assessment work of the prima facie effect of the filing.

Robinson v. Griest (California), 173 Pacific 88, p. 89.

PROOF IMMATERIAL IN SUIT ON ADVERSE CLAIM.

In an action by an adverse claimant to sustain his claim and to have determined the relative possessory rights to portions of the ground sought to be obtained, where there is no question of forfeiture or abandonment, the question of whether the required assessment work has been done or the required amount of improvement been made is immaterial.

Roberts v. Oechsli (Montana), 172 Pacific 1037, p. 1038.

PERFORMANCE BY COOWNER—RECORD OF SERVICE OF NOTICE.

The United States statutes (section 2324) provides that on the failure of a coowner of a mining claim to perform or to contribute to the performance of his part of the assessment work, the interest of the delinquent coowner may, upon proper notice, be forfeited. The statute of California (Civil Code, sec. 14260) provides that when the notice required by section 2324 of the United States statutes is given it may within 90 days be recorded in the office of the proper county recorder and it shall be prima facie evidence that the delinquent coowner has failed to contribute to the assessment work.

Robinson v. Griest (California), 173 Pacific 88, p. 89.

PERFORMANCE BY COOWNER—BURDEN OF PROOF.

A coowner of certain mining claims began an action against the other coowner to remove or to prevent a cloud on the title. He alleged in his complaint that he had paid his full share for work done upon the mines. The answer of the defendant, the other coowner, denied the allegations of the complaint and it devolved upon the complainant to prove the fact of the payment of his full share of the assessment work, and in the absence of such proof the court erred in denying the defendant's motion for nonsuit.

Robinson v. Griest (California), 173 Pacific 88, p. 89.

NOTICE TO COOWNER TO CONTRIBUTE—TIME OF FILING—INJUNCTION TO
PREVENT FILING.

The statute of California (sec. 1426o, Civil Code) provides that
a notice to a coowner to contribute to the assessment work on a mining claim with the proof of service on the coowner may be filed with
the county recorder within 90 days after the service of the notice.
The filing of the notice and proof within the time shows prima facie
that the interest of the noncontributing coowner is subject to forfeiture. The alleged delinquent coowner can not maintain an action
to enjoin the other coowner from filing the proof of the notice of delinquency after the expiration of the 90 days from the date of service,
as the filing of the proof after the expiration of the time provided
casts no cloud upon the title and is not then prima facie evidence of
a right of forfeiture.

Robinson v. Griest (California), 178 Pacific 88, p. 89.
See California Mining Statutes, Annotated, 140.

FAILURE OF COOWNER TO CONTRIBUTE—FORFEITURE—QUIETING TITLE--
PRIMA FACIE PROOF.

A coowner of mining claims sought by cross-complaint to quiet
title to the mining claim in himself as against his coowner on the
ground that the coowner had failed to contribute his share to the
performance of the assessment work and that notice of such failure
had been duly recorded in the office of the proper county recorded.
The proof shows that the cross complainant failed to file the notice
served upon the coowner within the 90 days as required by the California statute and therefore the record does not constitute prima
facie evidence under the statute and he was not entitled to a decree
quieting his title in the absence of actual proof of the failure of his
coowner to perform his part of the assessment work.

Robinson v. Griest (California), 173 Pacific 88, p. 89.

ABANDONMENT.

CONDITIONAL ABANDONMENT.

The owners of a valid mining claim consented to the location of a
second claim upon and over a part of their claim upon the condition
and pursuant to an agreement that the second locator and his claim
should be charged with a one-fourth interest in a tunnel site in
favor of the owners of the first location. The consent of the owners
of the first location to the second location upon and over a part of
their claim was equivalent to an abandonment by them of that part
of the claim; and the acceptance of the conditions pursuant to the
agreement and the location by the second location subjected the sec-

ond locator and his claim to the charge of a one-fourth interest in the tunnel site and the second location when made stood charged accordingly.

Walsh v. Klienschmidt (Montana), 173 Pacific 548, 549.

RELOCATION.

VALIDITY OF JUNIOR LOCATION.

It is not correct to say that a prior mining location, which subsequently fails, so aboslutely withdraws the land from entry as to defeat a junior location of the same ground otherwise valid.

Walsh v. Klienschmidt (Montana), 173 Pacific 548, 549.

INVALID AS TO EXISTING LOCATION.

Where there has been a valid location of a mining claim followed by the performance of the annual assesment work, there can be no relocation of any part of the ground embraced in an existing valid location. The attempted relocation of any part of the ground embraced in an existing valid location is void and confers no rights whatever upon the relocator; and a purchaser of ore from such a relocator, though in actual possession of a part of the ground embraced in the original location, is not an innocent purchaser and may be liable to the original locator for the value of the ore so purchased.

Kelvin Lumber & Supply Co. v. Copper State Min. Co. (Texas Civil Appeals), 203 Southwestern 68, p. 70.

POSSESSORY RIGHTS.

ADVERSE POSSESSION—INSTRUCTIONS BY COURT.

In an action by an adverse claimant, brought pursuant to the statute, on an adverse claim filed by him against the granting of a patent to the applicant, the defendant, the applicant for the patent, pleaded the statute of limitations of the State of Nevada and introduced evidence tending to show not only that there was a discovery within the boundaries of each of his lode claims in controversy years before the entries thereon and the placer locations claimed by the adverse claimant, but that for many years, more than are required by the Nevada statute of limitations, the lode claims had been possessed, worked, and claimed as lode mining ground adversely to all the world except the Government. Under these facts it was error for the court to refuse to instruct the jury on the subject of the adverse holding by the defendant and his predecessors in interest.

Ralph v. Cole, 249 Federal 81, p. 92.

In an action on an adverse claim as required by the statute, a finding by the court to the effect that the defendant, the applicant for patent to the disputed ground, had been in the actual, exclusive, and uninterrupted occupation and possession of all the mining ground in dispute, claiming title thereto adversely to the plaintiff, the adverse claimant, for a period greater than the statute of limitations of the State in which the claim is situated, prior to the commencement of the suit on the adverse claim, constitutes a complete bar to the action.

Ralph v. Cole, 249 Federal 81, p. 93.

POSSESSORY ACTIONS.

SUIT BY ADVERSE CLAIMANT—RIGHT TO POSSESSION.

In a suit by an adverse claimant to quiet his title the plaintiff's right of ownership and possession is in no way dependent upon a prescriptive title, and he is not required to prove possession of the claim for more than five years prior to the commencement of the action; but it is sufficient to prove and for the court to find that at the commencement of the action the plaintiff was the owner and entitled to the possession of the mining claim.

Batt v. Steadman (California Appeals), 173 Pacific 99, p. 100.

PROOF TO SUSTAIN.

To maintain an action for the possession of a mining claim in the State of Nevada and under the statutes of that State, the plaintiff must show by proper evidence that he, or those through or from whom he claims " were seized or possessed of such mining claim, or were the owners thereof according to the law and customs of the district embracing the same within two years next before the commencement of such action."

Ralph v. Cole, 249 Federal 81, p. 93.

ADVERSE POSSESSION—TRIAL OF RIGHTS—VENUE.

An action involving the title and ownership of minerals alleged to have been wrongfully mined and removed from the mining claim of the complainant, can not be maintained in the court of a foreign State to recover the value of the minerals so mined and taken from an innocent purchaser of the original trespasser where it appears that the alleged trespasser was in the peaceable adverse possession of the mining claim at the time the minerals were so mined and removed, as this involves the title to the mining claim as between the original parties and it is the policy of the law that such an issue should be tried in the local court where the property is situated.

Kelvin Lumber & Supply Co. v. Copper State Min. Co. (Teaxs Civil Appeals)'. 203 Southwestern 68, p. 70.

DESCRIPTION.

DESCRIPTION OF CLAIM—SURVEYOR'S MAP.

A map made by a surveyor showing a description and location of a mining claim in controversy is sufficiently supported where the surveyor testifies that he found fixed monuments on certain corners and on one side line of the claim and that in surveying he considered both the data on the ground as well as that given in the notice of location.

Batt v. Steadman (California Appeals), 173 Pacific 99, p. 101.

STATE STATUTES.

EFFECT ON MINING LOCATIONS—STATUTE OF LIMITATIONS.

The statute of Nevada provides that no action for the recovery of mining claims shall be maintained unless the plaintiff was possessed of the same within two years before the commencement of the action. It also provides that occupation and adverse possession shall consist in holding and working a mining claim in the usual and customary mode. Proof that a claimant has been in possession of the disputed ground either by himself or his predecessors in interest for more than the period prescribed by the statute of limitations, together with proof of working the same as required by law, is sufficient to entitle him to the right of possession, where his recorded location notice recited the fact of a discovery as such recital " creates a presumption of discovery of mineral and of a valid location."

Ralph v. Cole, 249 Federal 81, p. 92.

MINERS' RULES AND REGULATIONS.

EFFECT ON MINING LOCATION.

Prior to the passage of any mining law by Congress, both the Land Department and the courts always acted upon the rule that all mining locations were to be governed by the local laws, rules, and customs in force at the time of the location. Such practice was clearly intended by Congress as shown by the provisions of section 2 of the original act of July 26, 1866 (14 Stats. 251), as well as by section 2332 (United States Revised Statutes 1878).

Ralph v. Cole, 249 Federal 81, p. 87.

PLACER CLAIMS.

USE OF WATER.

A placer mining claim without water is only a site for a mine, and is in fact a mere prospect.

Roseburg National Bank v. Camp (Oregon), 173 Pacific 313, p. 315.

ADVERSE CLAIM.

FILING—CONTEST TRANSFERRED TO COURT.

On the filing of an adverse claim to an application for a patent as required by section 2326 (Revised Statutes 1878), the adverse claimant is required, within 30 days, to commence proceedings in a court of competent jurisdiction to determine the right of possession; and he must prosecute his action with reasonable diligence to final judgment. The question thus to be transferred from the Land Office to a court of competent jurisdiction for decision is that of the right of possession of the mining ground respecting which the contest arose in the Land Office; but the proceedings in the Land Office are suspended, the title to the ground remaining in the Government for disposal in accordance with the judgment of the court, and on compliance with all requirements of the statute. No form of action is prescribed by the statute and no court other than one of competent jurisdiction is designated.

Ralph v. Cole, 249 Federal 81, p. 87.
See United States Mining Statutes, Annotated, 447.

AMENDMENT OF STATUTE—PRACTICE.

The act of July 26, 1866 (14 Stats. 251), in which provision was made for contests between rival claimants of a mining location by the filing of adverse claims in the Land Office and their subsequent trial in a court of competent jurisdiction, was made more specific by the act of May 12, 1872 (17 Stats. 91). On the revision of the United States statutes in 1878, the provisions for application for patent and filing and determining of adverse claims were carried into the revision as sections 2325 and 2326.

Ralph v. Cole, 249 Federal 81, p. 87.

PROCEDURE AND TRIAL—JURISDICTION OF FEDERAL AND STATE COURTS.

Section 2326 (Revised Statutes 1878) requires an adverse claimant, within 30 days after filing his claim, to commence proceedings in a court of competent jurisdiction, to determine the relative rights of the parties to the ground in controversy. The adverse suit to determine the right of possession may not involve any question as to the construction and effect of the Constitution or the laws of the United States, but may present simply a question of fact as to the time of the discovery of mineral, the location of the claim on the ground, or a determination of the meaning and effect of certain local rules and customs, or the effect of State statutes, and is not necessarily a suit that must be brought in a Federal court. Such suits may sometimes so present questions arising under the Constitution and laws of the United States that the Federal courts will have jurisdiction, regardless of the citizenship of the parties, but the mere fact that such an adverse suit is authorized by the statutes of the United States is not in and of itself sufficient to vest the jurisdiction in the Federal courts.

Ralph v. Cole, 249 Federal 81, p. 90.

ACTION TO DETERMINE RIGHTS—ACTION AT LAW—JURY TRIAL.

The pleadings in an action commenced on an adverse claim, as required by section 2326 (Revised Statutes 1878), show the action to be one at law where the sole issue is the right of possession of the respective mining claims or of the ground in controversy. In such a case either party is entitled to a trial by jury, or the parties may stipulate for such a trial in a Federal court having jurisdiction because of diverse citizenship.

Ralph v. Cole, 249 Federal 81, p. 90.

TRIAL AND FINDING OF JURY.

Section 2326 (Revised Statutes 1878) was amended by an act of March 3, 1881 (21 Stats. 505), to the effect that in the trial of an action brought by a claimant to determine his adverse claim, if title to the ground in controversy should not be established by either party, the jury should so find and judgment must be entered accordingly.

Ralph v. Cole, 249 Federal 81, p. 87.

ACTION BY ADVERSE CLAIMANT—NATURE—TRIAL BY JURY.

It was not the intention of section 2326 (United States Revised Statutes 1878), providing for a trial on the filing of an adverse claim, to change the method of trial, but its object was to provide for an adjudication of the right of possession and to prevent the applicant from obtaining a patent if the adverse claimant failed to sustain his claim and where the applicant himself failed to prove his right of possession. The duty was imposed on the court trying the case to enter such judgment or decree as would evidence that the applicant had not established his right of possession and was therefore not entitled to a patent.

Ralph v. Cole, 249 Federal 81, p. 87.

TRIAL BY COURT OR JURY.

If Congress had intended to provide that litigation as provided in the amendatory act of March 3, 1881 (21 Stats. 505), to section 2326 (Revised Statutes 1878) must be an action at law, or must invariably be tried by jury, it would have said so in plain terms. But there is nothing to indicate the intention thus to circumscribe a resort to the accustomed modes of procedure or prevent the parties from submitting the determination of their controversies to the court.

Ralph v. Cole, 249 Federal 81, p. 87.
See United States Mining Statutes, Annotated, 452.

QUESTIONS DETERMINED IN SUIT.

In an action brought by an adverse claimant after filing his adverse claim, his right to a patent for the land is not settled by the judgment of the court beyond the right of inquiry by the Government; and the judgment does not necessarily give the claimant a right to the land in controversy.

Lane v. Cameron, 45 Appeal Cases (D. C.) 404, p. 410.
See United States Mining Statutes, Annotated, 483.

PROOF OF ASSESSMENT WORK IMMATERIAL.

An adverse claimant brought an action to have determined the relative rights of himself and the applicant for patent to portions of the ground claimed by each. No question of forfeiture or abandonment was involved, and the question whether the applicant for patent had performed the assessment work or had made improvements to the value of $500 can not be material on the trial of an action between an adverse claimant and the applicant for patent.

Roberts v. Oechsli (Montana), 172 Pacific 1087, p. 1088.

PATENTS.

PROCEEDINGS FOR OBTAINING—CONCLUSIVENESS.

The law intended in every instance where there was a possibility of a conflict of claims to give an opportunity to have such conflict decided by a judicial tribunal before the rights of the parties were foreclosed or embarrassed by the issuance of a patent to either claimant. The wisdom of the provision is apparent when its effect is considered upon the value of the patent which is thereby rendered conclusive as to all rights that could have been asserted in the proceedings. An opportunity is given to determine such conflict in the form of an action in a court of the viscinage, where witnesses can be produced and a jury, of miners if desired, can pass upon the rights of the parties, under instructions as to the law from the court. The parties and the Land Office are governed by the decision of the court, and if the court decides for one party or the other or against both parties, the Land Department is bound by the decision.

Ralph v. Cole, 249 Federal 81, p. 89.

PROCEDURE AFTER JUDGMENT ON ADVERSE CLAIM.

After the judgment of a court on an adverse claim on the question of the right of possession, the Land Department may still pass upon the sufficiency of the proofs to ascertain the character of the land, and to determine whether the conditions of the law have been complied with in good faith.

Lane v. Cameron, 45 Appeal Cases (D. C.) 404, p. 410.

MINERAL CHARACTER OF LAND—CANCELLATION OF ENTRY—NOTICE.

Two locators of a placer claim made joint application for patent and on investigation of the mineral character of the land by the Land Department an entry was authorized. Subsequently on application by a homestead entryman for patent and on investigation by the Land Department, his application was passed to entry. Before either entry was passed to patent a rehearing was ordered to determine the true character of the land. Notice of the rehearing was given to the homestead entryman and to one of the placer entrymen, but not to the other. On the rehearing, the land was found to be valuable only for agriculture and to have no value for placer mining, and the homestead entry was passed to patent. In an action to quiet title by the homestead patentee against the placer entryman not notified of the rehearing, he was entitled to assert and show that the tract was valuable for placer mining as originally found by the land officers, as the finding of the Land Department in his absence was not conclusive of the character of the tract, as against him.

Kirk v. Olson, 38 Supreme Court Rep. 114, p. 115.

CANCELLATION—MINERAL CHARACTER OF LAND—SUBSEQUENT DISCOVERY.

To justify the cancellation of a patent under a railroad grant as wrongfully covering mineral lands, it must appear that at the time of the order granting the patent the land was known to be valuable for minerals; that is, it must appear that the then known conditions were plainly such as to engender the belief that the land contained mineral deposits of such quality and in such quantity as would render their extraction profitable and justify expenditures to that end. But if the land was not then known to be valuable for minerals, subsequent discoveries can not affect the validity of the patent.

Southern Pacific Co. v. United States, 249 Federal 785, p. 804.

VALIDITY—COLLATERAL ATTACK.

A collateral attack upon a patent is authorized where the basis of the attack is that the patent shows upon its face that it is void.

Meade v. Steele Coal Co. (Kentucky), 203 Southwestern 1061.

ADVERSE CLAIM—NEITHER PARTY ENTITLED TO PATENT.

The proceedings provided for under section 2326 United States Revised Statutes, as supplemented by the act of March 3, 1881 (21 Stats. 505), were merely in aid of the Land Department, and the object of the amendment was to secure that aid as much in cases where both parties failed to establish title or right of possession, as where judgment was rendered in favor of either. If the adverse claimant chooses to proceed by a bill to quiet title, and if neither party is entitled by the proof to relief the court can render a decree to that effect, the same as if judgment was rendered on a verdict in an action at law.

Ralph v. Cole, 249 Federal 81, p. 87.

COAL LOCATIONS.

VALIDITY OF COAL LOCATION.

Sections 2347 to 2351, Revised Statutes of United States, impose certain restrictions in making coal locations on the public lands, and expressly forbid individuals and associations from acquiring coal land in excess of the quantities prescribed, whether directly by entries in their own names or indirectly by entries made for their benefit in the names of others. One person can not lawfully make an entry in the interest of another who has had the benefit of the law, or in the interest of an association where it or any of its members has had the benefit thereof, or in the interest of a person or association where he or it has not had such benefit but is seeking through entries made, or to be made by others in his or its interest, to acquire a greater quantity of land than is permitted by law.

United States v. Kirk, 248 Federal 80, p. 35.

CANCELLATION OF PATENT—NOTICE TO OFFICER OF CORPORATION.

The United States brought suit to cancel a patent for coal land on the ground of fraud. The proof showed that the vice president of the defendant coal company procured the entries to be made in the name of a third person for the benefit of the coal company, paying all the expenses and the purchase price of the land from the funds of the company. The coal company, under this proof, is chargeable with the knowledge possessed by its vice president and agent, and although it paid a valuable consideration, it took with notice of the fraud and can not hold the title.

United States v. Routt County Coal Co., 248 Federal 485, p. 486.

OIL LOCATIONS.

WORK LEADING TO DISCOVERY—GROUP CLAIMS.

The statute of the United States called the Pickett Act of June 25, 1910 (36 Stats. 847), provides that the rights of a bona fide occupant or claimant of oil or gas bearing lands who at the date of a withdrawal order is in diligent prosecution of work leading to a discovery of oil or gas shall not be affected by such order, if he shall continue in diligent prosecution of the work. A locator of an oil location had in June, 1909, encountered large quantities of gas in drilling operations on one of a group of oil claims. As soon as the gas was under control drilling continued and oil was discovered on February 1, 1910. On September 27, 1909, the land by presidential order was withdrawn from oil discovery and development. During the year prior to the order of withdrawal a cabin and skeleton derricks had been constructed on different quarter sections and in that manner and in the building of roads, making surveys, piping water, and otherwise providing for the general development of the property, the locator had spent considerable sums of money. Pending application for patents the United States brought suit to enjoin the locator from drilling further wells on the disputed tract and for the appointment of a receiver for the wells in operation. The question is, under such conditions, whether or not the group system of development can be applied. The testimony was sufficient to warrant the conclusion that future operations on the unpatented sections depended entirely upon the results obtained from the drilling on a particular section. But the drilling of a well on one location can not, where the group theory is inapplicable to the several locations, be treated as work leading to discovery on the other locations.

United States v. Honolulu Consolidated Oil Co., 249 Federal 167, p. 168.
United States Mining Statutes, Annotated, 1044.

DISCOVERY—SUFFICIENCY—PROBABILITY NOT SUFFICIENT.

Conditions which suggest the possibility of oil in paying quantities and sufficient to induce the more venturesome such as are willing to take chances to prospect the field is not sufficient to constitute discovery where such conditions were not plainly such as to engender the belief that any given section or other legal subdivision of the land in controversy contains oil of such quality and quantity and at such depth as would render its extraction profitable.

Southern Pacific Co. v. United States, 249 Federal 785, p. 804.

OPERATIONS ENJOINED—RECEIVERSHIP.

In a suit by the United States Government to enjoin the further drilling of wells on oil locations a receiver may be appointed for a limited period and with limited authority where little harm or inconvenience will result to the locator, and where the product of a completed well is practically impounded under an existing contract, and where it is not clear that further drilling was intended until pending controversies between the Government and the locator were settled.

United States v. Honolulu Consolidated Oil Co., 249 Federal 167, p. 168.

TRESPASS.

INNOCENT PURCHASER OF ORE—RECOVERY BY ORIGINAL OWNER.

A person who has made a valid location of a mining claim on the public land and maintained the same by the performance of the annual assessment work, may recover ore or its value in an action against the purchaser thereof from a person who wrongfully entered upon the mining claim and without right dug and removed the ore therefrom, where the owner of the claim did not consent to the mining and removal of the ore; and he does not by his mere silence lose his right to recover the ore or its value from a purchaser.

Kelvin Lumber & Supply Co. v. Copper State Min. Co. (Texas Civil Appeals), 203 Southwestern 68, p. 70.

LIABILITY FOR ORE MINED AND APPROPRIATED—DAMAGES.

Trespassers upon a valid location, who mined and removed ore therefrom, are not entitled to recover for expenses in taking out the ore; nor can a purchaser of ore from such trespassers, in an action against him by the original locator and owner for the value of the ore so taken, claim any credit or allowance for the expenses in the mining and taking of the ore by the trespassers.

Kelvin Lumber & Supply Co. v. Copper State Min. Co. (Texas Civil Appeals), 203 Southwestern 68, p. 70.

STATUTES RELATING TO MINING OPERATIONS.

RELATIVE DUTIES OF OPERATOR AND MINE FOREMAN—SAFE PLACE.

The question as to liability of a mine owner for injuries caused by the position of the post along a mine tract has not been settled by the Pennsylvania court. But the rule now stated is that where the act imposes upon the mine owner a specific duty as to the arrangement and management of the mine, the mine owner can not protect himself from liability by showing that the structure in question was put up by the direction of the mine foreman, at least when the fact of its existence become known to the owner. In such case it would be the duty of the owner to remove the dangerous structure. On the other hand, where the act does not call for any particular arrangement or structure, then the arrangement or structures ordered by the mine foreman for the purpose of supporting the roof of the mine are obligatory on the mine owner and he is not responsible for injuries caused thereby.

Whittaker v. Valley Camp Coal Co. (Pennsylvania), 103 Atlantic 594, p. 595.

APPLICATION OF STATUTE—EFFECT OF CUSTOM.

The statute of Kentucky on the question of propping the roof of a mine applies only to those miners or workmen charged with the duty of propping, and does not in terms impose upon every miner the duty of complying with the statute. Accordingly the custom of a mine may be proved for the purpose of showing upon whom the duty of propping devolved; but the effect is not to change the statutory requirement by custom. A mine operator's rules may impose upon every miner the duty of doing his own propping, and yet the proof may show that such rules had been waived and that it was the custom of the mine, since beginning its operations, to have a timberman whose duty it was to do the propping at such places as the miners were required to work. Under these circumstances the fact that the miner failed to prop his roof and was killed by a fall of slate from the roof will not prevent a recovery because of the alleged failure of the decedent to comply with the statute and the rules of the mine operator.

Borderland Coal Co. v. Kirk (Kentucky), 203 Southwestern 534, p. 535.

DUTIES IMPOSED ON OPERATOR.

DUTY TO FURNISH SAFE PLACE.

A mine operator may be liable for a failure to use ordinary care to furnish a miner a reasonably safe place for work; and the exception to the safe place doctrine is not applicable because the miner himself was creating the danger in the progress of his work, does not apply, where the duty of propping devolved upon the mine operator and not upon the miner; nor can the mine operator escape liability on the ground that the miner was engaged in removing the coal at the time of the accident and the operator had no opportunity to prop the roof. But the fact was that the coal had been removed some time before and the company's timberman had been sent to do the propping and had abundance of time in which to do so. Under these circumstances the mine operator was charged with the duty of using ordinary care to make the working place reasonably safe, and in an action for the death of the miner, caused by a fall from the roof, the question of the exercise of such care was to be determined by the jury.

Borderland Coal Co. v. Kirk (Kentucky), 203 Southwestern 534, p. 536.

SAFE PLACE—DUTY AS TO ENTRIES.

The bituminous mine act of 1893 (P. L. 52 (78)), provides that a mine owner shall see that the entries where braking is necessary shall have a clear level width of not less than 2½ feet between the sides of the cars and the rib to allow the driver to pass his trip safely and keep clear of the cars. But whether this requires a space of 2½ feet to be left between the cars on a switch and a row of posts, and another way of 2½ feet on the other side of the posts between the posts on the main track, is left in doubt.

Whittaker v. Valley Camp Coal Co. (Pennsylvania), 103 Atlantic 594, p. 595.

VENTILATION—DUTY NONDELEGABLE.

Section 40 of the Alabama statute (Acts 1911, p. 515) makes no practical change in section 1016 of the Code of 1907 except to include explosives as well as noxious gases and also to provide that the minimum amount of air to be supplied shall be 100 cubic feet per minute per man and 500 cubic feet per mule or horse. It was the intention of the legislature to protect the miner from the danger of noxious and explosive gases generated in mines.

Segrest v. Roden Coal Co. (Alabama), 78 Southern 756, p. 757.

KEEPING MINE FREE FROM GAS—STANDARD OF DUTY.

Section 1016 of the Alabama Code (1907) makes it the imperative duty of a mine owner or superintendent to keep the mine swept out and free from noxious gases generated therein; and this is a non-delegable duty, and the mere furnishing of the means is not sufficient.

Segrest v. Roden Coal Co. (Alabama), 78 Southern 756, p. 757.

VENTILATION—REQUIREMENT AS TO AIR.

The statute of Alabama (sec. 40, act of 1911, p. 515) requires a mine operator to keep his mine free from noxious and explosive gases, and that the minimum amount of air to be supplied shall be 100 cubic feet per minute per man and 500 cubic feet per mule or horse. Under this statute it is not sufficient to prove the means for supplying the air, but the operator is required to provide and maintain the same, the air or ventilation, to the extent requird by the statute, and to supply the air to the extent of accomplishing the purpose; and there would be a violation if the purpose was not accomplished whether it was due to the insufficiency of the means or of a failure to operate and maintain the same even if amply and sufficiently provided for.

Segrest v. Roden Coal Co. (Alabama), 78 Southern 756, p. 757.

DUTY TO FURNISH PROPS.

The statute of Alabama (General Acts 1911, 500, p. 514) requires a mine operator to keep at a convenient place at or near the main entrance to the mine, or in the mine, a sufficient supply of props and other timbers of suitable lengths and sizes to be furnished on request by the miners. But the mine operator is only required to deliver the props or timbers of proper length only when needed and requested and designated by the miner.

Clark v. Choctaw Min. Co. (Alabama), 78 Southern 372.

FURNISHING PROPS—NOTICE OF PLACE OF DELIVERY—PROOF.

The statute of Alabama (General Acts 1911, 500, p. 514) requires a mine operator to keep ready for use a sufficient supply of props and other timbers of suitable lengths and sizes to be used in the mine. The statute makes it the duty of the miner who needs props or other timbers to select and mark the same when needed, designating on such props or timbers the place at which the same are to be delivered, or give notice to the person whose duty it is to deliver the props of the kind and number of props needed and of the place at which they are to be delivered. In an action by a miner for damages for injuries occasioned by the alleged' failure of the mine operator to furnish props it was alleged that the miner designated on a blackboard used for that purpose the number and kind of props and timbers needed in his working place and designated the place at which they were to be delivered; but there can be no recovery under such an allegation where the evidence failed to show that notice was given to the mine operator of the place at which the props and timbers to be used in the miner's working place should be delivered.

Clark v. Choctaw Min. Co. (Alabama), 78 Southern 872, p. 873.

AGREEMENT AND CUSTOMS—DUTIES RELATING TO PROPPING.

The former statute of Kentucky did not in terms impose upon either the mine operator or miner himself the duty of propping, and whether the duty was imposed upon the one or the other could be shown by the agreement of the parties or the custom of the mine. But the rules of a mine are not conclusive evidence of the terms of a contract of employment or the custom of a mine, as these may be waived.

Borderland Coal Co. v. Kirk (Kentucky), 208 Southwestern 534, p. 535.

VIOLATION BY OPERATOR.

FAILURE TO EMPLOY MINE FOREMAN—FOREMAN FOR SEPARATE MINES.

The placing of a single mine foreman and his assistants in charge of several separate mines without underground connection, the mines being separate and distinct working places, is not a compliance with the terms of the Pennsylvania statute (P. L. 1891, p. 176). This statute contemplates a foreman in each mine, including all underground workings and excavations connected below the surface and operated by one general system of ventilation and haulage, but it does not include distinct and separate systems.

Kolalsky v. Delaware & Hudson Co. (Pennsylvania), 103 Atlantic 721, p. 723.

VIOLATION OF REQUIREMENT—QUESTION OF FACT.

The statute of Pennsylvania (Laws 1893, p. 52 (78)) requires the mine owner to see that entries where tracks are laid and braking is necessary shall have a clear level width of not less than $2\frac{1}{4}$ feet between the sides of the cars and the rib to allow the driver or brakeman to pass his trip safely and keep clear of the cars. The main hauling tracks and the switch in an entry were laid about 5 feet apart; between the track and the switch there was set by order of the mine foreman a row of posts necessary to support the roof. These posts left a space of not more than 3 or 4 inches between them and the cars standing or running on the switch, but leaving a clear space of something more than 2½ feet between the main track and the line of posts. The posts were from 13 to 15 feet apart. A brakeman operating the brakes on the cars on the switch was caught between a car and one of the posts and injured. In an action by the brakeman for damages for the injuries so occasioned, the court held that in view of the doubt whether the statute positively required a space of 2½ feet between the cars on the siding or switch and the posts, as well as 2½ feet between the posts and the cars on the main track, the question of the negligence of the mine operator in permitting the setting of the posts and in operating the cars on the siding with a space of not to exceed 4 inches between the cars and the posts was one of fact to be determined by the jury, and was not under the statute a question of law for the court.

Whittaker v. Valley Camp Coal Co. (Pennsylvania), 103 Atlantic 594, p. 595.

FAILURE TO FURNISH SAFE PLACE—LIABILITY. .

The complaint of a miner for injuries caused by a fall of the roof alleged a defect in the ways, works and plant of the mine operator in that the roof of that part of the mine where the complainant was set to work was cracked, and had not sufficient strength or cohesive power to hold itself together and was not properly supported and that the defect arose from or had not been discovered or remedied owing to the negligence of the mine operator or his superintendent. The proofs showed that the miner on entering his working place sounded the roof and found that it sounded all right; that he examined the rock in the roof to see whether it was dangerous and whether it was safe for him to work and made what he considered a careful examination and thought it was safe enough to work under. The miner himself was experienced and was as competent to sound roofs and tell when a roof is safe as any other person. Under such circumstances the mine operator was not liable for negligence in failing to inspect or discover the defect.

Red Eagle Coal Co. v. Thrasher (Alabama), 78 Southern 718.

STRUCTURES ERECTED BY MINE FOREMAN—DUTY AND LIABILITY OF OPERATOR.

The statute of Pennsylvania imposes upon a mine owner or operator a specific duty as to the arrangement and management of his mine, and the owner or operator can not protect himself from liability by showing that the structure in question was put up by the direction of the mine foreman, at least where the fact of its existence became known to the owner, and if dangerous or if it rendered the mine an unsafe place for miners to work, it would be his duty to remove such structure.

Whittaker v. Valley Camp Coal Co. (Pennsylvania), 103 Atlantic 594, p. 595.

DUTIES IMPOSED ON MINE FOREMAN.

DUTY AND LIABILITY OF MINE OPERATOR.

Where the statute of Pennsylvania does not call for any particular arrangement or structure as to the method of underground operations in a mine, the arrangement or structures ordered by the mine foreman for the purpose of supporting thereof are within the statutory powers of the mine foreman and over which the mine operator has no authority or jurisdiction and the operator can not be held liable for an injury to a miner caused by defective or unsafe structures or posts placed or set by the direction of the mine foreman.

Whittaker v. Valley Camp Coal Co. (Pennsylvania), 103 Atlantic 594, p. 595.

CONSTRUCTION OF ENTRIES—AUTHORITY OF MINE FOREMAN—KNOWLEDGE OF DANGER.

Under the statute of Pennsylvania, the mine operator is not responsible for the work done in the course of the construction of passageways and the mine operator can not be held liable for an injury to a miner caused by a fall of rock from the roof of a new and incomplete entry where the underground operations were under the control of a statutory mine foreman and where the mine foreman had directed the miners to pass through the new entry in going to and from their work, and where the mine foreman knew that the roof was dangerous and where it did not appear that either the mine operator or the mine superintendent had any knowledge of the new entry or that it was being used by the miners as a passageway and where the use by the miners of the new entry as a passageway prior to the accident was of such short duration that knowledge of its condition could not be imputed to the mine operator or his superintendent.

Cossette v. Paulton Coal Min. Co. (Pennsylvania), 103 Atlantic 346, p. 347.

CONSTRUCTION OF PASSAGEWAYS.

Under the general provisions of the Pennsylvania statute (P. L. 1911, p. 756), placing the workings of a mine under the mine foreman's charge and supervision, the mine foreman is responsible for all work in the course of construction of passageways, and a mine operator can not be held liable in damages for injuries to a miner caused by a fall of rock from the roof of an entry that was in the course of construction and incomplete and where the mine foreman permitted the injured miner and others to pass through such incomplete entry.

Cossette v. Paulton Coal Min. Co. (Pennsylvania), 103 Atlantic 346, p. 347.

FAILURE TO PERFORM—LIABILITY OF OPERATOR—FAILURE TO MAKE REFUGE HOLES.

The duty to see that refuge holes along motor roads in coal mines are maintained as required by the statute of West Virginia (Code 1916, ch. 15H, sec. 36d (2)) rests on the mine foreman and not on the mine operator, and for an injury to a miner resulting from defects therein the mine operator is not liable.

Kinder v. Boomer Coal & Coke Co. (West Virginia), 95 Southeastern 580, p. 584.

Strother v. United States Coal & Coke Co. (West Virginia), 95 Southeastern 806.

FAILURE TO MAKE OVERCAST—LIABILITY OF MINE OPERATOR.

The statute of West Virginia (Barnes' Code 1916, ch. 15H, sec. 63d (2)) requires the mine foreman to make or direct excavations for overcasts in the roofs of haulways or air courses in coal mines as the operations therein progress, for properly ventilating the mine and make the places to work therein safe. His negligence in the performance of these duties, unless he is authorized by the master to represent him in other ways incompatible with his statutory duties, can not be imputed to the mine owner or operator and render him liable for prsonal injuries sustained that are due to such negligence.

Strother v. United States Coal & Coke Co. (West Virginia), 95 Southeastern 806.

PRESUMPTION AS AGENT OF MINE OPERATOR.

The mere fact that a person was a mine foreman employed in a mine, without more, does npt authorize any one to assume that he represents the mine operator in any other capacity than as a statutory officer. It seems that the presumption should be that the mine foreman has no authority to act for the mine operator, especially where his duties as a mine foreman are incompatible with his duties as a representative of the mine operator.

Strother v. United States Coal & Coke Co. (West Virginia), 95 Southeastern 806, p. 808.

PROOF OF RELATION—FOREMAN OR SUPERINTENDENT—QUESTION OF FACT.

The Pennsylvania statute (P. L. 1891, p. 176) defines a superintendent as a person who shall have on behalf of the owner general supervision of one or more mines or collieries. The fact that an employee had authority to hire and discharge men and direct the manner in which entries should be driven and the mine developed was not inconsistent with his position as a mine foreman, and the fact that he was called superintendent was not sufficient to make him one, but his position as either superintendent or mine foreman was one of fact.

Kolalski v. Delaware & Hudson Co. (Pennsylvania), 103 Atlantic 721, p. 722.

VIOLATION BY MINE FOREMAN.

FAILURE TO PERFORM STATUTORY DUTY—LIABILITY OF MINE OPERATOR.

If a mine operator has employed a competent and qualified mine foreman, as required by the statute of West Virginia, the negligence of the mine foreman in the discharge of the duties imposed upon him by statute in the operation of the mine is not chargeable to the mine operator.

Kinder v. Boomer Coal & Coke Co. (West Virginia), 95 Southeastern 580, p. 584.

See Williams v. Thacker Coal & Coke Co., 44 West Virginia 599, 30 Southeastern, 107, 40 L. R. A. 812.

Jaggie v. Davis Collieries Co., 75 West Virginia 370, 84 Southeastern 941.

FAILURE TO PROP ROOF—OPERATOR LIABLE.

A mine superintendent marked out a space 15 feet wide and directed a miner to excavate to the depth of 30 feet for the purpose of installing a pump to remove water that had accumulated in the mine. The miner called the superintendent's attention to a rock projecting some distance into or over the space so marked. The superintendent after striking the rock with a pick informed the miner that it was "all right," and directed him to "go to work." The miner subsequently requested props for the rock and the superintendent promised to provide the same. After an absence the miner returned to know if the props had been supplied and as he stepped to one side of the track to permit a car to pass the rock in question fell inflicting the injuries for which he sued.' Under such circumstances the mine owner's duty is governed by the statute (P. L. 1891, p. 176), which requires a mine owner, superintendent, or foreman to provide props necessary for the safe mining of coal and protection of the lives of the workmen, and makes the failure to do so negligence per se in an action for damages for injuries due to insufficient propping.

Kolalsky v. Delakare & Hudson Co. (Pennsylvania), 103 Atlantic 721, p. 723.

DIRECTING MINERS TO USE DANGEROUS ENTRY—LIABILITY OF OPERATOR.

A mine foreman in charge of a mine under the statute of Pennsylvania (P. L. 1911, pp. 756, 762) knew of an unfinished entry and the defective condition of the roof and that miners were passing through such defective entry to and from their work, but posted no danger signals and nothing to guard and protect the miners against or from the existing danger, but permitted them, and even suggested to them that they might pass through such incomplete entry. A miner while passing through the entry was injured by a fall of slate from the roof. Under such circumstances the mine operator can not be charged with negligence and held liable for the injury, but the injury must be attributed to the negligence of the mine foreman who was, by the statute, given full charge of the inside workings.

Cossette v. Paulton Coal Min. Co. (Pennsylvania), 108 Atlantic 346, p. 347.

DUTIES IMPOSED ON MINER.

REQUEST FOR PROPS—STATUTORY METHOD FOR MAKING.

The statute of Kentucky (sec. 2726, subsec. 5) provides that miners in need of props and timbers shall notify the mine foreman at least one day in advance, giving the number, size, and length of props, cap pieces, and timbers required. But in case of emergency props may be ordered immediately upon the discovery of danger. Where the duty of propping devolves upon a miner he must request props in the manner pointed out by the statute, and the statutory requirement can not be changed by the custom of a mine so as to impose upon the mine operator a liability for failure to furnish props when the statute itself was not complied with.

Borderland Coal Co. v. Kirk (Kentucky), 203 Southwestern 534.

REQUEST FOR PROPS—DEMAND ON INDEPENDENT CONTRACTOR.

An independent contractor had six or seven rooms in a mine which he was having mined under his contract with the mine operator. One of the miners employed by such independent contractor placed upon the blackboard opposite the contractor's name a request for props and timbers, but did not designate the place at which the props and timbers were to be delivered. In an action by the miner for damages for injuries caused by the failure of the mine operator to supply the props and timbers, there was no evidence that the independent contractor was the representative of the mine operator to accept or receive the notice required by the statute. Proof that it was the contractor's duty to set the timbers up did not serve to prove compliance with the prerequisite of the statute requiring the mine operator to furnish the props and there could be no recovery against the mine operator in any event where the miner fails to designate the place at which the props and timbers were to be delivered.

Clark v. Choctaw Min. Co. (Alabama), 78 Southern 372, p. 373.

RIGHT TO STATUTORY PROTECTION.

APPLICATION OF STATUTE TO EMPLOYEE OF INDEPENDENT CONTRACTOR.

The statute of Alabama (General Acts, 1911, 500, p. 514) requires a mine operator to keep at hand a sufficient supply of props and other timbers and supply the same to a miner on request. Whether this statute and the duty imposed has application otherwise than as between the master and servant, the mine operator and his miner only, is a question, the Alabama court says, of doubtful solution.

Clark v. Choctaw Min. Co. (Alabama), 78 Southern 372.

FELLOW SERVANT RULE—DEFENSE ABROGATED.

ASSUMPTION OF RISK ELIMINATED.

The Indiana employers' liability act of 1911 (Burns 1914 Stats. sec. 8020) eliminates the defense of dangers or hazards inherent or apparent in the employment and abrogates the assumed risk rule as to the particular risk of a fellow servant.

Oolitic Stone Mills Co. v. Cain (Indiana Appeals), 119 Northeastern 1005, p. 1007.

EFFECT ON ASSUMPTION OF RISK.

KNOWLEDGE OF DANGER—ASSURANCE OF SAFETY—CONTINUING WORK.

Where a miner requested a timberman to prop his roof and where the mine foreman had directed the timberman to prop the roof, and the timberman after examining the roof assured the miner of its safety, the miner by remaining in the place and continuing work does not assume the risk of the danger from the roof unless the danger was so obvious that a person of ordinary prudence in his situation would have refused to do so.

Borderland Coal Co. v. Kirk (Kentucky), 203 Southwestern 534, p. 536.

EMPLOYERS' LIABILITY ACT.

ALABAMA EMPLOYERS' LIABILITY ACT—PURPOSE AND CONSTRUCTION— COMPETENCY OF WITNESS.

The purpose of the Alabama employers' liability act (Code 1907, sec. 3910) was to repeal the general inhibition against persons testifying who were interested in the issue to be tried except in transactions between the estate of a deceased person and a person whose testimony was intended to be used. But the purpose was not to protect a defendant, a mine operator, charged with negligence, from the testimony of a witness, because the other party interested in the act in controversy was dead. Under the terms of the statute the exception is confined to inhibiting the testimony where the transaction is with "the deceased person whose estate is interested in the result of the suit." But an action for the wrongful death of a miner is not for the benefit of his estate but for the next of kin under the statute; and a mine superintendent, who negligently directed the miner to work in a dangerous place, is a competent witness to testify as to the alleged dangerous condition surrounding the place where the miner was so directed to work.

Central Iron & Coal Co. v. Hammacher, 248 Federal 50, p. 52.

ACCEPTANCE OF PROVISIONS BY EMPLOYER—NOTICE TO EMPLOYEE.

The employers' liability act of Texas permits the employer to accept the provisions of the act and to take a policy in the Texas Employers' Insurance Association and thereby become a "subscriber" according to the provisions of the act; but the act requires the "subscriber" to give written or printed notice to the employee of the fact in order to avoid liability in a civil suit for damages. The provision of the act requiring such notice is absolutely necessary in order to create the relationship of subscriber between the employer and employee and without such notice the relation can not exist and the employer may be liable in a civil suit for damages.

Farmers' Petroleum Co. v. Shelton (Texas), 202 Southwestern 194, p. 109.

INDUSTRIAL INSURANCE LAW.

INDEMNITY INSURANCE—MINING COMPANY'S CONTRACT OF INDEMNITY—VALIDITY.

By an indemnity insurace policy a mining company as lessee of a mine agreed to indemnify the beneficiaries of the insurance policy against all cost, loss, and damages by reason of legal liability of the assured for and on account of bodily injury or death suffered through an accident by any employee of the assured while working for him in his mine. Such a contract is valid and binding on the mining company, although the injuries to the employees were caused by the negligence of the assured.

Harden v. Southern Surety Co. (Missouri Appeals), 204 Southwestern 84, p. 85.

WORKMEN'S COMPENSATION ACT.

ALASKA WORKMEN'S COMPENSATION ACT—VALIDITY—PROVISIONS FOR INDUSTRIAL INSURANCE.

The workmen's compensation act of Alaska does not contain any provision for industrial insurance, but it does contain regulations for securing payment for the compensation for injuries. A bond or cash deposit by a mining company is provided for where beneficiaries of the deceased persons are concerned and out of which to meet the compensation to which they are entitled; and in an action for the stipulated compensation the employee has his attachment for securing the demand and it can not therefore be said that the employee is without provision looking to the eventual payment of his claim. The district court is constituted a tribunal for ascertaining the legitimacy of the claims and the amount, and the fact that the insurance feature is wanting does not render the law invalid.

Johnston v. Connecott Copper Corp., 248 Federal 407, p. 410.

WORKMEN'S COMPENSATION ACT OF ALASKA—CONSTITUTIONALITY—CLASS LEGISLATION.

Classification of subjects for regulation by law is a function belonging to the legislative department of the Government. Generally, class legislation is prohibited, but legislation which is limited in its application, if within the sphere of its operation it affects alike all persons similarly situated is not within the prohibition. The legislature possesses a wide scope of discretion in the exercise of its functions of classification, and such legislation can be condemned as vicious only when it is without any reasonable basis and therefore purely arbitrary; but when legislative classification is called in question if any state of facts can be reasonably conceived that would sustain the law the existence of that state of facts at the time it was enacted must be assumed. Mining is the one great industry of Alaska and is attended by many hazards and complexities and it is not strange that the legislature should make of the single industry a single classification for adjustment of workmen's compensation. Such a classification does not render the act unconstitutional, and the fact that "mining operations" include all work performed on or for the benefit of any mine or mining claim and may include persons but remotely connected with the working of mines is within the legislative discretion and can not affect the validity of the act.

Johnston v. Connecott Copper Corp., 248 Fed. 407, p. 418.

WORKMEN'S COMPENSATION ACT OF ALASKA—VALIDITY—APPLICATION TO MINORS.

The workmen's compensation act of Alaska is not invalid because it makes no provision respecting workmen under the age of majority for accepting or rejecting the provisions of the act. The legislature may have assumed that a minor having the capacity to contract or to be contracted with has the capacity to reject or waive the provisions of the act. But minors are not denied the interposition of a guardian or next friend in doing the act for them.

Johnston v. Connecott Copper Corp., 248 Fed. 407, p. 414.

BENEFICIAL CONSTRUCTION ADOPTED.

In construing and applying the provisions of the Utah workmen's compensation act a court must keep in mind the purpose the legislature had in view in adopting the act. If the act is susceptible to two constructions, one of which in a large measure would make it useless and of no material benefit and the other construction would manifestly make it effective and beneficial and would subserve the public welfare, a court should adopt the latter construction if it can be done according to sound principles of law and rules of construction.

Industrial Commission v. Daly Min. Co. (Utah), 172 Pacific 301, p. 306.

WORKMEN'S COMPENSATION ACT OF ALASKA—SUPERBURDEN ON MINER.

The fact that the workmen's compensation act of Alaska makes it more difficult for the workmen or miners to make their election to accept the provisions and to waive them when the election is once made than for the employer, and that it is burdensome for the miners to pay the expense pertaining to verification and recording, is only a minor inequality, if any, and is without the indicia of arbitrary discrimination.

Johnston v. Connecott Copper Co., 248 Federal 407, p. 414.

COMPULSORY CHARACTER OF ACT.

The workmen's compensation act of Utah (laws 1917, ch. 100) is compulsory in its provisions and a mining corporation may be required, by mandamus proceedings instituted in the supreme court by the industrial commission, to conform to and comply with an order of the industrial commission requiring it to furnish security to the satisfaction of the commission where the mining company elected to pay the compensation direct to its employees.

Industrial Commission v. Daly Min. Co. (Utah), 172 Pacific 301, p. 306.

ELECTIVE FEATURE.

The provisions of the workmen's compensation act of Utah (laws 1913, ch. 100) can not be construed to be elective merely upon the part of a mining corporation if such a construction would practically nullify the provisions of the act, and where a different construction would render it effective.

Industrial Commission v. Daly Min. Co. (Utah), 172 Pacific 301, p. 306.

WORKMEN'S COMPENSATION ACT OF ALASKA—INSURANCE—METHOD OF
SECURING FUND.

The insurance feature of a workmen's compensation act is designed
to afford an employee adequate security for his compensation, but in
the workmen's compensation act of Alaska another scheme is evolved
intended to accomplish the same purpose. The particular method
for accomplishing the' purpose is mainly one of legislative choice,
and so long as a method selected is reasonably adapted for the purpose and is not arbitrary and without proper regard to cause and
effect it is beyond the scope of judicial function to disturb the choice.
The workmen's compensation act of Alaska of 1915 is reasonably
adapted to secure to an employee the compensation provided for in
the act and a court can not say that the legislation is arbitrary and
not based upon sufficient reason for its adoption.

Johston v. Connecott Copper Corp., 248 Federal 407, p. 413.

SECURING COMPENSATION TO EMPLOYEES.

The workmen's compensation act of Utah provides that mine operators, as one class of employers, shall secure compensation to their
employes either (1) by insuring payment of compensation in the
State insurance fund, (2) by insuring with some company engaged
in indemnity insurance, or (3) by proof of financial ability to pay
compensation direct to their employees. But if a mine operator elects
to pay direct without obtaining insurance, the industrial commission
may nevertheless require security in the manner provided.

Industrial Commission v. Daly Min. Co. (Utah), 172 Pacific 301, p. 304.

INJURY ARISING OUT OF EMPLOYMENT.

The death of a miner was caused by the bursting of the aorta,
which was in a diseased condition. Where the death was not caused
solely by the disease progressing naturally, but was due to the strain
of the work concurring with the disease, then it may be said that the
miner suffered an injury by an accident arising out of his employment, within the meaning of the Indiana workmen's compensation act.

Indian Creek Coal & Min. Co. v. Calvert (Indiana Appeals), 119 Northeastern
519, p. 521.

BASIS OF COMPENSATION—CHANGE FROM REGULAR TO TEMPORARY WORK.

Under the workmen's compensation act of Kansas an employee engaged for a specified employment at a fixed amount is entitled to compensation in case of injury, based upon the earnings of himself and

other persons in that grade of service, though the injury was received while he was working for less wages in a different grade to which he had been temporarily assigned for a short time by reason of a brief cessation of the work for which he was employed.

Bundy v. Petroleum Products Co. (Kansas), 172 Pacific 1020, p. 1021.

CLAIM FOR COMPENSATION—WHAT CONSTITUTES—FILING.

The workmen's compensation act of Michigan requires that a claim for compensation must be made within six months of the time of the injury. But a mere statement on the part of an injured miner to the effect "that I would have to make a claim for compensation if I didn't get along better," is not sufficient to constitute even an oral claim; and a subsequent written claim not made within six months can not be considered.

Basse v. Banner Coal Co. (Michigan), 167 Northwestern 954, p. 955.

DEATH DUE TO DISEASE—LIABILITY.

Where an employee afflicted with a disease received a personal injury under such circumstances as that he might have applied to the workmen's compensation act for relief on account of the injury if there had been no disease, but where the disease as it in fact existed was by the injury materially aggravated or accelerated resulting in disability or death earlier than would have otherwise occurred and the disability or death did not result from the disease alone progressing naturally, but the injury aggravating or accelerating its progress materially contributed to hasten its culmination in disability or death, then there may be an award under the workmen's compensation act of Indiana (Acts 1915, p. 392).

Indian Creek Coal & Min. Co. v. Calvert (Indiana Appeals), 119 Northeastern 519, p. 521.

PERSONAL INJURY.

A "personal injury," as the term is used in the Indiana workmen's compensation act (Acts 1915, p. 392), has reference not to some break in some part of the body or some wound thereon or the like but rather to the consequences or disability that results therefrom.

Indian Creek Coal & Min. Co. v. Calvert (Indiana Appeals), 119 Northeastern 519, p. 525.

ACCIDENT—MEANING.

The word "accident," as it occurs in the Indiana workmen's compensation act (Acts 1915, p. 392), is used in its popular sense and means any unlooked for mishap or untoward event not expected or designed. The meaning assigned to the word as used in compensation acts must be distinguished from its meaning as used in accident insurance policies.

Indian Creek Coal & Min. Co. v. Calvert (Indiana Appeals), 119 Northeastern 519, p. 521.

EXTRAHAZARDOUS BUSINESS.

A retail coal dealer delivering coal with his own teams from his bins, or when out of coal at his yards, obtaining and delivering coal from local mines direct to his customers, is not engaged as a common carrier of persons or property or a private carrier for hire; and the business or occupation as a retail coal dealer does not come within any of the various businesses, enterprises, or occupations classified in the Illinois workmen's compensation act (Hurd's Rev. Stats. 1917, ch. 48, sec. 129) as extrahazardous.

Fruit v. Industrial Board, etc. (Illinois), 119 Northeastern 931, p. 932.

RETAIL COAL DEALER—CARRIAGE BY LAND.

A retail coal dealer hauling coal from his own bins to his customers and when out of coal at his coal yards sending his teams to the local mines and having his deliveries of coal made from the mines to his customers direct is not engaged in "carriage by land" within the meaning of the Illinois workmen's compensation act (Hurd's Rev. Stats. 1917, ch. 48, sec. 129), and a driver of one of the teams of such dealer injured while hauling material other than coal is not entitled to compensation under the workmen's compensation act, especially where neither the employer nor the injured employe had elected to be bound by the act.

Fruit v. Industrial Board, etc. (Illinois), 119 Northeastern 931, p. 932.

RETAIL COAL DEALER—ELECTION NOT TO COME UNDER ACT.

A retail coal dealer purchased coal by the car load and sold it by the ton both at his bins and by delivering to his customers. When out of coal he would send his teams to the local mines and have his deliveries of coal made direct from the mine to his customers. The dealer did other hauling with his teams. A driver of a team while engaged in hauling material other than coal was injured. The workmen's compensation act of Illinois (Hurd's Rev. Stats. 1917, ch. 48, sec. 128) does not apply because the injured employee was not engaged in any extrahazardous occupation within the meaning of section 3 of that act, and neither the employee nor the coal dealer had filed an election to be bound by the act.

Fruit v. Industrial Board, etc. (Illinois), 119 Northeastern 931, p. 932.

See Sugar Valley Coal Co. v. Drake (Indiana Appeals), 117 Northeastern 937, p. 938.

DEATH OF SON—DEPENDENCY—COST OF BOARD.

In case of the death of a son who had been contributing to the support of his father and mother, in estimating the amount to which the dependents were entitled under the workmen's compensation act of Michigan, where the son lived with the father and mother, the cost of the board and maintenance of the deceased son should be deducted from the amounts that the son contributed to the support of the dependents.

Engberd v. Victoria Copper Min. Co. (Michigan), 167 Northwestern 840, p. 841.

DEPENDENTS—CHANGE OF DEPENDENT'S WORK.

A minor 19 years of age and unmarried was killed while working in a mine in the course of his employment. His father and mother were dependents and claimed compensation under the workmen's compensation law. The proof showed that after the death of the minor the father, because of poor health, changed his work from underground at a wage of $3.25 a day and did surface work at a wage of $2.65 a day. The change of work or compensation by the father after the death of the son was immaterial. Questions as to who constitute dependents and the extent of their dependency are to be determined as of the date of the accident to the employee and their right to any death benefit becomes fixed at such time irrespective of any subsequent change of conditions.

Engberd v. Victoria Copper Min. Co. (Michigan), 167 Northwestern 840, p. 841.

ALLOWANCE FOR HOSPITAL CHARGES.

The Kansas workmen's compensation act (Gen. Stats. 1915, sec. 5906) provides that in fixing the amount of compensation in case of injury, allowances shall be made for any payment or benefit which the injured employe may receive from the employer during the period of incapacity, and authorizes an allowance for reasonable hospital charges necessarily incurred and paid by the employer.

Bundy v. Petroleum Products Co. (Kansas), 172 Pacific 1020, p. 1021.

INDUSTRIAL COMMISSION—AUTHORITY AND DISCRETION.

The workmen's compensation act of Utah (Laws 1917, ch. 100, sec. 53) gives a mining corporation, with other employers, the option to pay compensation direct to its employes, but to obtain this option it must furnish satisfactory proof to the industrial commission of its financial ability to do so. The power is conferred upon the industrial commission to decide whether or not a mining corporation that has elected to pay the compensation direct to its employes is able to do so, and if the commission decides that it is not, it may order the corporation to secure the payment of the compensation as the statute provides, as in its discretion it may do. Under such circumstances insurance is not therefore forced upon the corporation " contrary to law," but the corporation is merely required to do what the law enjoins.   -

Industrial Commission v. Daly Min. Co. (Utah), 172 Pacific 301, p. 306.

INDUSTRIAL COMMISSION MAY REQUIRE SECURITY.

A mining corporation made application to the industrial commission for the privilege of paying compensation to its employees direct and the privilege was granted upon the express condition that the mining company would file with the commission a surety bond or liquid collateral in a stated sum and a formal order to that effect was entered by the commission. Employers are given the right to elect whether they will insure the payment of compensation or furnish proof of their financial ability to make prompt payment; but the legislature intended that all employers shall in advance secure the payment of compensation to which any one of their employees may become entitled. This feature of the workmen's compensation act is compulsory and is valid and enforceable and is not repugnant to any constitutional provision.

Industrial Commission v. Daly Min. Co. (Utah), 172 Pacific 301, p. 304.

INDUSTRIAL COMMISSION—MANDAMUS TO ENFORCE COMPLIANCE WITH STATUTE.

The workmen's compensation act of Utah imposes upon a mining corporation as upon other employers, the duty of complying with its provisions relating to the insuring of the payment of compensation provided for. The industrial commission is empowered to enforce obedience to these provisions and to enforce the payment of the compensation when it becomes payable in accordance with the terms of the statute; but no power is conferred upon the commission to sue to recover the amount employers are required to pay for insurance unless and until they have obligated themselves to pay. Where a mining corporation has elected to pay compensation direct to its employees but has refused to provide security for such payment, the only remedy the industrial commission has is to compel by mandamus the delinquent corporation to comply with the provisions of the statute relating to the insuring of the payment of the compensation provided by the statute. The fact that a mining corporation may be required to pay money for insurance without receiving anything in return is not a defense to an action by the industrial commission to compel it to furnish the security required for the reason that such statutes result in the public welfare and individual employers receive benefits therefrom in that they are saved from the expense of litigation and the beneficial provisions are usually sufficient to induce an employer to avail himself of the provisions of the act.

Industrial Commission v. Daly Min. Co. (Utah), 172 Pacific 301, p. 305.

MINES AND MINING OPERATIONS.

RELATION OF MASTER AND SERVANT.

FAILURE TO PROVE RELATION.

In an action by a miner for damages for injuries received by the failure of a mine operator to furnish props, certain counts proceed on the theory of the existence of the relation of master and servant between the injured miner and the mine operator; but there can be no recovery on such counts where the undisputed evidence disclosed no such relation.

Clark v. Choctaw Min. Co. (Alabama), 78 Southern 372, p. 373.

ACTIONS—PLEADING AND PROOF OF NEGLIGENCE.

PLEADING NEGLIGENCE—SUFFICIENCY OF COMPLAINT.

A complaint in an action by an injured miner against a mine operator for damages for injuries is sufficient where it avers that the complainant was in the defendant's mine as a laborer employed by an independent contractor by and through whom the defendant was operating his mine; that as such laborer the plaintiff was being transported on a tram car operated by the defendant for the transportation of ore and laborers, and that the defendant so negligently conducted itself in and about transporting the plaintiff that the tram car was derailed and the plaintiff was thereby injured.

Sloss Sheffield Steel & Iron Co. v. Crosby (Alabama), 78 Southern 898.

A complaint or declaration is sufficient and states a cause of action against a mine owner where it avers that the plaintiff, a minor, a poor boy of tender years without experience or means of support other than his daily wages as a common laborer, was engaged with his father in digging and loading coal into mine cars to be hauled to the drift mouth and tipple for shipment; that such work was not inherently dangerous or beyond his ability to perform while working with his father; that the coal mine operator through its superintendent and other agents ordered, required and coerced the plaintiff to perform a different and more hazardous work, and that he was injured while so engaged in performing such extrahazardous work.

Kinder v. Boomer Coal & Coke Co. (West Virginia), 95 Southeastern 580, p. 581.

DEFENSE—INSUFFICIENT PLEA.

In an action by a miner for damages for injuries caused by a fall of rock from the roof of the mine a plea is insufficient that averred that the plaintiff was hurt while working under or near a rock or place in the roof not securely propped and that the plaintiff knew the danger thereof and was sent by the mine foreman to remedy the defect and that he continued negligently to work under or dangerously near the rock until it fell and injured him, as it does not aver that it was the plaintiff's duty to remedy the defect or that in compliance with the order he undertook that duty.

Choctaw Coal & Mining Co. v. Dodd (Alabama), 79 Southern 54.

A plea to a complaint in an action for damages for injuries sustained by the plaintiff by a fall of rock from the roof of an entry is insufficient where it failed to aver that an examination of the roof of the mine at the place of the injury would have disclosed the defect as well as the danger of going to work at the place in question.

Choctaw Coal & Mining Co. v. Dodd (Alabama), 79 Southern 54.

In an action by a miner for damages for injuries caused by a fall of rock from the roof a plea is insufficient which avers that the plaintiff was at the time he suffered the injuries complained of the agent of the mine operator and entrusted with the duty of seeing that the roof was in a proper condition and that the plaintiff undertook to perform such duties, is insufficient, as it is no more than a denial of the superintendence of the person entrusted with the superintendence of the mine, and a denial that such superintendent negligently ordered the plaintiff in the entry and in dangerous proximity to the rock which fell and caused his injury.

Choctaw Coal & Mining Co. v. Dodd (Alabama), 79 Southern 54.

STATUTE OF LIMITATIONS—ISSUES INCOMPLETE.

A boy 16 years of age was injured while working in a mine. Within the year after he became 21 years of age he brought an action against the mine operator for damages for the injuries sustained by him during minority. The mine operator pleaded the one year statute of limitations as a bar to the plaintiff's action. In such case the plaintiff must file a special replication, in order to make an issue on the special plea of the statute, as a judgment is erroneous if obtained without such issue; and it is essential in order to render admissible proof of the age of the complainant to remove the statutory bar.

Kinder v. Boomer Coal & Coke Co. (West Virginia), 95 Southeastern 580, p. 584.

PROOF AND PRESUMPTION.

On the trial of an action by an injured miner for damages, the miner, as plaintiff caused to be subpoenaed as a witness a fellow miner who saw the accident and knew of the circumstances of the injury, but the witness was not called upon to testify on behalf of the plaintiff. On the failure of the plaintiff to call the witness the presumption arises that the witness if examined would not support the claim of the party upon whom the duty devolved to prove the facts of the injury.

Kinder v. Boomer Coal & Coke Co. (West Virginia), 95 Southeastern 580, p. 585.

See Producers' Coal Co. v. Mifflin Coal Co. (West Virginia), 95 Southeastern 949, p. 951.

PROOF OF INJURY—INFERENCE DRAWN FROM CIRCUMSTANCES.

It is one thing to say that the presumption of negligence arises from proof of injury, but it is a different matter to say that the jury may not in its consideration of the facts of a case reach the conclusion or draw the inference that an injury was occasioned by some happening which was not one of the ordinary risks of the employment and to conclude further that some negligent act of omission or commission caused the injury complained of.

Yellow Rose Min. Co. v. Strait (Arkansas), 202 Southwestern 691, p. 692.

PROOF AND PRESUMPTION—USE OF PROPER CARE.

Proof of the fact that the thing or instrumentality which caused an injury to an employee was under the management of a company producing commercial asphaltum from crude petroleum and that in the ordinary course of things the accident would not have happened if those having the management of the machinery and appliances had used proper care, is sufficient to justify a verdict in favor of an injured employee.

Hallawell v. Union Oil Co. (California Appeals), 173 Pacific 177, p. 180.

See Graves v. Union Oil Co. (California Appeals), 173 Pacific 618, p. 619.

PROOF—INSTRUCTIONS BY COURT.

In an action for damages for injuries to a miner the liability of the mine operator depended on whether the injuries complained of resulted from a fall of rock from the roof, and where the operator's defense was based on the claim that the injuries complained of were occasioned by the falling of material from the working face of the room and not from the roof, it was proper for the court to instruct the jury that in order to justify a verdict for the plaintiff he must prove by a preponderance of evidence that the injuries complained of were sustained by a fall of rock from the roof of the room as a result of negligence upon the part of the defendant in failing to supply the plaintiff with timbers and that the failure to so supply the timbers was the proximate cause of the injuries.

Victor American Fuel Co. v. Melkusch (New Mexico), 178 Pacific 198, p. 199.

ACCIDENT AS PROOF OF NEGLIGENCE—RES IPSA LOQUITUR.

A collision of a truck or handcar used by a mine operator for carrying express matter to his mine and on which persons were permitted to ride, with a train on the track, is prima facie proof of negligence, and if unexplained is conclusive, and the rule is applied in the case of a train or car operated by a private carrier. A collision in either case is not one of those accidents which is liable to happen if proper care is observed, and the doctrine or maxim res ipsa loquitur must be applied.

Hodge v. Sycamore Coal Co. (West Virginia), 95 Southeastern 808, p. 809.

TRIAL—INSTRUCTION REQUESTED BY COMPLAINANT.

On the trial of an action by a miner for damages for personal injuries occasioned by the alleged negligence of the mine operator, the complainant is only required to present an instruction which covers every element essential to be found in order to make a case for him and permits a recovery under such instruction; and he is not required to go beyond this and to present in his instruction the theory presented in the evidence of the defendant, the mine operator, which, if believed by the jury, would destroy the complainant's case.

Lawbaugh v. McDonnell Min. Co. (Missouri Appeals), 202 Southwestern 617, p. 619.

TRIAL—INSTRUCTION—THEORIES OF CASE.

In an action by a miner for damages for personal injuries it is sufficient if an instruction by the court to the jury trying the case does not include the defendant's theory, but merely puts the case to the jury on the evidence which the plaintiff relied upon for a recovery. Such an instruction does not necessarily exclude the consideration of other facts asserted by the defendant.

Lawbaugh v. McDonnell Min. Co. (Missouri Appeals), 202 Southwestern 617, p. 619.

TRIAL—INCOMPLETE INSTRUCTION—REFUSAL.

A miner brought an action against a mine operator for damages for injuries caused by the alleged negligence of the mine operator in not taking reasonable care to inspect certain appliances, the defective condition of which caused the injury complained of. The mine operator defended the action on the theory that the injury was the result of an unforeseen accident. Under such circumstances the court was justified in refusing an instruction to the effect that if the jury believed from the evidence that the injury to the complainant was due to accident or mischance not reasonably to be foreseen then a verdict should be for the defendant, in view of the fact that the instruction requested failed to define the words "accident" and "mischance."

Lawbaugh v. McDonnell Min. Co. (Missouri Appeals), 202 Southwestern 617, p. 619.

FAILURE TO INSPECT APPLIANCE—QUESTION OF FACT.

Whether or not a mine operator was negligent in failing to inspect the bail of a tub used in hoisting earth in a mine shaft where the tub or bucket was purchased from a reputable dealer is a question of fact to be determined by the jury in an action by an employee for damages for injuries caused by the fall of a tub due to the defective condition of the bail.

Cole v. D. C. & E. Mining Co. (Missouri Appeals), 204 Southwestern 197, p. 200.

MINOR'S IGNORANCE OF DANGER—PLEADING EXCUSED—CHANGE OF WORK.

A complaint in an action by an inexperienced boy working in a coal mine is sufficient where it avers that the coal-mine operator by its superintendent and other managing agents required and coerced the boy to undertake a dangerous work, one for which he did not possess the capacity or experience to perform with safety and that his injury occurred while so performing such work. In such case it is not necessary to aver that the plaintiff obeyed the commands without knowledge of the danger, as the averment of coercion excused the averment of want of knowledge.

Kinder v. Boomer Coal & Coke Co. (West Virginia), 95 Southeastern 580, p. 582.

NEGLIGENCE OF OPERATOR.

WHAT CONSTITUTES—USAGES OF BUSINESS.

The ground of liability under a charge of negligence is not danger but negligence, and the test of negligence in methods, machinery, and appliances is the ordinary usage of the business; and a usage of machinery and appliances that is inherently and necessarily dangerous and practically abandoned because of its danger may be sufficient to establish negligence as against a usage that is general and comparatively free from danger.

Ramage v. Producers' & Refiners' Oil Co. (Pennsylvania), 108 Atlantic 336, p. 337.

DEATH OF MINER—CAUSE UNKNOWN.

There can be no recovery for the death of a miner due to his fall from a bucket or tub while being lowered into a mine on the ground of the alleged negligence of the mine operator caused by leaving timbers in the shaft in such position that the tub or bucket striking against one such timber thereby throwing the deceased out of the bucket, where the plaintiff's evidence left in doubt the question of whether the miner was killed by the fall from the bucket or whether he was struck and killed by the falling of a piece of timber from the top of the shaft and where there was no averment of any negligence in connection with the falling of the piece of timber. Where the evidence did not show whether or not the miner was struck by the timber or whether his death was due to the fall from the bucket and where the evidence failed to show that the fall was due to the bucket striking against the timbers in the shaft, there could be no recovery under the allegations of the complaint.

Kidd v. Coahuila Lead & Zinc Co. (Missouri Appeals), 204 Southwestern 284, p. 285.

DAMAGES BY FIRE—LIABILITY.

The generally recognized and practically admitted danger attending the use of what is known as the hot tube method of ignition in power used in pumping oils in immediate connection with the gaseous and dangerous fluid and the further fact of the occasional and exceptional use of such method, are proper matters for consideration in determining the question of the negligence of an operator in pumping oil by such method.

Ramage v. Producers' & Refiners' Oil Co. (Pennsylvania), 108 Atlantic 336, p. 337.

DAMAGES BY FIRE—PROOF OF NEGLIGENCE.

In an action by a property owner for damages for loss of property occasioned by a fire alleged to have been caused by the negligent operation of an oil pumping station, proof that the oil company operated its pump by the hot tube method of ignition, in connection with the conceded liability to the escape of oil and gas in and around the open flame used in the hot tube, might be sufficient for a reasonable inference by a jury for the origin of the fire, as the fact that oil and gas coming in contact with the open flame would ordinarily produce an explosion and fire is so manifest as not to require proof.

Ramage v. Producers' & Refiners' Oil Co. (Pennsylvania), 108 Atlantic 836, p. 837.

DISCHARGE OF INFLAMMABLE FLUID INTO FURNACE FIRE.

It is negligence on the part of a manager or master mechanic of an oil company engaged in producing commercial asphaltum from crude petroleum to discharge the highly inflammable fluid in a way to come in contact with the intense furnace fires and to spread the fire quickly to the immediate vicinity of an open trap the surface of which carried more or less inflammable material and by reason of which an employee was burned to death.

Hallawell v. Union Oil Co. (California Appeals), 173 Pacific 177, p. 180.
See Graves v. Union Oil Co. (California Appeals), 173 Pacific 618, p. 619.

OIL OPERATIONS—METHODS OF PUMPING.

In an action for damages caused by fire from an oil pumping station the evidence showed that 95 per cent of the pumping powers used throughout the particular oil producing territory were owned by a single transit company and the gas engines and powers used by it in pumping were operated by the magneto instead of the hot tube method by reason of its greater safety and the dangers attending the use of the latter method. Of the remaining 5 per cent of the pumping powers there was a very general use of the magneto for like reasons and because of its attendant dangers the occasional use only of the hot tube method, and such occasional uses were limited to times when the power was not being applied to the propulsion of oil through the lines or when light pressure only was required. These are proper matters for consideration in an action against an oil pumping company for damages caused by a fire that started from the burning oil.

Ramage v. Producers' & Refiners' Oil Co. (Pennsylvania), 108 Atlantic 836, p. 837.

DANGEROUS GRADE OF DYNAMITE.

A mine operator may be liable on the ground of negligence for furnishing a miner a higher and more sensitive grade of dynamite for blasting purposes than he had been accustomed to using and for furnishing him a higher and more sensitive grade than that which he requested, and by reason of the negligence in supplying the higher and more sensitive grade the miner while in the act of properly and carefully tamping the same was injured by an explosion.

Terleski v. Carr Coal Min. Etc. Co. (Kansas), 173 Pacific 8, p. 9.

THREATENED INJURY TO REAL ESTATE—NO RECOVERY FOR ANTICIPATED
INJURY.

An iron mining company prepared its mine for operation by a
system known as open-pit mining. The company stripped and removed the overlaying earth and material, consisting of soft earth,
quicksand and gravel and negligently deposited such material on
ground that was low and spongy and that sloped toward the premises
of the complainant and other residences in a village on the opposite
side of a highway. By reason of the negligent manner of depositing and piling the material a large mass of the dump broke
loose, flowed and moved across the street, caught and removed and
injured the houses and buildings; but the residence of the complainant was not touched or damaged by the flow from the dump.
The condition of the dump continued dangerous from its negligent
construction and the complainant was apprehensive of injury to
his residence and premises and by reason of the situation the market
value of his premises was entirely taken away. Courts can not allow
damages for depreciation in the market value of real estate, due to
the apprehension of future injury by negligence for an act which has
not and may never happen. Damages based upon mere apprehension
of future injury to real property by an act yet to happen is too remote and speculative.

Johnson v. Rouchleau-Ray Iron Land Co. (Minnesota), 168 Northwestern 1.

LIABILITY FOR INJURY TO PERSONS RIDING ON CAR.

A person riding gratuitously on a truck or handcar maintained and
operated by a coal mining company and used to haul its express matter over a spur track of a railroad company leading from the main
line to its coal line with the consent of its general manager having
control of the operations of such car, is not a trespasser or a mere
licensee but is a passenger and the coal company owes him the duty
to use reasonable care for his safety. Such person is entitled to the
protection of a passenger although he has paid no fare.

Hodge v. Sycamore Coal Co. (West Virginia), 95 Southeastern 808, p. 809.

DIRECTING MINER TO WORK OUTSIDE OF EMPLOYMENT—FAILURE TO
WARN.

A miner inexperienced in general mining operations directed by
the superintendent to do work outside of that he is engaged to do, is
not presumed to be aware of its peculiar risks, and if the mine operator or his superintendent does not fully explain the risks to the
miner before putting him at such new work, he may assume that it

has no greater risks than those which attach to his regular work, either in the nature of the work itself or in the habits of fellow servants with whom it brings him into contact.

Kinder v. Boomer Coal & Coke Co. (West Virginia), 95 Southeastern 580, p. 582.

INSTRUCTIONS AS TO NEGLIGENCE.

A miner working in a mine was injured by a truck falling upon him at the bottom of a raise due to the alleged negligence of the mine operator in failing to give the miner sufficient time to properly adjust a rope on the truck in order to raise it. The mine operator sought to show that the plaintiff was negligent in the manner in which he tied the rope and in not stepping out of the way of the truck when he knew it had not been securely tied. On the trial of the case it was not error for the court to instruct the jury to the effect that if they·found certain facts to exist then such facts as a matter of law would constitute negligence and establish the defendant's liability.

Picino v. Utah-Apex Min. Co. (Utah), 173 Pacific 900.

DUTY OF OPERATOR TO FURNISH SAFE PLACE.

LIABILITY FOR FAILURE—INSTRUCTION AS TO DUTY.

In the trial of an action by a miner for damages for injuries caused by the alleged failure of a mine operator to furnish a miner a safe working place the court instructed the jury that it was the operator's duty to use ordinary care in furnishing a miner a reasonably safe place in which to work and to use ordinary care and prudence to prevent injury to the miner while engaged in such work, and a failure to discharge such duty constitutes negligence on the part of the mine operator and would permit the injured miner to recover damages by reason of any injuries resulting therefrom. Such an instruction is not erroneous on the ground that it excludes the doctrine of assumption of risk, where the court gave other instructions on the effect of the assumption of risk, and expressly told the jury that all instructions were to be considered as a whole.

Yellow Rose Min. Co. v. Strait (Arkansas), 202 Southwestern 691, p. 693.

DANGERS FROM FIRE—INSUFFICIENT MEANS OF ESCAPE.

A company engaged in producing commercial asphaltum from crude petroleum is liable for the death of an employee caused by fire, for failing to furnish the employee with a safe working place, where the oil-impregnated soot had been allowed to accumulate on the timbers and sides of the building and the room in which the employee was required to work was crowded with barrels and material and no sufficient means of escape were provided, and the fire was carried rapidly to the employee's working place by reason of the oil-impregnated soot.

Hallawell v. Union Oil Co. (California Appeals), 173 Pacific 177, p. 181.

DUTY TO PROVIDE SAFE APPLIANCES.

DUTY TO FURNISH MATERIALS AND AGENCIES NEEDED FOR REMOVING DEFECTS.

Where danger exists in a mine it is the duty of the operator or of the mine foreman acting for him, to furnish an employee ordered to remedy the defect the materials, agencies, and facilities needful for the remedying or removal of the dangerous defect without unnecessarily imperiling the employee. This primary duty of the operator carries with it a proper and reasonable exercise of the superior judgment of the operator as to the materials and facilities needful and as to the reasonable safety with which the employee, charged with no duty to remedy the defect, may comply with the operator's orders to remedy the same. In such case the employer's default suspends the employee's duty to remedy the defect, though he be ordered to do so by the mine operator. A mine operator can not escape responsibility for his negligent order, or failure of duty, that unnecessarily exposed to peril the employee complying therewith or acting thereon when it was not the employee's primary duty to do the thing commanded and it was perfectly obvious to the employer that to comply with the command was dangerous.

Choctaw Coal & Mining Co. v. Dodd (Alabama), 79 Southern 54, p. 55.

PURCHASE FROM DEALER—HAZARDOUS USE—CARE REQUIRED.

The trend of recent authorities is away from the old rule that where a master purchased a standard appliance from a reputable dealer he has discharged his duty in the first instance. Regard should always be had for the character of the work and its hazards and no invariable just rule can be conceived or stated. In the purchase of simple tools and appliances to be used in nonhazardous occupations the master has met the requirements of ordinary care where he has supplied the servant with an instrument and appliance of a reputable dealer or manufacturer, but the modern tendency seems to be in favor of holding that the mere purchase of an instrument or appliance from a reputable dealer is not alone sufficient to excuse the employer where the appliance is to be used in a hazardous occupation such as a bucket for hoisting dirt in lots of three-quarters of a ton up a shaft over the heads of the employees.

Cole v. D. C. & E. Mining Co. (Missouri Appeals), 204 Southwestern 197, p. 199.

HOIST—DEFECTIVE CLUTCH BOLT.

Under the rule that it is the duty of a mine operator or employer to furnish his employees with safe appliances it was the duty of the mine operator to tighten the screw or clutch bolt for properly adjusting the clutch band to a drum so that the hoist for lowering and hoisting men in the mine could be safely operated. The failure to readjust the clutch bolt before attempting to use the hoist for miners descending into the shaft may be regarded as the proximate cause of the death of a miner caused by the too rapid descent of the bucket due to the defective conditon of the screw or clutch bolt.

Consolidated Interstate-Callahan Min. Co. v. Witkouski, 249 Federal 833, p. 837.

REPAIRS—NONDELEGABLE DUTY.

It is the duty of an employer, including a coal mine operator, to furnish sufficient and safe materials, machinery, or other means by which service is to be performed and to keep them in repair and order. This duty can not be delegated to a servant or other person so as to exempt the employer from liability for injuries caused to another employee by its omission.

Consolidated Interstate-Callahan Min. Co. v. Witkouski, 249 Federal 833, p. 836.

DUTY TO INSPECT.

INSPECTION—QUESTION OF FACT.

In an action by a miner for damages for injuries caused by the giving way of certain defective clamps alleged to be one of the efficient causes of the injury, the defective condition of which was due to the alleged failure of the mine operator to make proper inspection, it was for the jury to say whether under the facts the injury would have occurred had a reasonable inspection been made of the condition of the clamps.

Lawbaugh v. McDonnell Min. Co. (Missouri Appeals), 202 Southwestern 617, p. 619.

TUB FOR HOISTING—PURCHASE FROM DEALER—INSPECTION.

A mine operator in the construction of a shaft used iron tubs with iron bails for the purpose of raising the earth and materials in the mine shaft. The tubs were designed to hold 1,650 pounds of earth material. The operator purchased a new tub from a reputable dealer and the bail of the tub after some two months' use broke while being raised causing the tub to fall and injure a miner. The operator was not justified in relying on the safety of the tub and the bail because he purchased it from a reputable dealer, but he owed his employees the duty of inspecting the tub and the bail where the tub was to be used in hazardous work like lifting large loads of earth material in the mine shaft over the heads of the employees.

Cole v. D. C. & E. Mining Co. (Missouri Appeals), 204 Southwestern 197, p. 199.

DUTY TO WARN OR INSTRUCT.

QUESTION OF FACT.

Whether an employee, either an adult or a minor, is within the classification which imposes upon a mine operator the duty and obligation to warn, usually is a question of fact for the jury to solve.

Kinder v. Boomer Coal & Coke Co. (West Virginia), 95 Southeastern 580, p. 582.

CHANGE OF EMPLOYMENT—INCREASED DANGER.

Where an inexperienced boy working in a mine was ordered and required to perform new work more hazardous, and had not the capacity to discern, appreciate, and apprehend the increased peril, the duty devolved upon the mine operator to warn and instruct him, and failing to do so the operator must respond in damages for injuries sustained by the boy while engaged in the new employment.

Kinder v. Boomer Coal & Coke Co. (West Virginia), 95 Southeastern 580, p. 582.

DELEGATION OF DUTY.

DUTY NONDELEGABLE—PRESENCE OF EMPLOYER.

Where a coal mine operator delegates to an agent or employee the performance of duties which the operator is legally bound to perform for the protection of his employees, the agent occupies the place of the employer by representation and the latter is deemed to be present and liable for the manner in which the duties are performed.

Kinder v. Boomer Coal & Coke Co. (West Virginia), 95 Southeastern 580, p. 581.

CHANGE OF EMPLOYMENT—DUTY NOT TO COERCE.

Where a superior employee has authority to change an inferior employee from one form of employment to another, the duty not to coerce is nondelegable and the delinquent superior servant will be held to be a vice principal. But the employer's nondelegable duties must be limited to those to whom he entrusts authority to change the employment of other servants.

Kinder v. Boomer Coal & Coke Co. (West Virginia), 95 Southeastern 580, p. 584.

OPERATOR'S ASSURANCE OF SAFETY.

WHAT CONSTITUTES ASSURANCE.

A timberman appointed by a mine operator to do the propping in the mine for the miners was requested by a miner and directed by the assistant mine foreman to prop the roof where the miner was at work. In response he went to the working place of the miner for that purpose and after examining the roof and upon leaving the working place he said, "All right!" "So long." After the timberman left the roof fell and killed the miner. In an action for damages for the death evidence was introduced to show that the words "all right" mean, in mining parlance, that everything was all right. Considering the circumstances under which the language was used it was a question for the jury whether it amounted to an assurance of safety and if it was an assurance of safety the miner had the right to continue his work and the mine operator would be liable in damages for his death.

Borderland Coal Co. v. Kirk (Kentucky), 208 Southwestern 534, p. 536.

DANGER NOT OBVIOUS AND IMMINENT—RELIANCE BY MINER.

A miner was directed by the superintendent of a mine to work under an overhanging rock that appeared to be dangerous. The superintendent after testing the rock with a pick directed the miner to proceed with the work and assured him that it was "all right." In the absence of evidence tending to show that the danger was obvious and imminent the miner was justified in relying upon the superior judgment of the superintendent and upon his assurance that the place was safe.

Kolalsky v. Delaware & Hudson Co. (Pennsylvania), 108 Atlantic 721, p. 723.

TIMBERMAN—ASSURANCE OF SAFETY.

A timberman to whom a coal mine operator has delegated the duty of propping is a vice principal and his assurance of safety to a miner is in effect an assurance by the mine operator. But it is not necessary that a mine operator or his timberman shall assure a miner in terms that the place is safe. It is sufficient if under the circumstances the acts and words of the mine operator, or his representative, are in effect an assurance of safety.

Borderland Coal Co. v. Kirk (Kentucky), 208 Southwestern 534, p. 536.

CONTRACTS RELATING TO OPERATIONS.

AGREEMENT TO OPERATE CONTINUOUSLY—BREACH—FORFEITURE.

A contract for the operation of a mine contained a covenant to the effect that the contractee or his assignee would operate the mine continuously, keeping the machinery in as good working order as when he took possession, less the natural wear and tear of the same, and on failure to operate the mine for 60 days without the written consent of the owner, the contract is to become null and void. The contractee did not operate the mine continuously and some five months after the date of the contract the mill used in operating the mine was destroyed by a flood and no effort was made on the part of the contractee to rebuild the mill, or to equip the mine with the necessary machinery to operate it. Neither the contractee nor his assignee was financially able to rebuild the mill, supply the machinery and operate the mine. In order to avoid a forfeiture of the contract the burden was on the contractee or his assignee to show that he intended to rebuild the mill and could have restored the status quo of the property within a reasonable time after its destruction by the flood and on failure so to do a forfeiture may be declared.

Bradley v. Holliman (Arkansas), 202 Southwestern 469, p. 471.

MODIFICATION AS A NEW CONTRACT.

The modification of a written agreement for a mining lease in so material a matter as fixing a different rate of rent or royalty is in effect the making of a new contract.

Last Chance Min. Co. v. Tuckahoe Min. Co. (Missouri Appeals), 202 Southwestern 287, p. 288.

MODIFICATION OF CONTRACT—STATUTE OF FRAUDS—SPECIFIC PERFORMANCE.

The parties to a written contract and agreement providing for the execution of a mining lease for certain mineral lands, subsequently made a verbal modification fixing a different rate of rent or royalty to be stated in the lease when executed. When a contract is one that must be in writing in order to be valid under the statute of frauds then a modification of the contract must be also evidenced by writing or it is void, and specific performance can not be enforced of such a contract with the oral modification.

Last Chance Min. Co. v. Tuckahoe Min. Co. (Missouri Appeals), 202 Southwestern 287, p. 288.

SPECIFIC PERFORMANCE—STATUTE OF FRAUDS—PART PERFORMANCE.

A contract for mining operations provided for the payment of royalty at 20 per cent of the proceeds of the operations. Conditions were encountered that rendered the operations unprofitable on the basis of the royalty named and mining operations could not be continued without additional investment and the contractee proposed to cease operations and abandon the contract. The contractor's tenure of the land depended upon the continuation of the mining operations. It was thereupon orally agreed to change the royalty from 20 to 15 per cent, and under this modification the contractee continued to mine, sunk shafts to lower levels, drained the land, and thereby discovered and opened up a rich run of ore which was mined for more than six months and the contractor accepted royalty at the modified price of 15 per cent. This was such part performance as to take the contract out of the statute of frauds and permit its specific enforcement by the contractee.

Last Chance Min. Co. v. Tuckahoe Min. Co. (Missouri Appeals), 202 Southwestern 287, p. 289.

METHOD OF OPERATING.

OBSTRUCTION OF HIGHWAY—RIGHT OF INDIVIDUAL TO SUE.

A company operating a zinc mining plant permitted a discharge of water therefrom which contained a murky sediment called "sludge," and which was emptied into a small, dry branch that crossed a private road used by certain landowners and which afforded them a way to the public road. The water containing the sludge was caused to run over and upon complainant's field and near to and into a spring house and into a pond on the complainant's land used for watering stock. In times of high water or floods the sludge would wash down upon the complainant's land, thereby injuring the land, contaminating the spring and stock pond, to the annoyance and damage of the complainant. Under such facts the complainant had the right to maintain an action for damages and to have the nuisance abated not only with reference to his private property but also with reference to the highway as interfering with the ingress and egress, and for the further reason that the highway was shown to be a private way.

Dickensheet v. Chateau Min. Co. (Missouri Appeals), 202 Southwestern 624, p. 625.

THREATENED INJURY TO ADJOINING PREMISES—RECOVERY.

An iron company prepared its mine for operation by the system known as open pit mining. In conducting the operation it removed or stripped the dirt and surface above the minerals and placed these in a large dump and pile on adjoining land and near to the real estate and residence of the complainant. Subsequently a large mass of soft, wet, mortarlike material gushed from the bottom of the dump, swept across a highway down a village street, carrying with it several houses. The complainant's residence was not touched or injured by this flowing mass, but by reason of the negligent manner in which the dump was constructed another slide was likely to happen at any time and would strike and injure his property, and that as the result of such apprehension the market value of his property had been greatly diminished. The owner of real estate can not recover damages for mere apprehension of injury, but he must wait until the injury or damage has actually happened. It is the damage, and not the anticipation thereof, that gives rise to the cause of action.

Johnson v. Rouchleau-Ray Iron Land Co. (Minnesota), 168 Northwestern 1.

MINE FOREMAN, VICE PRINCIPAL, SUPERINTENDENT—RELATION AND NEGLIGENCE.

PROOF OF RELATION—QUESTION OF FACT.

The fact that an employee in a mine had authority to hire and discharge men and direct the manner in which the entry should be driven and the mine developed was not inconsistent with his position as mine foreman nor was the fact that he was called superintendent sufficient to make him one. Where the evidence showed that such employee had full charge of a mine and the breakers of a colliery and that he had been known as superintendent of the mine, and where many of his duties and powers were consistent with those of a superintendent, the question of his relation as superintendent or foreman became one of fact to be determined by a jury.

Kolalsky v. Delaware & Hudson Co. (Pennsylvania), 103 Atlantic 721, p. 722.

SUPERIOR SERVANT—RELATION.

Whether a superior servant other than the superintendent or general manager is a vice principal depends upon the nature of his negligent act and not upon his grade or rank.

Kinder v. Boomer Coal & Coke Co. (West Virginia), 95 Southeastern 580, p. 584.

WHAT CONSTITUTES—VICE PRINCIPAL.

An employee of superior rank with authority to control and by whose negligent acts or orders injury to a servant of a lower grade is caused, is, as to such injured servant, a vice principal and not a fellow servant and his negligence is that of the employer.

Kinder v. Boomer Coal & Coke Co. (West Virginia), 95 Southeastern 580, p. 582.

WHAT CONSTITUTES—AUTHORITY TO EMPLOY AND DISCHARGE.

Mere authority to employ and discharge workmen is not of itself sufficient to constitute one employee a vice principal, as power to control the inferior servant in addition to superiority of rank is essential and when these two elements are present the superior is regarded as directly representing the master and his negligence in giving orders and directions is that of the master and not that of a mere fellow servant.

Kinder v. Boomer Coal & Coke Co. (West Virginia), 95 Southeastern 580, p. 583.

MINE FOREMAN AS AGENT OF OPERATOR—PROOF.

The authority of a mine foreman to represent the mine operator must be proven either by express authority given or by open, apparent, and notorious acts and conduct on the part of the mine foreman; but proof of one or two isolated instances thereof is not sufficient to establish such authority of the mine foreman and to render the mine operator liable in damages for injuries due to the negligence of the mine foreman in the discharge of his statutory duties.

Strother v. United States Coal & Coke Co. (West Virginia), 95 Southeastern 806.

DIFFERENCE IN RANK—LIABILITY OF MINE OPERATOR.

To render a mine operator liable for the consequences of a negligent act or omission to act of one servant where injury ensues therefrom to another servant of the operator in a common employment, the act or omission must be in respect to some duty incumbent upon the mine operator himself, some nondelegable duty. If a mine operator delegates to another the performance of nonassignable duties, thus making him a representative of the employer, the latter must respond for the results of the negligence of the former.

Kinder v. Boomer Coal & Coke Co. (West Virginia), 95 Southeastern 580, p. 583.

NEGLIGENCE OF VICE PRINCIPAL—LIABILITY OF OPERATOR.

An attempt on the part of a mine operator to delegate to a person in any rank or employment the duty of furnishing employees sufficient and safe materials, machinery, or other means by which the services are to be performed has the effect to make such person a vice principal. Such person discharges the employer's services and can not be reckoned as a fellow servant with a common employee.

Consolidated Interstate-Callahan Min. Co. v. Witkouski, 249 Federal 833, p. 836.

CHANGE OF WORK—INCREASED HAZARD—LIABILITY OF OPERATOR.

A coal-mine operator is liable for the negligence of his superintendent who ordered and coerced an inexperienced boy to leave the work he was employed to perform and go to a distant part of the mine and perform more hazardous work, work of which he did not possess the capacity to appreciate the increased risk, and where the superintendent failed to warn the boy of the risks of the new employment and gave no instructions as to the method of performing the work.

Kinder v. Boomer Coal & Coke Co. (West Virginia), 95 Southeastern 580, p. 582.

SERVANT EXCEEDING AUTHORITY—COERCING MINOR EMPLOYEE.

Where a foreman of mine drivers exceeded the authority conferred upon him by the mine operator and coerced a minor employee engaged at work in a different department, loading coal, to undertake the task of driving, a more dangerous and hazardous form of employment, and injury results to the minor in the latter employment, the foreman is, nevertheless, only a fellow servant, unless his negligence relates to some nondelegable duty which the employer owed to the injured employee.

Kinder v. Boomer Coal & Coke Co. (West Virginia), 95 Southeastern 580, p. 583.

FELLOW SERVANTS—RELATION—NEGLIGENCE.

WHO ARE—TEST.

The test as to whether one servant is a fellow servant of another is not the particular rank he sustains to that of others in the service, but the specific character of the act performed. If the act is one done in the discharge of some positive duty of the master to the servant then the negligence of the servant in the act is the negligence of the employer.

Consolidated Interstate-Callahan Min. Co. v. Witkouski, 249 Federal 833, p. 836.

Under the law as held and applied in Kentucky a miner pushing a car of dirt and refuse out of a mine to a dumping place, is not a fellow servant of other employees engaged in pushing other cars loaded with like material on the same track, and there may be a recovery against the mine operator for injuries occasioned by a car being pushed upon and over the miner engaged in pushing another car.

Elkhorn Min. Corp. v. Paradise (Kentucky), 203 Southwestern 291.

Two miners were engaged in tramming ore and the loaded cars for a part of the distance ran down a steep grade. One of the miners after having ridden the car several times insisted that the other miner ride a particular car. He did so without any direction or assurance of safety from the mine operator, with knowledge of the condition of the car and the track and of the danger in so doing. The sugges-

82478°—19——8

tion of the fellow miner was not binding upon the mine operator nor could it fix any responsibility upon him for the resulting injury to the miner who chose to act upon the suggestion of his fellow miner.

Dilley v. Primos Chemical Co. (Colorado), 171 Pacific 1146, p. 1147.

NEGLIGENCE OF FELLOW SERVANT.

In an action by a miner for damages for injuries caused by the fall of a heavy iron bucket used by the mine operator on his cars carrying ore to the tipple, an instruction by the court was not erroneous where it stated that if the mine operator negligently let fall a large can upon the complainant and that by reason thereof he was injured, that the carelessness, negligence, or recklessness on the part of any of the employes would be the negligence of the mine operator in respect to the acts charged. The instruction does not eliminate from the consideration of the jury the questions of assumption of risk and contributory negligence, but it merely states a condition under which a finding of negligence could be made against the mine operator.

Yellow Rose Min. Co. v. Strait (Arkansas), 202 Southwestern 691, p. 693.

NEGLIGENCE OF COEMPLOYEES—RECOVERY—PLEADING.

A miner while pushing a loaded car was injured by reason of another car being pushed against and upon him. In an action against the operator for damages for the injuries he alleged that the injury was occasioned by the negligence of the employer in failing to provide a safe working place in that the track over which the cars were pushed was defective and unsafe. Under such a complaint he can not recover where the proof showed that the injuries complained of were caused by the negligence of other employees, not fellow servants, in pushing another car against and upon him causing the injuries complained of.

Elkhorn Min. Corp. v. Paradise (Kentucky), 203 Southwestern 291.

MINER'S WORKING PLACE—SAFE PLACE.

INCREASED DANGER FROM DYNAMITE.

A miner was accustomed to use 40 per cent dynamite and knew how much force could be used in tamping the 40 per cent grade. He requested the mine operator's storekeeper to furnish him 40 per cent dynamite, but instead thereof and without any mark to indicate and without knowledge on his part the storekeeper furnished him 60 per cent dynamite, a higher and more sensitive grade and one that would explode with much less force and with the use of which the miner was not familiar. The miner without knowledge of the increased danger placed the higher and more sensitive grade in a drill hole prepared for it and while tamping it lightly with a broom stock, the dynamite, by reason of the sensitive grade, exploded and caused the injury complained of. Under such circumstances he can not be charged with a failure to keep his working place safe.

Terleski v. Carr Coal Min. etc. Co. (Kansas), 173 Pacific 8, p. 9.

CONTRIBUTORY NEGLIGENCE OF MINER.

EFFECT ON RECOVERY.

Under the rule as established in the State of Arkansas, the fact that a miner was guilty of negligence contributing to an injury only reduces the amount of the recovery proportionately and will not necessarily defeat an action for personal injuries.

Yellow Rose Min. Co. v. Strait (Arkansas), 202 Southwestern 691, p. 693.

CHOICE OF TWO METHODS—DUTY.

Where a miner has the choice of two methods of performing his work, one safe and the other dangerous, and he is aware of this fact, it is his duty to choose the safe method, but if he chooses the method which necessarily exposes him to danger and which would have been avoided had he chose the other and he is injured, he can not recover.

Dilley v. Primos Chemical Co. (Colorado), 171 Pacific 1146, p. 1147.

MINER SELECTING DANGEROUS METHOD.

A miner whose duty it was to push loaded cars of ore from the mine to the dumping place and who instead of pushing or following the loaded car got upon the car to ride down a steep grade and who by reason of the velocity attained by the car was injured by reason of selecting the dangerous method of tramming the ore, was guilty of such contributory negligence as would prevent a recovery.

Dilley v. Primos Chemical Co. (Colorado), 171 Pacific 1146, p. 1147.

INSTINCT OF SELF-PRESERVATION.

The instinct of self-preservation is sufficient to overcome the imputation of contributory negligence where an employee of a company, producing asphaltum from crude petroleum, was burned to death by reason of the negligence of the master mechanic of the oil company, and where the remains of the employee were found some distance from his working place, thereby indicating an effort on his part to escape.

Hallawell v. Union Oil Co. (California Appeals), 173 Pacific 177, p. 181.
See Graves v. Union Oil Co. (California Appeals), 173 Pacific 618, p. 619.

EMPLOYEE DOING WHAT HE WAS EXPECTED TO DO.

An employee in a mine was riding at the time of his injury on a trip or train of cars known in the parlence of the mine as the "slow drag," which was run by the operator for the purpose of carrying the employees out of the mine. Under such circumstances an injured employee can not be convicted of contributory negligence by reason of the fact that he did what he was expected to, where it appeared that the trip was run for the use and convenience of employees in the situation of the plaintiff.

Sloss-Sheffield Steel & Iron Co. v. Crosby (Alabama), 78 Southern 898.

RIDING ON TRAM CAR WITHOUT BRAKES.

A miner of several years' experience and of usual intelligence was with another workman engaged in tramming ore. The two miners conducted the loaded car from the breast of the tunnel to the ore bin by pushing. The car as the men knew had no brakes and a part of the way it passed down a steep grade. The miner without any direction from the mine operator and with knowledge that the car had no brakes and of the steep grade in the track voluntarily got upon the car, rode down the grade, and by reason of the velocity of the car was thrown off and injured. The negligence of the miner directly contributed to his injury and in the absence of any negligence on the part of the operator, he can not recover.

Dilley v. Primos Chemical Co. (Colorado), 171 Pacific 1146, p. 1147.

TRIP DRIVER—INJURY.

A mule driver hauling a trip of loaded cars out of a mine is guilty of such contributory negligence as will prevent a recovery for injuries, who while standing on the front of his car with his back to the side of the entry or tunnel and intending to step off at a room neck or entry to get and put into his trip a loaded car, but who without looking or making any effort to ascertain the proper place for alighting from his trip stepped off the car backward after it had passed the room neck and struck a rib or side of the entry and was then caught by the car and injured.

Maize v. Big Creek Coal Co. (Missouri Appeals), 203 Southwestern 683, p. 634.

INJURY FROM DYNAMITE—USE OF HIGH GRADE EXPLOSIVES.

A miner's action for damages for injuries caused by a premature explosion of dynamite used by him in his mining operations is not to be defeated on the ground of contributory negligence, where the miner had been accustomed to use and was familiar with a 40 per cent grade of dynamite but the operator furnished and delivered to him a higher and more sensitive grade of dynamite without his knowledge and without any knowledge on his part of its being the higher or more sensitive grade and without knowledge on his part that the dynamite furnished would explode with a less degree of force used in tamping it and where the tamping was proper for the grade of dynamite the miner believed he was using.

Terleski v. Carr Coal Min. etc. Co. (Kansas), 173 Pacific 8, p. 9.

DRIVER JUMPING FROM TRIP.

A trip driver was riding between the front of his car and the mule. It was his duty to step off the car, enter each of the rooms and get the loaded cars of coal therein, push them out to the main track and hitch them into the trip. At the time of the injury complained of he intended to let his trip pass the room neck and push the loaded car out of the room and attach it to the rear of the trip. To this end he attempted to step off the car in the room neck as the trip passed, but without looking, his face being in the opposite direction, he stepped off backward but did not step off into the room neck but stepped off after he had passed the room neck, his shoulders striking a hump or ridge in the side of the entry causing him to stagger forward, the car striking and crushing him against the side of the entry. Under the circumstances as disclosed by the proof, the driver was guilty of contributory negligence as a matter of law and was not entitled to recover for the injuries sustained.

Maize v. Big Creek Coal Co. (Missouri Appeals), 203 Southwestern 688, p. 684.

FREEDOM FROM CONTRIBUTORY NEGLIGENCE.

USE OF NEW ENTRY.

An old entry in a mine used by the miners going to and from their work became flooded and was partially unfit for use by the miners. The mine operator was engaged in the construction of a new entry, which at the time of the injury complained of, was not entirely completed, but had been opened sufficiently for the miners to pass through, and a miner while passing through the new entry was injured by a fall of slate from the roof. Under such conditions, the miner could not be charged with such contributory negligence as would defeat his action for damages.

Cossette v. Paulton Coal Min. Co. (Pennsylvania), 103 Atlantic 846, p. 847.

MINER ACTING UNDER SUDDEN IMPULSE.

A miner had complained to the mine superintendent of an apparently dangerous rock projecting on one side of an entry. The superintendent, after testing it, assured the miner that it was safe and directed him to proceed with the work. On the opposite side of the entry there was a wire carrying an electric current into the mine used as a motive power for the mine cars. The miner, seeking to avoid a rapidly approaching car, stepped to the side of the entry under the rock and was injured by the fall of the rock. He considered the opposite side of the entry dangerous and required as he was to act in an emergency, he can not be held responsible for an error in judgment or charged with contributory negligence.

Kolalsky v. Delaware & Hudson Co. (Pennsylvania), 103 Atlantic 721, p. 723.

RIDING ON UNGUARDED TRUCK.

A person permitted to ride gratuitously on a truck or handcar and injured by reason of a collision of the handcar with a train on the tracks is not guilty of negligence in riding on the truck because it is unprovided with seats or rails when it appears that the collision of the truck with the train of cars was the only proximate cause of his injury.

Hodge v. Sycamore Coal Co. (West Virginia), 95 Southeastern 808, p. 810.

ASSUMPTION OF RISK.

RISKS ASSUMED.

DANGEROUS METHOD OF UNLOADING COAL.

The ore from the operation of a mine was hauled in large iron cans on cars run by gravity from the mine to the tipple, striking the tipple with sufficient force to throw the cans over and empty the ore into the tipple. The empty can was replaced on the car and the car pushed back to the mine by an employee engaged in that particular duty. The employee knew of the dangerous method of taking the ore to the tipple and of emptying the can and the danger from the falling ore and the risk of injury from this method of unloading was assumed by him as one of the risks of the employment. But this assumption of risk did not extend to a danger in no wise connected with the method of unloading and that was wholly unknown to him.

Yellow Rose Min. Co. v. Strait (Arkansas), 202 Southwestern 691, p. 692.

CHANGE OF EMPLOYMENT—KNOWLEDGE OF DANGER.

An employee in a mine, either adult or minor, who is transferred by the mine superintendent to a new branch of employment, and who accepts the change voluntarily with full knowledge of its dangerous character, is regarded as assuming the risks ordinarily incident thereto.

Kinder v. Boomer Coal & Coke Co. (West Virginia), 95 Southeastern 580, p. 582.

PERSON RIDING ON COAL CAR.

A person riding on a truck or hand car used by a mining company to haul its express matter to its mine and not designed for carrying passengers assumes only such risks as are incident to the particular mode of conveyance.

Hodge v. Sycamore Coal Co. (West Virginia), 95 Southeastern 808, p. 809.

RISKS NOT ASSUMED.

INJURY FROM INSECURE COAL BUCKETS.

In mining operations carried on by a mine operator the ore was carried on cars in large 200-pound metal cans, and iron bars were placed in and across the tipple in such position as to keep the cans from falling through when dumped into it from the cars. The loaded cars on which the cans sat were run to the tipple by gravity but the cars with the empty cans were pushed back by an employee. The method of emptying the loaded cans was dangerous and the danger was known to the employee whose duty it was to run the cars back with the empty cans, but the employee did not assume the risk of the danger of an empty can falling upon him after it was placed on a car by reason of an insecure method of holding the can on the car or by reason of the failure of an employee to properly place the empty can on the car.

Yellow Rose Min. Co. v. Strait (Arkansas), 202 Southwestern 691, p. 692.

DEFECTIVE MACHINERY AND APPLIANCES.

An employee in a mine does not undertake to incur the risks attendant upon the use of defective machinery or other instruments with which to do his work unless reasonable care and precaution has been exercised by the mine operator in supplying such as are safe for the purpose and use to which they are adapted.

Consolidated Interstate-Callahan Min. Co. v. Witkouski, 249 Federal 833, p. 836.

SUPERINTENDENT'S ASSURANCE OF SAFETY.

A miner was directed to make an excavation under or near an overhanging rock that appeared dangerous. He was assured by the mine superintendent, after a test, that the rock was safe. Subsequently props were promised and the miner ceased work until the props should be set. Later on returning to see if the props had been set and while inspecting the place he stepped under the rock in order to avoid a passing car. The danger the miner feared was the fall of the rock as a result of working around or near it, but he had no reason to anticipate the rock would fall in the absence of a disturbance of the surrounding coal and he had been assured by the superintendent that there was no danger. In the absence of evidence tending to show that the danger was obvious and imminent he was justified in relying upon the superior judgment of the superintendent and can not be charged with the assumption of the risk of danger.

Kolalsky v. Delaware & Hudson Co. (Pennsylvania), 103 Atlantic 721, p. 723.

COERCION BY SUPERINTENDENT.

The doctrine of assumption of risks rests upon voluntary action and if there is coercion by an agent representing the master and acting in his behalf and upon his authority, express or implied, the law does not regard the servant as assuming the risk.

Kinder v. Boomer Coal & Coke Co. (West Virginia), 95 Southeastern 580, p. 583.

COERCION—FEAR OF DISMISSAL.

Ordinarily the fear of dismissal or loss of position is not such coercion as will prevent the application of the doctrine of assumption of risk, where the dangers of the work are apparent and known to the plaintiff. However this may be, when the threat is against an adult employee, but when the employee is a minor rendering obedience to a superior servant, great allowance should be made in his favor with respect to the voluntariness of the assumption of the risks and if coercion is used the risks will not be assumed. Coercion will be more readily inferred in case of a minor than in the case of an adult under similar circumstances.

Kinder v. Boomer Coal & Coke Co. (West Virginia), 95 Southeastern 580, p. 583.

CHANGE OF EMPLOYMENT.

A miner who is directed by a mine superintendent to change his employment and to engage in different work in another part of the mine is not presumed to have knowledge of the dangers or the peculiar risks of the new employment and may assume that it has no greater risks than those which attach to his regular work.

Kinder v. Boomer Coal & Coke Co. (West Virginia), 95 Southeastern 580, p. 582.

PROXIMATE CAUSE OF INJURY.

PROXIMATE CAUSE—PROOF.

In an action for damages for the death of an employee of a company producing commercial asphaltum from crude petroleum, the plaintiff is not required to show with absolute certainty that a fire negligently started was communicated to the asphaltum shaft and was the proximate cause of the injuries and death complained of; but it was sufficient to show that the burning oil was in close proximity to a trap; that the smoke and flames noticeably rose 2 or 3 feet from the ground; that the smoke contained more or less inflammable gases which might readily be communicated to the gases arising from the trap or the oil on it by the eddying currents of a strong wind, and that the fire burst forth almost simultaneously at the asphaltum shaft where there had been no fire from any other source.

Hallawell v. Union Oil Co. (California Appeals), 173 Pacific 177, p. 180.

See Graves v. Union Oil Co. (California Appeals), 173 Pacific 618, p. 619.

CONTRACTS RELATING TO OPERATIONS.

WRONGFUL DISCHARGE OF SUPERINTENDENT—LIABILITY.

The wrong doing or delinquency of a superintendent of a mine which justifies his discharge by the mine operator in violation of a contract of employment must have some relation to the duties imposed by the contract. The test is, when taking the nature of the employment into account, do the acts complained of render the superintendent unfit to perform the duties which he had undertaken.

Burch v. Conklin (Missouri Appeals), 204 Southwestern 47, p. 50.

See Consumers' Lignite Co. v. James (Texas Civil Appeals), 204 Southwestern 719.

INDEPENDENT CONTRACTOR.

INJURY—LIABILITY OF MINE OPERATOR.

An independent contractor mining coal as such and prior to the injury complained of was working and driving a place in the coal and in doing the work he hired the laborers and paid them out of his own money and the mine operator paid him for shooting it down by the ton. The defect causing the injury and for which he sued for damages was not such that arose from or had not been discovered or remedied owing to the negligence of the mine operator; but was in fact caused by the complainant himself at a time when he was working as an independent contractor and the defect could not have been discovered by an examination made by an expert and competent miner, at least not until shortly before the accident on account of the coal and rock that had been shot down by the complainant himself prior to the accident.

Red Eagle Coal Co. v. Thrasher (Alabama), 78 Southern 718, p. 719.

GENERAL INSTRUCTIONS—INDEPENDENT CONTRACTOR.

An independent contractor mining out coal in a mine, was given general instructions to "go to work on that side track and put up timber." A general instruction to a person who was undertaking to do a piece of work involving judgment, care, and skill to make it conform to the requirements for the safe operation of a mine, is a far different thing from ordering a dependent servant to do a particular act of service. A mine owner can not be held liable in damages for injuries to an independent contractor by a fall of rock from the roof where the contractor himself was an experienced miner and had sounded and examined the roof and thought it was safe to work under, and where, until shortly before the accident, nothing would have been discovered indicative of loose rock or specific danger therefrom.

Red Eagle Coal Co. v. Thrasher (Alabama), 78 Southern 718, p. 719.

RIGHT TO SURFACE SUPPORT.

RIGHT TO MINE SERVIENT TO SURFACE OWNERSHIP.

The right to mine is servient to the right of the owner of the surface to have it perpetually sustained in its natural state.

Faught v. Leth (Alabama), 78 Southern 830, p. 831.

INJUNCTION—DISSOLUTION.

A court on appeal will not disturb the finding of a trial court on the dissolution of a temporary injunction heard on affidavits and where the trial judge in person inspected the mining operations and on the affidavits and such personal inspection determined that the temporary injunction theretofore granted at the suit of the surface owner against the mining operations should be dissolved.

Faught v. Leth (Alabama), 78 Southern 830, p. 831.

NUISANCE.

MINE REFUSE—POLLUTION OF WATERS.

It is a nuisance justifying a recovery of damages for a company operating a lead and zinc mining plant to permit the mine refuse, a murky sediment, to run from its mining plant over the fields of an adjoining landowner, into a spring house and stock pond on his land, and over and upon a private road used by the landowner, to the injury of his land and to his annoyance and discomfort in polluting the waters of his spring and stock pond and in obstructing the private way used by him for ingress and egress to his premises.

Dickensheet v. Chateau Min. Co. (Missouri Appeals), 202 Southwestern 624, p. 625.

MINING LEASES.

LEASES GENERALLY—CONSTRUCTION.

CONSTRUCTION—MATURITY OF DEFERRED PAYMENT.

A lease or conveyance of minerals bound the grantee to pay the sum of $300 within six months after quarrying or mining should be begun on the premises, the lease stating: "The unpaid $300 is payable at any time second party sees fit to begin work on said land to remove stone." The grantee transferred his right to a stone company and the stone company began quarrying and removing stone from the premises. Whether the lease gave the company this right is not the question, but it does provide for the payment of the $300 upon that contingency.

Kennedy v. Hicks (Kentucky), 208 Southwestern 318, p. 320.

CONSTRUCTION—COMPROMISE AND SETTLEMENT—CONSIDERATION FOR DRILLING.

An oil company entered into a contract with an oil-well driller in settlement of an existing liability and agreed to pay the driller $3,000 if a well proposed to be drilled upon leased lands held by the oil company proved to be a commercially paying oil or gas well. Shortly after the execution of the contract, the oil company sold the lease on which the well was to be drilled and the purchaser, the assignee of the lease, thereupon developed a commercially paying gas well upon the leased land. Under the contract and according to its terms the oil company on the completion of the well became liable to the driller for the payment of the $3,000.

Gem Oil Co. v. Callendar (Oklahoma), 173 Pacific 820, p. 823.

FAILURE OF LESSEE TO OPERATE—FORFEITURE.

A mining lease obligated the lessee to operate the mine continuously, "keeping the machinery in as good working order as when he takes possession 10 days hence, less the natural wear and tear of same." The lease also provided that on the failure of the lessee or his assignees to operate the mine for 60 days without the written consent of the lessor, the lease should become null and void. The lessee did not operate the mine continuously and about five months after the execution of the lease the mill and machinery used in operating the mine were destroyed by an unusual flood. Some two months after the property was destroyed the lessee assigned the lease and his assignee made some effort to operate the mine, but did nothing toward rebuilding the mill. Under a covenant to repair, a lessee's liability is not confined to cases of ordinary and gradual decay, but extends to accidental injuries to the property, and if the improvements are entirely destroyed, the assignee is bound to repair within a reasonable time. The failure of the lessee or his assignee to rebuild the mill and to operate the mine according to the terms of the lease gave the lessor the right to treat the lease as canceled and to take possession of the property.

Bradley v. Holliman (Arkansas), 202 Southwestern 469, p. 471.

DIAMOND LEASE—CONTINUOUS OPERATIONS.

A mining lease recited that the lessor desired to have the property developed and worked as a diamond mine; that the lessee was a practical mining engineer with experience in testing deposits of Kimberlite or diamond-bearing dirt, and that he was associated with men of large means who had become interested in developing the mine. The lease provided that the lessee would diligently and faithfully prosecute the work as outlined in a scientific and practical manner and would commence operations within 30 days and would erect and install a modern washing and concentration plant of African type within one year and would treat and wash for the recovery and extraction of diamonds and other precious metals a minimum of 10,000 loads of material annually from the land described and as much more as could be reasonably done. The lease provided that the lessee should in no event cease work for a longer period than three months unless a necessity therefor should arise from contingencies beyond the control of the lessee or from physical or other conditions, not the fault of the lessee; but this clause was not to release the lessee from washing and treating for diamonds the 10,000 loads of dirt every year. One year later the lessor instituted proceedings to cancel the lease on the ground of fraud in its procurement, but on the final hearing the complaint was dismissed for want of equity. A year later another suit was instituted by the lessor to recover possession of diamonds which had been mined from the land, the complaint alleging in addition that the lease had been procured with the fraudulent intention of not complying with it. On the trial a jury found that the lease was not entered into with a fraudulent purpose and that the lessee had not failed to comply with the terms of the lease and judgment was entered in favor of the lessee. Each of these cases constituted an adjudication of the question of fraud in the execution of the lease and also to the effect that the lease had not been broken by the lessee, and in a subsequent action by the heirs of the lessor the right of action on these questions extended no further than the last of the adjudications, and if any recovery could be had it must be for a breach of the terms of the lease subsequent to the date of the last action.

Mauney v. Millar (Arkansas), 203 Southwestern 10, p. 11.

LEASE TO MINE DIAMONDS—DILIGENCE IN DEVELOPING.

By the terms of a lease to mine diamonds, the lessee obligated himself to treat and wash for the recovery and extraction of diamonds a minimum of 10,000 loads of material annually and that the work should be carried on with diligence and as much more than the amount stated as could reasonably be done. The lease contained no express provision for forfeiture but a forfeiture may be decreed where there is an abandonment of the lease. But such a lease will not be canceled where the proof shows that the lessee had expended $100,000 in developing and starting operations on the mine and that the year preceding the commencement of the action the lessee had taken out and treated more than 11,000 loads, and the proof was sufficient to show that the lessee had not abandoned operations and had not failed to comply with its terms.

Mauney v. Millar (Arkansas), 203 Southwestern 10, p. 12.

POWER OF EXECUTRIX TO LEASE—DISPOSITION OF ROYALTIES.

The will of a testator stated that certain mining properties in different States, each of an estimated value, should be held at such values until a certain named daughter should arrive at the age of 21 years. The will provided that the money received after the testator's death from each of the mining properties named should be divided and paid in a certain stated manner. Under these provisions the executrix could not lease the mines and receive and merge the royalties into what the will termed the residue of the estate, although the will did authorize the executrix to lease any part of the estate without an order of court.

Campbell's Estate, In re (Utah), 173 Pacific 688, p. 695.

LESSEE NOT PURCHASER OF ORE IN PLACE.

The lessee of an ordinary lease granting the privilege of mining ore from certain described land and providing for the payment of a stipulated royalty per ton on the ore mined is not a purchaser of the ore in place.

United States v. Biwabik Min. Co., 38 Supreme Court Rep. 462, p. 463.

ABANDONMENT—RIGHT TO CANCELLATION.

Where one party to a lease has completely abandoned performance a court of equity will give relief by canceling the lease. This principle applies particularly to mining leases where the sole benefit is to result from continued performance such as one to develop a mine, to pay royalty, or divide the proceeds. For partial breach of such a lease the parties will be remitted to their remedy at law, but in case of an abandonment equity will afford relief by rescission and cancellation.

Mauney v. Millar (Arkansas), 208 Southwestern 10, p. 12.

FORFEITURE—VIOLATION OF STATUTORY PROVISION.

The statute of Texas (Acts, 1913, p. 409) declares that the failure of the owner of a mining permit to comply with certain requirements shall work a revocation of such a permit and the termination of the right to the owner. The legislative purpose seems to have been to require an official ascertainment and record of a forfeiture rather than to leave open indefinitely the issue of the forfeiture. It is to be presumed that in the enactment of this material right into a statute the legislature intended that the decisions of the courts and the settled public policy against forfeitures should be read into it, and it should not be construed as contemplating ipso facto forfeiture.

Underwood v. Robinson (Texas), 204 Southwestern 314.

COAL LEASES.

CONSTRUCTION—REMOVAL OF ALL COAL—TERMINATION OF LEASE.

A coal lease was to continue for 20 years unless all the coal should be sooner removed. "All coal," as used in the lease, is the equivalent to all minable coal and the purpose of such a lease is that the coal should be removed by the lessee with diligence and within the shortest practical time. It is not within the power of the lessee to terminate such a lease as long as he exercises or claims the right to exercise the privileges granted in the lease. When the lessee desired absolution from further payment of a stipulated minimum royalty, it devolved upon him to take an irrevocable position to that end and to so notify the lessor; but so long as the lessee claimed any of the benefits of the lease or exercised the privileges granted thereby he is subjected to the burdens and liabilities imposed by the lease.

Fisher v. Maple Block Coal Co. (Iowa), 168 Northwestern 110, p. 111.

MINIMUM ROYALTY—LIABILITY OF LESSEE—TERMINATION OF LEASE.

A coal lease was to run for 20 years unless all the coal should be sooner removed. It provided for a royalty of 10 cents per ton and for a minimum royalty of $1,000 for each year, such sum to be paid whether sufficient coal was mined to produce the same or not. The lessee after 10 years of operation refused to pay the minimum royalty for the reason that the mine became unworkable on account of flooding; which he was unable after good faith efforts to overcome. The lessee did not abandon the mine but continued to exercise his rights under the lease. In order to avoid liability for the minimum royalty it was incumbent upon the lessee to show that he had a right to terminate the lease and that he did terminate the same by some form of surrender and by some manner of notice to the lessor.

Fisher v. Maple Block Coal Co. (Iowa), 168 Northwestern 110, p. 111.

RECOVERY OF MINIMUM ROYALTY—DEFENSE.

A coal lease was to run for 20 years unless the coal was all mined out before that date. The lessee was to pay a royalty of 10 cents per ton monthly and a minimum royalty of $1,000 for each year payable semiannually. This minimum royalty was to be paid whether sufficient coal was mined to produce the same or not. In an action by the lessor to recover the minimum royalty, the lessee offered to prove that the mine became unworkable because of flooding by reason of which he was unable to operate the mine with profit and had made a good faith offer by the expenditure of large sums to make the mine workable, but offered no proof that he had surrendered the lease or relinquished his rights thereunder. The lessee continued to exercise his privileges under the lease for the greater part of the term for which the minimum royalty was claimed and as long as he claimed any of the benefits of the lease he must be subjected to its burdens.

Fisher v. Maple Block Coal Co. (Iowa), 168 Northwestern 110.

RIGHT OF LESSEE TO REMOVE MINING MACHINERY.

A lease of coal land and a coal mine for 21 years provided that the "lessees shall have the right to remove all and any machinery placed upon said lands under this contract upon the termination of same." The lessee became insolvent, the property was sold at a receiver's sale, and the purchaser operated the mine for some three years. Under the terms of the lease the purchaser from the former lessee or at the receivers' sale had the right to remove the machinery within a reasonable time after the lease was canceled or abandoned. The provision giving the lessee the right to remove the machinery upon the termination of the lease did not mean upon the exact moment the lease was terminated, but it did mean that the property should be removed within a reasonable time after the termination of the lease. But the purchaser from the lessee, after waiting an unreasonable time and permitting the original owner and his second lessee to repair and improve the machinery, can not then claim the right to remove the machinery from the land.

Helm v. Brock (Arkansas), 202 Southwestern 36, p. 37.

LEASE BY GUARDIAN—DURATION—VALIDITY OF STATUTE.

The statute of Kentucky (Laws 1916, chap. 99) authorizes the guardian of an infant or of an insane person to lease the real estate of such infant or insane person for the purpose of mining and removing all or part of the coal, oil, or gas or any or all other minerals or mineral substances therein. " Such lease may be of such length of time as the guardian, curator, or committee may approve without being limited to the time at which the disability of such infant or person of unsound mind may be removed." Under this statute a guardian of three minor children, ages, respectively, 11, 16, and 18 years, by order of the proper court leased the lands of his wards for mining purposes and for the mining of all coal in and under the land for a period of 40 years with the privilege of renewal for an additional term of 40 years, at a stipulated royalty of 10 cents for each ton of coal mined and 15 cents for each ton of coke manufactured on the premises with a minimum royalty of $5,000 for the first year, $10,000 for the second year, and $15,000 for each year thereafter. Evidence was heard, and it was adjudged by the court authorizing the lease to the effect that the conditions of the lease were reasonable and satisfactory, and that the royalty was the best obtainable. On appeal it was held that the statute in so far as it attempted to authorize the leasing during the infancy of the owners of the coal, oil, gas and other mineral or mineral substances and products for a period beyond the minority of the infants was void.

Lawrence E. Tierney Coal Co. v. Smith (Kentucky), 203 Southwestern 731, pp. 732–738.

Note.—This decision is contrary to the rule adopted in other States under statutes similar if not identical:

McCreary v. Billing, 176 Alabama 314, 58 Southern 311.

Beauchamp v. Bertig, 90 Arkansas 350, 119 Southwestern 75, 23 L. R. A. (N. S.) 659.

Cabin Valley Min. Co. v. Mary Hall, —— Oklahoma ——, 155 Pacific 570.

Ricardi v. Gabouri, 115 Tennessee 485, 89 Southwestern 98.

Mallen v. Ruth Oil Co., 231 Federal 845.

Instances suggest themselves in all classes of mining, and especially in oil and gas, that would be ruinous to minors if their guardians were not permitted to make leases beyond the term of the majority of the wards. It would appear to be more reasonable and more consistent with the general rule of law and with sound business principles, for a court to hold invalid such a mining lease for an unreasonably long period rather than to hold invalid a statute authorizing guardians to make such leases.

ABANDONMENT—RIGHT OF LESSOR TO TAKE POSSESSION.

The owner of coal lands leased a mine thereon for a period of 21 years. The lease required the lessee to construct and equip a coal mine plant of 400 tons daily capacity and to maintain the plant and machinery in good repair. The lease provided that the lessee retained the right to remove the mining machinery placed upon the land upon the termination of the lease. The lessee became insolvent and the entire plant was sold by a receiver, the purchaser took possession and operated the property for some three years and finally abandoned the lease. The lessor on such abandonment had the right to take possession of and operate the mine included in the lease.

Helm v. Brock (Arkansas), 202 Southwestern 36.

CANCELLATION OF PATENT—BONA FIDE PURCHASER.

A patent for certain coal lands can not be canceled on the ground of fraud in its procurement as against a bona fide lessee who took a lease for a term of years for a valuable consideration, the greater part of which had been paid before the suit to cancel the patent was instituted and without notice of any fraudulent practice.

United States v. Routt County Coal Co., 248 Federal 485, p. 486.

OIL AND GAS LEASES.

CONSTRUCTION—LAND OWNER PROTECTED.

It is the purpose of courts in the construction of oil and mineral leases to give them when the lease will permit, such a construction as will protect the rights of the land owner and enable him to derive from the lease the benefit he had the reasonable right to expect when it was entered into.

Hughes v. Busseyville Oil & Gas Co. (Kentucky), 203 Southwestern 515, p. 517.

CONSTRUCTION—EFFECT GIVEN TO ALL PARTS.

An oil and gas lease was executed in 1902. In 1903, the parties entered into an agreement modifying the lease by providing that after the first four years of the first term of the lease one-half of the 15 per cent of the net proceeds, derived from the business, should be paid to the lessor. It was expressly stated in the modifying agreement that it was intended to modify clauses with which it was in conflict. The Civil Code of California (sec. 1641) requires that in the construction of a contract the whole of it is to be taken together so as to give effect to every part, if practicable, and each clause must help interpret the other. In the light of these provisions a reasonable effect must, if possible, be given not only the original lease as a whole, but as to the meaning and effect of the modifying clauses as well.

Nathan v. Porter (California Appeals), 172 Pacific 170, p. 173.
See Witherington v. Gypsy Oil Co. (Oklahoma), 172 Pacific 634.

CONSTRUCTION—REASONABLE CONSTRUCTION ADOPTED.

Where the meaning of the language of an oil and gas lease is doubtful, and is fairly susceptible of two constructions, that construction must be given which makes it fair and such as prudent men would naturally execute in preference to a construction that would make it inequitable or such as reasonably prudent men would not be likely to enter into.

Witherington v. Gypsy Oil Co. (Oklahoma), 172 Pacific 634, p. 637.

CONSTRUCTION—AMBIGUOUS TERMS—INTENTION OF PARTIES.

The object of the interpretation and construction of an oil and gas lease is to arrive at and give effect to the mutual intent of the parties as expressed in the lease, and where a lease is ambiguous, the true intention, if it can be ascertained from the contract, must prevail over verbal inaccuracies, inapt expressions, and dry words of the stipulations. It is the duty of a court to place itself as far as possible in the position of the parties at the time the lease was executed and to consider the instrument itself as drawn, its purpose and the circumstances surrounding the transaction and from a consideration of all of these elements to determine upon what sense and meaning of the terms used their minds actually met.

Witherington v. Gypsy Oil Co. (Oklahoma), 172 Pacific 634, p. 635.

CONSTRUCTION AND CONSIDERATION.

In the construction of an oil lease and in determining whether a consideration is sufficient in amount to be classed as adequate and valuable, a court must look to the subject matter of the contract and its value to the parties concerned. The value of an exclusive option to explore and mine oil and gas is governed largely by the prospects of their discovery in the given territory and their worth when discovered. It is easy to imagine territory in which this privilege would be worth much and other territory in which it would have no practical value. The sum of $70 cash actually paid must be regarded as a sum sufficient to make the issue of valuable and adequate consideration as one of fact to be passed upon by a court or jury. Such a consideration is sufficient to prevent the cancelation of an otherwise unilateral contract.

Griffen v. Bell (Texas Civil Appeals), 202 Southwestern 1084, p. 1087.

CONSTRUCTION—SUCCESSIVE EXTENSIONS OF TIME—CONSIDERATION.

An oil and gas lease stipulated that unless the lessee began drilling a well within six months from its date the lease should expire, but the lease gave the lessee the right of having successive extensions upon the payment of stipulated payments of money and the payment of a stated sum in advance was not only the consideration for the privilege extending over the first term of six months but for all the rights and privileges, conditional and unconditional, which the contract conferred, including the conditional right of claiming an extension of the option. The payment of the sum stated was what induced the execution of the contract and was the consideration for all the rights and privileges therein granted. The sum paid can not be treated as a payment of the rental for the first term of six months; but when that term expired the lessee could have demanded an extension of time upon the tender of the sum agreed upon. This was a valuable right which emanated from the contract and for the grant of which the sum paid was the consideration.

Griffen v. Bell (Texas Civil Appeals), 202 Southwestern 1034, p. 1037.

CONSTRUCTION—PERIOD OF DURATION.

An oil and gas lease provided that it should be for a period of two years and as long thereafter as the lessee may produce oil and gas therefrom in paying quantities. The lease bound the lessee to complete a well within two years and contained a provision to the effect that the extension of the lease after the two-year term should depend upon productions at the time and thereafter of oil and gas in paying quantities as "herein provided." A well was drilled on the premises as provided and was at the expiration of two years and thereafter producing oil in paying quantities and the lessor's royalty therefrom amounted to $180 a year. This was sufficient compliance with the requirements and sufficient to extend the period of the lease.

Hughes v. Busseyville Oil & Gas Co. (Kentucky), 203 Southwestern 515, p. 516.

CONSTRUCTION—PERIOD OF EXISTENCE—PAYMENT OF RENTALS.

An oil and gas lease provided that it should be for a term and period of two years and as much longer, not exceeding 50 years, as coal, oil, water, mineral water, gas, or other minerals were found in paying quantities thereon. If no mineral should be discovered within two years the lease was to become null and void unless the lessee should pay $1 per acre per year at or before the end of each year thereafter. By these provisions the lease was to extend 50 years if

gas or oil was found in paying quantities or if the annual rental of
$1 an acre was paid before the end of each year until a well should
be found on the premises. So long as the payments are made according to the terms of the lease, the lease is valid and not subject to forfeiture or cancellation, nor can the lessee be enjoined from drilling
upon the premises.

Arnold v. Garnett Light & Fuel Co. (Kansas), 172 Pacific 1012, p. 1013.

CONSTRUCTION—USE OF TECHNICAL WORDS—PAROL EVIDENCE TO EXPLAIN.

An oil and gas lease contained a provision to the effect that when
a well was commenced the drilling should be prosecuted with due
diligence until such well was completed. The word " completed " in
the lease is a technical word relating to oil and gas drilling operations; and technical words in a contract are to be interpreted as
usually understood by persons in the profession or business to which
they relate and expert testimony is admissible, where there is a disagreement as to the meaning and purport of the word "completed" to
explain the sense in which the term was understood generally by those
engaged in the business of drilling oil or gas wells.

Frost v. Martin (Texas Civil Appeals), 203 Southwestern 72, p. 74.

CONSTRUCTION—NATURE OF ESTATE GRANTED.

An oil and gas lease is not a grant of the oil and gas that is in the
ground, but of such part thereof as the lessee may find, and passes no
estate that can be the subject of ejectment or other real action.

Brennan v. Hunter (Oklahoma), 172 Pacific 49, p. 50.

NATURE—NO ESTATE GRANTED.

A lease or contract granting the right to the lessee to enter on the
land described to explore for oil and gas and if oil and gas be found
in paying quantities to operate and produce the same, is not a lease
in the strict sense; but the term " lease " is applied to such instruments merely through habit and for convenience. Such an instrument creates no estate in land but merely a kind of license. It creates an incorporeal hereditament, a right growing out of or concerning or annexed to a corporeal thing but not the substance of the
thing itself.

Huston v. Cox (Kansas), 172 Pacific 992, p. 993.

CONTRACT FOR EXPLORATION OF LAND FOR OIL AND GAS.

A contract or lease for the exploration of land for minerals, oil,
and gas, although designated a sale by the parties, was the grant of
an exclusive right to search for, take and appropriate the minerals

mentioned in the contract, and is in effect a lease of the land described for mining purposes.

De Moss v. Sample, 143 Louisiana ———, 78 Southern 482, p. 483.

CONSTRUCTION—DRILLING WELL—MEANING OF COMPLETED WELL.

An oil and gas lease required the lessee to begin operations within 12 months and provided that " when a well is once begun the drilling thereof shall be prosecuted with due diligence until same is completed." The lease also provided that a failure to perform the obligations of the lease should render it null and void. The lessee, pursuant to the terms of the lease began drilling a well that was carried on continuously night and day until a depth of over 2,100 feet was reached, but no oil or gas was discovered. If the lessee began in good faith the drilling of a well within the time stipulated he released himself from the alternative obligation to pay rental and from the penalty of forfeiture for a failure to do so. The term " completed " as used in the lease means finished or sunk to the depth necessary to find oil or gas in paying quantities, or to such a depth as in the absence of such oil or gas would reasonably preclude the probability of finding oil or gas at a further depth. It can not be construed to mean that the lessee bound himself, under penalty of forfeiture to sink a producing well or in the absence of oil or gas to bore through to China.

Frost v. Martin (Texas Civil Appeals), 203 Southwestern 72, p. 73.

CONSTRUCTION—COMPLETED WELL—MEANING.

A completed oil well, as used in a lease requiring the completion of a well in a certain time or right to forfeit the lease means the drilling of a well deep enough to discover oil or gas or deep enough according to the experience of well drillers in the same or adjoining fields to determine with reasonable certainty that there is no oil or gas bearing sand or rock in the particular field. And if a well has been drilled to a depth regarded by experienced drillers as sufficient to test the field, it is a completed well, although no oil or gas has been discovered.

Frost v. Martin (Texas Civil Appeals), 203 Southwestern 72, p. 74.

CONSTRUCTION—SUSPENSION OF DRILLING—FORFEITURE.

An oil and gas lease provided that if operations were not begun on or before a stated date the lease should terminate unless the lessee should pay to the lessor a certain stated amount. The payment of the stipulated amount should operate to confer on the lessee the privilege of deferring the time limit for six months from the stated date. It also provided that after drilling operations had begun, the payment of the sum should not be required if the period of suspension of drilling was less than 30 days. The lessee began drilling operations as required prior to May 29, 1916, and completed a producing oil well on August 7, 1916; but no further drilling operations were begun or attempted within 30 days from the completion of such well. The completion of a producing well did not terminate or avoid the provision for continued drilling, or the payment of the stated amount, and the lease was subject to forfeiture on account of the suspension of drilling following the completion of the well; but this forfeiture could have been prevented by the lessee by the payment of the stipulated amount in lieu of drilling operations for each six-month period following the suspension until resumption of operations.

Stein v. Producers' Oil Co. (Texas Civil Appeals), 203 Southwestern 126, p. 128,

CONSTRUCTION—EFFECT OF SURRENDER CLAUSE—RIGHTS OF LESSEE.

The owner of land under an existing oil and gas lease executed a second lease that contained a clause by the terms of which the lessee could at any time upon the payment of $1 surrender the premises and relieve himself from any obligation under the lease. This provision makes such a lease unalateral, and is such a one as a court of equity will refuse to enforce, and it will not furnish the basis for an action in ejectment or other real action. The lessee in such a lease has no standing to question the validity of the first lease or to maintain ejectment against the original lessee.

Brennan v. Hunter (Oklahoma), 172 Pacific 49, p. 50.

CONSTRUCTION—RIGHT OF POSSESSION—INJUNCTION.

The lessee of an oil and gas lease sought to enjoin a subsequent lessee of the same land from interfering with the peaceful possession of the complainant. The proofs showed that the complainant had never entered upon or taken possession of the premises under its lease, but that the second lessee had taken and was in possession of the premises under its lease and had been in the possession of the land for more than seven months at the commencement of the suit. A preliminary injunction may issue to maintain a complainant in possession, but it should not be allowed to oust one already in possession of property.

Pure Oil Operating Co. v. Gulf Refining Co., 143 Louisiana ———, 78 Southern 560.

CONSIDERATION—UNILATERAL CONTRACT—WHO MAY QUESTION.

The lessor in an oil and gas lease is the only person who can urge that the lease is unilateral by reason of the presence of a certain clause therein and claim a cancellation of the lease. A lessee in a subsequent lease of the same premises can not urge the invalidity of a prior lease on the ground that it is unilateral.

Brennan v. Hunter (Oklahoma), 172 Pacific 49, p. 50.

CONSTRUCTION—NET PROFITS AND NET PROCEEDS.

An oil and gas lease provided that the net profits were to be determined by deducting from the gross income only the royalties and operating expenses, as distinguished and considered apart from "capital expenses." A modifying clause provided for a change in the payment of royalty based on the "net proceeds" derived from the business. The "net proceeds" provided for in the modifying clause was not dependent upon the cost of capitalization but only upon the sum total of royalties and operating expenses and in estimating the net proceeds the lessee could not deduct capital expenses in addition to operating expenses.

Nathan v. Porter (California Appeals), 172 Pacific 107, p. 173.

CONSTRUCTION—RIGHT OF LESSOR TO GAS FROM OIL WELLS.

An oil and gas lease contained two separate provisions to the effect: (1) That if gas is found the lessor is to have sufficient gas for domestic purposes free of charge and the remainder, with all gas from oil wells to go to the lessee; (2) if the lessee shall market any gas from any well producing gas the lessor shall receive therefor at the rate of one-fourth of all the gas marketed. If the first provision stood alone the lessor would only be entitled to gas for domestic purposes free of charge and would not be entitled to royalty from gas produced from an oil well. But the first provision must be considered as affected or modified by the second, which provides that the lessor shall receive at the rate of one-fourth of all the gas marketed or sold from any well producing gas. The second provision entitles the lessor to an account for one-fourth part of the proceeds of gasoline produced from casing-head gas as well as the one-fourth part of all gas from oil wells that was marketed.

Witherington v. Gypsy Oil Co. (Oklahoma), 172 Pacific 634, p. 635.

NATURE AND CONSTRUCTION—LESSEE'S RIGHT TO POSSESSION.

An oil and gas lease gave the lessee the right to go upon the premises and search for oil and gas and to commence operations within the initial period and continue the same with reasonable diligence until oil or gas was found in paying quantities, or until it was demonstrated that neither existed in the land. The lessee by the lease acquired no vested estate in the premises, but he had the right to the possession of the land to the extent reasonably necessary to perform the obligations imposed by the terms of the lease.

Brennan v. Hunter (Oklahoma), 172 Pacific 49, p. 50.

CONSTRUCTION—IMPLIED COVENANT TO DEVELOP.

The development of mineral, oil and gas lands is too expensive for the land owner himself to undertake and requires skill and capital ordinarily beyond the reach of the owner of the soil and that these resources be developed and profit realized therefrom it is necessary that the privilege be granted to persons who are engaged in the business of exploring and developing oil and gas and mineral fields. The fluctuating and uncertain character and value of this class of property renders it necessary for the protection of the land owners that the properties should be developed as speedily as possible, and the lessee for such purpose will not be permitted to hold the land for speculative or other purposes an unreasonable length of time for a mere nominal rent when a royalty on the product is the chief object for the execution of the lease.

Hughes v. Busseyville Oil & Gas Co. (Kentucky), 203 Southwestern 515, p. 517.

IMPLIED OBLIGATION TO DEVELOP—INTENTION OF PARTIES.

The application of the rule to the effect that an oil lease contains an implied covenant on the part of the lessee to develop the leased premises, depends on circumstances and on the intention of the parties. An implied covenant to develop can not be read into a lease of land for oil and gas where the territory had not before been developed and its productive value was not known; where the lessee was a local corporation, with small capital and only willing to undertake the venture and the risk involved under the term of a lease that the land owner was willing to make in order to secure some development and profit from the land and where it does not appear that either of the parties at the time thought of putting down the probable number of wells that it would take to fully test the oil properties of the land, as to do so would require the expenditure of a much larger sum than the lessee had or could get or desired to invest in the enterprise.

Hughes v. Busseyville Oil & Gas Co. (Kentucky), 203 Southwestern 515, p. 517.

COVENANT TO DEVELOP NOT IMPLIED.

A covenant to develop thoroughly or completely the premises leased for oil purposes will not be implied where it appears that it costs about $4,000 to drill a well and that a well drilled, and other wells in the same field produce not more than 2 barrels a day, and where it appears that the lessee is not holding the land for speculative purposes and that wells drilled on neighboring or adjoining tracts are not draining the oil from the leased tract. A lessee may be compelled to develop and protect the lines where it appears that wells on adjoining lands are drawing from the leased lands.

Hughes v. Busseyville Oil & Gas Co. (Kentucky), 208 Southwestern 515, p. 517.

CONSIDERATION AND VALIDITY—IMPLIED COVENANT.

An oil and gas lease granted in consideration of $1 actually paid and by which the lessee covenanted to complete a well within one year or to pay 10 cents per acre yearly in advance for each additional year that such completion is delayed, and by which he covenanted to pay 'to the lessor one-eighth of all oil produced, is not invalid during the first year or within a reasonable time during which an implied covenant to commence operations under penalty of forfeiture may be enforced, either by reason of the smallness of the consideration or the reservation of the right of the lessee on payment of $1 to surrender the lease for cancellation.

Raydure v. Lindley, 249 Federal 675, p. 676.

FAILURE TO DEVELOP—RIGHT OF SECOND LESSEE TO TERMINATE FIRST LEASE.

An oil and gas lease dated February 18, 1914, required the lessee to commence development within 90 days. The lease contained no provision for delay in development by the payment of rentals. Development was commenced within the period stated and a well completed, in which gas was found in paying quantities and the lessee paid the lessor the stipulated rental for a period of 12 months. Under such a state of facts a second lessee of the premises can not maintain an action in ejectment against the first lessee upon the ground either for a failure to develop or that the lease is unilateral.

Brennan v. Hunter (Oklahoma), 172 Pacific 49, p. 50.

FAILURE TO DRILL—REMEDY OF LESSOR.

An oil and gas lease provided that if the lessee failed to commence operations by a certain date the lessee should pay the lessor a stated amount per month for each and every month in which default should be made in the commencement of operations. The lease also provided that a failure on the part of the lessee to comply with its conditions, or on failure to diligently prosecute the work of drilling the lease should be void. This latter provision constitutes an option given to the lessor which in lieu of insisting on the pay of the $100 per month he might or might not at his election cancel the lease. But the lessee after failing to commence operations before the date stated could not insist that his failure to perform the covenant should be equivalent to a performance. Under the terms of the lease the lessor may, instead of exercising the right which he has to terminate the lease, insist upon the payment of the sum stipulated to be paid for failure to comply with the covenant to commence operations at the date stated.

Allen v. Marver (California), 172 Pacific 980.

OPTION TO DEVELOP OIL LAND—FAILURE TO DEVELOP.

A lessee acquired from a land owner by the payment of a stated sum in advance an option to develop the lessor's land for oil. The lease was acquired for speculative purposes and without any intention on the part of the lessee of drilling a well on the land and these facts were known to the lessor and he accepted the cash payment with the knowledge of such intention. In such case the absence of an intention to develop the land is not such a fraud as will entitle the lessor to a rescission of the contract. The lessee did not bind himself to sink wells and violated no covenant of the agreement by a failure to do so.

Griffen v. Bell (Texas Civil Appeals), 202 Southwestern 1034, p. 1037.

CONTRACT TO DRILL OR PAY RENTAL—LESSOR'S OPTION.

An oil and gas lease required the lessee to begin drilling within a stated time or pay a certain stated sum per month for failure to commence drilling and also provided that a failure on the part of the lessee to comply with the conditions of the lease would render the lease void. These provisions give the lessor the option as to his remedy and he may elect to put an end to the lease or he may elect to have the lease continued in force to the end of the term and enforce the payment of the amount due each month.

Allen v. Marver (California), 172 Pacific 980, p. 981.

BREACH OF CONTRACT TO DRILL—MEASURE OF DAMAGES.

The lessee in an oil and gas lease of certain lands entered into a contract with an oil-well driller to drill on the leased land an oil well to the depth of 1,260 feet at 85 cents per foot. Pursuant to the contract the driller moved his drilling machinery and equipment onto the land, commenced drilling and drilled to the depth of 10 feet. A dispute arising between the lessor and the lessee as to certain terms of the lease the lessee directed the well driller to suspend drilling until the dispute over the lease should be settled, but requested him to remain on the premises until the dispute was settled, but finally stopped the drilling altogether and prevented the completion of the well. In an action for damages for a breach of the contract the measure of damages which the well driller was entitled to recover was the expense necessarily incurred in hauling his drilling rig and machinery from where they were to the well that he began drilling, the expense necessarily incurred in rigging up and drilling to the point where drilling was stopped, and reasonable compensation for services in removing the rigging and drilling machinery, reasonable compensation for the enforced idleness of the rig and machinery, and the reasonable value of the well-driller's services lost during the time he remained on the premises at the request of the lessee of the land.

Letcher v. Maloney (Oklahoma), 172 Pacific 972, p. 973.

BREACH—LIQUIDATED DAMAGES OR PENALTY.

An oil and gas lease provided that the lessee should begin drilling operations by a certain date and on failure to do so should pay the lessor $100 per month for each and every month in which he fails to commence operations. The covenant to pay the $100 a month is not a penalty but is liquidated damages. In an action on such a lease to recover the amount of the monthly payments, proof of the amount of damages is unnecessary as the amount is fixed by the terms of the lease. Damages for breaches of contract touching future interests in oil wells of unknown value are of such remote and speculative value as to bring them peculiarly within the rule that the parties should have the right to fix them by mutual agreement. It would be impossible to calculate with any degree of certainty the amount of damage sustained by a lessor by reason of the breach of the covenant of such a lease by the lessee.

Allen v. Marver (California), 172 Pacific 980, p 981.

LEASE OF MINERALS—RIGHT TO USE OF SURFACE.

Ordinarily the leasing of minerals, by implication, carries with it the right to use as much of the surface as may be actually necessary to mine from the stratum conveyed. The same rule applies to the grantor who excepts a stratum from the soil of his land conveyed to another.

De Moss v. Sample, 143 Louisiana ———, 78 Southern 482, p. 484.

CONSTRUCTION—CONSENT TO DRILL—CONSENT OF ASSIGNEE NOT REQUIRED.

An oil and gas lease conveyed all the oil and other minerals under a certain tract of land on the fixed payment of royalties. The lease gave the lessee the right to subdivide and sell portions thereof and prohibited the lessee from drilling wells within 300 feet of the residence of the lessor except with the consent of "both parties hereto, their heirs and assigns." This provision was intended to prevent the lessee from drilling within the excepted distance without the consent of the lessor. It was also intended to prevent the lessor from leasing to third persons for drilling purposes the tract within the prohibited distance without the consent of the lessee. But it can not be construed to mean that where the leased premises had been divided up into small areas and proportionate parts of the lease assigned to each purchaser that the consent of each of the assignees should be obtained, where the original lessor consented to the drilling within the excepted distance by either the original lessee or any one of his assignees. Under such circumstances one assignee of a part of the lease can not enjoin another assignee from drilling within the prohibited distance.

McFarlane v. Gulf Production Co. (Texas Civil Appeals), 204 Southwestern 460, p. 461.

DRILLING WITH CONSENT OF BOTH PARTIES—COVENANT RUNNING WITH LAND.

A stipulation in an oil lease to the effect that no wells should be drilled within 300 feet of a dwelling house unless with the consent of both parties, indicates that the parties in making the lease did not intend to burden the property and this intention will prevail as against an effort to make the provision a covenant running with the land.

McFarlane v. Gulf Production Co. (Texas Civil Appeals), 204 Southwestern 460, p. 461.

RIGHT OF RENEWAL—DISPOSITION OF DEPOSIT.

An oil and gas lease provided that it might be renewed for another year by the lessee paying or depositing in a certain stated bank a certain stated amount on or before the expiration of the lease. On failure of the lessee to make the required deposit within the stated time, he could not claim the right of renewal because the lessor failed to return the deposits as the lessor was under no obligation to do more than inform the bank of its willingness to accept the deposits.

Pure Oil Operating Co. v. Gulf Refining Co., 143 Louisiana ——, 78 Southern 560, p. 561.

RIGHT OF RENEWAL—PAYMENT OR DEPOSIT OF RENTALS.

An oil and gas lease expired on March 29, 1913. It contained a provision to the effect that the lessee could renew the lease for another year by a payment of 10 cents per acre and provided that " such payment for such renewal may be made directly to the party of the first part or deposited to his credit." The required amount for renewal was $8. Six dollars of this sum was deposited, but was refused by the lessor, and the lessee subsequently sent a check for the additional $2. The question as to whether this second deposit was made in time was one of fact to be determined in the trial court, and the burden of proof was on the lessee to show that it had been so made.

Pure Oil Operating Co. v. Gulf Refining Co., 143 Louisiana ———, 78 Southern 560, p. 561.

ACCEPTANCE OF DELAY RENTALS—FORFEITURE—VALIDITY OF SECOND LEASE.

An oil and gas lease provided for the payment of a certain sum quarterly as delay rentals. After its execution, on a survey of the land the acreage was found to be less than that stated in the lease, and the amount of the quarterly delay rentals was reduced accordingly. The land owner, standing in the place of the original lessor, accepted the smaller amounts in payment of the quarterly delay rentals. The land owner can not after accepting such payments arbitrarily refuse to receive them and demand the full amount as stipulated in the lease and on default of payment declare a forfeiture and execute a valid second lease on the land.

Monarch Gas Co. v. Roy (West Virginia), 95 Southeastern 789, p. 791.

ACCEPTANCES OF RENTALS BY MARRIED WOMAN—RIGHT TO FORFEITURE—ESTOPPEL.

A married woman acquired lands subject to an oil and gas lease. The assignee of the lease inquired of her if there were any unpaid delay rentals and was informed that there were none. It was a matter peculiarly within her knowledge and which directly concerned her and her reply was equivalent to an affirmation or representation of payment and she can not afterwards be permitted to take advantage of her own dereliction, but she is concluded by the information imparted. She can not afterwards be heard to assert a forfeiture which she failed to declare at the proper time when by her representations she induced the assignee of the lease to predicate his action upon the assumption of a previous satisfactory compliance with the requirements of the lease by the lessee.

Monarch Gas Co. v. Roy (West Virginia), 95 Southeastern 789, p. 790.

POWER OF LESSEES TO MORTGAGE.

The legislature of Louisiana (Acts 1910, p. 393) authorizes the lessees or owners of contracts granting the right to explore and develop land for oil, gas, and other minerals to mortgage their leases.

De Moss v. Sample, 143 Louisiana ——, 78 Southern 482, p. 484.

PROOF OF VALUE OF LEASE—OPINION OF WITNESS.

In proving the value of an oil lease on a 10-acre tract of land with a flowing well thereon, it is competent to prove the value of an oil lease on similar adjoining tracts and it was proper to prove what an owner of an adjoining lease was offered for his lease and also to prove the sum for which he sold part of his tract. It was also proper for the owner of the lease of the adjoining tract, who had sold part of his lease and had offers for the remaining part, to give his opinion of the fair and reasonable market value of the lease in controversy.

Peden Iron & Steel Co. v. Jenkins (Texas Civil Appeals), 203 Southwestern 180, p. 184.

LEASE BY ONE TENANT IN COMMON—FRAUD.

The owner of the surface of a tract of land who was also a tenant in common with others in the ownership of the oil and gas under the land, agreed for a sufficient consideration to lease the land for oil and gas purposes, the royalties to be divided according to the relative ownership of the oil and gas. The owner of the surface was to receive in person all sums paid in the way of delayed rentals. The surface owner after leasing the land for oil and gas purposes and after gas had been discovered in the immediate vicinity and wells drilled on other lands owned by him in such position that they would drain the land in controversy, extended the lease for a long period with an agreement to accept delayed rentals in lieu of drilling. Such a lease under such circumstances was a fraud on the rights of his cotenant. The defrauded cotenant could by suit have the lease canceled or he could sue and recover his proportion of the sums received as delayed rentals.

McMillan v. Connor (West Virginia), 95 Southeastern 642, p. 643.

OWNERSHIP OF OIL WELL—PRESUMPTION.

An oil well was located near a disputed boundary line and neither the lessee of the land on one side nor the owner of the land on the other side of the line knew that the well was on his land, but each claimed the well. In a subsequent lease by the landowner the chances that the owner of the land was the owner of the oil well, and that it would be included in the lease, may reasonably be presumed to have been in the contemplation of the parties and as between the lessor and the lessee the presumption is that the well was included in the lease.

Russell v. Producers Oil Co., 143 Louisiana ———, 78 Southern 473, p. 476.

PURCHASE PENDING SUIT—CONCLUSIVENESS OF JUDGMENT.

A person who, pending a suit by a lessor to cancel a lease, himself made a contract with the lessor to purchase the leased premises at a certain stated price per acre if the lease in the pending suit was declared void, otherwise he should be discharged from the obligation to purchase, became from that time a privy with the lessor and was bound by every act of his in the suit and by the judgment against the lessor, the complainant in the action, and can not in a subsequent action between the same parties relitigate the same questions.

Miller v. Belvy Oil Co., 248 Federal 83, p. 92.

APPOINTMENT OF RECEIVER—JURISDICTION.

One partner brought suit for a dissolution of a partnership and for an accounting by reason of the operations of the partnership under an oil and gas lease and for a disposition of the partnership property. The appointment of a receiver was asked by way of provisional and auxiliary relief. The real estate under lease was not within the jurisdiction of the court. A court having jurisdiction of the parties may appoint a receiver for an oil and gas lease of land beyond the jurisdiction of the court; and especially where the lease creates only an incorporeal hereditament.

Huston v. Cox (Kansas), 172 Pacific 992, p. 993.

LIABILITY OF AGENT—UNDISCLOSED PRINCIPAL.

The president of an oil and gas company entered into an oral contract with a well driller to drill an oil well on a certain tract of land. The president acting as agent did not disclose to the well driller his principal, the oil and gas company, and the well driller had no knowledge that the president was acting for the oil and gas company and did not have such information until after a breach of the contract by the agent. Under such facts the president of the company was personally liable under the rule that where an agent enters into a contract with a third person and does not disclose to such third person his principal, the agent is liable upon the contract, if the contracting party has no knowledge that the agent is acting for an undisclosed principal.

Letcher v. Maloney (Oklahoma), 172 Pacific 972, p. 973.

CANCELLATION—RECOVERY OF LIEU PAYMENTS.

An oil and gas lease provided that drilling should begin on or before a certain date or the lease terminate unless the lessee should pay at the beginning of each 6-month period a certain stated sum. The lease also recited that it should not obligate the lessee to make any such payment or to drill or otherwise carry on operations against its wish or option. Under these provisions the lease is not to be constructed as imposing an absolute obligation on the lessee either to drill or pay the stated sum, but as simply providing that the doing of the one or the other of these things is necessary to prevent a termination of the lease and the intent is to provide that the failure to drill or to make the payments in lieu of drilling merely terminates the contract and that is the only contractual liability that can be enforced against the lessee for a failure to do either of these things; and the lessor in an action to recover the unpaid sums stated in the lease and to cancel the lease is not entitled to a judgment both for the cancellation and for the cash payment. He can not invoke both the forfeiture for nonpayment and then collect the payment on account of the nonpayment of which the forfeiture was sought.

Stein v. Producers' Oil Co. (Texas Civil Appeals), 203 Southwestern 126, p. 128.

FORFEITURE—LESSOR.

The lessor in an oil and gas lease is ordinarily the only person who can take advantage of a condition in a lease providing for a forfeiture for failure of the lessee to comply with its terms, " unless there is an express stipulation in the lease that same shall be null and void upon failure of the lessee to comply with its terms."

Brennan v. Hunter (Oklahoma), 172 Pacific 49.

FORFEITURE—REASONABLE OPPORTUNITY FOR PREVENTING.

An oil and gas lease required the lessee to begin drilling operations on or before a stated date. It also provided that on payment of a certain stated sum drilling operations might be deferred for a period of six months. The lease also provided that if after operations were begun they should be suspended for 30 days or more without the payment of the stipulated sum the lease should be subject to forfeiture. The lease also provided that after oil was discovered in paying quantities the lessee should be exempt from loss or forfeiture except after judicial ascertainment and a reasonable opportunity thereafter to prevent forfeiture. The lessee began operations within the stated time and drilled a well producing oil in paying quantities and drilling operations were thereupon suspended for more than 30 days without the payment or tender of the stipulated amount. The lessor subsequently brought suit for a cancelation of the lease. Under these circumstances it was the duty of the court to enter judgment canceling the lease unless the lessee should pay the stipulated sum in lieu of drilling operations for each six-month period beginning with the completion of the producing well to the date of the judgment, and permitting the lessee 30 days after judgment in which to make payment and renew drilling operations.

Stein v. Producers' Oil Co. (Texas Civil Appeals), 203 Southwestern 126, p. 128.

FORFEITURE—EXCEPTION TO RULE.

Forfeitures are not generally favored by law but forfeitures which arise in oil and gas leases by reason of the neglect of a lessee to develop or operate the leased premises are rather favored because of the peculiar character of the product to be produced. It has been found necessary to guard the rights of land owners by numerous covenants, some of the most stringent kind, to prevent their land being burdened with unexecuted and profitless leases incompatible with the rights of alienation and the use of the land. Forfeiture for nondevelopment or delay is essential to profit and public interest in relation to the use and alienation of property. Perhaps in no other class of leases is prompt performance of contract so essential to the rights of the parties, or delay by one party likely to prove so injurious to the other.

Hughes v. Busseyville Oil & Gas Co. (Kentucky), 203 Southwestern 515, p. 517.

82478°—19——10

MINING PROPERTIES.

TAXATION.

CLASSIFICATION OF PROPERTY FOR TAXATION.

A State may classify the property within its borders and impose unequal taxation provided the taxes are uniform on all property in each class. Such a classification does not violate the provisions of the fourteenth amendment to the Constitution of the United States. A State statute for the taxation of sulphur mines, or persons engaged in mining sulphur, is not void because the person complaining is the only person in the State engaged in sulphur mining. This can not affect the classification.

Union Sulphur Co. v. Reed, 249 Federal 172, p. 174.

MINING INTERESTS TAXABLE.

As different elements of land are capable of being severed and separately owned the State of Louisiana now authorizes separate assessments against the owners of the different interests; and if different parts or interests of land have been severed and are held by different persons, the land may attach to one and the coal or oil, or mineral rights to another.

De Moss v. Sample, 143 Louisiana ——, 78 Southern 482, p. 484.

VALUATION OF MINERAL LANDS—ERRORS OF JUDGMENT.

Errors of judgment by taxing officers will not support a claim of discrimination in valuations between mineral lands and other lands. The good faith of such officers and the validity of their actions is presumed and when assailed the burden of proof is on the complaining party.

Sunday Lake Iron Co. v. Wakefield Township, 38 Supreme Court Rep. 495.

VALUATION FOR ASSESSMENT—MINERAL INTEREST.

Section 39, chapter 29, Barnes' Code (Code 1913, sec. 923), construed with reference to other provisions of the code does not require or authorize the valuation and entry on the land books for taxation of undivided interests in timber, coal, oil, and gas or other minerals or mineral substances but only the whole of such estates or interests may be separately entered and taxed to the owner or owners thereof.

State v. Guffey (West Virginia), 95 Southeastern 1048, p. 1049.

EQUALITY IN VALUATION—CONSTITUTIONAL PROTECTION.

The owner of a mine or of mineral lands is denied the equal protection clause of the fourteenth amendment of the Constitution where his mine or mineral lands is assessed for taxation at the full cash value and there is an intentional systematic undervaluation by State officers of other taxable property in the same class.

Sunday Lake Iron Co. v. Wakefield Township, 38 Supreme Court Rep. 495.

INCREASED VALUATION—VALIDITY OF ASSESSMENT.

Local taxing officers valued certain mineral lands for the purpose of taxation at $65,000. The State board raised the assessment to $1,071,000. At the time the owner of the mineral lands first challenged the valuation placed on the mineral lands of others as compared with that placed upon his mineral lands, no adequate time remained for detailed consideration and there was not sufficient evidence before the board to justify immediate and general revaluation. In refusing to hear evidence of the value of the other property of the same class, the owner of the mineral lands was not deprived of the equal protection of the law, where it is made to appear that the next year a diligent and successful effort was made to rectify any inequality.

Sunday Lake Iron Co. v. Wakefield Township, 38 Supreme Court Rep. 495.

VALUATION OF SEPARATE ESTATES—ASSESSMENT AS A UNIT.

The statute of 1913 (ch. 29) provides for annual valuations of lands and interests or estates therein and the assessor on a segregation of any such estate is required to assess such severed estates to the owners thereof and therein; but the assessor is not now limited as formerly to an apportionment of former values of the whole between the estates so segregated. There is nothing in the act of 1905 even impliedly sanctioning the assessment of undivided interests in such segregated estates. The statute contemplates the whole of any estate, not undivided interests therein.

State v. Guffey (West Virginia), 95 Southeastern 1048, p. 1049.

SEPARATE ASSESSMENT OR SALE OF UNDIVIDED INTEREST—VALIDITY.

The whole of any estate in land whether surface, timber, coal, oil or gas, after severance may be the subject of separate assessment, but the separate assessments or sale for taxes of undivided interest therein are void.

State v. Guffey (West Virginia), 95 Southeastern 1048, p 1049.

AUTHORITY OF COUNTY TREASURER—OMITTED PROPERTY.

The right and duty of a county treasurer under the statute of Oklahoma (Rev. Laws 1910, secs 1732, 7449, as amended by Laws 1915, ch. 189) do not cease until he has exhausted all the means for this purpose placed at his disposal by the legislature. As it is his duty and right to act for the State in giving the original notice, so it continues to be his duty and right to pursue to the end, with the advice and counsel of the law officers of the State all the remedies afforded by the statute for the collection of these taxes. In the case of omitted property a tax ferret is employed to assist the treasurer in collecting taxes and not to collect such taxes.

Kramer v. Gypsy Oil Co (Oklahoma), 173 Pacific 802, p. 805.

ASSESSMENT OF OMITTED PROPERTY—NATURE OF PROCEEDING.

A proceeding for the assessment and collection of taxes due on omitted property is not a civil action but is a remedial proceeding granted by the legislature conferring upon the county treasurer a remedial right and duty not heretofore existing for the collection of taxes due on omitted property. The legislature as a matter of grace gave the tax payer the right to appeal to the county court where summary proceeding may be heard de novo.

Kramer v. Gypsy Oil Co. (Oklahoma), 173 Pacific 802, p. 805.

FRANCHISE TAX—ANNUAL LICENSE FEE.

The annual license fee or franchise tax imposed upon corporations by the New Jersey statute (Laws 1906, p. 31) is payable each year in advance, the year beginning with the first Tuesday of May.

Old Dominion Copper Min. & Smelting Co. v. State Board of Taxes, etc. (New Jersey), 103 Atlantic 690.

LICENSE TAX ON MINING.

The statute of Louisiana (Act 145 of 1916) required a corporation engaged in mining sulphur to pay a tax of ten cents per ton for the privilege of continuing operations during the ensuing three months.

No lien is imposed on the amount of sulphur previously mined and no direct seizure of the sulphur is provided for. If the tax is not paid it must be collected in the same manner as all other license taxes by suit in a court of competent jurisdiction. This tax can be classed as a license tax and the fact that the operator of the sulphur mine can not pay the tax and can not stop operations without destroying his mine can not affect the validity of the tax.

Union Sulphur Co. v. Reed, 249 Federal 172, p. 178.

LICENSE TAX ON OIL AND GAS BUSINESS.

The Legislature of Louisiana (Acts 1912, p. 437; Acts 1916, p. 37) requires an annual license tax to be paid by persons engaged in "severing natural products, including all forms of timber, turpentine and minerals, including oil and gas, sulphur and sale from the soil."

De Moss v. Sample, 143 Louisiana ———, 78 Southern 482, p. 484.

CORPORATION EXCISE TAX—DEDUCTIONS FROM INCOME.

In 1898, a mining company by assignment of a lease acquired a leasehold estate in certain ore producing properties from which it mined ore from that date to and including the year 1910. The company in the latter year made a return to the collector of internal revenue of its gross income and from the total deducted, to cover realization of unearned increment, something over $256,000. The amount was found by multiplying the number of tons of ore mined during the year by 48¾ cents, which was the market value of the ore in place on January 1, 1909, the date upon which the returns for taxation were commenced. The deduction was made in good faith upon the claim that it was a reasonable allowance for depreciation of the property for the year. The assignee of a lease is in no sense a purchaser of ore in place, but he takes from the property the ore mined paying for the privilege so much per ton for each ton removed. He has this right or privilege under his lease so long as he sees fit to hold the same without exercising the privilege of cancellation therein contained. The mining company, as assignee of the lease, was not entitled to deduct $265,000, the market value of the ore in place from its gross income derived from a sale of the ore mined during the year as so much depletion of capital assets.

United States v. Biwabik Min. Co., 38 Supreme Court Rep. 462, p. 463.

CORPORATION EXCISE TAX—INCOME FROM SALE OF STOCK—METHOD OF COMPUTING.

A mining corporation, in 1902, purchased the shares of stock of another mining corporation for $800,000. In 1911 the company sold the same shares of stock for $1,010,000, a gain of $210,000. The proportionate part of the $210,000 that is income and subject to an assessment of 1 per cent under the corporation tax act of 1909 must be ascertained, under the Treasury regulations, by that proportion thereof represented by the ratio of 1,019 days that elapsed between January 1, 1909, when the corporation excise tax act became effective, and October 16, 1911, the date of the sale, to the 3,233 days that elapsed between the date of the purchase and the date of the sale. This proportionate part constitutes the income of the corporation for the year 1911 within the meaning of the corporation tax act.

Hays v. Gauley Mountain Coal Co., 38 Supreme Court Rep. 470, p. 471.

CORPORATION EXCISE TAX—SALE OF STOCK—INCOME.

In December, 1902, a coal mining corporation purchased certain shares of stock of another mining corporation for $800,000. In 1911 the mining corporation sold the same shares of stock for $1,010,000, being a gain of $210,000 over the purchase price, omitting the item of interest. The corporation tax act measures the tax by the income received during the year for which the assessment was levied, whether it accrued within that year or some preceding year, but it excludes all income that accrued prior to January 1, 1909, although received after the corporation tax act was in effect. It results that so much of the $210,000 of profits as may be deemed to have accrued subsequent to December 31, 1908, must be treated as a part of the gross income of the mining company and is subject to the excise tax provided for by the corporation tax act.

Hays v. Gauley Mountain Coal Co., 38 Supreme Court Rep. 470, p. 471.

FORFEITED LANDS—SUIT BY STATE TO SELL—PLEADING.

The commissioner of school lands of a county instituted a suit to sell for the benefit of the school fund certain lands and the undivided interest or estates in oil and gas under various tracts of land alleged to have been forfeited to the State for nonentry on the land books for taxation for five successive years. It is not sufficient to allege in the bill merely that upon the segregation of such undivided estates they were not thereafter entered on the land books. The presumption is that when the whole of such segregated estates have been separately valued and entered for taxation they continued to be entered and valued along with the other estate or estates in the land as an entirety, and the bill must contain allegations sufficient to overcome this presumption. The allegation that the tracts out of which such undivided mineral interests are reserved or granted were entered on the land books as "surface" and that such mineral estates were not subsequently valued or entered or taxed is not sufficient to overcome a presumption that such mineral estates continued to be entered and taxed as before severance.

State v. Guffey (West Virginia), 95 Southeastern 1048, p. 1049.

TAXATION PREVENTS FORFEITURE.

Taxation of mineral lands in the name of any person claiming the land in whole or in part under any given title prevents forfeiture not only as to him and the portion of land claimed by him but as to all other lands so taxed and all persons claiming the same or any part thereof under the same title.

State v. Guffey (West Virginia), 95 Southeastern 1048, p. 1049.

LIENS.

SALE OR RENTAL OF MINERALS—LIEN FOR ROYALTIES—PLACE OF ENFORCING.

Royalties upon minerals usually, if not always, become due and are a lien on the minerals after separation and consequently as chattels. But if by a conveyance the title to the minerals vests, with a lien reserved for unpaid purchase money payable before separation, then the lien attaches to the minerals as real estate. It follows, therefore, that an action to enforce a lien for the payment of royalties for minerals after separation is transitory, but if the conveyance vests the title to the minerals in the vendee and retains a lien for unpaid purchase money, due and payable before the severance, then the action would be local to the county where the land is located under the Statute of Kentucky.

Roseburg National Bank v. Camp (Oregon), 173 Pacific 313, p. 315.
Kennedy v. Hicks (Kentucky), 208 Southwestern 818, p. 320.

FIXTURES.

PIPES USED IN HYDRAULIC MINING.

Sections of pipe made into a pipe line for carrying water and giants or monitors purchased by the owner of a placer mining plant and placed upon the ground with the intention of using them for the sole purpose of operating the placer mining ground and without any expectation of removing them from the premises, but where it was necessary to make frequent changes in the pipe line, extending the same to keep pace with the progress made in washing the ground, or otherwise changing and separating the joints and removing and reuniting the same as convenience might require, are not fixtures in the sense that they can be sold on execution with the mining claim.

Roseburg National Bank v. Camp (Oregon), 173 Pacific 313, p. 315.

JUDICIAL SALE OF FIXTURES—VALIDITY.

Mining machinery, tools, and implements used in the ordinary process of hydraulic mining on a placer mining claim can not under the statute of Oregon be sold in connection with and as a part of the real estate and mining claim. Neither can a pipe line and the giants or monitors used for applying the water in the operation of placer mining be sold as fixtures where the pipe line was constructed of a series of joints put together in the manner of a stovepipe and constantly removed from place to place, lengthened or shortened as convenience required in the mining operations.

Roseburg National Bank v. Camp (Oregon), 173 Pacific 313, p. 316.

INSURANCE AND LOSS.

LIABILITY AND RECOVERY.

Certain mining property was shipped by boat from Seattle, Wash., to Uyak, Alaska. The property was covered by a policy of marine insurance "in the power schooner Harold Bleakum at and from Seattle, Wash., to Uyak, Alaska." The vessel encountered a storm near the entrance to Uyak Bay which carried away some of her sails, rendered her unmanageable, and after being abandoned by the crew and the master, she was stranded, within the meaning of that term, upon the east shore of the bay. The storm encountered near the entrance to the bay was the proximate cause of the loss of the property insured, as all the subsequent events depended and flowed from the injury to the vessel caused by that storm.

Amok Gold Min. Co. v. Canton Insurance Office (California Appeals), 171 Pacific 1098, p. 1099–1102.

TRUST.

DIVERSION OF TRUST FUND.

The fact that an executrix of a will improperly mingled the royalties received from mining properties with the residuary estate instead of paying the proper proportion to a legatee, gives the legatee an interest in the residue equal to the sum of money that was due her from the funds received as royalties; and a court in a State where ancillary administration proceedings are pending may compel the executrix to pay over to the legatee the amount due without regard to the domiciliary administration proceedings.

Campbell's Estate, In re (Utah), 173 Pacific 688, p. 692.

QUARRY OPERATIONS.

An employee injured while working in a quarry is not to be
charged with contributory negligence and his action for damages
defeated on the ground that he voluntarily chose a hazardous way
of doing his work when another and a safer way was open to him,
where the evidence shows he was doing his work in the usual and customary way and that he was at the time obeying an order of the
superintendent, who was present and saw and knew how the work
was being done, made no objections thereto, and gave no warning to
appellee of danger because of the way he was doing his work.

Oolitic Stone Mills Co. v. Cain (Indiana Appeals), 119 Northeastern 1005,
p. 1007.

NEGLIGENCE—EXCESSIVE BLAST.

In the operations of a quarry an explosion of dynamite threw a
rock one and one-half squares distant through a window and injured
a woman in a dwelling. Such a violent and unusual result amounts
in itself to evidence from which a jury might infer a lack of proper
care in the amount of explosive used for the blast.

Rafferty v. Davis (Pennsylvania), 108 Atlantic 951, p. 952.

USE OF EXPLOSIVES—INJURY FROM BLASTING—INFERENCE OF NEGLIGENCE.

In quarry operations where the blasting in which the accident
occurred was in the exclusive management of the quarry operator
all the elements of the occurrence were within his control. The
result was so far out of the usual course that no fair inference that
it could have been produced from any other cause than negligence
and no other cause is apparent to which the injury may with equal
fairness be attributed, an inference of negligence may be drawn.
Negligence in such case is not presumed, but the circumstances
amount to evidence from which it may be inferred by a jury.

Rafferty v. Davis (Pennsylvania), 108 Atlantic 951, p. 952.

BLASTING—CARE REQUIRED—LIABILITY—PROOF.

Where the work of blasting is done in a thickly settled portion of
a city by reason of which a person is injured, damages are recover-

able without proof of negligence and notwithstanding proof that the person firing the blast employed skillful and experienced persons and exercised the highest degree of care. Such work is so inherently dangerous that the doing of it, no matter how careful, is of itself negligence and no amount of care in doing a negligent act will excuse the actor from the responsibility of the consequences.

Rafferty v. Davis (Pennsylvania), 103 Atlantic 591, p. 952.

USE OF DYNAMITE—CARE REQUIRED.

The general rule in cases of the explosion of dynamite where third parties having no relation to the person having it in possession are injured is that the highest degree of care must be exercised.

Rafferty v. Davis (Pennsylvania), 103 Atlantic 591, p 952.

DAMAGES FOR INJURIES TO MINERS.

ELEMENTS OF DAMAGES.

DAMAGES FOR FUTURE PAIN AND SUFFERING.

In an action by a miner for damages for injuries caused by the alleged negligence of the mine operator, the court may properly instruct the jury that they have the right and should take into consideration among other things "the physical pain and mental anguish suffered and endured, and that he will probably hereafter endure by reason and on account of said injury."

Picino v. Utah-Apex Min. Co. (Utah), 173 Pacific 900.

DAMAGES NOT EXCESSIVE.

INSTANCES.

A judgment for $15,000 in favor of an oil-well driller, a healthy man 29 years of age, earning $90 a month, is not excessive where his right hand and arm below the elbow were crushed and rendered worthless and after the accident he was confined to a hospital for four months undergoing three operations at a total expense of $1,000.

Farmers' Petroleum Co. v. Shelton (Texas), 202 Southwestern 194, p. 198.

DAMAGES EXCESSIVE.

INSTANCES.

A miner caught between a coal car and a prop suffered a fracture of the pelvic bone and other injuries. The pelvic bone healed and he was able to work after about one month. There was some ossification of the muscle 'that prevented as free use of his legs as formerly and some pain and inconvenience suffered in that respect, but there was no prospect of any future loss of earnings and he was entitled to no damages except the one month's wages and compensation for pain suffered and the inconvenience. A verdict of $3,750 was regarded as excessive and a remitter of $1,250 was ordered and if refused a new trial to be granted.

Whittaker v. Valley Camp Coal Co. (Pennsylvania), 108 Atlantic 594, p. 595.

INTERSTATE COMMERCE.

TAXATION OF PRODUCT OF MINE.

A license tax imposed upon that part of the product of a sulphur mine that is sold in interstate and foreign commerce is invalid.

Union Sulphur Co. v. Reed, 249 Federal 172, p. 174.

Bulletin 175

DEPARTMENT OF THE INTERIOR
FRANKLIN K. LANE, Secretary
BUREAU OF MINES
VAN. H. MANNING, Director

EXPERIMENT STATIONS OF THE BUREAU OF MINES

BY

VAN. H. MANNING

WASHINGTON
GOVERNMENT PRINTING OFFICE
1919

First edition. August, 1919.

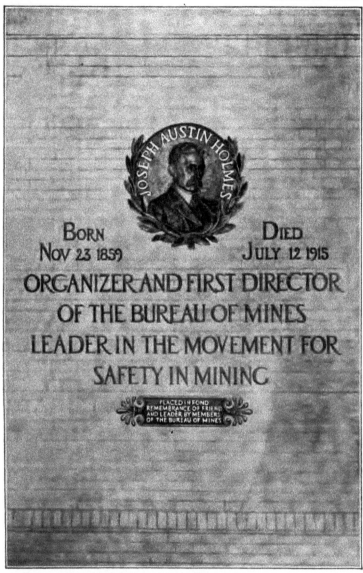

JOSEPH A. HOLMES MEMORIAL, AT THE PITTSBURGH STATION OF THE BUREAU
OF MINES.

CONTENTS.

ILLUSTRATIONS.

EXPERIMENT STATIONS OF THE BUREAU OF MINES.

By Van. H. Manning.

FOREWORD.

During the nine years that have elapsed since the Bureau of Mines was established in 1910, the work of the bureau has included many investigations that have proved of high value to the Nation. Eleven mining experiment stations established by act of Congress have greatly increased the bureau's usefulness; being situated in the mining fields, these stations are closely in touch with industrial needs, and the results of their research and educational work are more directly available to those engaged in the mining, metallurgical, and mineral industries.

In the United States mining is second in importance to agriculture only. The Federal Government annually appropriates large sums in the interest of agriculture, whereas the appropriations for mining are relatively small. But the nation-wide popularity of Government assistance to agriculture should not overshadow the need of Federal assistance to the mineral industry. With scientific management farms can be made to produce indefinitely without any depletion of the potential resources of the soil, but the mineral wealth of a country is a fixed quantity, and every year's production brings nearer its ultimate exhaustion.

As the deposits of the richer or more readily available ores, especially those in the West, have been depleted, development of methods for mining and treating the leaner or more inaccessible ores has lagged, so that meeting the increasing demands for certain metals is becoming more and more difficult. In some branches of mining the small independent operator is greatly handicapped. He has to rely on custom mills and smelters for the profit or loss of his product, and because of relatively inefficient methods of milling or smelting he may receive no return for much of the ore he mines. Even though the small operator thinks he knows of improved methods of obtaining the metals from the ores, he can not afford to install the necessary machinery or to carry on experiments that may or may not prove successful.

In view of the increasing need of metals and the impossibility of renewing the ore supplies, efforts should be directed toward develop-

7

ing the most economical and efficient methods in mining and metallurgy. The great mining companies carry on independent research
work, some of them on a large scale; but as such work is usually
undertaken for gainful purposes, these companies do not feel under
any obligation to give the results of their investigations to the world.
Clearly the solution of the difficulty is to place in the mining districts
trained mining and metallurgical engineers who will investigate and
solve the problems for the benefit of the whole industry. It is with
this purpose in view that experiment stations of the Bureau of Mines
have been established to investigate economy, efficiency, and safety
in mining, and to make public the results of the investigations.

The Federal Government can be of most help in developing the
mineral resources of the country when its forces are brought in direct
contact with these resources. Mining is as old as civilization, but
only within a generation has this country busied itself with any other
problem of mining than the amount of mineral products that could
be put upon the market. Little concern was felt if much of the
metal in an ore was wasted, if the miner lived in unsanitary surround-
ings and died of preventable disease, or if accidents occurred frequently and from causes that could easily be controlled. Now we
realize that the waste of any resource is inexcusable. The mineral
wealth of the country can be mined but once. The task before the
Nation is to recover as much metal as possible from the ore in the
ground, and to do this in the most economical way without waste of
human life. This task is of such magnitude as to warrant aid from
the Government through experiment stations. In its efforts to make
the work of these stations of most benefit to the mineral industry the
Bureau of Mines seeks the cooperation of State authorities, of mine
operators, of miners' organizations, and of every person interested.

These experiment stations also serve an educational purpose. Generally speaking, mining is not a preferred occupation. The miner
should have reason to take greater interest in his work; he should feel
that the experiments the Government is making affect him vitally by
enabling him to increase his production and by making his work safer.
The mine operator should realize that the Government stands ready
to aid him, and he should know that he can bring his own problems
to a Government organization that is trained and equipped to solve
them. The surest method of accomplishing these results is to carry
the work of the Bureau of Mines into the mining regions through the
establishment of experiment stations easily accessible both to the
mine operator and to the miner.

Many miners in this country do not read English, but they can
understand motion pictures of mining methods; they can learn how
lives can be saved in case of accident; and they can appreciate the
larger results gained by modern methods and equipment. In every

mining region there should be some agency for the disseminating of knowledge that will increase the efficiency of the miner and safeguard his life. Such work has been started in certain districts now served by experiment stations of the Bureau of Mines. The aim is ultimately to cover all mining areas, and Congress has authorized the opening of additional mining experiment stations from time to time.

The war has emphasized the need for governmental research into all matters that in any way concern the common welfare. The Federal Government has the funds and the power to conduct investigations in the most efficient way. It can avoid the duplication of effort that is apt to occur if such work is carried on by the States alone. By working in conjunction with the State authorities the Government can include the special phases of research work necessary in any particular State. The stations of the Bureau of Mines already established have demonstrated their value in solving unusual problems.

It is difficult to measure and appreciate the contribution of these stations to the winning of the war. They helped to find more economical and efficient methods of production and to stimulate the production of minerals necessary to the proper equipment of our Army and Navy. The successful results of their work can be traced in almost every branch of the Army, from certain steels needed for the big guns of the Ordnance Department to the hundreds of soldiers trained in the use of oxygen breathing apparatus in the mining and sapping regiments of the Engineer Corps.

The experiment station at the American University on the outskirts of Washington, D. C., solved new problems in the testing and manufacture of war gases so efficiently that long before the end of the war the production of poison gas was far ahead of the supply of shells. This station also helped to develop highly efficient gas masks, smoke screens, and other defensive devices.

Although the work at this station was taken over by the War Department on July 1, 1918, in accordance with the program of uniting the gas investigations under the Chemical Warfare Service, work is still being carried on there, and the valuable equipment and laboratories form a nucleus around which a great mining experiment, station system could be developed.

The station at Columbus, Ohio, situated at a clay-working center is employed mostly on ceramic problems. In this country there are about 4,000 firms manufacturing clay products, including brick, tile, sewer pipe, conduits, hollow blocks, architectural terra cotta, porcelain, earthenware, china, and art pottery. The amount invested in these industries is approximately $375,000,000 and the value of the products exceeds $208,000,000 annually.

The station at Bartlesville, Okla., is investigating problems that arise in the proper utilization of oil and gas resources, such as elimination of waste of oil and natural gas, improvements in drilling and casing wells, prevention of water troubles at wells, and of waste in storing and refining petroleum, and the recovery of gasoline from natural gas.

What the Bureau of Mines has done for the great coal-mining industry, chiefly through investigations at the experiment station at Pittsburgh, Pa., has been published in numerous reports issued by the bureau. Some of the more important accomplishments have been the development and introduction of permissible explosives for use in gaseous mines, the training of thousands of coal miners in mine-rescue and first-aid work, and the conducting of combustion investigations, aimed at increased efficiency in the burning of coal and the effective utilization of our vast deposits of lignite and low-grade coal.

How vast are the deposits of low-grade ores being made available through the experiment stations is shown by the work assigned to the station at Minneapolis, Minn. The primary purpose of this station is to devise methods of utilizing low-grade iron ores. It has been estimated that the reserves of low-grade magnetic iron ores in the State of Minnesota alone amounts to some forty billion tons, but until recently these ores have been untouched because no process of treating them profitably had been devised. Even now only one company is attempting to utilize them. The Minneapolis station has already demonstrated that one process for utilizing the great deposits of manganiferous iron ore on the Cuyuna Range is metallurgically possible.

Work such as this not only stimulates mineral production and helps to make available tremendous resources that are now unused, but it increases the total wealth of the Nation and ultimately benefits every citizen.

The mining industry is so related to commerce and manufacture that the importance of publishing the results of technical investigations of mining problems is becoming more and more evident. The miner and the mine operator are integral parts of the industrial system of the country, and each needs to keep in close touch with what the Government is doing through the mining experiment stations of the Bureau of Mines.

In the West vast quantities of low-grade complex ores will become available as soon as commercially feasible processes are devised. Many of the problems involved, which are being attacked at the Golden, Colo., station, the Salt Lake City, Utah, station, the Seattle, Wash., station, the Tucson, Ariz., station, and the Berkeley, Calif., station, are of such a nature that the small operator can not afford to attack them, and the large operator finds them outside his field.

Yet the solution of any one of them may add greatly to the available resources of the country and may result in establishing a new industry that will build up the district in which it is situated. Already these stations, although young, have witnessed such results.

To-day, through the efforts of men at the Golden station, there is an American radium industry. Formerly the low-grade radium-bearing ore was wasted, the best of the ore was bought by foreign concerns at ruinously low prices, and the radium was shipped back to this country at excessively high prices.

The Salt Lake City station has devised novel methods of treating certain low-grade and complex ores of lead and zinc. These methods show a large saving of metal over methods hitherto employed, and have made available ores that other methods could not treat profitably.

The Seattle station is busy with the beneficiation of the low-grade ores of the Northwest, and the mining and utilization of the coals of the Pacific States; the Tucson station is working on the beneficiation of low-grade copper ores; and the Berkeley station has shown how losses may be reduced at quicksilver plants and how methods at those plants can be improved.

In the conduct of these investigations the bureau seeks and is obtaining the cooperation of the mine operators. At more than a dozen mills in the West engineers from the stations are working directly with the mill men on various problems, and the results they already have obtained more than warrant the existence of the stations. Success in solving one problem may easily be worth millions to the country. Mining men are using these stations more and more freely as they realize that the Government maintains these stations to help them, and that the difficulties of the operators, both large and small, will receive sympathetic consideration and such aid as the stations can give.

ACKNOWLEDGEMENTS.

Dorsey A. Lyon, supervisor of stations, H. E. Tufft, of the editorial staff of the bureau, and Prof. W. C. Thayer, of Lehigh University, temporarily detailed for service with the Bureau of Mines, assisted in the preparation of this bulletin. Also, acknowledgement is due H. Foster Bain, formerly assistant director of the bureau, for criticism and correction of the manuscript.

COOPERATIVE WORKING AGREEMENTS WITH STATE INSTITUTIONS.

In order to insure the most effective action with State agencies working along similar lines to the bureau, to harmonize activities and avoid needless duplication of effort, and to have the advantage of the use of the facilities provided by such institutions, the Bureau of Mines endeavor to cooperate with State organizations and with State universities and mining schools. The wisdom of this policy has been amply demonstrated and the number of such institutions cooperating with the bureau is constantly being added to each year.

As has been stated, a number of the mining experiment stations of the bureau are established at or housed in buildings provided by State universities or mining schools.

Thus the station at Urbana, Ill., situated at the engineering experiment station of the University of Illinois, works in close cooperation with that institution and with the Illinois Geological Survey. The station at Tucson, Ariz., is housed in a building on the campus of the University of Arizona; the station at Berkeley, Calif., occupies a building of the University of California, and works in cooperation with that institution; the station at Moscow, Idaho, is situated at the University of Idaho; the station at Golden, Colo., is at the Colorado School of Mines; the station at Minneapolis, Minn., is at the University of Minnesota; the station at Columbus, Ohio, at Ohio State University; the station at Salt Lake City, Utah, at the University of Utah; and the station at Seattle, Wash., at the University of Washington. In addition to cooperating with the institutions mentioned, the bureau also does cooperative work with the Industrial Accidents Commission of California, the Oregon Bureau of Mines and Geology, the University of North Dakota, and Cornell University.

SAFETY AND RESCUE STATIONS.

The mine-rescue and safety work is under the supervision of George S. Rice, chief mining engineer of the Bureau of Mines, and is in charge of D. J. Parker, mine safety engineer, with headquarters at Pittsburgh, Pa.

For the purpose of facilitating the mine-rescue and safety work of the bureau the country is divided into districts, the safety and training stations and the headquarters for mine rescue cars being selected with regard to convenience and effective effort. The seven safety stations are at Pittsburgh, Pa.; Norton, Va.; McAlester, Okla.; Birmingham, Ala.; Jellico, Tenn.; Seattle, Wash.; and Vincennes, Ind. The eight mine rescue cars, three of which are specially designed all-steel cars, have headquarters at Pittsburgh, Pa.; Huntington, W. Va.; Ironwood, Mich.; Evansville, Ind.; Pittsburg, Kans.; Raton, N. Mex.; Butte, Mont.; and Reno, Nev. In addition to these eight cars, three all-steel cars similar to those now in use are under construction and will be put in service as soon as completed. These new steel cars will probably replace three of the five wooden cars now in use. Also three rescue trucks for general training work are maintained, one each at the Pittsburgh, Birmingham, and Seattle stations. An additional rescue truck is now under construction for the Vincennes station.

Each station is in charge of a foreman miner, who gives both mine-rescue and first-aid training. The stations are equipped with emergency sets of rescue apparatus and first-aid supplies for rendering assistance at near-by mines in the event of a disaster. The mine-rescue cars are completely equipped for giving aid at such disasters and for training work, and have accommodations for about 12 persons. Each car when fully manned has a foreman miner who trains men in safety and rescue methods, a first-aid miner who gives training in first aid, a mining engineer, and a surgeon. Plate II shows a mine-rescue car. The interior of such a car is shown in Plate III.

WORK OF THE CARS AND STATIONS.

In 1917 and 1918 a special feature of the work of the cars and stations, in addition to the rendering of aid at mine disasters, the training of miners in first-aid and mine rescue work and the organizing of field contests in first aid and mine rescue, has been the training of soldiers at military camps in rescue and safety methods and in the use of breathing apparatus in poisonous or irrespirable gases.

An important duty of the men engaged in this work is inspecting rescue apparatus for mining companies and giving advice as to its condition and the repairs needed.

During the fiscal year ending June 30, 1918, fully 8,851 miners were trained in mine rescue and first aid at more than 70 towns in the 19 States where such training was given; the lectures and safety demonstrations were attended by more than 33,000 miners. Since its establishment the bureau has given first aid or rescue training, or both, to more than 55,000 miners.

An important feature in stimulating the interest of miners in rescue and safety work is the holding of mine-rescue and first-aid contests at which teams from the different mines compete. During the fiscal year 1918 Bureau of Mines employees participated in 13 field contests. Some of these were company affairs, others were intercompany or interstate. The bureau's men helped to arrange and supervise some of the contests and gave competing teams special instruction and training. Handbooks on rescue and first-aid methods, published by the bureau in 1916 and 1917 have proved of especial value in training men at mines.

RESCUE WORK AT MINE DISASTERS.

Members of the Bureau of Mines investigated 38 mine accidents during the fiscal year ended June 30, 1918. Thirty of these were in coal mines and eight in metal mines.

The value of oxygen breathing apparatus in exploring mines after fires and explosions has been demonstrated many times. As the possibilities and limitations in the use of breathing apparatus are becoming better understood, and as recovery work at mines is being placed on a more systematic basis, success in the employment of such apparatus is more assured, and the possibility of the wearer coming to grief through lack of proper knowledge of the device becomes decreasingly small. A technical paper describing the principal types of breathing apparatus and the proper precautions in their care and use was published by the bureau in 1917.

INVESTIGATIONS OF MINE SANITATION AND HYGIENE.

Through a cooperative agreement with the United States Public Health Service, experienced surgeons are detailed for service with the Bureau of Mines rescue cars. These surgeons investigate sanitary conditions at the mines and camps visited by the cars, and give illustrated lectures on how to keep the home and the mine sanitary, and how to prevent the spread of communicable diseases. Health hazards and conditions tending to favor occupational diseases in mines, mills, and industrial plants are studied and methods of improving or alleviating such conditions are suggested. Nearly always this work has received hearty cooperation from the mine operators, and the workers and their families show keen interest in the lectures. The surgeons also examine applicants for rescue training to determine whether they are physically qualified for mine rescue work.

Special investigations relative to silicosis, or miners' consumption, and the general health of miners in the Butte, Mont., district are being made by a mining engineer of the Bureau of Mines and a surgeon of the Public Health Service. These investigations will show whether the disease is prevalent in Montana mines and the precautionary measures that will have to be taken where working conditions tend to injure the miners' health.

The complete results of a similar investigation in the Joplin district, which had been underway for some years, was published as Bulletin 117 of the bureau.

THE MINING EXPERIMENT STATIONS.

The work of the Bureau of Mines in connection with the 11 mining experiment stations that have been established since 1908, the needs that led to the establishment of each station, and the problems that confront the bureau are discussed in the following pages. There were general reasons for placing each of these stations in the territory where it is situated, as well as such local reasons as position near mining fields and convenience of transportation.

BUREAU OF MINES RESCUE CAR AND CREW.

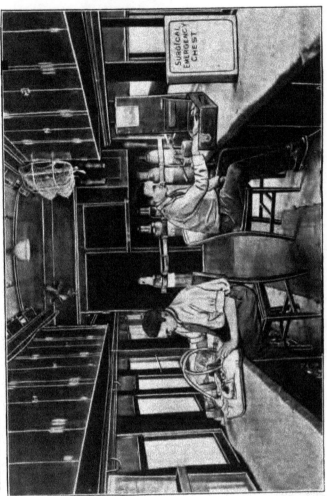

INTERIOR OF BUREAU OF MINES RESCUE CAR.

General oversight of all the mining experiment stations is intrusted to Dorsey A. Lyon, supervisor of stations, with headquarters at Washington, D. C.

The superintendents of the stations are selected because of their having broad technical knowledge, as well as skill in handling economic problems. Solution of a mining or milling problem is often complicated by the fact that the mine or plant under consideration may be situated in a new or undeveloped district, so that the availability of fuel or supplies and the transportation facilities must be taken into account in devising methods of extracting or treating the ore. Each station looks after the interests of a diversified mining population with its many problems of safety and health. Without police powers or other means of enforcing its views, each station must depend upon the good will of the region that it is established to serve. It must consider problems of transportation and supply, study markets, and devise feasible methods of production and treatment.

The list of stations, with the dates of their establishment and the general character of the minerals or industries that they are to study, is as follows:

Experiment stations of the Bureau of Mines.

Station.	Date.	Chief minerals or industries.
Station at Pittsburgh, Pa..........	1908	Coal mining in all its phases, including mine safety, first-aid, and mine rescue work; testing of explosives; also metallurgy of iron and steel.
Station at Urbana, Ill..............	1908	Coals of the Middle West fields, and general coal preparation.
Station at Salt Lake City, Utah.....	1910	Complex and low-grade ores, especially those of zinc and lead.
Station at Golden, Colo............	1910	Rare metals and problems connected with the mining and metallurgical treatment of other ores commonly found in Colorado and adjoining States.
Station at Berkeley, Calif..........	1911	Metallurgical treatment of the ores common to California and adjacent States.
Station at Tucson, Ariz............	1916	Metallurgy of the ores common to the Southwest and especially the treatment of low-grade copper ores.
Station at Seattle, Wash...........	1916	Problems connected with the mining and utilization of the coals of Alaska and the Pacific Northwest; also problems connected with the treatment of the nonferrous ores of those districts.
Station at Fairbanks, Alaska.......	1916	Problems connected with the mining industry in Alaska.
Station at Columbus, Ohio.........	1917	Clay industries and ceramics.
Station at Minneapolis, Minn.......	1917	Problems connected with the mining and beneficiation of low-grade iron ores, manganese ores, and other ores of the Lake Superior district.
Station at Bartlesville, Okla........	1917	Problems connected with the production of petroleum and the conservation and utilization of petroleum and natural gas.

Each station makes a monthly report to the director of the Bureau of Mines, through the supervisor of stations, on the progress of the investigations that are being conducted. Also, suggestions and

117615°—19——2

requests for the maintenance and extension of the work are made to the director, through the supervisor, who determines what problems call for immediate solution.   Thus the work of the stations is coordinated; each station pursues its own specific duties; the director and his assistants decide the broader questions of relative importance; and the supervisor sees that the work at each place is kept steadily advancing in accord with the general plan.   Each station acts independently, but interests and investigations are closely coordinated; an earnest purpose to regulate official action in the correlated interests of science and of industrial progress, with due regard to public happiness and prosperity, is the end sought.

THE PITTSBURGH STATION.

ORIGIN OF THE WORK ON FUELS.

The work now being done by the Bureau of Mines had its beginning in the testing and analyzing of fuels, as authorized by Congress, at the Louisiana Purchase Exposition in St. Louis, Mo., in 1904—a duty then under the supervision of the United States Geological Survey. Early in 1905 the work at St. Louis was placed under the direction of Joseph A. Holmes, as expert in charge. Analyses of coals, steaming and coking tests, and various related tests were made, and after 1905, under appropriation made by Congress, the testing of structural materials was added. By order of the Secretary of the Interior, in April, 1907, these two allied groups of investigation—testing of fuels and testing of structural materials—were made a separate branch of the United States Geological Survey, with Dr. Holmes, as chief technologist, in charge. Early in the same year all the fuel-testing equipment, except that for coking and washing tests, was moved to the Jamestown Exposition at Norfolk, Va., where tests of coals used in the Navy were made during the years 1907 and 1908.

ESTABLISHMENT OF THE PITTSBURGH STATION.

In 1908, before the Federal Bureau of Mines was established, a beginning of Government investigations relating to coal mining was made at Pittsburgh, Pa. By a temporary arrangement with the War Department, part of the grounds of the old Arsenal on Butler Street was leased to the Department of the Interior. There material and equipment were assembled, the nucleus being formed from exhibits and machinery used in the investigative work at the St. Louis and Jamestown expositions. The chief motive for the establishment of this station was the desire to help and guard the miner, to prevent mining accidents, to rescue the victims of mine disasters, and to experiment on different means of meeting and eliminating the ordinary risks of the miner's life.

The first efforts at this station related to the testing of explosives for use in coal mines and to gas-producer, briquetting, and steaming tests of coal. Large-scale tests to determine the explosibility of coal dust and the best methods of preventing dust explosions began in 1910 at the experimental mine established at Bruceton, about 13 miles south of Pittsburgh. In 1911 Dr. Holmes issued his first annual report as director of the new bureau. This report presents not only the practical results achieved, but also the patience and

17

reading room. The top floor accommodates the division of technical service, with rooms for computing, drafting, photographic, and motion-picture work, with dark rooms and appliances for printing

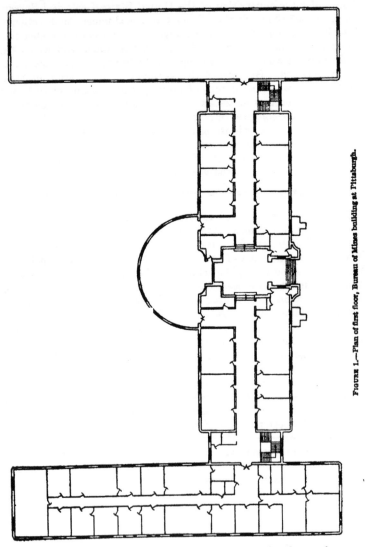

FIGURE 1.—Plan of first floor, Bureau of Mines building at Pittsburgh.

and development, and storerooms where thousands of negatives are kept. On this floor there is also a kitchen and a large room adjoining which is used as a cafeteria for the employees of the station.

NEW BUILDING OF THE PITTSBURGH EXPERIMENT STATION.

WEST END OF CORRIDOR, PITTSBURGH STATION.

The ground floor holds the instrument shop, a supply and store-room, a smoke room for testing oxygen rescue-apparatus and for training men in their use, rooms for rescue apparatus, for shower baths and toilets. The entire east wing is occupied by the offices, workrooms, and laboratories of the chemical division. The west wing houses the engineering interests and the investigations of fuels. The mechanical engineering laboratory on the ground floor runs the whole depth of the wing and is two floors high.

THE POWER HOUSE.

At the old site, as already noted, the power plant has been dismantled. The new power house on the Forbes Street site has been completed, and the machinery is installed. Advantage has been taken of the slope to place the power house on a lower level, behind the main building and out of sight from the front. (See Pl. VI.) The structure is 55 by 220 feet, built of reinforced concrete, and divided into three parts. A short spur from the railroad tracks enters the grounds at the southwest corner. Coal is lifted from the cars by means of an electric hoist to a large pocket, from which, on the inside, fuel may be taken into the boiler house. This is at the west end of the building, where three boilers are installed—two water-tube boilers of the Babcock-Wilcox type, formerly in the Capitol at Washington, and one Parker boiler. The center space in the building is occupied by the engine room. There are now installed three simple engines of the Nordberg type, connected with 200 k. v. a. three-phase, 440-volt generators. The engines can be run independently or any two can be run in parallel. The motor-generator set will furnish direct current at 110 and 220 volts.

The space at the east end of the building is intended for several kinds of apparatus. The apparatus for the study of heat transmission through boiler tubes is being set up here; also special furnaces for the study of the combustion process within the fuel bed. House-heating boilers, four in number—perhaps six if there is room enough—will aid in the study of problems of efficiency in heating houses by both steam and hot-water systems.

A De Laval turbine has been brought over from the old arsenal site and set up in the engine room for the use of the electrical laboratory in special testing.

PROPOSED SHOP.

Among the buildings projected there may be a structure known as the service building, to be between the main building and the power house, running lengthwise from east to west, for storing tools and supplies, and available for the manifold needs of a large plant employing many men. It should be a long and narrow building, two stories in height, with the probable dimensions of 200 by 40 feet,

of an architectural style to conform to the general appearance of
the main building.

As grading and filling goes on the property will be developed on
an orderly and well-conceived general plan, making this site a digni-
fied and acceptable home for the Pittsburgh station.

DIVISIONS OF THE BUREAU REPRESENTED AT PITTSBURGH STATION.

The following divisions of the bureau are represented at the Pittsburgh station: Mining, fuels and mechanical equipment, explosives,
petroleum, metallurgy, chemical, administrative.

Below is given a summation of the character of the work of each of
these divisions:

MINING DIVISION.

The principal work of the mining division at Pittsburgh has to
do with coal mining. Two phases of this work are emphasized—
(1) Mine rescue and first-aid work; and (2) the prevention of mineral
waste.

MINE RESCUE SECTION.

The mine rescue section of this division is engaged in mine rescue
and first-aid work, chiefly the rendering of aid at mine disasters,
the training of miners in first-aid and mine rescue, and the conducting
of first-aid and mine rescue contests.

A mine safety motor truck fully equipped with mine rescue apparatus and mechanical resuscitating devices, inhalator, safety lamps,
oxygen cylinders, life-line reel, stretchers, and first-aid cabinets,
complete with sterilized bandages, compresses, bandages, etc., is
held in readiness to be dispatched to the scene of disaster on short
notice. The truck is in charge of a crew expert in mine rescue
and first-aid work, and is sent to all mine fires, explosions, and disasters in the Pittsburgh district.

All work relating to the testing of rescue apparatus and of mechanical resuscitating devices is in charge of this section. Both the
Gibbs and the Fleuss machines are being tested for efficiency. All
apparatus sent to the mine rescue cars and field stations is tested at
Pittsburgh before being shipped.

A schedule of tests has been prepared for establishing a list of
permissible mine rescue breathing apparatus, giving fees, character
of tests, and conditions under which apparatus will be tested.

Names of men trained by the bureau between July 1, 1914, and
June 30, 1916, were published in Technical Paper 167,[a] and those of

---

a Parker, D. J., Men who received Bureau of Mines certificates of mine rescue training, July 1, 1914, to
June 30, 1916: Tech. Paper 167, Bureau of Mines, 1917, 66 pp.

REAR VIEW OF PITTSBURGH STATION, SHOWING POWER HOUSE.

men trained between July 1, 1916, and June 30, 1918, are printed in Technical Paper 226.[a] This list will supplement that published as Technical Paper 167, which gives the men trained by the bureau during the period July 1, 1914, to June 30, 1916. The publication of the names and addresses of men so trained enables State and mine officials to obtain with minimum delay the services of the nearest available trained men in the event of a mine disaster.

THE EXPERIMENTAL MINE.

The primary purpose of the bureau's experimental mine at Bruceton, Pa., 13 miles south of Pittsburgh, is to enable explosions of coal dust and gas to be made on a scale comparable with ordinary mine explosions. The walls, roof, and timbering of a mine present conditions that can not be duplicated in a gallery of wood, steel, or concrete. Hence preventive measures that are quite successful under artificial conditions in a surface gallery are not necessarily successful in a mine. Arrangements of the experimental mine make possible the duplication of any kind of explosion that may occur in a coal mine. In this mine, moreover, other investigations relating to safety can be made, such as tests of gasoline locomotives, of mining or cutting machinery, of electrical equipment, and of ventilation methods. Gas-producer tests and physical and chemical tests of explosives can well be associated with a plant of this sort.

COAL-DUST EXPLOSIBILITY TESTS.

Dusts of various coals have been tested for explosibility in the experimental mine. The coals tested range from anthracite, containing 5 or 6 per cent volatile matter, to subbituminous coals containing more than 40 per cent. Four of the coals tested by the standard method [b] are: One from Clearfield County, Pa.; one from near Benham, Ky.; one from the Crows Nest Pass field, British Columbia; and one from Vancouver Island, British Columbia. Each coal was tested for explosibility in sizes corresponding to the average fineness of road dusts obtained from the mine in which the sample was taken, and also in a size of standard fineness for comparison with other coals.

Also, tests of various sizes of dust of Pittsburgh coal from the experimental mine have been made. This series, when completed, gave the explosibility of four sizes of coal dust under three conditions—without gas, with 1 per cent, and with 2 per cent of gas in the ventilating current.

A series of tests was made to determine the effect of free moisture, when mixed with coal dust and with rock dust, upon the explosibility of the four standard sizes of Pittsburgh coal dust.

---

[a] Parker, D. J., Men who received Bureau of Mines certificates of mine rescue training, July 1, 1916, to June 30, 1918: Tech. Paper 226, Bureau of Mines, 1919. 72 pp.
[b] Manning, Van H., Yearbook of the Bureau of Mines for 1916: Bull. 141, 1917, pp. 26–27.

All of these tests were made in the standard test zone in the experimental mine, which is lined with "gunnite," or blown concrete. From these tests three charts or sets of curves, have been prepared, which show how the composition of the coal, the size of the dust particles, and the moisture content effect the explosibility of the dust. These charts have been found useful in interpreting the results of explosion tests and in determining what dust mixture is necessary at any particular mine to render the coal dust inert.

A view of a manometer, for measuring the pressures developed by an explosion, is shown in Plate VIII, *B* (p. 30).

EXPLOSION BARRIERS.

In this mine, explosion tests of barriers containing rock dust in shelves and troughs which tip with the force of the initial explosion, have been repeatedly made. Rock-dust barriers of various types which have been devised by bureau engineers and successfully applied in commercial mines are described in detail in Technical Paper 84.[a] Recently much success has been had with troughs that discharge a thin sheet of water in fine spray through a ½-inch aperture. In tests made no explosion has penetrated this barrier.

EFFECTS OF MINE STOPPINGS ON EXPLOSIONS.

A series of explosion tests was made in the two butt entries in the experimental mine to determine the effect of mine stoppings in checking or propagating explosions. Explosive limits for two sizes of coal dust were determined in one of the butts, with heavy stoppings placed in the cut-throughs and with no stoppings. A similar series of tests was made in the other butt, from which five rooms have been turned. These tests show that while the rooms did not change the explosive limits much, they did cause lower pressures and velocities. The series indicates that the danger of explosions being propagated is practically no greater with strong stoppings than with light stoppings or no stoppings at all. Resistance and confinement, however, within certain limits increase the dynamic thrust of the explosion. The effect of an explosion on a concrete stopping, which was completely destroyed, is illustrated in Bulletin 141.[b]

A series of tests was made to determine how near the mouth of the entry an explosion could be started that would travel inby and through the mine. It was found that a blown-out shot of 2 pounds of black blasting powder would propagate an explosion through pure coal dust when the source of ignition was 180 feet from the mine mouth. When the point of ignition was moved to within 112 feet

---

[a] Rice, G. S., and Jones, L. M., Methods of preventing and limiting explosions in coal mines: Tech. Paper 84, 1915, 44 pp.

[b] Manning, Van H., Yearbook of the Bureau of Mines for 1916. Bull. 141, Bureau of Mines, 1917, p. 30.

of the mine mouth, propagation was not obtained even when the igniting charge was increased to 4 pounds of powder. In all of these tests the flame traveled outby to the mouth of the entry.

STENCHES FOR WARNING MINERS OF DANGER.

A novel method of warning miners of danger by turning a foul-smelling gas into compressed-air lines has been developed at the experimental mine. One of the most promising stenches tried is a preparation of butylmercaptan or isomercaptan; its pungent and disagreeable odor gives timely warning of an unsafe condition and allows the miner time to escape. Its manufacture involves the replacement of the oxygen of alcohols by sulphur, compounds known as hydrosulphides or mercaptans being formed. Thus butyl alcohol (C_4H_9OH) is converted to butylmercaptan (C_4H_9SH).

In the event of sudden danger, such as a fire on the surface near the mouth of the mine, or in one of the upper levels with a large number of men in the next level below, or any condition necessitating hasty escape of the men, the engineer at the surface can at once, by means of a simple device, introduce this pungent substance into the compressed-air pipes. Its odor is unmistakable, and is instantly detected by the miner, who has been told to regard the odor as a danger signal. The odor reaches him through the drill that he is using, the hoists where he is working, the pump that he is controlling, and through all forms of machinery using compressed air. The method is simple and direct; it replaces expensive telephone equipment, and is more dependable than electric bells and signals.

COAL-MINE ENGINEERING.

One of the chief efforts of the mining engineering division is to increase the output of coal by advising the use of mining methods that insure the recovery of a larger proportion of the coal mined and a general saving in expenses. A progressive operator using improved methods and fewer men will mine an acre thoroughly, whereas a non-progressive competitor will mine two acres with less production, greater cost, and more discomfort to the miner. A bigger production per acre decreases the expense for mine props, for haulage, for pipes, and for electric systems. Thus, the small mine, intensively developed, may prove a far better venture than a larger one poorly worked.

Better light is an immediate benefit, as a miner working by good light can do more work with greater safety and comfort, and will give more willing and efficient service than one working in a dimly lighted place. It is obvious that in a mine worked intensively much better lighting facilities can be had for less cost than in a poorly worked mine covering a much larger area.

ELECTRICITY IN MINES.

The work of the electrical section of this division at the Pittsburgh station has largely to do with the testing of electric mining equipment submitted by manufacturers to determine its safety for use in gaseous coal mines. The danger of an explosion being started by an electrical machine, lamp, or installation sparking or becoming overheated when gas is present is so obvious that the importance of this work is evident to the most casual observer. Equipment that passes the tests prescribed by the bureau is approved by the bureau, this approval being attested by a plate bearing the seal of the Department of the Interior. As a result of this work, mining has been made immeasurably safer and the use of equipment approved by the bureau is being rapidly extended.

Apparatus is tested under schedules issued by the bureau from time to time, as the need for a particular class of apparatus develops. The types of apparatus for which schedules have been issued to date include the following: Permissible electric motors for mines, permissible portable electric mine lamps, permissible miners' safety lamps, permissible gas detectors for mines, permissible mine-locomotive headlights, permissible electric lamps for mine rescue service, permissible flash lamps for use in explosive mixtures of methane and air, permissible single-shot blasting units, and permissible self-contained oxygen breathing apparatus.

These schedules are becoming widely known throughout the mining industry, and a large proportion of the mines are using apparatus approved by the bureau.

Besides extending the approval system manufacturers are encouraged to develop safe apparatus, and the bureau is constantly seeking to improve laws and rules covering the installation and maintenance of electrical equipment in mines.

TESTING OF MINE MOTORS.

The coal-mine operator requires a motor that can be used in a mine where fire damp may occur. In the bureau's tests of explosion-proof apparatus, assumption is made that such explosive mixtures are present. The motor or other explosion-proof piece of apparatus to be tested is placed in a gallery (Pl. VII, A), where it is not only surrounded by an explosive mixture of gas but the motor casing is filled with it. Plate VII, B, shows a motor piped so that an explosive mixture of gas can be placed within the casing. The gas within the motor is ignited by a spark plug or from the commutator. If no explosion of the gas outside the motor occurs after repeated tests of this kind, the motor passes the permissibility tests and is given the bureau's approval for use in gaseous mines.

A. MOTOR TESTER.

B. MOTOR PLACED IN GALLERY FOR TESTING IN EXPLOSIVE GAS.

PHYSICAL TESTS OF EXPLOSIVES.

As previously stated in this bulletin, all work relating to the physical testing of explosives is now conducted at the experimental mine. Congress appropriated $17,000 for moving the equipment, and for building foundations, instrument shelters, and service lines (water, gas, steam, electric, sewerage, and drainage). The equipment, including two galleries each 100 feet long, the ballistic pendulum, the Mettegang recorder, small impact machine, large impact machine, pendulum friction machine, flame-testing apparatus, Bichel gages, calorimeter, and a variety of lead-block expansion and compression-test equipment, were moved. These apparatus and the methods of using them have all been described in Bulletin 66 [a] of the bureau.

The moving was planned and so conducted as to interfere the minimum length of time with any given test, so that there was practically no interruption of the testing work during the time of moving. This was accomplished by preparing the foundations for a given apparatus and then dismantling, moving, and setting up promptly. The moving began in July, 1917, and continued intermittently for two months.

The new installation (Pl. VIII, *A*) includes four instrument shelters 20 by 30 feet, and one story high, of hollow-tile construction and concrete floors; three "bomb-proofs"—one very large, within which several pounds of explosive may be detonated, one within which an hydraulic press is installed, and one for apparatus for mixing explosives; separate magazines for the storage of high explosives, including permissible explosives, black blasting powder, and other easily ignited powders, detonators, electric detonators, and samples sent to the station to be tested. Especially heavy foundations were required for the ballistic pendulum, the galleries (Pl. IX, *A*), and the large bomb-proof (see Pl. IX, *B*.) Several hundred feet of concrete walks and concrete steps were built in order that explosives and dangerous blasting supplies could be transferred with safety. The magazine for high explosives is built in accordance with the principles laid down in Bureau of Mines Technical Paper 18, "Magazines and Thaw Houses for Explosives."

The work of this division consists of studying the physical characteristics of the various classes and grades of explosives used in mining and also the physical characteristics of different kinds of blasting supplies; the character and quantity of poisonous and inflammable gases produced by the explosives used in close or inadequately ventilated places are also studied, the purpose being to classify explosives

---

[a] Hall, Clarence, and Howell, S. P., Tests of permissible explosives: Bull. 66, Bureau of Mines, 1913, pp. 4-17.

on this basis so that the user may have information on one of the dangers attending the use of explosives in mines. Exhaustive tests are also made of coal-mining explosives to determine their permissibility for use in dusty or gaseous coal mines.

CHEMICAL TESTS OF EXPLOSIVES.

In the explosives chemical laboratory of the bureau at Pittsburgh examinations are made of samples of explosives submitted to the bureau for test, to determine the permissibility of the explosives for use in dusty or gaseous coal mines or the suitability of the explosives for use in metal mines, tunnels, and quarries. Under this head is included the examination of samples of permissible explosives collected in the field from time to time in order to determine whether the permissible explosives on the market maintain the composition of the samples originally tested by the bureau.

In addition to the chemical examinations of mining explosives to determine their permissibility for use in dusty and gaseous coal mines, and for other industrial purposes, much work in connection with the analysis and inspection of explosives has been conducted for the War Department and other departments of the Government, including the inspection of dynamite and detonators for the Panama Canal. The cooperative work for the War Department included tests of grenade fillers, shell fillers, and gun greases. On account of the excellent facilities possessed by the bureau for research work in explosives, the cooperative work with the War Department on explosives and explosive materials for use in munitions developed rapidly during 1917 and 1918.

An investigation is being made of different methods of analyzing and of testing explosives, and of the standardization of all methods with a view to preparing a manual describing the best methods of procedure to follow in such work. Also a study was made of the preparation, properties, and uses of explosives. A technical paper describing the methods used in the explosives laboratory has been published by the bureau.

Work on a propylene glycol dinitrate, to be used as a commercial explosive, has been completed. The process involved the nitration of propylene glycol, the determination of propylene glycol nitrate thus made, and the preparation, analysis, and tests of different types of commercial explosives in which this material is to be substituted for the nitroglycerin now commonly in use.

Further work has been done on the determination of explosive materials or their products of combustion in coal samples from boreholes or blown-out shots. A paper on this subject is in course of publication.

A. THREE OF THE NEW INSTRUMENT SHELTERS, EXPERIMENTAL MINE.

B. MANOMETER FOR MEASURING THE PRESSURES DEVELOPED BY AN EXPLOSION.

A. GAS-AND-DUST GALLERY, EXPERIMENTAL MINE.

B. THE "BOMB-PROOF," AT EXPERIMENTAL MINE, FOR TESTING RATE OF
DETONATION OF EXPLOSIONS.

Representatives of this laboratory assisted in the investigation of explosions at places where explosives were being made for military use.

Experiments have been made to determine the suitability of flax and the down of the milk weed, cat-tail, and white smoke-root as substitutes for cotton in the manufacture of nitrocellulose for nitroglycerin.

As nitro substitution compounds are used in all classes of explosives, a knowledge of the nitrogen content of such compounds is useful in identifying them and determining their purity. The bureau has issued a technical paper describing the various methods used for the determination of nitrogen in substances used in explosives.

"SAND TEST" FOR DETONATORS.

The sand test for detonators, devised by the bureau, has been modified to give results more nearly proportional to the quantities of priming employed. The sand test [a] has been thoroughly described in former publications of the bureau. The detonator whose strength is to be determined is fired by electricity in the center of a definite quantity of standard 30-mesh quartz sand contained in a block of steel with cylindrical bore. Then the sand is sifted and the amount of crushing of the sand measures the strength of the priming. A set of such tests soon leads to a general standard for the detonating force of all explosives so tested. It is also possible with this device to determine whether the detonation of the explosive was partial or complete. This test has proved to be a distinct aid to manufacturers and users of detonators.

PETROLEUM DIVISION.

The routine work of the petroleum laboratory of the Pittsburgh station consists largely of the inspection of fuel oil purchased for the Government. Other work has included the analysis of samples of crude oil, tests of lubricating oils, tests of fuels for airplane motors, and various products of cracking processes, so-called "gasoline improvers," gun greases, and coal-tar derivatives. Practically all of this work was of military nature and is mentioned in Bulletin 178. [b]

PITCH FORMATION IN THE INTAKE MANIFOLDS OF ENGINES.

The formation of pitch sometimes occurs when gasoline is carbureted. Physical conditions controlling this phenomenon have been determined and the chemical factors involved are to be studied.

---

a Storm, C. G., and Cope, W. C., The sand test for determining the strength of detonators: Tech. Paper 125, Bureau of Mines, 1916, 68 pp.

b Manning, Van H., War work of the Bureau of Mines: Bull. 178, Bureau of Mines, 1919, 106 pp.

DETERMINATION OF UNSATURATED HYDROCARBONS IN GASOLINE.

The results of a study of methods of determining the unsaturated hydrocarbons in gasoline has been published as Technical Paper 181.[a] In the manufacture of motor fuel from heavy petroleum products by cracking processes, and in the testing of such fuel, a knowledge of the percentage of olefins or unsaturated hydrocarbons in the gasoline is desirable. The various methods for making such determinations were studied and experiments made to determine their relative value and develop them to a maximum of convenience for laboratory use.

MISCELLANEOUS WORK.

Apparatus for the making of dynamometer tests of gasoline engines is now being set up and a series of experiments is to be started at an early date.

Much work has been done in developing and perfecting laboratory apparatus used in testing petroleum. Particular attention has been given to the problem of determining the calorific value of liquid fuels.

METALLURGICAL DIVISION.

The technical staff of this division at the Pittsburgh station has recently been doing research work on nickel, tungsten, and molybdenum steel, especially as regards their use for military purposes.

METALLOGRAPHIC LABORATORY.

The work of this laboratory is devoted entirely to the examination of iron and steel for industrial and other purposes. Since May, 1918, when this laboratory was organized, something like 1,500 samples of metal have been prepared, studied, and photographed for the Ordnance Department.

CHEMICAL DIVISION.

WORK OF THE FUEL-ANALYSIS LABORATORY.

The analysis of fuels belonging to or for the use of the United States Government is one of the chief duties of the analytical laboratory and the large number of determinations involved occupy most of its time. In addition, there are analyzed in this laboratory samples of coal, coke, ores, and various materials collected in the course of investigations being made by the Bureau of Mines and other branches of the Federal Government and by State geological surveys and experiment stations. These include the War Department, Navy Department, Panama Canal, United States Geological Survey, Indian Office, and the Illinois State Geological Survey.

---

[a] Dean, E. W., and Hill, H. H., Determination of unsaturated hydrocarbons in gasoline: Tech. Paper 181, Bureau of Mines, 1917, 25 pp.

Sample of coal, mine-road, mine-rib, stone, and coke dusts taken by mining engineers of the bureau during investigations of explosions and fires in coal mines are analyzed in this laboratory. From these analyses and the information obtained in the examination of the mine, it is sometimes possible to form conclusions as to the general nature and causes of the explosion, which can be applied in making recommendations for the prevention of future explosions or fires.

Problems of military importance included the analysis of samples of pyrite taken in connection with the investigation of sources of pyrite for manufacturing sulphuric acid; analyses, for the fuels and mechanical division, of samples of coal, ash, clinker, and residual fuel taken in connection with steaming tests; and examination of materials collected in connection with chemical warfare investigations at the American University.

FUSIBILITY AND CLINKERING OF COAL ASH.

During 1918 the investigations of the fusibility of coal ash from the coals of Pennsylvania, Virginia, West Virginia, Indiana, Illinois, and Maryland were practically completed. The method employed was developed by the bureau and is described in Bulletin 129 [a]. In the laboratory furnace used for fusing the ash a reducing atmosphere is maintained by using much more gas than is needed for complete combustion.

This investigation is of value because the fusibility or softening temperature of the ash of a coal is an indication of the liability of that coal to form clinkers. For most uses the more easily a coal clinkers the less valuable it is for fuel. When the data obtained from these tests has been classified and arranged, every fuel user will be able to judge better the type of coal best adapted to his plant, and to guard against the purchase of coal that may give excessive trouble from clinkering.

PHYSICAL LABORATORY.

The physical laboratory at the Pittsburgh station is equipped for testing and calibrating physical apparatus and measuring instruments used in the work of the bureau and for making a number of tests that require apparatus not usually found in an ordinary laboratory. The calibrations and tests made include the calibration of potentiometers, millivoltmeters, sling meters, and mercury thermometers.

THE GAS LABORATORY.

Immediately after this country entered the war the splendid facilities of the gas laboratory at the Pittsburgh station and the services of the bureau's experienced chemists were offered to the War Depart-

---

[a] Fieldner, A. C., Hall, A. C., and Feild, A. L., The fusibility of coal ash and the determination of the softening temperature: Bull. 129, Bureau of Mines, 1918, 146 pp.

ment for experimental work in connection with gas warfare. This offer was cordially accepted. Many of the testing methods used in grading and inspecting gas masks and absorbents for gas masks were developed here. This testing work and considerable research work was carried on until October, 1917, when most of the war work was transferred to the American University experiment station at Washington, D. C.

On account of the pressing importance of the military investigations, only the more urgent and necessary routine mine-gas investigations, such as are required in the investigations of fires and explosions in mines, were conducted during 1918.

A new method and a portable apparatus have been developed for the determining of small percentages of sulphur dioxide in the air of metal mines. By this method mining engineers can make tests quickly and easily in the mines where the samples of mine air are taken. The apparatus is cheap and all broken parts are easily replaced.

The work on the employment of stenches in compressed air lines as a danger warning to miners, described on page 25, was largely done in this laboratory.

MICROSCOPIC INVESTIGATIONS OF COAL AND OTHER MATERIALS.

A bulletin entitled "Structures in Paleozoic Coals" is in course of publication. The purpose of this report is to clear up some of the confusion that exists as to the character of the components of coal. Conceptions as to the origin, composition, and general nature of the coal substance vary widely, so that it has been difficult for the chemist to attack many coal problems intelligently, and the engineer has not had a broad chemical basis for studies of combustion and other processes relating to the use of coal. Hence the efficient utilization of coal in industrial plants has suffered from this lack of knowledge as to the composition of coal itself. The bulletin now being published contains the result of a microscopic study of coals from different parts of the United States.

The work of this laboratory includes microscopic examinations of coal dust, rock dust, charcoal, and of miscellaneous materials submitted by the various divisions of the bureau. A microscopic study was also made of the structure of charcoal to determine the relation of its structure (both physical and gross) to its efficiency as an absorbent for use in gas masks.

GLASS BLOWER.

The extensive research carried on by this division demands many special scientific instruments of a degree of accuracy not procurable in the open market. All such requirements are taken care of by an expert glass blower.

ADMINISTRATIVE DIVISION.

The administrative division of the Pittsburgh station comprises six sections, as follows: Technical service, clerical, purchases and supplies, library and translations, shops and power plant, care of buildings and grounds.

The chief clerk of the station has general supervision of all clerical work, purchasing, accounting, receiving, shipping, and the supply room. Supervision of shops and power plant is in charge of a mechanical superintendent. The custodian of the buildings and grounds has general supervision of the labor force, the janitors, and the watchmen.

THE LIBRARY OF THE PITTSBURGH STATION.

The largest of the seven branch libraries of the bureau—parts of the main library at Washington—is at the Pittsburgh station. Although these branches will ultimately be duplicates of one another, they differ greatly at present in equipment. The Pittsburgh branch has some 6,600 books and received about 190 periodicals. The books are chiefly technical works on physics, chemistry, geology, and mechanical and electrical engineering, mining, metallurgy, chemical technology, industrial safety and hygiene for miners and metal workers, and miners' diseases.

TECHNICAL SERVICE SECTION.

In order to eliminate as far as possible the doing of routine work by engineers, chemists, and other investigators, the station has a corps of draftsmen, photographers, computers, and clerks who cooperate in designing and drafting special apparatus; in reducing laboratory and test observations; and in preparing data and illustrations for reports, and frequently assist in making test observations.

An important phase of the work from April, 1917, to July, 1918, was the computing, plotting, drawing, and the making of large numbers of blue prints of photographic reproductions of the results of observations relating to the work on war gases at the Pittsburgh station and the American University experiment station.

MOTION PICTURES.

Taking, developing, and exhibiting moving pictures has become an important feature of the bureau's educational campaign for promoting safety and health among miners. The large set of records at the Pittsburgh station representing various subjects is constantly being increased. The bureau has a complete equipment of camera and electric lamps for taking films either on the surface or underground. Usually the taking of pictures has been done in cooperation with the owners of the mine or plant to be pictured, the bureau furnishing the photographer, films, and camera, the operator supplying the workmen, making all electric connections, and sometimes paying the expenses of the bureau's representative.

The films are stored in fire-proof safes which have a large flue connecting directly with the outside of the building. When needed for use these films are placed in special metallic shipping cases holding four to six rolls each.

By a recent arrangement a large film corporation cooperates with the bureau in obtaining motion pictures of mining events of interest to the general public, these pictures are shown at theaters throughout the country in connection with the weekly news items supplied by the corporation.

THE URBANA STATION.

ESTABLISHMENT.

The first mine rescue station west of Pittsburgh was established by the technological branch of the United States Geological Survey, at Urbana, Ill., in 1908, in cooperation with the State geological survey. Later a more direct connection with the University of Illinois led to an investigation of the coals of the interior field of Illinois and Indiana, with reference to methods of mining and utilization. At first limited in scope to mine-safety work, the Urbana station soon became extremely useful in the investigation of certain special problems whose solution has been of general interest and wide application.

On July 1, 1911, the Federal Bureau of Mines, the University of Illinois, and the State geological survey entered into a formal cooperative agreement jointly to investigate methods of mining coal in Illinois, particularly with reference to safety measures and appliances for preventing accidents, to improvement of working conditions, to the use of explosives and electricity in mining, and other inquiries pertinent to the mining and utilization of coal. On July 1, 1917, a new agreement was entered into whereby the scope of the cooperative investigations was broadened to include, besides coal mining, quarrying, metallurgical and other mineral industries. The Bureau of Mines agreed to maintain a mining experiment station at the university, to be devoted to the purposes mentioned and to such other investigations as should be assigned to the station by the director. The university agreed to furnish free of charge to the bureau a laboratory, offices, and the use of its library. In addition the State geological survey and the university set aside certain funds to be used in connection with the cooperative work.

The results of this work in Illinois are contained in 28 bulletins. Of these the University of Illinois, through its engineering experiment station, has issued 12 bulletins dealing principally with coal-mining practices and with coal preparation. The Illinois geological survey has issued 10 bulletins covering the coal resources of the State, a chemical study of Illinois coal, clay materials in coal mines, and surface subsidence resulting from coal mining. The Federal Bureau of Mines has issued seven bulletins, which deal particularly with Illinois mining conditions, as follows: Bulletin 72, "Occurrence of Explosive Gases in Coal Mines"; Bulletin 83, "The Humidity of

Mine Air"; Bulletin 99, "Mine Ventilation Stoppings"; Bulletin 102, "The Inflammability of Illinois Coal Dusts"; Bulletin 137, "Use of Permissible Explosives in the Coal Mines of Illinois"; Bulletin 138, "Coking of Illinois Coals"; and Technical Paper 190, "Methane Accumulations Due to Interrupted Ventilation."

SCOPE OF THE STATION.

Under the new cooperative agreement the Urbana station became a full-fledged mining experiment station of the bureau. Thus it has been possible not only to continue the cooperative investigations in Illinois, but also to extend the work of the station over the contiguous coal-mining fields in Indiana, western Kentucky, Iowa, and Missouri.

Illinois, with more than 36,000 square miles underlain by coal, is the center of the coal-mining industry of the interior coal basin. In 1917 there were mined in Illinois more than 80,000,000 tons of bituminous coal, and the industry employed more than 80,000 men. Of the adjacent States, in Indiana there were mined about 22,000,000 tons; in west Kentucky, 13,000,000 tons; in Iowa, 9,000,000 tons; and in Missouri, 5,000,000 tons—a total for the district of more than 129,000,000 tons, or nearly one-fourth of the entire bituminous production of the United States.

Within the district served by the station are large cement mines and quarries, perhaps the largest limestone quarries in the country, the lead and zinc mines of Missouri, Oklahoma, Kansas, and southern Wisconsin, and practically all the fluorspar mines of the country. In addition Illinois is third among the petroleum-producing States of the Union. About 5,000 underground metal miners are employed in the district.

As regards metallurgical industries the State of Illinois alone contains nearly 40 per cent of all the zinc retorts of the country. Chicago has become a center of iron and steel manufacture second only to Pittsburgh.

To this tremendous mining and metallurgical activity the station at Urbana is centrally situated, as it is midway between Chicago and St. Louis, and nearly every mining district in the interior coal basin lies within a radius of 200 miles. Railroad facilities are convenient in every direction.

THE STATION QUARTERS.

The station occupies five offices on the third floor of the north wing of the ceramics building of the university. Plate X is a view of the building, the position of the offices being indicated by the flag. The rest of the third floor is occupied by the State geological survey.

Directly in the rear of the ceramics building are the mining laboratories (Pl. XI), which are thoroughly equipped for coal washing and

BUILDING HOUSING THE OFFICES OF THE URBANA EXPERIMENT STATION OF THE BUREAU OF MINES.

MINING LABORATORY BUILDING, URBANA STATION.

general coal preparation. Plates XII and XIII are interior views of the laboratories. The analysis laboratory is completely equipped for coal sampling and analysis, mine gas analysis, and investigational work. All these laboratories are being used by the bureau representatives. The work at the station has been in charge of E. A. Holbrook, superintendent. Bureau of Mines rescue car 3, with headquarters at Evansville, Ind., is under general charge of the mining engineer of the station.

WORK OF THE STATION.

RECOVERY OF PYRITE FROM COAL.

Much of the work of the station during 1918 was in connection with the recovery of pyrite from the coals of the central basin, for use in making sulphuric acid. Field trips and extensive concentration experiments carried out in the mining laboratory on machinery of commercial size lead to the conclusion that from many coal-mining districts in the central basin, pyrite now mined with the coal and rejected as worthless, can under most market conditions be sent to a central concentrating plant and there converted into commercially pure pyrite, and usually the coal accompanying the pyrite can be recovered as a by-product. A report describing the results of the tests has been prepared.

The station has cooperated with the geological surveys of the States of West Virginia, Pennsylvania, Ohio, Indiana, Tennessee, Illinois, Iowa, and Missouri in investigating the possible recovery of pyrite from coal mines in these States. Tests and analyses of methods of treatments have been worked out at the station and indicated that a large quantity of commercial pyrite could be obtained from these sources.

MINE ACCIDENTS AND ACCIDENT PREVENTION.

Many of the fatal accidents in the central coal field result from local explosions of gas being propagated by coal dust. During the year a number of such explosions in Indiana and Illinois have been investigated, and at one mine a thorough study of the explosion hazards was made.

On July 1, 1917, the various State agencies in Illinois charged with the inspection and supervision of mines were consolidated as a single State department of mines and minerals under the directorship of Evan D. John. This department has lent its assistance freely in promoting the special work of the station in the Illinois field. On its part the station was able to assist the State department in the recovery of a large coal mine in southern Illinois, after an explosion disaster. Later the mining engineer of the station sampled the air in a number of mines under investigation in southern Illinois by a State mining commission, and the results of the analyses were submitted to the commission. The station also cooperates with the State department in training mine rescue and first-aid teams throughout the State.

SURFACE SUBSIDENCE AT MINES.

Three years ago the Bureau of Mines, as a part of its cooperative agreement, established a series of permanent surface monuments above four coal mines in different parts of Illinois where mining was conducted under different systems and conditions. As mining has progressed beneath these monuments, careful surveys have been made both inside and outside the mine, to determine the effect on the surface of such mining. The question of surface movement or subsidence at mines has become of great importance, especially in Illinois, where much of the coal underlies valuable farm land or building property.

COKING AND CARBONIZING ILLINOIS COALS.

The work on the coking of coal and on the carbonization of coal in inclined gas retorts has aroused interest in the commercial possibilities of the central-district coals for these uses. The wastefulness of present methods of using coal is well known. When coal is burned in furnaces to generate steam, the possible by-products that might be obtained are lost, and only the heat generated by the fuel is utilized. The volatile matter, containing tar and hydrocarbon gases, is hard to burn completely, and the result is often a heavy black smoke issuing from the stacks which, in many cities, constitutes a nuisance. When coked in a by-product oven this coal gives not only a valuable coke, but may be made a source of benzol, toluol, gas, tar, ammonia, and cyanogen, which are all valuable by-products. The modern economic methods of coking the bituminous coals of Illinois and Indiana have been a source of considerable revenue.

The Urbana station has investigated and published the results of past efforts to coke Illinois coal and has cooperated in by-product coke-oven tests in an effort to produce a successful metallurgical coke from this coal.

Considerable headway has been made in the substitution of Illinois coal for eastern gas coal in coal-gas retorts and in substitution of Illinois coal for eastern coke in the water-gas sets used in the manufacture of illuminating gas. This work is being carried on by the Illinois cooperation with the added cooperation of the Illinois gas association. The solution of these problems would give an added demand of considerable commercial importance for Illinois coal.

BUREAU OF MINES

INTERIOR OF MINING LABORATORIES, URBANA STATION.

COAL JIGS, URBANA STATION.

FUTURE WORK.

Great interest is being manifested by mine operators and others in new methods of mining that will permit the recovery of a greater proportion of the coal. At no time has the subject of proper ventilation and the overcoming of dangers from gas and dust in the mines been of so great interest to the practical mining engineer as at present. The campaign of education, among such a multitude of mines and miners, to be effective, must be practical, continuous, and personal. The Urbana station should take, in cooperation with the State and other agencies, a leading part in this work.

A wide field open to the station deals with the technical problems in mining, on which little information is available. Among these are studies of mine layouts, mine haulage, mine hoisting, and mine fires.

The cooperative work with the State geological survey, the university, and the gas association should be vigorously prosecuted. The State survey has appointed a gas engineer to act with the bureau's engineer, and through the other agencies a bench of gas retorts has been secured at Pontiac, Ill., where large-scale tests of the central basin coals can be conducted and operating difficulties overcome. These investigations are being extended to the use of Illinois coal and coke in water-gas sets and in gas producers.

The State agencies of the gas investigation can work in Illinois only, but the bureau's representatives may act in an advisory capacity to the plants in the surrounding States in which Illinois coal can be used.

The investigations of methods in the preparation of coal, of which the saving of pyrite from coal is one phase, are to be continued. Washing and other means of removal of impurities and preparing coal for the market have not received the same technical recognition in the United States as in France, England, and Germany, on account of the great quantity of clean coal here and the low price of the prepared product. Under the present fuel conditions an awakening of interest is taking place along the general line of preparation of coals for the market. The Urbana station, because of its location and unexcelled laboratory facilities for washing and other treatment of coal, should be able to do much work in this general technical field and be prepared to be of real assistance through testing work in specific cases.

Due to the Illinois cooperative agreement and to the relative importance of the Illinois coal field, the station has confined much of its coal investigation work in the central west to mining problems in Illinois. This work should be broadened so as to include the contiguous coal fields of adjacent States, with particular regard to their local conditions. In Illinois the various State agencies have made more progress than the other States in this field in the study of mining conditions and in maintaining a centralized agency for the State functions connected with mining. The direct effect of such agencies must always necessarily be limited by the State boundaries. It should be a particular function of this station, by reason of its ability to work independently of State boundaries, to disseminate information on improved safety practices and more efficient technical methods and be of direct aid in time of mine fires or other unusual occurrences. On the other hand, the bureau's engineers engaged in this work should be able to receive and compile the kind of information that has been of such value to the Illinois mining industry. All this work implies a closer cooperation between the station and the agencies having supervision of mine inspection in the various States of the district.

In sum, the chief work of this station must always relate to the mining, preparation, and utilization of the central basin coals. Each of these divisions offers fields for research that hardly have been entered.

THE SALT LAKE CITY STATION.

ESTABLISHMENT.

Utah's standing as a mining State is due to large bodies of low-grade ore, such as the mammoth deposits at Bingham. Successful treatment of these ores presents a problem not yet fully solved. This was the reason for the establishment of a department of metallurgical research at the State School of Mines. As stated in the act (Laws of Utah, 1913) the purposes of this department were to conduct experiments and research "with a view to finding ways and means of treating profitably low-grade ores, and of obtaining other information for the benefit of mining industry and the utilization and conservation of the mineral resources of the State." During the latter part of 1913 this research work was directed by the head of the department of metallurgy of the University of Utah.

In January, 1914, an agreement was entered into with the Federal Bureau of Mines under which the work of this research department was to be under the direction of metallurgists of the Bureau of Mines assigned to this station, the university to provide quarters and assist in the investigations. Other State institutions and mining and metallurgical companies have freely cooperated in this work.

PRELIMINARY SURVEY OF UTAH ORES.

One of the first steps undertaken was to determine the location and extent of the various low-grade and complex ores in the State. The principal mining districts were visited to determine whether any considerable tonnage of ore had been, or could be, developed which was not amenable to profitable treatment by present day milling and metallurgical processes. The results of this preliminary survey were published in 1915 by the bureau as Technical Paper 90.[a] Since then other districts in the State have been visited and examined from time to time, and these investigations are being continued.

Samples of complex and low-grade ores were collected, especially those offering distinct metallurgical problems, such as oxidized ores of copper, lead, zinc, silver, or gold that were too lean to be smelted direct. These ores were analyzed and the results for ores which up to that time could not be treated economically were tabulated. Immediate need of metallurgical experiment and research for the handling of all such ores was made apparent.

---

[a] Lyon, D. A., and others, Metallurgical treatment of the low-grade ores of Utah, a preliminary report: Tech. Paper 90, Bureau of Mines, 1915, 40 pp.

NEED OF METALLURGICAL RESEARCH.

Mechanical concentrating processes have proved to be unsuited to the treatment of low-grade oxidized ores; therefore attention had to be given to the development of new processes for the treatment of such ores. Research work carried on by the principal copper companies has developed flotation processes for the treatment of low-grade copper ores, but the flotation plants do not save all the copper in these ores, because a large part of the ore treated is a mixture of sulphides and oxides of copper. The problem presenting itself in some places was the recovery of mineral which escaped the concentrators and went into the tailings, with the added problem of finding a suitable process for recovering the copper from the capping of oxidized ore overlying the ores being treated. It was not possible to erect such plants and operate them with any prospect of success until much research and experimental work had been done. A little later much attention was given to hydrometallurgical treatment of zinc ores, especially by the process of roasting the ore, leaching the zinc with sulphuric acid, and then precipitating the zinc. Such work did not solve the zinc problem as a whole, nor was it undertaken for that purpose, but to solve the special problem of each particular ore. It was soon found that there were five distinct metallurgical problems confronting the metal workers of Utah. These five problems were as follows: Treatment of (1) low-grade zinc ores, (2) complex lead-zinc ores, (3) oxidized lead ores, (4) low-grade copper ores, (5) ores carrying gold and silver.

THE ZINC PROBLEM.

The chemistry of zinc is so different from that of other metals that the retort process, until the recent introduction of the electrolytic process, was the only one used to produce the metal commercially. The losses occurring in the zinc industry in mining, milling, and smelting are very great; probably less than 50 per cent of the zinc mined actually reaches the form of spelter. Losses of zinc at the mine amount to a large tonnage. In the average western lead mine zinc generally occurs as a shell in the stopes from which oxidized ores have been taken. To fill the stopes with waste makes these shells an absolute loss. In indiscriminate dumping of low-grade ore and waste on the surface much zinc is also lost. A vast tonnage is tied up in low-grade zinc ores; most of this is oxidized zinc.

The retort process, the method generally used for recovering zinc from its ore, has seemed until recently the only one suited to the requirements of this metal. However, zinc losses in retort smelting are large, and only those ores which have a high zinc content can be treated at a profit. Research work at the station has shown that other methods, such as igneous concentration or volatilization and leaching, have great promise for the successful treatment of low-grade ores of zinc.

THE LEAD-ZINC ORE PROBLEM.

As with zinc, a great deal of lead is lost in mining lead ores from leaving in the mine ores of too low a grade to be treated, and from indiscriminate dumping and waste on the surface. The problem in Utah was twofold. Lead smelters naturally prefer ores containing a high percentage of lead and a minimum of zinc; a point is therefore reached where the zinc present prohibits profitable smelting of the lead. Lead blast furnaces have to accept ores containing more or less zinc until the problem of removing the zinc from the ore before smelting it for lead can be solved. If a cheap and efficient method for separating lead and zinc and saving both could be devised, the miner would receive pay for both metals, whereas now he is penalized for whatever excess of either lead or zinc his ore contains above a specified percentage. The principle involved is that in smelting ores containing lead and zinc the latter volatilizes at a much lower temperature than the lead, makes heavy fumes, and clogs the passages, and also forms accretions on the sides of the furnace. Hence ore containing more than 5 per cent of zinc is penalized, and if it carries as high as 15 per cent is rejected. The saving of both the lead and the zinc would be a distinct advancement in economic conservation of our mineral wealth.

THE OXIDIZED LEAD ORE PROBLEM.

Utah has large bodies of low-grade lead ore which had been oxidized by contact with the air, for the level of ground water in Utah is so low that often the mines will be entirely dry and the lead ores weathered to a condition not permitting concentration by mechanical methods. Some of the ores also are contaminated with zinc, making them complex. The problem in all these ores is one of concentration. The amount of such ores available is hard to estimate, yet hardly a county in Utah fails to have deposits.

THE COPPER PROBLEM.

The problem of treating low-grade copper ores, like many others, is twofold—to concentrate to a point justifying smelting at a profit and to remove copper from ores of lead and zinc by some process other than smelting. The copper ores of Utah were generally treated by the former process. From ore which is a mixture of sulphides and oxides of copper a poor saving is made, oxides not being recovered by ordinary processes of concentration and escaping with the tailings.

The problem is to find a suitable process for the treatment of oxidized ores. Distance from railroads and lack of water complicate the problem.

THE PRECIOUS METALS PROBLEM.

Utah has in the past produced some free gold ores, but seldom very rich and never in large quantities. Some gravels yielded placer gold for a time, and some mines yielded finely disseminated free gold ore rich enough to pay handsomely when reduced by the cyanide process. Mercur gave by far the greatest returns as a strictly gold camp. Its ore was of low grade and could not have been treated but for the advent of the cyanide process just when old processes had failed.

Utah seems destined to produce and handle only low-grade ores. Silver ore rich in the metal was common in most of the early camps. Later developments showed increased tonnage but distinctly lower grade. From the first mining in Utah the State has stood high in silver production; it still awaits some cheap method of treating plenty of "ore in sight" carrying silver only in small quantities. The problem of treating this ore has always been due to the presence of some other metal, generally copper, preventing treatment by the cyanide process, or some other method equally applicable.

Distance from railroads, or lack of water, has made treatment by ordinary metallurgical methods impracticable; in many places even impossible. In connection with these researches the station at Salt Lake City has become an extremely important center for mining and metallurgy.

WORK OF THE STATION.

Under the cooperative agreement with the University of Utah, the university furnishes the offices and part of the funds for the station. Five or six "fellows," graduate students in either mining or metallurgy and candidates for a master's degree, work under the direction of the Bureau of Mines staff on mining or metallurgical problems. The university has recently constructed additions to the metallurgical building for the purpose of accommodating the bureau in this cooperative work. These additions, with their equipment, cost $32,000.

The building housing the offices is shown in Plate XIV.

Laboratory work is for the most part carried on in the metallurgy building, therefore the station is in close touch with the mining and metallurgical staff of the university. Without their help and the untiring efforts of the director of the State School of Mines, this station could not have accomplished the work it has done. Also, outside cooperating agencies are spending much money for specific work carried on by them at the station. The station is at present in charge of Thomas Varley, superintendent, assisted by F. G. Moses and J. C. Morgan.

BUILDING HOUSING OFFICES OF SALT LAKE CITY STATION.

PROBLEMS ATTACKED.

The station has succeeded in working out several processes having to do with some of the important problems connected with the mining industry of the State, and has, it is believed, succeeded in making possible the treatment of ores that could not otherwise be handled at a profit. One of the most important lines of work has been that of assisting the popularizing of the flotation process in the inter-mountain region.

A great deal of research work dealing with brine leaching of oxidized lead ores [a] was done at this station with results so favorable that it was taken up by various mining companies and worked out on a much larger scale than had been possible for the Bureau of Mines. The advent of flotation retarded the progress of the process as a means of treating lead-sulphide mill slimes, for which it was originally developed, but the large tonnage of oxidized ores in different parts of the intermountain region leaves the method a large and fertile field for development.

The chloride volatilization process for oxidized and sulphide complex ores, described in Bulletin 157, [b] was worked out in this station. The process promises to have a wide application and to become one of the most important factors in lead metallurgy. At first the opinion was held that this process could be applied only to the oxide ores of lead and possibly of silver, but later developments have proved that it can be applied to the complex lead sulphide ores as well, and it seems to be admirably suited to the treatment of certain classes of copper ores.

One company has constructed and is operating a plant to treat oxide lead ores by this process; another plant is to be built to treat complex lead-silver sulphide ores by the volatilization process; and another to treat copper ores. All of these plants are the direct results of experimental research at the Salt Lake City station.

The results of the work on lead ores have been incorporated in Bulletin 157, "Innovations in the Metallurgy of Lead," previously mentioned. The results of the experiments in the concentration and hydrometallurgy of zinc ores has recently been published as Bulletin 168, "Recovery of Zinc from Low-Grade and Complex Ores."

COOPERATIVE WORK WITH MINING COMPANIES.

Close cooperation between the Bureau of Mines and the mining interests of the State is made possible by an arrangement through which mining men or companies may enter into cooperative agreements with the department of metallurgical research of the Uni-

---

[a] For further details regarding this work the reader is referred to Bureau of Mines Bull. 157, Innovations in the metallurgy of lead, by D. A. Lyon and O. C. Ralston, 1918, pp. 58-74.
[b] Lyon, D. A., and Ralston, O. C., Work cited, pp. 74-84.

versity of Utah (with which the bureau is cooperating), and have its aid and the use of its equipment for their special problems. Any such person or company pays the university a certain fee for the use of the equipment and the power and supplies that they actually use. The bureau's engineers assist in directing the work and keep fully informed of the progress made. When the work is terminated a complete report is made to the university and to the bureau. If so desired these reports are considered confidential for a period not to exceed two years, after which time the bureau or the university has the right to publish them. To take out patents on any discoveries made in connection with this work is not permitted.

On the other hand, the university and the bureau agree to assist cooperating parties in every way possible, to allow them the full use of the university's and the bureau's equipment, and to give them the benefit of the technical knowledge of the university and of the bureau.

This arrangement has made it possible for the station to be of effective service to the mining people of the State, and has also brought to the files of the station a mass of valuable data. Several cooperative agreements have been made and some companies have been able to work out problems of interest to them in a way that would have been impossible except by such an arrangement. Among the companies that have taken advantage of this agreement have been the General Engineering Co., of Salt Lake City; the Mineral Products Co., of Marysvale, Utah; the Combined Metals Co., of Pioche, Nev.; the Utah-Apex Mining Co., of Bingham Canyon, Utah; and the Emma Consolidated Mines Co., of Alta, Utah. Successful results were obtained in most instances, and commercial mills have been or will be built as a result of these cooperative agreements.

VOLATILIZATION AND ROASTING PLANTS.

Some of the most important advances made at the station were in the volatilization of metal chlorides from ores. The results of the small-scale tests were so successful that a plant which could handle 100-pound samples was built.

This was replaced by a new volatilization plant at the pyrometallurgical laboratory, which consists of a rotary kiln with a maximum capacity of 300 pounds per hour and an electrical precipitator with a capacity of 3,000 cubic feet of gas per minute.

The kiln has a length of 20 feet with an inside diameter for 15 feet of 13 inches, the remaining 5 feet having a diameter of 20 inches. It is driven by an electric motor through a series of cone pulleys so that the speed may be varied. The kiln is heated with a Hauck high-pressure oil burner, burning a gas oil of 25° Baumé gravity.

The material is fed into the kiln through a screw feed with a water-cooling jacket on the outside. This feeder is arranged with variable speed pulleys so that the material can be fed at various rates from 100 to 300 pounds per hour. The gases from the kiln first enter a dust chamber where the heaviest material being carried is settled out. From the dust chamber they are carried through 15-inch diameter pipe up through vertical stacks to the top of the treaters. In these vertical stacks more of the solid material is settled out from the gases and collected in a dust box at the bottom.

The treaters have a capacity of 1,500 cubic feet in each unit, there being two units in the installation. Each unit consists of twenty 6-inch pipes 8 feet long. The gases enter the top of the treaters, go down through the treater tubes, leave the bottom hopper of the treaters, and are exhausted through a fan to the stack.

The electrical equipment for the treaters consists of a 5 k. v. a., 80,000-volt precipitator transformer, and a rectifier driven by a three-horsepower synchronous motor and a controlled switchboard. The switchboard contains switches for the main control, motor control, and transformer control. The regulation of the primary voltage of the transformer is obtained by means of an auto transformer and a theater dimmer. Meters are mounted on the board for the measurement of the voltage current and wattage. A milliammeter is placed in the treater circuit for measurement of the high-tension direct current.

The supply circuit for the equipment is over a separate line from that used for the other motors in the building. This gives a more even operation of the equipment than could be otherwise obtained, as the voltage does not fluctuate with the change of load on the other motors.

Under normal operation the treaters usually take about 65,000 volts, this voltage varying somewhat with the atmospheric conditions and with the nature of the fume being precipitated.

Many tests have been made in this plant with the most gratifying results. The tests have proved conclusively that it is possible to treat many types of oxide and sulphide ores, and that the electrical precipitator is a successful device for recovering the valuable metal constituents; also, enough fume has been recovered from the operation of this plant to show that the problems of fume treatment can be solved successfully.

The equipment in this laboratory also comprises a 2-foot revolving-hearth roaster of the Wedge type. With certain types of high-sulphur ores part of the sulphur must be removed from the ore before further treatment. For this purpose the Wedge type of furnace has proved satisfactory.

VOLATILIZATION OF LEAD AND SILVER FROM COMPLEX SULPHIDE ORES.

The most important of the major problems investigated at the station was the recovery of lead and silver from complex sulphide ores containing lead and silver. Utah has an enormous amount of this class of ore. Heretofore, it has been impossible to treat such ore successfully, owing to its physical makeup, by any known method except direct smelting, and the lead and silver content of much of the ore is so low that smelting yields little or no profit. Exceptionally gratifying results were obtained in experiments with this class of ores, and a cheap and satisfactory method of treatment devised. By adding the correct amount of sodium chloride or calcium chloride to the ore and heating the mixture to between 850° and 950° C. in a revolving cement kiln, 95 to 99 per cent of the lead and 80 to 85 per cent of the silver can be easily driven from the ore in the fumes. This fume can then be easily recovered from the furnace gases with an electrical precipitator. The precipitated fume is mixed with a small amount of coal, say 15 per cent, and the equivalent proportion of limestone, and fused. The resulting products are lead-silver bullion and calcium chloride slag. The slag can be used in the first roast to replace the sodium chloride that was used in the previous operation.

This process can be used for either oxide or sulphide ores of lead and silver, even those containing zinc, as the zinc is difficult to volatilize and a good separation of the lead and zinc is obtained. Such separation has always been difficult under the older methods. Enough work has been done on the process to prove that it will be a commercial success and will make possible the treatment of great quantities of ore that up to the present time has been difficult or impossible to handle. It is expected that in the near future a large mining company of Utah will build a plant to test further the possibilities of the process.

COMPLEX LEAD-ZINC ORES.

One of the most important of the problems attacked to date has been the treatment of the lead-zinc sulphide ores. As is well known, lead in a zinc ore is not paid for by the zinc smelter, while zinc in lead ore is always penalized by the lead smelter. Thus if the two metals are not separated before they are shipped, one of them is a total loss.

Metallurgists of the Bureau of Mines have suggested several schemes for the treatment of such ore, some of which have proved very promising. The most encouraging method tried to date is light roasting of the ore with a small amount of common salt. The sodium chloride converts nearly all of the lead into lead compounds that are soluble in a saturated brine solution, by which the lead can be leached from the ores and recovered, leaving the zinc unaffected and in a condition for recovery by gravity concentration or flotation. This process can be applied to all ores in which the lead and zinc are not so closely combined that they can not be separated by crushing.

OXIDIZED ORES.

Another important problem attacked by the station is the treatment of oxidized ores that carry both lead and zinc. These ores are difficult to sell because the zinc causes trouble in lead smelters and the lead gives trouble in zinc smelters. A satisfactory method of handling this type of ore has been evolved. This depends on the fact that the chloride of lead can be driven from the ore by heating, leaving the zinc in suitable condition for leaching with sulphuric or sulphurous acid solution. The process has proved satisfactory for the types of ores that are adapted to it. Fortunately many of the oxidized ores are of this type.

OXIDIZED LEAD-ARSENIC ORES.

Treatment of oxidized lead-arsenic ores, of which there are large tonnages in Utah and adjoining States, was one of the minor problems taken up at the station. At first the intent was to make, if possible, a salable product of both the lead and arsenic in the ore, but a study of the arsenic situation revealed that this source of arsenic was not important enough to justify further experiment. It was found, however, that high recoveries of the lead by leaching with saturated brine were possible, and that volatilization would recover the lead and more than half of the arsenic.

MANUFACTURE OF AMMONIUM PHOSPHATE.

Nearly all of the ammonium phosphate consumed in the manufacture of fertilizer has been made in the southeastern part of the United States near supplies of phosphate rock. The present method is to ship ammonium sulphate from the by-product coke plants, the producers of most of the ammonia, to the phosphate fields. The sulphuric acid necessary for the manufacture of the ammonium sulphate is bought in the open market.

A vast quantity of sulphuric acid can be made from waste gases at western smelters. Near these smelters are large supplies of phosphate rock. It was thought that possibly this phosphate could be treated in the West to make phosphoric acid, which could then be shipped to the ammonia makers, who could use it to make the ammonium phosphate direct from the phosphoric acid.

Work at the Salt Lake City station to determine whether it would be feasible to put the idea into commercial practice, without too much change in the existing plants, showed that chemically the method was feasible, but that the mechanical difficulties at present would make the venture unattractive.

OIL-SHALE INVESTIGATIONS.

Utah is, geographically, the center of the large oil shale deposits of the Inter-Mountain region, and on account of its position with respect to these deposits the Salt Lake City experiment station is at present the center of the oil-shale investigations. The station has begun a series of investigations to determine the best methods of treating shales with a view to determining their present economic status. Undoubtedly these shales will be an important source of hydrocarbon oils, perhaps in the near future.

FLOTATION WORK.

Experiments made at the Salt Lake City station to determine the possibility of floating the lead or zinc from some of the oxidized ores of Utah have shown that some of the lead ores could be sulphidized and good recoveries obtained, but the zinc ores could not be handled satisfactorily in this way.

In connection with flotation work, some important results were obtained in cooperative tests of wood oils for the Forest Service. That bureau wished to have some of the wood-oil products from its laboratories tested for their possible value as flotation oils. Samples of these different products were sent to the station at Salt Lake City, where they were thoroughly tested to determine their efficiency. Many of them proved to be efficient and could be used to good advantage in commercial flotation plants.

POTASH INVESTIGATIONS.

As is well known, potash salts are important constituents of the commercial fertilizers that are vital to our agricultural welfare. Before the war practically all of the potash used in this country was imported from Germany. This foreign potash was so cheap that no attempt was made to obtain potash from local sources. Hence, when the German supply was cut off, the situation became serious, and an investigation of potash resources and possible sources of supply was immediately begun.

Potash salts exist in nature in enormous quantities, but unfortunately in such a condition as to be useless until converted to soluble form. The principal natural source of potash is the silicates of potassium, and the development of a cheap method of recovering the potash from such silicates would solve the problem. The mill dumps of the Cripple Creek district in Colorado, those of the Utah Copper Co. at Garfield, Utah, and of the Inspiration Consolidated Copper Co. at Miami, Ariz., all contain large quantities of potash salts. The recovery of the potash from the materials named is attractive, as they have been mined and are already fine enough for direct treatment.

During the past few years many schemes have been suggested for the recovery of potash from silicates, hence experiments were begun at the station to discover whether any of the suggested methods could be applied commercially to the mill tailings mentioned.

The numerous methods tried fall naturally under three heads: (1) Decomposition with sulphuric acid, (2) treatment with phosphoric acid and calcium hydroxide, (3) chloridizing roasting and precipitation. The first and second were found to be impracticable, but the third might be applied where gold or some other valuable material is present in addition to the potash. Although the tests can not be considered a success in demonstrating a means of recovering potash from mill tailings on a commercial scale, there are indications that under more favorable conditions the tailings might easily be the source of much American potash.

MICROSCOPIC EXAMINATIONS OF ORES AND METALS.

The microscopic laboratory of the Salt Lake City station has a broad field of usefulness. This laboratory is equipped with everything necessary to make microscopical studies of ores, metallic substances, alloys, or other materials. Each year hundreds of specimens of ores and minerals submitted by mine operators, prospectors, and others are examined, and reports rendered as to the character of the samples and the probable value of the deposits.

GRAPHITE MILLING TESTS.

In an investigation of methods of purifying graphite concentrates, two methods of concentration were worked out with satisfactory results. The first, a combination treatment in an electrostatic machine and a finishing treatment in a burr mill, gives a finished product containing about 90 per cent carbon, with a recovery of 85 per cent. The other method is a combination treatment in a ball mill, flotation machine, and burr mill. The finished product contains about 90 per cent carbon and the recovery is about 80 per cent.

THE GOLDEN STATION.

The station in Colorado was the second of the two established in 1910, its special province being the investigation of rare metals. At first this station was at Denver, but about two years ago much of the equipment was transferred to Golden, where the station is now housed in one of the buildings of the State School of Mines. A view of this building is shown in Plate XV.

The work of this station has, so far, related chiefly to investigations of uranium, radium, molybdenum, tungsten, vanadium, and the methods of saving rare metals occurring in minute quantities in commercially valuable ores.

WORK AT DENVER.

Shortly after Congress made its first metal-mining appropriation for the Bureau of Mines, the director felt that a part of this money could be used to no better advantage than in investigating the rare metal resources of the United States, with a view to increased production, higher efficiency, and the elimination of waste. It was known that there were in the United States deposits of carnotite, pitchblende, molybdenum ore, vanadium ore, etc., from which a small output was obtained, but that there was room for considerable improvement in methods of concentration and in the treatment of the concentrates. It was, therefore, decided to establish a branch of the Bureau of Mines in Denver, Colo., to investigate rare metals, including the mining and concentration of the low-grade ores, their metallurgy, and any chemical problems of commercial or scientific interest.

The minerals that have been of particular interest to the Colorado station of the Bureau of Mines are those containing radium and uranium, thorium, vanadium, molybdenum, nickel, tungsten, and manganese.

COOPERATION WITH NATIONAL RADIUM INSTITUTE.

When the Colorado station was established, large quantities of radium-bearing ores from Colorado were being exported for treatment in foreign countries, and excessive prices were being paid for that part of the manufactured product returned to this country. In addition, much of the low-grade ore was being thrown on the dump and lost, owing to the lack of satisfactory concentration methods. As the carnotite deposits of Colorado and Utah represent the largest

BUILDING HOUSING THE OFFICES AND LABORATORIES OF THE GOLDEN STATION.

source of radium-bearing ores of the world, the need of investigation was obvious. Under a cooperative arrangement between the Bureau of Mines and the National Radium Institute, founded by Dr. Howard A. Kelly and Dr. James Douglas, the institute was to furnish the necessary funds for experimental work and receive most of the radium produced. The problems studied were the methods of recovering radium from carnotite, purification of the radium salts, methods of concentrating the low-grade ores, and the recovery of uranium and vanadium at the same time as the radium. Another agreement, made between the National Radium Institute and the Crucible Steel Co. of America, provided for leasing 16 carnotite claims in Montrose County, Colo., thus insuring a supply of ore.

An experimental plant was built in Denver and subsequently enlarged. This plant was operated for about two years, and during this time produced 8,543 milligrams of radium element at a cost of a little over $40,000 per gram. A considerable quantity of uranium oxide and vanadate of iron was recovered at the same time.

As methods of radium determinations ordinarily used in scientific work were not applicable to plant conditions, it was necessary to develop new methods for plant control. The electroscopes developed during the course of the work are now more or less standard in this country.

At that time practically nothing had been published on the fractionation of radium salts, and this work had to be developed from the start. Full details were worked out, showing the best conditions for fractionating radium from barium salts, and the losses that are likely to be met. The results of this work have been published and have been of great use to others who have contemplated going into the business.

Incidentally, a new method of converting sodium uranate cheaply and efficiently into a high-grade uranium oxide product was developed. A concentrating mill was built at Long Park, near the leased mines. Here, concentrating methods were studied and ultimately more than 300 tons of concentrates were produced from low-grade ore that otherwise would have been left on the dump.

OTHER WORK ON RARE METALS.

An investigation of the ratio of radium to uranium in carnotite ore was also made and the results published in Technical Paper 88 [a] of the bureau. It was ascertained that whereas in some secondary uranium minerals of recent origin the ratio was low, this was not true of large bodies of carnotite ore. In all places where 1 ton or more of the ore was thoroughly sampled, the ratio was practically

---

a Lind, S. C., and Whittemore, C. F., The radium-uranium ratio in carnotites: Tech. Paper 88, Bureau of Mines, 1915, 29 pp.

the same as that of pitchblende, although in some small samples of carnotite the ratio was below the normal.

The molybdenum deposits of the country were investigated, and the possibilities of the molybdenum industry in this country were thoroughly studied. A considerable amount of preliminary work was also done on both molybdenite and wulfenite, as well as on methods of separating copper minerals from molybdenite.

A preliminary investigation of some of the complex lead and zinc ores of Colorado, particularly those of Leadville, was started. It was known that in many ores the minerals could not be separated by concentration processes. The reasons for this were determined, and valuable information was obtained as to the physical composition of these ores.

REMOVAL TO GOLDEN.

The station was moved from Denver to Golden in July, 1916, under a cooperative arrangement with the Colorado School of Mines. This arrangement aimed at the promotion of close relations and general cooperation in mining, ore dressing, and metallurgical research.

The school of mines transferred its Physics Building (see pl. XV) for the use of the station. This building is of brick, has two stories and a full basement, and is equipped with high-pressure steam, compressed air, and other facilities. The upper story is mainly occupied by two chemical laboratories, each of which can accommodate about six men. The first story contains offices and the station's library. In the basement are the ore-grinding equipment, laboratory for the treatment of rare-metal ores, two other small laboratories, tool shop, and storerooms. The station has full use of the experimental ore-dressing and metallurgical plant of the School of Mines. This plant is about one-half mile from the main buildings and is completely equipped for large scale experiments in sampling, mechanical concentration, cyanidation, magnetic and electrostatic concentration, and flotation. Views in the metallurgical laboratory are shown in Plates XVI and XVII. The station is in charge of Richard B. Moore, superintendent.

EQUIPMENT.

The laboratories of the station are equipped with all the chemical apparatus necessary for analytical and research work in connection with the rare metals. A special room is reserved for electroscopes for radium determinations. A Hilger constant deviation type of spectroscope is available for ordinary spectroscopic work, and a Hilger spectrograph for special research work in which great accuracy is desired.

METALLURGICAL LABORATORY, COLORADO SCHOOL OF MINES, USED BY GOLDEN STATION.

ANOTHER VIEW IN METALLURGICAL LABORATORY, GOLDEN STATION.

One room is used for work in connection with large quantities of radium emanation. Three hundred milligrams of radium element in the form of bromide is held constantly in solution, and to this is adapted a Duane apparatus (Pl. XVIII) for pumping off and purifying the emanation from the solution. In this manner a supply of purified emanation can be obtained at any time for experimental work.

The room in the basement for grinding ores is equipped with a crusher, ball mill, coffee mill, disk grinder, Newaygo screen, and Rotap sizer, all power driven. An adjoining room is completely fitted up with apparatus for experimental work in the treatment of rare-metal ores on a semicommercial scale. The equipment was installed with the idea of making it as flexible as possible; and any rare-metal ore can be given any kind of an acid leach, alkali leach, roast, reduction, or fusion; also any combination of the above treatments can be readily effected. Ores can be handled in quantities of 10 to 500 pounds at a run.

RECENT WORK OF THE STATION.

RARE METALS.

SOLUBILITY OF PURE RADIUM SULPHATE.

The solubility of pure radium sulphate in water and in aqueous solutions of sulphuric acid of various concentrations was determined at 25°, 35°, and 45° C. The results showed that radium sulphate has a low solubility, the solubility increasing somewhat at the higher temperatures and with concentrations of acid in excess of 50 per cent. The industrial recovery of radium as sulphate depends, fortunately, not on its own solubility, but on the sulphate being precipitated with barium in much greater proportion than alone. Were it not for this circumstance the practical recovery of radium would be almost impossible. Further investigation is being made of the action of mixtures of radium and barium, for the purpose of enabling one to predict how much barium should be added to a given radium ore in order to obtain a maximum recovery of radium.

RECOVERY OF MESOTHORIUM IN THE MANUFACTURE OF THORIUM NITRATE FROM MONAZITE SAND.

Although the carnotite deposits in Colorado and Utah constitute the largest source of radium-bearing ores in the world, most of this carnotite will be mined within five or six years at present rates of production. As most of the radium produced is being disseminated, through its wide use in luminous paint, the finding of a substitute for radium is important. Such an element is mesothorium, found in all thorium ores. The life of mesothorium is not nearly as long as that of radium, but is long enough for use in luminous paint and for some of the uses of radium in therapeutic work. In the past no attempt has been made to recover mesothorium in this country, all of the material having gone on the dump.

The Bureau of Mines made a cooperative arrangement with the Welsbach Co. to investigate the best methods of recovering mesothorium in the manufacture of thorium nitrate from monazite sand. This work involved making a number of plant runs on approximately 5 tons of ore, and tracing the losses and final recovery throughout the whole process. It also involved methods of quantitative determinations of mesothorium to find what the losses were, as well as methods of fractionation and purification of the mesothorium salts. Satisfactory results were obtained which will be applied in the plant of the Welsbach Co., Gloucester, N. J., and ultimately published.

EFFECT OF RADIUM EMANATION ON MIXTURES OF HYDROGEN AND OXYGEN.

Part of the radium that has been reserved by the Bureau of Mines for scientific use has been put into solution for the purpose of collecting the gas, radium emanation, by means of the Duane apparatus (see Pl. XVIII). The emanation thus collected has been used for studying the laws governing the combination of hydrogen and oxygen gases under the influence of the alpha rays when these gases are mixed with radium emanation. These investigations, to be extended to other gases, may throw some light on the chemical reactions that take place during electrical discharges in gases. Some of these reactions are of considerable industrial importance.

TREATMENT OF PITCHBLENDE FOR THE RECOVERY OF RADIUM AND URANIUM.

Under a cooperative arrangement with Mr. Alfred I. Du Pont and Mr. William Wright, of the Colorado-Gilpin Gold & Radium Mining Co., about 100 tons of low-grade pitchblende ore and 6 tons of high-grade ore were furnished the Bureau of Mines for experimental treatment. Work on this ore was begun in 1917 and completed in 1918. Methods of concentrating the low-grade ore were worked out and applied. The high-grade ore and the concentrates, after a thorough investigation of the most efficient methods that might be used, were treated in the plant of the National Radium Institute, and the radium recovered was returned to the company. It was found that the original ore could be treated much more easily than the concentrate, which consisted almost entirely of pyrite. However, methods were developed that could be applied to either product. One of the results of this work was that the plant of the institute was finally equipped for the treatment of both pitchblende and carnotite.

APPARATUS FOR PURIFICATION OF RADIUM EMANATION.

FRACTIONATION OF RADIUM SALTS.

In connection with its cooperative work with the National Radium Institute, the Bureau of Mines has fractionated about 9 grams of radium element, starting with barium chloride containing only one part per million of radium chloride. After seven or eight fractionations in aqueous solutions containing hydrochloric acid, the radium reached a concentration of about 25 parts per million, at which point it is converted into bromide and fractionation continued in aqueous solutions containing hydrobromic acid until the concentration of radium bromide reaches about 1 per cent. At this stage the salt is collected until several hundred milligrams of radium element are obtained. Fractionation is then continued on a small scale to any desired degree, and on two occasions the bromide has been brought to 100 per cent purity, as verified by gamma ray measurements of the Bureau of Standards.

When large quantities of salt are being handled in a continuous system it is not difficult to crystallize radium to 100 per cent purity at an expense nearly as small as for the 1 per cent material; but in treating a small quantity of radium, crystallization to a high degree of purity becomes much more difficult, requiring many repetitions of the fractionation.

RATIO OF RADIUM TO URANIUM.

The determination of the ratio of radium to uranium in primary minerals has been made by several competent observers. Rutherford obtained the figure 3.4×10^{-7}, but afterwards found that his radium was impure. Correcting for this impurity, his final figure was 3.2×10^{-7}. Hyman and Marckwald obtained the figure 3.328×10^{-7}; while Becker and Jannasch found 3.415×10^{-7} when the emanation was obtained by melting the radium salt, and 3.383×10^{-7} when the emanation was obtained from a solution.

A knowledge of the exact ratio bears directly upon the question of radium recovery in commercial work, besides having pertinent scientific value. This ratio must be used in determining the recovery of radium from the original ore in a commercial plant; and as the figures given vary as much as 6 per cent, it is important that the true ratio be determined beyond question. The radium obtained by the Bureau of Mines in its cooperative work with the National Radium Institute has given an opportunity to purify some radium bromide up to 100 per cent; and with this as a standard, a start has been made on the redetermination of the ratio of radium to uranium.

RADIUM LUMINOUS PAINT.

Some work was done to determine the best methods of producing phosphorescent zinc sulphide, which is mixed with radium salts in making radium luminous paint, and to determine the effects of certain physical and chemical factors on luminosity. This investigation had a decided bearing on war work, as the luminous paint is used on all of the instruments carried by airplanes, such as clocks, manometers, and compasses, and is also used to a considerable extent for gun-sights and other military equipment.

ACTINIUM AND IONIUM.

Two other elements associated with radium in all of the uranium ores are actinium and ionium. These elements are long-lived and active enough to be useful as substitutes for radium in the manufacture of luminous paint. An investigation was made of the best methods of recovering these elements from radium-bearing ores during the manufacture of thorium. A considerable quantity of concentrates was obtained during the operations of the National Radium Institute, and these concentrates with other material were used for this work.

POLONIUM.

Some work has been started on the recovery of polonium from lead residues and also from some old glass tubes that contained radium. Some of this recovered material has already been purified and used for experimental purposes in making luminous paint.

TUNGSTEN, NICKEL, MOLYBDENUM, AND VANADIUM.

The conditions under which tungstic acid may be efficiently reduced to metallic tungsten have been studied. Metallic tungsten is a commercial product almost always obtained when low-grade and medium-grade tungsten ores are treated chemically with a formation of tungstic acid. It is largely used in place of ferrotungsten in making tungsten steel.

Some work has been done on the metallurgy of wulfenite or lead molybdate with the object of recovering not only the lead, but also the molybdenum in a convenient concentrate free from arsenic and sulphur. Some experiments have been undertaken on the separation of wulfenite from vanadinite, as the two minerals frequently occur together.

Experimental work was started on the metallurgy of several vanadium minerals, particularly vanadinite, or lead vanadate, and cuprodescloisite, as these minerals offered an additional supply of vanadium for war purposes. The metallurgy of two nickel ores has been studied in detail, namely, a garnerite from North Carolina and an ore consisting mostly of sulphides from Alaska. Some success has been obtained and the work is being continued.

COOPERATIVE INVESTIGATION OF THE CARON PROCESS FOR THE TREATMENT OF MANGANESE SILVER ORES.

M. H. Caron, of the Netherlands East Indies Bureau of Mines, is the inventor of a process for recovering silver from oxidized manganiferous ores. This process was investigated and developed at the Colorado station at Golden under a cooperative agreement between the Netherlands East Indies Bureau of Mines, the United States Bureau of Mines, and the Research Corporation. Each of these interests furnished engineers and funds for the work.

For a long time the recovery of silver from many oxidized manganiferous ores has proved a difficult metallurgical problem. Such ores can not be treated directly by hydrometallurgical methods, and many of them are too low grade for profitable smelting. The Netherlands Government is particularly interested in the process on account of the large deposits of such ore in certain government-owned lands in Sumatra. Also, the mining industry of the United States would be benefited through the development of a satisfactory process for treating refractory silver ores. Mr. Caron has assigned his original patent to the Research Corporation of New York, which is acting as an administrative agency for handling the original patent, as well as such other patents as may develop in the course of the work.

Briefly, the process consists of heating the ore to approximately 800° or 900° C., and bringing the heated ore into intimate contact with reducing gases, such, for example, as producer gas. The manganese dioxide is reduced to manganese monoxide, and simultaneously with this reaction the silver is converted into such a form as to be readily recovered by cyanidation.

Although the process appears to be simple, considerable investigation has been required to establish its limitations and the best conditions for its operation. Among the many points covered might be mentioned the problem of preventing reoxidation. If the hot reduced ore comes in contact with air, reoxidation takes place and refractory silver compounds are formed; therefore, methods of cooling under reducing conditions must be developed. Also, if cooling is not done under the proper conditions, this reoxidation may even take place slowly with a corresponding loss during the subsequent cyanide treatment. The problems that arose in this connection have now been definitely solved. To effect satisfactory reduction on a commercial scale a specially modified direct-fired furnace had to be developed. This furnace does away entirely with the disadvantage of having to work in closed tubes or retorts. Lots of 500 to 1,000 pounds of ore have been put through the experimental equipment with entire satisfaction. Development of the process has now proceeded to the point where it can be introduced commercially with assurance of its success.

Similar manganese-silver ores from widely separated localities are amenable to treatment by the process. The value of the process is shown by a few of the results obtained. The recovery of silver from some Sumatra ores has been raised from 11 to 96 per cent. Ore from a deposit in Colorado gives only 30 per cent by ordinary cyanidation but readily yields more than 90 per cent when treated by this process. A sample from Arizona containing 62 per cent of manganese dioxide yields only 6.9 per cent by direct cyanidation and 93.1 per cent by the new method. A Mexican silver ore yielding 40 to 50 per cent by ordinary cyanidation gives over 90 per cent extraction by the Caron process.

An interesting feature of the process is that the manganous oxide resulting after reduction is readily soluble in acids, hence the manganese can be recovered as a by-product. The possibility of recovering marketable manganese product from the residue by simple mechanical means is being investigated.

THE ITHACA BRANCH OFFICE.

For some years past the Bureau of Mines has maintained a branch office at Cornell University, Ithaca, N. Y. The Bureau of Mines, in cooperation with the American Institute of Metals and Cornell University, undertook a study of brass and nonferrous alloys manufacture, and has maintained a small staff of alloy chemists, who have worked in cooperation with the university staff and used its laboratories in their experiments. This work is in charge of H. W. Gillett, alloy chemist.

The work has had especial reference to the relative efficiencies of different types of furnaces and the improvement or development of furnaces with a view to reducing metal losses.

A bulletin on "Brass Furnace Practice in the United States" was published by the bureau in 1914, and another bulletin on "Melting Aluminum Chips," which discusses the loss of aluminum in melting scrap and suggests improved methods, was published in 1916. More recently, Dr. Gillett and his associates, working in cooperation with brass companies of Detroit, Mich., have evolved a new type of rocking electric brass furnace that has many superior features and is now in commercial use by a number of firms. During 1917 and 1918 most of the work at this laboratory was on alloys used in steel making, including ferrotungsten and ferro-uranium, this work being in cooperation with the Bureau of Ordnance of the War Department. A special study was made of the possible value of zirconium for alloying steel, and a series of samples of zirconium steels was prepared. These samples are to be tested by the Bureau of Standards.

THE BERKELEY STATION.

In 1911 the Bureau of Mines established an office in San Francisco, Calif., for conducting investigations of smelter-fume problems and various problems relating to petroleum. Later, when the mining experiment station at Berkeley was established, the metallurgical work was transferred to that station, the office at San Francisco being maintained for the study of petroleum problems.

In 1914, at the Panama-Pacific Exposition in San Francisco, the Government board allotted to the Bureau of Mines $7,000 and the space for an exhibit. By a cooperative arrangement between the director and various representatives of the mining and metallurgical industries, exhibits were made of devices to increase safety and efficiency in mineral industries of typical mining machinery, of experiments in metallurgy, and of petroleum machinery and methods. Much machinery and equipment thus assembled was given to the Bureau of Mines at the close of the exposition and was transferred to the station at Berkeley.

EQUIPMENT.

The station is housed in the Mining Building (Pl. XIX) of the University of California. The laboratory and offices are on the second floor; a well-equipped machine shop and storerooms are in the basement. One section of the laboratory is equipped for assaying and analytical work. Another section is devoted to research work on the chemical problems of metallurgy. The equipment includes special apparatus for studying chemical changes at high temperature, and for the determination of fundamental physical and chemical constants involved in metallurgical operations. As the chemical department of the university is large and active and the scientific library is extensive, the station is particularly well situated for conducting highly specialized research work on problems of the mineral industry.

SCOPE OF THE STATION'S WORK.

The importance of San Francisco as a mining center, the facilities for scientific research in its technical schools, and the two great universities near it, were strong reasons but not the chief reason for establishing an experiment station at Berkeley. The territory tributary to this station, which includes California, western Nevada, and the Klamath region of southern Oregon, is remarkable for the

BUILDING HOUSING THE OFFICES AND LABORATORIES OF THE BERKELEY EXPERIMENT STATION.

A CALIFORNIA GOLD DREDGE.

extent and variety of its mineral resources. The mineral production of California alone in 1916 was worth, roughly, $130,000,000, and included the following materials, the production of which was valued at $100,000 or more: Borax, brick, tile, cement, chromite, pottery clay, copper, gold, lead, lime, limestone, magnetite, manganese ore, mineral water, natural gas, petroleum, potash salts, pyrite, quicksilver, salt, silver, soda, building stone, gravel and road materials, tungsten concentrates, and zinc.

The mining industry in California employs over 13,000 men, which places the State among the first seven as regards the number of men employed in mineral industries.

FIELD OF RESEARCH WORK.

As the list of minerals shows, the mining industry of this region has a considerable future. Moreover, some parts of California—notably the desert areas—have not been thoroughly prospected, and discovery of other mineral deposits is to be expected. The rate at which the mineral deposits of California and adjoining territory can be developed is limited by certain economic factors, chiefly transportation and market. At present, and probably for many years to come, the principal points of consumption for the mineral products of California must be at a considerable distance, either in the central and eastern States of this country, or in foreign countries bordering the Pacific. Hence, these products must have some advantage, such as exceptionally high quality or low cost of production, to offset the high freight charges.

Fortunately the disadvantages mentioned are in a measure offset by certain natural advantages, such as undeveloped water power and abundant oil fuel. The further development of these mineral resources, especially the complex and low-grade ores, will involve a variety of problems. In such a field as this, it is believed, the station can be particularly helpful.

GOLD.

One can not mention the gold output of California without recalling the early placer mining. The picturesque side of gold mining in the State has largely disappeared, but production on a large scale continues, and California still leads the Union in this respect. The annual yield is roughly one-fourth of the total amount produced by the United States and Alaska together, and almost 5 per cent of the world's production.

In recent years the growing scarcity of high-grade ore has brought into the foreground the problem of treating the low-grade and refractory ores. Many unsolved metallurgical problems in this field demand early attention. The claim of the gold-mining industry is particularly strong because of the unique position of gold as a monetary standard. The price of gold is fixed, and the rising cost of labor and supplies can not be met, as in all other industries, by an increase in the price of the product. The alternative is to improve recovery methods and to reduce the cost as much as possible. Thus there is a big field for investigation in the gold industry of California and the adjoining region.

MANGANESE, CHROME, AND QUICKSILVER.

The war markedly stimulated the production of manganese, chrome, and quicksilver in California, as is illustrated by the following figures:

Production of manganese and chrome ores and of quicksilver in California.

	1914	1917
Manganese ore..tons..	150	15,500
Chrome ore..do....	1,500	52,400
Quicksilver...flasks..	11,373	24,251

MANGANESE.

The endeavor to stimulate production of the so-called war minerals brought up for immediate solution several important metallurgical problems. Manganese may be taken as an example. Nearly all of the manganese ore in California and the adjoining region is high in silica, and much of it is so low-grade that concentration is necessary before shipment to the smelters. Moreover, in some of these siliceous ores the manganese and silica minerals are so finely divided and intimately mixed that separation is extremely difficult. In attacking this problem the station cooperated with the State Council of Defense for California and with the mining department of the University of California.

CHROME.

The chrome situation is somewhat similar to manganese, except that the low-grade ores are in general more amenable to concentration. The important tonnage of chrome ore produced in California in 1917, amounting to about one-third of our domestic requirements, was obtained from a large number of deposits. Chromite in California occurs chiefly in small pockets scattered over a considerable area, many of them on ranch land which had not been thought to contain valuable minerals. These scattered deposits attracted the attention of many people unfamiliar with prospecting or mining but who had heard that minerals previously unsalable were now necessary for war purposes and could be mined to advantage. Soon there arose an active demand for information in regard to these minerals and for assistance in classifying them and determining their value. The Berkeley station has been able to assist many people thus interested.

QUICKSILVER.

Quicksilver production was stimulated by the war, but the history of quicksilver in California goes back to the Spanish régime. Reminders of that period are found in the names of some of the mines, the technical terms still current, and the numerous Spanish and Mexican workers in the industry. Some 40 years ago the output of quicksilver in California was greater than that from any other quicksilver district in the world. One of the old mine records contains an illuminating comment of a mine foreman to the effect that the mine was evidently failing, as it had been necessary to drop the grade of ore from 30 to 28 per cent. The change that the industry has undergone since then is illustrated by the fact that the average grade of ore treated in California in 1917 contained probably not far from one-half of 1 per cent of mercury. In fact, before the war, quicksilver mining in California and in other sections of the West was on the decline. The greatly increased demand for this metal for military purposes gave the industry a new lease of life.

In treating such lean ores as those containing one-half of 1 per cent, low mining and reduction costs and a high recovery are essential. In the early days, with high-grade ore, an inefficient process yielded a handsome return. To-day 40 to 60 tons of ore and frequently more have to be treated in order to obtain the amount of quicksilver yielded by 1 ton of the ore mined at New Almaden in the old days.

WORK OF THE STATION.

INVESTIGATIONS OF LOSSES IN QUICKSILVER PLANTS.

Having the military importance of quicksilver in view, and realizing the difficult problems before the industry, the Bureau of Mines, early in 1917, through its station at Berkeley, entered into a cooperative agreement with several California quicksilver operators for the purpose of studying the metallurgy of mercury. Little technical or scientific work had been done in this field for many years, and such data as were available were neither accurate nor capable of direct application to present conditions. Although the methods of treatment in common use differed little from those in vogue many years ago, it was necessary to consider whether some of the new processes or appliances for ores of other metals might not be applied to advantage in the treatment of quicksilver.

The furnace in common use for reducing quicksilver ore is known as the Huttner-Scott, or more commonly the Scott, furnace (Pl. XXI, A). In this type of furnace the ore passes slowly over a series of inclined tiles in direct contact with the hot gases from the combustion of fuel.

The mercury distilled from the ore is taken up by the stream of gas. With the low-grade ores treated, the volume of the mercury vapor is vanishingly small as compared with the volume of gas into which it passes, being frequently less than 0.1 to 0.2 per cent. It is obviously not an easy matter to recover the quicksilver from such dilute gases, and the problem has received much attention from quicksilver operators.

The efficiency of existing methods of recovery was, therefore, one of the first questions taken up by the station in its study of the metallurgy of quicksilver. The conventional condenser system for the recovery of quicksilver from furnace gases consists of a series of brick or stone chambers, followed frequently by some wooden chambers, and then a side-hill flue leading to a short stack. In view of the large volume of gas through which the quicksilver vapor was disseminated, it was thought that important amounts might be carried through the condenser system and escape from the stack. No direct determination of the magnitude of such losses had ever been made. In studying this point it was found that the method developed for sampling other smelter gases could not be applied to quicksilver plants, owing to the low temperature and large water content of the condenser gases. After several trials a suitable type of apparatus was developed (Pl. XXI, B). Several determinations at two different plants showed that the stack losses are extremely small, not exceeding 1 to 2 per cent of the quicksilver output.

Several other possible sources of loss in condensers were also investigated. As a result of this work, the conditions for efficient condensation of quicksilver vapor have been determined and several modifications in present condenser practice have been suggested. Some of these undoubtedly will be tested soon on a practical scale at commercial plants.

The application of new types of furnaces to the treatment of quicksilver ore and of various concentrating procedures to the treatment of low-grade ores is being studied.

Another investigation dealt with the methods used in assaying quicksilver ores and furnace products. Some of these were not accurate and others were too tedious. Moreover, certain types of ore, high in sulphur or bituminous matter, were found exceedingly difficult to assay by any known method. As a result a method that is both rapid and accurate has been devised.

In studying the changes taking place within the quicksilver furnace information was needed with regard to the volatility of cinnabar, the common quicksilver mineral in California ores. As the boiling point of metallic mercury is 357° C., it has been commonly supposed that heating quicksilver ore to this temperature will expel the mercury. The experiments showed that cinnabar sublimes at about 500° C., and that when cinnabar ore is heated in the absence

A. TYPICAL SCOTT QUICKSILVER FURNACE. PHOTO BY W. W. BRADLEY,
CALIFORNIA STATE MINING BUREAU.

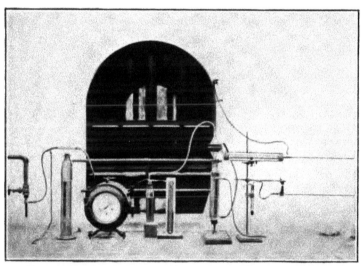

B. FUME-SAMPLING APPARATUS FOR QUICKSILVER DETERMINATIONS.

of air it does not give up all its quicksilver until the temperature is somewhat higher. The information obtained from this study has proved to be of value in interpreting the process taking place within quicksilver furnaces and retorts.

POTASH.

The station at Berkeley has cooperated with the scientific research committee of the State Council of Defense for California in several chemical and metallurgical matters. The demand for a domestic supply of potash brought about by the war, has stimulated activity at the various saline deposits in the West, among others, Searles Lake in southern California.

The bittern from the manufacture of salt from sea water is also a possible source of potassium chloride, magnesium chloride, and bromine, for all of which there was strong. demand. In devising processes for the recovery of these valuable materials, considerable saving of time, money, and effort can be effected through cooperation between various private individuals and Government institutions having information bearing on the question at hand. Through the Federal Bureau of Mines and the State Council of Defense some developments in the recovery of saline products were made immediately available to those engaged in the industry, and at the same time information desired by the Federal Government for military purposes was obtained.

PROSPECTING FOR WAR MINERALS.

The war stimulated many people to search for new sources of material for military purposes and to develop new processes for the winning of these materials. Many projects have originated in the territory tributary to the station at Berkeley, and it has been important to investigate them rapidly and thoroughly in order that nothing of value might be overlooked. The station was asked to make a number of investigations of this sort and among the many propositions of doubtful merit some have appeared to be of real value. In this work the advantage of a local station representing the scientific and technical activities of one of the branches of the Government has been particularly evident.

FULLER'S EARTH.

The possible use of California fuller's earth for purifying lubricating and edible oils offered a problem calling for immediate solution. At present all fuller's earth used by California oil refiners is shipped from Georgia and Florida, although there are thousands of tons of fuller's earth in the State. An investigation to determine whether California earths can be used for filtering purposes has been completed.

SAN FRANCISCO FIELD OFFICE.

The field office at San Francisco, Calif., is on the fifth floor of the United States Customhouse, its laboratories being in the Treasury Building close by. Until the establishment of the Bartlesville station, the work on petroleum technology was centered at San Francisco. From this office field investigations of various problems relating to petroleum engineering and technology are conducted throughout the United States. The chemical and technological laboratories of the station are well equipped.

PROBLEMS UNDER INVESTIGATION.

PROTECTING OIL SANDS AGAINST WATER.

Many oil fields in the United States have had their potential yield much reduced by water entering the oil sands. In California great advance has been made in methods of excluding underground waters, and the Bureau of Mines through its investigations in that State has demonstrated how such methods can be applied in other fields. The principal method advocated is the use of cement.

INCREASING THE ULTIMATE PRODUCTION OF WELLS.

Investigations have disclosed that in many fields far more of the oil originally in the oil sand is not recovered than is brought to the surface by the wells. Investigations have been conducted throughout the United States to ascertain the quantity of oil remaining in the oil sands, and methods are being devised for increasing this recovery by the use of compressed air and other means.

STUDY OF THE MIGRATION AND ACCUMULATION OF OIL AND WATER.

The relation between the accumulation of oil and of water in oil-bearing formations is of great importance to the oil industry. Experiments are being conducted in the San Francisco office to ascertain the underlying causes and effects of the migration of oil and of water through oil sands. Plate XXII shows apparatus used in such pl. XXII experiments.

VALUATION OF OIL PROPERTIES.

Although the proper valuation of oil properties is a matter of fundamental importance to the industry, rational engineering principles have not been applied to it in the past. Recently methods for

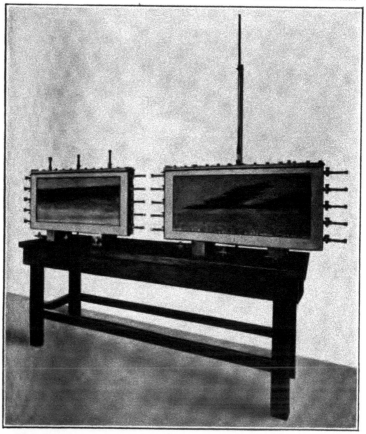

APPARATUS FOR STUDYING CAUSES AND EFFECTS OF MIGRATION OF OIL AND
WATER THROUGH OIL SANDS.

estimating future production and thus determining a just basis for the appraisal of oil properties have been devised. These methods are being employed by the Bureau of Internal Revenue in making valuation allowances under the taxation acts passed by Congress. The San Francisco office has been asked to appraise Government oil lands in California and elsewhere, and to advise on the royalties involved in leasing oil and gas lands.

DRILLING METHODS.

The bureau has begun a study of well-drilling methods in the United States. Various factors that have determined the methods used in different fields have been investigated and many problems that call for the use of special methods have been solved.

FISHING PROBLEMS.

Troubles encountered in placing and in carrying casing through caving formations are being investigated. Methods of overcoming some of the troubles encountered are being demonstrated.

PUMPING OF OIL WELLS.

Investigations have been made with reference to the effect of methods of perforating casing and screen pipe in oil wells on the production of oil. The deposition in an oil sand of minerals that tend to exclude oil from a well and to decrease the total amount of oil recovered from the sands has also been investigated.

SUPERVISION OF GOVERNMENT OIL LANDS.

The San Francisco office of the Bureau of Mines has been called on to advise with the Land Office on the operation of certain oil lands in the California oil fields.

TRANSPORTATION OF HEAVY OIL THROUGH PIPE LINES.

Some of the heavy viscous oils in the California oil fields move through pipe lines with difficulty. Investigations of the engineering features involved in the transportation of these oils, and the cracking of such heavy oils, in order to reduce their viscosity and increase their fluidity, are under way.

A STUDY OF TOPPING PLANTS.

Plants for the continuous distillation of the lighter fractions of crude oil, known as topping plants, are much used in California fields and to a less extent in the Gulf coast fields and in the fields of Oklahoma. These plants have been investigated.

STORAGE OF OIL.

Investigations throughout the United States have dealt with the different types of storage tanks and reservoirs for crude oil—their cost, construction, maintenance, the evaporation losses, and the other factors that have to be considered in the storage of petroleum.

Investigations are now being made to determine the actual evaporation losses as related to climate, type of tank, and character of oil.

OIL AND GAS WELL FIRES.

The causes, means of prevention, and methods of quenching fires at oil and gas wells and of oil in storage have been investigated.

OIL-CAMP SANITATION.

Most oil and gas fields are developed rapidly under boom conditions that lead to highly insanitary conditions surrounding the living quarters of the men and their families. The bureau has investigated oil-town sanitation, with special regard to the peculiar conditions that menace health and life.

BIBLIOGRAPHY OF PETROLEUM.

A bibliography of petroleum, containing thousands of entries, is maintained at the San Francisco office. Although primarily for the use of the members of the petroleum division, it is available for the benefit of the petroleum industry.

STATIONS ESTABLISHED IN 1916.

The act of Congress approved March 3, 1916 (40 Stat. 969), authorized 10 new mining experiment stations to be under the Bureau of Mines, not more than 3 of which were to be established in any one year. The three stations established in 1916 were placed at Fairbanks, Alaska, at Tucson, Ariz., and at Seattle, Wash. Each of these had its particular field of service. The station at Tucson is to study problems connected with the mining, concentration, and metallurgical treatment of low-grade copper, gold, and silver ores; the station at Seattle is to investigate similar problems in connection with the coal and ore deposits of the Northwest and of the coast of Alaska; and the station at Fairbanks will take up the problems connected with the development of the mineral resources of Alaska.

THE SEATTLE STATION.

The station at Seattle was established specifically to investigate the problems connected with the mining and treatment of coal and other mineral deposits of the Northwest. Its first task was to make a thorough survey of the needs of the territory it was intended to serve and to get into close touch with local mining and metallurgical interests, in order to learn how it could be of most help.

In 1916 the division of metallurgy was created in the bureau, and the Seattle station was one of the first stations to profit by the change. The work of the station thus includes mining and metallurgy. Chief accomplishments to date have been in the study of scientific methods in connection with those subjects and in suggesting the best means of applying such methods to commercial conditions. Also careful attention has been given to some of the more theoretical questions involved. The station is housed in a building (Pl. XXIII) provided by the University of Washington. The work of the station is in charge of F. K. Ovitz, superintendent.

EQUIPMENT OF ORE-DRESSING LABORATORY.

The laboratories at the University of Washington are well equipped for ore-dressing tests. The equipment includes jaw and gyratory crushers, rolls, disk crushers, ball mills, tube mills, concentration tables, and almost every kind of flotation machine, both mechanical and pneumatic. The laboratory contains also a Dings electromagnetic separator, which was lent jointly to the Bureau and the College of Mines by the American Smelting and Refining Co.

ELECTROMETALLURGICAL LABORATORIES.

The Seattle station has two electrometallurgical laboratories, electrolytic and thermoelectric, situated in the south end of the College of Mines building. The electrolytic laboratory, of dust-proof construction, is fitted with a motor-generator set and other equipment.

The thermoelectric laboratory is separated from the rest of the building by cement partitions. The transformers and switchboards are separated from the furnace room proper by cement partitions, fitted with large glass panels. At the top of the transformer room are the voltage regulators, the regulator platform serving also as a charging floor for the electric furnaces. The furnace room is of fire-proof construction throughout. Fumes from the furnaces are carried through a hood and flue to a 90-foot brick stack at the end of the building. The electrical equipment includes two 140 k. v. a. transformers; two 77.4 k. v. a. duplex induction regulators, complete with motor and necessary shafting and gears for coupling them. Bus bars convey the current from the switchboard to the furnaces.

WORK AT THE SEATTLE STATION.

The work of the station may be grouped as follows: Studies of methods of mining and preparing western coals, research work in flotation, mining and concentration of ores of metals, especially those that were needed during the war, and electrometallurgical investigations. Much work was done in cooperation with the University of Idaho at Moscow in obtaining data for a report on the mining districts of that State. Other work having a military bearing included investigations of the tin deposits of the Black Hills; concentration tests of quicksilver and chrome ores, in cooperation with the station of the bureau at Berkeley, and investigations of manganese and other ores in cooperation with the Oregon Bureau of Mines and Geology.

COAL INVESTIGATIONS.

MINING METHODS IN WASHINGTON.

The station is making a detailed study of coal-mining methods in the State of Washington, taking into account the thickness and dip of the beds; the characters of the walls; systems of mining, and methods of loading, haulage, hoisting, and ventilation.

Some of the important problems are a study of roof pressures under great depths, the control of roofs at varying depths with different systems of mining, and the gases produced at the various mines, with particular reference to the outbursts of gas in the mines of the Pierce County coal field.

WASHING TESTS OF WASHINGTON COALS.

A study is being made of the specific gravities of the various coals of the State and the impurities that may accompany the coal from the working face to the cleaning plant.

BUILDING HOUSING THE OFFICES OF THE SEATTLE STATION.

During the field season of 1918 a detailed study was begun of the coal-washing plants in the State. Efficiency tests were made of the various types of washeries, and representative samples of the various products were analyzed. It is expected that these results will show the efficiencies of the different plants and also show the grades of coal produced.

Studies will be made of the lignitic coals of Washington in the attempt to find more efficient ways and means of utilizing them. Also fuel tests of powdered coal will be made with the coals of this State, either independently of the bureau or in cooperation with the companies that are now experimenting with the utilization of powdered coal.

IDAHO COALS.

The coal-mining engineer of the station at Seattle made a preliminary examination of the Teton Basin coal field in Idaho. A report of the results of this investigation was included in the bulletin on the mining districts of that State.

OREGON COALS.

Two visits have been made to the Coos Bay coal field of Oregon for the purpose of examining the mining conditions and for investigating the possibility of finding improved methods of using this coal. These two phases of work are still under consideration. Also the coal field east of Medford, Oreg., was visited.

GOVERNMENT MINES IN ALASKA.

The coal-mining engineer of the northwest station, acting as consulting mining engineer to the Alaskan Engineering Commission in the development of coal mines in Alaska, made a field examination of two coal mines being opened by the Government in the Matanuska field, and suggested plans for the development of the mines.

WASHING AND COKING TESTS OF ALASKA COAL.

At the close of the season of 1917, 40 tons of coal were shipped to Seattle, for the purpose of making a washing test of two of the coals and a blacksmithing and coking test of the other. The 40 tons comprised approximately 20 tons of Chickaloon coal and 20 tons of Eska Creek coal. Although both coals could be improved by washing, a great loss of combustible material results when a suitable product is made of the Eska Creek coal, which had a large inherent ash content. The washed Chickaloon coal makes a satisfactory blacksmith coal and will also make satisfactory coke.

Float-and-sink determinations were made on a thousand-pound sample of coal from the Baxter prospect on Upper Moose Creek.

CONCENTRATION AND FLOTATION OF ALASKA NICKEL ORES.

Some work was done, in cooperation with the station at Golden, on the concentration of a low-grade nickel ore from Alaska. This ore is massive sulphide and is composed of sulphides, chiefly of pyrrhotite; the copper content by assay is a little less than 2 per cent and the nickel content about 4 per cent. Tests were made with a view to removing the copper as a high-grade copper concentrate and then separating the nickel as a high-grade commercial concentrate. The ore-dressing tests were followed by hydrometallurgical treatment. Possibly the electrometallurgy of the nickel products will be investigated.

CONCENTRATION TESTS OF TIN ORES.

Past methods of concentrating the tin ore in the Black Hills district of South Dakota were investigated and typical samples of the ores were sent to Seattle for concentration tests. These tests showed what could be expected in the recovery of tin from the low-grade ores of that district and the tin content of the products.

CONCENTRATION TESTS OF QUICKSILVER ORES.

In cooperation with the station at Berkeley, some concentration tests of typical samples of quicksilver ores from California were made to determine what recovery could be effected by concentration, and whether the method would have any advantages over direct smelting.

CONCENTRATION TESTS OF CHROME ORES.

Concentration tests of chrome ores from beach sands of southern Oregon were made in cooperation with the Oregon Bureau of Mines and Geology, in order to determine what could be expected in the way of recoveries under commercial practice. These tests were by wet concentration and electromagnetic separation of the concentrates. Concentration tests have also been made of chrome ores from different parts of California and Oregon, with the purpose of ascertaining how the chrome content of low-grade deposits could be made to yield commercial products. So far the results have been encouraging.

CONCENTRATION TESTS OF MANGANESE ORES.

Metallurgical analyses of the ore and concentrates at a concentrating mill near Lakecreek, Jackson County, Oreg., showed poor recoveries of the manganese. Tests were made in an attempt to help the operators in recovering mineral that is now being wasted.

ELECTROMETALLURGICAL INVESTIGATIONS.

Field investigations include a review of the present status of electrometallurgical plants in the Northwest, showing the general character of the work, equipment, and operating conditions. Particular attention has been paid to plants engaged in the production of ferro-alloys, on account of the importance of these alloys for special steels for military uses. It was found that such plants need much encouragement and attention. In the past there has been a tendency to start electric furnaces without much regard to basic technical considerations. With the intent of supplying some of this much-needed information, the station at Seattle has endeavored to collect such data as is needed by those concerns engaged or embarking in such work. Frequent personal visits have been made to the plants of the district.

INVESTIGATION OF HYDROELECTRIC POWER IN NORTHWEST.

Data have been compiled on hydroelectric power in the Pacific Northwest and Southeastern Alaska with reference to the potential power available and the situation of proposed sites. Data have also been tabulated, as far as possible, showing the power now available at existing plants, the maximum power available with present equipment, details of installations, and operating costs.

UTILIZATION OF METALS FROM WASTE TIN CANS.

An investigation was made of the probable metal waste of tin and iron in the Northwest and some experimental work was carried on having for its object the elimination of this waste. However, the work had to be stopped because of the lack or nonarrival of apparatus that had been ordered for trying out the small-scale experiments on a semicommercial scale.

FERROMANGANESE.

Pending the receipt of thermoelectric apparatus, a study was made of the manganese situation with reference to plants operating on the Pacific coast and in the Pacific northwest. Some of the data has been gathered directly, some was supplied by field engineers of the bureau.

FERROCHROME.

Small-scale experiments have been made in the electric reduction of chrome-bearing beach-sand concentrates to ferrochrome. Concentrates from low-grade chrome ores will be given similar treatment.

OTHER FERROALLOYS.

From the inquiries and samples submitted there seem to be numerous tungsten and molybdenum prospects in the Northwest, and it is planned to do some electric furnace work on these ores in order to determine their suitability for the production of ferromolybdenum and ferrotungsten.

SMELTING OF TIN CONCENTRATES.

- Because of the natural advantages of the ports of Puget Sound and their proximity to large salmon canneries, one of these ports is probably the logical place for a tin smelter for smelting tin concentrates from Alaska and Bolivia, as well as the by-product tin oxide which, it is found, may be obtained from detinning old tin cans. This problem will receive early attention.

ELECTROLYTIC TREATMENT OF COMPLEX LEAD-ZINC ORES.

The proposed treatment of the complex lead-zinc ores suggest a combined thermoelectric and electrolytic process. In the Cœur d'Alene district, Idaho, some large-scale experiments have been conducted for the recovery of lead by roasting the ores to lead sulphate, leaching with brine solution, and then electrolyzing. To date, the drawback has been the lack of coherent deposition. This problem, together with the recovery of lead and lead-chloride fume, obtained by volatilization of lead as lead chloride after a chloridizing roast with sodium chloride, will receive early attention in the electrolytic laboratory.

ANTIMONY.

More antimony ore (from Alaska) passes through Seattle than any other one place in the United States except New York City, when imports from Bolivia are particularly large. In addition, antimony prospects in all three northwestern States might possibly become commercial mines if there should be an outlet nearby. For these reasons, the station has taken a more than passing interest in antimony. The metallurgical treatment has already been pretty well worked out.

AMMONIUM NITRATE.

It is proposed to do some work at this station along the lines of making ammonium nitrate compound by electrolyzing nitric acid. The results of some small-scale experiments seem to indicate that it may be possible to devise such a process that would be commercially feasible.

MANUFACTURE OF METALLURGICAL CHEMICALS.

The use of the electric furnace in the manufacture of chemicals, such as hydrochloric acid and phosphoric acid has been advocated and it is proposed to try out some of these suggestions at an early date.

THERMOELECTRIC SMELTING OF COPPER ORES.

Some of the deposits of low-grade copper ore in southeastern Alaska seems to be situated so as to favor the development of thermoelectric plants for smelting. Ample power is available, and when the problem of a supply of charcoal or coke is solved much interest should be manifested in the development of such a process.

PHYSICAL PRINCIPLES OF FLOTATION.

The development of the flotation process has been rapid, and practical knowledge of the process is in many ways in advance of the science. Although many details of methods of operation, the use of oils, and other phases of mill work have been worked out, adequate explanation of the underlying theories has been lacking.

The results of studies made at the Seattle station of the physical principles of flotation have been illuminating. Surface tension, the angle of contact between liquid surfaces and solids, the spreading of oils on water, the flotability of different shaped particles, and the effects of colloids on flotation have received special attention.

The modifying effects of oil on the surface tension of water proved to be less then had been supposed. An apparatus (see Pl. XXIV, A) for the measurement of surface tension was developed, based on the drop-weight method. A mathematical formula expressing the relation of the angle of contact between liquids and solids to flotability was worked out. In studying the flotability of various shapes of particles, a surface was designed that illustrates film suspension remarkably. Plate XXIII, B, shows a piece of window glass cut in this design floating on pure water by film suspension. The interfering effects of colloids on flotation were clearly demonstrated.

The results of these tests were published in Technical Papers 182, "Flotation of Chalcopyrite in Chalcopyrite-Pyrrohotite Ores of Southern Oregon," and Technical Paper 200, "Colloids and Flotation."

COOPERATIVE WORK IN IDAHO.

In cooperation with the University of Idaho, at Moscow, an examination of the mining districts of Idaho was begun late in May, 1917, and continued during the summer. E. K. Soper, then dean of the mining department of the University of Idaho, visited the mining districts in the central and southwestern parts of the State and also some of the districts in the northern part. D. C. Livingston, professor of geology at the University of Idaho, visited some of the mining districts in central Idaho during the summer, and prepared maps covering the State. Thomas Varley, then superintendent of the station at Seattle, spent a short time in the field, principally in the mining districts in central and northern Idaho, and made a brief visit to the southwestern districts with Mr. Soper. The metallurgical engineer of the station spent most of his time in the Cœur d'Alene region, where he visited nearly all of the mining districts and devoted most of his time to examining milling methods.

The results of these field investigations have been published by the bureau as Bulletin 166. This bulletin aims to give the location of the various mining districts and the nature of the present operations as well as of those that have been carried on. Field work is to be continued and the results are to be published in more detail, with greater emphasis on the metallurgical treatment of the ores.

FLOTATION TESTS AT IDAHO UNIVERSITY.

The experimental work done at the University of Idaho under the direction of the Bureau of Mines has been chiefly on the flotation of mill feeds and tailings of lead and zinc ores from mills of the Cœur d'Alene district. The purpose of these flotation tests was to find the most suitable flotation mixture, for separating the lead minerals from the zinc minerals, what proportions of the different reagents are needed, and what operating conditions give the highest recovery of both minerals.

All of the tests have been conducted in a Varley mechanical type flotation cell. Air at low pressure can be used with all the machines.

In general, the reagents that have been given the highest differential selection of galena from sphalerite are common salt (NaCl) sodium carbonate (Na₂CO₃), coal-tar creosote oils, and a few heavy coal-tar oils. Special laboratory distillates from mixtures of different coal-tar creosotes and alcohol and coal-tar creosotes dissolved in alcohol offer promise for these ores. Seemingly, the coal-tar creosotes and distillates with alcohol used in an alkaline solution give the best results.

As regards recovery it has not been difficult to obtain a high selection of the metallic sulphides from the gangue, in fact 95 per cent recovery has easily been made. The problem is to make a high recovery of each metallic sulphide from the first differential separation. Some of the tests will be applied on a large scale in mills of the Cœur d'Alene district, as soon as the laboratory results warrant such trial.

THE TUCSON STATION.

FIELD OF THE STATION.

The territory served by the station at Tucson, Ariz., includes Arizona, southwestern New Mexico, southwestern Texas, and that part of southeastern California which borders on the Colorado River. The special task of this station is the investigation of methods of recovering copper and associated metals from low-grade ores, and in this respect its activities are not limited to the area named. Although the principal metal produced in this area is copper, there is an important production of lead and zinc, of the precious metals, and of some of the rare metals such as molybdenum, tungsten, and manganese. Views at manganese mines are shown in Plate XXV, A and XXV, B. Promising deposits of quicksilver, of vanadium, and of nickel have been found, but are only partly developed. On account of the importance of these metals, the station has an important duty to fulfill in studying the problems that arise in connection with their mining and treatment.

A. DROP-WEIGHT APPARATUS FOR DETERMINING SURFACE
AND INTERFACIAL TENSION.

Left, apparatus as arranged for surface-tension test; middle, interfacial tension, dropping downward; right, interfacial tension, dropping upward.

B. GROOVED GLASS PLATE LOADED WITH SAND AND RAFTED ON WATER BY FILM
SUSPENSION.

A. MINING AND SORTING MANGANESE ORE AT A MINE IN ARIZONA.

B. WINCH AND SORTING BOX AT A MANGANESE MINE IN ARIZONA.

The mineral production of the area tributary to the station is large and increasing. In 1917 the output was in excess of a quarter of a billion dollars. The value of the output of copper, lead, zinc, gold, and silver in Arizona for 1917 was roughly $214,000,000, and that of New Mexico, $33,700,000, as follows:

Approximate output and estimated value of principal metals produced in Arizona and New Mexico in 1917.

	Arizona.		New Mexico.	
	Production.	Value, roughly.	Production.	Value, roughly.
Copper	*Pounds.* 688,000,000	$200,000,000	*Pounds.* 104,500,000	$28,000,000
Lead	18,000,000	1,865,000	8,340,000	717,000
Zinc	20,700,000	2,000,000	27,900,000	2,790,000
Gold	a 192,800	3,988,000	a 48,389	1,000,000
Silver	a 6,354,000	5,900,000	a 1,262,000	1,045,000

a Troy ounces.

The number of men employed in mining in Arizona during 1917 was approximately 23,000; employment statistics for New Mexico are not available. Among the largest employers in the metal-mining field are the Chino Copper Co., with mines at Santa Rita, N. Mex., and mills and smelter at Hurley, N. Mex.; the Copper Queen Mining Co. and the Calumet & Arizona Mining Co., at Bisbee, Ariz.; and the Burro Mountain Copper Co., at Tyrone, N. Mex.

PROBLEMS CONNECTED WITH THE MINING OF LOW-GRADE COPPER ORES.

On account of the large scale on which mining of the porphyry copper ores is conducted, the tonnages ranging from 5,000 to 20,000 tons a day for a single property, problems of development, mining, transportation, ventilation, and the prevention of mine fires and their control are more than ordinarily difficult. As the low grade of much of the ore compels careful attention to economy in mining, the method of mining is of great importance. Many experiments have been made with different variations of underhand stoping. Two large mines have resorted to open-pit mining with steam shovels. One mine after many years of underground mining contemplates changing to open-pit methods.

MILLING AND LIXIVIATION PROBLEMS.

On account of the large tonnages involved, the location of concentration and leaching plants must be carefully studied with reference to water supply and to tailings storage, because of possible re-treatment and of pollution of streams. The mine and the mill are usually connected by a broad-gage railroad, and the ore trains are hauled by steam.

Milling of porphyry copper ores is in a transition stage. The ores may be roughly classified as follows:

1. Sulphide ores containing some oxidized copper, where table concentration followed by flotation—or flotation alone—recovers most of the sulphide copper and some of the oxidized copper, leaving for re-treatment a tailing of relatively high grade which contains a high percentage of the oxidized copper. A large proportion of the Chino, Ray, Miami, Inspiration, and Utah Copper ores is of this character.

2. Clearly defined bodies of oxidized ore with sulphide content low enough to warrant leaching with sulphuric acid and wasting the sulphides. Such bodies occur at the Utah Copper and the New Cornelia properties.

3. Mixed oxidized and sulphide ore for which straight leaching or straight concentration would be too wasteful. Ore bodies of this class are found at all of the properties.

4. Accumulations of mixed oxidized and sulphide tailings which offer much the same problem as the ores of class 3.

TREATMENT OF SULPHIDE ORES CONTAINING SOME OXIDIZED COPPER.

A brief description of the Miami and Inspiration mills will suffice to illustrate the treatment of sulphide ores containing some oxidized copper.

MIAMI MILL.

The Miami mill was originally designed to treat, by gravity concentration only, some 6,000 tons daily of ore having an assay value of 2.25 to 2.5 per cent copper, and containing perhaps 0.5 per cent of copper in oxidized form. The mill began work early in 1911 and was in full operation the latter part of the year. Since then the original design has been modified in many details. The ore is crushed with gyratory crushers and rolls, separated on Callow screens, ground in Chilean mills, which are fast being replaced by Hardinge mills, and concentrated on Deister tables and Deister slime tables, with the necessary intermediate apparatus. In 1914 flotation of tailings was tried and a flotation system installed. The mill is now treating 10,000 tons daily of ore averaging about 2.25 per cent copper content, and is making concentrates that average around 24 per cent copper. The mill recovers 90 per cent of the sulphide copper and less than 25 per cent of the oxidized copper.

INSPIRATION MILL.

The Inspiration mill as first designed was based largely on experience gained at the Miami, as the same engineer designed both plants. When the Inspiration mill was developed the flotation process was developing rapidly in the United States, so that opportunity favored the testing and comparison of gravity concentration and flotation.

Flotation was first considered for the treatment of tailings from gravity concentration. The results obtained from experimental work, however, soon led to the conclusion that flotation would play a much more important part. In January, 1914, a 600-ton experimental plant was built in which flotation was the principal method. This plant solved many questions and led to the final adoption of the simple flowsheet of the new mill of the Inspiration Company, which includes Marcy mills with Dorr classifiers, Callow or Inspiration flotation cells, followed by classification and secondary flotation treatment, with final table concentration of the tails from secondary flotation. It is to be noted that this latter concentration could be effected as well by additional flotation equipment, but the cost would be greater, because of the special oil needed to save this particular product.

The Inspiration mill treats 20,000 tons daily, making a 29.5 per cent concentrate from a mill feed carrying about 1.4 per cent copper, of which 0.3 per cent is oxidized. The mill recovery of sulphides is 90 per cent; the cost of crushing and concentration is about 63 cents a ton, including royalty on flotation patents.

Thus it happens that within a short distance of each other are two magnificent plants, one representing the highest development of the old practice, the other representing the highest development of the new practice.

It would seem that the flotation practice developed at the Inspiration mill, with such modifications and betterments as will naturally result from constant experiment, has solved the sulphide recovery problem.

TREATMENT OF OXIDIZED ORE LOW IN SULPHIDE COPPER.

The New Cornelia Copper Co., at Ajo, has standardized a sulphuric acid leaching practice for ores of the second class. The Ajo deposit comprises a low-grade sulphide ore, which must be mined by open-pit methods, overlain by a capping which may be regarded as a well-defined body of 12,000,000 tons, of oxidized ore, containing 1.65 per cent oxidized copper and a negligible quantity of sulphide copper.

The ore is crushed in tandem gyratory crushers to 3-inch size, then in tandem Symons horizontal disk crushers to ¼-inch size. This product goes to the leaching vats.

The leaching plant consists of eight lead-lined concrete vats, with a 15-foot ore column, and 1,500,000 gallons of leaching solution in circulation, having a rate of flow of 1,000 gallons per minute. The solution contains .3 per cent sulphuric acid, 2.5 per cent copper, and 2 per cent iron as sulphate. It is introduced at the bottom of the last vat which is receiving its final (eighth day) treatment, percolates upward through this charge and then through the charge that is receiving its seventh-day treatment, and so on successively to the first vat, whence it emerges as a practically neutral solution containing 3 per cent copper and ferric sulphate. The latter is reduced to ferrous sulphate because ferric sulphate redissolves copper precipitated by electrolysis. This reduction is accomplished by passing the solutions through absorption towers against ascending gas, containing 7 per cent sulphur dioxide. The resulting 3 per cent copper solution passes to the electrolytic tanks where one-sixth of the copper is removed. The leaching solution, with the wash solutions from the charges, is then built up for refuse. Besides dissolving the copper the leaching solutions take up various constituents from the ore, such as alumina, iron, etc., which would accumulate in succeeding cycles, and the copper after each cycle if reduced to 2.5 per cent. Therefore, after each cycle one-fourth of the solution is run to waste and is replaced with fresh acid and water. The copper in the wasted solution is precipitated as cement copper on scrap iron.

Sulphuric acid for leaching is made at the new acid plant recently put in operation at the Calumet & Arizona smelter. The output of this plant is 210 tons of 60° B. acid, most of which is consumed at the New Cornelia plant. Sulphur dioxide gas is obtained by roasting a high-grade copper-bearing pyrite produced in the Warren district. The calcined ore is returned to Douglas to be smelted. Experiments are now under way looking to metallizing the calcined iron by heating with coal in a closed rotating kiln, thereby making sponge iron for the precipitation of cement copper.

In all its mechanical details the new Cornelia plant is a model of construction; it operates smoothly and satisfactorily. It is treating daily 5,000 tons of oxidized ores containing 1.65 per cent copper and is making an 80 per cent extraction.

TREATMENT OF MIXED OXIDIZED AND SULPHIDE ORES AND TAILINGS.

Mining operations in the various districts, in so far as possible, are confined to ores of the first and second classes. On account of the larger demand and increased price of copper in recent years, many of the mills have worked far above their rated capacity, with resultant imperfect crushing and high sulphide tailings that carry varying, but always important, percentages of oxidized copper. At all of the large plants the treatment of such tailings and of the mixed oxidized-sulphide ores is a most important problem. In some of the older camps, where selective mining is no longer possible, and where ores containing a large proportion of oxidized copper must be milled, the situation is acute. This problem is being attacked vigorously from different sides and all the companies are working at it.

That the outstanding problem in the copper industry of the Southwest was the recovery of copper from mixed oxidized-sulphide ores became evident some time ago. Several properties have distinct bodies of mixed ore high in oxidized copper, these bodies ranging from a couple of million to perhaps ten million tons.

An examination of the various large tailings dumps showed that sulphide losses in tailings are due to the following causes: Insufficient crushing; deliberate overloading of mills to obtain a large output of copper, and an oxidized coating of film on sulphide particles, which must be removed in order to insure efficient recovery of the remaining sulphide by flotation.

There is very little sulphide ore that does not contain enough oxidized copper to warrant retreatment of mill tailings. Therefore, practically the entire tonnage milled and to be milled involves the retreatment of this mixed oxidized and sulphide material.

GENERAL CONCLUSIONS AS TO COPPER-ORE PROBLEMS.

A general survey of the situation by the technical staff of the bureau's station at Tucson has led to the following conclusions:

Flotation would make satisfactory recoveries of clean sulphides.

Sulphuric acid leaching seemed to offer a satisfactory process for oxidized copper ores containing negligible quantities of sulphides.

Certain operators felt that the solution of the problem of mixed oxidized and sulphide ores, as regards their particular ore, lay in sulphidizing or "filming" the oxidized mineral and subsequently floating it with the sulphides in one operation. Other operators felt that their ores were more amenable to an adaption of sulphuric acid leaching, followed by precipitation of the dissolved copper on metallic iron and recovery of both sulphide and cement copper by flotation, thus doing away with some rather objectionable steps in the recovery of copper from sulphate solutions. It seemed at first that the field had been rather completely covered by the well-organized experiment staffs of the large copper companies, but further investigation developed the fact that the companies were somewhat disappointed with the net results of their large-scale experimental work; also that a very promising field—leaching and sulphur dioxide gas—had been left entirely untouched.

It was evident that the first requisite to solving the mixed ore problem was a dependable analytical method for determining the exact proportions of sulphide and nonsulphide, or oxidized, copper present. The known methods were faulty and the one generally in use (the sulphuric acid method) gave erroneous results, part of the sulphide copper in the tails being reported as oxidized copper on account of the solubility of chalcocite in cold dilute sulphuric acid.

Therefore, the station gave attention first to the development of a reliable analytical method for the selective determination of oxidized and sulphide copper.

SMELTING PROBLEMS.

There are many important smelting plants in the area served by the station at Tucson, and all of them do a custom business. The larger plants are at El Paso, Tex.; Hurley, N. Mex.; Clifton, Morenci, Douglas, Miami, Globe, Ray, Jerome, and Humboldt, Ariz. Among the problems of smelter practice are those mentioned below.

The mining public regards the sampling of mill products and mine ore at the smelter with undisguised suspicion. Standardization of the practice and a description written in nontechnical language for the information of the public would be a benefit.

The basis of smelter schedules is not understood. The producers do not appreciate the fact that most of the smelters were built to furnish an outlet for the ores of one or more groups of mines, nor do they comprehend the difficulties under which the smelters are laboring. Neither do the smelters appear to be always solicitous of the problems of producers. Many misunderstandings that arise through the producer not knowing the difficulties and losses encountered in smelting practice could be cleared away. Similarily the attitude of smelting corporations toward the producer and his problems might be improved.

A study and correlation of practice in the following departments of the smelters might be helpful: Dust losses; preparation of fine material destined for the blast furnace; converter practice; and utilization of waste products, such as gases and slags.

BUILDINGS AND EQUIPMENT.

The station at Tucson, Ariz., was established in the summer of 1916, and active work was begun in January, 1917. The first six months were devoted to a general survey of the area to be served by the station; the second half of the year to the ordering and installation of equipment.

The station is housed in buildings of the University of Arizona and works under a cooperative agreement with the Arizona State Bureau of Mines and with the college of mines and engineering of the State University. The State Bureau of Mines furnishes the funds for two or more metallurgical fellowships, and recipients of these fellowships do research work in the laboratories of the station. The station is in charge of Charles E. Van Barneveld, superintendent.

EQUIPMENT.

During 1917 the station occupied temporary office and laboratory quarters on the university campus pending the construction of the new mines and engineering building. Provision was made in the new building for offices, for a chemical laboratory, an electrical laboratory, an assay-furnace room, a general metallurgical laboratory, and an overflow laboratory. The equipment selected includes metallurgical apparatus to handle roasting, leaching, and concentration tests of ores in lots of 100 pounds to several tons. Much of this equipment was temporarily erected and in use several months before it was installed in the new building.

INVESTIGATIONS.

SELECTIVE DETERMINATION OF COPPER MINERALS IN PARTLY OXIDIZED ORES.

The first problem undertaken was the development of a dependable analytical method for the selective determination of the sulphide and nonsulphide copper minerals in partly oxidized copper ores. A satisfactory method was worked out, the details of which have been published in Technical Paper 198.[a]

SULPHUR DIOXIDE METHOD FOR COPPER ORES.

In connection with this work a careful study was made of the action of sulphur dioxide gas on copper minerals, and an apparatus was constructed for the use of hot sulphur dioxide gas, direct from the roasting furnace, as a solvent for nonsulphide copper in any form. Small-scale laboratory tests gave most encouraging results on materials of the mixed oxidized-sulphide class, and as a result a plant was constructed to test ores and tailings in quantity ranging from 1 ton upward with a maximum capacity on tailings of 10 tons per 8-hour day.

The work undertaken requires experimentation in different processes, as follows: Crushing, gravity concentration, flotation, leaching, and precipitation of ores and tailings from the five principal districts in the territory. The field is broad and the tonnage involved is enormous.

THE FAIRBANKS STATION.

The station at Fairbanks, Alaska, was established primarily to aid in solving problems arising in connection with the development of the gold, copper, and other mineral resources of the Territory, and expressly those that relate to the prevention of waste in the mining and milling of low-grade ores and the application of hydroelectric power in the mineral industries. This station is housed in temporary quarters (Pl. XXVI) furnished by the Commercial Club of Fairbanks. The equipment of the station includes an ore-testing laboratory, where milling and concentrating tests on a small scale can be made, and a laboratory for chemical analyses, assaying, and microscopic examinations of ores and minerals.

---

a Van Barneveld, C. E., and Leaver, E. S., Sulphur dioxide method for determining copper minerals in partly oxidized ores: Tech. Paper 198, Bureau of Mines, 1918, 14 pp.

Fairbanks is on the Tenana River, a tributary of the Yukon, and has many of the conveniences of a modern city, including an electric light plant, a telegraph system connecting it with the mining camps in the vicinity, and a wireless station. Transportation from points on the Pacific coast is chiefly by way of Dawson and the Yukon River. A railroad up the Tenana Valley makes travel in the area immediately tributary to Fairbanks comparatively easy, but outside that area travel is difficult. The new Government railroad, in course of construction, will undoubtedly prove a powerful stimulus to mining and to commercial ventures.

In Alaskan mining the cost of power is a factor of growing importance. As the shallower deposits are worked out and the mines become deeper, more elaborate pumping and hoisting equipment must be used. Power for running machinery at mines and mills is now chiefly obtained from wood and gasoline. Timber near the mines is scarce and is being rapidly cut off; gasoline is brought from the Pacific States, and its cost is accordingly high. Utilization of Alaska coal and development of hydroelectric power would be of advantage. A study of the possible application of hydroelectric power to mining, milling, and smelting will be a special feature of the work of the station.

Another feature of the work is the supplying of information to prospectors and others regarding economic minerals submitted for identification. In connection with this work a collection of minerals is being formed and information is being compiled on the appearance, mode of occurrence, value, and uses of economic minerals found in Alaska.

In the metallurgical laboratory studies are made of the best methods of treating low-grade ores. Samples of ores submitted by parties desirous of the bureau's help and ores selected as typical of certain districts are studied under the microscope and by screen analyses and by laboratory methods. Then test runs are made on lots of 1 to 5 tons to determine the type of concentrating machinery or the process best suited to the ore studied.

Crushing and milling tests have been made on a few ores from certain properties and recommendations made to the owners.

BUILDING HOUSING THE FAIRBANKS STATION.

STATIONS ESTABLISHED IN 1917.

The three mining experiment stations established in 1917 are at Columbus, Ohio, Minneapolis, Minn., and Bartlesville, Okla. The chief function of the station at Columbus is the study of problems of the ceramic industries; that of the station at Minneapolis, the beneficiation and utilization of low-grade iron ores, including hematites, magnetites, titaniferous magnetites, and manganiferous ores; and that of the station at Bartlesville, investigations relating to petroleum and natural gas.

THE COLUMBUS STATION.

As the mining and manufacture of clay products is one of the chief industries in the United States, one of the stations established in 1917 deals with ceramic problems exclusively. The main reasons for selecting Columbus, Ohio, as the site of this station were that Columbus is the center of important clay-working industries, Ohio being the leading State in manufacture of clay products, and that the University of Ohio, which has a school of ceramics, offered to provide quarters for the station and, as soon as possible, furnish funds for active cooperative research work. The specific purposes of the station are to conduct research work and carry on investigations on problems arising in connection with the ceramic industry, to encourage the development of ceramic raw materials, and to devise improved methods for the refining, manufacture, and utilization of domestic clays, with especial regard to the prevention of waste and the improvement of the quality of the products.

IMPORTANCE OF CERAMIC INDUSTRIES IN THE UNITED STATES.

Ceramic industries in the United States are many and varied. They include the mining, refining, and utilization of the nonmetallic minerals used in the manufacture of products that require baking, hardening, or fusing in kilns and furnaces.

In the manufacture of clay products the three materials used in greatest quantities are clay, sand, and limestone. These are widely distributed and exist in vast quantities.

The most important ceramic industries are those utilizing clays. Plants for making clays are in operation in every State and Territory of the Union. The most important products made are brick, tile, sewer pipe, flue lining, conduits, hollow blocks, fireproofing, architectural terra cotta, stoneware, porcelain, earthenware, china, and art pottery.

The number of firms manufacturing clay products in the United States is about 4,000, representing an investment of approximately

89

$375,000,000. The value of the Nation's clay products exceeds
$200,000,000 per annum. Approximately 75 per cent of this sum
represents brick, tile, and other structural material, and the remainder higher grade material classed as pottery.

The six leading States in order of production of clay products are
Ohio, Pennsylvania, New Jersey, Illinois, New York, and Indiana.
The value of their production exceeds $135,000,000, or is about 68
per cent of the total domestic production. Ohio, the leading State,
produces annually clay products valued at more than $45,000,000, or
approximately 22 per cent of the Nation's output. More than 600
plants in the State are engaged in clay-working operations representing an investment of approximately $50,000,000.

In the manufacture of structural materials—such as brick, terra
cotta, tile, and sewer pipe—a wide variety of clays is utilized.

The principal materials used in the manufacture of pottery are
kaolin, ball clay, potter's flint, whiting, and feldspar. In the manufacture of glazes, enamels, and colors for decorative purposes over
250 different minerals and chemically prepared materials are required.

Refactories for furnace linings are of the greatest importance to
the metallurgical industries, for without suitable refractories most
of the metallurgical processes would be impossible. Where the best
quality of clay fire brick will not withstand the severe furnace treatment often necessary, it is necessary to use fire brick made from other
mineral substances, such as ganister, magnesite, chromite, and
zirconia. Graphite is important in the manufacture of graphite
crucibles used in making crucible steel, brasses, and other alloys.

In the manufacture of Portland cement, large quantities of clay,
shale, slag, limestone, and marl are needed. In the manufacture of
quicklime, used in mortars and plasters and for certain chemical
purposes, a pure grade of limestone is required. Gypsum is a mineral
of considerable importance. From it wall plaster, plaster of Paris,
fireproofing and Keen cement are made. Common window and
plate glass, bottle glass, optical glass, tableware and cut glass, chemical glassware, and special glasses for scientific purposes utilize large
quantities of high-grade sand as well as pure limestone.

Industries manufacturing grinding wheels and other abrasive
products from natural corundum, as well as from the electric furnace
products, such as silicon carbide and fused bauxite, are important
ceramic industries.

In the sheet-metal and cast-iron enameling industries, feldspar,
fluorspar, cryolite, kaolin, silica, borax, boric acid, red lead, white
lead, whiting, tin oxide, and antimony oxide are used as ingredients
of the enamel for such products as bathtubs, lavatories, kitchen
sinks, laundry tubs, chemical ware, stove parts, granite ware, automobile tags, signs, and enameled jewelry.

NEED OF SCIENTIFIC RESEARCH IN CERAMICS.

Although clay working was one of the first arts toward which savage man turned his attention, the ceramic industries are among the last to receive much scientific attention. The scientific study of ceramics began in Europe about 1869. In the United States the first degree in ceramic engineering was conferred in 1900, and only four universities and one college now offer instruction in ceramic chemistry and ceramic engineering.

Technical assistance is needed especially in the clay industries. Clay plants are widely distributed and, in general, comparatively small, representing an average capitalization of less than $100,000 per plant. In the bottle-glass industry over $500,000 was spent in the development of the modern automatic bottle-blowing machine, but no clay-working firm feels that it can afford to spend a fiftieth part of that sum for experimental purposes. The general methods employed in the manufacture of clay products are substantially the same as those employed half a century ago. Advance has been made in kilns and machinery, but has not kept pace with that made in many other industries.

A more thorough acquaintance with the raw materials, a more extended development and improvement in automatic machinery for molding and handling the products, and a wider adoption and further improvement of the continuous kiln and the tunnel kiln are needed in order to reduce losses and to conserve both labor and fuel.

EQUIPMENT OF THE STATION.

The laboratories and offices of the station are housed in Lord Hall (see Pl. XXVII), a building containing the mining engineering, metallurgical, and ceramic laboratories of the University of Ohio. A two-story addition and a kiln house have been erected for the use of the station (see fig. 2). The first floor contains two general laboratories, a machine shop, and a pottery laboratory. On the second floor are offices, a drafting room, an optical and chemical laboratory, and an electric furnace laboratory.

In the laboratory in the west basement are a dry pan, pug mill, and molding machine, bins for storing crude and screened clay, a screen and two bucket elevators for conveying the crushed clay from the dry pan to the screen and the screened clay from the bin to the pug mill. Provision has been made for driving this machinery with two 30-horsepower motors. One drives the dry pan, elevators, and pug mill, and the other is connected to the molding machine. This arrangement gives flexibility and permits tests being made on the molding machine independently, as, for example, in determining the power required for molding stiff mud clay products, a subject on which no reliable data exist. This machinery, with the tunnel

FIGURE 2.—Addition to Lord Hall, Ohio University, for the ceramics station of the Bureau of Mines.

BUILDING HOUSING THE COLUMBUS STATION.

driers and large kiln, also permits tests of clays on a scale comparable to that of small commercial plants.

The two large laboratories of the new addition have been equipped for general experiment work. The machine shop contains a turning lathe, drill press, shaper, double grinder, and workbench. In ceramic experiment work it is sometimes necessary to make apparatus that can not be purchased in open market.

The pottery laboratory contains a blunger, filter press, vertical pug mill, potter's jolly, a battery of six porcelain-lined ball mills, and a disk grinder. This equipment is used for the preparation of clays for casting and jiggering, for pressing test pieces, and for preparing glazes and enamels.

The optical laboratory is equipped for the study of optical properties of ceramic materials and for making refined physical measurements.

The electrical furnace laboratory contains two carbon-resistance furnaces, two nichrome-resistance furnaces, a platinum-resistance furnace, and a small arc furnace.

WORK OF THE STATION.

KAOLIN INVESTIGATIONS.

During the war an unusual demand for chemical stoneware arose from the manufacture of dyestuffs, of ingredients for poisonous gases and explosives, and of other chemicals. A study of the properties of stoneware clays of Ohio and Pennsylvania was begun, chiefly to call attention to valuable stoneware clays not widely used and to show how their qualities could be improved by inexpensive physical and chemical treatment. Over 50 per cent of the china clay or kaolin used in the pottery, paper, and oilcloth industries in normal times has been imported from England. Although there are large quantities of high-grade kaolin in the United States, consumers of this commodity claim that the kaolin miners are unable to deliver a product that is comparable to the English.

In the manufacture of the better grades of pottery, the domestic kaolins either do not give ware of good color or else cause high bisque losses. The principal objections to the use of domestic kaolins as fillers in the paper and oilcloth industries are that the domestic kaolins do not give good spreading qualities and wear out the machinery rapidly.

The workable kaolin deposits east of the Mississippi River are being investigated by engineers of the bureau. At each deposit data have been collected on its extent and structure, its location and accessibility, and the methods of mining and refining used. Samples have been collected and sent to the Columbus station, where experiments in refining and blending will be conducted with the object of producing a clay equal to imported varieties.

In connection with the work on kaolins, two investigations are underway. These are the use of sulphuric acid and the sedimentation of clays, and the use of American clays as fillers for oilcloth.

REFRACTORIES.

The station is investigating a number of problems in refractories. Some of these problems are: The use of domestic magnesite as a substitute for imported varieties for the manufacture of refractories; the use of dolomite for furnace linings; the use of American graphite in the manufacture of crucibles; and the utilization of American clays for bond clay in the manufacture of graphite crucibles.

OTHER PROBLEMS.

Other problems in connection with the clay industries are the improvement of methods of handling and burning clays so as to obtain greater efficiency and less waste of fuel, and a study of the effects of dust on the lungs of workers in mines and shops.

THE MINNEAPOLIS STATION.

The Lake Superior iron district, which includes the iron ranges in Minnesota, Wisconsin, and Michigan, produces about 88 per cent of the total iron ore mined in the United States. Much of this production comes from ore of good grade, but the district has vast deposits of low-grade ore. The supply of high-grade ore may be exhausted in 30 years, or even less, and the problem of the beneficiation and utilization of the leaner ores is of great importance. As Minnesota ranks first in production of iron ore, and the State university had offered to cooperate with the Federal Bureau of Mines in furnishing quarters for the station and the use of its laboratories, the station was established at Minneapolis. In a city as large as Minneapolis supplies and material can be promptly obtained, and the numerous foundries and shops offer excellent facilities for the construction of special apparatus and machinery. Because of railroad connection, large samples of ores for testing can readily be assembled from the ranges of the Lake Superior district, and also from the West and Southwest. The problems to be investigated at this station include the concentration of sandy and cherty hematites, of lean magnetites and titaniferous magnetites, and the concentration of manganiferous ores; a study is also to be made of new processes for the reduction of iron ores.

PRODUCTION OF IRON ORES IN THE LAKE SUPERIOR DISTRICT.

The preeminence of the Lake Superior district in production of iron ores is shown by the following table, which gives the production of Minnesota, Michigan, Wisconsin, New York, Alabama, and the other States for the years 1908 to 1917.

BUILDING HOUSING OFFICES OF MINNEAPOLIS STATION.

plemented by special investigations, and then published so that it will be available to all. As the Minneapolis station works in cooperation with the mine operators, and with the State School of Mines, which maintains a mining experiment station for the study of Minnesota ores, it is expected that this station will offer unusual facilities for research and be able to assist the industry by coordinating and disseminating data bearing on the efficient utilization of iron ores.

WORK OF THE STATION.

The Minneapolis station is housed in temporary quarters in the old laboratory of the Minnesota School of Mines, and in an adjacent office (Pl. XXVIII) built for the use of the station. The experiment station of the School of Mines is elaborately equipped for experiments in ore dressing and metallurgy, and the Federal station will have the advantage of using the laboratories and equipment. An apparatus for magnetic concentration tests is shown in Plate XXIX.

As the steel industry during the war needed large amounts of manganese, and as imports of foreign manganese ore were curtailed by the necessity of transporting troops and supplies to Europe, the obtaining of enough domestic ore to meet the needs of our blast furnaces became an imperative necessity. Therefore, it was decided that as long as should be necessary the station should devote its energies to manganese, and the first work of the station was a survey of the manganese situation.

During the first year the work of the station was in charge of Edmond Newton, superintendent, who was succeeded by C. E. Julihn, the present superintendent.

During the summer of 1918, in cooperation with the State School of Mines, a comprehensive examination was made of the manganiferous mines of the Cuyuna Range, and manganiferous deposits on the other ranges, including the Mesabi, were examined.

Extensive research work was conducted on the use of manganese alloys in open-hearth steel practice, and the extent to which low-grade manganiferous domestic ores could be utilized. A study was made of the production of ferromanganese in blast furnaces, every furnace in blast on manganese alloys being visited. A similar investigation was made of the production of spiegeleisen in blast furnaces.

At the station a large amount of experimental work was conducted on the concentration and beneficiation of low-grade ores of manganese. These tests were conducted in the ore-dressing laboratories of the State mining experiment station.

Two processes for the direct reduction of manganese from its ores, known as the Jones process and the Bourcoud process, were investigated.

Reports prepared on all of the manganese investigations mentioned will be assembled and published as a bulletin of the bureau.

BUILDING HOUSING OFFICES OF MINNEAPOLIS STATION.

MAGNETIC CONCENTRATOR.

THE STATION AT BARTLESVILLE.

The petroleum experiment station, established in 1917, was placed at Bartlesville, Okla., because of that city being in the Mid-Continent field of Kansas and Oklahoma, which is the center of the petroleum industry in the United States and produces more than one-third of , the country's output.

Authorities estimate that there are in the United States 300 refineries, valued at approximately $430,000,000. The total value of the output of crude petroleum and natural gas in 1916 amounted to $450,000,000; the copper output had a value of $474,000,000, and gold not more than $193,000,000, but these prices are for mine metal.

The Mid-Continent field, being connected by pipe line with the Atlantic and Gulf coasts, supplies oil to some of the largest refineries in the world, which are on the Atlantic seaboard. Moreover, many large refineries in the Mid-West, which supply the consumers of the Mississippi and Missouri valleys, depend entirely on Oklahoma and Kansas for supplies of crude oil.

BUILDINGS AND EQUIPMENT.

The Chamber of Commerce of Bartlesville donated $50,000 for the station and four acres of land as a site were given by Mr. G. B. Keeler. This sum was used to construct buildings and purchase equipment. The contract for two buildings, one for offices and one for a laboratory, was let June 3, 1917. Pending the construction and equipping of these buildings, work was conducted from temporary quarters supplied by the chamber of commerce.

PETROLEUM PROBLEMS. .

The tremendous development of the oil industry in Oklahoma, Texas, and adjoining States makes this station a natural center for investigation, experiment, and report, as well as for helpful practical suggestions to operators. Situated in the oil fields, its position insures close cooperation with the producer and investor.

Petroleum engineers of the bureau, working from this station as a center, will conduct field investigations aimed to prevent waste in the production of oil and natural gas. Studies will be made in the improvement of methods of drilling and casing wells, the causes of water troubles and their abatement, prevention of waste at refinery plants, and the saving of valuable by-products, such as gasoline present in natural gas or in "casing head" gas.

The station endeavors to cooperate in every way with producers and refiners and aid them in solving their problems. Conferences and correspondence with various oil company officials about definite problems have awakened much interest among operators and have shown that they stand ready to aid the station to the fullest extent. For example, a request for the aid of certain officials in a campaign to shut off water from oil wells and prevent damage by infiltration has evoked a hearty response.

In the laboratory of the station research work on practical problems in the recovery of gasoline, the refining of petroleum, and similar work will be conducted. The investigations will not be confined to any one branch of the industry, nor to any one part of the country, but will extend to all districts having interests that are similar. The problems studied will be followed from the preliminary stages in the laboratory to final adjustment to practical conditions in the field.

Among the development and production problems suggested are the dehydration of emulsified oil and the prevention of its formation in oil wells. Experience has shown that probably all emulsions in wells are formed after the oil has left the sand, and their formation can often be prevented. Studies of methods for excluding water from oil and gas wells, of the effect of water on the production of oil wells, and of casing and tubing problems, would be of value. Work on the use of safety devices in the oil fields, the possible substitution of explosives safer than nitroglycerin for shooting oil wells, and the effects of shooting on oil wells, would be of interest. Investigations for determining the capacities and characteristics of oil and gas sands, the proportions of oil not being recovered by present methods of production, and the laws governing the expulsion of oil from the oil sands will be started at the station. There is little information on this subject, and it will not be possible to interpret data and to judge the value of various methods of producing until the underlying scientific laws are better understood. Methods for stimulating production and increasing the recoveries of oil from the sands, including the effects of vacuum pumping, use of compressed air or gas, and use of water flooding, present a fertile field for research. In this connection, pumping problems and equipment, including the use of automatic air and gas pumps, electrical equipment, and air lifts would receive attention.

Storage and transportation problems include losses of oil in storage, prevention of fires at tanks and wells, the flow of oils in pipe lines, and the construction of storage tanks and reservoirs.

At refineries the problems are many and varied. The efficiencies of condensers and heat exchangers, reduction of wastes and losses of the lighter fractions, methods for effecting a cleaner separation of gasoline from the heavier cuts, recovery of acid, especially in small refineries, and utilization of sludge and acid tar are some phases demanding attention. Collection and dissemination of data on properties of petroleum for refinery engineering, such as latent and specific heats of oils and various fractions and other properties that will influence designs of stills, heat interchangers, condensers, would be of value to refinery engineers.

Investigations on the construction and operation of compression, refrigeration, and absorption plants for recovering gasoline from natural gas, with regard to the recoveries being made, and improvements in the plants or methods employed are being conducted by bureau engineers. It is believed that many of the existing plants are not recovering all the gasoline, and that gasoline plants could be profitably installed for the treatment of much gas now going to waste at oil wells or being turned into pipe lines untreated.

INDEX.

○

Bulletin 176   Petroleum Technology 50

DEPARTMENT OF THE INTERIOR
FRANKLIN K. LANE, Secretary

BUREAU OF MINES
VAN. H. MANNING, Director

RECENT DEVELOPMENTS IN THE ABSORPTION PROCESS FOR RECOVERING GASOLINE FROM NATURAL GAS

BY

W. P. DYKEMA

WASHINGTON
GOVERNMENT PRINTING OFFICE
1919

CONTENTS.

ILLUSTRATIONS.

TABLES.

ILLUSTRATIONS.

RECENT DEVELOPMENTS IN THE ABSORPTION PROCESS FOR RECOVERING GASOLINE FROM NATURAL GAS.

By W. P. Dykema.

INTRODUCTION.

This report gives the results of a study conducted by the Bureau of Mines for the purpose of informing the petroleum industry on the recent progress in the development and application of the absorption process for recovering gasoline from natural gas. The work has been undertaken entirely with regard to plant practice and for the purpose of describing the features and the operation of units of the plants now in use, the improvements that have been made, and the reasons for any changes contemplated by the engineers in charge of the plants studied.

Throughout the United States the improvements in the absorption process that are under consideration by engineers not only widen practice so as to include gases at all pressures and percentages of gasoline content, but seem to be developing definite standards as to dimensions of towers, velocities of flow, saturation of oils, areas of cooling surfaces, and quantities of menstruum to be circulated. Although perfect practice has by no means been attained, the writer believes that a description of the current practice at various plants, including those obviously more efficient features that each plant has developed and uses at one or another stage of treatment, will interest and possibly help all persons operating or designing absorption plants.

ACKNOWLEDGMENTS.

The possibilities and the usefulness of a study of the absorption process by the Bureau of Mines were recognized and suggested by W. A. Williams, when chief of the petroleum division of the bureau. Chester Naramore, the present chief of the division, has shown untiring interest in this study and has given his hearty support and helpful guidance. Particular thanks are due Messrs. P. M. Biddison and Thomas R. Weymouth for much data regarding plant operation and construction. The writer acknowledges the courtesies extended and the information given him by Messrs. Guy L. Goodwin, J. J. Allison, W. P. Gage, W. R. Hamilton, J. D. Creveling, W. R. Finney, E. Hepner, E. T. Buckley, Carl Holt, D. L. Newton, and

A. E. Hurley, all of whom furnished original data on both experimental work and plant operation.

J. O. Lewis, R. O. Neal, C. P. Bowie, F. B. Tough, and J. M. Wadsworth, all members of the Bureau of Mines, rendered efficient cooperation in matters pertaining to the particular phases of the oil industry in which each is a specialist. J. G. Shumate, of the bureau, gave aid in making drawings from notes and sketches. The writer is indebted to J. H. Wiggins for a number of the illustrations used.

Much use has been made of Bureau of Mines publications[a] on this and related subjects, also the publications of the Bureau of Standards.[b]

PRODUCTION OF GASOLINE FROM NATURAL GAS.

The marketed production of gasoline recovered from natural gas sold in the United States during the years 1911 to 1917 is shown in Table 1 following.

TABLE 1.—*Gasoline from natural gas marketed in the United States, 1911–1917.*[c]

1911.

State.	Number of operators.	Plants.		Gasoline produced.			Gas used.		Average yield in gasoline per M cubic feet.
		Number.	Daily capacity.	Quantity.	Value.	Price per gallon.	Estimated quantity.	Value.	
			Gallons.	*Gallons.*		*Cents.*	*Cubic feet.*		*Gallons.*
West Virginia.....	47	72	16,819	3,660,165	$262,661	7.18	1,252,900,600	$76,074	2.92
Ohio...............	26	39	6,454	1,678,985	118,161	7.04	469,672,000	37,574	3.57
Pennsylvania.....	43	50	5,669	1,467,043	109,649	7.47	526,152,663	52,615	2.79
Oklahoma.........	8	8	4,800	388,058	20,975	5.40	144,629,000	4,378	2.68
California.........									
Colorado..........									
Illinois............	8	7	3,358	d 231,588	20,258	8.75	82,343,000	6,320	2.81
New York.........									
Kentucky.........									
	132	176	37,100	7,425,839	531,704	7.16	2,475,697,263	176,961	3.00

[a] Allen, I. C., and Burrell, G. A., Liquified products from natural gas, Tech. Paper 10, 1912, 23 pp.; Calvert, W. R., A preliminary report on the utilization of petroleum and natural gas in Wyoming, with a discussion of the suitability of natural gas for making gasoline by G. A. Burrell, Tech. Paper 57, 1913, 23 pp.; Burrell, G. A., and Jones, G. W., Methods of testing natural gas for gasoline content, Tech. Paper 37, 1916, 26 pp.; Burrell, G. A., and Robertson, I. W., The compressibility of natural gas at high pressures, Tech. Paper 131, 1916, 11 pp.; Dean, E. W., Motor gasoline; properties, laboratory methods of testing, and practical specifications, Tech. Paper 166, 1917, 27 pp.; Burrell, G. A., Seibert, F. M., and Oberfell, G. G., The condensation of gasoline from natural gas, Bull. 88, 1915, 106 pp.; Burrell, G. A., Biddison, P. M., and Oberfell, G. G., Extraction of gasoline from natural gas by absorption methods, Bull. 120, 1917, 71 pp.

[b] Edwards, J. D., Effusion method of determining gas density, Technologic Paper 94, Bureau of Standards, 1917, 30 pp.; United States standard tables for petroleum oils, Circular 57, Bureau of Standards, 1916, 64 pp.

[c] Figures for 1911 to 1916 from Mineral Resources of United States for 1916, U. S. Geol. Survey, 1917, pp. 648–649; figures for 1917 supplied by U. S. Geol. Survey.

[d] Includes gasoline produced in Kentucky which came from natural condensation in gas mains.

TABLE 1.—*Gasoline from natural gas marketed in the United States, etc.*—Contd.

1912.

State.	Number of operators.	Plants.		Gasoline produced.		Price per gallon.	Gas used.		Average yield in gasoline per M cubic feet.
		Number.	Daily capacity.	Quantity.	Value.		Estimated quantity.	Value.	
			Gallons.	*Gallons.*		*Cents.*	*Cubic feet.*		*Gallons*
West Virginia.....	66	97	22,366	5,318,136	$513,116	9.6	1,972,882,212	$163,749	2.8
Pennsylvania.....	69	83	10,524	2,041,109	217,016	10.6	722,730,117	62,010	2.8
Ohio..............	25	43	7,791	1,718,719	173,421	10.1	576,123,700	46,090	2.98
Oklahoma.........	11	13	11,910	1,575,644	99,626	6.3	701,044,300	24,901	2.25
California........	7	7	6,609	1,040,695	112,502	10.8	600,743,000	25,573	1.7
Illinois...........	4	4							
Colorado..........	2	2	2,008	a 386,876	41,795	10.8	114,273,000	9,662	3.4
New York.........	1	1							
Kentucky.........	1								
	186	250	61,268	12,061,179	1,157,476	9.6	4,687,796,329	331,985	2.6

1913.

State.	Number of operators.	Number.	Daily capacity.	Quantity.	Value.	Price per gallon.	Estimated quantity.	Value.	Average yield.
West Virginia.....	63	115	31,930	7,662,493	$807,406	10.54	2,961,119,000	$181,337	2.57
Oklahoma.........	19	40	61,633	6,462,968	577,944	8.94	2,152,503,000	82,742	3.00
Pennsylvania.....	100	113	22,207	3,680,696	405,186	11.01	1,372,056,000	114,783	2.68
California........	12	14	21,135	3,460,747	376,227	10.87	2,436,445,000	106,539	1.42
Ohio.............	25	41	8,142	2,072,687	212,404	10.25	744,226,000	63,233	2.79
Illinois...........	6	12							
Colorado..........	2	2							
New York.........	3	3	7,368	a 721,826	79,276	10.98	203,092,500	17,590	3.55
Kansas............	1	1							
Kentucky.........	1								
	232	341	152,415	24,060,817	2,458,443	10.22	9,889,441,500	566,224	2.43

1914.

State.	Number of operators.	Number.	Daily capacity.	Quantity.	Value.	Price per gallon.	Estimated quantity.	Value.	Average yield.
Oklahoma.........	35	58	74,798	17,277,555	$1,113,059	6.44	5,738,549,000	$273,940	3.01
West Virginia.....	65	121	34,460	9,278,108	691,899	7.45	3,005,292,000	172,396	2.58
California........	17	19	32,360	7,581,309	633,517	8.36	5,129,709,000	197,056	1.48
Pennsylvania.....	96	119	21,456	4,611,738	359,402	7.79	1,560,064,000	125,690	2.89
Ohio.............	25	47	9,319	2,440,171	184,097	7.54	832,277,000	68,935	2.86
Illinois...........	7	14	5,300	1,164,178	100,331	8.62	462,321,000	43,017	2.52
Kansas............	3	3							
New York.........	3	3	1,665	a 299,573	23,604	7.88	146,345,000	8,862	2.03
Colorado..........	2	2							
Kentucky.........	1								
	254	386	179,353	42,652,632	3,105,909	7.28	16,894,557,000	889,906	2.43

1915.

State.	Number of operators.	Number.	Daily capacity.	Quantity.	Value.	Price per gallon.	Estimated quantity.	Value.	Average yield.
Oklahoma.........	36	63	111,463	31,665,991	$2,361,029	7.46	8,791,881,000	$435,512	3.60
California........	18	20	40,755	12,835,126	975,397	7.60	8,006,888,000	288,669	1.60
West Virginia b ...	66	114	34,422	10,853,608	927,079	8.54	3,526,575,000	150,918	2.30
Pennsylvania b	116	139	22,754	5,898,597	569,873	9.66	1,838,034,000	186,325	2.73
Ohio.............	29	50	8,995	2,198,715	167,138	7.60	785,041,000	77,767	2.80
Illinois...........	8	16	8,500	1,035,204	80,049	7.73	451,663,000	34,405	2.29
Texas.............	1	1							
New York.........	4	4							
Louisiana.........	2	2	3,447	877,424	70,258	8.01	664,309,000	28,959	1.32
Kansas............	3	2							
Colorado..........	2	2							
Kentucky b........	2	1							
	287	414	232,336	65,364,665	5,130,823	7.88	24,064,391,000	1,202,555	2.57

a Includes gasoline produced in Kentucky which came from natural condensation in gas mains.
b Includes gasoline resulting from natural condensation in gas mains.

TABLE 1.—*Gasoline from natural gas marketed in the United States, etc.*—Contd.

1916.

State.	Num-ber of opera-tors.	Plants.		Gasoline produced.			Estimated quantity of gas treated.	Average yield of gasoline per thousand cubic feet of gas.
		Num-ber.	Daily capacity.	Quantity.	Value.	Price per gallon.		
			Gallons.	*Gallons.*		*Cents.*	*M cubic ft.*	*Gallons.*
Oklahoma	77	116	233,077	48,359,602	$5,865,145	12.13	24,749,454	1.954
West Virginia	105	147	98,659	18,765,056	3,025,293	16.12	104,664,536	.179
California	28	26	54,060	17,158,754	2,298,822	13.37	24,836,354	.691
Pennsylvania	167	195	46,487	9,714,926	1,726,173	17.77	38,490,621	.252
Ohio	40	55	18,391	2,638,571	804	17.84	5,435,750	.485
Illinois	17	32	12,070	2,260,288	269,664	11.58	1,338,504	1.688
Louisiana	7	7	10,661	2,113,159	269,564	12.76	907,153	2.329
Texas	3	4	6,688	1,292,811	201,023	15.55	948,485	1.363
Kentucky	5	5	11,300	725,467	141,347	19.48	5,614,613	.129
Kansas	4	3	3,080	215,000	35,030	16.29	1,626,635	.132
New York Colorado }	7	6	1,025	249,055	40,283	16.17	102,819	2.422
	460	596	495,448	103,492,689	14,331,148	13.85	208,705,023	.496

1917.

State.	Num-ber of opera-tors.	Plants.		Gasoline produced.			Estimated quantity of gas treated.	Average yield of gasoline per thousand cubic feet of gas.
		Num-ber.	Daily capacity.	Quantity.	Value.	Price per gallon.		
Oklahoma	167	234	492,436	115,123,424	21,541,905	18.71	84,719,941	1.359
West Virginia	128	188	135,663	32,666,647	6,511,813	19.93	167,771,351	.195
California	45	49	99,761	28,817,604	4,438,022	15.40	45,351,247	.635
Pennsylvania	287	251	59,164	13,826,250	2,778,098	20.01	49,487,056	.279
Texas	10	11	32,550	6,920,405	1,149,441	16.61	12,677,216	.546
Ohio	49	61	25,137	5,439,560	1,051,376	19.33	30,062,141	.181
Louisiana	15	20	20,118	4,977,764	814,747	16.36	2,233,511	2.229
Illinois	33	55	17,392	4,937,009	866,033	17.55	2,685,805	1.837
Kentucky	5	5	13,400	3,818,209	753,186	19.99	24,915,946	.153
Kansas	4	6	4,642	1,174,980	241,219	20.53	9,315,339	.126
New York Colorado }	7	6	2,122	181,262	33,116	18.27	68,154	2.659
	750	886	902,385	217,884,104	40,188,956	18.45	429,287,797	.508

The production of gasoline from natural gas by compression and vacuum methods and by the absorption method for the years 1916 and 1917 is shown in Tables 2 and 3. The tables show that the quantity of gas treated in 1917 was more than double that of 1916, and that most of this increase was from gas treated in absorption plants.

TABLE 2.—*Production of gasoline from natural gas in 1916 by principal methods of manufacture.*[a]

Gasoline produced by compression and by vacuum methods.

State.	Plants.		Gasoline produced.			Gas used.	
	Number.	Daily capacity.	Quantity.	Value.	Price per gallon.	Estimated quantity.	Average yield in gasoline per M cubic feet.
		Gallons.	*Gallons.*		*Cents.*	*M. cubic ft.*	*Gallons.*
Oklahoma	101	215,377	45,827,325	$5,471,307	11.94	14,018,757	3.280
West Virginia	133	39,276	9,280,624	1,642,031	17.67	3,550,523	2.616
Pennsylvania	185	30,287	6,722,370	1,216,717	18.10	2,693,215	2.496
Louisiana	7	10,661	213,159	269,564	12.76	907,153	2.329
Texas	4	6,688	1,292,811	201,023	15.55	948,485	1.363
New York	5	1,025	249,055	40,283	16.17	102,819	2.422
Colorado	1						
California	24						
Ohio	53						
Illinois	29	72,251	19,428,443	2,652,776	13.65	14,492,463	....
Kentucky	3						
Kansas	2						
	550	375,365	84,922,787	11,493,701	13.53	36,713,415	....

Gasoline produced by absorption methods.[b]

West Virginia	14	59,383	9,475,432	1,383,262	14.60	101,114,013	0.094
Pennsylvania	10	16,200	2,992,556	509,456	17.02	35,797,406	.084
Oklahoma	12	17,700	2,532,277	393,838	15.55	10,730,697	.236
California	2						
Kentucky	2						
Illinois	3	26,600	3,569,637	550,891	15.43	24,349,492	....
Ohio	2						
Kansas	1						
	46	119,883	18,569,902	2,837,447	15.28	171,991,608	....
Grand total	596	495,448	103,492,689	14,331,148	13.85	208,705,023	.496

[a] Mineral Resources of United States for 1916, U. S. Geol. Survey, 1917, pt. 2, p. 650.
[b] Includes drips.

TABLE 3.—*Production of gasoline from natural gas in 1917 by principal methods of manufacture.*[a]

Gasoline produced by compression and by vacuum pumps.

State.	Plants.		Gasoline produced.			Gas used.	
	Number.	Daily capacity.	Quantity.	Value.	Price per gallon.	Estimated volume.	Average yield of gasoline per M cubic feet.
		Gallons.	*Gallons.*		*Cents.*	*M. cubic ft.*	*Gallons.*
Oklahoma	207	456,632	108,728,213	$20,321,067	18.68	36,399,280	2.987
California	40	82,092	23,478,521	3,637,827	15.49	27,477,443	.854
West Virginia	159	44,348	12,276,784	2,211,494	18.01	4,845,648	2.534
Pennsylvania	234	32,564	9,011,199	1,792,430	19.89	3,572,356	2.522
Louisiana	18	17,915	4,459,920	719,758	16.14	1,558,346	2.862
Illinois	54	15,392	4,268,158	756,344	17.72	2,020,044	2.113
Texas	8	10,900	3,942,337	664,543	16.86	2,666,983	1.478
Ohio	54	8,337	2,331,498	423,106	18.15	836,639	2.787
New York	5						
Kansas	1						
Kentucky	3	3,322	369,925	70,361	19.02	150,784	2.133
Colorado	1						
	784	671,502	168,866,555	30,596,930	18.12	79,527,523	2.123

[a] Figures supplied by United States Geological Survey.

TABLE 3.—*Production of gasoline from natural gas in 1917, etc.*—Continued.

Gasoline produced by absorption.[a]

State.	Plants.		Gasoline produced.			Gas used.	
	Number.	Daily capacity.	Quantity.	Value.	Price per gallon.	Estimated volume.	Average yield of gasoline per M cubic feet.
		Gallons.	*Gallons.*		*Cents.*	*M. cubic ft.*	*Gallons.*
West Virginia.........	29	91,315	20,391,863	$4,300,319	21.09	162,925.703	0.125
Oklahoma.............	27	35,804	6,395,211	1,220,838	19.09	48,320,661	.132
California............	9	17,569	5,339,083	800,195	14.99	17,873,804	.299
Pennsylvania.........	17	26,600	4,815,051	985,668	20.47	45,914,700	.105
Kentucky [a]...........	2	[b] 13,000	3,725,893	745,210	20.00	21,871,390	.150
Ohio.................	7	16,800	3,108,062	628,270	20.21	20,225,502	.108
Texas [c].............	3	21,650	2,978,068	484,898	16.28	10,010,233	.298
Kansas..............	5	3,842	1,071,033	220,550	20.58	9,274,389	.116
Illinois [a]...........	1	2,000	665,851	109,689	16.47	665,851	1.000
Louisiana [a].........	2	2,203	519,834	94,989	18.27	675,185	.770
New York [d].........			7,000	1,400	20.00	2,776	
Colorado............							
	102	230,883	49,017,549	9,592,026	19.57	349,760,274	.140
Grand total...........	886	902,385	217,884,104	40,188,956	18.45	429,287,797	.508

[a] Includes drip gasoline.
[b] Includes gasoline produced in Kentucky from West Virginia gas.
[c] Includes some gasoline produced by compression.
[d] Drips only.

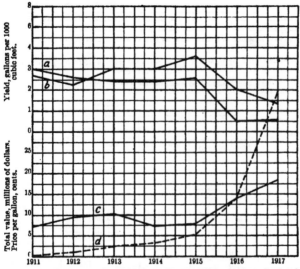

FIGURE 1.—Curves showing yield of gasoline per 1,000 cubic feet of natural gas treated and value of gasoline produced, 1911–1917. Curve *a*, yield in United States; curve *b*, yield in Oklahoma; curve *c*, price of gasoline in United States in cents per gallon; curve *d*, total value of gasoline from natural gas in United States in millions of dollars.

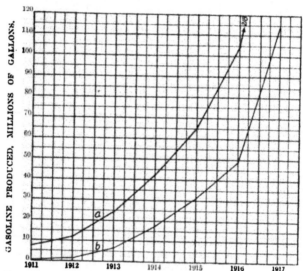

FIGURE 2.—Curves showing annual production of gasoline from natural gas in United States (curve *a*), and in Oklahoma (curve *b*), for the years 1911 to 1917, in millions of gallons.

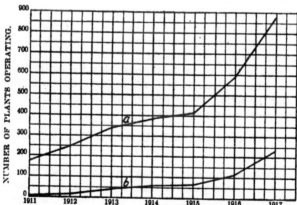

FIGURE 3.—Curves showing number of plants treating natural gas for gasoline in United States (curve *a*), and in Oklahoma (curve *b*), 1911-1917.

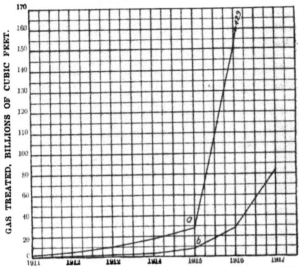

FIGURE 4.—Curves showing quantity of natural gas treated in United States (curve a), and in Oklahoma (curve b), 1911–1917.

The curves of figures 1, 2, 3, and 4 show graphically the output of gasoline from natural gas since 1910. In the years 1916 and 1917 construction was exceedingly active, many improvements being made in both compression and absorption plants. The output of gasoline from natural gas in Oklahoma in July, 1918, was probably more than 300.000 gallons a day, and in California 75,000 gallons a day from the 30 plants in operation. Eastern as well as western districts will undoubtedly show a steady increase in the production of gasoline from gas, because of the larger volume of gas being treated and the increased efficiency of the plants.

From what he has seen in the various fields, the writer thinks that the number of new plants in 1918 will be less than the number built in 1917, for the ratio between gas production and number of plants has become more nearly equalized.

CHARACTER OF NATURAL GAS TREATED BY AB-SORPTION.

The first absorption plants treating natural gas were built only for the treatment of lean gases at high pressure and commonly were placed on the gas transmission lines after the gas had been compressed and passed through the cooling coils usually employed in gas pumping. The coils are used to remove the heat generated by compression and to precipitate and collect from the gas as much as possible of the condensable vapors. Otherwise these vapors would cause rapid deterioration of rubber gaskets, which are used on pipe lines connected by couplings, and would entail additional expense through the construction and maintenance of a large number of pipe-line drips. In commercial gas pumping an absorption plant becomes a drying equipment that is of enough value for this purpose alone to warrant installation. In fact, drying the gas was one of the primary incentives in the development of absorption processes.

The gas being treated by large commercial gas companies in the eastern United States usually originates in gas fields that yield no oil, the gas being known as high-pressure "dry" gas. There are, however, in the East a few compression gasoline companies that after extracting as much condensate as possible, deliver the residual gas to the pipe lines of commercial gas companies, this gas receiving a second treatment by an absorption plant at the gas-pumping station.

In the Mid-Continent fields, besides gas from "dry" gas wells in gas pools, gas produced with the oil under high pressures (gushers) is separated from the oil in traps*, admitted under its own pressure to the pipe lines of gas companies, and treated by absorption at the first pumping station through which it passes. Large quantities of such gas have carried 0.28 gallon of condensable vapor per 1,000 cubic feet, even when the pressure ranged between 300 and 500 pounds at the casing head. For such gas the percentage of condensable vapors carried invariably increases as the pressure drops.

In both the Mid-Continent and California fields the gas originating in oil wells at all pressures and vacuums is now being treated successfully by the absorption process; much of the gas contains quantities of vapor too small for treatment by compression alone. Practice seems to indicate that natural gas containing three-fourths of a gallon of gasoline per 1,000 cubic feet of gas is the driest gas yielding profitable results by compression, whereas absorption is used successfully for gas yielding from less than one pint to three and one-half gallons per 1,000 cubic feet, and also for uncondensed still vapors in refineries and topping plants and the vapors rising from flow tanks and storage tanks, as well as the residue gases discharged from casing-head gasoline plants.

In the study of both the compression and absorption processes in the various fields of the United States the writer found marked differences in the characteristics of the condensate recovered from gases yielding equal quantities of gasoline. This variation seems to depend on the proportion of each of the hydrocarbons in the gas; its existence is proved by the varying characteristics of the conden-

---

* Hamilton, W. R., Traps for separating gas at oil wells: Tech. Paper 209, 1918, 86 pp.

sates produced as well as by the varying quantities obtained at the different pressures and temperatures at which the condensable vapors precipitate. Unfortunately, no method of analysis has been devised by which the percentages of the different condensable hydrocarbon members in natural gas can be even approximately estimated.

It seems that the character of the oil with which gas is, or has been, associated and the factors developed and used in compression plants would afford a good indication of the character of the gas and would indicate roughly the treatment necessary to obtain the desired quality of product by absorption.

PHYSICAL LAWS CONTROLLING ABSORPTION OF GASES IN LIQUIDS.

Inasmuch as the physical laws of gases apply closely to the absorption process, the more important are given here:

Henry's law is as follows: " The weight of any gas absorbed in the unit volume of a liquid, at any pressure, will be equal to the weight absorbed in that volume at one atmosphere pressure, times that pressure (in atmospheres)." This statement of the law presupposes that the temperatures are always equal. The coefficient of absorption of a liquid equals the volume of gas (at $0°$ C., $32°$ F.), which is absorbed in the unit of volume of a liquid at one atmosphere pressure. To illustrate how the absorption of gases by liquids varies with temperature the following example is given: Water absorbs 0.0498 volume of oxygen at $0°$ C. ($32°$ F.) and 0.0244 volume at $35°$ C. ($95°$ F.); at the boiling point $100°$ C. ($212°$ F.) practically no oxygen will be absorbed or remain in solution.

Dalton's law of partial pressures is as follows: " Every portion of a mass of gas in a vessel contributes to the pressure against the sides of a vessel the same amount that it would have exerted by itself had no other gas been present. "

Partial pressures have direct effects in the absorbing action in the towers, in the condensing of the distilled vapors, and in the separating of the gasoline from the oil in the still. As partial pressures affect practically all of the major functions of absorption operations, plant designers should completely understand the physical laws involved.

VAPOR PRESSURES.

When water is brought into a space that it does not fill, evaporation takes place until either all the water has been converted into vapor or the pressure of the water vapor has reached a certain definite limit, and a state of equilibrium exists between water vapor and liquid water. The limit of vapor pressure varies with the temperature and is entirely independent of the size and shape of the vessel or the relative volume of the liquid or of the water vapor. The pressure at which the water vapor is in equilibrium with the liquid water is called the vapor pressure of the liquid.

These laws help to explain the absorption process but can not be mathematically applied to the problems met in plant designing except in a general way, because it is impossible to learn the percentages of the many hydrocarbon compounds that constitute the gas, and therefore the partial pressure of each at a given pressure, or to know the absorption coefficient of the absorbing medium for each of the fractions of condensable vapor, or the temperatures that will be developed in the plant at different seasons.

These facts explain why the many plants installed were designed in accord with results of experimental tests and small units operated as nearly as possible under plant conditions.

FACTORS CONTROLLING PLANT DESIGN.

The controlling factors of plant design are the pressures, temperatures and gasoline content of the gas to be treated and the time and intimacy of contact between the gas and the absorbing medium necessary to absorb the gasoline fractions that are at present commercially valuable.

PRESSURE.

Pressure affects the absorption process mechanically and physically. Mechanically it determines the weight and strength of materials used for pipe lines, towers, traps, and valves. Physically, it controls the actual volumes of gas, and thereby affects the speed of gas flow or flow velocity and is a factor in determining the percentage of saturation to which the absorption oil may be raised while removing the maximum quantities of condensate, that is, the quantity of oil to be circulated per 1,000 cubic feet of gas.

TEMPERATURE.

The temperatures of gas and oil during treatment affect gas volume, percentage of saturation, extraction efficiency, and also bear directly upon the character of hydrocarbons removed, because of the rise in vapor pressure of the condensable members with higher temperatures.

GASOLINE CONTENT.

The gasoline content of the gas being treated, together with the temperature and pressure of the gas and oil and the time of treatment, will determine the quantities of oil to be circulated.

CONTACT OF GAS AND OIL.

The intimacy and duration of contact between the oil and the gas must be regulated by the height of towers used as absorbers, and by the system of baffles together with the other factors controlling plant design in order to expose as large a surface of oil as is possible without too great constriction of the cross-sectional area of the tower.

HORIZONTAL ABSORBERS WHERE INSTALLED BEING REPLACED BY VERTICAL UNITS.

Although horizontal absorbers have been used at many plants and are at present included in the designs of some plants under construction, the trend of development is toward vertical units. This trend was so evident at the plants studied by the writer that horizontal types of absorbers are not included in the general discussion presented on the pages following. Many operators are rebuilding plants that originally had absorbers of the horizontal types, and are installing vertical units instead. The reasons given for this change are the difficulty of exposing large surfaces of oil to the gas being treated, excessive loss of pressure through friction in the treating units, and the low efficiency of extraction.

TESTING NATURAL GAS FOR GASOLINE CONTENT.

The methods of testing natural gas for gasoline have been described in Bureau of Mines publications[a] and many articles on the subject have been published in the technical press. Only such methods as are not already well known and understood and are of particular interest and use in connection with absorption practice are given here.

LINE-DRIP MEASUREMENTS.

A company pumping between 30 and 40 million cubic feet of gas a day, using the ordinary cooling coils after compression, made tests preliminary to the designing of an absorption plant and found that 4.600 gallons of condensate were being blown from the

[a] Burrell, G. A., Biddison, P. M., and Oberfell, G. G., Extraction of natural gas by absorption methods: Bull. 120, Bureau of Mines, 1917, 71 pp.; Dykema, W. P., Recovery of gasoline from natural gas by compression and refrigeration: Bull. 151, Bureau of Mines, 1917, 151 pp.; Burrell, G. A., Seibert, F. M., and Oberfell, G. G., The condensation of gasoline from natural gas: Bull. 88, Bureau of Mines, 1915, 106 pp.; Burrell, G. A., and Jones, G. W., Methods of testing natural gas for gasoline content: Tech. Paper 87, Bureau of Mines, 1916, 27 pp.; Calvert, W. R., A preliminary report on the utilization of petroleum and natural gas in Wyoming, with a discussion of the suitability of natural gas for making gasoline, by G. A. Burrell: Tech. Paper 57, Bureau of Mines, 1913, 23 pp.

line drips and wasted in one day. This single test proved the advisability of building a plant. The test was made during the coldest part of the year when condensation in the pipe lines was at a maximum. However, the results are reliable, as they indicate the minimum quantity of gasoline that an absorption plant would be able to produce at all periods of the year. The proposed plant was built and it now produces between two and three times the amount of condensate recovered during the drip measurements.

The tests described were made on pipe lines conveying gas under high pressure. The same method may be used for lines carrying gas under partial vacuum, but the percentage of condensable content will be much lower on account of the reduced pressure. The usual arrangement of piping for a drip on a vacuum line shown in figure 5 is self-explanatory. Mr. G. L. Goodwin says that from his experience, " it is always advisable to connect an equalizing line from

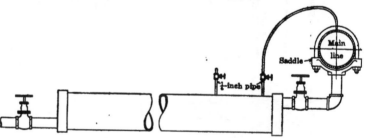

FIGURE 5.—Construction of line drip.

the highest point on the receiver to the top of the gas main line, as this greatly increases the efficiency of the drip collector."

The drips are always placed at low points in pipe lines and are built of standard pipe and fittings. A 2-inch pipe carries the condensed liquid through a valve and into a storage pipe made of one, two, or three lengths of large (6, 8, or 10-inch) pipe screwed together to form an accumulator or reservoir. This storage pipe slopes so as to drain into the 2-inch discharge pipe placed at the lower end and fitted with a valve to allow the liquid to be blown out.

TYPES OF TESTERS USED.

Numerous testing devices have been developed by the chemists and engineers of the natural-gas gasoline industry. They may be divided into two classes, those for lean gas, and those for rich gas, but no line of demarcation can be drawn, as by proper manipulation many of them may be used for gas of any gasoline content.

TESTERS USED IN PRACTICE.

SINGLE-CELL TESTERS.

The type of testing apparatus most often found in use is that designed by P. M. Biddison[a] and described by G. A. Burrell as follows:

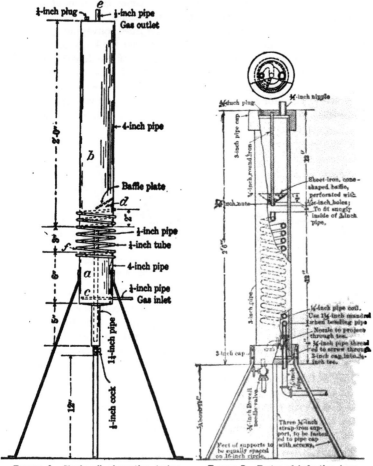

FIGURE 6.—Single-coil absorption tester.   FIGURE 7.—Tester of induction type.

The absorber, shown in figure 6 and Plate I, *A*, which is built on the principle of a laboratory gas-washing bottle of the Friedric type, is of iron pipe thoroughly welded. It is about 3 feet high, and the two main barrels

---

[a] Chief engineer, Ohio Fuel Supply Co., pipe-line department.

A. SINGLE-COIL ABSORPTION TESTER
SET UP FOR TEST.

B. QUADRUPLE-COIL, SINGLE-
CELL TESTER.

C. QUADRUPLE-COIL, FOUR-CELL TESTER.

a and *b* (fig. 6) are of 4-inch pipe. The entering gas bubbles up through the absorbing oil at *c* and passes through the coil of pipe *f* and into the part *d* and out at *e*. This apparatus was designed as a very efficient absorber, to show the maximum possible yield of gasoline from natural gas.

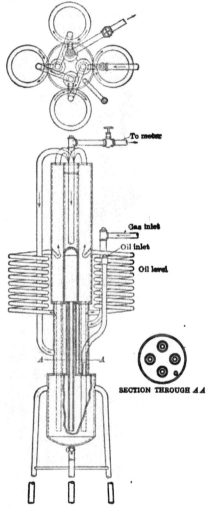

For making preliminary tests a single absorption unit of this type gives satisfactory results as long as the percentage of saturation is kept below 4 per cent and the gas is passed through the oil and tube slowly. When the amount of gas passed through allows the oil to become saturated to 6 or 7 per cent, the results of the test will be 15 to 25 per cent or more below the true value. Gas under atmospheric or low pressures, because of larger volume, should be passed through the unit more slowly than gas at high pressures, as time of contact is one of the factors controlling absorption. Figure 7 shows the detail of a single treatment tester in which a small nozzle on the pipe supplying the gas acts like an injector to force and mix the oil in the tubing coil. This type of absorption tester is usually termed an induction tester because of the mechanical principle involved.

FIGURE 8.—Multiple-coil tester with single oil chamber.

In final testing, three, four, or five units of the single cell type are connected in series, the gas is passed through each in turn, the gasoline is extracted from the oil in each unit separately by distillation, and the total extraction is

computed. The oil in the last unit should show very small extraction
or none at all. This proves the test.

A large company in Oklahoma uses four units of this type as a
check on plant operation, tests being run once or twice weekly. The
results are considered as 100 per cent extraction, and the daily plant efficiency is computed on that basis.

Mr. Biddison improved the original single cell, single coil tester by using multiple coils and larger quantities of absorbing oil and increases the number and the duration of the contacts between oil and gas. (Fig. 8, and Pl. I, B.)

The method of operation is essentially the same as that of the unit shown in figure 6, although the quantity of gas can be increased because of the large amount of oil used, which is three to five times as much as that in the bowl of the single coil type. Because of the successive contacts the flow velocity may be increased. This type of tester is used on rich or lean gas at

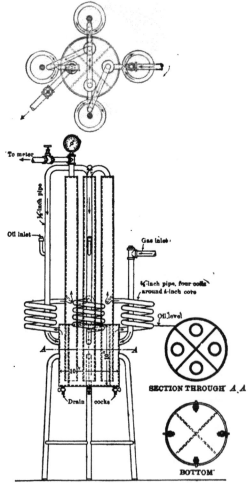

FIGURE 9.—Multiple coil tester with quadruple cell chamber.

any pressure by varying the quantity of gas treated and the rate of gas
flow. It gives the gas four treatments with the oil, uses only one oil
charge and requires but one distilling to determine the total extraction.

MULTIPLE-CELL TESTERS.

A combination of the two types which is shown in figure 6 (single) and figure 8 (quadruple-coil single-cell tester) has been designed and uesd by Mr. J. D. Creveling, chief gasoline engineer of the Logan Natural Gas Co., Columbus, Ohio. Figure 9 and Plate I, C, show this type of tester. The bowl is divided into four compartments, each holding approximately 3,000 c. c. of absorption medium, and each filled through a short, upturned arm welded to the gas pipe that leads to each oil chamber.

When in use each chamber of the bowl is charged with an accurately measured quantity of absorption oil (approximately 3,000 c. c.), and the charging pipe is securely capped. The gas passes into the first section of the bowl, then upward through the first coil to the separating pipe chamber; the oil carried up with the gas descends to the bottom of the chamber from which it came, and the gas passes up and into the pipe leading to the second oil-filled chamber where the second contact with oil is obtained. The gas flows through each of the four chambers in turn and out at the top of the fourth separating pipe to the meter.

The oil is drained from each chamber, measured, tested for gravity, distilled separately, and the total extraction is computed. According to the rate of flow, gasoline content, the total quantity of gas treated, and some other small factors, the variation in the percentage of extraction in each of the four chambers will be as follows:

Gasoline absorbed in each chamber.

	Per cent.
First compartment	60 to 90
Second compartment	10 to 25
Third compartment	1 to 15
Fourth compartment	0 to 6.5

The controlling functions of the tester are shown in the tables below.

Controlling functions of quadruple tester.

Maximum rates of flow of gas, cubic feet per hour:	Pounds absolute pressure.
400	300
200	150
100	75
50	40
20	Atmospheric pressure.

Maximum gas capacity.

Gas, cubic feet:	Gasoline content, gallons per 1,000 cu. ft.
800	.125
400	.25
200	.50

	Gasoline content, gallons per 1,000 cu. ft.
Gas, cubic feet—Continued.	
150 _____	0. 75
100 _____	1. 00
66 _____	1. 50
50 _____	2. 00
35 _____	3. 00
25 _____	4. 00

If the maximum quantities of gas are passed through the tester the oil in the fourth chamber will probably contain a small fraction of gasoline and should always be tested.

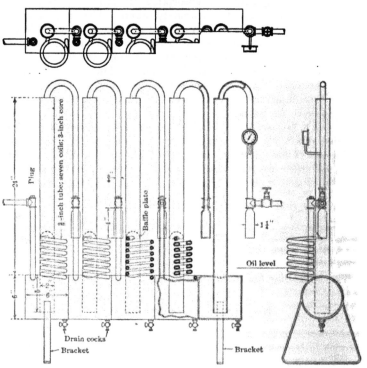

FIGURE 10.—Multiple tester used by Bureau of Mines.

In figure 10 and Plate II, *A*, is shown an absorption tester designed by the writer to embody the most recent improvements; it provides multiple treatment in a rigid unit. Large gas companies give gas four or five treatments as a check on plant extraction. The gas is put through the tester under pressures and temperatures like

B. TOWERS 20 INCHES BY 25 FEET. DE-
SIGNED FOR CAPACITY OF 1,000,000
CUBIC FEET EACH, AT 100-POUND
PRESSURE.

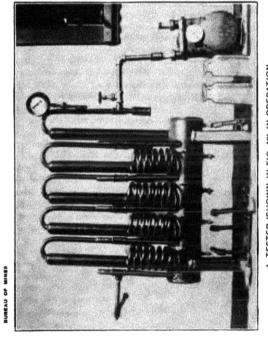

A. TESTER (SHOWN IN FIG. 10) IN OPERATION.

those in the plant proper. The oil in the last chamber is tested for gasoline but the chief purpose of testing is to check the extraction in the other chambers to prove that the extraction is maximum.

Although these types of testers, all based on the same principle, are most commonly used throughout the United States, there are other types whose value is indicated by their continued use. The autoclave tester is a type frequently used.

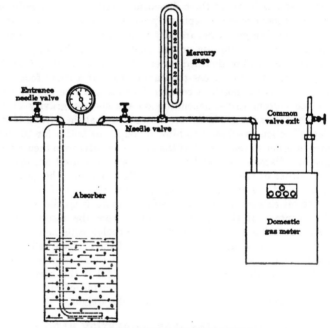

FIGURE 11.—Autoclave absorption tester.

AUTOCLAVE ABSORPTION TESTER.

Figure 11, taken from California State Mining Bureau Bulletin 73,[a] shows the tester used by a large California company. Two or more of these absorbers are often coupled in series in the same manner as those previously described, and for the same purpose. This tester treats both lean and rich gas. It is especially useful in tests with crude oil as an absorbent when low saturation and large quantities of oil are customary.

---

[a] McLaughlin, R. P., First annual report of the State oil and gas supervisor of California for the fiscal year 1915–16: Cal. State Mining Bull. 73, 1917, p. 281.

THE CUBIC FOOT BOTTLE.

Oberfell[a] has designed and used an absorption tester for gas of high gasoline content. A cylindrical tank of galvanized iron approximately one cubic foot in capacity is built, 10 inches in diameter and 24 inches in length, with slightly conical heads to allow drainage of oil. In the center of each head is welded or soldered a small (½-inch) valve for charging and discharging both oil and gas. Although a tank or bottle of these dimensions will hold practically one cubic foot of gas under atmospheric pressure, in order to be accurate enough for testing gas it is carefully measured by the displacement and measurement of the air or gas contained and by weighing it empty and when full of distilled water. The bottles used by Oberfell are gaged carefully to the one-thousandth part of a cubic foot.

In testing gas the bottle is connected to the gas line at either the top or the bottom valve, depending on the specific gravity of the gas—at the top if the gas is lighter than air, and at the bottom if it is heavier. The gas is allowed to blow through the bottle for 30 to 60 minutes in order to remove all of the air. The valves are then closed, the gas line disconnected, and 850 c. c. of absorption oil are injected through either of the valves into the bottle containing the gas. The bottle and contents are violently agitated for a period of 20 minutes while the oil extracts all of the gasoline vapors from the gas. The gasoline content of the gas is computed from the results of distilling about 800 c. c. of the oil, using ice-cooled coils and collector to condense distilled-off vapors. Because of the extreme refinements required and the comparative simplicity and equal or greater accuracy of other methods of testing, it is doubtful if this method will be useful for final testing. In preliminary testing of rich gas it is practical because of its portability and light weight.

DISTILLING GASOLINE FROM ABSORPTION MEDIUM.

In completing the tests, the absorbed gasoline is separated from the absorption oil in fire stills or steam stills. The vaporized fractions pass through coils usually cooled by water and ice, and are condensed and discharged into a graduated vessel also cooled by water and ice. A steam still and coil of this type are shown in figure 12, taken from Bulletin 73, of the California State Mining Bureau.[b] Many condensers do not depend on a flow of ice water as the vessel through which the vertical worm condenser passes is large enough to allow the ice to be added directly to the water surrounding the

---

[a] Oberfell, G. G., Testing natural gas for gasoline: Jour. Ind. and Eng. Chem., vol. 10, March, 1918, p. 211.

[b] McLaughlin, R. P., First annual report of the State gas and oil supervisor of California for the fiscal year 1915-16: Cal. State Mining Bull. 73, 1917, p. 232.

coil. This simplifies the operation and gives equally accurate results.[a]

GENERAL USE OF TESTING APPARATUS.

Testing apparatus is now used not only to determine the quality of gas before a plant is designed, but also at large plants to show

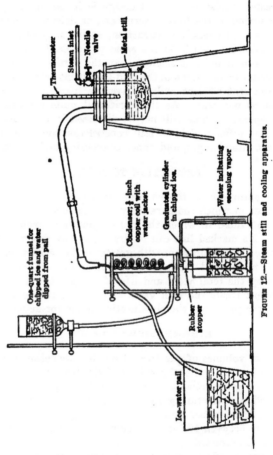

FIGURE 12.—Steam still and cooling apparatus.

from day to day the variations in the quantity of gasoline content and as a check on operation and efficiency. Such tests should be made frequently at all plants and the results filed with the records of plant pressures, production, and other essential information. In

[a] See Dean, E. W., Motor gasoline; properties, laboratory methods of testing, and practical specifications: Tech. Paper 166, Bureau of Mines, 1917, p. 20.

some fields, Osage and Cushing for example, tests are made at regular intervals to determine the gasoline content of the dry gas, since royalties are paid on the gasoline content and figured on a 50 per cent extraction basis.

PLANT DESIGN AND OPERATION.

The greatest variation in plant design is found in the type, size, and construction of the towers, scrubbers, or driers, as they are variously called.   In gasoline extraction, as in many other scrubber systems designed to work on the same principle, horizontal scrubbers have been found less efficient than the vertical or tower type. The horizontal installations in the Mid-Continent fields are not more than 75 per cent efficient and the average plant of this type not more than 60 per cent.   As the horizontal units are being abandoned by operators, they will not be discussed in this paper.

Plants also differ widely in still construction and operation, design of heat interchanger, and vapor condensing and treatment.

ABSORPTION TOWERS.

Vertical absorbers or towers may be divided into two general classes—those in which comparatively lean gases are treated at high pressure, and those in which the richer gases are treated at lower pressures.   No marked line can be drawn, as any gas, rich or lean, can be treated at any pressure with greater or less efficiency, depending on the mechanical features of the plant unit and the expense warranted by the total yield and life of the plant.   As the industry at present treats the lean gases at high pressure and the rich gases at lower pressures, the classification holds for actual practice.

HIGH-PRESSURE TOWERS.

The largest volumes of gas treated by the absorption process are lean gas (0.1 to 0.3 gallons per 1,000 feet) from gas fields, which is pumped by commercial gas companies for distribution and use in cities and manufacturing plants.   In order to reduce the size of pipe and the number and size of pumping stations, the gas is compressed to between 100 and 350 pounds in a pumping plant, and the gasoline is extracted by absorption.

High pressures affect tower design in two ways—first, by reducing the gas volume; and, second, by requiring heavy construction. Increased pressure allows towers of smaller cross section and, according to Henry's law, is a factor in determining the quantity of oil necessary to extract the vapors.

DIMENSIONS AND CAPACITIES OF HIGH-PRESSURE TOWERS.

Towers made of single lengths of pipe vary in outside diameter from 18 to 36 inches and in height from 18 to 40 feet. The thickness of metal is one-fourth to five-eighths inches, depending upon the diameter of the pipe and the working pressure.

Plate III, *A*, *B*, and Plate IV, *A*, show three views of a battery of absorption towers. Each unit is designed to treat a maximum of 6,000,000 cubic feet of gas daily at 275 pounds pressure. Figure 13 shows details of this type of tower. The double gas intake and discharge lines shown in Plate III, *B*, allow any or all of the towers to be cut into either of the two pipe lines into which the plant is pumping gas at different pressures.

Plate IV, *B*, shows towers of the same general construction as the one shown in detail in figure 13; the diameter is 30 inches outside and the height 32 feet, the pressure to be withstood is about the same. These towers are handling 7,000,000 cubic feet of gas a day, at a pressure between 225 and 275 pounds.

In comparison to these larger towers, Plate II, *B* (p. 20), shows eight towers 20 inches by 25 feet, each having a capacity of 1,000,000 cubic feet at 100 pounds, or 2,000,000 cubic feet at 200 pounds pressure. Figure 14 shows in detail the construction of towers similar in design and capacity, also the baffles in the form of a cartridge, mentioned under wood

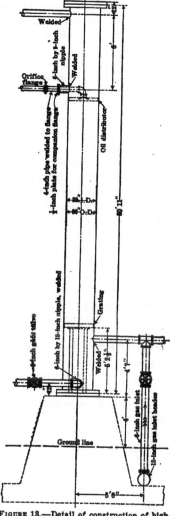

FIGURE 13.—Detail of construction of high-pressure tower.

baffle construction. Plate V, *A* and *B*, shows 20-inch by 30-foot towers that handle from 150,000 to 200,000 cubic feet of ¼ gallon gas each per day, at 25 to 35 pounds pressure.

FLOW VELOCITY OF GAS.

The diameters of towers vary, and also the pressures and volumes of gas at different plants; however, when the actual volumes of gas at the given pressures and the velocities of flow through the tower are computed, the velocity of gas through the towers is calculated to be from 30 to 75 feet a minute, disregarding the area reductions caused by baffling. The different systems of baffling, however, have " porosities "[a] varying between 3 and 75 per cent that cause the actual flow velocities to increase to as much as 200 more feet a minute in some plants; this is far too rapid for good extractions.

A company in the Mid-Continent field, using 30-inch towers 32 feet high and 200 pounds gas pressure puts seven million feet a day through each tower, using baffling with 50 per cent " porosity " that causes a flow velocity of 150 feet a minute in the baffled part, or 75 feet a minute in the open, or unbaffled part, of the tower. They find that higher velocities carry oil out of the towers.

The gas before leaving the tower through the outlet passes through a length either of open towers or of a series of baffles constructed to separate entrained oil from the gas. It is usual practice to place an oil trap or separator of some kind between the tower discharge and the gas main intake to induce a complete separation of oil and gas in an emergency, such as flooding, or excessive velocity of gas in the absorption tower, which will often carry oil over with it. Under normal conditions of running the open length of tower completely separates the oil and gas when the velocity of gas flow is not greater than 30 to 75 feet a minute in unbaffled parts of the tower. The extremes of flow velocity found in the plants visited were 30 feet a minute minimum, and 260 feet a minute maximum. At the latter plant oil was sometimes carried through the separating traps placed in the upper part of the tower and into the oil traps on the main gas-lines, and the extraction of gasoline from the gas was not satisfactory.

At the flow velocities given above any particle or molecule of gas passing through a tower will be in contact with the oil not over 50 seconds. In order to allow a great number of contacts, or contact of longer duration between gas or vapor particles and the oil, many engineers are studying results with towers of greater height.

One company is increasing its towers from 20 to 30 feet, and another, that has used 30-inch by 30-foot towers, plans to install 30 or 36 inches by 40-foot towers, claiming that by reducing the flow velocity and increasing the time of contact a greater extraction can

---

[a] The expression porosity as here used refers to the ratio of the area of open space to that of the solid parts of the baffle.

B. SAME BATTERY OF TOWERS, VIEWED FROM END.
EACH TOWER IS 30 INCHES BY 30 FEET 11
INCHES; CAPACITY 6,000,000 CUBIC FEET OF GAS
DAILY, AT 275-POUND PRESSURE.

A. BATTERY OF ABSORPTION TOWERS. NOTE PILE OF
BUILDING TILE IN FOREGROUND AT EXTREME
RIGHT, USED AS TOWER FILLING OR BAFFLING;
ALSO NOTE OIL-DISCHARGE TRAPS AT FOOT OF
EACH TOWER.

B. TOWERS, 30 INCHES BY 32 FEET, IN COURSE OF CONSTRUCTION. DESIGNED FOR CAPACITY OF 7,000,000 CUBIC FEET EACH, AT 250 POUNDS.

A. GAS-INTAKE SIDE OF TOWERS SHOWN IN PLATE III.

be obtained and less oil circulated per 1,000 cubic feet of gas. As a result the boiler horsepower required will be less for both pumps and stills, and the recovery of gasoline will be greater. The total increase in the height of a tower can be used for baffling space, as the open portions of towers for intake and discharge of gas need not be increased. The addition of 10 feet to the height of a tower 30 feet high will increase by 50 per cent that portion in which the extraction is accomplished.

As the pressure at which gas is to be treated is usually fixed, the number of towers necessary to treat the gas will be controlled by the actual volume of gas at the given pressure, and the diameter of pipe chosen for tower construction.

In designing plants it is well to choose towers of such diameter as to require a number rather than one tower of cross-section area large enough to treat all of the gas. As in the practice of extracting gasoline from gas by compression, the number of units should be such that if one unit is for any reason cut out, the number left in the gas circuit will for a short time, at least, be able to treat all of the gas without overtaxing their capacity or lowering the plant efficiency unduly. A designer in Oklahoma, having to treat daily an average of 6,000,000 feet of gas at 250 pounds pressure, put in three 20-inch towers rather than one 36-inch tower, although the latter would have had capacity enough for all of the gas at approximately the same flow velocity. Two 20-inch towers will handle all the gas by working at full capacity at maximum pressure. The third tower is to be used in parallel with the other two while the plant runs normally, but it is expected to pay for itself during breakdowns by treating gas that would be by-passed around the towers and the gasoline lost if only two towers were installed.

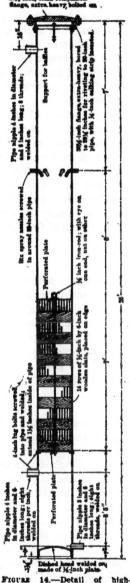

FIGURE 14.—Detail of high-pressure tower, 20 inches by 25 feet in size.

The largest welded towers of line pipe that were treating high-pressure gas were 36 inches in diameter, and the largest built-up tower was 12 feet in diameter. The latter was used in treating rich gas at less than 10 pounds pressure.

The height to which a tower must be built to yield the maximum economical extraction with the least oil, has not yet been determined, but as the present tendency is to increase this dimension in towers for high and for low pressure, the limit may soon be found. In the absorption process developed for scrubbing coke-oven gas the height limit is 75 feet.

Many methods of treatment have been devised, such as scrubbing the gas in sprays of oil before sending it to the usual towers, passing the gas through two or more towers, and bringing the oil and gas in contact in seven different units at different pressures or temperatures. As the form of single towers shown is simple, inexpensive, and perfect in counter-current flow treatment, it seems improbable that, except for changes in height, any other system of treatment equally efficient and simple in construction and operation will be invented for lean high-pressure gas.

MECHANICAL AGITATION.

Machines to agitate oil in the presence of gas are being tested at present, and experimental plants of small capacity using such machines have made good extractions and consumed little power. It is too soon to predict to what ultimate use mechanical agitation may be put, but indications are that it will prove useful. The process will be thoroughly tested, particularly in connection with small quantities of rich gases and still vapors and with flow-tank and weathering-tank vapors.

LOW-PRESSURE TOWERS.

The same principles hold true for low pressures (up to 30 pounds) as for high pressures; the important difference is the effect on tower construction. Because of the low pressures towers may be constructed of light boiler plates riveted together and, owing to the increased volume of gas, a greater cross-section area of tower is needed in order to maintain the minimum flow velocity and maximum time of treatment.

Plate VI (A and B) shows low-pressure towers built of light boiler plates riveted together, 12 feet in diameter by 48 feet high. The cross-section area is approximately 110 square feet, and the capacity in the unbaffled portions of the tower is 2,500,000 cubic feet of 2.5-gallon gas daily, with a flow velocity of about 30 feet a minute, at atmospheric to 5 pounds pressure. The towers are filled or baffled

B. SIDE VIEW OF SAME TOWERS SHOWING OIL INLET.

. TOWERS, 20 INCHES BY 30 FEET. CAPACITY 150,000 TO 200,000 CUBIC FEET EACH, AT 25 TO 35 POUNDS.

A. TOWERS, 12 BY 48 FEET; WORK AT ZERO TO 10-POUND
PRESSURE, AVERAGE CAPACITY 2,500,000 CUBIC FEET
EACH. NOTE TRAP ON GAS-DISCHARGE LINE.

B. ANOTHER VIEW OF SAME INSTALLATION.

with 1 by 4 inch boards set 1 inch apart in each grating, and one grating set on top of the other with its strips placed at a small angle to the one below. The gratings are set in 12-foot tiers, with open spaces of 2 feet above each tier. Plate VI, *B*, shows the manholes in the side of the tower entering the 2-foot open section above each of these tiers of baffles. The piping is so arranged that the two towers can be used either in series or parallel.

Multiple units of smaller diameters are also used for absorptions at the lower pressure. The towers in Plate VII, *A*, are 18 inches in diameter by 33 feet high, filled with cobbles about $2\frac{1}{2}$ inches in diameter. The pressure used is approximately two atmospheres, and the total quantity of gas is two to three million feet a day. When 3,000,000 feet are passing through the three towers at two atmospheres pressure, the flow velocity in the unbaffled portion of the tower is 190 feet a minute and in the baffled part (if the "porosity" is 50 per cent) the velocity will be twice that, or much higher than should be used in treating $\frac{1}{2}$-gallon gas. Higher pressures, or more or larger towers, would overcome this difficulty. These towers are about the proper size to treat 250,000 feet of gas a day each at 25 pounds pressure.

After the gas has passed through the towers and traps and is completely separated from the oil, it is delivered into mains and distributed in the same manner as it was before the absorption or drying units were installed.

LOSS IN VOLUME AND HEATING VALUE OF GAS DUE TO ABSORPTION TREATMENT.

The loss in gas volume owing to the extraction of condensable fractions is represented, first, by the liquid products recovered and counted as net gasoline production, and, second, by the part of the gas that is carried from the towers as dissolved gas but can not be condensed after passing through the distilling process, or can not be held as liquid in the condensed still-vapors during subsequent handling and shipping. Of the first portion each gallon of product represents between 30 and 35 cubic feet of gas, or each gallon of condensate from 1,000 cubic feet reduces the original volume of the gas 3.5 per cent.

The second portion is more difficult to estimate, its volume depending on temperatures, pressures, and methods used at different plants as well as on the vapor tensions of the condensable fractions contained in the gas treated. In most plants this second portion is not to be computed as loss, as the vapors from still coils, accumulators, and weathering tanks are collected and used as fuel for power for the plant or are pumped back to the gas lines. The practice of blowing into the air the gases from stills, weathering tanks, and coils or any other point is to be condemned as dangerous and wasteful.

As to the loss of heat units, through the extraction of the condensable vapors, Biddison [*] has published the statement following:

The reduction in heating value due to the recovery of the gasoline content can be closely estimated. Suppose a given gas has a gasoline content of 100 gallons per million feet of gas, or 0.0001 gallon per cubic foot of gas. The gasoline made by the absorption process will cause a shrinkage in volume of the gas treated, each gallon of gasoline being equal to about 35 feet of vapors. Then 0.0001 gallon of gasoline extracted from one cubic foot of gas would cause a shrinkage of 0.0001 × 35, or 0.0035 cubic foot, leaving, out of each cubic foot treated, 0.9965 cubic feet available for marketing. Now, gasoline of the nature made by this process has a heating value of about 20,400 B. t. u. per pound and is about 80 Baumé gravity, or 5.549 pounds per gallon. The heating value per gallon is then 5.549 × 20,400, or 113,200 B. t. u. per gallon. The 0.0001 gallon extracted from 1 cubic foot of gas could contain 113,200 × 0.0001, or 11.3 B. t. u.

If the gas contained before treatment 1,000 B. t. u. per cubic foot, the residue of 0.9965 cubic foot would contain 1,000 minus 11.3, or 988.7 B. t. u. If 0.9965 cubic foot contained 988.7 B. t. u. the heating value per cubic foot is 988.7 divided by 0.9965, or 992.17 B. t. u. per cubic foot. Thus the extraction of 100 gallons per million of gasoline from a gas of 1,000 B. t. u. heating value results in a reduction to 992 B. t. u., or 0.8 of 1 per cent. The effect on a gas of any other heating value of gasoline yields can be similarly calculated.

I submit herewith the result of some calorimeter tests of gases before and after removal of their gasoline content.

Inlet gas $\left\{\begin{array}{l}1066\\1070\\1060\end{array}\right.$	
Average	1065.3
Outlet gas $\left\{\begin{array}{l}1058\\1058\\1058\end{array}\right.$	
Average	1058
Loss due to gasoline extraction	7.3
Inlet gas $\left\{\begin{array}{l}1068\\1070\\1070\end{array}\right.$	
Average	1069.3
Outlet gas $\left\{\begin{array}{l}1042\\1046\\1042\end{array}\right.$	
Average	1043.3
Loss due to gasoline extraction	26
Inlet gas $\left\{\begin{array}{l}1094\\1092\\1091\end{array}\right.$	
Average	1092.3

* Biddison, P. M., The reduction in heating value due to recovery of gasoline content very slight: Natural Gas and Gasoline Journal, August, 1917, p. 213.

B. OIL-COOLING COILS DURING CONSTRUCTION. NOTE WIDE RETURN BENDS.

. EIGHTEEN-INCH BY 33-FOOT TOWERS; WORKING AT 30-POUND PRESSURE; SCAFFOLDING NOT YET REMOVED.

Outlet gas	1067
	1062
	1060
Average	1063

Loss due to gasoline extraction	29.3

Inlet gas	1094
	1097
	1100
Average	1097

Outlet gas	1063
	1064
	1059
Average	1062

Loss due to gasoline extraction	35

Inlet gas	1086
	1087
	1084
Average	1085.66

Outlet gas	1071
	1066
	1071
Average	1069.33

Loss due to gasoline extraction	16.33

The above tests are all from the same plant at different periods and the gasoline yield very closely the same at the time of each test. Calorimeter tests are not accurate to a degree closer than 4 B. t. u. This error, which may be a plus or minus error, makes the determination of the loss in heating value determinable only approximately by the calorimeter, since the error in measurement is so large a fraction of the quantity to be measured. However, the average of a large number of tests would show closely the loss due to extraction. The average of all the above tests shows a loss of 22.98 B. t. u. per cubic foot in heating value. For other data on this subject reference may be had to paper by G. A. Burrell, G. G. Oberfell, and P. M. Biddison before the Natural Gas Association of America, Pittsburgh, 1916; and Bureau of Mines Bulletin 88 (p. 60) and to articles appearing in the trade papers and journals.

OIL CIRCUIT AND TREATMENT.

The oil used as an absorption medium passes through the entire course of plant operation in a closed circuit of pumps, coils, interchangers, weathering tanks, and towers. Theoretically, no loss of this oil is necessary, but through leakage, breaks, and other irregularities in operations some oil is invariably lost, although in regular operation not enough to have much effect on costs.

ABSORPTION OILS.

The oils used as the absorbing medium in many plants are known as "mineral seal oil," "straw oil," and "gas oil," these being the names given by refineries to products distilled after the distillation of all the water-white products, including heavy kerosene or stove distillate, as it is called in California practice. The light lubricating or spindle-oil stocks are at times used and have the same characteristics as those of the oils mentioned above, so far as regards their use in absorption plants.

The names of such stocks are a poor guide to use in selecting an oil for an absorption plant, since oil of the same name varies at different refineries, depending on the number of redistillations or treatments through which it has been put, and also the character of the crude treated. The properties to be considered in selecting oil are vapor tension, initial boiling point, and the viscosity at different temperatures.

Vapor tension is important because, as the vapor tension of the absorbing medium and the vapor tension of the liquid to be absorbed approach the same point, the capacity of the oil to absorb the liquid decreases and this limits the total percentage of vapor that will be absorbed and held by the oil. Temperature affects vapor tension, but will be considered later.

The initial boiling point of the oil used should be considerably higher than that of the heaviest of the gasoline fractions extracted from the gas, so as to leave a range of temperature between the final boiling point of the lowest of the absorbed fractions and the initial boiling point of the absorbing oil wide enough to insure a minimum distillation of the heavy yellow oil along with the gasoline vapor. This oil, when condensed, discolors the gasoline and makes it unfit for market without some treatment to remove these impurities.

The viscosity of absorption oil, while of less importance than other factors, may affect results in two ways—first, by making pumping and equal distribution to all towers or units more difficult, especially in cold weather; and, second, by the oil not spreading in thin films and running freely over the baffles in the tower.

In regard to emulsification of absorption oil, R. O. Neal,[a] chemist of the Bureau of Mines, says:

Information regarding the emulsification characteristics of the absorption oil is helpful, although not necessary in choosing an absorption medium. Some oils have a tendency to form emulsions with water when they are agitated in passing through pumps into the flow tanks, stills, and absorbers (more especially horizontal absorbers). In one plant where the difficulty had resulted in low plant extraction, it was discovered that the trouble was due to the large per cent of emulsion (35 per cent), that was contained in the oil and would not separate out in the still.

---

[a] Neal, R. O., Personal communication.

· The Baumé gravity of the oil, although an indication of the characteristics, is too indefinite for a standard by which to be guided. The oils found in use range from California crude having a gravity of 19° B. to light-yellow mineral seal oil having a gravity of 38° B. A plant in California found that a refined oil with a gravity of 36° B. had too low an initial boiling point and too high a vapor tension for use even in a steam still, and is now using successfully a narrow cut of oil (27° B.) from Fullerton crude. A narrow cut of oil, that is, one with initial and final boiling points not too widely separated, is recommended with an initial boiling point not lower than 450° F. The Baumé gravity of the refined absorption oils in use ranges from 37° to 27° B, the tendency being toward the lower gravities and toward narrow cuts. After having passed through the plant a few times the mineral seal oil gains about one degree Baumé gravity, through a small fraction of gasoline that is not extracted by a single steam distillation. At plants where gas oil is used a decrease in Baumé gravity often occurs after the first passage of the oil through the still, because gas oil is not entirely freed from the kerosene fractions in refining. This small fraction of kerosene will give the gasoline made a high end point, especially the gasoline made just after the addition of fresh oil to the circuit. An operator of a low-pressure plant in the Mid-Continent field uses oil complying with the following specifications:

	° F.
Initial boiling point	418
90 per cent over at	679
96 per cent over at	720

Viscosity factor, 45 at 100° F. Saybolt.

Such specifications are unusual because, first, the percentage of oil distilled at temperatures above that at which this oil will begin to crack is large, and, second, the initial boiling point is lower than that most engineers desire. The following specifications define an oil which is better suited to this work and which is much used for this purpose:

Specifications for an absorption oil.

Flash point	247° F.
Fire test	262° F.
Gravity	37° B.
Over point	450° F.
End point	695° F.

Viscosity, 1.6 E. D. (at 20° C.)

Per cent.	Temperature, ° F.
10	528
20	540
30	556
40	565
50	572
60	572
70	595
80	608
90	633
100	695

OIL PUMPS.

Two sets of pumps are usually installed, one to pump oil against the pressure in the towers, and the other, called transfer pumps, for handling the oil at low pressure to and from the stills, interchangers, coils, and storage.

In many plants duplex steam pumps have been used, the exhaust steam going to the steam still, though this practice is sometimes condemned as wasteful of gas used for fuel under boilers. In later plants are gas engines directly driving either plunger or duplex geared pumps, or driving generators that supply electricity to run the pumps and compressors. In these plants the boilers generate steam merely for still use. The fuel economy is obvious, for gas engines use about 12 cubic feet of gas per horsepower-hour, whereas boilers use 60 to 120 cubic feet.

In plants where the oil circulated in the towers must be pumped against pressures of 100 pounds or more, the oil from the heat exchangers is cooled, in water-cooled coils, after passing through the high-pressure pumps, because pumping the cold oil at high pressures raises its temperature 6° to 7° F. The plant that noted this rise in temperature used steam duplex pumps, so part of the heat entering the oil was probably transmitted by conduction from the steam end.

COIL AREAS USED IN COOLING ABSORPTION OIL.

Oil from the pumps under the full tower pressure passes into a header at the low end of continuous return-bend coils, through which it passes upward to a header connecting with the top pipes of the different water-cooled coils. The number of coils and number of pipes in each coil depend on the total area necessary to accomplish the cooling and on the size of pipe used. Pipes of 2, 3, and 4 inch diameters are used. The smallest pipes, 2-inch pipes, expose the greatest radiating area in proportion to their oil capacity and are generally used in coils cooled by water and air in towers and similar cooling apparatus.

Figure 15 shows a coil of this character having 13 pipes 4 inches in diameter. The coils exposed an area of 7.64 square feet of radiating surface per gallon per minute when the plant was working at 60 per cent capacity. When the plant ran at full capacity the coils exposed an area of approximately 4 square feet per gallon per minute.

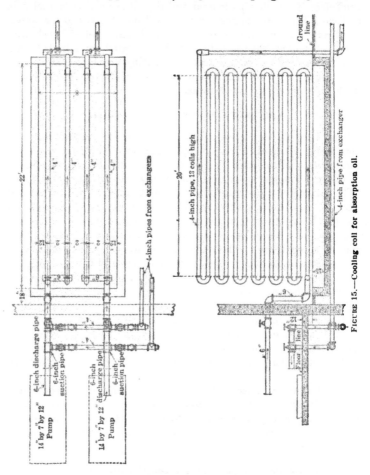

FIGURE 15.—Cooling coil for absorption oil.

At the time of inspection the temperature reduction between inlet and outlet was 70° F. (from 150° to 80° F.). The water used came directly from wells on the property, and was circulated without being cooled in towers. In dry hot climates the tower cooling of the water is necessary.

In most oil fields coils of 2-inch pipe 20 or 40 feet long with return bends placed in cooling towers are preferred. The practice is similar to that for cooling gas from compression plants, as described in Bureau of Mines publications.[a]

Between 3 and 4 square feet of radiating area per gallon per minute is sufficient if 2-inch pipe is used and the oil is previously cooled in an efficient heat interchanger on its way from the still to the water-cooled coil. An operator in an eastern field uses 6 feet of 2-inch pipe, or about 3 square feet per gallon of oil circulated per minute.[b] Another factor used is 2 square feet of radiating area per 1,000 gallons of oil a day; that is, 2.88 square feet, or about 5.50 linear feet of 2-inch pipe per gallon per minute.

CONSTRUCTION OF COOLING COILS.

Plate VII, *B*, shows oil-cooling coils during construction, and Plate VIII, *A*, shows the coils completed, with a louver tower used as a protection and to direct air over the coils, and not as a water-cooling tower. The louver extends only as high as the top of the pipes.

The spacing of pipes and coils that depend on the action of air and water for their cooling effect on oil or vapor has caused trouble, yet is seldom brought to the attention of the plant operators. Although conditions may force closer spacing, it is well when possible to place pipes in a continuous coil so that an air space not less than twice the diameter of the pipe will be left open vertically between pipes of the same coil, and a space 8 to 10 times the diameter of the pipe will be left open horizontally between sets of coils. This rule is entirely empirical, but is based on installations that have proved satisfactory. Another rule is to flow hot oil upward whenever possible, especially during heating, to avoid the formation of gas pockets. These rise against the flow of oil if the oil is flowing downward, lodge at high points, cause "bumping," and reduce the flow by constricting the area.

From the water-cooled coils the oil goes to the pipe header from which it is distributed to the absorption towers. The temperatures of the oil in the pipe header is between that of the atmosphere and 40° F. below atmospheric, and the pressure is 5 to 100 pounds above that of the gas to be treated.

FLOW OF OIL THROUGH ABSORPTION TOWERS.

Where two or more absorption units are connected in parallel, the oil from the high-pressure pumps and the cooling coils is divided as

---

[a] Dykema, W. P., The recovery of gasoline from natural gas by compression and refrigeration: Bull. 151, Bureau of Mines, 1918, 123 pp.

[b] Radiating areas throughout this bulletin are computed on the nominal diameters of the various sizes of pipe used.

B. SINGLE SECTION OF WOOD BAFFLE OR FILLER.

A. COILS SHOWN IN PLATE VII, B, COMPLETE WITH LOUVER TOWER.

it flows from a header into smaller lines that carry it into the distributing device inside of towers above the baffles. In order to divide the oil equally among the different units of baffles, an orifice may be inserted in the line at some point between the oil header and the tower connection. In figure 13 (p. 25) this orifice is shown, as are the methods of connecting the oil line with the tower and distributor.

Figure 16 shows the detail of this distributor. The pressure used is 5 to 10 pounds above that of the gas.

Other devices for dividing the oil equally over the baffles area are perforated pipes either circular or with 4, 6, or 8 radiating arms, sprays, slotted pipes, and troughs. There seems little choice among these types, and no trouble with any of them was reported to the writer.

METHODS OF EXPOSING OIL TO GAS.

Obviously, the best practice is to expose the largest possible surface of oil to the gas with least constriction of the cross-sectional area of the pipe or tower. As it is only those oil surfaces exposed to the gas that can absorb vapor, a large surface of oil should be exposed so that each molecule of condensable vapor in the gas may at some period of transmission through the tower be in actual contact with the oil and be absorbed by it.

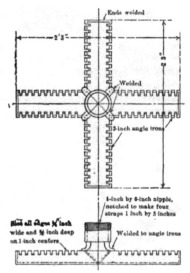

FIGURE 16.—Oil distributor.

OIL SPRAYS.

Oil sprays are being used at tanks or towers through which gas is passing, either as a preliminary treatment before the gas passes to the towers or final extractors, or as the final treatment at the top of towers. In figure 14 (p. 27) the oil is sprayed into the tower at a pressure of 100 pounds above that of the gas. Around the tower wall six spray nozzles are set 60° apart, and inclined downward 30° from the horizontal, so that a fine mist of oil opposes the rising current of gas and the oil surface exposed is the maximum. The fresh

oil comes in contact with gas that has passed through the wood
baffles shown below, and has been relieved of part of its content of
condensable vapor. Oil from spray nozzles settles on the baffles and
continues downward, still in contact with the gas and being enriched
by the gasoline vapors, until it reaches the bottom of the tower.

Some operators object to sprays, contending that with many oils
the water in the oil coming from the still will form an emulsion when
forced through a small orifice or a spray nozzle.

BUBBLING.

Gas is at times bubbled through oil either in the bottom of towers
after the oil has passed over the baffles, or in tanks. The testing
apparatus shown in figure 11 (p. 21) shows the principle of bubbling
gas through oil. In large tanks the contact obtained is not satisfactory because of the formation of large bubbles that, rising through
the oil, allow only a relatively small proportion of the surface of the
contained gas to come in contact with the oil, and when the gas
velocity is high such bubbles tend to carry the oil through the tower
and into the gas lines.

At one plant the towers are filled two-thirds of the way to the top
with brickbats, and the oil level is held half way up. The gas bubbles upward through the oil and bricks to the top of the oil and then
passes through brick baffles counter-current to the descending oil,
receiving a second scrubbing action, and finally ascends through an
unbaffled spray to the gas discharge. Counterflowing over baffles
has proved more efficient in practice, whereas bubbling through oil
has limited application, and only a few engineers are willing to install this type of apparatus.

BAFFLES.

Many materials have been used for spreading oil into thin surfaces in the presence of gas. Among them are stone, tile, and wood
and steel shavings.

In the experiments by Biddison and Burrell, on which the design
of the plant of the Ohio Fuel Supply Co. at Homer, Ohio, was based,
round cobblestones were used in the vertical absorbers to spread the
oil in thin sheets while in contact with the gas, and worked admirably, but constricted unduly the area of the tower as compared
with the actual surface exposed. When the plant was built, 4 by 5
by 6 inch hollow wall tile was put in the place of stones; these
tiles have given satisfaction as baffles, but have crushed or decomposed somewhat and particles of the burnt clay follow the absorption oil to low points in its circuit, from which they have to be

drained. Vitrified tile or pipe would be stronger and would not disintegrate. Fragments of vitrified sewer pipe have been used satisfactorily to expose large surfaces of liquids to vapors in towers.

One plant uses successfully two sizes of carefully screened crushed rock, packed in towers of 36-inch diameter. However, the great weight of such filling must be considered, as well as the cost and time of cleaning if the towers should be clogged with dust, dirt, or other impurities, and the constriction of the cross-section area, which increases the flow velocity. A cobble-filled tower showed by test that only 32 per cent voids were left for the passage of gas. Another plant having towers 20 inches in diameter is using cobbles 3 or 4 inches in diameter, in the manner described by Burrell[a] and used in his experimental work.

When stone, tile, or other heavy baffling is used, the weight is supported on a frame set on the tower bottom as shown in figure 13 (p. 25).

Wood baffling has been proved satisfactory in the treatment of coke-oven gas and is being widely adopted in the treatment of natural gas, because of its lightness and the ease of making different shapes and thicknesses. It can be worked into baffles that expose a large surface yet constrict the area but moderately. In one plant the writer saw lath used for baffling in such a way as to give 75 per cent voids. Metal strips would seem to offer even greater advantages than wood, but their use has not been observed.

Steel shavings or cuttings are being tried by several companies. The quantities used are 15 pounds to 30 pounds per cubic foot, giving a high porosity and a large area on which to expose the oil.

WOOD BAFFLES.

Wood baffles are made of different widths and thicknesses of wood strips, built up as gratings. The spaces between the strips are usually two or three times the thickness of the strips. Widths and thicknesses range from house lath to 1-inch by 6-inch boards. The gratings are set in the towers so that each supports the one above it, each set of strips being placed at an angle with those below. In some towers the angle is 90°; in others, it is less and gives a spiral effect. The relative efficiency of the various angles of setting has not been determined.

Figure 14 (p. 27) shows a unique method of inserting baffles. An iron bolt and perforated plates at the top and bottom hold the baffle cartridge together, allowing the cartridge to be handled as one piece. The baffles, as shown, are strips of wood, $\frac{1}{2}$ inch by 6 inches, spaced

---

[a] Burrell, G. A., Biddison, P. M., and Oberfell, G. G., Extraction of gasoline from natural gas by absorption methods: Bull. 120, Bureau of Mines, 1917, 71 pp.

¼ inch apart. A section of wood baffle, or filler, having wedge-shaped slats placed so that the thin edges alternate is shown in Plate VIII, B (p. 36).

In many towers of large diameter, wood gratings, or sets of baffles, are placed one on top of the other all through that part of the pipe used in baffling; in others, 4 or 5 feet of baffle gratings is supported on lugs, or angle-iron ribs, and an open space of about one foot is left between sets of baffles to allow the gas currents to mix before passing through the next baffle above. The angle-iron ribs act as oil baffles, distributing the oil deposited on the tower wall over the wood gratings below.

Wood parts always should be made of dressed lumber, as the use of rough pieces has caused trouble from splinters and fibers and consequent cleaning of valves, tanks, stills, and other places offering lodgment.

In the opinion of the writer neither the most useful material for baffles nor the most efficient percentage of voids has been definitely determined. Some operators have made tests that indicate the most efficient voidage for a given material; none, so far as the writer knows, has tested more than one or two materials in an endeavor to determine the factors for plant construction. One company has adopted as its standard form of tower filling a perforated round tile that gives a voidage of 50 per cent. This tile has proved satisfactory for use with the pressures, flow velocity, and gasoline content of the gas treated.

The problem seems to be one of combining the factors of open space and exposed surface in such a way as to obtain the maximum of each without undue friction or reduction of pressure.

SEPARATING OIL AND GAS IN TOWERS.

At both the top and the bottom of a tower open unbaffled lengths are left to permit the separation of gas and oil. At the bottom the gas inlet is placed between 3 and 5 feet above the tower base, and the oil outlet as near the bottom as possible. More or less oil from the baffles collects in the lower chamber before being discharged from the tower, so that the gas has a brief time to separate from the oil before going to the weathering tank.

The gas outlet is placed close to the top of the tower, and the oil inlet 5 to 7 feet below the top, so that there is left 5 to 7 feet of open tower. A much shorter space might be left here, as well as at the bottom, and the final separation of the oil from the gas might be completed in a trap in the vertical gas-discharge line, thus allowing a greater length of tower to be used for baffles. A trap of the type used to separate the oil and gas outside of the tower is shown in Plate

VI, *A* (p. 29). The trap discharges the oil collected from the gas into the oil-outlet line at the bottom of the tower. This practice is consistent with the tendency to increase the height of tower to gain greater length in which baffles may be placed, for it has the same effect, by permitting, as it does, a greater part of the tower to be used for bringing the oil and gas into intimate contact. At present only one-half to two-thirds of the tower is filled with baffles. The remaining space at top and bottom is used to separate the oil and gas. Usually the space at the top of the tower is left as an open separating chamber. Figure 17 shows an oil trap for placing inside the tower, as used by an eastern company. This trap is placed above the oil inlet and below the gas outlet, being suspended from the top of the tower by a bail bolted to an angle iron. The action of the trap is self-explanatory.

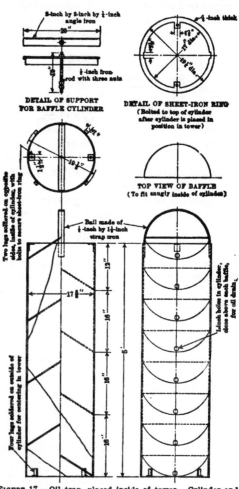

FIGURE 17.—Oil trap, placed inside of tower. Cylinder and baffles made of 16-gage galvanized sheet iron; baffles soldered to cylinder.

OIL TRAPS OUTSIDE OF TOWERS.

The traps most used embody the principle of enlarged diameter, either of pipe or a tank, so that the velocity of the passing stream of gas is reduced, and any entrained particles of oil have opportunity

to separate from the gas and collect. Many traps are so designed as
to insure a change in direction of flow, the oil settling to the lower
part of the trap. The oil flows from the bottom of the trap, controlled by back pressure, or is automatically discharged.

Water traps used in steam lines work satisfactorily, if large
enough, but are more expensive and are no more efficient than traps
built on the ground of va-
rious standard sizes of
pipes. Plate IX, *A*, shows
a trap built of standard
pipe and fittings through
which 40,000,000 cubic feet
of gas passes daily at 250
pounds pressure. There
are many designs of effi-
cient and inexpensive traps. Plate **X**, *A* and *B*, and figure 18 show
traps with prostrate tank.

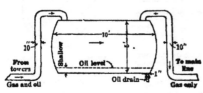

FIGURE 18.—Oil trap placed in discharge line.
Not drawn to scale.

Regardless of the separating devices in the towers, the gas always
passes through some form of trap before it goes to the gas mains.

QUANTITIES OF OIL.

The quantity of oil used is computed in the same terms as the condensable content of the gas, that is, in gallons per 1,000 cubic feet,
so that all primary operations can be referred to the same base, the
volume of gas treated.

SATURATION.

The table following shows the relation of pressures, saturation, oil
circulated, and gasoline content, as found at the plants listed:

TABLE 4.—*Saturation of oil and recovery of gasoline at plants visited.*

Plant No.	Pressures.	Oil circu- lated.	Satura- tion.	Gasoline recovered.
	Pounds per sq. in.	*Gallons per 1,000 cu. ft.*	*Per cent.* 2.55	*Gallons per 1,000 cu. ft.*
1...............	175	5.95	5.35	0.152
2...............	172	3.80	6.30	.202
3...............	185	4.96	6.60	.264
4...............	142	2.61	2.50	.173
5...............	30	20.00	2.10	.500
6...............	25	18.00	4.00	.375
7...............	250	5.00	6.50	.200
8...............	250	5.00	5.00	.325
9...............	5	70.00		3.500

This table shows far too few plants to serve as a basis of final conclusions or for the development of an empirical formula for calculating the quantity of oil necessary for the complete extraction of gaso-

A. PIPE OIL TRAP, USED ON GAS LINES FROM TOWERS TO COMPLETE THE OIL AND GAS SEPARATION.

B. TWO AUTOMATIC TRAPS.

C. WEATHERING TANK MOUNTED ON HIGH FRAME, TO INSURE GRAVITY FLOW TO STILL AGAINST PRESSURE OF 2 TO 5 POUNDS IN STILL.

A. HIGH-PRESSURE, PROSTRATE TANK USED AS
OIL TRAP.

B. LOW-PRESSURE, PROSTRATE TANK USED AS
OIL TRAP.

line from gas of varying gasoline content at any given pressure. However, the table shows that the limit of saturation may be more than 4 per cent, which was once considered the maximum for high-pressure treatment. Although oil begins to absorb gasoline imperfectly when the saturation is 4 per cent, the oil can still absorb large amounts of gasoline vapors from fresh untreated gas.

This point is demonstrated in multiple-coil absorbers using separate volumes of oil in each coil or unit. In experiments and tests with such absorbers as shown in figure 10 (p. 20) and Plate I, *B* (p. 16) the oil in the first chamber has been saturated to 10 per cent, the oil in the second chamber became saturated to 4 per cent, and that in the third chamber about 1 per cent of its volume. In the fourth and fifth chambers the volume increased and the Baumé gravity became higher, but the amount of gasoline was so small as to be difficult or impossible to distill out and measure with the usual laboratory equipment. The increase in volume and the increase in Baumé gravity are due to the absorption of gas, the absorption of vapors too light to be condensed as they issue from the still, or to both. Although the plants installed do not treat the gas in several separate units each with its circulation of fresh oil, the countercurrent flow of oil and gas in a tower is quite similar, the richest gas meeting first the oil that has been exposed to the gas in the upper part of the tower. This brings up again the point previously mentioned, that of height of towers, but in regard to saturation rather than to construction.

HEIGHT OF TOWER AS A FACTOR IN OIL CIRCULATION.

Operators now are planning to increase the height of towers with the object of reducing the amount of oil circulated and increasing the time of contact between oil and gas, thus obtaining equal recovery with less power for pumping and less heat in the stills. Decreasing the amount of oil circulated means that the saturation must be increased. and this will be accomplished by longer exposures of oil to gas in the counterflow system of treatment.

The amount of oil circulated, besides varying with the pressure, height of tower, and condensable content recovered from the gas, varies with the temperatures of oil and gas, and with the characteristics of the hydrocarbons recovered. At the plants the writer visited the greatest quantity of oil circulated was 70 gallons and the least was 2.61 gallons per 1,000 feet.

PUMP CAPACITY.

When an absorption plant is being designed from data obtained in tests, it is well to have the capacity of pump and still somewhat larger than the maximum indicated by the tests, to cover the possibility of the chosen height of tower not being efficient enough to give the total extraction desired. This deficiency in the towers may develop because of insufficient height, the type of baffling used, the high temperature of the oil, and gas or some other factor, and the quickest and cheapest method of overcoming this fault, at least temporarily, will be by increasing the flow of oil. The most flexible and easily adjusted part of plant operation is the oil circulation, and therefore it is the part where control should be provided. Hence the size and reliability of the pumps should be carefully considered before a choice is made.

EFFECT OF TEMPERATURE.

The temperature of the oil and gas during the period of absorption decidedly affects the quantity and quality of the vapors absorbed and consequently the efficiency of the process. Lowering the temperatures lowers the vapor tensions of the condensable fractions of the gas and thereby renders them more easily absorbed. Moreover, the absorbent, as is true of all liquids that absorb gases selectively, has a greater capacity for taking in and holding vapors and gases. Hence it is clear that lowering the temperatures of the oil and gas permits a reduction in the quantity of oil circulated or an increase in the percentage of saturation.

However, it is evident that greatly reducing the temperature and maintaining the same flow of oil would increase the production only a little, as the lightest fractions absorbed at extremely low temperatures, if they could be condensed after distillation, would be too volatile to ship. In distilling and cooling the light uncondensable vapors would have a tendency to carry valuable fractions through the coils. Just how much the flow of oil could be reduced with lowering of temperature has not been determined, but there are decided advantages in keeping the temperatures in the towers well below the overpoint of the gasoline produced, which is usually about 80° F. In order to keep the gas at the same temperature as the oil, many operators cool the gas just before it enters the absorbers by passing it through a coil under water sprays in a louver tower. This should always be done when the gas lines lie on or near the surface of the ground for any distance before they enter the absorption units.

In Oklahoma, Texas, and California a well-designed louver tower with enough coil area should cool the gas 20° to 40° F. below outdoor temperature during the hot dry season.

The temperature of the oil and gas should be nearly equal and as low as the cooling system used will permit. In an Eastern plant the gas entering the tower at 60° F. was treated with oil at 84° F. and temperature of the gas was raised 10° to 70° F., in passing through the tower. Such increase of temperature is undesirable. The temperature of the oil and gas can not be lowered so far by atmospheric and water cooling as to cause the absorption of undue quantities of uncondensable fractions. Such low temperatures can be developed only by some process of refrigeration. To recapitulate, the quantity of oil circulated through the absorption tower may be reduced without lowering the extraction if the oil and gas temperatures are lowered or a greater percentage of saturation is used.

The following table gives an idea of the temperature changes of both oil and gas in passing through the absorption towers:

TABLE 5.—*Oil and gas temperatures at inlet and outlet of absorption towers. Average for 24 hours.*

	Plant No. —				
	1	2	3	4	5
	°F.	°F.	°F.	°F.	°F.
Temperature of gas entering tower	70	68	39	117	116
Temperature of gas leaving tower	73	72	50		
Temperature of oil entering tower	95	97	81	101	96
Temperature of oil leaving tower	66	64	42	106	101
Temperature of atmosphere	70	72	30		

Figures for plants 1, 2, and 3 were taken in winter, and those for plants 4 and 5 were taken in summer. The temperature of the oil leaving the towers in plants 4 and 5 show that cooling systems used were entirely inadequate and could easily have been improved. As the height of the towers and the gas pressure are fixed, the temperature of oil and gas will be one of the factors determining the quantity of oil that must be circulated, this becoming less as the temperature is lowered. This factor of temperature seldom becomes evident, as the quantity of oil to be circulated is generally determined at commercial plants under the usual working conditions, or at experimental plants under practically uniform conditions.

FLOW OF OIL FROM TOWERS.

Two distinct methods of discharging oil from towers have been developed. In one the oil flows through a self-dumping trap that automatically reduces the pressure. In the other the oil flows from the towers into a collecting header from which it goes to a receiving tank under the full tower pressures.

AUTOMATIC TRAPS.

In the first method the automatic self-dumping traps shown in Plate IX, *B* (p. 42) immediately reduce the pressure to that in the weathering tank, which ranges from 2 to 40 pounds per square inch. Plate IX, *C* shows a weathering tank on a high frame; the object of placing the tank on high ground and a frame is to obtain enough pressure to force the oil through the interchangers and into the still without the use of pumps.

The number of traps required depends on the quantity of oil passing through the towers, and on the capacity of the traps used. Many plants include a trap for each tower and some, as shown in Plate IX, *B*, take the oil from all the towers and discharge it through a manifold header to traps of the required capacity. Oil-level regulators are also being used to obtain automatic discharge from towers.

FLOW TANKS.

In the second method, using a collecting header, the oil flows directly from the tower under the full gas pressure and is discharged to the weathering tank through an automatic level-regulator and valve of the type used on stills and other apparatus. This method requires but one automatic device for the discharge of oil from any number of towers, and has the advantage of requiring less attention, lower upkeep charges, and smaller first cost. The tank or receiver held under tower pressure may be made of standard pipe and fittings. One such unit installed in an eastern plant was built of a 20-foot length of 30-inch pipe set horizontally.

OIL WEATHERING TANKS.

Weathering the absorption oil in tanks at a reduced pressure before heating and distilling has become the general practice. The object of weathering is to relieve the oil of gas taken up during the scrubbing process, as this gas can not increase the gasoline yield and is troublesome and detrimental in the distilling and cooling. Regardless of the higher hydrocarbon selectively absorbed from the gas treated, in making a complete extraction of those higher fractions that after distilling can be marketed as gasoline, different proportions of all of the other hydrocarbon constituents of the gas, will also be in the oil, even after the oil has been weathered at pressures lower than those in the towers. As soon as the pressure on the oil is reduced, these lighter fractions begin to distill from the gas. Some operators believe these fractions carry with them valuable quantities of the condensable fractions. The quantities will vary in different plants, and whether they are always large enough to be worth extra units for their recovery has not been determined, but the belief is becoming general that these gases are too lean for profitable treatment by compression.

Some operators treat the gases discharged from the weathering tank either in the compressor or in an absorption tower that treats also any still vapors not condensed in the water-cooled coils. Other engineers doubt the usefulness of treating these gases, claiming that the recovery is too small to justify the added cost of equipment and operation. At some plants, the treated gas is passed back to the outgoing gas mains, as its heating value is higher than the other treated dry gas. In this way the absorption plant is relieved of part of the charges for smaller volume and lowered heating value of gas, consequent on the removal of the gasoline. In some plants this gas, with the uncondensed still vapors, is used in gas engines or under the boilers instead of gas from the outgoing mains, thus saving fuel and reducing costs. To allow any vapors of gas to flow into the atmosphere is dangerous and needlessly wasteful. If the gases that leave the oil in the weathering tank were held in the oil while it is being heated in interchangers and stills, gas pockets might form in the interchanger pipes and there might be an excessive liberation of gas in the still; this gas when coiled later in the vapor coils would hinder the precipitation of the condensable fractions.

As stated, the pressure held on the weathering tanks ranges from 2 to 40 pounds, according to the pressures used in the tower and the method of flowing the oil through the heat interchangers. The weathering tank is usually a horizontal cylinder constructed of boiler plates. A valve set to open at the desired pressure controls the gas outlet which discharges into a gas line to the compressor or the absorption unit, or to the boilers.

If the oil level in the weathering tank is not held automatically at a given level, the tank may be, and often is, used as a receiver or temporary reservoir for oil when the quantities of gas being treated or of oil circulated vary. This tank is the only unit in the system where such storage or variation of content is allowable or possible, for the oil must move continually through the other units and usually to their full capacity under the working conditions. If the weathering tank can not be used in this way there must be an extra tank to care for irregularities in the quantity of oil circulated.

From the weathering tanks the oil is forced through the heat interchangers to the still, either by transfer pumps working at a pressure of 15 to 25 pounds to overcome friction in the interchangers and the head due to the height of the still inlet, or by the pressure in the weathering tank. The rate of flow is regulated by the speed of the pump, or if the oil flows under tank pressure, by the discharge-valve regulator or a hand valve in the line to the still inlet.

HEAT INTERCHANGERS.

The primary function of heat interchangers is to save heat and fuel. In the absorption process the interchangers also cool the residual oil flowing from the still. Otherwise this cooling would have to be done entirely in water and air cooled, or double-pipe water-cooled coils. At a refinery where the residual oil is to be pumped through a long pipe line, cooling below a certain temperature is neither necessary nor desirable; but in absorption practice lowering the temperature of the oil from the still lessens the cooling to be done in the water tower, and raising the temperature of the saturated oil saves much boiler horsepower.

No radically new designs of heat interchangers have been developed in absorption practice, the types being of the same general plans as those used in refineries.

In absorption plants the conditions for developing efficiency in the heat interchangers are ideal. The amount of oil going to the still is only 6 or 7 per cent more than that from the still, the viscosity of the two oils is little changed, and the oil contains no fraction, such as wax, that will be precipitated or solidified by any decrease of temperature. In refinery practice an interchange of 22 B. t. u. per square foot per degree of mean temperature difference per hour is considered exceptionally good practice and is far above the average efficiency obtained in topping plants.

One absorption plant claimed an interchange much higher than mentioned above, but complete data could not be obtained by the writer.

Lucke,[a] gives 50 to 75 B. t. u. per hour per square foot per degree of mean temperature difference as the average practice for the transfer of heat between brine and ammonia in ammonia absorber cooling coils. The difference, in heat transfer, between refinery or absorption practice and ammonia absorbers is due to the specific heats of the liquids concerned being different, and to the fact that in ammonia absorbers latent heat has to be considered. Also, the difference is partly due to the oil circulated in refineries being more viscous, and therefore being not only more difficult to pump at the high velocities necessary for maximum heat transfer, but requiring pipes of large diameter in which the rate of transfer is not so rapid as in smaller pipes.

The rate of heat transmission depends on the velocity of the fluids along the separating walls of the pipes as well as on the difference in temperature of the two liquids. Lucke[b] says of the velocity function:

---

[a] Lucke, C. E., Engineering thermodynamics, 1914, p. 550.
[b] Lucke, C. E., Work cited, p. 539.

"The quantity of heat transmitted is found to be proportional to the velocity of the fluid, to some power, when one fluid is in motion, and to some power of each, not necessarily the same, when both move."

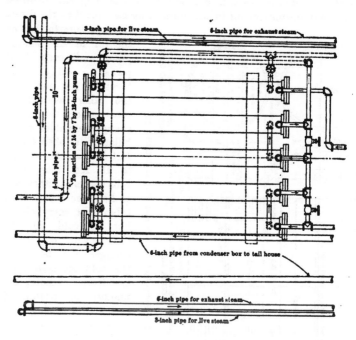

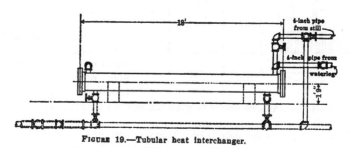

Figure 19.—Tubular heat interchanger.

TYPES OF INTERCHANGERS.

Two distinct types of interchangers are being used in absorption plants, the tubular and double pipe, or the jacketed line. Each type has a number of practical applications.

TUBULAR INTERCHANGERS.

Two methods are used in constructing heat interchangers of the tubular type. In one a number of tubes 1 to 2¼ inches in diameter are placed between heads in a pipe 9 to 20 feet long and 18 to 36 inches in diameter, the oil in the tubes being circulated in the direc-

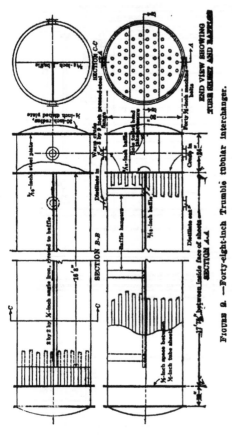

FIGURE 2.—Forty-eight-inch Trumble tubular interchanger.

tion opposite to that of the oil in the outside pipe, and each fluid passing only once through the unit. In the other method of construction the interchange resembles a tubular boiler shell 4 to 6 feet in diameter and 10 to 18 feet long with tube sheets at either end to hold the 2 or 3 inch tubes that carry either hot or cold oil. In this type of interchanger the hot and the cold oil each pass two times or more, the number of passes of each medium being governed by the number and position of the baffle sheets.

After the size, length, and number of tubes for each interchanger have been determined, the total heat-transfer area necessary is obtained by installing enough of the units in series to give the desired or maximum heating effect.

Figure 19 shows the details of a tubular interchanger made of an 18-inch pipe, 10 feet long, with thirty-one 2-inch pipes 16 feet between tube sheets. As each medium passes only once through each unit, there are no baffles to direct return flows. Plate XI, A shows such interchanger before being insulated, placed horizontally. Plate XI, B, shows the same size and type of interchanger placed vertically, seven in series, with the insulation in place.

A. HEAT INTERCHANGERS, FIVE IN SERIES, SET HORIZONTALLY; SHOWS PIPING BEFORE INSULATION HAD BEEN APPLIED.

B. HEAT INTERCHANGERS SET VERTICALLY, SEVEN IN SERIES, WITH INSULATION IN PLACE. OUTSIDE PIPE IS 18 INCHES BY 18 FEET; CONTAINS 31 TWO-INCH TUBES, 16 FEET LONG.

C. END VIEW OF INTERCHANGER IN-
STALLATION SHOWN IN PLATE XII, B.

B. VERTICAL TUBULAR INTERCHANGERS, 36 INCHES
BY 19 FEET HIGH, CONTAINING 127 TWO-INCH
TUBES, 16 FEET LONG.

A. TWO VERTICAL INTERCHANGERS
IN SERIES, 8 INCHES BY 20 FEET
IN SIZE; 1-INCH TUBES, 18 FEET
LONG. NOTE SECTION WITHOUT
INSULATION.

Plate XII, *A*, shows two interchangers in series; they are made of 8-inch pipe 20 feet long with 1-inch tubes set vertically. Plates XII, *B*, and XII, *C*, which show two vertical tubular interchangers 36 inches in diameter, placed in series and uninsulated, gives a good idea of the piping and connections.

Tubular interchangers of the second type, such as the one shown in figure 20 are used somewhat, the units being so constructed and baffled that the oils make two or more passes through or over the different sets of tubes. In figure 20 the tubes shown are so baffled as to cause six passes of the oil through them and the drum is so baffled as to cause two passes of the oil over the tubes. In this unit the heat exchange obtained is between 7 and 10 B. t. u. from each square foot of surface per hour per degree of mean difference in temperature. In either of the two types of tubular interchangers described, it is difficult to obtain the maximum exchange of heat from a given area of tube, because of the large diameter of the tubes and the consequent low velocity of flow. Although this type may give the same outlet temperature as one that is more efficient, the surface required will be larger than in interchangers of the double pipe or jacketed line construction, and construction and maintenance will be much more expensive.

JACKETED LINE OR DOUBLE-PIPE INTERCHANGERS.

Two types of double-pipe or jacketed line interchangers have been developed. One has a number of small pipes within a large pipe and resembles the tubular type first discussed, except in length and in flow velocities. The other has a single pipe inside of another pipe large enough to give the space desired for the flow of oil.

An exceedingly good example of the first type is shown in Plate XIII and figure 21; in it crude oil is being heated on its way to the stills and residuum from the stills is being cooled on its way to storage. The necessary area of heat-exchange surface is obtained by using as many units as the quantity of oil being treated requires. The rate of heat transfer is 20 to 22 B. t. u. per square foot, per hour per degree of mean temperature difference.

As an example of the effect of velocity of flow, J. M. Wadsworth, petroleum engineer of the Bureau of Mines, says that "in an installation similar to this the quantity of oil circulated in both directions was doubled, thus doubling the velocity, while the temperature of the oil discharged in each direction remained practically the same. The rate of transfer must necessarily have been increased 100

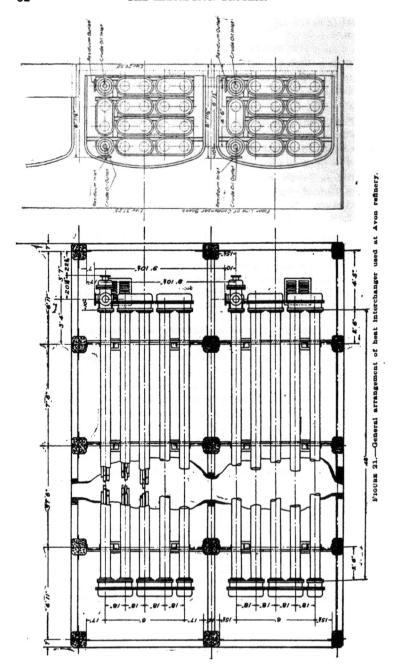

FIGURE 21.—General arrangement of heat interchanger used at Avon refinery.

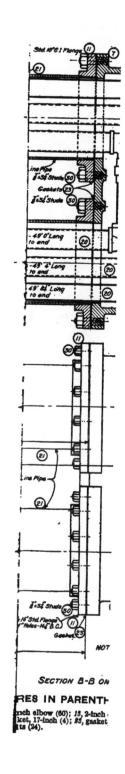

Std. 10"G I Flange — ⑪ ⑦

㉑

ine Pipe
⅝"×5⅛" Studs ㊿
Gaskets ㉓
⅝"×5⅛" Studs ㊿

- 49' 0" Long
to end ⑳

-49' 4" Long
to end ⑳

49' 8½" Long
to end ⑳

⑪

㊿

㉑

ine Pipe

㉑

⅝"×5⅛" Studs ㊿

14" Std. Flange
1" Holes-14⅝" B.C. ⑪
Gaskets ㉓

NOT

SECTION B-B ON

per cent to allow all temperature to remain the same while the flow of oil was doubled."

Interchangers constructed of two pipes, one within the other, give a high rate of heat exchange. One built of 3-inch pipe inside of 6-inch pipe, shown in figure 22, having a radiating area of 455 square feet, or 0.1 square foot per gallon per hour, raised the temperature of the oil 80° F. Another interchanger using 1,600 feet of 2½-inch pipe inside of 3¾-inch casing, a radiating area of 0.28 square foot per gallon per hour, raised the oil 60° F.

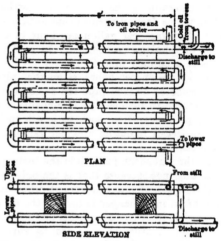

FIGURE 22.—Continuous double-pipe counter-flow interchanger.

Table 4 following gives the sizes, capacities, and area of the different types of heat interchangers used in absorption plants for oils of practically the same character from stills at nearly the same temperatures.

TABLE 6.—*Transfer area per gallon per hour in heat exchangers.*

Plant No.	Oil flow per hour.	Type.	Interchanger.						Total area.	Transfer area.
			Outside tube.		Inside tubes.				
			Diameter.	Length.	Number.	Diameter.	Length.		
	Gallons.		*Inches.*	*Feet.*		*Inches.*	*Feet.*		*Sq. ft. per gal. per hr.*
1	13,300	Tubular............	30	9	357	1	7½	a 4,362	0.328
2	8,500	.....do.............	18	18	31	2	16	a 2,594	.326
3	3,750	Double pipe........	4	1,520	....	2½	1,600	1,050	.28
4	1,250	Tubular...........	36	19	127	2	16	2,120	1.696
5	4,941	Double pipe........	b 6¼	504	....	....	....	....	....

a Ten units, five in series; later changed to 14, 0.42 square feet per gallon.　　　　b Casing.

CONCENTRIC PIPE HEAT EXCHANGERS.

An unusual but entirely practical exchanger in use is made by fitting pipe concentrically within larger pipe, the pipe diameters selected being such that the area of the cross section through which oil flows in each pipe being as nearly equal as possible. As the oil being cooled and that being heated occupy alternate pipes, each of the fluid streams is effective toward both the inside and outside of the coil, so that loss of heat by radiation to the air is lessened.

In determining the number of interchangers to be erected, operators use as factors various constants, such as 0.324 to 0.375 square feet of surface per gallon per hour in the unbaffled or single-pass type of tubular interchanger. In a plant using 1-inch tubing inside of a 30-inch pipe the factor used is 1.4 linear feet of 1-inch pipe per gallon per hour; at another plant which uses 2-inch tubing inside of 18-inch pipe the factor is 0.68 linear feet per gallon per hour. In double-pipe interchangers of both single and multiple pipe construction, such as 2-inch inside of 3-inch, 0.25 to 0.30 square feet of radiating area per gallon per hour is giving satisfactory results.

In the statement just made the main point has been the rate of heat exchange, because that is the only criterion for comparing the various types of interchangers at different plants, whereas the function of an interchanger is to heat the oil going to the still as much as possible in order to save fuel and to cool correspondingly the outgoing oil in order to save cooling surface and water. The rate of heat transfer is a factor of first cost rather than of overall efficiency, for the incoming and the outgoing oil may be brought to the same temperature regardless of the rate of transfer by allowing enough transfer surface. First cost should be considered, however. The writer has seen an exchanger having a transfer of 1 to 2 B. t. u. that was estimated to have cost $50,000, whereas another having the same oil capacity but a transfer rate of about 20 B. t. u cost $4,000.

If the hot oil coming from the still is cooled as much as possible by radiation and by transfer in reducing the temperature of the oil going to the still to within 40° F. of the still temperature, the work done by the interchanger is satisfactory, regardless of the required number of feet of radiating surface. With equal results, the factors controlling design are first cost and cost of upkeep.

Rather than use hot oil pumps to obtain high velocities in the exchangers, some operators prefer to allow the oil to flow by gravity, and in order to get the desired temperatures at both ends they increase the areas of the contact surface or use more interchanger units. This method has the advantage of relieving the pump that forces oil to the still, of overcoming the friction due to high velocity, and of lifting the oil into the still or the stone tower.

Refinery operators who have had broad experience and technical training say without reservation that heat interchangers of the type and general construction shown in figures 20, 21, and 23 are the most efficient and the cheapest to build, operate, and maintain. The coils require only standard fittings and can be kept in repair by refinery employees.

If the oil being cooled is in the outside pipe, as shown·in the figures, there is no need of insulation, for the radiation loss makes practically no difference in the final temperature of the oil being heated and tends to lower the temperature of the oil being cooled.

In absorption plants the problems of heat exchange are less complicated than in refining practice because the temperatures are more uniform, the quantities of oil circulated in each direction are nearly equal, differing only 6 or 7 per cent, and the specific heats and viscosities are also nearly equal. In fact, the factors named are so nearly equal that the inequality may be disregarded and the variations in heat-exchange areas and in sizes of pipe can be compensated by using a reasonable factor of safety.

Daily records of interchanger temperatures from six plants. in different parts of the country. and taken at different seasons, follow:

TABLE 7.—*Oil temperatures to and from heat exchangers.*

Points at which temperatures were taken.	Temperatures in plant No.—					
	1	2	3	4	5	6
	°F.	°F.	°F.	°F.	°F.	°F.
Unsaturated oil from interchangers.	144	150	163	147	138	130
Saturated oil entering interchanger.	a 64	a 12	a C6	76	102	101
Unsaturated oil from still.	214	208	194	216	208	205
Saturated oil to still.	150	97	140	160	170	170

a Temperatures taken at tower discharge, from which the oil passed through a steam pump before reaching the heat exchanger.

The temperatures are not exactly comparable, as they were not taken at corresponding points in any two plants. Some plant operators take the temperatures of the oil before it reaches the dephlegmating columns; others take the temperature afterward. Moreover, temperatures in the table were read at different seasons, the figures in the first three columns being taken in winter and those in the last three in summer. At plant No. 6 the condensed water from the still passed through one unit of the interchanger, replacing oil. This method undoubtely economizes heat, as water has twice the heating effect of oil for an equal change of temperature. Although the data in the table were taken from daily report sheets, and show figures necessary in calculating overall efficiency. they can not be used for computing the actual transfer of heat from the saturated to the unsaturated oils in the interchanger.

However, it is clear that the exchange of heat lessens fuel cost and boiler capacity, so that it is far too valuable a feature to be omitted at any plant. Besides conserving fuel interchangers reduce decidedly the amount of water required for cooling and the amount lost by evaporation. The latter item deserves careful attention in many plants, especially in California where water costs 3 cents a barrel.

The heat saved by interchangers at the plants shown in Table 5 ranges between 22 and 77 boiler horsepower. Using 50 cubic feet of gas per horsepower hour the interchangers save between 26,000 and 95,000 cubic feet of gas a day.

RADIATOR INTERCHANGERS.

Although the radiator type of heat-interchanger has not proved a success a short description of it may be of interest. On its way to or from the still the oil is passed in series through the coils of a steam radiator made of cast iron or pressed steel. These radiators are placed in light-weight, tightly closed steel boxes, the oil in the steel boxes flowing over and under baffles countercurrent to the flow in the coils. The greatest fault found with this type of coil is that the radiators are not built to withstand pressures above 15 pounds. At one plant a coil of this type burst and the oil was ignited, causing the death of one of the operators. The unit was dismantled and a double-pipe coil used in its place.

In general, it has been found that about 0.30 square feet of radiating surface per gallon per hour is enough in interchangers of the double-pipe continuous-coil type and about 0.40 in the multiple-pipe units. As velocity of flow is important in heat exchange, the continuous double-pipe coil appears to have the greater advantages in both construction and operation.

STILLS.

Stills of a number of different types have been installed at absorption plants, most of the types being those used in refineries for somewhat similar operations. The action desired is practically the same as that in topping plants, for only two products are made— the gasoline fraction, which is vaporized, and the residuum, or absorption oil, which remains in the still.

Usually the oil used for absorbing has an initial boiling point of 450° F., or more, and the end point of the gasoline absorbed is 300° F., so that the gap between the two temperatures is not less than 100° F. Although this condition greatly simplifies the operation of the still, it is impossible, even in a steam still, to remove all of the gasoline from the oil without the gasoline vapors carrying a small fraction of the heavier oil from the still. This fraction is taken up by the action of partial pressures and of other physical principles; moreover, small particles of the oil are carried mechanically with the gasoline vapors and the steam rising from the surface of the oil in the still.

Separation of this small content of heavy oil from the vapors is accomplished by fractional cooling. To determine and maintain the proper temperatures in still, still towers and cooling coils is a most vital matter, and one with which many operators have the most difficulty. Steam stills with either closed or perforated coils, or both, are the simplest and the most used.

STEAM STILLS.

Figure 23 gives details of a steam still that has proved satisfactory to many operators. From the heat interchangers the oil with a temperature of 160° F. to 170° F. rises through the 4-inch line to the top of the still tower, and is discharged over the stones, tile, or baffles, with which the tower is filled. As the figure shows, the tower is set over the still; it is connected to the still by two large vapor lines that enter near the bottom, and by an oil outlet from the bottom of the tower to the top of the still.

Oil from the intake at the top divides over the baffling of wall-tile, or stone, and flows downward through the tower countercurrent to the vapors rising from the still. The vapors pass from the top of the tower to the cooling coils through the inclined 8-inch pipe.

The object of the tower is to condense and carry back into the still the fractions of heavy oil that are vaporized or are mechanically entrained with the gasoline and steam vapors; to liberate the vapors that come from the oil at the temperature it has acquired in the interchange; and also to collect the heat produced by the condensation of these heavy vapors, thus reducing the work required of the steam in the still in the same way as the interchanger reduces the necessary amount of heat.

This type of tower is justly criticized by California refiners on the ground that the heat interchange itself is too small to be of decided value. Moreover, their long experience with topping plants has proved that the cooler incoming oil has a strong tendency to absorb the light vapors as well as to condense the heavy one; hence the practice of refiners, in order to eliminate the absorbing action, to remove the vapors from contact with the the heavy or incoming oil as soon as possible.

These refiners suggest that any heat interchange between vapor and oil should be made with the oil in double-pipe coils or in tubes and out of contact with the vapors. Also, if it is desired as is considered the best practice, to remove the vapors from the hot oil in a tower, to spread the oil in thin sheets in some type of apparatus, allow the vapors to come off, and remove them as soon as possible from contact with the oil. This practice is followed most perfectly in

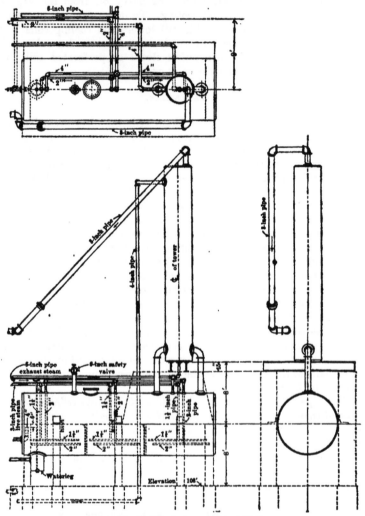

FIGURE 23.—Detail of steam-still installation.

the Trumble type of fire-still tower. A dephlegmating action, such as is used in rerunning " tops " to obtain narrow cuts of the different fractions, is neither necessary nor desirable in absorption practice.

Plate XIV, *A* shows this type of stone tower. When this tower was full of stone the absorbing action proved much too great, consequently more than half of the filler was removed.

Following the example set by one of the largest and most successful absorbtion plants, many operators have placed towers of this character over their stills, and the practice is now general. The writer believes stone towers of this type are not only unnecessary but are prejudicial to the action desired and that a careful study will lead to many changes in the design and operation of stills in absorption plants. Plates XIV, *B*, XV, *A*, and XV, *B* show stills and towers of the type described above.

In these stills the temperature of the incoming oil is often as high as 160° or 170° F. and at times 190° F., depending on the heating effect obtained in the interchangers. As the oil passes through the stone tower on its way to the still it may become warmer, so that at times it may enter the still at a temperature 10° F. higher than that at which it entered the top of the tower. Oil enters a still of this type at the end opposite the discharge, passing over and under vertical baffle plates, if baffles are used, and meeting steam, from open or perforated pipes, which completes the separation of the volatile hydrocarbons.

The pressure in the stills is 2 to 5 pounds, the stills being fitted with safety valves set at this pressure, and the temperature ranges from 206° to 218° F. As an aid in heating and temperature control, many stills have closed steam coils at the bottom; through these coils live steam at any pressure and temperature may be passed to raise temperature quickly. In coils that are open or perforated exhaust steam from the pumps is often used, being discharged against the still pressure of 2 to 5 pounds at temperatures corresponding to the steam pressure. Using the exhaust steam from the oil pumps is economical when steam pumps have to be used, but when pumps driven by gas engines are used, and the steam for heating is taken directly from the boilers at the desired pressures and temperatures, the saving would appear in the quantity of gas used for power. This saving would be an object to companies that sell gas.

In stills of this type the oil level is maintained at about half of the height of the still by an open discharge pipe, or by any one of the fluid-level regulators now obtainable, the oil flowing to the heat interchangers.

In stills baffled as those shown by figure 23 (p. 58) the oil is forced to flow over and under each baffle in series on its way to the discharge, and provision must be made for water, condensed from the steam in the still, to flow along the bottom of the shell and to be drained away. Drainage is usually effected through perforations in the baffles near the still bottom. In the still shown by figure 24 the water leg is used to trap the water and discharge it to one of the series of seven heat interchangers; in other plants the water goes to the hot well and is used to feed the boilers.

Stills of this type are of such diameter and length that when in full operation the oil they contain and treat, is approximately the amount circulated through the system in one hour; in other words, the heat treatment in the still takes one hour.

VERTICAL STILL.

A large operator in western Pennsylvania uses a vertical steam still of quite different type, with notable success. The still has no tower, but instead a vapor and oil heat interchanger. Oil from the interchanger at a temperature of about 160° F. passes through the tubes of a preheater of the type used in heating boiler feed-water with exhaust steam or hot gas. The still vapor at 212° F. passes from the top of the still to the heater drums which surround the tubes carrying the oil; the oil is warmed and the vapors cooled, a large part of the absorption oil and steam carried over with the vapors is condensed, and the particles of oil carried mechanically can settle. Thus the interchanger acts as an auxiliary condenser and separator or "knock-out box." The condensed oil and water are separated from the vapors by a trap similar to a steam trap and are discharged to the still, while the vapors are conducted directly from the top of the preheater to coils cooled by air and water. This arrangement eliminates the auxiliary water-cooled coil, used in many plants, which is maintained at a temperature of approximately 185° F. to separate the heavy oil and water.

The steam still used is a boiler shell set vertically and filled with banks of steam radiators; in flowing over these the oil is heated and gives up the gasoline vapor. An open coil at the bottom introduces steam to simplify the heat control and aids the separation of the vapors. The steam in the radiators has a pressure of 10 to 16 pounds and a corresponding temperature. In a day the still handles easily 126,000 gallons of oil containing about 2 per cent of gasoline.

HORIZONTAL-PAN STILL.

Figure 24 shows a still of the horizontal cylindrical type, 5 feet in diameter and 16 feet long. Instead of the oil being allowed to lie in the still at a height of about half the diameter, it is discharged into a shallow, slightly inclined pan, 5 inches deep; in this the oil runs slowly toward the opposite end of the still where it is discharged through an overflow pipe 1 or 2 inches above the pan bottom to the next pan below which slopes in the opposite direction. Four or five such pans are placed in the still shell. On the bottom of each pan is a continuous return-bend coil of 1-inch pipe 210 feet long,

A. STILL, AND STONE TOWER.

B. STEAM STILLS WITH TOWER SUPPORTED ON SEPA-
RATE BASE; TREAT 10,000 GALLONS PER HOUR.

A. STEAM STILLS, 8 BY 24 FEET; TREAT 5,000 GALLONS PER HOUR; ALSO SHOWS AUXILIARY SUBMERGED COIL AND "FINAL" COIL.

B. STEAM STILLS, 6 BY 12 FEET IN SIZE; CAPACITY 1,000 GALLONS PER HOUR.

over which the oil travels. Steam intakes and discharges for each coil are connected to headers outside of the still shell, and are controlled by valves. In the bottom of the still is a perforated steam

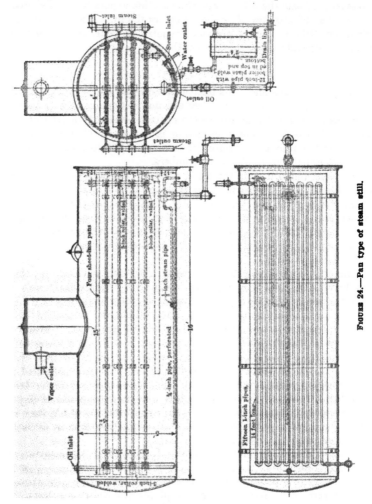

FIGURE 24.—Pan type of steam still.

line that is kept covered with oil to a depth of 6 to 12 inches by a float, as shown, or by a fluid-level regulator. The steam discharged into the oil removes the last gasoline vapors and helps to carry all of the vapors out of the still to the vapor lines.

Such a still is simple to erect and to run, and has an exceedingly large capacity in this service. The still shown in figure 24 has a capacity of 5,000 gallons per hour when using steam at 90 pounds pressure; it heats the oil quickly, the gasoline vapors being liberated in less than 20 minutes after the oil enters the first pan. The large-diameter stills previously described require an hour, and often two hours, to separate the gasoline from the oil.

FIRE STILLS.

Fire heated stills and stills heated by both fire and steam are used at the absorption plants, but they are not common and at present are considered good practice, as the control necessary in separating the two factions without overheating is not easily obtained where a still or the piping containing the oil is subject to open flame or to direct heating in combustion chambers.

In steam and fire stills the oil is placed in the still shell over an open fire or over the combustion chamber of a fire box. The hot gases strike directly against the still shell, heating the oil inside. A perforated steam coil is placed within the still near the bottom, to heat the oil and help the escape of vapors, and also to agitate the oil so as to keep the temperature near the bottom sheets of the still from rising unduly.

One still heated only by fire was seen by the writer, the heating being done in 4-inch pipes in continuous return-bend coils set near the top of the combustion chamber of the open fire box. The oil always was circulated through the coils in an upward direction in order to avoid the formation of gas pockets. From the last pipe of the top coil the hot oil went to the top of a tower called a separating tower, or evaporator, where it passed in thin sheets over a series of baffles, so that the gasoline vapor had unlimited opportunity to escape from the oil. The vapors collect under hoods and flow through perforated pipes to a header, from which they go to the water-cooled condensers. At the time of the writer's visit, this plant was having difficulty in controlling temperatures so as to prevent overheating and loss of production.

Because the steam induces the vaporized light hydrocarbons to separate from the oil, and because the temperature of steam at low pressures (2 to 5 pounds) is that best suited to distilling in the open or perforated coils used in absorption plants, the writer believes that fire stills will never become general in absorption practice.

HEAT NECESSARY FOR DISTILLING.

Probably the most common of the serious errors, in plant design hitherto has been the calculation of the heat or steam requirements for distilling.

By using the constants given below, or those that are similar, the theoretical heat requirements can be calculated, but the efficiency of the still in using heat is given as 80 per cent, whereas in practice the efficiency is practically never more than 40 per cent, and 30 per cent is a better figure to use in plant design. This figure allows some leeway for seasonal changes in temperature and for changes in quantity of oil circulated on account of variations in the quantity of gas treated.

Constants sometimes used in calculating heat for distilling.

One boiler horsepower_____B. t. u___	34,000
Latent heat of gasoline_____do___	100
Specific heat of " mineral seal oil "_____	0.50
Specific heat of gasoline_____	0.58
Weight of " mineral seal oil " per gallon_____pounds___	7.00
Weight of gasoline per gallon_____do___	5.5

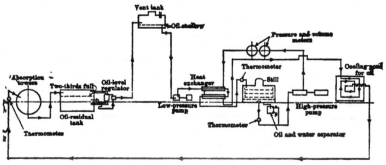

FIGURE 25.—Oil path in an eastern refining plant.

From refineries treating somewhat similar oils in steam stills the heat used was reported as 1 boiler horsepower per barrel of distilate produced per hour with open steam stills and 2¼ to 4 boiler horse-power for each barrel of product per hour with closed steam coils inside the stills.

From the stills and interchangers the oil at 80° to 140° F., goes to the high-pressure pumps, thence to the water-cooled coils, which reduce the temperatures as low weather conditions permit, and thence to the absorption towers. Figure 25 shows the complete circuit of the oil in one of the large Eastern plants.

TIME OF CIRCULATION OF OIL.

The total time required for any given particle of oil to make a complete circuit of the absorption plant varies greatly, but depends mostly on the time required to pass through the stills, interchangers, and weathering tanks because there are the points where oil is held in considerable volume and its velocity slow.   Three to four hours is probably a fair estimate of the average time of a circuit in usual practice, being more at some plants and less at others.   The total quantity of oil in use at any one time is controlled in the following way: Assume that a plant treating a given quantity of gas requires 4 gallons of oil per 1,000 cubic feet of the gas and 204,000 gallons for the gas passing through the plant in 24 hours.   Then the oil pumped through the circuit would be 8,500 gallons per hour, and if this oil required 4 hours to travel through the circuit, 4 times 8,500 or 34,000 gallons of oil would always be in active circulation. If three hours were required, 25,000 gallons of oil would be in use, the oil making 8 complete circuits each 24 hours.

Usually more or less tankage is provided for oil stock not in actual use, to allow buying oil in car lots and having enough always on hand to make up any loss or to meet changes in plant functions, such as an increased flow of gas.

TREATMENT OF STILL VAPORS.

The vapor, whether coming from the stone towers, the vapor-heat interchangers, or directly from the still with no auxiliary cooling at all, contains not only the gasoline vapors, but also gases liberated from the oil by the heat and steam in the still, water vapor from the steam used, and small portions of the absorption oil itself, the latter probably both as vapor and as finely divided particles of liquid.

At this point a clean separation of these four constituents is made by fractional cooling; that is, by reducing the temperature of the vapors in steps, or stages, and removing the products precipitated at the temperature in each stage.   The order of separation is naturally reverse to that in distilling, the constituent of highest boiling points separating first.

As in distilling, however, any one cut of distillate, or any one member of the mixture of vapors, can not be perfectly separated at any given temperature, but portions of fractions of both higher and lower boiling points will come down with it.   The liquid precipitated, or condensed, from the vapor depends on the vapor pressure of that particular component under the existing conditions of temperature and pressure.

SEPARATING OIL, WATER, AND GASOLINE.

RERUN METHOD.

Two methods are used to separate such mixtures, rerunning and fractional cooling.   The rerunning method consists of vaporizing and condensing from the mixture a fraction large enough to carry with it all of the desired cut and those portions of lower boiling fractions that can not be held as liquids, drawing off the portion left, or the still "bottom," and repeating this process with the condensed portion until the product from the coils has the character desired.

FRACTIONAL COOLING.

Fractional cooling is of greater advantage in absorption practice, because the gasoline cut is the only one that must be perfectly clean. Owing to their mutual insolubility water and gasoline and water and oil separate from each other by gravity. Water is the overlapping fraction precipitated throughout fractional cooling with both the oil and the gasoline, and in each instance it is separated after condensation by settling, the oil or gasoline being floated away.

COOLING OF THE VAPORS.

PRIMARY CONDENSERS.

The vapors from the stills have temperatures between 200° and 220° F. In order to completely reciprocate the absorption oil vapors, they must be reduced to between 180 and 190° F., which is the first step in fractional cooling. This cooling and separation is obtained by one of several methods. An Eastern plant reduces the vapor temperature by heat interchange with the incoming oil and by air cooling. Air cooling is in a large uninsulated pipe exposed to the atmosphere, the oil and water condensed being separated in a steam trap and returned to the still. Air cooling when conducted in the open has the disadvantage of being uncontrollable, because of the atmospheric temperature and the wind velocity changing, causing diverse rates of heat radiation, under dissimilar temperature difference, and varying dissipation by air currents. This disadvantage may be partly overcome by housing the pipe within a still building, engine or pump room, where temperature and air currents vary less than outside and are to some degree controllable. Its successful operation speaks for itself. The size and length of pipe are the factors to be determined and these depend on the temperature necessary to make the separation and the volume of vapor being treated.

Other plants use a submerged water box coil of the usual refinery construction, the water being kept at 185° F. by an electric thermal element, as shown in figure 26, at the discharge of the smaller of the coil boxes. The condensed products, oil and water, are separated from the vapors and each other by the overflow trap, the vapors from the separating chamber flowing into the pipes leading to the second coil. Although this construction and operation are both simple and efficient, it is a question whether the cost is warranted, although the water discharged can be used as boiler water, the condenser taking the place of a pre-heater. If gas engines were used to drive the pumps the same saving could

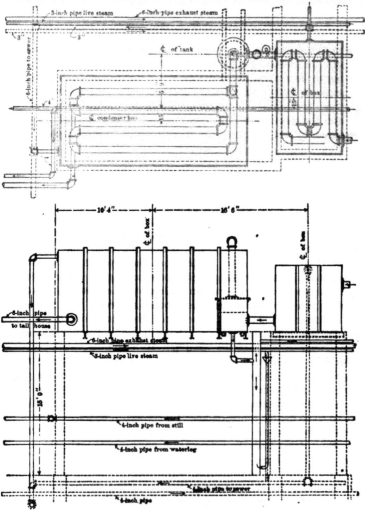

FIGURE 26.—Submerged cooling coils. Small coil is the primary condenser; electric thermal control is placed in the connection between the two boxes.

easily be made by passing the exhaust gases through the feed-water preheaters of the tubular type, but care must be taken not to have the back pressure on the exhaust too great.

At the plants visited, the areas of the first-stage submerged cooling coil of 8-inch pipe was between 2 and 2.5 square feet for each gallon of gasoline extracted from the gas per hour. Although this coil precipitates no gasoline, heat is extracted from the vapors to liquify the heavy oil and part of the water vapor at this point, and the amount will be about the same for each gallon of gasoline distilled. The factor is therefore given in terms of gallons per hour, rather than in British thermal unit equivalents.

If smaller pipes are used the area can no doubt be considerably reduced, the velocity of flow and the consequent rate of transfer becoming more rapid.

Cooling the vapors by interchange of heat with the incoming saturated oil has been discussed elsewhere. To this point cooling has precipitated all of the Mineral Seal oil and a large part of the water vapors coming from the still.

FINAL COOLING COILS.

Vapors from the first stage, or auxiliary cooling coils, whether submerged coils, heat exchanger, or air-cooled pipes, are led directly to where the second reduction of temperature, precipitating gasoline and water, takes place. Here two methods, singly or in combination, are used to cool the vapors—submerging the coils or cooling them by air and water in a louver tower. With either, the vapors at the coil intake have a temperature between 170° and 190° as they come from the auxiliary coolers.

A typical submerged coil is shown in figure 26 (p. 66). It has a radiating area of between 3.5 and 4 square feet per gallon of gasoline per hour. The reader should note that this coil, besides condensing and cooling the gasoline vapors and liquids, must also cool varying volumes of gas or vapors that do not condense, and also condense and cool any water carried as steam or water vapor past the first cooling. These last duties vary greatly with the cooling required at different times in the same plant, and are not at all comparable in different plants, as they depend on other functions and operations through the entire plant and on what hydrocarbons the gas carries to the absorbers.

In Plate XVI, A, a combination submerged coil and louver tower is shown. The coils are placed in both the box and louver. Little can be said for a cumbersome hybrid of this type.

COOLING COILS IN LOUVER TOWERS.

The second method of cooling and condensing vapors is by coils exposed to water and air in louver towers, as discussed on a preceding page in connection with the cooling of oil and in another publication[a] of the Bureau of Mines. From 2 to 8 square feet of radiating surface is exposed per gallon of condensate per hour. An area of 2.5 square feet per gallon per hour has been adopted by a number of engineers for return-bend coils made .of 2-inch pipe. This type of cooling is used in an absorption plant in California, where submerged coils proved an utter failure because of the hot, dry climate. Plate XVI, *B*, shows a louver tower of modern design in which compressed gas, still vapors, and absorption oil from the interchanger are cooled in small (1-inch) pipe coils in headers. Temperatures 40° F. below atmospheric are obtained in summer.

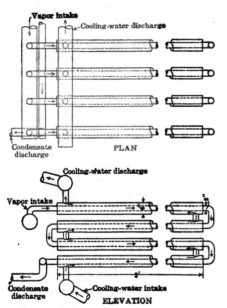

Figure 27.—Double-pipe vapor condenser.

It is generally true that in hot, and especially hot, dry climates the use of a louver tower for cooling water is far the best practice. A tower may be used in two ways—either with coils supported in the tower under the dripping water, or cooling the water in the tower and then circulating it over submerged coils, the first method being the more efficient, as well as much more convenient, as all of the coils are at times subject to inspection, and repairs can be made quickly without stopping other plant operations.

Refineries, absorption plants and compression plants in all parts of the country are using cooling coils in louver towers more and more.

DOUBLE-PIPE CONDENSER.

Figure 27 and Plates XVII, *A* and XVIII, *A* (p. 72) show an entirely different system of cooling and condensing vapors. This system uses four return-bend, continuous double-pipe coils, 36 feet long and four pipes high, with a total radiating area of 450 square feet, or between 3 and 4 square feet per gallon per hour of condensed gasoline. The vapor-intake header acts as a trap for vapors

---

[a] Dykema, W. P., Recovery of gasoline from natural gas by compression and refrigeration, Bull. 151, 1918, 123 pp.

A. COMBINATION LOUVER AND SUBMERGED COIL.

B. A LOUVER TOWER OF MODERN DESIGN, ERECTED IN HOT, DRY CLIMATES.

B. TRAP FOR SEPARATING OIL AND GAS.

A. END CONNECTIONS OF DOUBLE-PIPE CONDENSER.

condensed in the line from the still. The vapors and condensate ·
flow parallel and not countercurrent, and water flows countercurrent
to them. From the coils the con-
densed vapors—gasoline and water—
flow to the water separators and to
inspection or look boxes. The cooling
water flows from a header into the
lowest outside pipe of each coil, flows
upward, and discharges to an upper
header.

SEPARATION OF GASOLINE, WATER, AND GASES.

Water, gasoline, and the uncondensed vapors and gases coming from
the final coil discharge may be separated into two products—liquids and
gases—at the coil outlet by connecting the end of the coil pipe to
a pipe set vertically in a T connection. The pipe rising from the T
carries away the vapors and the horizontal pipe leads the liquids to the
water separator.

Figure 28 shows a system in which
all the products are carried together
through one pipe to the separating
chamber, which allows the water
to settle into the leg below the horizontal cylinder. This water over-
flows through a differential column
to a waste pipe, which is fitted at
the top with a funnel; the overflow
pipe ends about 2 inches above the
funnel to permit inspection and
sampling of the passing liquid. The
gasoline flows off continuously through
a horizontal pipe into the inspection
box, from which it goes to storage
or a blending tank. The gravity of
product obtained at this point ranges
rom 72° to 84° B.; the vapor tension

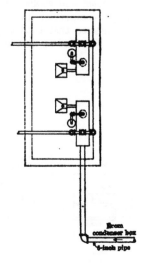
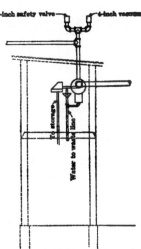

FIGURE 28.—Connections for "look
box," showing method of separating
vapor and water.

is well within the shipping regulations and is often as low as 3
pounds at 100° F.

As the gases rising from the separator contain considerable condensable vapor they are put through a small absorption unit or are compressed and cooled in coils. In figure 29, which shows a plant using a compressor on these vapors, the pipe carrying the gases is so connected that both a pressure and a vacuum valve may be put on the line leading to the compressor. The use of these valves, set

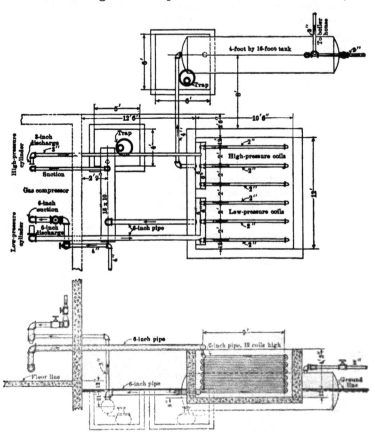

FIGURE 29.—Connections for compression coils.

to relieve pressures over 2 or 3 pounds or vacuums over 4 to 6 inches, is advisable as a safety measure. The lines are directly connected to the still which would be damaged if a high vacuum were held on it by the compressor suction at the other end of the line, and the safety valve keeps the pressure in the entire system from building up if the compressor is stopped or is unable to take the vapors fast enough.

The condensate from the compressed gas is a "wild" product unfit for shipment unblended, and is, so far as the writer knows, added to the still product in such amounts as to bring the whole to a point just within the legal requirements for shipment in tank cars. From 15 to 30 per cent of the plant product is derived from the compressed still vapors and from gas from the weathering tanks. The compression and cooling of uncondensed gases is the same as that described in other Bureau of Mines bulletins.[a] To this gas is often added the gas coming from the weathering tank in which the absorption oil is first run from the towers, but as stated before, it is questionable whether these gases add materially to the net output of gasoline, whereas the compression of the gases from the still alone is known to add a valuable amount.

Figure 29 shows the coils and connections of two-stage compression units. The 18-inch by 10-inch pipe forms the accumulator of the low-pressure coil. The high-pressure product is run with the gas to the 4-foot by 16-foot high-pressure accumulator, which is a 4-foot by 16-foot riveted horizontal tank.

COST OF PLANTS.

Under conditions such as those prevailing during the war estimates of the cost of material or labor necessary for the construction of gasoline plants are necessarily high. In 1916 a plant to treat a daily flow of 60,000,000 cubic feet of gas at 250 pounds pressure was built at a cost of about $1.75 per each 1,000 feet of capacity. A plant for which material was ordered late in 1917 cost $2.25 per 1,000 feet capacity. Both of these plants were large and did not include gas pumping or cooling systems, but did include compressors to be used on still and on weathering tanks vapors. An installation being erected in August, 1918, to treat 6,000,000 to 8,000,000 cubic feet of gas at 250 pounds pressure will cost between $25,000 and $30,000, or about $3.75 per 1,000 cubic feet of capacity.

Construction costs vary not only with prices for labor and steel but with the pressure and richness of the gas. A tower that will treat 3,000,000 feet a day at 300 pounds being capable of treating not more than 500,000 cubic feet of the same gas at 50 pounds pressure; or if the plant capacity in feet remain the same, the towers for gas treatment will cost six times as much.

The increase because of the greater richness of the gas results from the increased pump, boiler, still, and coil capacity necessary, all increasing with the greater volume of oil circulated to extract the

[a] Burrell, G. A., Seibert, F. M., and Oberfell, G. G., The condensation of gasoline from natural gas, Bull. 88, Bureau of Mines, 1915, 106 pp. Dykema, W. P., Recovery of gasoline from natural gas by compression and refrigeration, Bull. 151, Bureau of Mines, 1918, 133 pp.

larger quantities of gasoline. Absorption plants designed to treat casing-head gas under vacuum require the added expense of the pumps, and often of the compressors installed to develop the desired working pressures so that the cost of such plants runs as high as $40 or $50 per 1,000 feet of capacity.

ABSORPTION OF GASOLINE VAPOR IN CRUDE OIL.

Here it may be of interest to describe a plant that embodies a unique adaptation of the absorption process as developed and used by several of the large California oil companies that produce large quantities of both casinghead and dry gas containing gasoline vapor to make treatment profitable. Instead of using mineral seal oil or other refinery product as the absorbing medium, these companies use the crude oil just as it comes from wells, choosing the heaviest oil produced in the immediate vicinity of the gas pumping station. The gas from the wells is compressed to line pressure and then sent directly through a series of coils cooled with air and water in a louver tower of the usual construction. The coils are made of 2-inch pipe—13 pipes high, 40 feet long—set in headers. From the coils the gas goes to drips or accumulator tanks that separate the small amount of gasoline vapors condensed from the gas by compression and cooling. Then the gas goes to a header connected by short nipples to each of the four towers. These towers are 42 inches in diameter and 20 feet high, are made of riveted ¾-inch to 1-inch plate, and are tested at 300 pounds pressure. In Plate XVIII, B, a side view of the battery of towers is shown. The gas from the header connection bubbles up in the towers through the crude oil, which is held by an automatic control (in Plate XVIII, B, the 12-inch vertical pipe in the foreground) at approximately 10 feet (one-half the total height) above the bottom of the tower. The tower is two-thirds filled with bricks to break up bubbles as they rise through the oil, the top third of the tower being an open chamber in which the oil and gas can separate. The gas from each tower flows through a short vertical connection to a horizontal header directly above the tower this header collects all the gas from all of the towers, which work in parallel at the same pressure, and from it the gas passes downward through a vertical pipe into the oil trap shown in Plate XVII, B. This trap is simple, has no baffles or moving parts, and its design is shown by the illustration. The gas passes out at the left end through the vertical connection, which is connected to the main-line pipe.

OIL CIRCUIT.

The incoming crude oil is handled by a pump shown in Plate XIX, A, from whence it goes to a number of cooling coils set in the louver tower. It leaves these coils at a temperature of 75° F., goes

A. VIEW OF DOUBLE-PIPE CONDENSER.

B. ABSORPTION TOWERS FOR CRUDE OIL.

A. PUMP FOR ABSORPTION OIL.

B. WELDED FITTINGS AND PIPE; WORKING PRESSURE, 500 POUNDS PER
SQUARE INCH.

directly to the open chamber in the top of towers, passes downward through the towers and back to the storage tanks of the oil pipe-line pumping station. At one of these plants 14,000,000 cubic feet of gas is treated daily, this volume including 6,000,000 cubic feet of "dry" gas, or gas from gas wells, which is compressed to 260 pounds pressure, and 8,000,000 cubic feet of casing-head gas, which is compressed to 150 pounds pressure.

As is the usual practice in the San Joaquin Valley field, the casing-head gas and oil together are permitted to flow under pressure into a trap, when the gas and oil separate. The pressure is maintained to induce the crude to absorb as much as possible of the gasoline fractions during such treatment. A description of gas traps is given in another publication.[a]

During a given period 21,300 barrels of crude oil (gravity 21.6° B.) were pumped into the towers, and during the same period 22,729 barrels of oil (gravity 24.7° B.) were taken from the towers, showing an increase of oil of 1,400 barrels, and of gravity of 3.1° B. During a period at a different plant, 12,600 barrels of oil of 19.1° B. gravity were put through the towers, and 14,000 barrels of oil of 23.3° B. gravity were produced. The increase in volume and in gravity is due to dissolved fractions of hydrocarbons contained in gasoline, but as in all other absorption processes, there is probably an increase in Baumé gravity due to dissolved gas and to hydrocarbons, such as propane, that are too light to be held in any great proportion in ordinary motor fuels.

An eastern plant using oil with a gravity of 41° B. as the absorption medium raises the gravity in the absorption towers, under 140 pounds pressure, to 43.9° and actually produces from the oil by distillation about 2 per cent of gasoline, showing that not more than one-half of the 2.9° increase in gravity is due to recoverable gasoline. The light hydrocarbons not recovered and the dissolved gas must account for the balance of the increase.

Companies using this system of drying gas gain a double advantage; besides recovering the gasoline extracted from the natural gas they get the benefit of pumping a more fluid oil through their pipe lines. The exact value of the latter is difficult to compute, but nevertheless it is of decided importance. The operation is continuous, simple, and requires little expense for material and labor, as all of the distilling is done at the companies' refineries several hundred miles from the oil and gas fields. The cost of compressing and cooling the gas can not be charged to the absorption account, as it is a necessary part of the charge for transporting and marketing the gas.

---

[a] Hamilton, W. R., Gas traps for oil wells: Tech. Paper 209, Bureau of Mines, p. 86.

The drying of gas or the recovery of gasoline at plants operated in this way is claimed to be entirely satisfactory, and, whether
or not the method is the most efficient that could be used, it undoubtedly shows exceptionally large returns on the investment because of the low first cost and the small expense for operation.

ABSORPTION GASOLINE.

The gasoline obtained through absorbing the lighter hydrocarbons from natural gas and then distilling the absorbent has a high
gravity, 72° to 84° B. and a low vapor tension (2 to 5 pounds at
100° F.), so that it commands a premium over the usual "straight" .

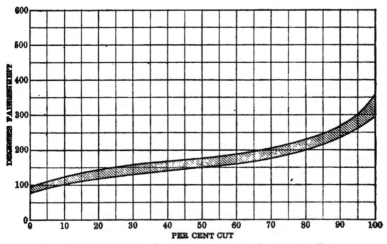

FIGURE 30.—Fractionation diagram for eight absorption-plant gasolines.

run or blended casinghead gasoline. Although it contains extremely
light hydrocarbons, it has practically none of the dissolved gases
such as make gasoline from compression plants so volatile and practically none of the heavy ends that are often found in "straight" run
gasoline as marketed at present. An equal product can easily be
made from crude oil by regular distillation, but this would require
the heavier fractions now sold as motor fuel to be sold for other use.
In fact, it is the general opinion that the greatest part of the absorption gasoline made is used for blending with the heavy fractions
from stills, to make the latter a suitable motor fuel.

The gasoline made by the compression of the uncondensed vapors
from the coils has practically the same characteristics as the product
of plants that compress casing-head gas. It is always blended as fast

as produced, either with naphtha or the still product from the same gas.

The pressures and temperatures used in the treatment of the uncondensed still vapors determine to a large extent the quantity and gravity of the compression product.

Figure 30, which shows a fractionation diagram of eight absorption-plant gasolines, chosen at random from different plants, gives a clear representation of average absorption-plant gasoline.

PLANT RECORDS.

All large gas companies that have operated gas-pumping plants for years keep complete records of all important data. Smaller operators with less experience may find the form below of value as indicating the pressure, temperature, and production records that large operators have found necessary and valuable. Besides its usefulness in checking, records of plant function of this kind holds the interest of men running the plants each tour. By a few recording pressure gauges and thermometers the entire record can be checked; this will show whether recorded figures were read from the instruments or were guesses by the men on tour.

In the accompanying form the daily report sheet used by a large eastern company the entire set of temperatures and pressures is shown as taken at 2-hour intervals, and the total daily records are placed below the hourly readings. Besides the data shown, it is usual to give the number of feet of gas used as fuel in the boilers or for power per gallon of gasoline produced, a figure obtained by dividing the total gas consumed by the total production of gasoline in gallons.

DAILY ABSORPTION GASOLINE REPORT.

From .... gasoline plant for 24 hours ending 8 a. m. ...., 191..

Nature of observation.	Averages.	No. 1 tour.					No. 2 tour.			No. 3 tour.			
		10 a. m.	Noon.	2 p. m.	4 p. m.	6 p. m.	8 p. m.	10 p. m.	Midnight.	2 a. m.	4 a. m.	6 a. m.	8 a. m.
Air temperature (in shade)													
Gas pressure, entering absorber													
Gas temperature, entering absorber													
Gas temperature, leaving absorber													
Oil temperature, entering absorber													
Oil temperature, leaving absorber													
Oil temperature, entering interchanger (sat.)													
Oil temperature, entering dephlegmator (sat.)													
Oil temperature, entering still													
Oil temperature, leaving still													
Oil temperature, leaving interchanger (unsat.)													
Oil temperature, leaving cooler (unsat.)													
Water temperature, entering condenser													
Water temperature, leaving condenser													
Gasoline temperature, vapor leaving still													
Gasoline temperature, vapor leaving dephlegmator													
Gasoline gravity, leaving condenser													
Gasoline temperature, receiving tank													

Gas used for fuel under boiler .... ft. for 24 hours.
Gas passed through absorbers .... ft. for 24 hours.
Gasoline produced .... gal. in 24 hours.
Gasoline produced .... gal. per million cu. ft. of gas treated.
Gravity gasoline, before pumping into storage; run of tank No. 1, ... ° B; No. 2, ... ° B.
Total absorbent oil on hand in storage tanks this morning .... gal.

Oil pressure entering absorber .... lb.
Oil pumped in 24 hrs. .... gal.
Ratio of oil flow .... gal. per 1,000 cu. ft.
Pressure on run tank .... lb.
Number of absorbers used ....
Vel. of gas in absorbers .... ft. per min.

.... Engineer No. 1 tour. .... Engineer No. 2 tour.

.... Engineer No. 3 tour. .... Chief Engineer.

A. WELDED HORIZONTAL ABSORBERS.

B. EIGHTEEN-INCH PIPE REDUCED TO 4-INCH NIPPLE BY
CUTTING AND WELDING.

OXY-ACETYLENE WELDING.

In this paper are illustrations that show many units which are partly or wholly of welded construction. The testing apparatus is almost universally welded, welding being especially suitable for small work of that character.

Many of the more progressive operators have carried welding much further than simply as an alternative for small fittings, using it to replace practically all fittings except valves. Plates XIX, B (p. 73), and XX, A show how this process may be used in complicated units. The plant illustrated worked at pressures ranging up to 500 pounds per square inch, yet had no trouble with any of the welded joints. Plate XX, B shows a near view of an 18-inch standard pipe reduced to take a 4-inch nipple which is also welded in.

Some operators claim that a welded pipe line will not hold high-gravity compression gasoline because the welding burns the carbon out of the iron in the weld, leaving it so porous that " wild " gasoline under line pressures would penetrate every joint. In regard to this claim Mr. J. J. Allison, chief engineer of the Empire Gas & Fuel Co., says: " The strength and tightness of a weld depends entirely upon the welder; competent, skilled men can make welded joints that are as strong and tight as joints made with fittings, and the welded work is often better and less expensive."

SUMMARY.

USES OF TESTS.

In the foregoing pages the writer has described the absorption-plant operations in successful use at the many plants visited. Although many changes and improvements have been made since the first horizontal units were started, yet the industry has not reached a standard of units, for many experiments are being made at almost every plant to advance efficiency, to improve the product, and to lower costs.

Teaching plant foremen and shift men to use testing apparatus for both absorption efficiency and specific gravity is one of the first steps toward improvement. These operations are simple and easy; they hold the interest of the force, and also give a continuous record of the performance of the various units. Not only should both absorption and specific gravity tests be made of the incoming and treated gas, but also of the still vapors from the cooling coils and from the last unit, whether a compressor or another small absorption tower.

GAS TREATMENT.

QUALITY OF GAS.

Any gas containing the hydrocarbons that constitute the gasoline content can be treated by absorption for the recovery of these hydrocarbons. Rich casing-head gas and still vapor containing as much as 4 gallons of gasoline per 1,000 cubic feet are being successfully and efficiently treated, the gasoline content merely being a factor in the quantity of oil circulated. The process started, however, in the drying plants of companies pumping lean or "dry" gas which yields only a pint or less per 1,000 cubic feet and this fact led to the general impression that richer gases were not amenable to the process.

PRESSURE NECESSARY.

The pressures being used in the absorption process range from atmospheric to 500 pounds per square inch. Although it is true that gases containing all proportions of gasoline may be treated at all pressures, there are practical limitations of construction and size of units.

78

EFFECT OF TEMPERATURE.

Gas should always be cooled to as low a temperature as is practicable without an undue expense for an artificial cooling plant. The primary reason for cooling is that at low temperatures a given quantity of oil will absorb greater quantities of gasoline; moreover gasoline of a low boiling point will not be recovered by oil at temperatures higher than that point. Thus low temperature not only increases the total production but also permits a higher percentage of saturation of the oil thereby reducing the total oil circulated. Consequently the pumps, still, and other units have to do less work and fuel is saved. The oil sent to the towers should have a correspondingly low temperature for the same reasons.

EFFECT OF PRESSURE.

Although any pressure may be used in conjunction with the quantity of oil in towers, the pressure determines the actual volume of the gas, and this volume is a function of the velocity of flow through a tower or a given cross section. Thus the pressure controls the cross section or diameter of the towers to be used, since if rate of flow is too great, oil is carried over mechanically and the oil and gas are in contact too brief a time to obtain the desired extraction. Flow velocities over 75 feet per minute in the unbaffled or open part of the tower are too rapid, so that with such velocities the gas will usually carry oil with it from the tower.

The diameter of the towers used ranges from 18 inches to 12 feet and the pressures from a few ounces to 500 pounds, the low-pressure towers being the larger. For high pressure (100 pounds and over) the strength of pipe or metal necessary limits the diameter. The largest high-pressure tower seen in operation was 36 inches; the pressure used was less than 300 pounds.

HEIGHT OF TOWER.

The percentage of gasoline that is absorbed by the oil is, within the limit of total saturation, controlled by the duration and intimacy of the contact between the oil and gas and the temperature of the two. The duration is controlled by the velocity of the gas and by the height of the effective or baffled part of the tower. This fixes the height of the tower as a function of the saturation desired or possible. Thirty-foot towers are yielding saturations as high as 6 per cent, and aggressive operators are planning 40-foot towers in the hopes of increasing this figure to 10 per cent with the aid of more efficient baffling materials. If this increase can be had without hindering the complete recovery of gasoline, the saving of fuel will be well worth the cost and effort.

BAFFLING.

Many substances have and are being tried for baffling in towers, the object being to obtain material that will leave the greatest open space for the flow of gas and still break the gas into many fine streams that are continually divided and mixed, and in addition will expose a maximum surface of the absorbent to the gasoline vapors.

Wood gratings of many dimensions have been used in different plants and while the work done by them is satisfactory they constrict the tower section unduly and prevent the passage of the greatest volume of gas; that is, they cut down the capacity. Stone filling has the same effect but to a greater degree.

Chemical stoneware, blocks or tile, are being used successfully, but they also have a large bulk (50 per cent), which is absolutely inactive as far as concerns useful work. Hollow building tile are also used; they give satisfactory results but lessen the openings considerably. Steel cuttings are being tried, and, so far as the experiments have been carried, seem to give better promise than other material. One company is using 15 pounds of steel cuttings per cubic foot of tower, a porosity of 97 per cent. Another operator uses 30 pounds of the cuttings per cubic foot and has a porosity of 94 per cent. In addition to this high percentage of voids, the material offers an exceptionally large surface for exposing the oil to the gas.

The gas after passing out of the towers is taken through a trap, where any oil carried out of the tower with the gas is separated and run back to the weathering tanks.

OIL CIRCUIT.

COOLING COILS.

Oil entering the towers should be as cool as is economically possible. To obtain this temperature the cooling is done in towers under water sprays. An area of 3 square feet of radiating surface per gallon per minute is abundant, provided the pipe used is 2 inches in diameter and is placed in a well-designed and properly constructed cooling tower.

QUANTITY OF OIL CIRCULATED.

The quantity of oil circulated per 1,000 cubic feet of gas depends upon the gasoline content of the gas, the pressure, the temperature of both the oil and gas, and the vapor tension of the absorbing medium. Vapor tension is largely determined by temperature. As an interesting example of this fact, a company is using gasoline as the absorbent, tue vapor tension of the gasoline being reduced by the use of an ammonia cooling plant to the point that permits absorption. In this plant the product from the tower is a marketable motor fuel and, consequently, redistilling is unnecessary. It is questionable, however, whether equipment for ammonia cooling can be erected and operated at nearly as low a cost as a still, which, in this instance, it replaces.

Obviously, gases rich in gasoline vapors require large quantities of oil; and as the absorbing power of an oil increases with the pressure, the higher the pressure used the less oil need be sent through the tower.

Plants using pressures of 100 pounds and higher run the saturation as high as 6.5 per cent, and many operators believe that by increasing the height of the towers 50 to 100 per cent this figure may be increased to 10 per cent. In plants using lower pressures (atmospheric to 30 pounds) the saturation percentage can not be brought as high, 2 to 4 per cent being good practice under present conditions. The richness of the gas in gasoline vapor is of little moment as regards saturation percentages.

From the towers the oil is run through automatic traps to a weathering tank. In this tank, which is held at low pressures (atmospheric to 30 pounds) the dissolved gases are allowed to separate from the oil so that the following units do not have to treat this gas, a consideration of importance in large, high-pressure plants. The gas would also dilute the vapors from the still, causing a lower percentage of precipitation in the coils. These gases with the uncondensed vapors from the still are at many plants treated by compression or in a small absorption unit working under low pressure.

After half an hour to 2 hours weathering the oil is either pumped through the heat interchanger or allowed to flow through it under the pressure held on the weathering tank.

HEAT INTERCHANGERS.

Although many types of heat interchangers have been installed, the industry at present appears to have discarded all types except the jacketed line. Such interchangers are either made of two pipes, one within the other, each containing oil flowing in opposite directions, and connected in series; or of a large pipe with a number (five to eight) of smaller (1-inch to 2½-inch) pipes inside of it so arranged as to give a continuous counter flow of oil.

On its way to the still the oil is heated as high as 170° F. whereas the oil from the still is cooled to 130° F. The interchanger effects two savings, first in the boiler power for heating the oil, and second in the water pumped and the coil area for cooling the unsaturated oil going to the towers.

In coils of the type most used in radiating area of 0.25 to 0.35 square feet is allowed for each gallon per hour passing through the unit on its way to the still.

STILLS.

With the exception of experiments being made in baffling material for towers, no part of absorption plants is now receiving more attention than the still. The older plants all used the type steam still used in refineries, but this has proved inefficient in absorption practice. Such stills are of large (8 to 12 feet) diameter and 16 to 30 feet long. They are held one-third to two-thirds full of oil, which is all partly treated and continuously dilutes the freshly saturated entering oil. As it is the last 1 or 1½ per cent of gasoline contained in the oil that is difficult to separate and send out as a vapor, the continual dilution keeps the saturation of the oil in the still at the point where the vapors are most resistant; hence the distilling action requires excessive time as well as an undue amount of heat.

The pan type of still seems to overcome at least part of these objections, as the oil flows through in a small stream while being heated by the steam coils and being treated with live steam in the bottom of the still. The entire operation requires only a few minutes and the separation of the gasoline fractions is accomplished more easily and thoroughly.

One operator is trying a vacuum still to operate at low temperatures, while another is experimenting with a combination fire and steam still. The results are not complete but show promise of interesting results.

VAPOR TREATMENT.

In distilling an absorption oil with open steam coils (which are necessary if the gasoline is to be vaporized at the temperatures used) not only the gasoline vapors, but vapors of the heavier fractions are driven from the absorption oil itself and these with the uncondensed steam leave the still together.

It is now the usual practice to pass this mixture of vapors through some apparatus by which their temperature is reduced through a transfer of heat to the oil entering the still. This transfer of heat is obtained either by direct contact between the oil and vapors or with a heat exchanger of the preheater type, in which the oil and vapors are separated by the walls of tubing.

The object of this interchange is to precipitate as much as possible of the heavy oil and water vapors without condensing the gasoline vapors and also to recover a part of the heat from these vapors.

From an open contact, or stone tower, which seemingly is not very efficient, the vapor mixture has to be cooled further in water-cooled coils held at temperatures between 180° and 190° F. in order to remove completely the heavy absorption-oil fraction carried over. This "knock-out" coil has not been used in plants with a preheater type of unit at this point in the circuit. None of the absorption oil should be allowed to pass this point, as it would condense with and discolor the gasoline. A part of the water vapor will pass this point, but when this condenses with the gasoline it is easily separated by the gravity or the overflow method.

After this point, whether a water-cooled coil or an interchanger be used to precipitate the absorption oil, the gases (gas, gasoline, and water vapors) are further cooled in water-cooled coils.

FINAL COOLING COILS.

The vapors are often cooled in submerged coils of 2-inch, 4-inch, 6-inch, or 8-inch pipe, having an area of from 2 to 4 square feet' per gallon of gasoline per hour. As previously stated, this type of coil is rapidly being replaced by the spray-cooled coils in louver towers, which in most respects are far better. Spray-cooled coils are almost universally of the continuous return-bend type, made of 2-inch pipe, enough coils being taken out of a header to care for the volume of gas and vapor to be cooled. The total length of pipe or number of coils of given length is computed on a basis of 2 to 3 square feet of radiating area per gallon of gasoline condensed per hour.

The reduction in temperature should cool the vapor as much as the absorption oil and gas entering the towers are cooled; that is, as low as is possible with cooling by water sprays.

Gasoline and water are condensed to liquids in these coils and are separated together from the gas. Later the gasoline and water are separated by gravity in any one of the simple apparatus for this purpose or the water is drawn from the bottom of a "make" tank.

The gas from these coils carries some condensable gasoline vapors, and is further treated by compression or with a small (8 inches or 10 inches by 20 feet high) absorption tower. The writer favors the latter method. Oil for this tower is taken from the high-pressure oil line leading to the larger towers, and the saturated oil from the towers is sent directly to the still. Between 10 and 25 per cent of the total plant production of gasoline is usually recovered by this treatment. Some operators mix the weathering tank gas with the cooling-coil gases before this final treatment, but the value of this practice is doubtful.

THE PRODUCT.

Gasoline made by the absorption process is especially sweet, has a high gravity (70° to 85° B.), and as it comes from the still has a low (3 to 6 pounds) vapor tension.

Commanding through these properties a premium in the open market, absorption gasoline is a more profitable product than either raw or blended compression condensate. For this reason and because absorption plants cost less to install per 1,000 feet of capacity and are more efficient than any others in use, absorption plants seem likely to supersede compression and refrigeration plants for recovering gasoline from natural gas.

PUBLICATIONS ON NATURAL GAS AND INDUSTRIAL GASES.

A limited supply of the following publications of the Bureau of Mines has been printed and is available for free distribution until the edition is exhausted. Requests for all publications can not be granted, and to insure equitable distribution applicants are requested to limit their selection to publications that may be of especial interest to them. Requests for publications should be addressed to the Director Bureau of Mines.

The Bureau of Mines issues a list showing all its publications available for free distribution as well as those obtainable only from the Superintendent of Documents, Government Printing Office, on payment of the price of printing. Interested persons should apply to the Director Bureau of Mines for a copy of the latest list.

PUBLICATIONS AVAILABLE FOR FREE DISTRIBUTION.

BULLETIN 6. Coals available for the manufacture of illuminating gas, by A. H. White and Perry Barker, compiled and revised by H. M. Wilson. 1911. 77 pp., 4 pls., 12 figs.

BULLETIN 55. The commercial trend of the producer-gas power plant, by R. H. Fernald. 1913. 93 pp., 1 pl., 4 figs.

BULLETIN 89. Economic methods of utilizing western lignites, by E. J. Babcock. 1915. 74 pp., 5 pls., 5 figs.

TECHNICAL PAPER 9. The status of the gas producer and of the internal-combustion engine in the utilization of fuels, by R. H. Fernald. 1912. 42 pp., 6 figs.

TECHNICAL PAPER 38. Wastes in the production and utilization of natural gas, and methods for their prevention, by Ralph Arnold and F. G. Clapp. 1913. 29 pp.

TECHNICAL PAPER 106. Asphyxiation from blast-furnace gas, by F. H. Willcox. 1916. 79 pp., 8 pls., 11 figs.

TECHNICAL PAPER 112. The explosibility of acetylene, by G. A. Burrell and G. G. Oberfell. 1915. 15 pp.

TECHNICAL PAPER 131. The compressibility of natural gas at high pressures, by G. A. Burrell and I. W. Robertson. 1916. 11 pp., 2 figs.

TECHNICAL PAPER 158. Compressibility of natural gas and its constituents with analyses of natural gas from 31 cities in the United States, by G. A. Burrell and I. W. Robertson. 1917. 16 pp., 9 figs.

PUBLICATIONS THAT MAY BE OBTAINED ONLY THROUGH THE SUPERINTENDENT OF DOCUMENTS.

BULLETIN 4. Features of producer-gas power-plant development in Europe, by R. H. Fernald. 1910. 27 pp., 4 pls., 7 figs. 10 cents.

BULLETIN 7. Essential factors in the formation of producer gas, by J. K. Clement, L. H. Adams, and C. N. Haskins. 1911. 58 pp., 1 pl., 16 figs. 10 cents.

BULLETIN 9. Recent development of the producer-gas power plant in the United States, by R. H. Fernald. 1910. 82 pp., 2 pls. 15 cents.

BULLETIN 13. Résumé of producer-gas investigations, October 1, 1904, to June 30, 1910, by R. H. Fernald and C. D. Smith. 1911. 393 pp., 12 pls., 250 figs. 65 cents.

BULLETIN 19. Physical and chemical properties of the petroleums of the San Joaquin Valley, Cal., by I. C. Allen and W. A. Jacobs, with a chapter on analyses of natural gas from the southern California oil fields, by G. A. Burrell. 1911. 60 pp., 2 pls., 10 figs. 10 cents.

BULLETIN 31. Incidental problems in gas-producer tests, by R. H. Fernald, C. D. Smith, J. K. Clement, and H. A. Grine. 1911. 29 pp., 8 figs. 5 cents.

BULLETIN 42. The sampling and examination of mine gases and natural gas, by G. A. Burrell and F. M. Seibert. 1913. 116 pp., 2 pls., 23 figs. 20 cents.

BULLETIN 88. The condensation of gasoline from natural gas, by G. A. Burrell, F. M. Seibert, and G. G. Oberfell. 1915. 106 pp., 6 pls., 18 figs. 15 cents.

BULLETIN 109. Operating details of gas producers, by R. H. Fernald. 1916. 74 pp. 10 cents.

TECHNICAL PAPER 3. Specifications for the purchase of fuel oil for the Government, with directions for sampling oil and natural gas, by I. C. Allen. 1911. 13 pp. 5 cents.

TECHNICAL PAPER 10. Liquefied products of natural gas, their properties and uses, by I. C. Allen and G. A. Burrell. 1912. 23 pp. 5 cents.

TECHNICAL PAPER 20. The slagging type of gas producer, with a brief report of preliminary tests, by C. D. Smith. 1912. 14 pp., 1 pl. 5 cents.

TECHNICAL PAPER 54. Errors in gas analysis due to the assumption that the molecular volumes of all gases are alike, by G. A. Burrell and F. M. Seibert. 1913. 16 pp., 1 fig. 5 cents.

TECHNICAL PAPER 57. A preliminary report on the utilization of petroleum and natural gas in Wyoming, by W. R. Calvert, with a discussion of the suitability of natural gas for making gasoline, by G. A. Burrell, 1913. 23 pp. 5 cents.

TECHNICAL PAPER 104. Analysis of natural gas and illuminating gas by fractional distillation in a vacuum at low temperatures and pressures, by G. A. Burrell, F. M. Seibert, and I. W. Robertson. 1915. 41 pp., 7 figs. 5 cents.

TECHNICAL PAPER 109. Composition of the natural gas used in 25 cities, with a discussion of the properties of natural gas, by G. A. Burrell and G. G. Oberfell. 1915. 22 pp. 5 cents.

TECHNICAL PAPER 120. A bibliography of the chemistry of gas manufacture, by W. F. Rittman and M. C. Whittaker, compiled and arranged by M. S. Howard. 1915. 30 pp. 5 cents.

INDEX.

Bulletin 177   Petroleum Technology 51

DEPARTMENT OF THE INTERIOR
FRANKLIN K. LANE, Secretary
BUREAU OF MINES
VAN H. MANNING, Director

THE DECLINE AND ULTIMATE PRODUCTION OF OIL WELLS, WITH NOTES ON THE VALUATION OF OIL PROPERTIES

BY

CARL H. BEAL

WASHINGTON
GOVERNMENT PRINTING OFFICE
1919

The Bureau of Mines, in carrying out one of the provisions of its organic act—to disseminate information concerning investigations made—prints a limited free edition of each of its publications.

When this edition is exhausted, copies may be obtained at cost price only through the Superintendent of Documents, Government Printing Office, Washington, D. C.

The Superintendent of Documents *is not an official of the Bureau of Mines.* His is an entirely separate office and he should be addressed:

SUPERINTENDENT OF DOCUMENTS,

Government Printing Office,

Washington, D. C.

The general law under which publications are distributed prohibits the giving of more than one copy of a publication to one person. The price of this publication is 30 cents.

Persons desiring for lecture purposes the use, free of charge, of lantern slides of the illustrations in this publication, should make request of the Director of the Bureau of Mines, Washington, D. C.

Reprint, September, 1919.
First edition issued in April, 1919.

CONTENTS.

CONTENTS.

TABLES.

ILLUSTRATIONS.

PREFACE.

One of the problems that confronts petroleum producers and petro-
ium engineers is the estimation of the total amount of crude oil that
iay be obtained from oil lands. A producer with some means of
etermining, even approximately, the total amount of oil that a given
rea of ground will produce in the future, is able to estimate, within
irtain limits, the future gross receipts from his property—a bit of
ttremely valuable information. The engineer or geologist engaged
. determining the value of an oil property must know first approxi-
ately the amount of oil that can be expected from the area, and,
cond, the probable yearly rate at which it will be obtained under a
ecified program of drilling. Furthermore, the determination of
xable income involves an estimate of the total recoverable oil from
e lands belonging to oil companies with income enough to place
em in the taxable class, for the redemption of invested capital, by
owing yearly deductions from gross income, and is accomplished
deducting each year from gross income the same proportion of in-
sted capital as the yearly oil production bears to the amount ulti-
itely recoverable. Only by estimating the ultimate amount of oil
overable can invested capital be written off at the proper rate,
d that depends on the rate at which the total oil resource is de-
ted.

[n this bulletin Mr. Beal furnishes several new methods for
imating the future output of oil lands and gives curves and other
a that should assist oil producers and engineers in determining
probable amount of oil that oil properties in the different fields
:he United States will yield. The material is, therefore, presented
wo parts. The first part explains in considerable detail the meth-
that should be used in estimating the future production of oil
l the manner of applying those methods to the valuation of oil
ds; whereas the second gives in detail such information as was
ilable on the ultimate oil recovered in different fields and the usual
i at which the average oil well in each field will produce.

VAN H. MANNING,
Director.

3

THE DECLINE AND ULTIMATE PRODUCTION OF OIL WELLS, WITH NOTES ON THE VALUATION OF OIL PROPERTIES.

By Carl H. Beal.

GENERAL STATEMENT.

INCREASING PRODUCTION OF OIL IN THE UNITED STATES.

The oil industry in the United States is further advanced than in any other country, because of American initiative and the development of industries dependent in some way on petroleum or its products. For this reason the output has constantly increased (fig. 1), and as a result this country has produced more than half of the total output of the world. The total past output of the world is approximately 7,000,000,000 barrels of 42 gallons each; of this the United States produced about 4,000,000,000 barrels, or about 57 per cent. The limit of production in this country is being approached, however, and although new fields undoubtedly await discovery, the yearly output must inevitably decline, because the maintenance of a given output each year necessitates the drilling of an increasing number of wells. Such an increase becomes impossible after a certain point is reached, not only because of a lack of acreage to be drilled, but because of the great number of wells that will ultimately have to be drilled. According to figure 1, the daily production per well each year has increased during the last few years. However, this increase is abnormal, being caused by the new pools brought in. Although such a condition may continue for several years, the average production per well will finally begin to decrease on account of the lack of new pools to make up for the normal decline in production of the old ones.

WORK OF THE BUREAU OF MINES.

At present the country is facing a serious shortage of petroleum. By way of preparing for the best way to meet this emergency, the Bureau of Mines has been carrying on technical investigations in the petroleum industry for several years. The results of these investigations have been published in technical papers and bulletins that have discussed many such topics as methods of drilling, more efficient recovery of oil, exclusion of water from oil wells, storage and protection of oil in reservoirs, manufacture of gasoline from natural gas, and utilization of petroleum fuels. In addition, these publications have covered many chemical problems. Little information has been published, however, and very few investigations have been

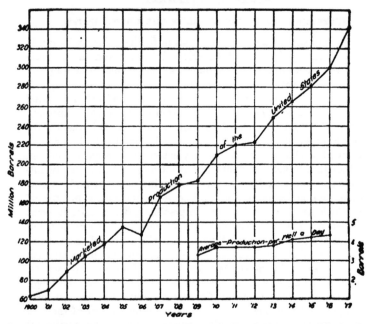

Figure 1.—Chart showing annual oil production of the United States since 1900 and the average daily production per well for each year since 1909 of the oil fields of the United States. (Statistics from the United States Geological Survey.)

made dealing with the ultimate amount of oil that a well or a property can be expected to produce under certain conditions or the rate at which the oil can be obtained.

INFORMATION NEEDED BY PRODUCERS.

Formerly the producer did not deem such information essential to the operation of his oil lands with most success, for until the last few years he has not found it necessary to work on as narrow a margin of profit as is common in other kinds of business. Undrilled prospective acreage has been plentiful, so the producer directed his attention mainly, and with considerable success, to the development of new pools. During the last few years, however, the oil industry has been forced to adopt a more conservative basis. Favorable territory has become scarcer, competition has increased, and the demand for petroleum and its products has created a market that can not be adequately supplied. As a result, small holdings have been consolidated; men of broader vision and greater ability have directed the proper investment of capital; advantage has been taken more generally of technical knowledge that tends to reduce the risk involved in the search for oil, such as the application of geological principles; and business methods have been applied to the industry as its proper development has demanded.

These changes have resulted in the introduction of conservative methods and a growing demand during the last few years for more complete knowledge on the probable future production of oil wells and of oil lands. How much oil will they produce in the future? At what rate will the oil be obtained? These questions occur to oil operators rather often nowadays. If such queries could be answered with even approximate exactness, the producer would be able to obtain a fairly trustworthy idea of his future yearly income and the worth of his property, as well as that of other oil lands he may contemplate purchasing.

Such information is valuable not only to oil operators, but also to engineers engaged in the valuation of oil lands. Unfortunately, those technical men engaged in valuing oil property have access to data, needed in their work, covering small areas only. To determine accurately the weight to be given certain factors controlling the production of oil, the valuer should possess not only a broad knowledge of oil-field conditions, but should also have access to much statistical information that heretofore has been inaccessible to him because of its being in the hands of competing companies. Undoubtedly such information can best be collected, analyzed, and published through the medium of the Federal Government.

INFORMATION NEEDED BY PETROLEUM ENGINEERS.

The training of a large proportion of petroleum engineers has not provided them with a knowledge of the fundamental principles of valuation. This lack of knowledge is not surprising in view of the newness of such work and the recent requirements of the oil industry; but is none the less to be deplored as valuation obviously should be done by the petroleum engineer, because his familiarity with the origin and accumulation of oil and gas and their relation to geologic structure is prerequisite for the scientific valuation of producing and prospective oil lands. Probably, as fewer and fewer large oil fields are discovered, the oil industry will seek the aid of science more and more, and the technical man will find his services needed in every branch of the industry, from the selection of probable new territory and the determination of the value of producing and prospective oil lands to the utilization of the refined products. For these reasons the writer believes that the petroleum engineer should strive to meet the probable requirements of this specialized branch of petroleum technology.

SCOPE OF THE INFORMATION PRESENTED.

The information in the following pages is offered with the knowledge that some of the conclusions are tentative and have not been proved, but such presentation is necessary in supplying data on a subject of which so little is known. Moreover, some of the methods set forth are not proposed for final adoption by other persons interested in the estimation of future production, but are given with a twofold purpose: (1) To supply such information on the estimation of future production as heretofore the engineer or geologist in commercial work could not obtain, and (2) to stimulate discussion and thought on the subject, for it is only by the painstaking accumulation and publication of such data and the discussion of any new methods evolved from new data that there can be permanent advance. The author, therefore, expects criticism of some of the methods described. Mistakes also may be found, for the subject is new and no precedent was available for compilation of the data. However, the author welcomes any discussion or criticism that will result in advancing the practice of oil-land valuation. Some of the material has been repeated for the sake of clarity.

The scientific appraisal of oil lands is practically new work; surprisingly little has been published on it, and on the methods of estimating the future production of oil wells. But the proper valuation of oil lands must be based on trustworthy estimates of future production, and it is with methods of making such estimates and with the practical application of these methods to appraisals, that this paper is chiefly concerned. Therefore many seemingly elementary features are discussed in detail. If the publication of this bulletin results in stimulating others to increased effort or in leading them to amplify the basic principles presented, the author will feel that a start has been made in the proper direction.

PURPOSE OF THE REPORT.

The present investigation was undertaken for the purpose of supplying some of the material so badly needed in the valuation and more efficient operation of oil properties. Because of the lack of time for further studies, the amount of data already accumulated and the policy of the Bureau of Mines of publishing as soon as possible any information that may be of use to an industry, the bulletin is issued now. As all the fields of the United States have not been studied in detail, it has been deemed wise to publish the information gathered and later to collect and publish more detailed information on the fields not yet studied.

As already stated, the information given in this bulletin has been roughly divided into two parts, as follows: •

Part 1. Methods for estimating the future and ultimate production of oil properties and the application of these methods to oil land valuation.

Part 2. The decline and ultimate production of different oil fields in the United States, including curves showing the actual rate of decline and the ultimate production of various properties and fields; discussions of the conditions existing in the fields represented; and the influence of these different conditions on the rate of production and the amount obtained.

Curves in considerable number are presented; some of them are based on scanty data. It was thought advisable to publish practically all the available information that would be of use because of its being valuable even though incomplete. Many statistics on ultimate and total past production per acre appear in part 2 of the text. Some of these statistics will be of little use to many readers, but of great use to a few. Moreover, the publishing of such detailed information will give other investigators the benefit of practically everything that has been available to the author, and may thus aid them in the solution of other and more complex problems.

COLLECTION AND COMPILATION OF THE DATA.

The collection of data was begun in the Oklahoma fields during the first part of 1916 while the writer was Federal oil and gas inspector for the Five Civilized Tribes. The other chief fields of the country were visited and information regarding production was collected during the next 18 months. The report was put in final form for publication during the spring and summer of 1918.

The information given represents the results of more than a year's work in compiling figures transferred from the books of various oil companies and in making the thousands of calculations necessary for preparing the charts and statistics that cover the annual yield of about 20,000 wells representing, as a rule, the most typical examples of production decline.

More information was available in some fields than in others; fur-thermore, the need for detailed study was more pressing in some fields. The Oklahoma fields were studied in greater detail than any of the others, information of varying fullness having been obtained from the Nowata-Chelsea district in northern Oklahoma, the Bartlesville field, the Osage Indian Reservation, the Glenn pool, the Cushing pool, the Ponca City pool, and the Healdton field.

Several production records of properties in the shallow oil fields of southeastern Kansas were available for study. The Augusta and El Dorado fields, in Butler and Montgomery Counties, at present furnish the bulk of the Kansas production, but it was not possible to obtain any records of production long enough to be of much use in the present work.

The fields in northern Texas and Louisiana and the salt-dome fields of the Gulf coast were studied in some detail. Present knowledge of the underground geology of the Illinois field makes possible the obtaining of data that will probably be of much use to operators and technical men in that State.

Little information was collected regarding the Lima-Indiana field in northwestern Ohio and Indiana and the Appalachian field, which extends over parts of New York, Pennsylvania, West Virginia, Ohio, and Kentucky. The productive areas of the Appalachian fields are of such extent, the oil and gas occur under such a wide range of conditions, and the production is from so many different sands that it was not deemed advisable to spend as much time in this field as in some of the younger fields. The data can be collected in more detail later and be presented in other publications.

The California fields were not studied in the detail they deserved, because of the variable conditions under which oil is produced and also because of a lack of time enough for collecting and compiling the wealth of information available in that State. Already some work of this kind has been done in California. In 1915 the appraisal committee of the Independent Oil Producers' Agency compiled composite decline curves for several of the most important California fields. These curves were published by Requa [a] and some of them are reproduced in this bulletin for the purpose of comparison with similar curves constructed by the author.

In collecting the material, statistics on production were usually deemed of first importance. Other information, such as the relative location of the productive oil wells, their depth, the thickness of the productive sand, the initial production, the viscosity of the oil, and the closed pressures of the oil wells, was included when available. It was rarely possible to obtain production records of individual wells except in California. In most of the other fields the records

---

[a] Requa, M. L., Methods of valuing oil lands: Am. Inst. Min. Eng., Bull. 134, February, 1918, pp. 409–428.

of production are and have been kept only by properties; that is, the production of all the wells on each property was combined.

ACKNOWLEDGMENTS.

The author gratefully acknowledges his indebtedness to W. A. Williams, former chief of the petroleum division of the Bureau of Mines, under whose direction the work resulting in the publication of this bulletin was done; to Chester Naramore, successor to Mr. Williams as chief of the petroleum division, who has offered many valuable suggestions; and to A. W. Ambrose, of the bureau, for suggestions and assistance in compiling the material in the office. He also acknowledges the aid rendered by several of the different State geologists in the United States, as well as by Dr. Inouyi, Director of the Japanese Geological Survey, who kindly furnished data on the Japanese oil fields. J. O. Lewis, who has done considerable work on the same subject, has courteously made many suggestions; in fact many of the ideas presented are the direct outgrowth of suggestions he has made. The manuscript was carefully read and numerous valuable suggestions were made by Roy E. Collom, Clarence G. Smith, E. D. Nolan, J. O. Lewis, R. B. Moran, and R. V. Mills. Acknowledgments are also due to scores of oil companies who have furnished information and allowed the author access to valuable production data, without which this could not have been prepared. The author desires to express his indebtedness especially to the various officials of these companies: In California, the Southern Pacific Co., the Standard Oil Co., the General Petroleum Corporation, the Santa Fe, and the Shell Co. of California; in Wyoming, the Midwest Oil Co.; in Oklahoma and Kansas, the Prairie Oil & Gas Co. and the Gypsy Oil Co.; in Texas, the Producers' Oil Co. and the Gulf Production Co.; in the eastern fields, the Chartiers Oil Co., the Ohio Oil Co., the National Transit Co., the Pure Oil Co., the Ohio Fuel Supply Co., the Ohio Cities Gas Co., and Brundred Bros., of Oil City, Pa. Many other companies have furnished valuable information. A. R. Elliott and J. G. Shumate, of the Bureau of Mines, drew the charts and figures.

DEFINITIONS.

In this bulletin several new terms have been used; these as well as some old terms that may be used in more than one sense are defined below:

Total production is the total amount of oil produced in the past by a well, property, or field.

Future production is the amount of oil that will be produced in the future by a well, property, or field.

Ultimate production (or the recoverable oil) is the amount of oil that a well, property, or field will ultimately produce. The ultimate production is therefore the sum of the past, or total production, and the future production.

The decline of a well is the decline or falling off in the production of a well.

The decline of a property means the falling off or reduction in yield based upon the average amount of oil each well makes. The actual daily production of the property may be increasing while the actual daily production per well may be decreasing. This decrease is the decline of a property.

A per cent decline curve is one showing the decline in the production of a well, property. or field, each year's production being expressed as percentages of the first year's production, which is taken as 100 per cent. For example, a well may produce 10,000 barrels the first year. 5,000 barrels the second year, and 3,000 barrels the third year. The percentages of this property for the first, second, and third years are successively 100, 50, and 30 per cent.

The cumulative percentage of a property is the sum of the percentages for all years. In the preceding example the cumulative percentage is 180.

The ultimate cumulative percentage is the sum of the past and the estimated future percentages; it is expressed as a percentage of the first year's production and may be further defined as the cumulative percentage of the property at the time of its exhaustion.

The expectation of a property or well may be defined as the amount of oil that is expected from a property or well—the future production.

Total production per acre is the average amount of oil produced in the past from each acre of a drilled tract of land.

Ultimate production per acre is the average amount of oil produced ultimately from each acre of a drilled tract of land.

Production per acre-foot is the total or ultimate production per acre divided by the average thickness of the productive sand.

Initial production is used in this bulletin in two different senses: (1) The production of a new well for the first 24 hours; and (2) the average daily production of a new well during the first year. Where the term is used in the latter sense attention is called to the fact.

Decline of initial yearly production is the decline in the daily production for the first year of wells drilled on a property for several consecutive years. For instance, if the average daily production of new wells drilled on a property for the years 1907, 1908, and 1909 were successively 50, 30, and 20 barrels, these figures are the decline, or decrease, of initial yearly production.

A composite decline curve is a curve showing the average per cent decline of many wells or properties.

An appraisal curve is a series of curves used to determine the amount of oil that under certain conditions will be produced in the future by the properties in the area for which the curve has been made.

Oil content is the amount of oil contained in a given porous reservoir or oil sand.

Recoverable oil is the amount of oil that ultimately may be recovered with profit from such a porous reservoir. It is necessarily a relative term, as the amount of oil taken from a field (the recoverable oil) varies with the price of oil, and with several other factors. More oil will be recovered when the price is high than when it is low. A distinction should be drawn between oil naturally recoverable and the extra amount possibly recoverable by suction, compressed air, and other stimulative processes.

Interference between wells signifies the interfering drainage areas of adjacent wells. In general, as a well becomes older its drainage area extends—the oil coming from greater distances—and often reaches and interferes with the drainage area of another near-by well.

PART 1.—METHODS FOR ESTIMATING THE FUTURE AND UL-TIMATE PRODUCTION OF OIL PROPERTIES AND THE APPLICATION OF THESE METHODS TO OIL-LAND VALUATION.

GENERAL STATEMENT.

Many workers in the petroleum industry do not realize that the most valuable and useful knowledge an oil operator can possess with reference to his property is (1) the ultimate amount of oil the property will probably produce, and (2) the rate at which the oil will be obtained under a specified drilling program. Generally, the pumping costs may be closely computed and the operator may be able to tell within narrow limits the outlay of capital required to drill a well to a given depth. Such data are, of course, invaluable, but if he knew, even within rather wide limits, the probable amount of oil he would be able to sell from his lease each year, the advantages would be manifold. He would be able to determine with fair accuracy the sale value of his property or the amount he could afford to pay for another, to calculate the proper rate of capital redemption, to estimate the proper rate of drilling to insure the greatest return in income, and to compute his probable future annual income.

Oil properties, however, differ so much in character, and the conditions affecting the recovery of oil vary so greatly, that it is impossible to lay down an absolute rule that can be applied with certainty to estimating the future production of all properties alike. However, by studying the conditions under which oil is being and has been produced and the factors that have governed the amounts properties have produced under certain conditions, as well as the rate at which oil has been obtained under those conditions, the estimation of future output is by no means so hopeless as at first it may seem. In fact, if data enough are collected, close estimates of the future production of oil properties can ordinarily be made.

PREVIOUS LITERATURE.

Publications dealing with estimating future production or with valuing oil land are few, because practically no intensive studies have been made. Such textbooks as those by Johnson and Huntley.[a]

---

[a] Johnson, R. H., and Huntley, L. G., Principles of oil and gas production, 1916. 871 pp.

14

Thompson,[a] and Bacon and Hamor[b] contain some valuable information on methods of estimating future output, and Arnold[c] gives a theoretical curve of final decrease. Washburne[d] gave a method of determining oil content by what is known as the "saturation method."

More recent publications are by Pack,[e] who gives an excellent résumé of previous methods, and by Lewis and the present author,[f] who present some new methods for making estimates.

One of the earlier papers on the valuation of oil lands was presented in 1915 by Lombardi.[g]

Hager[h] offers examples of the method of determining the amount that an operator could afford to pay for a property under certain assumed conditions. Requa,[i] in 1912, gave some information on this subject, and a more recent paper[j] by the same author presents the method adopted by the appraisal committee of the Independent Oil Producers' Agency of California for determining the future production and value of several properties in that State.

A bibliography containing the titles of several other papers dealing directly or indirectly with these subjects is given at the end of this bulletin.

REVIEW OF PREVIOUS METHODS.

Three general methods are commonly employed for the estimation of future output. These are (1) the saturation method, based on a calculation of the oil content of the productive sand, (2) the production-curve method, which consists of determining from the decline in production of a well in the past the amount of oil that probably will be produced in the future, and (3) the production per acre method, which estimates the future output by comparing actual recoveries per acre from similar properties in the same district or in one where the conditions are comparable.

[a] Thompson, A. B., Petroleum mining and oil-field development, 1910. 862 pp.

[b] Bacon, R. F., and Hamor, W. A., The American petroleum industry, 1916. 963 pp.

[c] Arnold, Ralph, The petroleum resources of the United States: Econ. Geol., December, 1915, pp. 695–712.

[d] Washburne, C. W., The estimation of oil reserves: Am. Inst. Min. Eng., Bull. 98, February, 1915, pp. 469–471.

[e] Pack, R. W., The estimation of petroleum reserves: Am. Inst. Min. Eng., Bull. 128, August, 1917, pp. 1121–1134.

[f] Lewis, J. O., and Beal, C. H., Some new methods for estimating the future production of oil wells: Am Inst. Min. Eng., Bull. 134, February, 1918, pp. 477–504.

[g] Lombardi, M. E., The valuation of oil lands and properties: International Eng. Cong., San Francisco, September, 1915. The same paper with a few changes was published in Western Eng., vol. 6, October, 1915, pp. 153–159.

[h] Hager, Dorsey, Valuation of oil properties: Eng. and Min. Jour., vol. 101, May 27, 1916, pp. 930–932.

[i] Requa, M. L., Present conditions in the California oil fields: Am. Inst. Min. Eng., Bull. 64, April, 1912, pp. 377–386.

[j] Requa, M. L., Method of valuing oil lands: Am. Inst. Min. Eng., Bull. 134, February, 1918, pp. 409–428.

SATURATION METHOD.

BASIS OF METHOD.

The saturation method is based on several factors the values of which are uncertain, these factors being the porosity, the thickness, the extent, and the saturation of the oil sand. From these the percentage of oil that may be recovered is estimated. The capacity of an oil sand of uniform thickness and porosity can be determined with fair accuracy from samples of the sand from different wells, especially if the wells are scattered over the area and several determinations are made from a number of representative samples. The most difficult determination in this method is the proportion of the total oil content that may be recovered from the sand. This is closely related to the error resulting from estimating the saturation of the oil sand.

OIL CONTENT OF SAND AND OIL RECOVERED.

Usually more oil is left underground than is brought to the surface; in fact, only a small percentage of the total oil content of a sand is ordinarily recovered. The so-called recovery factor varies widely according to the conditions controlling the production of oil and gas. Hence an arbitrary value nearly always has to be assigned to the recovery factor. As the accuracy of the whole procedure is thereby greatly reduced, the estimate becomes little more than a guess.

The difference between oil content and the amount of oil that may be recovered (the ultimate production) must be kept carefully in mind. The factors that govern the amount of oil present in a sand of course control the amount of oil that can be recovered from it. But other factors must be studied. The recoverable oil, or the ultimate production of a sand underlying an area, is the quantity that may actually be taken from the sand rather than the amount present in it. This recoverable oil is a percentage of the total oil content and varies with the conditions of occurrence of the oil and the conditions of production. Unquestionably a much larger percentage of the oil content can be recovered from a coarse porous sandstone subjected to a high gas pressure than from a fine-grained denser sandstone under no great pressure. The coarse porous sand offers little resistance to the flow of the oil toward the well, whereas, with a fine sand and a low gas pressure, the production is retarded, both by the greater frictional resistance and the lack of expulsive force. Undoubtedly the recovery factor varies with the conditions between these two extremes.

VARIABILITY OF RECOVERY FACTOR.

It should be borne in mind that the recovery factor is not the same for all pools nor even the same for all properties in a pool. The factor may differ for different parts of a single property, as it depends on the conditions that influence production and may be as variable as they. For instance, the recovery factor of a small area on the crest of a dome may be much greater than that of another

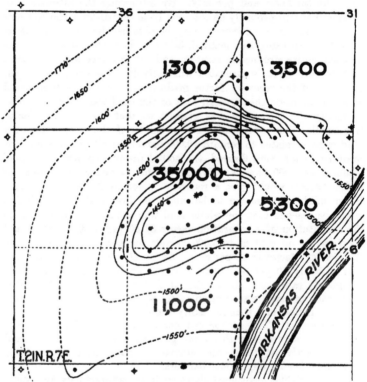

FIGURE 2.—Sketch showing the relation between total production per acre (in barrels) and geologic structure in the Boston pool, Okla.

area a few hundred feet nearer the edge of the pool. As much oil is contained in the sand in one place as in the other, but the conditions governing its expulsion are less favorable in the second tract than in the first, because of the smaller proportion of dissolved or compressed gas. Figure 2 shows this relation excellently; just as much oil was originally contained in the sand on the edge of the pool, but the ultimate production of those areas is much smaller. In the figure the dots are oil wells; the dots with four points are dry holes; the circles with four points are abandoned wells; the stars are gas wells.

ESTIMATION OF RECOVERY FACTOR.

The recovery factor may be estimated in several different ways. One method is by computing the oil content of the productive sand from the thickness and porosity. By estimating the ultimate production that will probably be obtained from this district, the recovery factor, or percentage recovery, may be obtained by dividing one by the other. In using this method, however, it is necessary to assume that the entire productive stratum and all the pore space, or some definite part of it, is saturated with petroleum or compressed gas; estimates of the recovery factor made in this way have been based on the assumption that a small part of the sand was saturated when in reality the saturation was greater, so that the recovery factor was smaller than estimated. For example, if the pore space of a sand were estimated to contain 10,000 barrels of oil per acre and the recovery factor were desired, the actual production per acre (here assumed as 2,000 barrels) divided by the estimated oil content of 10,000 barrels would give a recovery factor of 20 per cent. This factor might be assumed to be much too low, and the conclusion might be reached that the whole sand was not saturated, leading to the substitution of a saturation factor of 50 per cent, which would decrease the oil content to 5,000 barrels per acre and thereby raise the recovery factor to 40 per cent. As a matter of fact, the recovery was probably 20 per cent instead of 40 per cent, and the sand was completely saturated. It is believed that a similar error has often been made in using a saturation factor to determine oil content or a recovery factor, and that in reality the recovery factor was smaller than estimated. The author can not conceive a uniformly porous sand partly saturated with oil and gas under pressure. If the sand be only partly saturated, the saturated portion must be separated from the other by impervious barriers.

If the variables mentioned above could be satisfactorily and easily determined, this method of estimating the recovery factor would be easy to use and therefore of great value, as a single well on a property would make available some of the information on which future production could be based.

Lewis[a] gives other methods of estimating the recovery factor, as well as estimates made by various persons of the amount of oil left underground, these estimates ranging from 25 to 90 per cent; he also presents statistics that lead him to believe that only 10 to 20 per cent of the oil is ordinarily recovered. This is much lower than the usual estimate of approximately 50 per cent.

---

[a] Lewis, J. O., Oil-recovery methods: Bull. 148, Bureau of Mines, 1917, pp. 25–32.

DISADVANTAGES OF SATURATION METHOD.

The disadvantages of the method may be summed up as follows: (1) Impossibility of obtaining information as to the exact thickness of the "pay"; (2) difficulty of accurately determining the voidage, because of differences in porosity; (3) practical impossibility of determining the percentage of oil recovered from the sand, as the proportion varies with temperature, pressure, character of the oil, character of the sand, relation of edge-water to oil, encroachment of water, the space in the reservoir occupied by gas, and the underground migration of oil.

The author believes that the value of the saturation method has been overestimated, and is not willing to concede that it can be applied under ordinary circumstances without the making of hazardous assumptions. Hence it should be used only until data are available for the use of other methods. Conceivably the method might be used to obtain an approximate idea of the probable productivity of a newly discovered reservoir under ideal conditions, such as the accumulation of oil being controlled by a definite anticline or dome and the sand being probably of uniform thickness and porosity. However, such a combination of ideal conditions is seldom encountered. Furthermore, the existence of such conditions can not be known until several wells are drilled, and then information enough is available to allow the use of a more trustworthy method.

PRODUCTION-CURVE METHOD.

BASIS OF METHOD.

The production-curve method is based on the recorded production of the wells themselves; in other words, future recovery is based on past yield, and as the record of the actual output of a well is an index to the quantity of recoverable oil rather than to the total oil content, a determination of the recovery factor is unnecessary. In using this method the output of the well is plotted for time periods for the term of production, and to estimate the future output the curve thus determined is projected from the terminus to the point representing the minimum production at which the well can be pumped with profit. This method is most valuable for estimating the future output of individual wells or a group of wells on a property, but often the production records of individual wells are not available and it is necessary to construct a curve showing the decline in production of a property.

In this event it is a better practice to construct the decline curve of each property by computing and plotting the average daily output per well (total property production divided by the number of wells producing) for each time period. If all the wells were completed at the same time such a curve would show the decline in production of the property as accurately as the decline curve of one well would show its production. But all the wells on a property are rarely drilled at once, and as a result the average daily production per well may increase until the later wells can no longer offset the combined decline of the older wells. This time may be a year or two years after the initial well is drilled, but as a general rule it has been found that curves showing the average decline of wells on different properties in pools like those in the Mid-Continent field are little affected by slow or rapid rates of development (pp. 21-24).

MERITS OF METHOD.

The author believes that the estimating of future output, especially the future output of small areas, by the production-curve method, and by certain variations of it to be given later, is the most practical method. One of the great disadvantages of the method is the impossibility of constructing a decline curve for a property until the property has produced for some time. This defect, however, may be offset, if data enough are at hand, by applying the decline curve of one district or property to the probable production of another similar district or property. It is believed that if data enough are gathered and analyzed, the application of typical production curves of properties or wells producing under certain specified conditions can be applied with considerable certainty to other properties where approximately the same conditions will influence output.

PRODUCTION-PER-ACRE METHOD.

MANNER OF USE.

The production-per-acre method is used considerably with the production-curve method to estimate future and ultimate production. It consists in reducing the actual output of exhausted properties, or of properties so nearly exhausted that their ultimate production can be estimated with fair accuracy, to the amount of oil produced per acre, and in applying the values to other similar properties from which approximately the same ultimate production may reasonably be expected. Estimating ultimate production per acre from the recoveries obtained from similar older properties is possible, of course, only where records of past output are available; yet it is believed that with such information this method should be used much more than it has been, because usually most of the production from an acre of ground is obtained during the first two or three years and the proportion of the total obtained after that time is small compared with that already recovered. Hence fairly accurate estimates can be made of the ultimate production of an acre of ground after it has produced two or three years, and these data can then be applied to other districts where the conditions affecting output are similar and the properties are operated in a like manner.

FACTORS GOVERNING THE DECLINE OF OIL WELLS.

GENERAL STATEMENT.

Many factors influence the decline of oil wells. The effect of some is great and should be given due consideration; that of others is small and may be neglected. None of the factors will be discussed at length in this report, except the rate of development, which often is overlooked and sometimes is greatly overestimated, especially in the Mid-Continent field.

RATE OF DRILLING.

To determine the influence of the rate of drilling on the decline of a group of wells. several properties in the Glenn pool (Okla.) and in the Blue Creek field (W. Va.) were studied. In this study those properties in the Glenn pool that were more than one-half drilled the first year were separated from those that were less than one-half drilled in that time. The average daily production per well for each year for each property was determined for the two classes. They were averaged together and the decline plotted.

Figure 3 shows these curves. In this figure the solid line is the decline of those properties that were more than one-half drilled the first year; the dashed line is the decline of those properties that were less than one-half drilled the first year or, in other words, those on which the wells were drilled gradually through several years; and the dotted curve represents the composite decline of the properties entirely drilled the first year. The decline of these was considerably slower because of the wells being smaller. The figures along the three curves indicate the number of properties used in constructing the curves.

It will be observed that the decline of the properties developed gradually, was less during the first three years, and greater thereafter than that of the properties that were more than one-half drilled. An analysis of two concrete examples follows.

DECLINE OF SLOWLY DRILLED PROPERTIES.

Select one property, comprising 160 acres, from those that were gradually developed; assume that only 5 equidistant wells were drilled the first year; then approximately 30 acres would contribute to each

well, and for that reason the wells would " hold up " unusually well
and show a high average yield for the first year. Next, assume that
5 more wells were drilled the year following; then the area allowed

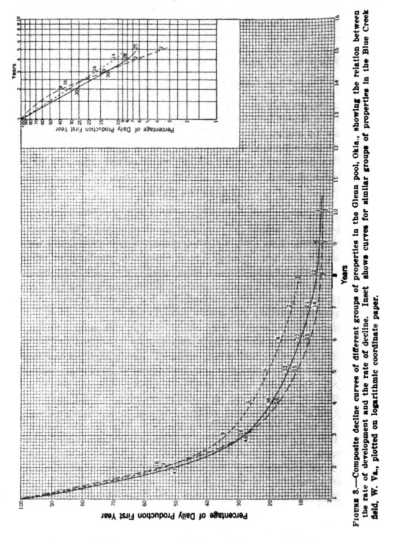

FIGURE 3.—Composite decline curves of different groups of properties in the Glenn pool, Okla., showing the relation between the rate of development and the rate of decline. Inset shows curves for similar groups of properties in the Blue Creek field, W. Va., plotted on logarithmic coordinate paper.

for each well, if all were practically equidistant, would be 16 acres.
As this acreage is much more than the average for the Glenn pool,
during the second year also the average daily production per well

would be larger than if the wells were more closely spaced. If during the next year 10 more wells were drilled, there would be a total of 20 wells producing on the 160 acres, an average of 8 acres per well. Thus the contributory area would be cut down more, and the new wells being adjacent to wells that have produced one or two years would stand little chance of being in areas unaffected by drainage. These new wells must be considered, however, in computing the average daily production per well for the third year and for succeeding years. If it were possible to obtain the production data for each of the wells it would be seen that the output of the first wells would decline slowly, but as soon as the wells drilled during the third year started producing the output would fall off sharply. The new wells would have a smaller initial production and in addition would probably decline more rapidly than if they had not been drilled into a partly drained sand in which the expulsive force of the gas had become much less. With heavy oil, tight sand, and wider spacing of the wells, the decline of the later wells would probably not be so rapid, but if the communication between all the wells on the property were easy, the interference would extend rapidly.

DECLINE OF RAPIDLY DRILLED PROPERTIES.

Consider a property of 160 acres in that class of properties that have been more than half drilled the first year. Suppose 16 wells were drilled the first year, giving 10 acres per well. Drilling 4 more wells the second year would reduce the area per well to 8 acres. The first 16 wells would show a much smaller average daily production for the first year than the wells drilled the first year in the slowly developed tracts, because of there being only 10 acres contributing to each well instead of about 30 acres, and because of interference starting and cutting down the average production per well. During the second year the effect of the new wells would be proportionately much less. On such a property, as compared with the other, there would be a smaller average production of the wells during the first year, but also a lower rate of decrease during the second year.

The same reasoning applies to the decline of properties completely drilled the first year.

STUDY OF PROPERTIES IN THE BLUE CREEK FIELD (W. VA.).

Fifty properties in the Blue Creek field (W. Va.) were studied in the same manner. The inset on figure 3 shows the curves, plotted on logarithmic coordinate paper, from the records of the three different classes of properties—that is, those fully drilled the first year, those more than one-half drilled the first year, and those one-half or less than one-half drilled the first year. The relationship of the curves is the same as that of the curves for the Glenn pool (Okla.).

One interesting similarity is that the curves representing the decline of slowly and rapidly drilled properties in both districts cross each other the third year.

METHOD OF SHOWING THE DECLINE OF OIL WELLS.

GENERAL STATEMENT.

The method adopted by the author for showing graphically the rate at which oil is obtained is not new and has been in rather general use among geologists, especially in the California oil fields, for several years. Briefly, the method consists in showing each year's production as a percentage of the first year's production; that is, where the production record of a whole property is involved, the average daily production per well for the property for each year is computed and the first year's average called 100 per cent. Next, the average daily output for each succeeding year is shown as a percentage of the first year's output and these percentages are plotted, the first year being called 100 per cent. For example, if the average daily production per well on a property the first year was 1,000 barrels, the second year 500 barrels, the third year 300 barrels, the decline of the property would be expressed in percentages by plotting the first year as 100 per cent, the second year as 50 per cent, the third year as 30 per cent, and so on to the exhaustion of the property. In the present bulletin the author has endeavored to show the average rate of decline for a whole field by averaging the decline curves of as many representative properties as possible. For some fields several hundred properties, involving probably 4,000 or 5,000 wells, were used to construct these composite decline curves.

EFFECT OF DIFFERENCES IN INITIAL OUTPUT.

One objection to the use of composite curves is that the decline curves of wells of different initial yearly production are employed in determining the average. In general, wells of small initial yearly production decline more slowly than those of large; hence, if the composite decline curve is used in estimating the future production of a well of larger output than the average the estimate will considerably exceed the actual production, and the estimates of the future production of small wells will be unduly low.

Composite or average curves, regardless of this fault, have much value, however, and can be used to good advantage if due allowances are made for the possibilities of error. Often, because of lack of in-estimate.

To show further the fundamental error in promiscuously using such curves without making allowances for differences in the rate of decline of large and small wells, figure 4 has been prepared. It shows the rates of decline of properties in the Bartlesville pool (Okla.), on which the average daily production per well the first year was different. In constructing these curves, all properties were separated into six classes according to the average daily output per well during the first year. These six classes were those in which the wells averaged the first year zero to 10 barrels, 11 to 20 barrels, 21 to 30 barrels, 31 to 40 barrels, 41 to 50 barrels, and in excess of 50 barrels a day. The first year's daily production for each property was called 100 per cent, and the production for succeeding years for each property was shown as a percentage of the first year's production. The average yearly percentages for each class thus obtained were plotted on logarithmic coordinate paper, because with this kind of paper the curves came out more nearly as straight lines.

The difference in decline for the different classes of wells is striking. For example, the wells which made less than 10 barrels daily the first year average during the second year 61 per cent of that; whereas wells that made more than 50 barrels a day the first year average during the second year about 29.5 per cent as much. The rate of decline of wells averaging between 10 and 50 barrels a day the first year varied regularly between these two extremes.

The dashed line shows the composite curve for all wells and clearly exemplifies the error of using the average curve for the whole field in estimating the future production of wells making less than 10 and more than 50 barrels a day. Had the average curve been projected at the end of the second year, as shown by the dotted line, it would have come close to giving an accurate prognostication of the future decline of the average well in that pool.

Average, or composite, curves are based on the law of averages. There are, of course, many properties that will not follow any of the decline curves shown in figure 4. Some properties produce less than they should because of natural and artificial causes that affect the yield. In fact, there are several factors that can cause the decline

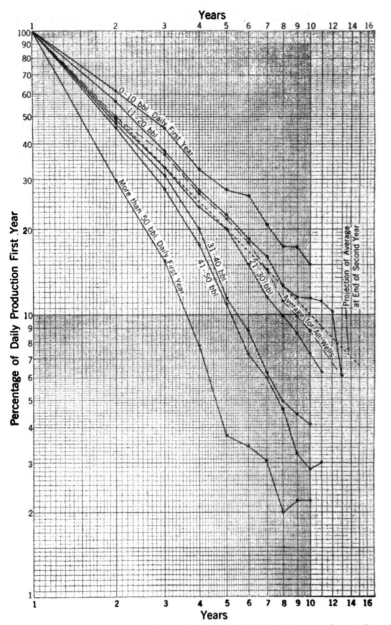

FIGURE 4.—Curves showing the difference in the rate of decline of groups of properties in the Bartlesville field, Okla., on which the initial yearly output was different.

curves of different properties to become irregular, but the chances are that the decline of a property will follow some one of the curves shown in figure 4.

As an example of the method by which averages are applied, the familiar instance of the probabilities of the stature of man may be cited. If 1,000 men are selected at random, fully 500 of them will be between 5 feet 5¼ inches and 5 feet 9¼ inches high, or their average height will be 5 feet 7¼ inches. Of the same 1,000 men probably 10 would be shorter than 5 feet, and three or four would be taller than 6 feet. But the chances are greater of selecting at random from the 1,000 men a man approximately 5 feet 7¼ inches tall than of selecting a man of any other height. The chances are slight that one would obtain in this random choice a man less than 5 feet tall, and still less that one would select a man more than 6 feet tall.

The same principle applies to the average decline of oil properties. The chances are that a property selected at random in the Bartlesville pool will approximately follow the average decline curve shown in figure 4, but if selection be limited to properties that during the first year averaged less than 10 barrels daily per well, there will be little likelihood of the property chosen deviating far from the average decline of 10-barrel wells.

Some of the curves rise toward the end of their lives because of the abandonment of some of the less productive wells on a property, thus raising the average production per well.

USE OF LOGARITHMIC COORDINATE PAPER.

Composite curves may be shown on either rectangular or logarithmic coordinate paper. The author has found the latter advantageous in studying production curves, as many production records of individual wells and also many composite decline curves can be made to approach straight lines when plotted on it. This fact has many advantages. For instance, the curve may be projected more easily and accurately, and in the later life of the well when the output of the well becomes small, the curve is projected into an area on the logarithmic paper where the scale is large and more easily read. The reduction of production curves to their algebraic equations is also much simplified by the use of such paper. Figure 4 is an excellent example of curves plotted on logarithmic coordinate paper. When plotted on such paper an equation of the form $y=cx^n$ will be represented by a straight line whose slope is n.

COMPOSITE DECLINE CURVES.

DATA NECESSARY FOR CONSTRUCTING CURVES.

In constructing composite decline curves one should use only the production data of those properties whose output is not materially affected by the rate of drilling. If production is upheld by drilling, the curve for that property is drawn out to a much greater length, and its decline should not be considered in constructing the average curve for the field. The effects of rate of drilling have already been discussed.

WHEN RATE OF DRILLING IS NEGLIGIBLE.

In fields like the Glenn pool where the wells are spaced closely, communication between wells is easy, and, if the sand is porous, the rate of drilling on different properties may be practically ignored. Thus all properties may be used in constructing the average, regardless of whether wells were drilled after the first year. In a field where spacing is not so close, however, or where the sand is thicker, or where any condition exists that materially reduces the rapidity of interference between wells, some consideration must be given the rate of drilling—as, for example, to records of properties in the San Joaquin Valley district (Cal.), where the initial productions are large and the sand is thick, so that the wells decline slower than in a district where conditions are not so favorable. However, records of the production of individual wells can be used. Some of the fields in Illinois are comparable to the Glenn pool. In fact, the author has found, in constructing composite decline curves of the different fields, that where production data are not available for individual wells, the rate of drilling can be ignored in practically every case, except in such fields as those of the Gulf coast and of California, and where wells are pumped down to a few gallons a day, as in the Appalachian region, where the production is so small a new well will materially increase the average daily production as compared with the first year's production.

PRODUCTION BY MONTHS NOT NECESSARY.

The author made no attempt to obtain production records by months except in the Gulf coast and the California fields, because constructing monthly decline curves for different fields would involve a tremendous amount of labor. Moreover, various irregularities in output, such as those caused by winter weather, variations in pipe-line runs, etc., are averaged out by taking the output for the whole year and determining the average produced during each day. Therefore, the annual production for each property was obtained when it was not possible to obtain the annual production for each well, and the average number of wells producing each year was also determined. Often this figure could be determined only by obtaining the dates of completion of these wells, computing the number of wells producing each month, and thus obtaining the average number producing during the year.

CONSTRUCTION OF CURVES.

CURVE FOR A SINGLE PROPERTY.

In making the computations, if a property began producing after July the first year, the output during that year was ignored and the next year's output was called 100 per cent. If, however, the production began during the first six months of the year, that year was called 100 per cent. Obviously a slight error is thus introduced, but it is believed that with a large number of properties the errors will balance. One may use monthly productions if he desires, or those for any other period that will serve best, to show the average decline of a group of properties. For instance, the average daily production could be computed for each half year or for periods of two or three years. The latter unit might be selected for some of the wells in the McKittrick field (Cal.), where the daily output of a well sometimes increases for two or three years after the completion of the well.

CURVE FOR A GROUP OF PROPERTIES.

To obtain the average decline for a group of properties the yearly percentage decline of the properties is determined. The average is then taken for the first year's percentage of all the properties (in this case 100 per cent) and then for the second year's percentage, and so on. This procedure naturally involves many more wells than properties. The first part of the composite, or average, curve thus obtained is usually more accurate than the last, because many of the properties began producing later than others, and therefore had not as long a decline. For example, a property that began producing in 1907 would show a record of 11 years if 1917 is counted, whereas another property that began producing in 1913 would have produced only five years. The percentage for these two properties would be averaged together for the first five years, but after that time the average decline curve would be the same as that of the property that began producing in 1907. This fact is exemplified in the following table, which also shows the method by which the mathematical averages were determined:

TABLE 1.—*Tabulation of statistics showing the method used in computing the percentages for a composite decline curve.*

Property.	Average daily production per well during first year.	First year.	Second year.	Third year.	Fourth year.	Fifth year.	Sixth year.	Seventh year.	Eighth year.	Ninth year.	Tenth year.	Eleventh year.
	Bbl.	Per cent.	Per cent.[a]	Per cent.[a]	Per cent.[a]	Per cent.[a]	Per cent.[a]	Per cent.[a]	Per cent.[a]	Per cent.[a]	Per cent.[a]	Per cent.[a]
A	22	100	64	34	22	14	9	5	3	2	1	......
B	3	100	68	54	42	28	36	22	......	......	......	......
C	5	100	68	46	30	17	14	9	......	......	......	......
D	6	100	66	30	13	12	9	......	......	......	......	......
E	2	100	65	40	30	......	16	14	13	13	......	......
F	3	100	57	24	16							
G	4	100	36	24	11							
H	13	100	78	34	15	10	5	4	3			
I	7	100	65	51	26	19	13	7	7	6		
J	13	100	69	45	21	19	17	12	11	10		
K	41	100	41	11	5	5	3	1	1	1		
L	118	100	15	12	4	2	2	1	1			
M	8	100	40	38	37	18	9	7	3	3		
N	8	100	49	39	24	13	12	10	8	4		
O	11	100	42	26	16	10	7	4	2			
P	28	100	53	36	13	10	8	6	5	6		
Q	5	100	63	53	37	21	15	10	8	6		
R	5	100	62	36	24	18	10	11	9			
S	7	100	63	38	25	20	18	14	10	7		
T	10	100	73	33	19	12	7	8	5	4	2	1
Average..	16	100	56.8	35.2	21.5	14.7	11.6	8.5	5.9	4.9	1.5	1

a Expressed as a percentage of the first year's average daily production per well.

Another source of inaccuracy in decline records is the abandonment of wells having a small output, for this increases the average daily production of those remaining. Because of this and other occasional irregularities many of the curves shown in part 2 are the result of drawing a smooth average curve through the plotted points instead of actually joining these points with lines. The reader should note the difference in the rate of decline of wells of large and of small yearly initial outputs in Table 1.

APPRAISAL CURVES.

GENERAL STATEMENT.

The term " appraisal curves " was first used by Lewis and the present author,[a] and was applied because of the use of the curves in determining the amount of oil that may be expected from a given area of land, which is one of the most important factors in appraising the monetary value of an oil property.

---

a Lewis, J. G., and Beal, C. H., Some new methods for estimating the future production of oil wells: Am. Inst. Min. Eng., Bull. 134, February, 1918, pp. 477–504.

From a review of the present methods of estimating future production one can see that the most profitable research has been the preparation of percentage curves showing the decline in output of a well or a property. The computed average daily production per well the first year is called 100 per cent, and the corresponding amounts for succeeding years are expressed as percentages of that amount. The future production is then estimated by the projection of the curves. In spite of the progress made there is obvious need of more exact and easily applied methods from which the probabilities and the limitations of the accuracy of the estimates may be determined.

The outline of the following method describing the preparation and use of appraisal curves was first published by Lewis and Beal.[a] The description of the method will be repeated for the sake of clarity, and several methods for the application of such curves will be presented in this paper. The appraisal curve used for illustration is for the Clark County and Crawford County fields (Ill.). Similar curves for other fields are given in part 2 of this bulletin.

DERIVATION AND CONSTRUCTION.

MAXIMUM, AVERAGE, AND MINIMUM CUMULATIVE PERCENTAGE CURVES.

Briefly, the principle of the appraisal curve is based on the difference in the rate of decline of wells of large and of small initial yearly output. The curves printed herein were constructed by using the percentage decline curves of as many properties as are available. Statistics for these decline curves were collected, plotted on prepared forms, and curves drawn through the plotted points. The curves were projected and cumulative percentage curves were constructed that gave the total percentage of oil produced to the end of any year. Thus, if the average daily production per well was, first year, 25 barrels; second year, 15 barrels; third year, 10 barrels; the percentage record would read 100 per cent, 60 per cent, and 40 per cent, respectively. These figures, when plotted, determine the percentage decline curve.

To obtain the cumulative percentage curve with the same figures, the cumulative percentage for the first year would be 100 per cent; for the second year, 160 per cent; for the third year, 200 per cent; and so on. The projection of this cumulative percentage curve to the point where the well reaches its minimum economic production gives the ultimate cumulative percentage; the percentage expresses the ultimate production as compared with the first year's production and is identical with the factor called " volumetric content " by Requa.[b]

---

[a] Lewis, J. O., and Beal, C. H., Some new methods for estimating the future production of oil wells :. Am. Inst. Min. Engr., Bull. 134, February, 1918, pp. 477–504.

[b] Requa, M. L., Methods of valuing oil lands : Am. Inst. Min. Eng., Bull. 134, February, 1918, p. 410.

Using this factor, the appraisal committee of the Independent Oil Producers Agency of California computed ultimate production by multiplying the first year's average daily production per well by 365—the number of days in a year—and then by the "volumetric content." In the present paper the term "ultimate cumulative percentage" is used instead of the somewhat inadequate term applied by Requa.

APPRAISAL CURVE FOR ROBINSON POOL (ILL.).

The appraisal curve in figure 5 was drawn by plotting on rectangular coordinate paper the ultimate cumulative percentage statistics of all the available properties in the Robinson pool, in Crawford County and Clark County (Ill.). The average daily production per well the first year is shown at the bottom of the figure, and the ultimate cumulative percentage is shown on the left margin of the figure. As the ultimate cumulative percentage was plotted against the average daily production per well the first year, each dot represents the ultimate cumulative percentage of a property having a certain average daily output per well the first year.

Then the maximum cumulative percentage curve (fig. 5) was drawn so that practically all properties represented lay below it, and the minimum cumulative percentage curve was drawn to bound the bottom of the area occupied by the dots. The average cumulative percentage curve was drawn as a mean between these two extremes, although an attempt was first made to determine this average by computing the mathematical average of all the properties in several different successive segments of the area bounded by the maximum and minimum cumulative percentage curves. However, the curve determined in this manner was so close to the actual mean between the maximum and minimum that in most of the other appraisal curves constructed the actual mean was taken. Possibly this may be a mistake. The conditions affecting production may be such that the actual average curve in some fields may be above or below the mean. In fact, in the appraised curve for the Osage Indian Reservation (Okla.), the numerical average was obtained and was found to be considerably below the mean. But this average curve should not be considered as a mean on the left side of the chart because all three curves approach the y-axis at infinity, and as the average cumulative percentage curve nears the left margin of the chart it approaches the minimum cumulative percentage curve.

Although the fact of the curves meeting at infinity along the vertical line representing zero production has no great practical importance, it nevertheless establishes the interesting deduction that the smaller the initial output of a well, the smaller will be its ultimate output and the larger its ultimate cumulative percentage. This deduction indicates that, on the average, the smaller the well the slower its rate of decline. As a matter of fact, a well whose initial

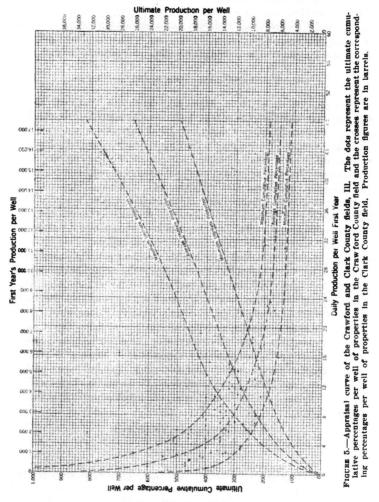

FIGURE 5.—Appraisal curve of the Crawford and Clark County fields, Ill. The dots represent the ultimate cumulative percentages per well of properties in the Crawford County field and the crosses represent the corresponding percentages per well of properties in the Clark County field. Production figures are in barrels.

production is below the minimum profitable production would not be pumped, so that the three curves indicated as approaching infinity would, in fact, reverse themselves at the line representing the minimum profitable production and join at zero. In the preparation of this and all other appraisal curves, however, the maximum, average, and minimum cumulative, percentage curves have been drawn to approach the y-axis at infinity. Although the ultimate production of a large well is greater, its ultimate cumulative percentage is much less than that of a small well; the larger wells, as the chart shows, tend to have small ultimate cumulative percentages, because of rapid declines, and conversely the small wells tend to have large ultimate cumulative percentages, because of gradual declines.

USE OF ULTIMATE CUMULATIVE PERCENTAGE CURVES.

The application of these curves may be shown by an example, as follows: Assume that the average well on a property during the first year produces 20 barrels daily, or 7,300 barrels for the year, and that the average future production of a 20-barrel well in that pool is desired. Follow the 20-barrel line vertically to the point where it intersects the average cumulative percentage curve, then trace from this intersection horizontally to the left. The reading is 197 per cent; that is, the average ultimate cumulative percentage of a 20-barrel well is 1.97 times its first year's production, or, in this case, about 14,400 barrels. But 7,300 barrels have been produced during the first year, therefore the future output of the well will be 7,100 barrels. Similarly, the maximum that this well will produce is obtained by following the 20-barrel line vertically to the maximum cumulative percentage curve and thence to the left, the figure thus obtained is 255 per cent. In other words, the maximum that such a well will produce is 2.55 times the first year's production, or about 18,600 barrels. Likewise, the minimum yield of the well may be obtained from the minimum cumulative percentage curve, which indicates 1.4 times its first year's production, or about 10,200 barrels. Therefore, the average 20-barrel well in this pool, after its first year will not make *more than* 11,300 barrels (18,600–7,300), will make *on an average* 7,100 barrels (14,400–7,300), and will make *at least* 2,900 barrels (10,200–7,300).

ULTIMATE PRODUCTION CURVES.

DERIVATION OF CURVES.

Although the ultimate cumulative percentages are less for the wells of larger output, the actual ultimate production of such wells is greater and usually varies with the initial yield, as is shown by the maximum, average, and minimum ultimate production curves in figure 5. These curves were plotted to bring out this relation and were derived directly from the ultimate cumulative percentage curves by multiplying the first year's production of wells of various initial yearly capacities by their respective ultimate cumulative percentages. In other words, these three curves were derived from the maximum, average, and minimum cumulative percentage curves by choosing different initial productions (daily per well the first year) and multiplying by 365 (days in a year) and by the respective ultimate cumulative percentages. The same curves might have been prepared by plotting the estimated ultimate production per well against its initial yearly output, basing the curves on actual output and estimated future production instead of deriving them from the percentage curves. By using the ultimate production curves instead of the ultimate cumulative percentage curves, an estimate of the ultimate and future production of a well may be obtained much more rapidly and easily.

APPLICATION OF CURVES.

Take the example already cited; that is, a well on a property averages 20 barrels a day the first year. What is its probable future production? Following the 20-barrel line to the points where it intersects the three different ultimate production curves and thence to the right margin, shows that the maximum, average, and minimum ultimate productions of such a well are 18,600, 14,400, and 10,200 barrels, respectively. These estimates are the same as those obtained by using the maximum, average, and minimum cumulative percentage curves. To determine the actual future production, the first year's production $(20 \times 365 = 7,300)$ is subtracted from the estimates of ultimate production; the differences are 11,300, 7,100, and 2,900 barrels, or the same values as those obtained by the previous method (p. 34). It will be seen that the percentage deviation of the extremes (or limits) from the average ultimate production is 29—that is, if this method of determining ultimate production were used, the possible deviation from the average, according to the histories of the wells upon which these curves are based, would be not more than 29 per cent. The percentage deviation from the average actual future production, however, would be greater. A study of the decline of the wells during the months of the first year would yield a closer estimate, for the monthly figures would indicate whether the wells would approach · the maximum or minimum curves.

Two different districts are represented in figure 5. Most of the properties shown lie in Crawford County (Ill.), but a few are in the shallower Clark County district, a few miles north. Because of this difference in the depth of the sands there is a slight difference in the decline of the wells, so that the ultimate cumulative percentage of a property in one of the fields differs a little from that of one in the other. The ultimate cumulative percentages of the properties in Clark

County are therefore shown by crosses and those of the Crawford County district by dots. The crosses tend to approach the upper limits of the area bounded by the maximum and minimum cumulative curves, hence a person using the chart for estimating the ultimate or future production of properties in Clark County should take this fact into consideration. The lower limit, defined by the minimum cumulative percentage curve, is the same for both districts. The average curve, like the maximum, is a trifle higher than that shown.

Curves may be prepared on this same chart to show the actual future production for wells of different initial production, the first year's production being deducted from the determined ultimate production and the remainder being plotted. By the preparation of such a curve one may read directly the actual future production of wells of any output.

USE OF APPRAISAL CURVES.

BASIS FOR USING.

The use of appraisal curves in determining future output is based on the average daily production per well on a property during the first year. However, if a property were, for example, four years old, it would be advantageous, because of fewer calculations or lack of data, to take the most recent year's output in determining future production. But the curves given can not be used thus unless the average future production of wells of equal output in the same district is approximately the same. This "law of equal expectations" has been shown to be true in a previous publication.[a] Data collected more recently confirm absolutely this law which may be restated as follows: "*If two wells under similar conditions produce equal amounts during any given year, the amounts they will produce thereafter, on the average, will be approximately equal regardless of their relative ages.*" The law applies particularly to the output of wells that have become "settled."

After an estimate of future output has been made as outlined above, it can be made more nearly accurate by determining from the first four years' production whether the property follows the average, the maximum, the minimum, or some combination curve. In other words, to determine the future production use the last year's production on the appraisal curve and modify the results thus obtained according to whether the action of the wells during the past four years indicates that the well ranks above or below the average well.

[a] Lewis, J. O., and Beal, C. H., Some new methods for estimating the future production of oil wells: Am. Inst. Min. Eng., Bull. 134, February, 1918, pp. 477–504.

ADVANTAGE IN USE.

Using the appraisal curve to determine future production makes unnecessary the employment of a composite decline curve which, from its being the average decline of wells of all sizes, as cited on page 25, gives an estimate too large for large wells and too small for small wells. By preparing appraisal curves applicable to a district the future or ultimate production of wells of any size or age can be determined at once.

Ability to determine the limitations of estimates of future or ultimate production is important. From a composite decline-percentage curve it is impossible to determine how large the maximum production will be or how small the minimum. By the use of appraisal curves, however, it is a simple matter to determine at once the maximum and the minimum amounts of oil that may be expected.

ACCURACY OF APPRAISAL CURVES.

From the appraisal curves, with the first year's production of a well or group of wells given, it is possible to determine both the minimum and the maximum amounts of oil that probably will be produced, and also the amount that the average well of a certain initial yearly output will ultimately yield.

CARE TAKEN IN COMPILING DATA.

Records of the actual performance of wells were used in preparing figure 5; enough trustworthy records having been taken to insure that all curves marked " average " represented the actual average performance of many wells or properties, unless the conditions were radically changed later, as drilling into a deeper sand.

Unusual wells were omitted in making this and other similar charts. There are, of course, extraordinary wells whose production would not correspond with that indicated on the chart. Some wells, for example, instead of a decreasing daily yield show a sustained or even an increasing daily output for several years. Other wells may cease producing suddenly and, perhaps, begin again after many months or years. A sudden increase in the production of old wells is not unusual, and the termination of the life of a well by accident or by the infiltration of water is rather common. The futility of estimating the future production of such freakish wells is obvious.

One great advantage of the method is that the ordinary irregularities, so common in wells on many properties, disappear by averaging. For instance, using yearly productions instead of daily productions and averaging several wells on a property, eliminate the irregularities of any well whose yield fluctuates rapidly. Thus,

92436°—19——4

figure 5 represents the production of 83 properties, including about 900 wells, so it is evident that the uncommon wells have little influence even though they were on a property that was used in preparing this chart.

RELIABILITY OF ESTIMATES MADE FROM CURVES.

In view of the derivation of the data and the systematic manner in which they arrange themselves when plotted to show how ultimate cumulative percentages compare with initial yearly productions, it is believed that much reliance may be placed on such curves. Not only can the appraisal curve in figure 5 be used with confidence, but estimates of future and ultimate production of properties in other fields may be made by the use of similar charts to be presented later (see figs. 24 to 70). In practically any field, where the conditions affecting production are not too diverse, such curves can be prepared and can be used confidently. The less diverse the conditions that affect production, the closer the maximum and the minimum limits will be to the average curve, so that future and ultimate production can be estimated much more closely than in fields where the conditions have a wide range. For instance, the table on page 205 shows the percentage of deviation above and below the average for the estimates of future production in some of the fields for which appraisal curves have been made.

APPRAISAL CURVES DIFFICULT TO PREPARE FOR SOME DISTRICTS.

For districts like the San Joaquin Valley fields in California the preparation of appraisal curves is difficult because the conditions affecting production vary decidedly. Figure 6 shows the wide variation of the ultimate cumulative percentages for several wells selected at random in the West Side Coalinga field (Calif.). Because of these variations, the wells used in preparing an appraisal curve must be selected with care and from areas where the conditions affecting production are similar.

RELATION OF INITIAL PRODUCTION TO ULTIMATE CUMULATIVE PERCENTAGE.

Figure 5 (p. 33) shows that the ultimate cumulative percentages vary considerably for wells of different daily production the first year, and that the smaller the output of the well during the first year the greater the variation of the ultimate cumulative percentages. For instance, for a well averaging 4 barrels daily the first year, the ultimate cumulative percentage will vary between 270 and 555, whereas, for a 40-barrel well the variation is only 118 to 193. Conversely, the larger the ultimate cumulative percentage of a property, the narrower will be the limits of the possible initial production. Thus, a property whose ultimate cumulative percentage is 550 will have a daily pro-

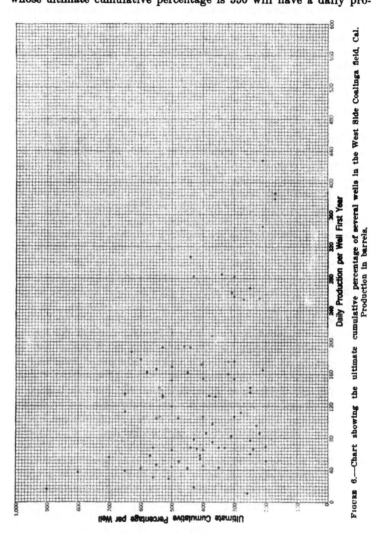

FIGURE 6.—Chart showing the ultimate cumulative percentage of several wells in the West Side Coalinga field, Cal. Production in barrels.

duction per well the first year of 0.2 to 4 barrels, but a property whose ultimate cumulative percentage is 200 will have a daily production per well the first year of 9 to 34 barrels.

POSSIBLE SOURCES OF ERROR IN CONSTRUCTING CURVES.

One source of error that occasionally may have to be considered, if records of individual wells are not available, or if all wells were not drilled at approximately the same time, is the rate of decline changing because of the rate of development of different properties. On some properties the wells drilled later retard, by their initial " flush " production, the actual rate of decline of all the wells. The errors introduced in this manner, however, are usually not as great as supposed, especially in such fields as the Oklahoma, northern Texas and Louisiana, West Virginia and the Illinois, as the initial production of new wells near old ones tends strongly, because of interference or drainage, to become constantly smaller as time goes on, so that the new wells drop into the general average soon after completion and may be ignored as far as they affect the rate of decline. This matter has been discussed in some detail previously (pp. 21 to 24).

The abandonment of wells of smallest yield also tends to keep up the average daily output of the other wells on a property, and often more than doubles their average production. Occasionally abandonment noticeably affects the production curves, as is shown in figure 4 (p. 26) by the composite curves during the eighth, ninth, and tenth years. Another source of error in constructing the curves is the minimum limit to which nearly exhausted wells can be pumped with profit. An increase in the price of oil may raise this limit so that wells of much smaller output may be profitably pumped. For wells of large initial yield this additional production will be so small compared with the first year's output that the error in the cumulative percentage is negligible, but for wells of small initial yield the error may be of some magnitude. Other sources of error are the alterable factors of recovery—such as manner of operation, " shooting," and the application of compressed air, vacuum pumps, or water drive—which may cause changes in the ultimate production and the rate at which the oil is recovered. These changes will increase the ultimate production as well as the rate at which the oil is obtained, but their effects, especially during the later life of a well, may usually be ignored as far as any great variation of ultimate production is concerned, although there may be a noticeable change in the rate at which the remaining recoverable oil is extracted.

The damaging effect of water invading the productive sand is to be noted. Occasionally a young field of much promise will be greatly damaged by water. The water may encroach naturally or be let into the sand by careless operators.

DETERMINING THE MAXIMUM, AVERAGE, AND MINIMUM RATES OF DECLINE OF WELLS.

VALUE OF KNOWING RATES OF DECLINE.

The author has discussed the use of appraisal curves in determining the ultimate and future production of oil properties producing under certain given conditions. A knowledge of the amount of oil that may be obtained in the future is important and valuable, but information on the rate at which this oil will probably be obtained—that is, the future annual production—is still more important. With this knowledge the operator can easily estimate his yearly income from the property by assuming a certain price per barrel for oil and a certain drilling program. Fortunately, to determine, through the use of appraisal curves, the future annual production of a property is not difficult.

Not only can one determine the probable average future production, but one can also obtain an excellent idea of the maximum and minimum rates of yield. The possibility of determining these maximum and minimum limits, as well as the average rate, greatly extends the use of the appraisal curves, for in such determinations lies the crux of the valuation of oil properties.

DECLINE OF A WELL IN THE CLARK COUNTY AND CRAWFORD COUNTY FIELDS, ILLINOIS.

Take, for example, a property in the Clark County and Crawford County fields (Ill.), on which the average well during the first year produced 10 barrels daily. To determine the average decline of such a well, follow the 10-barrel line in figure 5 (p. 33) to the average ultimate-production curve and thence to the right; the ultimate production is 10,100 barrels. This figure includes the first year's production of 3,650 barrels; deducting that leaves a future production of 6,450 barrels. Reversing the process and reading from the right margin horizontally to the left along the line representing 6,450 barrels to the point where this line intersects the average ultimate-production curve and thence reading downward, one finds that a well whose ultimate production is 6,450 barrels will produce during the first year 1,680 barrels, or 4.6 barrels a day. Deducting 1,680 from 6,450 gives a future of 4,770 barrels, which in turn is the ultimate production of a well producing 1,020 barrels the first year, or 2.8 barrels a day. These calculations may be continued until the original 10,100 barrels are exhausted. The yearly production of the average decline curve for a 10-barrel well is therefore, successively, 3,650, 1,680, and 1,020 barrels for the first three years.

In a similar manner maximum and minimum decline curves may be constructed by determining the intersection of the line, representing 10 barrels a day the first year, with the respective maximum and minimum ultimate production curves. In this way one may determine the average rate of decline of a 10-barrel well and also the maximum and minimum rates at which the oil may be obtained. There is little likelihood of a 10-barrel well in Clark and Crawford Counties (Ill.) declining more rapidly than the maximum decline curve or more slowly than the minimum decline curve.

If a property has been producing some years, and one wishes to determine the average rate of decline of the future production, the same process is used, but the future curve itself is modified in accordance with the rate of yield of the well during the first few years of its life. In other words, the average well on a property may decline slower or faster than the average decline curve would indicate for a well of that output; then the future annual production should be modified.

METHODS OF MAKING CLOSER ESTIMATES.

EFFECT OF DIFFERENCES IN INITIAL OUTPUT OF WELLS.

In the preparation of figure 5 (p. 33) the average daily production per well the first year was plotted against the ultimate cumulative percentage for each property of the district. It has been shown that by constructing the appraisal curves from these data certain definite limits to estimates of future and ultimate output can be obtained. Although the author has already found the chart of much value, refinements are desirable that will define more closely and with fewer data the probabilities of wells at earlier periods in their lives.

As has been shown, wells of the same initial yearly production may ultimately yield widely differing amounts of oil, and, conversely, a certain ultimate amount of oil may be produced by wells of widely varying initial yearly output. These variations are not without cause, and it is believed that a systematic study of the factors influencing production will result in materially reducing the limits of the estimates of future and ultimate yield. For instance, on figure 5 are shown the ultimate cumulative percentages of properties in two different fields in Illinois. The percentages of properties in the Clark County field, shown by crosses, generally lie higher than the ultimate cumulative percentages of properties in the Crawford County field.

Moreover, closer study may permit curves derived from the production data of one field to be applied with more confidence to a new field. Assembling enough data may show that in one pool a well making 20 barrels daily during the first year with a certain gas pressure would nearly always be above the average in that pool or, if the wells were spaced a certain distance apart, the tendency would be above or below the average for wells of that size. Undoubtedly, some of the factors that influence the production of many wells are of so little consequence that they may be neglected. The character of the production curve, however, is determined by the synthetic influence of several factors, and the effect of any one varies with the local conditions in each field.

Other factors, in addition to the initial output, that may profitably be studied are the area allotted each well and the depth and thickness of the producing sand. Other less important factors are gas pressure, the character of the oil sand, and the quality of the oil.

EFFECT OF ACREAGE PER WELL.

An attempt has been made to narrow the limits and thus enable closer estimates of future and ultimate production by using some of these factors. Enough information was obtained, however, for only such factors as acreage per well and depth and thickness of the sand. No attempt was made to utilize the scattered information on geologic structure, the character of the oil sands, or the quality of the oil. Figure 7 was prepared by selecting all those properties in the Crawford County field (Ill.) for which the acreage per well could be determined with fair accuracy. The same ultimate cumulative percentages for each property were used as in preparing figure 5. In figure 7 the dots that represent ultimate cumulative percentages for certain acreages per well seem to arrange themselves consistently in an area that ascends to the right. In other words, the greater the acreage each well drains the greater is the ultimate cumulative percentage of the well. This fact is so evident as to need no proof; a well draining a large area should decline slower and produce longer than one draining a small area.

To illustrate the use of figure 7, assume that a well in the Crawford County field drains 6 acres; the maximum, average, and minimum cumulative percentages of that well therefore will be, in order, 355, 275, and 200. Similarly the minimum ultimate cumulative percentage of a well that drains 9 acres of sand is 300 per cent.

EFFECT OF THICKNESS OF OIL SAND.

In the same manner the ultimate cumulative percentages of different properties in the Crawford County field were plotted against the average thickness of the oil sand under each property (fig. 8). The limits thus established were by no means so narrow as those obtained by plotting ultimate cumulative percentages against acreage per well,

for it is difficult to determine exactly the total thickness of a producing
sand. Also, it is difficult in this case to determine accurately where
the maximum and minimum limits should be drawn. However, the

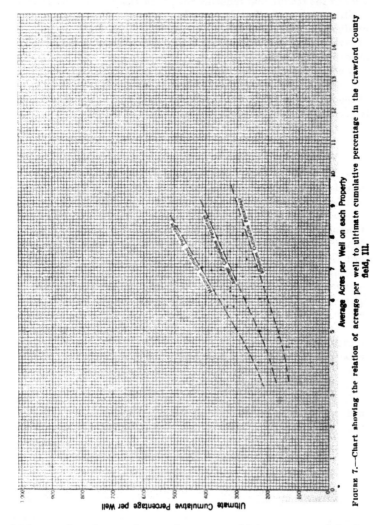

FIGURE 7.—Chart showing the relation of acreage per well to ultimate cumulative percentage in the Crawford County field, Ill.

thickness of sand was taken as the average thickness reported in the
different wells on each property. Examples to be given later show
that even these limits are occasionally of material aid in making
closer estimates of future and ultimate production.

Figure 8 demonstrates the well-known fact that the thicker the sand the slower the rate of decline of the wells draining it, as the three curves gradually rise to the right. In other words, the thicker the

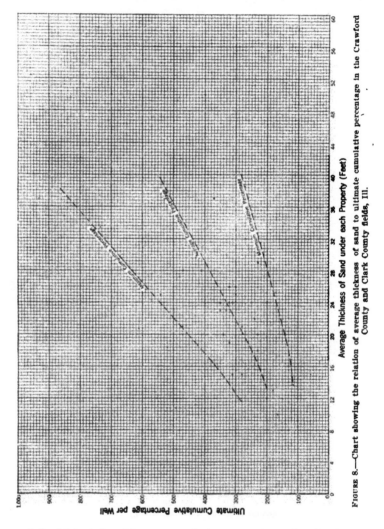

FIGURE 8.—Chart showing the relation of average thickness of sand to ultimate cumulative percentage in the Crawford County and Clark County fields, Ill.

sand the higher the ultimate cumulative percentage, for the decline of wells draining thick sands is ordinarily slower than that of wells drilled into thin sands.

EFFECT OF DEPTH OF OIL SAND.

The third method employed for reducing the limits is shown in figure 9, which gives the limits determined by plotting ultimate

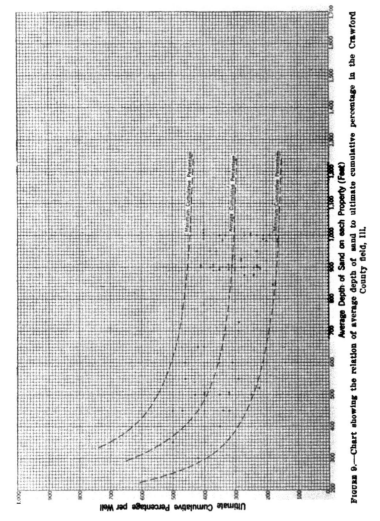

FIGURE 9.—Chart showing the relation of average depth of sand to ultimate cumulative percentage in the Crawford County field, Ill.

cumulative percentages against the average depth of the oil sand on each property. This figure proves the general rule that the deeper the sand the smaller the ultimate cumulative percentage.

NARROWING OF ESTIMATES IN CRAWFORD COUNTY FIELD, ILL.

In figure 10 the ultimate cumulative percentage of each of several properties in the Crawford County field were plotted against four conditions that control the ultimate production and decline of oil

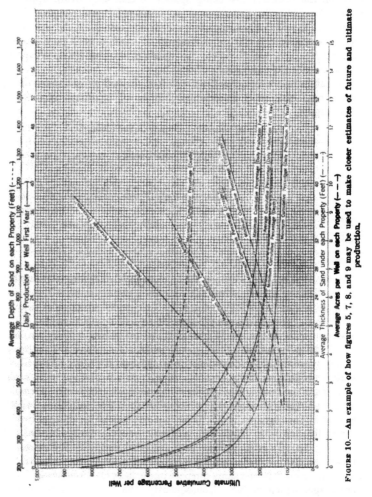

FIGURE 10.—An example of how figures 5, 7, 8, and 9 may be used to make closer estimates of future and ultimate production.

wells, as follows: (1) The average initial production per well (daily production per well the first year); (2) the average acreage per well; (3) the average thickness of sand; and (4) the average depth of the producing sand.

Let us now determine how this new composite chart may be used in making more accurate estimates of the future and the ultimate output of a well that, for example, has been producing one year. The average daily production during the first year is 14 barrels; the wells on the property are being spaced so that they drain approximately 5 acres each; the oil sand is 20 feet thick and 600 feet deep. How can the information prepared in figures, 5, 7, 8, and 9 be used to estimate the future production of such a well more closely than by figure 5, which shows the relation between daily yield the first year and ultimate production?

The table following gives the limits established by each controlling factor and the final determined limits of ultimate cumulative percentage:

Limits of ultimate cumulative percentage.

	Initial production, 14 barrels daily, first year.		Area drained, 5 acres.		Thickness of sand, 20 feet.		Depth of sand, 600 feet.	
	Maximum.	Minimum.	Maximum.	Minimum.	Maximum.	Minimum.	Maximum.	Minimum.
Limits of ultimate cumulative percentage established by each factor..	310	170	295	170	450	140	500	210
Determined limits of ultimate cumulative percentage..................			295					210

Maximum ultimate production, 15,000 barrels; minimum ultimate production, 10,700 barrels.

By the use of figure 10 one finds that a well making 14 barrels a day the first year will ultimately produce a minimum of 170 per cent and a maximum of 310 per cent of its first year's production (shown by arrows). A well draining 5 acres in this field will furnish an ultimate minimum output of 170 per cent and an ultimate maximum output of 295 per cent of its first year's production (shown by arrows). Similarly, a well in the same pool draining a sand 20 feet thick will ultimately produce on the average a minimum of 140 per cent and a maximum of 450 per cent of its first year's production (shown by arrows); and a well 600 feet deep will produce ultimately a minimum of 210 per cent and a maximum of 500 per cent of its first year's production (shown by arrows). Thus, four different minima are established, 170 per cent, 170 per cent, 140 per cent, and 210 per cent. Of these the highest, or 210 per cent, may be selected for making the estimate. In like manner and for the same purpose one selects the lowest maximum percentage, which is 295. Consequently, the final narrowed maximum and minimum limits are 295 and 210 per cent instead of the 310 and 170 per cent obtained by using figure 5 alone in determining the limits. In other words, a well that under the specified conditions of acreage, sand thickness, and depth makes 14 barrels a day the first year, will produce ultimately a minimum of 2.1 and a maximum of 2.95 times its first year's production.

Without the use of these additional factors for narrowing the limits, the deviation of the minimum percentage below the average is 27.5 per cent, and that of the maximum above the average is about 32 per cent; whereas with all four factors used the deviation of the minimum percentage below the average is about 17 per cent and that of the maximum percentage above the average is about the same. Hence, by using all four factors, the ultimate production of a well of this output can be determined within about 17 per cent. For the 14-barrel well cited the ultimate production is between 10,700 and 15,000 barrels.

By using these factors for narrowing the limits of the estimates, a man knows, judging from the past history of regularly operated wells in this field, that he should not expect a 14-barrel well to yield ultimately, under the most favorable conditions, more than 17 per cent above the average; and that, under the most unfavorable conditions, the ultimate production of the well will fall not more than 17 per cent below the estimated average ultimate output. On the other hand, if a man with no knowledge of these other factors has to estimate the ultimate production of a 14-barrel well, the actual final yield may exceed the estimated average by about 32 per cent or may fall as much as 27 per cent below it.

DETERMINING THE AVERAGE DAILY PRODUCTION FOR THE FIRST YEAR.

GENERAL STATEMENT.

The preceding discussion leads naturally to this question: If one has determined the influence of acreage, of depth, of sand thickness, and of initial yearly output on the ultimate or future production of a well in a certain district, why can he not determine the probable future initial yearly production of a well if he knows the acreage, sand thickness, and depth? These last three factors can readily be determined before the well is drilled, especially if it be situated in a proved district or in one where conditions are like those in a district where data can be obtained.

The ultimate production of a well depends on the initial production, and almost all estimates of the ultimate output of undrilled territory are based in some manner on the early action of the well. Estimating the future production by percentage decline curves rests entirely on the average daily production the first year. In fact, this unknown quantity has been decidedly puzzling and has been an element of uncertainty and error in estimating future output by production curves. Drainage or interference of producing wells is **an** important factor, and its extent, which is a function of time, distance, and gas pressure, greatly modifies the initial production of new wells. The lenticular structure of oil sands and differences in their porosity, as well as other geologic conditions, also affect initial production, and to determine the composite influence of all these factors is extremely difficult.

FACTORS DETERMINING VALUE OF OIL LANDS.

Among the more important factors determining the value of an oil property are (1) the amount of oil that will be ultimately recovered from it, and (2) the amount that will be produced each year. Obviously, if (2) can be determined, (1) may be obtained by adding the annual output during the life of the property. To make trustworthy estimates of the future annual output, one must have (1) some knowledge of the rate of decline of similar producing properties, and (2) either an estimate of the first year's production of each new well, or an estimate of the probable yield per acre of the undrilled area. Often it is more advantageous to estimate future initial yearly production and then by using the decline curve to compute ultimate output, but occasionally estimates of ultimate production are made and the initial yearly output computed, as explained later (pp. 53–55). Clearly, methods for determining the probable initial yearly production of undrilled wells are greatly to be desired.

THE DECREASE IN INITAL YEARLY PRODUCTION.

CAUSE OF DECREASE.

When a well is drilled into an oil sand the oil is expelled through the opening and the pressure in the sand decreases first at the well. This area of lowered pressure gradually extends in every direction at a rate determined by the thickness and porosity of the sand, the viscosity of the oil, the strength of the expulsive force, and other factors. The extent of the area varies directly with the time during which extraction takes place. Within the drainage area, in a sand partly drained in this manner the pressure is always lowered, and another well drilled within that area will have a smaller production, because of the expulsive forces being partly exhausted. In a field that is being drilled, therefore, the drainage areas of the producing wells finally begin to interfere with one another. This interference reduces the gas pressure over large areas and consequently reduces the initial production of new wells, but as long as the new wells are drilled in areas unaffected by drainage, their initial production will not be affected. As a general rule, however, the initial production of wells drilled during the later life of a field are decidedly less than the initial production of the first wells, so that this condition must be considered in estimating the future output of new wells, inasmuch as ultimate production is influenced by initial production.

IMPORTANCE OF RATE OF DECREASE.

The rate of decline in initial yearly production is of so much importance that it must be kept in mind in drilling to maintain production and in estimating the future output of producing oil properties and the ultimate production of undrilled oil properties. Frequently this decrease in the initial output of undrilled wells is not considered and the assumption is made that all the wells drilled on a certain tract will have approximately the same initial production.

Curves have been prepared showing the decrease in production of new wells drilled during several successive years of several hundred properties in different parts of the country. Some concrete examples (figs. 11, 12, and 13) show declines of the Lawrence County pool (Ill.), the Bartlesville field, the Bird Creek-Flatrock and Glenn pools (Okla.), and the Kurokawa field (Japan). In figure 55 (p. 157) the inset shows the decrease in initial yearly production of several wells in the Caddo field (La.). The figures on each curve denote the number of properties taken to obtain the average daily output per well during that year. In preparing these curves all properties beginning to produce in one year, for instance, 1908, were separated from those beginning in other years, and the average daily production per well for each property was computed for the first year. The curve represents this average for each year. The figures on the curves indicate the number of properties used to determine the average.

DETERMINING INITIAL YEARLY OUTPUT FROM COMPOSITE DECLINE CURVES.

When composite percentage decline curves are used for estimating future annual production, the average daily production of a well during the first year is called 100 per cent. The yearly output of properties after the first year is shown as a percentage of the first year's output, so that the production for any future year may be obtained by multiplying the average daily production for the first year by the yearly percentage indicated by the curve. In this way, if the first year's production of a well is known and if the production of a well will probably decline along a known curve, it is easy to determine the future annual output.

ESTIMATING FROM THE RECORDS OF OTHER WELLS.

In case a well has not produced one whole year, or an estimate of the future annual production of an undrilled well is desired, it is nec-

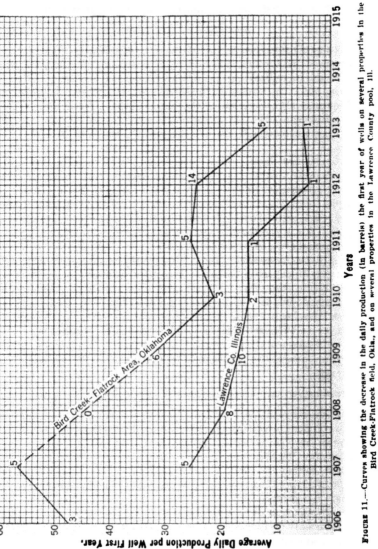

FIGURE 11.—Curves showing the decrease in the daily production (in barrels) the first year of wells on several properties in the Bird Creek-Flatrock field, Okla., and on several properties in the Lawrence County pool, Ill.

essary to assume a certain daily production the first year, unless in the case there is information available to show the relation between the initial output the first 24 hours and the average daily production of a well during the first year. (See p. 59.)

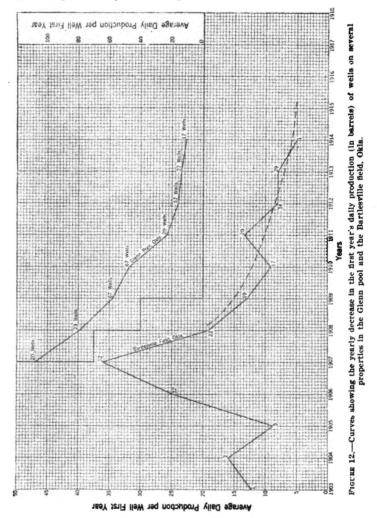

FIGURE 12.—Curves showing the yearly decrease in the first year's daily production (in barrels) of wells on several properties in the Glenn pool and the Bartlesville field, Okla.

ESTIMATING FROM ASSUMPTIONS OF ULTIMATE PRODUCTION.

The first year's daily output may be estimated by assuming a certain ultimate production per acre of the undrilled territory and then figuring backward. To illustrate the use of this method, assume that a composite decline curve is available, showing an ultimate cumulative percentage of 500; in other words, the ultimate production of a

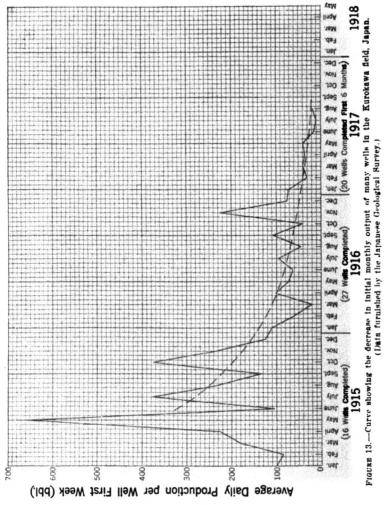

FIGURE 13.—Curve showing the decrease in initial monthly output of many wells in the Kurokawa field, Japan. (Data furnished by the Japanese Geological Survey.)

well following this curve may be estimated by multiplying the first year's production by 5. Assume also that the undrilled area in question will produce ultimately 18,250 barrels per acre and that the wells will be spaced so that each well will drain 10 acres. The average daily production the first year may be determined by the following equation:

$$\text{Daily production the first year} = \frac{18{,}250 \times 10}{365 \times 5.00} = 100 \text{ barrels.}$$

COMBINATION METHOD.

Sometimes a combination method is used, the initial yearly production being estimated directly and then being checked by estimating the ultimate production per acre and calculating the initial yearly production of the new wells by the method just given.

USE OF APPRAISAL CURVES.

The same kind of estimates may be had by using appraisal curves (fig. 5. p. 33), but these curves have the advantage of giving also the maximum and minimum initial yearly production that may be expected. Essentially, these curves are based on the relation between the ultimate and the first year's production, and with a certain first year's output the annual output can be computed by the method explained on page 34. The ultimate production on the right margin of figure 5 gives the total amount; that is, the average a well with a certain initial yearly production will make.

If the average acreage per well is known, and the ultimate production per acre has been estimated, the ultimate production of the well is determined by multiplying the estimated ultimate production per acre by the acres drained by the well. This product should be found on the right margin of figure 5. Then, by following the line representing this product to the left to where it cuts the average ultimate production curve and following the curve downward the first year's daily production is determined. In other words, by using this method one finds that, on an average, a well having a certain ultimate production will have a certain first year's daily production. But the chart gives additional information. By tracing the same line to the left and thence downward from where it intersects the minimum and the maximum ultimate production curves, one can determine theoretically the minimum and maximum daily output that a well on acreage of a certain productivity will yield the first year.

EXAMPLE OF USE OF APPRAISAL CURVES.

Take this example: Suppose the wells, to be on undeveloped lands, will drain an average of 4 acres each when the land is completely drilled, and that the area will ultimately yield 4,000 barrels per acre. The total amount of oil to be produced by each well, therefore, is 16,000 barrels. Follow the 16,000-barrel line from the right margin of figure 5 toward the left to where it intersects the minimum and the maximum ultimate production curves. The minimum limit thus determined is 13.6 barrels and the maximum is 36.6 barrels daily. In other words, the history of the properties upon which the appraisal curve is based indicates that a well with a daily output during the first year of 13.6 barrels may ultimately produce as much as a well that made daily during the first year 36.6 barrels; this leads to the conclusion that the former well produces under more favorable conditions.

This method gives some idea as to the probable range of initial yearly production, and is preferable to that of determining initial yearly production by the use of composite decline curves. A composite decline curve is based on the average decline of wells of all sizes, and the use of appraisal curves, which show the difference in the ultimate production of wells having different daily productions the first year, obviates the error introduced by averaging the declines of all wells.

To determine initial yearly output is certainly the most difficult problem in estimating the probable annual production of undrilled ground. If one can determine, even within rather wide limits, the amount a well will make during the first year, and if production curves are available that show the yearly decrease in production of a well under similar conditions, the making of the remainder of the estimate is comparatively simple. The approximate depth and, under certain circumstances, the approximate thickness of the sand, and the acreage to be allotted each well are usually known before drilling begins, but nothing as to the probable output of the well during the first year. As has been shown, close estimates of future and ultimate production may be made by plotting the ultimate cumulative percentages of properties against the daily output the first year, thickness of sand, depth of sand, and acreage per well. The limits thus determined are shown in figure 10 (p. 47). With the data given, figure 10 may be used to determine the reasonable limits of the initial yearly production of new wells, and by applying the law of probabilities to these estimates the first year's daily production of a well may be estimated much closer, theoretically, than by using any of the methods outlined above. Although the procedure may be of little practical value in actually estimating future production for the first year, an example is given by way of illustration.

APPLICATION OF METHOD TO UNDRILLED LAND IN CRAWFORD COUNTY FIELD (ILL.).

Suppose one desires to determine the annual rate of production of a well to be drilled on a large tract of undrilled land in the Crawford County field (Ill.). From the geologic data available and the records of surrounding wells it is fairly certain that the sand lies 400 feet below the surface and is approximately 30 feet thick: It is planned to space the wells so that each will drain approximately 6 acres. What will be the probable daily production the first year of one of the wells?

The conditions affecting the production of the new well are likely to be similar to those in the drilled part of the field, so that by knowing the acreage per well and the depth and thickness of the sand, one will have a fair start in making an accurate estimate of the annual production of new wells on the undrilled tract. The following table shows the statistics obtained from figure 10 in determining the probable future daily production the first year under the specified conditions of acreage, depth, and thickness:

Data obtained from figure 10 in determining probable daily production the first year.

	Limits of ultimate cumulative percentage.		Determined limits of ultimate cumulative percentage.		By applying determined percentage limits—					
					To maximum percentage curve for initial production.		To average percentage curve for initial production.		To minimum percentage curve for initial production.	
	Maxi-mum.	Mini-mum.	Maxi-mum.	Mini-mum.	Maxi-mum.	Mini-mum.	Maxi-mum.	Mini-mum.	Maxi-mum.	Mini-mum.
	Per ct.	Per ct.	Per ct.	Per ct.	Bbls	Bbls.	Bbls.	Bbls.	Bbls.	Bbls.
Acres per well, 6	355	200	355							
Depth of well, 400 feet	630	290		290						
Thickness of sand, 30 feet	680	210								
Determined daily production the first year					16	11	9	6	3	1

To determine the limits of the ultimate cumulative percentages as determined by acreage per well, the 6-acre line is followed upward to where it cuts the maximum and minimum percentage lines; then by reading on the left margin the respective values are found to be 355 and 200 per cent. To determine the same percentages for depth the line showing depth is followed downward to the maximum and minimum percentage lines; then readings on the left give 630 and 290 per cent. The maximum and minimum percentages as determined by sand thickness are 680 and 210 per cent. In other words, these are the limitations prescribed by the three different conditions that influence production in Crawford County, Ill.

As shown already (p. 48), the highest minimum percentage, 290, and the lowest maximum percentage, 355, are chosen. To find the most likely maximum daily production for the first year follow from the left margin the lines representing 290 and 355 per cent to where these lines cut the maximum cumulative percentage curve, and then read upward to the maximum (about 16 barrels) and the minimum (about 11 barrels) daily production. Similar procedure is used to determine the probable variation of wells that follow the average curve or the minimum curve, the same lines being followed toward the right to where they cut the average and the minimum percentage curves. The readings are 9 to 6 barrels daily the first year for the average and 3 to 1 barrels daily for the minimum. The respective limits determined, therefore, are 1 barrel and 16 barrels.

These limits are rather wide, but the chances are much greater that the new daily production will approach the average more nearly than it will the maximum or the minimum lines, and the law of probabilities can be applied to see what the chances are of its falling between 6 and 9 barrels. Obviously, if the probability is 0.5 that the first year's daily production of a new well will be between 6 and 9 barrels and that the chances for its being more or less will decrease with the distance from the 6 and 9 barrel lines, the operator has a definite basis upon which to work. If he knows that there is an even chance that the new well will yield 6 to 9 barrels a day the first year and that there is only a remote possibility of its producing less than 1 barrel or more than 16 barrels, he is in a fairly strong position. With enough data on the yield of different properties, he can apply the law of probabilities and determine the actual probability of the well making a certain initial yearly production. In practice, the wide use of such a method is doubtful because of the variations in the natural and artificial conditions affecting yield and of the lack of data with which to work.

USE OF CURVES SHOWING THE DECREASE IN THE INITIAL YEARLY OUTPUT OF WELLS DRILLED DURING CONSECUTIVE YEARS.

As explained above (p. 51), curves can be drawn to show the decrease in the initial yearly output of wells drilled during consecutive years. Often, if part of a field is uniformly drilled and an estimate of the average initial yearly production of the wells to be drilled in the rest of the field is desired. curves like those in figures 11, 12, and 13 may be prepared and the estimates made from them.

An interesting comparison may be made of the percentage decrease in these curves by designating the amount at the high point on the curve 100 per cent and the amounts during the following years as percentages of that amount. Except in the Kurokawa field (Japan) the percentage decrease in initial production is notably similar, despite the variety of conditions in the different fields.

RELATION BETWEEN INITIAL PRODUCTION OF A WELL THE FIRST 24 HOURS AND ITS DAILY PRODUCTION THE FIRST YEAR.

RELATION IN NEW STRAITSVILLE FIELD, OHIO, AND LAWRENCE COUNTY FIELD, ILL.

Frequently it is important to know the relation between the amount of oil a well yields during the first 24 hours and its average daily production during the first year. The composite decline curves and the appraisal curves prepared are based on the average daily production per well during the first year and often such data for so long a period are not available. Moreover, it is often desirable. after a well has produced a short time, to estimate approximately how much the daily production will be for the first year. Figures 14 and 15 have been prepared to show this relationship.

Figure 14 shows the relation for wells to the Clinton sand in the New Straitsville field. southeastern Ohio, and figure 15 is an example of the relation in the Lawrence County field (Ill.). The close upper and lower limits. shown by the dots (fig. 14), should be noted. The average line was determined by finding the average for the values indicated by the different dots in each vertical area of the chart. Then a smooth curve was drawn through the irregular line drawn from dot to dot. Similar curves can easily be made for any other field.

METHODS OF DETERMINING FUTURE AND ULTIMATE PRODUCTION.

USE OF COMPOSITE DECLINE CURVES.

Composite decline curves show the average decline for wells of different output. Therefore if one of these composite curves is used in figuring the future annual production and the ultimate production of a property whose daily output the first year is considerably below the average, the estimate of future annual production, and consequently of ultimate production, of that property will be less than it should be; whereas if the first year's daily output exceeds that of the average curve the estimates of future annual production and of ultimate production will be considerably higher than they should be.

However, such curves serve a certain purpose, and it was thought wise to give in this form the information collected for the use of those who prefer to use these curves, with their certain error, to other methods.

ESTIMATING THE FUTURE OUTPUT OF UNDRILLED LAND.

In estimating the future production of an undrilled property by means of composite decline curves it has been shown that the most

important and difficult problem is that of determining the prob-
ible production of each new well for the first year.   After the best
estimate possible has been made it is a fairly simple matter to deter-

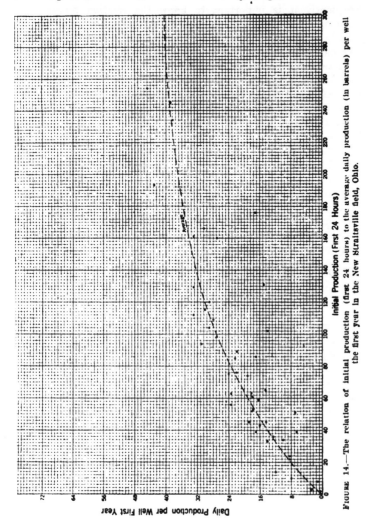

FIGURE 14.—The relation of initial production (first 24 hours) to the average daily production (in barrels) per well the first year in the New Straitsville field, Ohio.

nine the future annual output of the well if a composite decline curve
s available.   For instance, if such a curve shows the percentages for
he first two and three years to be 100, 60, and 30 per cent, and 500

barrels is a trustworthy estimate of the average daily production the first year, the annual output of this well for the first three years, if the well is not far above or below the average, will be, successively,

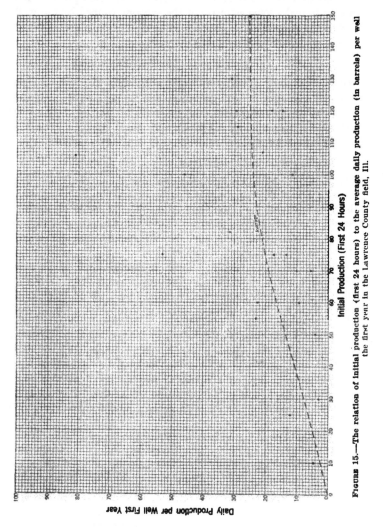

FIGURE 15.—The relation of initial production (first 24 hours) to the average daily production (in barrels) per well the first year in the Lawrence County field, Ill.

182,500, 109,500, and 54,750 barrels. To determine the ultimate production of the well the amounts obtained annually are added; or the yearly percentages can be added to make the ultimate cumulative percentage, and that can be multiplied by the first year's production. For example, in this instance, if the ultimate cumulative percentage is 300 the ultimate production of that well will be 500 × 365 × 3, or 547,500 barrels. To obtain the ultimate production per acre it is necessary to make some estimate of the acreage drained by the well; on the assumption that this well drains 15 acres, the ultimate production per acre would be about 36,500 barrels.

ESTIMATING THE FUTURE OUTPUT OF PARTLY DRILLED LAND.

When a property is being drilled the operator has not only the data for a composite decline curve, but also the actual past performance of the wells themselves; by these he is able to decrease or increase his estimate of future production according to whether the output follows a curve above or below that of the average well.

DETERMINATION OF THE AVERAGE AGE OF A BARREL OF PRODUCTION.

The method suggested by Washburne [a] for determining the average age of a barrel of production has been used occasionally by the author. It is especially valuable in estimating the future production of several wells of different ages when only a composite decline curve is available. Because of its importance, the salient points and certain modifications of the methods are given here.

Briefly, the method consists of determining the point on the composite decline curve at which the property is producing. For this purpose it is necessary to determine the average age of a barrel of the oil produced. The decline in output of a property that is being drilled is often distorted by the large initial production of new wells. An error would be introduced by using the average age *of the wells* in determining on the typical curve the point occupied by the production of that property, because the wells differ in output. To obtain the average age of a barrel of production on such a lease the daily output of each well is weighted by its age. For the sake of clarity, Washburne's table, with certain modifications, is given below:

Table for determining the average age of a barrel of production.

Well No. 1	Daily production, barrels. 2	Age, months. 3	Product of terms. 4
1...	10	18	180
2...	12	14	168
3...	15	8	120
Total.....	37	....	468

[a] Washburne, C. W., The estimation of petroleum reserves: Am. Inst. Min. Eng. Bull. 130, October, 1917, pp. 1866–1868; discussion of paper by R. W. Pack, in Am. Inst. Min. Eng. Bull. 128, August, 1917, pp. 1121–1134.

The average age of the wells is here shown to be 13⅓ months (40 divided by 3), whereas the average age of a barrel of production, determined by dividing the sum of the products (column 4) by the total daily production (column 2), is about 12¼ months. Hence to estimate the future production of these wells one finds the point on a curve 12¼ months after its beginning. In other words, the 37 barrels of production is the result of a decline for 12¼ months of some larger production which, during the first year, is called 100 per cent. If the curve at 12¼ months from the beginning reads 60 per cent of what it was 12¼ months earlier, then 37 barrels equals 60 per cent of the original production, or about 62 barrels. In other words, if one well made originally 62 barrels daily, 12¼ months later it would be making 37 barrels. This method, which has been used with certain modifications by the author and checked against other methods, has decided value under some circumstances. It is based, of course, on the assumption that a composite decline curve showing the yearly production of an average well may be interpreted in terms of months or fractions thereof.

ADVANTAGES OF COMPOSITE DECLINE CURVES.

Composite decline curves have many advantages and the finding of a better method for determining future or ultimate production would be difficult were it not for the basic difficulty of determining the limitations of the estimates made and for the curves representing the average decline of wells of all sizes. One advantage of such curves is that after the average daily production has been determined for the first year the production for any one year can be immediately determined by multiplying the first year's production by the percentage expressed on the decline curve. This can not be done with appraisal curves unless a generalized decline curve, explained on page 64, is used.

LIMITATIONS OF COMPOSITE DECLINE CURVES.

Composite decline curves should not be used in estimating yields in new territory unless the conditions that affect production there are approximately the same as those prevailing in the territory for which the curve has been prepared; in other words, the acreage per well, the operating conditions, the geologic conditions, and the thickness of the sand should be the same. In using a composite curve for estimating yields in a new district some distance from the field for which the curve was prepared considerable inaccuracy must be expected and a large factor of safety allowed for uncertainties of production, differences in geologic conditions, and so forth.

USE OF APPRAISAL CURVES.

As the use of appraisal curves for estimating ultimate and future production for wells already drilled has already been discussed on page 35, examples will not be repeated. The most important advantage in using appraisal curves is that the maximum and minimum limits of the decline of a well may be determined, as well as the same limits for the ultimate production. For determining the limits of the decline of a well, generalized decline curves are of great use.

USE OF GENERALIZED DECLINE CURVES.

Generalized decline curves showing the maximum, average, and minimum rates of decline of wells of any output may be constructed from the appraisal curves, so that one may obtain a ready idea as to the future annual production of a well of a certain size, especially if the well is two or three years old and the trend of its decline is known.

GENERALIZED DECLINE CURVES FOR WELL IN OSAGE NATION.

Generalized decline curves for a large well in the Osage Nation are shown in figure 16. These curves were derived from the appraisal curve of the Osage Nation (fig. 24, p. 107) by computing the decline from the curves of maximum, average, and minimum ultimate production, as explained on pages 41 and 42. Wells with large initial outputs usually follow along or near the minimum curve (fig. 16), whereas wells with small initial outputs usually follow the maximum curve. Production curves of other wells are intermediate, but will fall systematically within the extremes.

PLOTTING THE DECLINE.

To estimate the decline of a well, the daily production the first year may be plotted on the average decline curve at its intersection with the line indicating that production. The ensuing annual productions are then plotted at time intervals of one year each. If the latter points deviate from the average decline curve, the well is not an average well. Thus the plotted decline may approach more nearly the minimum decline curve or the maximum decline curve. Then the first year's output should be plotted on the minimum or maximum curve. Some wells may begin on one curve and then for a few years produce according to one of the other curves, but finally come back to the first curve. However, the actual future of a well can be determined within fairly narrow limits, even though only the first year's output is available.

MANNER OF USING THE CURVES.

The best way to use these curves in estimating the future of a property is to plot the production on a piece of tracing cloth at intervals of one year, using the same vertical scale as that on which the

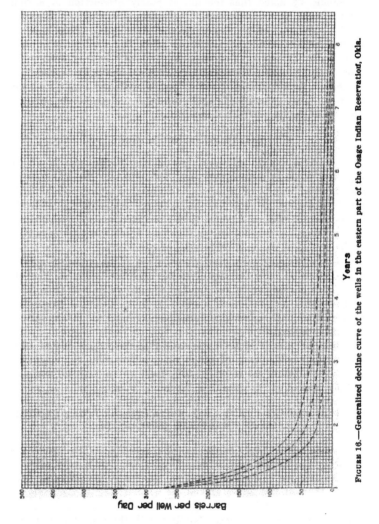

FIGURE 16.—Generalized decline curve of the wells in the eastern part of the Osage Indian Reservation, Okla.

generalized decline curves are plotted. The tracing cloth can then be shifted from right to left to make the production curve fit best one of the three generalized decline curves. The older the well the more accurate will be the estimate of future production. If a well is only a year old, however, it can be assumed to be an average well and its decline estimated to follow the average curve, with the same probabilities of inaccuracy as in the use of appraisal curves.

It should be noted that the initial yearly production of a well is not always an indication of its decline, for wells of different initial production follow curves of the same type. This is to be expected because different conditions affect the production of different wells. However, as all such variations have determinable causes, it is probable that when more data are available to show the way certain factors influence the output of wells the causes for such deviations from generalized curves can be found and more accurate estimates of future production made.

ADVANTAGES OF THE CURVES.

An operator should construct generalized decline curves of oil properties when he has enough data, for obviously the information thus obtained will be of great value to him. For instance, if individual production records for the wells are available the probable future of every well on a property can be determined approximately by inspecting the generalized decline curve and noting whether or not the decline is above or below the average shown.

In determining ultimate production by use of the appraisal curves, estimates can be made directly from the appraisal curve or generalized decline curves can be constructed for any large well in a field, and the actual production of average wells on different properties can be plotted on the generalized decline curve and the future annual production estimated.

Another advantage of these curves is that the future decline of a well can be readily seen, which is not possible with the appraisal curves. As pointed out later (p. 198) these curves may be used exclusively instead of the appraisal curves.

USE OF APPRAISAL CURVES FOR ESTIMATING FUTURE PRODUCTION OF UNDRILLED PROPERTIES.

Limitations that apply to the use of composite decline curves for estimating the future production of undrilled tracts must also be applied to the use of appraisal curves. All conditions affecting production of undrilled areas should be approximately the same as those for the district for which appraisal curves are available, or the estimator should be sufficiently familiar with the probable influence of the variable conditions.

METHODS OF DETERMINING THE FUTURE PRODUCTION OF FIELDS.

VALUE OF ESTIMATES OF FUTURE PRODUCTION.

Occasionally it is necessary to estimate the future output of an oil field in order to determine the probable amount of recoverable oil not yet produced. Several estimates of this kind have been made for the different fields of the United States, as well as for tracts of land on which production was expected. The sum of the estimates for the fields was taken as the total future production of the United States. The value of such knowledge lies in its giving an approximation of the country's oil resources.

In 1909, Day[a] estimated that the total yield of the petroleum fields of the United States would be not below 10,000,000,000 barrels, and not above 24,500,000,000 barrels. Estimates made in 1916 by members of the United States Geological Survey and the Bureau of Mines, published as a Senate Document,[b] indicated that Day's minimum estimate was more nearly correct than his maximum. In this estimate the ultimate production of the fields of the United States was estimated at about 11,000,000,000 barrels, of which 7,629,000,000 barrels remained to be produced. In general, estimates for large areas if made during early drilling have proved considerably above what the district finally produced. It is impossble to estimate with any degree of certainty the ultimate production of undrilled fields.

METHODS OF MAKING ESTIMATES.

Estimates of the future production of large areas may be made by the same general methods used in estimating the future production of individual tracts; that is, the future yield of the producing wells is estimated, and then the amount that ultimately will be produced per acre is computed. This acreage production is applied to the undrilled area and the ultimate output is estimated.

Another method is to estimate the initial yearly production of the undrilled wells and to find the future annual production by using a composite decline curve applicable to the district.

PLOTTING TOTAL OUTPUT BY TIME PERIODS.

A simple and fairly trustworthy method for estimating the future production of a field, if the output is declining and the field is nearly drilled up, is to plot a curve showing the total output of the field for

---

[a] Day, D. T., Papers on the Conservation of Mineral Resources: U. S. Geol. Survey Bull. 394, 1909, p. 35.

[b] Gasoline: Letter from the Secretary of the Interior transmitting certain information, in response to a Senate resolution of Jan. 5, '1916. relative to the production, consumption, and price of gasoline. 64th Cong., 1st sess., Senate Doc. 310, 17 pp.

each time period from the beginning of production and to project this curve to the point at which the wells in the field can no longer be operated at a profit. However, if the output of the field is not declining, some other method must be used.

PLOTTING DAILY PRODUCTION PER WELL.

Another simple method, one more nearly accurate than the preceding, is to plot the curve showing the production per well per day for any time period. By projecting this curve the average production per well per day can be estimated for any period. This method has not been used as much as it should be, because some persons believe that inasmuch as all the wells in the field have not been drilled the daily production per well will be upheld by the yield of new wells. This objection is not valid, especially if the production per well in a district has begun to decline on account of interference. After a field has attained a certain age the rate of decline in the daily production per well remains practically unchanged, regardless of the number of new wells drilled. Essentially, the reasons for the rate not changing are that the initial production of the new wells drilled is reduced by interference of older wells, and the actual number of new wells is so small a proportion of the total number of wells producing that the output of the new wells is correspondingly insignificant. Hence, the daily production or the monthly or yearly production, per well after a field has attained maturity and interference is fairly prevalent throughout, may be used in estimating future production without fear that the new wells drilled will materially change the rate of decline of the average well. It is necessary that the wells shall be drilled close enough to be affected by drainage. In a field where the productive sand is lenticular, or made up of several disconnected lenses, or if the wells are being widely spaced the statement will not always hold.

This method has often been used by the author in determining the normal decline of a field, in finding the number of wells to be drilled in an area to maintain or increase production, or in estimating the future production of the area. It is easy to apply and is sound in principle, as it is based entirely on the past performance of the wells. Several years ago McLaughlin[a] mentioned the possibility of using the method.

DETERMINATION OF NORMAL DECLINE OF MIDWAY-SUNSET FIELD, CAL.

To determine the normal decline of the field for any period the time when a constant number of wells were producing is selected, and the decline for the whole group from the beginning to the end

---

[a] McLaughlin, R. P., Petroleum Industry of California: California State Mining Bureau, Bull. 69, 1914, pp. 57–58.

of the period is determined. Figure 17 shows the curves used for this purpose in the Midway-Sunset field (Cal.). Here the production has declined to 40 to 50 barrels daily per well, and the influence

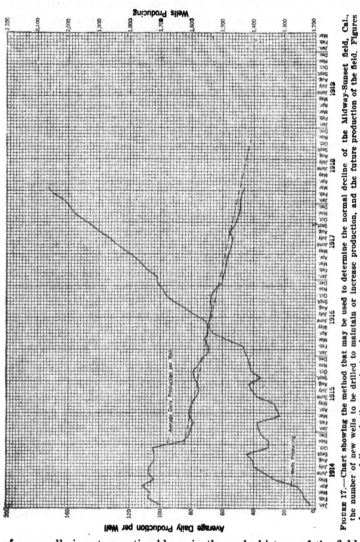

FIGURE 17.—Chart showing the method that may be used to determine the normal decline of the Midway-Sunset field, Cal., the number of new wells to be drilled to maintain or increase production, and the future production of the field. Figures for average daily production are in barrels.

of new wells is not as noticeable as in the early history of the field. The curve showing the daily production per well declines regularly for several years and can be projected into the future with small error. In this way the percentage decline for any period can be computed.

Although such a selection really was unnecessary, as the influence of new wells was negligible, a period was selected when little drilling was done. For instance, the daily production per well was 85.5 during November, 1914, and 71.5 barrels during February, 1916, a decline of 14.0 barrels daily, or 16.4 per cent in 1¼ years. The resulting average is about 13 per cent a year. This estimate was checked by taking the period from January, 1916, to January, 1918, and using the average line instead of the actual production figures. The computed decline was about 16 per cent. The two estimates are in fair accordance regardless of the fact that nearly 600 wells were drilled during the latter period and only 85 during the former. So that 15 per cent was taken as the normal decline in the Midway-Sunset field for the coming year.

DETERMINATION OF NUMBER OF NEW WELLS NEEDED TO MAINTAIN OUTPUT.

To determine the number of new wells to be drilled to maintain the production for March, 1918, to March, 1919, the following procedure was adopted: The daily production per well during March, 1918, was 46 barrels. By applying the normal decline of 15 per cent one finds that the daily production during March, 1919, would be 39 barrels per well. The daily production of the field during March, 1918, was 95,334 barrels, and if this production is to be maintained, the number of wells that must be producing during March, 1919, is determined by dividing 95,334 by 39, which gives 2,440. As the number of wells producing March, 1918, was 2,065, 375 wells (2,440– 2,065) must be drilled during the year between March, 1918, and March, 1919.

Suppose it was desired to increase the daily production of the field to 120,000 barrels daily during the ensuing year. The number of wells that must be producing in March, 1919, is determined by dividing 120,000 by 39. The quotient is 3,075. Subtracting 2,065—the number of wells producing in March, 1918—leaves 1,010 wells as the number to be drilled. This procedure not only affords a quick and trustworthy method of estimating the number of wells to be drilled to ·carry out any production program, but also shows what a large number of wells must be drilled in a field of that size to increase materially the daily production.

ESTIMATING FUTURE OUTPUT OF A FIELD.

To estimate the future output of the field requires only the determining of the number of wells that remain to be drilled, the plotting of these in a curve showing the rate at which they are to be drilled, the projecting of the curve showing the decline in the average daily output per well to a point at which the wells can not be economically operated, and then determining the total production for each month by multiplying the daily output per well by the days in a month and this product by the number of wells producing. The future output may be obtained by addition.

Some persons may think that this method is fundamentally wrong because it takes no account of the flush production of new wells. But in a field several years old the new wells have little influence on the average production per well a day, and drilling may be practically ignored. This generalization holds true in the other California fields, and unquestionably is true in the oil pools of the Mid-Continent district, where the sands are not as thick, interference between the wells is more rapid, and the wells decline more quickly.

ESTIMATED NORMAL DECLINE OF OTHER CALIFORNIA FIELDS.

By use of this method the author has determined the normal decline of other California fields. as follows: McKittrick, 9 per cent; Kern River, 8 per cent; Lost Hills-Belridge, 10 per cent; Coalinga, 10 per cent; and Lompoc-Santa Maria, 10 per cent. The normal decline for all of California was also determined, by combining the results of the individual fields, as between 11 and 12 per cent. The percentages of new wells required to maintain production as determined in this manner coincides rather closely with Lombardi's[a] estimates. For instance, Lombardi gave for the Coalinga field, 9.33 per cent, and for the Midway-Sunset field, 14.8 per cent of new wells, whereas the respective percentages, by using the method above, are 10.9 and about 18 per cent. For the whole State the author estimates that about 11 per cent more wells must be drilled during the next year to maintain production, whereas Lombardi estimated the increase as 8.22 per cent.

Pack[b] gives a modification of the method; by this the future production of a field at different times can be estimated. Most of the methods evolved were based on the maintenance of production and on the number of wells drilled during the periods in which production was maintained or was declining at a known rate.

ESTIMATING FUTURE OUTPUT BY DETERMINING AVERAGE AGE OF PRODUCTION.

As explained on page 62, a feasible method of estimating the future production of wells already drilled in a field is to determine

---

[a] Lombardi, M. E., The cost of maintaining production in California oil field : Am. Inst. Min. Eng., Bull. No. 105, September, 1915, pp. 2109–2114.
[b] Pack, R. W., The estimation of petroleum reserves : Am. Inst. Min. Eng., Bull. 128, August, 1917, pp. 1121–1134.

the average age of production. The only information required is: (1) a composite decline curve of several wells or properties in the field; (2) the present daily production of the wells for which the estimate is to be made; and (3) the total amount of oil already produced. The object is to determine the point on the composite decline curve at which the average barrel of production is declining at present—in other words, to determine the average age of a barrel of production. The total daily output for all the wells can then be considered as the actual output for one day for one great well of a certain age. With this age determined, the first year's daily production of this well may be found by dividing its present daily output by the percentage, on the decline curve, indicated by the age of the production. After this has been found the actual output in future may be obtained by applying the curve values for the future years to the first year's daily production. As the method has already been fully explained (p. 62), it is not discussed in detail here.

ESTIMATING BY USE OF APPRAISAL CURVE.

An appraisal curve, if it is available, may also be used for estimating the future or ultimate output of the producing wells in a field, the average production per well for the field being determined and the future or ultimate production of a well of that output being estimated. Then the future production of this average well should be multiplied by the number of wells in the field.

ESTIMATING BY " SATURATION METHOD."

If a field has just been discovered and little information is to be had on the probable action of the wells drilled, a possible way of estimating its future production is by the "saturation method." As an example of the application of this method during the early history of a field, the estimate made by Shaw as to the quantity of oil in the Carlyle field (Ill.) may be cited. Shaw says[a] that this estimate was made five months after the field was discovered and that it was proving better than estimates that could have been made by any other known method. As a general rule, however, estimates by this method should be made only after careful consideration of all available data and then only with the object of determining within wide limits the field's probable future production. Obviously, the unknown factors entering into an estimate of this kind are too many

---

[a] Shaw, E. W., Discussion of paper by A. W. Lauer on the Petrology of reservoir rocks and its influence on the accumulation of petroleum. Econ. Geol., vol. 13, No. 3, May, 1918, p. 214.

for the accurate determination of future output and the accuracy of successful estimates made by the method is due probably to the compensating error introduced by guessing at many unknown factors. For instance, no one can say for a certainty what percentage of the total oil stored in the rocks will be recovered. Furthermore, practically nothing is known as to the pore space in the reservoir, and information on the extent and thickness of the reservoir must be inferred almost entirely from geologic observation. Even though it were possible to determine closely, after the first well had been drilled, the total amount of oil in a field there still would be great chance of error in estimating the recovery factor.

OTHER POSSIBLE METHODS OF DETERMINING FUTURE PRODUCTION.

USE OF RECORDS OF COMPOSITION AND VOLUME OF GAS.

Profitable investigations on estimating future oil production can undoubtedly be made by studying the changes in the composition of the gases accompanying the oil as a well becomes older. The author has done no work of this kind, and owes the suggestion to Mr. J. O. Lewis, of the Bureau of Mines. More data should also be obtained as to the cubic feet of gas accompanying each barrel of oil produced for different periods of the life of a well. Everyone familiar with the drilling of oil fields knows that a well usually gives most gas when it begins producing. From that time onward, under normal conditions, the quantity of gas per barrel of oil produced constantly decreases; moreover, the composition of the gas changes. Hence, these two possibilities should be studied together. Very likely if oil producers noted the average quantity of gas accompanying each barrel of oil from their wells during each month, the data could be compiled so as to yield inferences of much interest, for it is believed that the decrease in the volume of gas accompanying each barrel of oil throughout any period corresponds closely to the decline in the output of oil. To collect this information in the oil fields of the United States is difficult, for, so far as the author knows, not many records of the gas that accompanies the oil are kept.

There seems much promise of obtaining profitable and interesting results by comparing the decline of well pressures with the decline of oil production, for the output of a well, especially during its flowing life, is beyond doubt closely related to the force that causes the well to flow. Moreover, it is fairly simple to ascertain and record the pressure in the well during its flowing life. Figures 18, 19, and 20 are excellent examples of the relation between the decline of well pressure and that of oil production. The reader should note how closely the curves showing well pressure and oil production correspond. Thus, in figure 20, after the month of November, 1914, the curve representing the well pressure seldom fluctuates without there being a noticeable similar fluctuation in the curve showing the average daily output of the well. For example, during January, 1915, there is a depression in the curve showing the pressure of the well. The production curve shows a similar depression. During

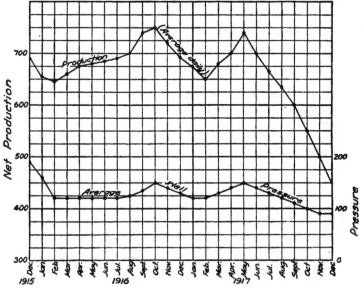

FIGURE 18.—Chart showing the decline in rock pressure (in pounds per square inch) and the corresponding decline in oil production (in barrels) of a well in the Midway field, Cal.

the next month the pressure went up and the production correspondingly increased. For the next two months the pressure was reduced somewhat and the production dropped off rapidly. From May, 1915, both production and pressure decreased rapidly to June, 1916, when the production increased a trifle, and during July, 1916—a month later—the pressure became greater. The increase of pressure and production continued until October, 1916, when both declined. The curve showing the increase in water should be noted also.

Some of the more progressive companies in the different fields of the United States are now taking similar observations, and it is

hoped that enough data will soon be available to prove or disprove
the value of such a method of estimating future production. One

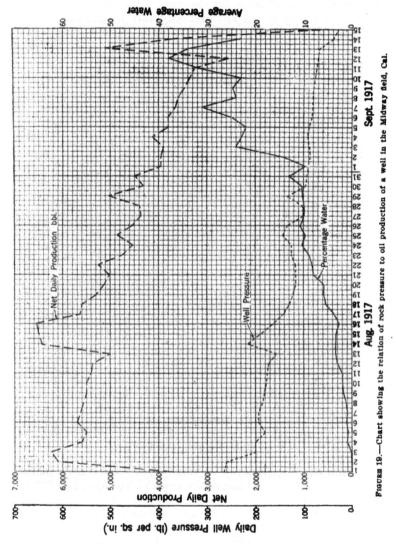

Figure 19.—Chart showing the relation of rock pressure to oil production of a well in the Midway field, Cal.

of the disadvantages is the difficulty in ascertaining the well pressure
in pumping wells.

MAKING HASTY ESTIMATES OF THE FUTURE PRODUCTION OF OIL WELLS.

In valuing drilled oil lands the appraiser often finds he has to make hasty estimates of the approximate future production of a property, though he may have few data at hand. Hence it is neces-

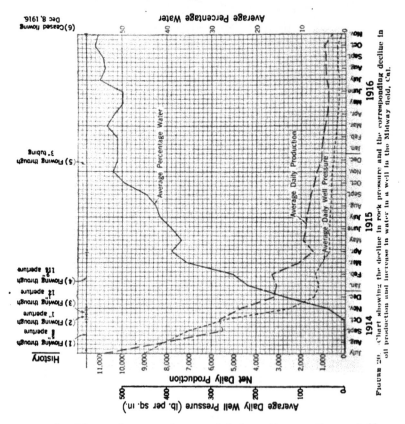

sary for him to have much general information on the probable action of oil properties under certain conditions, and, in addition, several rule-of-thumb methods that enable him, on short notice, to estimate roughly the future output under certain conditions. Although these methods may not insure exact results, nevertheless they afford a general idea of the probable worth of a property as shown by a rough determination of its future output.

About the only data immediately available, when estimates are required on short notice, are as follows: (1) The past production of the property; (2) the approximate age of production, especially if the property has been completely drilled at approximately the same time; (3) the present daily production per well; and, occasionally, (4) the output of the property during the first year, or the average daily production per well during that period. If considerable information can be collected the appraisal curve may be used, especially in making a rather detailed estimate of the future production. Otherwise. for rapid calculations, figure 21, which for the lack of a better name has been called an estimating chart, will prove useful.

USE OF ESTIMATING CHART.

BASIS OF CHART.

Briefly. this chart is based on the annual ratio of the future to the past production of wells of various sizes the first year, and has been computed from the appraisal curve. For example, if the annual production of a 10-barrel well on a property in the Crawford County field (Ill.) is known, the future production at the end of each year can be calculated in accordance with the following table:

Relation of past production to estimated future production.

Years. 1	Estimated annual production, barrels. 2	Total past production at end of year, barrels. 3	Estimated future production at end of year, barrels. 4	Ratio between future and past production, cols. 3 and 4. 5
1	3,650	3,650	6,450	1.77
2	1,680	5,330	4,770	.89
3	1,000	6,330	3,770	.60
4	730	7,060	3,040	.43
5	650	7,710	2,390	.31
6	390	8,100	2,000	.25
7	320	8,420	1,680	.20
8	280	8,700	1,400	.16
9	230	8,930	1,170	.13

CONSTRUCTION OF CHART.

In the table above the average ultimate production for a 10-barrel well in the Crawford County field is 10,100 barrels. After the well has produced one year at the rate of 10 barrels daily, its estimated future yield will be 6,450 barrels. The remainder of the table is computed in the same manner. Thus at the end of the first year there is an estimated future production of 6,450 barrels, and the well

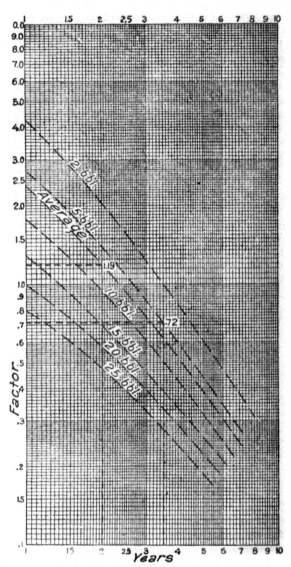

Figure 21.—Estimating chart for roughly calculating the future production of wells in the Crawford County and Clark County fields, Ill.

has produced 3,650 barrels, which gives a ratio of 1.77. In other words, with the past production given—3,650 barrels—the future can be determined by multiplying it by the " factor," or ratio, of 1.77. The ratios are then plotted on logarithmic coordinate paper against the ages of the well during successive years. This paper was used because, as has been said, the curves plotted on it more nearly approach straight lines. It will be noted that the same method of procedure as was used in determining the curve for the 10-barrel well was also used for determining curves for wells of other sizes and also for the average well. The last curve was computed so that if it would be impossible for the appraiser to determine the average daily production per well on a property during the first year, he could use the average curve with the same chance of error as is involved in the use of composite decline curves instead of appraisal curves for estimating future production.

MANNER OF USING CHART.

With such a chart available, if the appraiser can determine the average age of the production, and if he knows the average past production per well, the future output can be found by simple multiplication. For example, the production on a property in the Crawford County field is 3½ years old (fig. 21). The average past production per well is 4,000 barrels. Assume that the average daily production the first year was 5 barrels, then the average future production per well is determined by following the 3½-year line vertically to where it intersects the 5-barrel line, following thence horizontally to the left margin, and reading the factor, 0.72. The past production—4,000 barrels—multiplied by 0.72 gives the average future production for each well, or approximately 2,900 barrels.

If the average daily production the first year is not known the best method of procedure when only the average past production per well and the age are given, is to use the average curve. For example, assume the age is 2 years (fig. 21). The average well has produced 6,000 barrels. The average future production of each well is determined by following the 2-year line vertically to its intersection with the average line and thence to the left margin where the reading found is 1.19. The average past production, 6,000 barrels, multiplied by 1.19 gives the average future production per well, or 7,150 barrels.

One point that should be brought out is the compensating error introduced by the action of properties that are better or worse than the average. The curves in figure 21 are based on the average well. Now the output per well on a certain property may have been much above the average, so that multiplying the past production by the factor or ratio would make the estimate of future production too high. However, estimates of the future production of such wells have disproved this conclusion, for the future production of the better properties is higher than that of the average properties. The ratio between future and past production is, therefore, approximately the same, and practically no error is introduced by using the chart for estimating the future output of properties above or below the average in productiveness.

If a property is not completely drilled, the future production obtained by the method just given can be added to the past production and the ultimate production per acre of the drilled area can be determined. These values may be applied, with proper modification, to the undrilled proved acreage to determine its approximate ultimate production.

VALUE OF ESTIMATING CHARTS.

These charts for roughly estimating future output should not be used except where hasty estimates are to be made, for they represent the action of average wells of the indicated daily production for the first year, and do not show, as do the appraisal curves, the maximum and minimum limits, so that if enough time is available and the necessary information at hand, much closer estimates of ultimate and future production may be obtained by the use of appraisal curves and from the generalized decline curves. Moreover, the rate at which the oil is to be obtained may be computed.

Part 2 contains estimating charts for some of the fields of the United States, and although they may have little value except in exceptional cases, yet if used with due regard to their limitations, they may save considerable work in obtaining rough estimates of ultimate and future production.

NOTES ON THE VALUATION OF OIL LANDS.

GENERAL CONSIDERATIONS.

The valuation of oil properties demands much more than the mere determination of the possible yield of a property, although this and the rate at which the oil is to be obtained are the chief factors in making valuations.

No set rule can be laid down for determining future and ultimate production, because the procedure to be followed depends primarily on the character of the valuation required and the number and character of the properties to be valued. For instance, an approximate value may be desired of several hundred properties or a painstaking estimate of one property. The methods of making these two valuations are radically different. In making the first the appraiser, on account of the labor involved in determining even a rough value of several hundred properties, must adopt methods of shortening the time and of minimizing the labor. The variations in the manner of applying the methods given are many and the checking of estimates made by two or more different methods is always desirable.

CLASSIFICATION OF PROPERTIES TO BE VALUED.

Regardless of the accuracy of the estimates to be made, the appraiser is confronted at the start by the kinds of properties to be valued. These may be divided roughly into three different classes: (1) Properties practically all drilled; (2) properties partly drilled and on which more drilling is to be done, and (3) properties on which no drilling has been done.

In determining the value of properties that are practically all drilled, the most important task necessarily is to determine the future output of the drilled wells and the rate at which the oil is to be obtained. The drilled area of a partly developed property may be treated in the same way; but for the undrilled areas of the partly developed properties and for the undrilled properties, a different procedure must be followed. These undrilled tracts must be classified according to their probable productiveness, a drilling program must be postulated, and the future annual production then determined.

CLASSIFICATION OF UNDRILLED LAND.

The classification of undrilled tracts according to their probable productiveness is necessarily based on geologic inferences that may be more or less uncertain, so that the whole valuation becomes just that much more liable to error. Geology is not a mathematical science; it can be applied to oil-land valuations only in so far as it aids the making of an estimate that in itself is usually an approximation.

Undrilled acreage should be divided, if possible, into (1) proved oil land, (2) probable oil land, (3) prospective oil land, and (4) worthless territory, so far as the economical recovery of oil is concerned. Occasionally much more detailed classifications are made, but the writer believes this refinement incompatible with the uncertainty as to underground conditions.

The classification is given not as something that geologists engaged in oil-land valuation should adopt but as a suggested method of ascertaining the probable value of undrilled land. The boundaries between the different classes of land can not be easily determined, and no set rule can be made for any classification, as, in the final analysis, the segregation rests practically altogether on personal opinion, molded by the balancing of all available evidence.

PROVED OIL LAND.

The first class should include those areas in which drilling involves practically no risk. Just what constitutes proved oil land depends, it is true, upon local conditions. All of some quarter sections on which only one well has been drilled may be called proved oil land, even though it may not be surrounded by wells. Other tracts, on the contrary, before they could be considered proved, would require many tests; in drilling such land "every well is a wild-cat well," to use an oil-field saying.

The following definition of proved oil land has been modified from that given by the California State Mining Bureau [a]: "*Proved oil land* is that which has been shown by finished wells, supplemented by geologic data, to be such that other wells drilled thereon are practically certain to be commercial producers."

PROBABLE OIL LAND.

Probable oil land, as the name implies, includes those areas where from geologic inferences as to the structure and the continuity of the sands, and from the information obtained by drilling, the chances favor the finding of oil, although considerable uncertainty may attach to drilling on such land.

PROSPECTIVE OIL LAND.

Prospective oil lands include those areas, classified entirely by geologic observations, on which all available evidence indicates the possible presence of oil in commercial quantities.

WORTHLESS LAND.

Worthless territory is that which has been proved nonoil bearing by drilling, or that in which from geologic observations, the chance of obtaining a profitable yield of oil is remote. All land containing oil that can not be recovered with profit must also necessarily be placed in this class; hence, an area that may be rated worthless at the present time may become, with higher prices for oil or the development of more economical processes, proved oil land in the future. Similarly the drilling of one well on what is now prospective oil land may place the land in the probable or proved class.

---

[a] McLaughlin, R. P., Petroleum Industry of California : Cal. State Min. Bureau, Bull. 69, 1914, p. 69.

NEED OF ESTIMATING INITIAL YEARLY PRODUCTION.

In estimating the annual production of undrilled land, it is necessary to estimate the future initial yearly production of the wells to be drilled or to make estimates of the probable production per acre of the tracts of land in the different classes, so that composite decline or appraisal curves may be used to determine the future production for each year. Often it is a good plan in making such ultimate-production estimates to call the ultimate production of the proved oil land 100 per cent and to estimate the probable productiveness of the tracts in the remaining classes by assigning a factor to each class. For instance, probable oil land may be considered 75 per cent as valuable as proved land, and so on. The percentages assigned each class are necessarily arbtirary, as they depend on local conditions and on the judgment of the person making the estimate.

IMPORTANCE OF PROSPECTIVE RATE OF DRILLING.

A matter of primary importance in oil-property valuation is the rate at which the oil may be expected, or the future annual production. For properties not fully drilled, this production depends on the rate of development or the yearly drilling program; no accurate valuation of oil land can be made, nor should it be attempted, without considering this program. Usually the company desiring the valuation has a fair idea as to the rate at which the property is to be drilled. If such knowledge is not available the engineer should assume a drilling program that in his opinion will insure the most economical development of the property.

Estimates must be made of the probable future annual production of each property so that one may determine: (1) Future annual net income (apparent value), and (2) the present value of these deferred net receipts (actual value). The actual present value of a tract is the sum of the present value of the future net annual receipts. The present value of an income that is to be available, for example, 10 years hence, is the sum that when invested at an accumulative rate of interest, compounded for 10 years, will equal the income. "Accumulative rate of interest" is here defined as the rate of interest on capital on a practically secure investment, as distinguished from the "remunerative rate of interest," which is the rate the investor seeks when his capital is to be put in an uncertain venture.

RELATION OF PRESENT VALUE TO DEFERRED PROFITS.

To understand more clearly the application of present value to deferred profits one must consider that when an oil property is bought the purchase price is taken from where it may earn interest and invested in oil that brings no returns until recovered and sold. Thus there is a loss of interest pending the recovery and sale of the oil. Where oil deposits are relatively certain investments and the future annual production, and hence the annual net profit, can be estimated, the present value factor should be applied to the deferred profits. Occasionally, however, the investment is rendered so speculative by uncertainty as to the presence of oil in commercial quantities that the added refinement of discounting deferred profits is not warranted. Furthermore, the uncertainty of the future price of oil may not justify such refinement.

The annual net receipts for the future are controlled by (1) the future annual production; (2) the future cost of development; (3) the future cost of production; (4) the future price of oil; (5) the rate of amortization of capital; and (6) the salvage value of the equipment. Thus, to compute the value of an oil property, one must determine the present value of the deferred or anticipated profits, which will be the equivalent of a remunerative interest on the capital invested. The rate of interest demanded depends on the risk of the investment, and the return, or the amortization, of the capital invested, depends on the hazard of the investment and the probable life of the property.

To determine the influence of the different factors on annual net receipts is sometimes very difficult. For instance, although one may determine, with what is believed to be considerable accuracy, the future annual output of a property he may find great difficulty in predicting the price of oil, although that of course controls the annual net receipts. The future cost of development and of producing oil should give little trouble to the appraiser who has at hand data on the past cost of development and production.

Not only the interest demanded on the investment but the capital itself must be returned to the investor before the property is abandoned. This redemption of capital is necessarily based on the life of the field, but as the life of most oil properties is uncertain the time in which the invested capital is written off is in general considerably less than the probable actual life of the field.

After a field has been depleted so that pumping is no longer profitable, some of the equipment may be sold for a certain percentage of its original cost. This asset, however, is usually a very small factor in the actual value of the property, inasmuch as many properties have a life of 20 to 30 years and the material actually used in pumping to exhaustion has little salvage or scrap value. The usually small value of scrap at the time the lease is abandoned is further reduced by the necessity of computing its present value, as this sum, like the net receipts from the property during any year of productive life, is a deferred or anticipated return. Furthermore, the cost of abandoning a property is often greater than the value of the "junk."

METHODS OF PURCHASING OIL LANDS.

There are two general ways of purchasing oil lands. The first is by what is known as the "settled-production" method. In this a certain unit value per barrel of daily production, exclusive of the royalty, is given, this value being based on the gage of the output of a property for several days. The second method is by actually appraising a property, or, in other words, by determining the amount of money that the purchaser can afford to invest in the property under certain conditions.

"SETTLED-PRODUCTION" METHOD.

BASIS OF METHOD.

The first, or "settled-production" method of buying producing-oil properties is the one in common use east of the Rocky Mountains. This method was probably evolved for the dual purpose of rapidly valuing properties and of determining depletion for writing off the capital invested; at any rate it originated in the Appalachian region, where some rapid method of determining the relative value of different properties was necessary. The method is based upon the "settled production" of a well or property, the output of the property being gaged for several days to determine the actual daily yield. Then the royalty interest and the pipe-line deduction of 2 or 3 per cent (common in most of the fields east of the Rocky Mountains and made on account of the presence of sediment and water in the stock tanks) are subtracted from the daily output. A unit value per barrel is paid for the remainder of the daily production, or the working interest.

For example, if the daily output of a property as determined by a several days' gage is 100 barrels, the pipe-line deduction of 3 per cent leaves 97 barrels and the deduction of the royalty interest of $12\frac{1}{2}$ per cent from this leaves a working interest of 84.9 barrels daily. The value of a barrel of daily output depends on the number and value of the undrilled wells, on what percentage of the daily output is "flush" production—that only a few weeks or months old—on the present and the probable future price of oil, on operating costs, and on the productivity of the present wells. In the example given, if the value per barrel is $1,000, the value of the whole lease is $84,900.

An attempt has been made to discover what is the basis for determining the value of a property by this method. On what does the purchaser rest his estimate? Has it any scientific foundation or is it a mere guess? A method so widely used must be based on some easily ascertained condition that makes it of value and applicable in so many different districts and properties. It was found that the method probably has no scientific basis.

With a knowledge of the factors controlling the value of a barrel of production, the usual buyer proceeds much as he would in buying a horse, the value depending not so much upon the profit he can derive as upon the market value of the horse. This value—which is not the price of oil—seems to be the average of the opinions of the operators dealing in oil lands, the opinion of each operator being determined by the " feel " of conditions, and not, except for a few men, by actually computing the difference between probable total expenditure and probable gross income. This explanation does not signify that operators who employ this method in purchasing oil lands proceed blindly; the opposite is true.

In some districts an increase of 10 cents in the price paid the producer for oil arbitrarily increases the unit value per barrel of the oil by $100. For instance, if the settled daily production of a certain property is worth $1,000 per barrel with oil selling at $2.50, an increase in the price of oil to $2.60 would increase the unit value to $1,100 per barrel. This graded increase may not hold at the present time.

RETURN OF PURCHASE PRICE.

Frequently the purchaser demands that his money be returned within three years from date of purchase. For instance, one operator in Oklahoma expects 45 to 50 per cent of the invested capital returned the first year after he purchases the property, 30 per cent during the second year, and 20 to 25 per cent during the third year. The return of the purchase price may be expected from the production of the old wells, which of course are declining, or from the production of the wells to be drilled. The author believes that the demand for the return of invested capital in such a short time as three years is ultraconservative, except where the future yield of properties is highly uncertain.

INCOME OF PROPERTIES BOUGHT ON " SETTLED PRODUCTION " BASIS.

In studying this method of oil-land valuation the author collected data on the income derived by several companies from many properties purchased by the " settled production " method. The age and amount of the settled production when a property was purchased and the number of wells producing at that date were determined. Next the total investment was computed from the unit price paid per barrel, and the properties were classified according to the royalty paid, whether or not drilling was done after the purchase. Then all the different classes of properties were compared.

A most interesting table was prepared and comparisons made as to the length of time required for the properties on which new wells were drilled and for those on which no new wells were drilled to repay original cost. Such statistics have no great value, as conditions for different properties vary, and an increase in the amount received per barrel by the producer during a few months of the prolific part of a well's life will materially increase his net income and cause the property to " pay out " so much the sooner; but some of the conclusions obtained by studying one group of properties are given below because of their possible interest to the reader.

For instance, in Illinois, of 88 properties that changed hands in Clark, Crawford, and Lawrence Counties, at an average price of about $375 per barrel of settled production, 50 of the properties, or 57 per cent, "paid out" within 4 years and 10 months. At the time the information was collected—an average of about 6 years after the properties were purchased—the remaining 43 per cent were still in debt. On 39 of these properties no new wells were drilled after the date of purchase; and 19 of these 39 properties, or 49 per cent, "paid out" in the same length of time as the others— 4 years 10 months. New wells were drilled on 47 properties, of which 61 per cent "paid out" in the same length of time—4 years 10 months.

The properties were also classified according to the age of the production when the purchase was made. Those having a production less than 2 years old (1) were separated from those with production more than 2 years old (2). Likewise, the properties in each of these two groups were separated into two classes, (a) those on which no drilling was done after the date of purchase, and (b) those on which new wells had been drilled. The results of this classification and the conclusions obtained are shown in Table 2.

Sixty-eight per cent of those properties in group 1, class a (production more than 2 years old at the date of purchase and no drilling after that date), "paid out" in 5 years; and 100 per cent of the properties in group 1, class b (production more than 2 years old and new wells drilled), "paid out" in 4 years and 3 months. It is interesting to note that all the 9 properties "paid out" in 9 months less time than those properties whereon no drilling was done.

Of the 8 properties in group 2, class a (production less than 2 years old and no drilling done), 12 per cent "paid out" in 4 years and 5 months. The remaining 88 per cent had not "paid out" when the data were collected. Of the properties in group 2 (production less than 2 years old), 57 per cent of those in class b (drilling done after purchase) "paid out" in 5 years and 1 month; the remaining 43 per cent had not "paid out" at the time the data were collected, this being about 6 years on the average after the date of purchase.

TABLE 2.—*Classification of the oil properties purchased in Illinois by an oil company, showing the percentage of those that had returned their cost and the length of time required.*

(1) Production more than 2 years old at date of purchase.				(2) Production less than 2 years old at date of purchase.			
(a) No drilling after purchase.		(b) New wells drilled after purchase.		(a) No drilling after purchase.		(b) New wells drilled after purchase.	
Paid out.	Number not paid out.	Paid out.	Number not paid out.	Paid out.	Number not paid out.	Paid out.	Number not paid out.
Months.		Months.		Months.		Months.	
56	1	47		53	1	21	1
87	1	47			1	92	1
66	1	42			1	75	1
45	1	70			1	35	1
81	1	38			1	48	1
55	1	61			1	30	1
57		60				66	1
51		47				78	1
39		48				91	1
53						57	1
70						74	
57						66	
57						74	
774	6	460	0	53	7	798	10
Average, 5 yrs.	....	4 yrs., 3 mos.	....	4 yrs., 5 mos.	....	5 yrs., 1 mo.	....
Per cent paid out, 68.	....	100	....	12	....	57	....

Of interest is the accord in the length of time that most of these properties "paid out" and the lack of accord in the percentage of those that "paid out" under certain conditions. The average time of "paying out" ranged from 4 years and 3 months to 5 years and 1 month, whereas the percentage that "paid out" ranged from 100 per cent to 12 per cent.

PROPERTY THAT WAS SAFEST INVESTMENT.

It would seem from this analysis that under the circumstances prevailing at the time the safest property to buy was one on which the production was well "settled" and on which drilling could be done; also, that if at the date of purchase the production is comparatively young, it is a safer business venture to purchase those properties on which new wells can be drilled. Although the statistics given may not be applicable to other districts, and all the factors influencing the value of the properties have not been considered, yet at least the data furnish some interesting conclusions for that particular district.

A few years ago the author helped to make an interesting valuation of several partly developed oil properties in the Osage Nation (Okla.). These properties were to be leased at public auction by the Federal Government, acting as agent for the Osage Indians, at a royalty of one-sixth, to the person bidding the highest bonus. The bidding was on a "barrel basis," or the amount per barrel of daily production (the working interest) that would be paid as a bonus. But the valuation was made by estimating the future annual production of each property with a specified future drilling program, allowing a certain remunerative rate of interest on the capital invested, and computing the present value of the deferred profits. It was estimated that the future price of oil would be slightly greater than the then existing price. The object of the valuation was to determine the minimum price at which the Government, acting for the Osage Indians, could afford to dispose of the leases.

Thus it was possible to compare directly the probable minimum value, as determined by one method of valuation, with the highest price, a group of operators felt should be paid for a property. In general, the prices paid per barrel were considerably above, and often two and three times, the minimum value set by the Government. This difference was probably due to the optimism of the purchasers in regard to the future price of oil and to the intense competition for oil leases in that part of Oklahoma. Fortunately the price of oil since has increased considerably, even more than was then expected, so that the investments probably have proved lucrative.

APPRAISAL METHODS.

GENERAL STATEMENT.

An oil property is appraised on the basis of the net returns to be obtained annually from it. Whether or not the deferred profits are discounted depends in a measure on the accuracy with which the future annual production is estimated. The purchase value of a property is the sum that will be paid back with interest to the investor before the oil is exhausted, and it is governed by all the conditions that control oil production, only a few of which have been discussed in the preceding pages.

In appraising the value of oil lands it is imperative that the undrilled part of a producing property be segregated and appraised on a different basis from the drilled part. Slow drilling ordinarily reduces the ultimate amount of oil that may be obtained from a property because of the accompanying drainage of the sand and the gradual reduction of gas pressure; so that a property is worth more if it is to be drilled at once, for not only will the ultimate production probably be greater, but also the present value of deferred profits is greater, provided the price of oil does not change. Therefore, as a result of postponing the drilling of a property, a double factor militates against its value.

Many mathematical computations are involved in the valuation of oil lands, especially if a careful estimate is to be made of the value of many properties. Hence, there then may be great need of short cuts and rule-of-thumb methods. Such figures as the estimating chart (fig. 21) for determining future production quickly, would be of value if they could be so revised as to show the actual value of oil properties for different rates of drilling, for different future prices of oil, and the like.

METHOD OF APPRAISING SEVERAL SCORE PROPERTIES.

The author recently appraised several score oil properties for the Government, and because of the necessarily short time available the adoption of some system by which much of the tiresome mathematical work would be eliminated became imperative. The properties were in three different main districts, but the controlling conditions in each district differed radically. For instance, in district No. 1 were several wells; these partly proved the oil-bearing character of the geologic structure where they were drilled. In the second district several hundred wells were drilled which defined the probably productive area with fair accuracy. No wells had been drilled in the third district, and there a geologic report had to be used as the basis of the appraisal.

A part of the detailed work necessary for the appraisal of the properties in the second district is given here, because although they varied widely there was no great uncertainty, on account of the number of wells completed, as to the ultimate outcome of drilling. The properties to be appraised were all partly drilled, and the lack of detailed data made necessary the use of composite decline curves in estimating future annual production. All the properties lay on a large anticline and individual well-production records were available.

DATA PLOTTED OR COMPUTED.

In general the work consisted of (1) constructing a composite decline curve of the wells in the district; (2) projecting these composite decline curves and thus estimating the future annual production of the productive wells; (3) dividing the undeveloped territory into areas of varying productivity and estimating the probable ultimate production per acre of the different classes of undrilled land; (4) computing the first year's average daily production per well from the estimates of ultimate production per acre, as explained on page 54;

(5) estimating the average daily production per well the first year and checking against the computations made in (4) ; (6) assuming a certain drilling program to be instituted on each lease—if more than one class of acreage occurred on a lease, the richest was assumed to be drilled first; (7) computing the value of the oil obtained annually from one well of each class of acreage, deducting drilling and production costs, discounting the deferred profits, and allowing a certain remunerative rate of interest on the investment; and (8) determining the acreage per well and dividing with this figure the total undrilled acreage of a certain class. The number of wells to be drilled on each class of acreage multiplied by the net present value of the deferred profits from one well gave the present value of that class of acreage.

DETERMINING THE FUTURE PRODUCTION OF THE DRILLED AREA.

As already stated, the main object of the appraisal was to determine the total value of all the properties. A composite decline curve was computed for the producing wells. Seemingly the simplest method of determining the future annual production of the district was to determine at what point on the composite curve the field was producing; in other words, to determine the average age of a barrel of production in that field. Obviously it was impracticable to determine the average age of a barrel of production in the manner used by Washburne (p. 62), for several hundred wells were involved. Therefore it was decided to determine the average age by trial—by estimating the probable age of production and then applying the curve values for the past years and determining whether the total past production thus determined would equal the actual amount produced.

DETERMINATION OF PROBABLE AGE OF PRODUCTION.

The total past production of all the wells amounted to 72,000,000 barrels; the existing daily production of the district was 37,000 barrels, and the average age per barrel of production was at first estimated at about 4 years. The percentage readings on the composite decline curves for the first 4 years are in succession 100 per cent, 66 per cent, 51 per cent, and 41 per cent, so that if the daily production of the wells was approximately 37,000 barrels and the average age of production was 4 years, the first year's production of these wells would be 37,000 divided by 41 per cent, or approximately 90,300 barrels. This then was the average daily production for the first year if the present wells all began producing 4 years before. The first year's average daily output, or 100 per cent, was multiplied by the percentages for the second, third, and fourth years, and these products in turn by 365 (days in a year). The daily outputs obtained in this way were 90,300 barrels, 59,700 barrels, 46,300 barrels, and 37,000 barrels successively for the 4 years, or a total of 233,300 barrels; this sum multiplied by 365 days gave a total production to date of 85,200,000 barrels. But the actual total production was approximately 72,000,000 barrels, showing that the assumed average age for a barrel of production was too great. The same method, using an assumed age of 3 years, gave too small a total output. A third estimate sufficed to determine rather closely the average age, which proved to be 3½ years. This average age gave the first year's average daily production as 82,200 barrels. This multiplied by 66 per cent and 51 per cent gave an average daily production of 54,500 barrels the second year and 42,200 barrels the third year, or a total average daily production for the first three years of 178,900 barrels; this average multiplied by 365 days gave a total production of about 65,300,000 barrels for the first three years. Adding the amount obtained by multiplying the first year's production by one-half the percentage for the fourth year, made a total production of 72,050,000 barrels for 3½ years.

RESULTS OBTAINED.

Although this method is merely an approximation, it served as a check on the estimates made by other methods. Similarly the approximate future annual production of the present wells was obtained by applying the values on the composite decline curve for all years following three and one-half years. The results are shown in the following table:

Probable future output, apparent net value, and present value as determined from the composite decline curve.

Year.	Estimated annual production of present wells.	Apparent net value after deducting production costs.	Present value (at 6 per cent).
	Barrels.		
Fourth..	6,115,000	$4,280,000	$4,280,000
Fifth...	10,100,000	7,070,000	7,280,000
Sixth...	8,250,000	5,780,000	5,150,000
Seventh...	6,760,000	4,740,000	3,980,000
Eighth..	5,660,000	3,970,000	3,140,000
Ninth...	4,830,000	3,380,000	2,530,000
Tenth...	4,140,000	2,900,000	2,040,000
Eleventh..	3,630,000	2,540,000	1,690,000
Twelfth...	3,180,000	2,230,000	1,400,000
And so on to—			
Thirty-fourth.......................................	270,000	189,000	33,000
Total...	78,085,000	....	39,128,000

APPARENT NET VALUES.

The apparent net values shown in the third column are deferred receipts; they were obtained by multiplying the annual production shown in the second column by 70 cents, which was assumed as the net amount received, after deducting the costs, per barrel of oil produced. Although it was realized that this net value probably was low, a conservative estimate was desired. In view of the uncertainty of the price of oil at almost any future time because of increased production and drilling costs and of possible price fixing by the Government, the approximate net value per barrel the producers received at that time was used.

PRESENT VALUE OF DEFERRED RECEIPTS.

In the fourth column are the present values of these deferred receipts, so that the total of $39,128,000 represents the total present value of the oil to be produced by the existing wells on the assumptions of a certain future price per barrel of oil and certain production costs. This sum is what a prospective purchaser could afford to pay for the oil from the existing wells and yet have as much money at the end of the period as though he had invested otherwise in securities at 6 per cent. But oil-property investments are uncertain, and returns on them are speculative. Hence, the buyer demands a rate of interest on his capital commensurate with the risk of his investment, and this rate is high. Consequently, the prospective buyer of the property mentioned could not afford to pay the sum indicated.

DETERMINING THE FUTURE PRODUCTION OF THE UNDRILLED AREA.

The proved undrilled territory amounted to about 11,000 acres and the producing wells were scattered over the whole field, thus proving much undrilled ground. Hence, it was fairly safe to assign a certain ultimate production to each undrilled acre by comparing this value with the probable amount of oil to be obtained from the near-by drilled territory instead of classifying the acreage in accordance with the scheme given on page 81.

ASSUMPTIONS MADE.

Because it was necessary to use a single composite decline curve for all the wells to be drilled it was also necessary to assume that the output of all the wells would decline along this curve. This assumption probably introduced some error in the calculations, for the initial yearly output of the new wells would decline through interference, but the error was compensating. Now, if the ultimate production per acre of a tract of land is less than that of an adjoining tract, the initial yearly production of the new wells on the first tract will be less than that of those drilled on the second tract, so that different amounts of oil will be obtained annually from wells on tracts of different productiveness. Because of the depth at which some of the productive oil zones lie below the surface, it was estimated that only 7,840 of the undrilled 11,000 acres would repay drilling.

The new wells were allotted 10 acres each, the drilling cost was estimated at $50,000 per well, and a drilling program of eight wells yearly on each tract of 640 acres was assumed, although it was realized that actual drilling probably would not proceed so rapidly. The time taken to drill the wells would influence slightly the present value, but the added refinement of computing this difference was incommensurate with the accuracy of the other parts of the estimate.

To determine carefully the present value of the oil to be obtained from such a large number of tracts would have been very laborious, and a chart (fig. 22) was prepared that greatly reduced the work required. Briefly, the chart was prepared by determining the total present value of the output of wells drilled during different years on tracts of varying productiveness. The total present value of the output of a well decreases with the deferment of drilling; furthermore, the total present value is less for a well drilled on acreage that ultimately will produce 10,000 barrels than that of a well drilled on a tract that ultimately will produce more.

Figure 22 was prepared by computing, for example, the total present value of the annual output from one well drilled during the present year on an undrilled tract that ultimately will yield 11,000 barrels an acre, and again, to determine the total present value of an identical well drilled a year hence, two years hence, three years hence, and so on, for eight years. When plotted on semilogarithmic paper these total present values for wells drilled in different years lie on a straight line, as shown in figure 22. Similar computations were made for the total present value of the output of wells drilled during different years on lands that ultimately would produce 20,000 barrels and 60,000 barrels per acre, respectively. The inclination of these three lines proved to be the same, and the additional parallel lines were determined by interpolation.

EXAMPLES OF USE OF CHART.

The chart was used as follows: If eight new wells were to be drilled during the first year on a tract of land, the total present value of their output was determined by multiplying the total present value of the output of one well by 8. Furthermore, if the tract was large enough to support eight new wells the second year and five the third year, the total present values of the output of the wells drilled during the second and third years were obtained by multiplying the present values of the output of one well drilled the second year by 8 and of one well drilled the third year by 5.

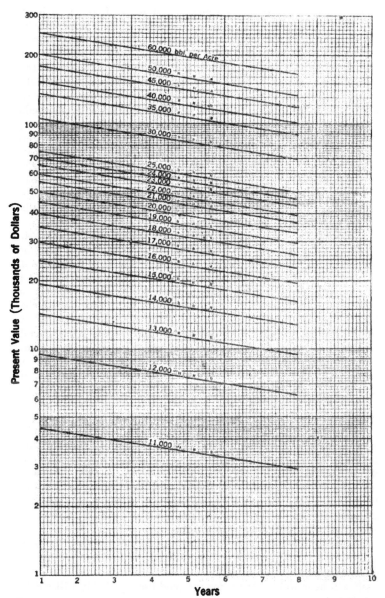

FIGURE 22.—Chart used in an appraisal of oil land for determining the total present value of the oil to be derived from wells drilled on acreages of different productiveness.

Again, suppose a property contains 330 acres of undrilled land, of which 230 acres, it is assumed, ultimately will produce 30,000 barrels per acre, and 100 acres ultimately will produce 20,000 barrels per acre. The best land will be drilled first, and it was assumed all the wells could be completed so that all would produce from the first of the year. If 10 acres are allowed for each well the 30,000,- barrel acreage will support 23 wells, and the remaining acreage 10 wells. With the drilling program of eight wells a year, eight wells will be drilled during the first year on the 30,000-barrel territory, eight wells the second year, and seven wells the third year. The total present value of eight wells drilled the first year on 30,000- barrel land is 8 times $105,000 (fig. 22), or $840,000; that of eight wells drilled the second year is 8 times $98,000, or $784.000; and that of seven wells drilled the third year is 7 times $92,000, or $644,000. This gives a total present value of $2,268,000 for the output of wells drilled on the 30,000-barrel acreage.

As only seven wells were drilled during the third year, one well may be drilled on the 20,000-barrel land to carry out the drilling program of eight wells a year. By following the same procedure it can be shown that the present value of one well drilled the third year on the 20,000-barrel acreage is $44,000. Ninety acres of this class of land remaining gives eight wells to be drilled during the fourth year and one during the fifth year. The present value of eight wells completed during the fourth year on the 20,000-barrel acreage is 8 times $41,000, or $328,000; and that for one well completed during the fifth year is $39,000. Hence the total present value of all the wells drilled on the land with an ultimate production of 20,000 barrels an acre is $411,000. This sum added to $2,268,000 gives $2,679,000, or the total present value of the undeveloped acreage.

The reader should remember that on account of the lack of specific information it was impossible to take into consideration the inevitable decline in initial yearly production. If the initial yearly production should decline during the latter part of the drilling of any one lease, the ultimate production recoverable would of course be smaller. However, as a factor of safety, the estimate of the future price of oil was made rather low, and the estimates of ultimate production were modified to take care of this factor.

The example just given shows one of a variety of methods that may be used in determining the present value of the deferred receipts in large-scale oil-property valuation. It is not expected that the chart itself will be useful to others, for the conditions probably will not be duplicated.

COMPUTING DEPLETION FOR PURPOSES OF TAXATION.

GENERAL STATEMENT.

A subject that bears on the estimation of the ultimate output of oil properties is the determining of the deductions from gross income, because of the depletion of the recoverable oil, for the purpose of computing the income tax authorized by the revenue act of September 8, 1916, as amended by the act of October 3, 1917. Individuals and corporations owning oil or gas properties are authorized by this act to deduct from gross income [a] " a reasonable allowance * * * for actual reduction in flow and production, * * * provided that when the allowance authorized * * * shall equal the capital originally invested, or in case of purchase made prior to March 1, 1913, the fair market value as of that date, no further allowance shall be made."

The reason for this provision is that capital, being returned out of profits, shall not be subject to tax. Return of capital comes out of gross profit. Hence a deduction must be made from the gross profit in order to find the true profit (taxable income) made over and above all costs. It is intended as a relief of tax upon such part of the gross profits as represents a return of the capital invested.

DEPRECIATION AND DEPLETION.

Depreciation and depletion should be clearly distinguished. Deductions on account of both are made from gross income to determine what proportion of the income is subject to tax.

Depreciation covers loss through exhaustion and wear and tear of physical property, such as machinery, and the decline in value of lease rights. The annual depreciation allowed is determined by the probable life of the physical property.

Depletion, on the other hand, covers the decrease in the amount of natural deposits through production. The production of oil entails depletion of the natural deposits, and as long as production continues depletion allowances should be made.

The Treasury Department encountered considerable difficulty in the administration of the act of September 8, 1916, because of the various methods used by operators in determining depletion, and because of the operators being uncertain as to what procedure the Treasury Department wished them to follow.

---

[a] Regulations No. 33 (revised), governing the collection of the income tax imposed by the act of Sept. 9, 1916, as amended by the act of Oct. 3, 1917, Bureau of Internal Revenue, 1918.

METHOD FIRST REQUIRED BY TREASURY DEPARTMENT.

The method first set forth[a] provided that the annual deduction authorized by the provision quoted above " must be reasonable and not
in excess of such a percentage of the cost or value, as the case may
be, and as herein defined, of the oil or gas producing properties
as is indicated by the reduction in the original flow or ' settled ' production of one year as compared with that of the preceding year."
An example of the method set forth in T. D. 2447 is as follows:

If the decline in the flow and production during the year of, say, 10 wells,
costing $100,000. has been 5 per cent. as compared with the production and
flow as indicated by a test made at the beginning of the period, then 5 per cent
of $100,000, or $5,000, will, for the year for which the compensation is made,
constitute an allowable depletion deduction in favor of the individual or corporation owning and operating the property.

FALLACIES INVOLVED.

There are two inherent fallacies in this method of determining depletion.   The first is that " settled " production is a comparative term
only and not susceptible to a specific definition; the second is that if
the aggregate flow of all the wells in a district for which depletion
is to be determined is greater than the aggregate flow of the wells
in the same district a year before (this often happens on account of
new drilling), the book value will show an *appreciation* instead of a
depreciation.   Obviously this is absurd, for production means depletion, and wherever there is production the operator should have
the privilege of deducting from his gross income a certain proportion
of his invested capital.   Thus, at the time when depletion is greatest no deductions from gross income are allowed.

METHOD USED BY SOME OIL COMPANIES.

BASIS OF METHOD.

Some of the larger oil companies operating east of the Rocky
Mountains adopt a somewhat different method for computing depletion, but the results obtained are practically the same.   The method
consists of applying a certain unit value per barrel of oil, as was
explained on page 85 in discussing the valuation of oil properties by
the " settled " production method.   For instance, a gage of the property to be purchased is taken for several days and the net daily production, exclusive of royalty and pipe-line deductions, is computed.
Then the property is purchased for several hundred dollars a barrel,
the sum paid per barrel depending on the probability of obtaining
more production on the undrilled part of the lease, the age of the
present production, the cost of development, and the like.

---

[a]T. D. 2447, Office of Comissioner of Internal Revenue, Feb. 8, 1917.

COMPUTING BOOK VALUE.

This same method is utilized in determining the book value of a property. If the property is undrilled the actual cost of obtaining the lease and the bonus paid per acre is, say, $5,000. The cost of developing oil is $5,000. The total cost is, therefore, $10,000. On the assumption that at the end of that year the production has declined to 25 barrels daily the unit value set per barrel becomes $10,000 divided by 25 barrels, or $400. If no drilling is done during the next year and production declines to 20 barrels daily, the book value of the lease will be $400 times 20 barrels, or $8,000. This value, as compared with the book value of the lease a year before, shows a depreciation of $2,000 which in this case includes depletion. On the other hand, suppose that by the expenditure of $10,000 the production increased during the second year to 60 barrels daily, then the total investment is $20,000, and the book value of the lease is $400 times 60, or $24,000. This method would make the book value of the lease show an appreciation of $4,000, although the oil deposits were depleted more during that year than if no wells had been drilled.

COMPUTING THE UNIT VALUE PER BARREL.

If a producing property is bought, the unit value per barrel paid would be adopted as the future unit value per barrel, and depreciation would be governed by that value. If the capital is not redeemed rapidly enough, the owners may find their property exhausted before the redemption fund is wholly realized; if the capital is redeemed too rapidly the owners are deferring the date at which they can obtain their profits.

Another point that should be brought out is the difficulty presented by the purchase of more than one property at different unit values per barrel of daily production. The value is computed in the following manner. Suppose 10 barrels of net daily production are purchased at a cost of $500 unit-value per barrel, a total value of $5,000, and that 20 barrels of production are bought for $800 a barrel, a total of $16,000. In the two properties the total investment becomes $21,000 for 30 barrels of production, or a unit value per barrel of $700. As properties are purchased at greater or less unit values per barrel, the average unit value per barrel is raised or lowered accordingly. Ordinarily, the same unit-value per barrel is maintained for each district where the operating conditions are similar, and if a company owns a large number of properties in different districts the depreciation is charged off for the properties as a whole in each district.

METHODS AT PRESENT REQUIRED BY THE TREASURY DEPARTMENT.

The regulations[a] recently issued authorize the use of either one of two different methods in computing depletion for the purpose of determining deductions from gross income. These methods apply: (1) Where the probable ultimate recovery of oil or gas is uncertain; (2) where the probable recoverable oil or gas can be estimated.

COMPUTING DEPLETION WHEN RECOVERY IS UNCERTAIN.

Under the first case the method required is that explained above, the percentage reduction in flow of wells on a property being computed at intervals of one year each. The objections to this method have been given.

COMPUTING DEPLETION WHERE PROBABLE RECOVERY CAN BE ESTIMATED.

Under the second case, the total amount of oil that will probably be recovered from a property is estimated, and such a percentage of the invested capital is deducted from the gross income as a year's production bears to the probable ultimate production. This has been called the " unit cost method," and obviously, this is the only method fundamentally sound for determining depletion. To illustrate its use, suppose a property will produce ultimately 100,000 barrels, and that during the first year of its life it produced 10,000 barrels, or one-tenth of the ultimate production, then the owner may charge off against gross income 10 per cent of the capital invested in the land.

To explain the designation " unit cost " we may use the same example. Suppose the property cost $20,000, and the gross income the first year was $30,000 from a production of 10,000 barrels. Because the property is estimated to be capable of producing 100,000 barrels ultimately, then the buyer has invested 20 cents in each barrel of recoverable oil underlying the property. This is the unit cost and the operator has the privilege of deducting this cost from the gross amount he derives from the sale. Here he derives $3 a barrel from each barrel produced. The same result is, of course, obtained by multiplying 10,000 by $0.20, which equals $2,000. This represents the amount of his deduction for depletion and he is allowed to subtract this amount from his gross receipts.

The amount of oil that a property may ultimately produce depends on the factors that govern the production of oil. Exact calculations of future or ultimate output can not be made, but the error introduced on this account is less than that involved in the use of the other

[a] Regulations No. 33 (revised), governing the collection of the income tax, 1918, imposed by the act of Sept. 9, 1916, as amended by the act of Oct. 3, 1917, Bureau of Internal Revenue, 1918. T. D. 2447, Office of Commissioner of Internal Revenue, Feb. 8, 1917.

method. All methods of computing depletion that are based on probable ultimate production can not be exact, although they are the only ones fundamentally sound and equitable to the oil producer.

The principal advantage of the method is that the rate at which invested capital is redeemed is in direct proportion to the rate at which oil is obtained, so that during the early life of a well when its production is nearly always largest, the depletion is greatest, and, consequently, the capital invested in the land is retired most rapidly.

The more speculative the production of oil, the more rapidly should the capital invested be retired. In other words, the estimates of ultimate recoverable oil should be conservative so that the production during any one year will be a larger percentage of the ultimate production, thereby allowing invested capital to be charged off more rapidly.

USE OF COMPOSITE DECLINE CURVES.

In using the second method of determining depletion, the operator must estimate the amount of oil he may ultimately recover from his property. Such an estimate is not difficult if the oil comes from only one well, for by projecting the curve showing the decline in output of that well, he is enabled to compute rather closely the ultimate production. The accuracy of his estimate increases as the well gets older. It is, therefore, advantageous when only a few wells are involved, to keep separate production records for each well, and to compute depletion for each. If a company operates several hundred or even several thousand wells such records are impracticable, especially if the daily production per well is small and the oil from several wells is run into one tank. Then some other method must be used.

One method is to estimate the probable ultimate amount of oil to be recovered by use of a composite decline curve showing the average rate of the decrease of production of wells in a certain district. The method of making such estimates from composite decline curves, and the advantages and disadvantages in their use have been fully explained.

USE OF APPRAISAL CURVES.

Appraisal curves may be of use for quickly estimating the ultimate output of oil from several hundred or several thousand wells in a district where conditions are similar, the average daily production per well on each property, or for the whole district, is determined and then the future production of the average well is multiplied by the number of wells. The ultimate production is obtained by adding the future and past production. In this way depletion may be computed for many wells and properties with no great difficulty and with considerable accuracy.

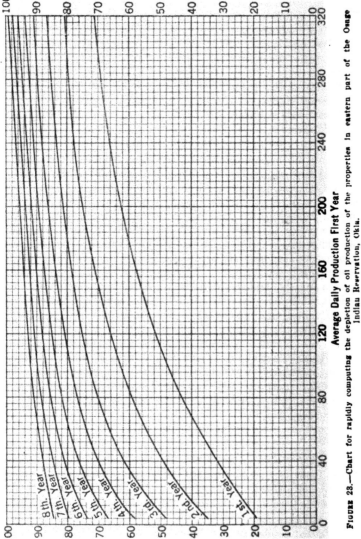

FIGURE 23.—Chart for rapidly computing the depletion of oil production of the properties in eastern part of the Osage Indian Reservation, Okla.

Figure 23 [a] has been prepared for the purpose of rapidly computing depletion. This chart shows the percentage of the probable ultimate production that will be obtained from wells of different output at the end of each year in the eastern part of the Osage Nation (Okla.). For instance, a well that yielded during its first year 100 barrels per day will have made at the end of that year about 48 per cent of its ultimate output. In other words, 48 per cent of the capital invested in the land drained by the average well of this size should be deducted from the gross income of the oil produced by the well during the first year. Similarly, the same well will have yielded at the end of the second year 63 per cent, and the third year 73 per cent of its ultimate production. Like charts for any other district may be computed directly if an appraisal curve is available. The ultimate production- of wells of different output are computed from the appraisal curve and the total amounts produced during succeeding years are expressed as percentages of the ultimate production. These points are plotted and curves are drawn, as shown in figure 23.

If a property yields oil from wells of widely different output and ages, the average age of a barrel of production may be computed, and figure 23 utilized by interpolating this average age between the curves showing the years and along the vertical lines showing average daily production the first year.

CURVES SHOWING THE PRODUCTION OF A PROPERTY.

One of the simplest methods of estimating the amount of oil recoverable from a property is to plot a curve, as explained on page 68 showing the average daily or yearly production per well. After the property is partly drilled and interference through drainage has shown itself, the curve begins to decline and regardless of the amount of new drilling the average production *per well* continues to fall off.

By projecting this curve to the minimum economic production, estimates of the property production for each time period may be made. One of the greatest advantages of using this method, at least for a check on other methods. is that by its use the making of estimates of the future initial yearly production of new wells becomes unnecessary, because for all practical purposes the initial yearly production of the new wells will be approximately the same as the average production per well of the old wells at any time. Theoretically, therefore, the postponement of drilling the remaining locations means the gradual reduction in the amount of oil that may be expected from these wells. This is known to be true, as a general rule, from experience.

---

[a] First published as figure 10 in " Some new methods for estimating the future production of oil wells by J. O. Lewis and C. H. Beal, Trans. Am. Inst. Min. Eng., vol 59, 1918, p. 516.

PART 2.—DECLINE AND ULTIMATE PRODUCTION OF DIFFERENT OIL FIELDS IN THE UNITED STATES.

INTRODUCTION.

Few of the numerous publications on petroleum in this country deal with the rate of decline and the ultimate production of oil lands. Information is plentiful, but it must be gathered and analyzed and put into a form of use to operators and engineers who desire to predict the future of wells or properties under certain conditions. The need of such information is becoming more evident to the operator as the margin of profit per barrel of oil decreases and he realizes that only through careful engineering supervision, long since recognized as essential in all important industries, can he conduct his business profitably.

This bulletin is to be regarded as only a start in the systematic compilation of such material and the methods it presents can be amplified and extended in studying data from other fields. All the data given are not of the same value, and some of the information is discussed much more thoroughly than the rest. For instance, all the fields of the country were not studied in the same detail as were the Oklahoma fields. This fact applies especially to the California fields, although they deserve careful study because of the variable conditions of production and because of the wealth of information available to the engineer who can spare the time to collect it.

In part 1 the author outlined a few methods for estimating the future and ultimate production of oil properties and the possible use of these methods for several allied purposes. In part 2 are composite curves, appraisal curves, estimating charts and other collected material on the decline and ultimate production of different properties and fields. Because of the lack of information, appraisal curves and estimating charts have not been given for many of the fields for which composite decline curves have been constructed.

THE OKLAHOMA-KANSAS DISTRICT.

GENERAL STATEMENT.

The fields of the Oklahoma-Kansas district lie chiefly in northeastern Oklahoma and southeastern Kansas. Their productive sands are mainly in rocks of Pennsylvanian age, on the great westward and northwestward dipping monocline formed in the uplifting of the

104

Ozark Mountains in southwestern Missouri and northwestern Arkansas. At the Kansas-Oklahoma line the beds dip west about 30 feet to the mile. The Ozark uplift probably caused much of the major folding in both Oklahoma and Kansas. Generally, these folds trend with the strike of the formations, the character of the folds seeming to change, however, as one approaches the center of uplift. For instance, in the Bartlesville and Nowata districts, Okla., the folding of the oil-producing formations is not so marked as it is farther west. The general geologic structure of the Oswego limestone in a part of northeastern Oklahoma is shown in Plate I.

Many of the folds, such as those in the Cushing, Blackwell, Augusta, and Eldorado fields, have been considerably modified by folding subsequent to the Ozark uplift. For instance, two separate series of cross folds along the north and south Cushing uplift have already been pointed out.[a]

The Healdton field, in southern Oklahoma, has not been studied in detail, but the conditions affecting the origin of the fold and the accumulation of the oil and gas are probably in no way related to those that govern the distribution of the oil pools in northeastern Oklahoma and southeastern Kansas. However, the Healdton field is included in the Oklahoma and Kansas district.

SPACING OF WELLS IN OKLAHOMA FIELDS.

In the Oklahoma oil fields 10 acres are ordinarily allotted each well, where the sands are more than 2,000 feet deep. If the sands are between 1,000 and 2,000 feet deep, the custom is to drill one well for every five or six acres, and for sands less than 1,000 feet deep, one well for approximately four or five acres. However, this practice is not strictly followed. In 1915 the average acreage per well on the restricted Indian leases of the following Indian nations in Oklohoma was determined by the author to be as follows:[b]

Average acres per well in the Cherokee, Chickasaw, and Creek Nations, Okla.

Indian nation.	Average acres per well.	Number of wells used in obtaining average.
Cherokee	5.1	4,500
Chickasaw	[c] 14.8	19
Creek [d]	7.4	917
Cushing field	7–10	544
Entire State	5.7	6,020

[a] Beal, C. H., Geologic structure in Cushing oil and gas field, Oklahoma, and its relation to oil, gas, and water: U. S. Geol. Survey Bull. 658, 1917, pp. 34–35.

[b] From a report of the author to the Superintendent of the Five Civilized Tribes, June 30, 1916.

[c] Drilling only begun at time of determining average.

[d] Excluding the Cushing field.

These averages were obtained without regarding the depth of the sand in different parts of the State. In the Cherokee Nation the wells ranged in depth from 300 to 400 feet to about 2,000 feet.

OSAGE INDIAN RESERVATION.

GENERAL DATA.

Some information on the Osage Indian Reservation has already been presented in a preliminary paper by Lewis and Beal[a] and some of this is republished here in revised form. Most of the oil and gas wells have been drilled in the eastern part of the reservation, and the figures compiled were for properties in that district. Records of the production of 68 of these properties, including more than 1,000 wells, were studied and analyzed. The properties are scattered over a district 60 miles long and 20 miles wide, so that the conditions affecting production differ greatly. Depths range between wide limits, and the geologic structures on which the oil has accumulated are rather diverse, so that a great variety of data was at hand.

Most of the oil comes from the Bartlesville sand in the lower part of the Pennsylvanian series. The depth of the sand ranges from 1,500 to 2,500 feet, and the average area allotted to each well is approximately 10 to 12 acres. Such differences in depth and the comparatively large acreage for each well cause considerable differences in the initial production and the rate at which the oil is obtained. The first year's average daily production per well, of the wells for which records were available, is about 35 barrels, the range being from three to four barrels to more than 300 barrels.

APPRAISAL CURVE.

Figure 24 shows the appraisal curve for this field. The construction and use of appraisal curves have been explained previously. The maximum and minimum ultimate production curves do not establish such narrow limits as some of the other appraisal curves because of the variety of conditions affecting output and because of the properties being scattered over a large area.

COMPOSITE AND GENERALIZED DECLINE CURVES.

Figure 25 represents the decline of the average well in the eastern part of the Osage Nation. If the first year's average daily production per well on a property is 100 barrels, during the second year the average output will be a little more than 63 barrels daily. The

[a] Lewis, J. O., and Beal, C. H., Some new methods for estimating the future production of oil wells: Am. Inst. Min. Eng. Bull. 134, February, 1918, pp. 477–504.

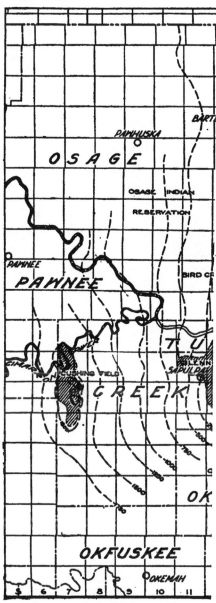

SKETCH MAP OF NORTHEASTERN OKLAHOMA, S
FIELDS AND THE GENERAL GEOLOGIC STRUC
LOGS.

number of properties used in determining this average curve is shown along the curve. The generalized decline curve for the Osage Nation (fig. 16) was discussed on page 64.

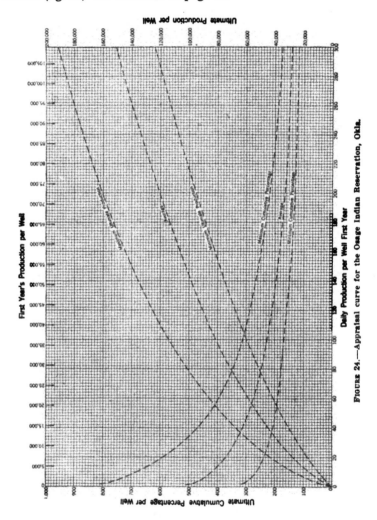

ESTIMATING CHART.

Figure 26 shows a chart prepared for rapidly estimating the approximate amount of oil that will be produced by wells that have produced a certain amount of oil and have a certain age. Such charts have been explained on pages 76 to 80.   Figure 26 should be used with caution because it applies to a large area where new pools may produce oil under different conditions.

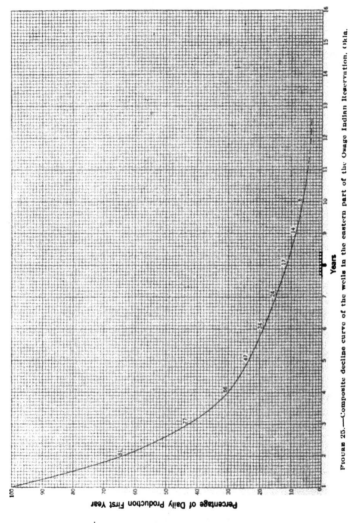

FIGURE 25.—Composite decline curve of the wells in the eastern part of the Osage Indian Reservation, Okla.

ACREAGE CHART.

Figure 27 shows the ultimate cumulative percentage for average wells on different properties plotted against the average acres per

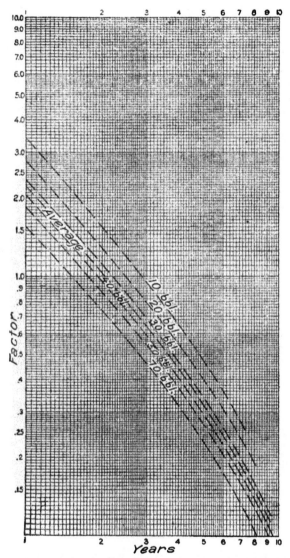

FIGURE 26.—Estimating chart for the properties in the eastern part of the Osage Indian
Reservation, Okla.

well.   The use of such charts for making closer estimates of the
probable future production has been explained on pages 56 to 58.

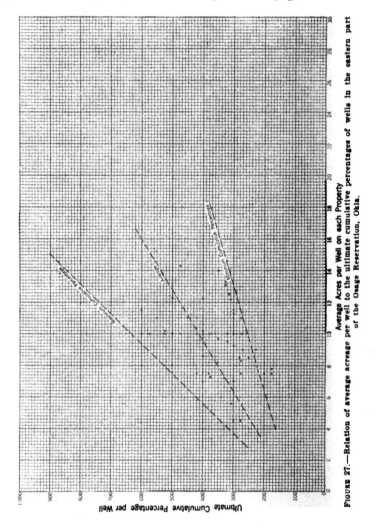

FIGURE 27.—Relation of average acreage per well to the ultimate cumulative percentages of wells in the eastern part of the Osage Reservation, Okla.

AVERAGE TOTAL PRODUCTION PER ACRE.

Prior to June 30, 1917, the Osage Nation produced more than
91,500,000 barrels of oil, the first recorded output having been in

1901. The area drained by productive wells is approximately 50,000 acres, so that the average total production per acre is approximately 3,000 barrels. It should be remembered that much of the production included in the total output is only a few years old, and that the wells producing this oil will yield more in the future, so that the ultimate production of the Osage Nation will be considerably more than 3,000 barrels per acre.

THE BARTLESVILLE FIELD.

DATA COLLECTED.

The Bartlesville field (Pl. I, p. 106) covers a large area, about 45,000 acres, and includes many different pools of oil lying on different geologic structures and occasionally in different sands. The field has been taken to include all wells between the Nowata-Chelsea field and the boundary line between the Osage and Cherokee Nations, and to extend as far north as the Kansas line. Because it is one of the oldest producing fields in Oklahoma, the field furnished much information, the records of more than 300 properties being available. As a whole, the district is divided into many small holdings because of the small size of the tracts allotted to the Cherokee Indians when the land was divided among the members of that tribe. Many of the leases are not larger than 10 acres, so that drilling has been rather close, and on many of the 10-acre tracts four wells have been drilled. However, the average area per well in the field is five to seven acres. Most of the oil comes from the Bartlesville sand, which lies near the base of the Pennsylvanian series, of Carboniferous age.

In the Bartlesville field the conditions affecting production are by no means so variable as those in the Osage Nation a few miles west. About the only factor common to the two pools is that both produce chiefly from the Bartlesville sand, which here is 500 to 1,700 feet deep. For several hundred wells in the Bartlesville pool the average daily production per well the first year is 17 barrels, ranging from 2 to about 150 barrels. A production of about 20,000,000 barrels has been obtained from 306 properties having a total productive territory of about 11,000 acres, or an average output of about 1,800 barrels per acre.

APPRAISAL CURVE.

Figure 28 represents the appraisal curve for the Bartlesville field. Because of the large number of properties available and the age of the production on most of the properties used, which allowed a more accurate estimate of ultimate cumulative percentages, the appraisal curve can be used without much fear of inaccuracy.

COMPOSITE DECLINE CURVE.

Figure 29 shows the average decline of wells of all sizes in the Bartlesville field. On account of the many properties used in its

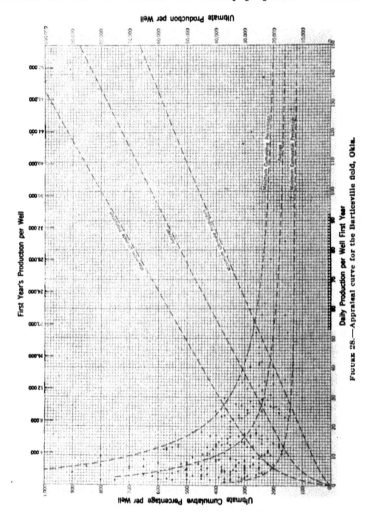

construction, this curve may be used with confidence, although, as already pointed out, the accuracy of estimates made with such average curves will vary considerably because of differences in the rates of decline of large and small wells (fig. 4, p. 26). Estimates of future and ultimate production will be in error in practically all curves for large wells and for small wells, but will be more nearly correct for average-sized wells. For such estimates an appraisal curve, if avail-

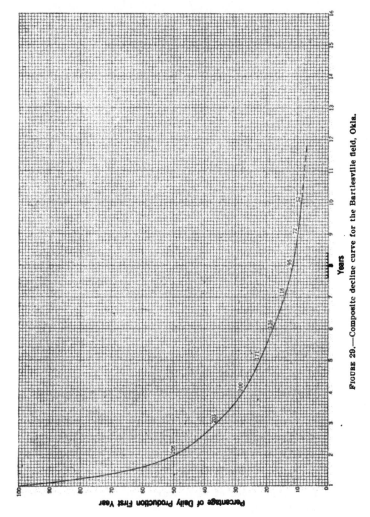

FIGURE 20.—Composite decline curve for the Bartlesville field, Okla.

able, is much better; a well yielding an average of two barrels daily the first year will decline more slowly than a well averaging 150 barrels daily the first year. These are the maximum and minimum

figures for the Bartlesville field. The numbers at yearly intervals on
the curve (fig. 29) represent the number of properties employed in

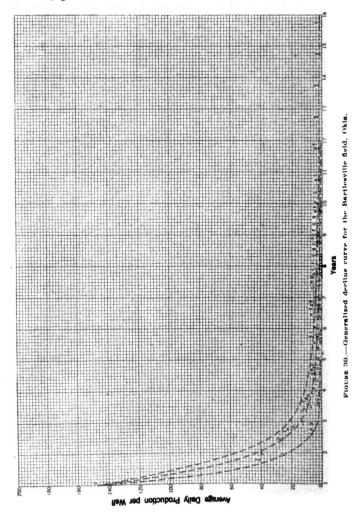

FIGURE 30.—Generalized decline curve for the Bartlesville field. Okla.

determining the averages of each year. No estimating chart was
prepared for the field.

GENERALIZED DECLINE CURVE.

The generalized decline curve shown in figure 30 has already been
explained. It is used in estimating the future production of wells

of various sizes and is based on the law (see page 36) that wells of equal output will have the same future output regardless of their age. Given the first year's average production of a well, one is justified by the law of averages in plotting the output on the average curve shown in figure 30 and assuming that the decline of the well will follow the average curve. Thus the future production can be estimated. However, after the average daily production for one or more succeeding years has been obtained, more accurate estimates may be made by fitting the production curve of the well to the type of curve it most likely will follow.

The lettered data represent the actual decline of several properties in the Bartlesville field, the first year's average daily production being plotted on the curve that it seemed to follow best. This is further proof of the law advanced by Lewis and Beal [a] that wells of the same settled output will produce on the average approximately the same amount, regardless of their ages.

CHART SHOWING VARIATION IN SAND THICKNESS.

Figure 31, which gives the average thickness of sand underlying different properties, is reproduced not so much for its value in limiting the estimates of future output of properties in the Bartlesville field as for its showing the great range of ultimate cumulative percentages for sands of the same thickness. For instance, a well producing from a sand 24 feet thick will ultimately yield a minimum of about 135 per cent and a maximum of about 615 per cent of its first year's production.

Such wide differences in the productivity of sands of the same thickness prevent the use of this chart, with a few possible exceptions, in estimating future and ultimate production. The differences shown undoubtedly result from the chart covering a large area. Moreover, in some parts of the field, oil is obtained from other sands than the Bartlesville. Probably these sands differ in porosity, in gas pressure, in thickness, and in character.

RELATION OF ACRES PER WELL TO ULTIMATE CUMULATIVE PERCENTAGE.

Figure 32 shows the relations of the average area per well on different properties to the ultimate cumulative percentages of the average well on these properties. These curves will very likely be of assistance in limiting estimates of future production from different properties.

---

[a] Lewis, J. O., and Beal, C. H., Some new methods for estimating the future production of oil wells: Am. Inst. Min. Eng. Bull. 134, February, 1918, pp. 477–504.

DATA ON TOTAL AND ULTIMATE PRODUCTION.

Table 3 shows the total production per acre of several hundred leases in Oklahoma, and gives considerable information on the total

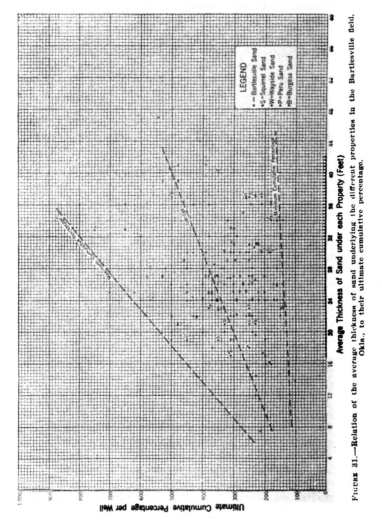

FIGURE 31.—Relation of the average thickness of sand underlying the different properties in the Bartlesville field, Okla., to their ultimate cumulative percentage.

yield per acre of different properties in the Bartlesville field. The reader should note that the average age in the fourth column of this table is the average obtained by determining the number of years the first productive well on a lease has produced. As a matter of fact, a

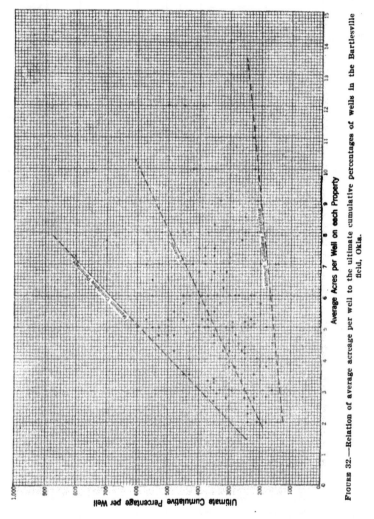

FIGURE 32.—Relation of average acreage per well to the ultimate cumulative percentages of wells in the Bartlesville field, Okla.

large part of the output from a lease may have come from wells drilled since, so that in general the age will be less than that given.

TABLE 3.—*Average total production per acre of leases in several hundred sections of land in Oklahoma.*

Township	Range	Section	Approximate average age.	Average total production per acre.	Number of leases used in determining average.	Township	Range	Section	Approximate average age.	Average total production per acre.	Number of leases used in determining average.
			Years.	Barrels.					Years.	Barrels.	
29 N.	12 E.	27	6	1,700	2	27 N.	13 E.	33	10	3,800	3
		34	4	2,000	2			34	8	2,700	4
	13 E.	31	8	700	3			35	4	800	1
		34	8	5,000	1			36	3	300	1
		35	5	700	1		14 E.	1	2	500	1
	14 E.	17	12	3,800	1			2	1	300	1
28 N.	12 E.	12	4	500	3			5	5	200	1
		25	6¾	700	3			7	9	2,500	4
		35	4	900	1			12	6	1,100	2
		36	7	3,200	2			17	3	900	1
	13 E.	1	9	2,000	1			18	8	1,900	7
		6	5	870	3			19	5	1,400	2
		7	4½	400	3			20	5	400	2
		10	6½	800	5			30	5	300	1
		11	11	11,000	1		15 E.	3	7	7,900	1
		14	7	400	1			7	6	3,900	1
		20	4	100	1			16	6	500	1
		21	10	1,900	2			17	7	5,200	4
		22	5¾	1,600	3			18	6	5,900	1
		26	6	340	1			20	7	4,000	1
		27	9	4,600	3			21	3	800	4
		28	5	700	1			22	7	1,500	4
		29	5	870	3			24	7	3,000	1
		30	4	100	1			26	7	900	2
		31	12	4,000	1	16 E.	19	8	1,000	1	
		32	4	300	1			25	10	2,700	1
		33	12	1,000	1			26	10	5,900	2
		34	5	700	2			27	8	5,400	1
	14 E.	17	6	300	1			28	8	2,400	1
		20	3	500	1			32	10	2,300	1
		29	5	400	2			34	10	8,500	2
		30	4	700	3			36	10	6,000	2
	15 E.	34	3	1,100	1		17 E.	31	10	4,400	2
		16	5	500	1	26 N.	12 E.	11	10	2,105	1
		17	5	400	1			12	14	3,600	4
27 N.	12 E.	1	4	1,200	1			13	10	2,570	1
		2	4	300	2			14	11	4,525	1
		12	7	800	2			23	8	2,645	1
		13	4	1,200	1			24	10	3,220	1
		14	7	3,400	2		13 E.	1	5	1,807	4
		23	13	1,900	2			2	7	1,380	1
		24	12	6,400	2			3	7	2,700	3
		25	11	3,600	2			4	10	2,900	4
		26	11	3,555	2			6	1	80	1
	13 E.	1	6	900	3			7	13	3,000	1
		2	4	900	3			8	9	800	2
		3	5	600	2			9	11	9,800	3
		4	6	1,600	2			11	6	440	1
		5	10	2,100	4			12	6	690	4
		6	9	3,700	2			13	4	2,800	1
		7	12	3,000	2			14	5	600	2
		8	7	1,500	2			17	7	550	1
		9	11	2,400	1			18	9	3,400	4
		11	8	2,100	1			19	5	200	4
		12	8	1,500	4			20	9	1,000	1
		14	10	3,500	1			21	9	1,900	2
		15	9	1,500	1			22	5	1,500	2
		16	5	1,400	2			23	7	1,700	2
		17	9	760	3			24	5	360	1
		18	12	2,300	4			25	7	1,400	3
		19	11	1,900	3			26	6	1,300	2
		20	9	1,200	3			27	6	950	2
		22	8	800	3			28	6	400	1
		23	5	400	2			29	9	1,200	3
		24	5	570	6			33	6	1,400	2
		25	11	2,000	2			34	7	80	2
		27	8	3,200	5			35	8	300	1
		29	3	400	1			36	5	3,000	1
		32	5	800	1		14 E.	5	6	1,500	1

TABLE 3.—*Average total production per acre of leases, etc.*—Continued.

Township.	Range.	Section.	Approximate average age.	Average total production per acre.	Number of leases used in determining average.	Township.	Range.	Section.	Approximate average age.	Average total production per acre.	Number of leases used in determining average.
			Years.	*Barrels.*					*Years.*	*Barrels.*	
26 N.	14 E.	6	5	600	3	25 N.	17 E.	4	4	900	1
		7	6	1,800	3			18	9	1,400	2
		16	4	2,800	1			20	10	1,200	2
		18	6	960	3			21	12	3,100	1
		19	10	1,500	2			28	11	2,300	3
		21	4	3,100	1			29	11	870	1
		27	4	1,500	1	24 N.	12 E.	2	9	1,600	2
		30	10	4,600	1			11	4	1,300	1
		31	9	2,100	5			14	9	1,600	1
		33	6	800	1			23	9	6,000	2
		35	4	300	1		16 E.	12	11	9,100	1
		36	5	2,100	5			13	6	1,100	1
	15 E.	1	9	840	2		17 E.	2	8	1,370	1
		2	8	3,400	1			5	11	2,900	3
		11	2	290	3			9	9	2,400	2
		13	6	600	2			16	10	1,500	7
		29	3	250	3			17	11	2,400	5
		30	4	500	1			20	10	4,000	3
		31	5	500	1			21	2	5,100	1
	16 E.	2	9	6,400	1			22	3	8,500	1
		3	8	600	2			27	2	1,800	2
		7	3	600	1			30	1	80	1
		12	10	2,400	2		18 E.	1	2	510	1
		13	9	1,700	1		19 E.	6	4	990	1
		14	9	1,000	1	22 N.	12 E.	26	7	3,900	2
		15	11	900	1		13 E.	31	5	5,400	2
		16	8	800	1			32	5	550	1
		18	7	1,300	1		14 E.	11	4	2,800	1
		19	7	500	1			14	4	840	1
		25	11	1,800	2		15 E.	31	3	790	1
		26	11	2,300	1	21 N.	12 E.	11	7	5,600	2
		27	4	580	1			12	4	980	3
		28	4	400	2			13	10	4,700	5
		35	6	900	6			14	11	6,400	2
	17 E.	18	11	3,000	1			25	3	1,600	4
		20	10	1,100	1			36	5	2,000	2
25 N.	12 E.	1	3	1,300	1		13 E.	5	6	3,600	4
		11	3	1,200	3			6	5	2,500	2
		12	4	400	1			7	7	2,600	3
		23	12	3,600	1			8	6	7,400	1
	13 E.	1	6	900	4			9	6	3,700	1
		2	4	1,100	1			16	9	1,600	1
		3	4	150	2			17	7	4,800	5
		4	7	940	3			18	8	4,300	5
		5	4	1,000	2			19	6	3,000	4
		11	3	1,400	1			20	5	6,400	4
		13	8	1,050	2			25	4	2,200	1
		14	4	200	1			30	4	820	1
		26	6	180	1			32	12	7,300	1
		34	4	80	2			33	11	5,600	2
		35	9	860	3	20 N.	12 E.	13	4	1,300	1
	14 E.	1	5	3,200	2		13 E.	5	4	300	1
		2	5	2,100	2		16 E.	18	8	7,100	3
		6	8	1,500	3			33	3	1,100	1
		7	10	2,500	2	19 N.	10 E.	8	7	1,200	1
		11	3	460	1			11	1	220	1
		13	4	2,000	2		11 E.	25	3	1,100	1
		14	5	500	1			36	1	450	1
		24	2	900	2		12 E.	16	2	16	1
	15 E.	19	3	70	1			18	2	1,300	1
		30	4	280	1			29	3	18	1
	16 E.	1	8	1,100	2			32	3	180	1
		2	9	870	2		13 E.	9	2	440	1
		5	4	630	1			16	4	650	1
		9	3	260	1			30	4	900	1
		11	10	4,600	1			31	3	570	1
		13	7	350	1		16 E.	3	3	400	1
		22	5	360	1			4	3	390	1
		24	11	2,200	3	18 N.	7 E.	5	4	10,300	2
		25	9	820	3			8	3	3,000	1
		36	9	1,400	3			9	2	8,700	3

TABLE 3.—*Average total production per acre of leases, etc.*—Continued.

Township.	Range.	Section.	Approximate average age.	Average total production per acre.	Number of leases used in determining average	Township.	Range.	Section.	Approximate average age.	Average total production per acre.	Number of leases used in determining average
			Years.	*Barrels.*					*Years.*	*Barrels*	
18 N.	7 E.	15	4	3,000	2	17 N.	10 E.	24	3	380	1
		16	3	6,500	2			35	1	360	1
		17	3	4,800	1		11 E.	7	5	930	1
		20	4	2,800	3			10	3	400	2
		21	2	500	1			11	3	640	3
		22	3	500	2			14	4	180	1
		28	4	600	2			26	3	110	1
		29	5	2,000	2		12 E.	1	4	520	1
		30	5	2,000	2			3	7	800	1
		31	4	2,500	2			4	1	230	1
		32	5	2,200	2			5	9	3,400	1
		34	3	3,700	2			7	10	9,800	3
	10 E.	35	3	60	1			12	5	1,600	1
	11 E.	1	6	7,100	4			20	10	9,600	1
		7	3	2,700	1			21	11	8,600	2
		8	3	1,700	4			27	1	3,000	2
		9	3	1,000	1		13 E.	9	2	380	1
		12	6	5,900	5			22	1	4,200	1
		13	7	5,100	3			28	1	2,000	1
		14	7	3,400	2		14 E.	27	1	1,300	1
		16	2	4,000	1			33	1	4,200	1
		17	2	2,100	2			34	1	2,000	1
		21	4	3,500	2			35	1	860	2
		23	7	2,300	2		15 E.	5	1	1,400	1
	12 E.	6	7	1,800	1			7	1	310	1
		7	4	720	3			12	3	1,300	1
		9	3	3,100	2		16 E.	7	1	880	1
		12	3	1,200	1	16 N.	7 E.	3	1	4,200	3
		15	1	900	3			4	1	7,000	3
		16	1	930	1			9	2	1,200	1
		17	7	2,500	2		10 E.	2	3	4,500	1
		18	8	4,000	2		11 E.	5	3	1,100	3
		19	5	3,300	2			8	2	95	1
		20	8	4,800	6			9	2	430	1
		21	7	2,400	1			10	4	900	1
		27	9	3,400	2		12 E.	2	3	1,500	1
		28	8	4,300	2			4	1	170	1
		30	6	570	1			5	3	730	2
		31	10	10,470	1		13 E.	10	3	140	1
		32	5	2,750	1			11	4	2,300	1
		33	9	2,200	1			26	4	1,500	1
		34	9	3,300	3			27	3	970	1
		36	5	220	1			35	4	790	2
	13 E.	4	4	740	1		14 E.	18	4	1,100	1
		7	6	2,500	1		15 E.	5	2	3,900	2
		17	3	980	1	15 N.	12 E.	35	7	3,000	1
		18	5	2,300	1		13 E.	23	1	80	1
		20	3	200	1		14 E.	2	1	1,900	1
		30	4	7,700	1			11	2	1,500	1
	14 E.	25	3	600	1			15	1	1,900	1
		26	2	200	1			16	2	2,100	1
		30	2	90	2			23	4	2,000	1
	15 E.	29	1	3,400	1			28	7	1,600	1
17 N.	7 E.	3	4	10,500	2			30	5	350	1
		4	3	11,700	3	14 N.	12 E.	3	7	5,400	1
		5	5	18,200	4			4	1	400	1
		6	4	8,400	3			9	7	26,000	1
		7	4	700	2			11	4	1,700	1
		8	4	11,200	5			12	5	2,400	1
		9	3	14,900	4			13	4	2,200	2
		10	3	43,500	2		13 E.	2	3	300	1
		11	3	10,500	3			9	1	200	1
		15	2	600	1			18	6	1,700	3
		16	3	16,900	3		17 E.	12	7	2,800	3
		17	4	2,000	3			13	9	2,000	1
		18	4	1,100	1		18 E.	1	1	200	1
		20	3	1,500	3			7	8	1,000	2
		21	3	1,800	2	13 N.	11 E.	26	3	450	1
		22	3	3,200	2		13 E.	12	4	4,900	2
		27	1	3,300	3			14	3	570	1
		33	1	5,100	2			15	5	1,100	1
		34	1	1,700	2			18	1	3,200	1

TABLE 3.—*Average total production per acre of leases, etc.*—Continued.

Township.	Range.	Section.	Approximate average age.	Average total production per acre.	Number of leases used in determining average.	Township.	Range.	Section.	Approximate average age.	Average total production per acre.	Number of leases used in determining average.
			Years.	*Barrels.*					*Years.*	*Barrels.*	
13 N.	13 E.	27	3	700	1	13 N.	16 E.	21	6	2,100	2
		29	3	900	1	12 N.	12 E.	9	4	1,600	1
	14 E.	3	3	200	1			10	4	500	1
		4	2	160	2			30		1,200	1
		13	3	2,700	1		13 E.	19		200	1
		18	1	1,300	1	11 N.	14 E.	6		6,400	1
		24	4	1,250	1	4 S.	3 W.	3		1,300	2
		25	4	1,450	2			4		6,500	4
		27	4	1,100	2			6		20,500	5
		28	10	2,600	2			9		1,700	1
		29	10	10,800	1			10		1,950	4
		32	10	7,000	1			15		1,500	5
		33	3	1,600	1	3 S.	3 W.	30		8,000	1
		36	6	1,300	1			31		12,000	2
	15 E.	7	3	600	1			32		6,100	2
		18	2	240	1						

BIRD CREEK-FLATROCK AREA.

GENERAL STATEMENT.

The Bird Creek-Flatrock area (Pl. I, p. 106) lies a few miles south of the southern end of the Bartlesville field, and is similar to it in depth of sand and in geologic conditions. Approximately 10,000 acres have been drilled. Most of the oil is obtained from the Bartlesville sand, although some comes from deeper sands, such as the Burgess and the Tucker. For many wells the average daily production per well the first year is about 30 barrels, the high and low limits being 125 and 3 barrels.

APPRAISAL CURVE.

Figure 33 shows the appraisal curve of this district. Because of the curves being based on data insufficient for locating the limiting curve accurately, the ultimate production estimates made by its use should carry a liberal factor of safety.

COMPOSITE DECLINE CURVE.

The curve showing the average decline in the production of wells in the area is printed as figure 34. It will be seen that from an average of 38 properties the second year's average daily production per well is an average of 60 per cent of the first year's average daily production. During the third year the average well makes about 37 per cent of the daily production of the first year. Although the curve shows the second year's production to be greater than that of the

TABLE 3.—Average total production per acre of leases, etc.—Continued.

Town-ship.	Range.	Section.	Approximate average age. (Years.)	Average total production per acre. (Barrels.)	Number of leases used in determining average
18 N.	7 E.	15	4	3,000	2
		16	3	6,500	2
		17	3	4,800	1
		20	4	2,800	3
		21	2	500	1
		22	3	500	2
		28	4	600	2
		29	5	2,000	2
		30	5	2,000	2
		31	4	2,500	2
		32	5	2,200	2
		34	3	3,700	2
	10 E.	35	3	60	1
	11 E.	1	6	7,100	4
		7	3	2,700	1
		8	3	1,700	4
		9	3	1,000	1
		12	6	5,900	5
		13	7	5,100	3
		14	7	3,400	2
		16	2	4,000	1
		17	2	2,100	2
		21	4	3,500	2
		23	7	2,300	2
	12 E.	6	7	1,800	1
		7	4	720	3
		9	3	3,100	2
		12	3	1,200	1
		15	1	900	3
		16	1	930	1
		17	7	2,500	2
		18	8	4,000	2
		19	5	3,300	2
		20	8	4,800	6
		21	7	2,400	1
		27	9	3,400	2
		28	8	4,300	2
		30	6	500	1
		31	10	10,370	1
		32	5	2,750	1
		33	9	2,200	1
		34	9	3,300	3
		36	5	220	1
	13 E.	4	4	740	1
		7	6	2,500	1
		17	3	980	1
		18	5	2,300	1
		20	3	200	1
		30	4	7,700	1
	14 E.	25	3	600	1
		26	2	290	1
		36	2	90	2
17 N.	15 E.	29	1	3,400	1
	7 E.	3	4	10,500	2
		1	3	11,700	3
		5	5	18,200	4
		6	4	8,400	3
		7	4	700	2
		8	4	11,200	5
		9	3	14,900	4
		10	3	43,500	2
		11	3	10,500	3
		15	2	600	1
		16	3	16,000	3
		17	4	2,000	3
		18	4	1,100	1
		20	3	1,500	3
		21	3	1,800	2
		22	3	3,200	2
		27	1	3,300	3
		33	1	5,100	2
		34	1	1,700	2

Town-ship.	Range.	Section.	Approximate average age. (Years.)	Average total production per acre. (Barrels.)	Number of leases used in determining average
17 N.	10 E.	24	3	380	1
		35	1	360	1
	11 E.	7	5	930	1
		10	3	400	2
		11	3	640	3
		14	4	180	1
		26	3	110	1
	12 E.	1	4	520	1
		3	7	800	1
		4	1	230	1
		5	9	3,400	1
		7	10	9,800	3
		12	5	1,600	1
		20	10	9,600	1
		21	11	8,600	2
		27	1	3,000	1
	13 E.	9	2	380	2
		22	1	4,200	1
		28	1	2,000	1
	14 E.	27	1	1,300	1
		33	1	4,200	1
		34	1	2,000	1
		35	1	860	2
	15 E.	5	1	1,400	1
		7	1	310	1
		12	3	1,300	1
	16 E.	7	1	880	1
16 N.	7 E.	3	1	4,200	3
		4	1	7,000	3
		9	2	1,200	1
	10 E.	2	3	4,500	1
	11 E.	5	3	1,100	3
		8	2	95	1
		9	2	430	1
		10	4	900	1
	12 E.	2	3	1,500	1
		4	1	170	1
		5	3	750	2
	13 E.	10	3	140	1
		11	4	2,300	1
		26	4	1,500	1
		27	3	970	1
		35	4	790	2
	14 E.	18	4	1,100	1
	15 E.	5	2	3,900	2
15 N.	12 E.	35	7	3,000	1
	13 E.	23	1	80	1
	14 E.	2	1	1,900	1
		14	2	1,500	1
		15	1	1,900	1
		16	2	1,100	1
		23	4	2,000	1
		28	7	1,600	1
		30	5	350	1
14 N.	12 E.	3	7	5,400	1
		4	1	400	1
		9	7	26,000	1
		11	4	1,700	1
		12	5	2,400	1
		13	4	2,200	2
	13 E.	2	3	300	1
		9	1	200	1
		18	6	1,700	3
	17 E.	12	7	2,800	3
		13	9	2,000	1
	18 E.	1	1	200	1
		7	8	1,000	2
13 N.	11 E.	26	3	450	1
	13 E.	12	4	4,900	2
		14	3	570	1
		15	5	1,100	1
		18	1	3,200	1

TABLE 3.—*Average total production per acre of leases, etc.*—Continued.

Township.	Range.	Section.	Approximate average age.	Average total production per acre.	Number of leases used in determining average	Township.	Range.	Section.	Approximate average age.	Average total production per acre.	Number of leases used in determining average.
			Years.	Barrels.					Years.	Barrels.	
13 N.	13 E.	27	3	700	1	13 N.	16 E.	21	6	2,100	2
	14 E.	29	3	900	1	12 N.	12 E.	9	4	1,600	1
		3	3	200	1			10	4	500	1
		4	2	160	2			30	4	1,200	1
		13	3	2,700	1		13 E.	19	2	200	1
		18	1	1,300	1	11 N.	14 E.	6	3	6,400	1
		24	4	1,250	1	4 S.	3 W.	3	2	1,300	1
		25	4	1,450	1			4	3	6,500	2
		27	4	1,100	2			6	3	20,500	4
		28	10	2,600	2			9	3	1,700	5
		29	10	10,800	1			10	2	1,950	1
		32	10	7,000	1			15	1	1,500	4
		33	3	1,600	1	3 S.	3 W.	30	1	8,000	5
		36	6	1,300	1			31	3	12,000	1
	15 E.	7	3	600	1			32	3	6,100	2
		18	2	240	1						

BIRD CREEK-FLATROCK AREA.

GENERAL STATEMENT.

The Bird Creek-Flatrock area (Pl. I, p. 106) lies a few miles south of the southern end of the Bartlesville field, and is similar to it in depth of sand and in geologic conditions. Approximately 10,000 acres have been drilled. Most of the oil is obtained from the Bartlesville sand, although some comes from deeper sands, such as the Burgess and the Tucker. For many wells the average daily production per well the first year is about 30 barrels, the high and low limits being 125 and 3 barrels.

APPRAISAL CURVE.

Figure 33 shows the appraisal curve of this district. Because of the curves being based on data insufficient for locating the limiting curve accurately, the ultimate production estimates made by its use should carry a liberal factor of safety.

COMPOSITE DECLINE CURVE.

The curve showing the average decline in the production of wells in the area is printed as figure 34. It will be seen that from an average of 38 properties the second year's average daily production per well is an average of 60 per cent of the first year's average daily production. During the third year the average well makes about 37 per cent of the daily production of the first year. Although the curve shows the second year's production to be greater than that of the other pools in Oklahoma the decline during the subsequent years is more rapid. However, this, as well as many other composite curves presented, are based on the records of only a few properties.

ESTIMATING CHART.

Figure 35 shows the chart, prepared from the appraisal curve, for use in making rapid estimates of the future output of producing tracts, as described on pages 76–80. The chart may be used for wells making 10 to 80 barrels daily.

GENERALIZED DECLINE CURVE.

On account of the duplication of work involved, it was not thought necessary to prepare a generalized decline curve for this area, especially as the data were not so complete as in other pools, for the reader, if he finds such curves will be of value, should have no difficulty in preparing them himself.

SAND THICKNESS AND ACREAGE PER WELL.

Figure 36 shows the curves determined by plotting the average sand thickness under each property against the ultimate cumulative percentage of that property, and also by plotting average acreage per well on each lease against its ultimate cumulative percentage. It should be noted that the limits of productivity as determined by the average thickness of sand are rather narrow. For instance, a well producing from a sand 24 feet thick will ultimately yield a minimum of 1.5 times and a maximum of 3.0 times its first year's output.

On account of the uniform depth of the wells on the properties in this area and because of the lack of records, it was not feasible to construct a curve showing the variation of ultimate cumulative percentages with the depth of the productive sand.

TOTAL PRODUCTION DATA.

Table 3 shows the average total production per acre of the properties on different sections.

THE NOWATA FIELD.

GENERAL STATEMENT.

The Nowata field (Pl. I, p. 106) is the shallowest in Oklahoma, for the productive sand in parts of the field lies 300 to 800 feet below the surface. Consequently most of the wells are drilled by portable rigs in a few days. The drilled area comprises about 48,000 acres. Production has declined so that the present average daily output per well is very low. Statistics prepared by the author in 1916 showed that during 1915 out of 4,500 producing wells on Indian land in the Cherokee Nation 1,079, or 24 per cent, averaged less than 1 barrel daily; 3,253, or 72 per cent, averaged between 1 and 10 barrels daily;

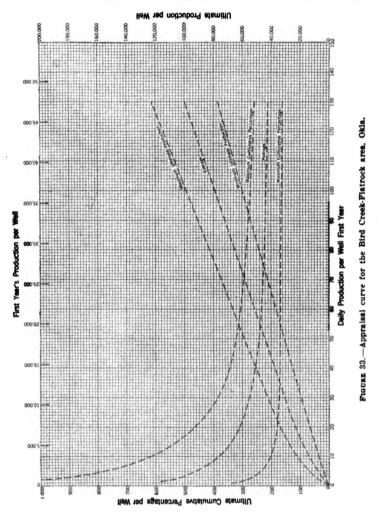

FIGURE 33.—Appraisal curve for the Bird Creek-Flatrock area, Okla.

and 168, or approximately 4 per cent, averaged between 10 and 100 barrels daily. The wells on no property in the Cherokee Nation were averaging more than 100 barrels daily. These figures of output, of course, include only the Indian acreage, which lies in all the fields in the Cherokee Nation, but may be taken as a fairly trustworthy index of the output of the rest of the drilled acreage there. The restricted Indian land that was producing oil at that time comprised between

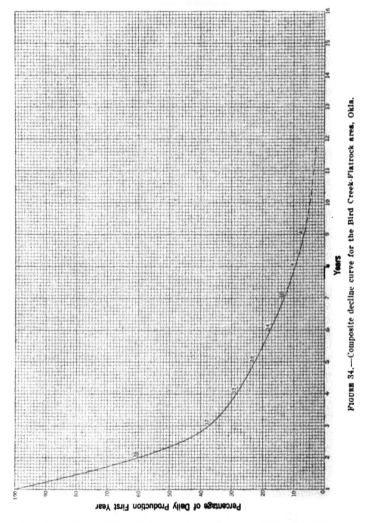

FIGURE 34.—Composite decline curve for the Bird Creek-Flatrock area, Okla.

one-fourth and one-fifth of the total productive oil land within the nation.

Inasmuch as other pools, such as the Bartlesville, were included in these calculations, the information can not be confined to the

Nowata field alone, but it is safe to say, because of the shallow depth of the productive sand in the Nowata field, that the average

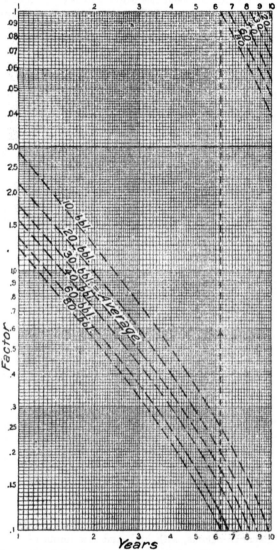

FIGURE 35.—Estimating chart for the Bird Creek-Flatrock area, Okla.

output per well is considerably below that determined to be the average for the wells on the restricted land in the whole Cherokee

Nation. Two principal sands are producing; the first, the Oswego. lies about 300 feet below the surface and is equivalent to the Oswego limestone. The Bartlesville sand lies below and is equivalent to the sand of the same name from which the bulk of the oil in Oklahoma

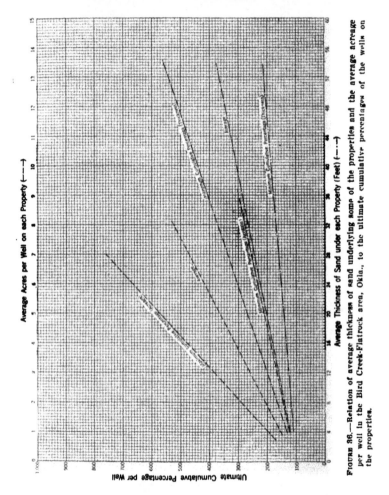

FIGURE 36.—Relation of average thickness of sand underlying some of the properties and the average acreage per well in the Bird Creek-Flatrock area, Okla., to the ultimate cumulative percentages of the wells on the properties.

is produced. The average daily production of wells the first year in this field is about 19 barrels, with high and low limits of 1 to 118 barrels.

The range of thickness of the productive sand is shown in figure 37, which was constructed by plotting the average thickness of the sand under each property against the ultimate cumulative percentage of that property. The spacing of the wells is close, presumably because of the low gas pressure, and undoubtedly is economically

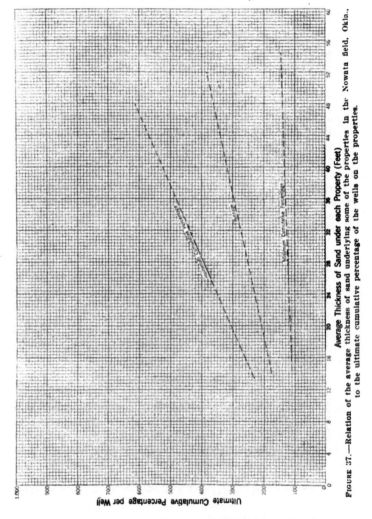

FIGURE 37.—Relation of the average thickness of the average thickness of sand underlying some of the properties in the Nowata field, Okla., to the ultimate cumulative percentage of the wells on the properties.

possible because of the small cost of drilling. Many 10-acre tracts have as many as five wells, and the average spacing is probably about three to five acres per well.

APPRAISAL CURVE.

Lewis and Beal[a] in a short preliminary paper published the appraisal curve for the Nowata field (fig. 38). This curve is based on the action of 69 regularly operated properties on which are a total of about 700 wells. The limits determined by plotting the ultimate cumulative percentages of these properties are very narrow, especially for the large wells, so that estimates of future and ultimate production made from this curve may be used with confidence.

COMPOSITE DECLINE CURVE.

Figure 39 gives the composite decline curve constructed from the properties in this district. Actual records were obtained for 11 years, so that the construction of this curve is a matter of certainty. During the first two years 68 properties were available for the average, but this number fell off, so that during the eleventh year only four properties were used. It is noteworthy that the average well in this field yields during its second year one-half the output of its first year, and during the third year yields only 30 per cent of its first year's output.

ESTIMATING CHART.

The chart prepared for making rapid estimates of future production of developed properties is shown in figure 40. These curves are somewhat generalized from those actually determined by calculations made from the appraisal curve, as outlined on pages 76-80. The reader is again warned that these charts are only for making hasty estimates; the appraisal curve itself should be used for more accurate estimates.

GENERALIZED DECLINE CURVE.

Figure 41, which shows the generalized decline curve, brings out the narrow limits of the actual decline of the type curves. The duration of the production of most of the wells used in constructing the generalized decline curve permits considerable reliance to be put on estimates made from this cause.

DATA ON THICKNESS OF SANDS.

Figure 37, already mentioned, shows the limitations of productiveness as determined by the thickness of the producing sand in this field.

---

[a] Lewis, J. O., and Beal, C. H., Some new methods for estimating the future production of oil wells: Am. Inst. Min. Eng. Bull. 134, February, 1918, pp. 477-504.

One can see by referring to this figure that the minimum ultimate production of the property underlain by a productive sand 28 feet

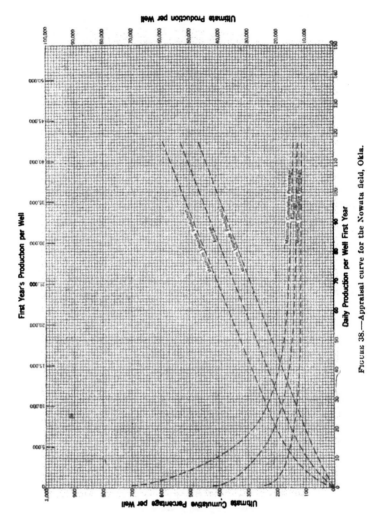

FIGURE 38.—Appraisal curve for the Nowata field, Okla.

thick is about 1.2 times and the maximum ultimate production of the same property is about 3.9 times its first year's output. Although the limits thus determined are by no means narrow, nevertheless they should often be of considerable aid in making closer estimates of future and ultimate output.

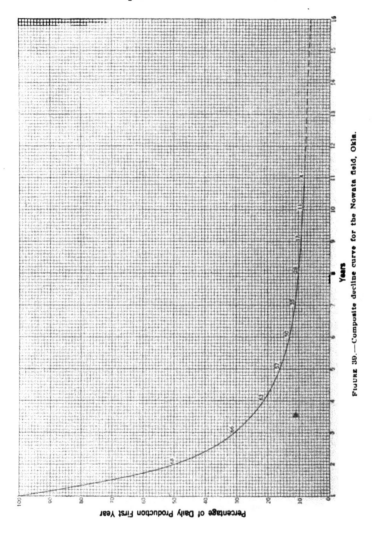

FIGURE 39.—Composite decline curve for the Nowata field, Okla.

ACREAGE OF WELLS.

Figure 42 shows the curves derived by plotting the average acres per well of different properties against the ultimate cumulative percentage of each property. The limits determined by this method of plotting were not so narrow as those shown in figure 37, although

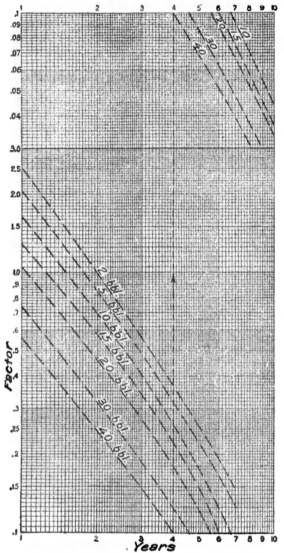

FIGURE 40.—Estimating chart for the Nowata field, Okla.

they undoubtedly will prove of aid in making closer estimates of future and ultimate production.

COMPARISON OF THE NOWATA FIELD WITH THE OSAGE DISTRICT.

As was pointed out in a paper by Lewis and Beal,[a] the appraisal curves of the Nowata and Osage fields differ considerably, and the

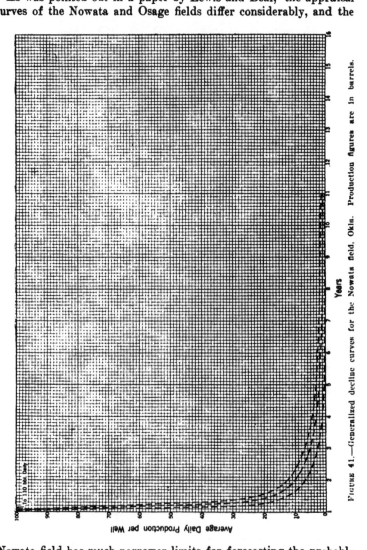

FIGURE 41.—Generalized decline curves for the Nowata field, Okla. Production figures are in barrels.

Nowata field has much narrower limits for forecasting the probable future production from the first year's production. The *maximum*

----

[a] Work cited, p. 490.

cumulative percentages of wells of 30 barrels in the Nowata field is about 225 per cent, which is about the same as the *minimum* cumula-

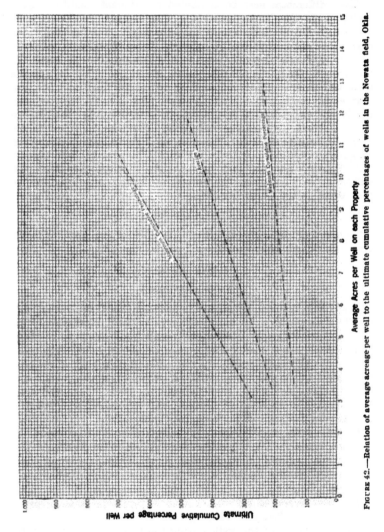

FIGURE 42.—Relation of average acreage per well to the ultimate cumulative percentages of wells in the Nowata field, Okla.

tive percentage of a well of the same size in the Osage Reservation. That is, a well of the same size in the Osage Nation will ultimately produce at least 2¼ times the first year's output, whereas one in the

Nowata field will make not more than 2¼ times the first year's production.

These differences may be attributed to different conditions governing production in the two fields. For instance, the Osage wells are scattered over a district 60 miles long and 20 miles wide; production comes mostly from the Bartlesville sand, which ranges from 1,500 to 2,500 feet deep; and each well is allotted 10 to 12 acres. In the Nowata field the production comes mostly from the Bartlesville sand and from a shallower sand at 300 to 800 feet, and the area drained by each well is 2½ to 5 acres. The Nowata field is practically in one pool and conditions are more uniform than in the Osage district. One of the underlying differences is, of course, the original gas pressure (in Oklahoma, the original gas pressures are roughly proportionate to the well depths); this greatly influences the initial production and thereby controls the ultimate production. In addition, the oil sands in the Nowata field are known to be thinner than those in the Osage district.

TOTAL AND ULTIMATE PRODUCTION.

The average total production per acre for several sections in the Nowata field are obtainable from Table 3, page 118.

GLENN POOL.

GENERAL STATEMENT.

The Glenn pool was discovered in 1906, and until the discovery of the Cushing pool was the most productive oil field in Oklahoma. Its area is approximately 19.000 acres, which happens to be approximately the same as that of the productive area of the Cushing field. Plate I (p. 106) and figure 50 (p. 147) show the situation of the pool. The depth of the pool varies with the geological structure and with the stratigraphic position of the producing sand. The Glenn sand, by far the most productive, is considered by some as equivalent to the Bartlesville sand of northeastern Oklahoma.

Because of its being divided into small tracts and because of its productiveness the Glenn pool has been drilled rather closely, that is the acreage per well is smaller than it should be, averaging about five to eight acres, although on some leases each well is allowed about 10 acres.

The average output per well on about 60 properties was 45 barrels daily during the first year, ranging from about 4 barrels to 200 barrels daily. Many of the wells during the first 24 hours were very productive, but declined rapidly, partly on account of the close spacing, so that the actual daily production during the first year does not average high.

One of the most puzzling characteristics of the Glenn pool is its structure, several poorly developed anticlines plunging west; but the contours showing the structure are hardly more than wavy lines, indicating that the strike is but little interrupted by folds. Plate I shows the generalized structure of the Fort Scott limestone of northeastern Oklahoma. The structure contours of this formation in and adjacent to the Glenn pool were determined by Smith.[a] The Fort Scott limestone lies above the productive oil sands, so that the structure contours in Plate I may not represent the structure of the deeper beds if there is an unconformity between them and the Fort Scott limestone, or if the underlying productive formations are not similarly folded; in fact, this possible lack of conformity may be one of the causes of the apparent anomaly of a phenomenal output from a slightly warped monocline. However, the lenticularity of the productive formations in the Glenn pool may greatly influence the production.

The curves following are based on data furnished by 60 representative properties in the Glenn pool.

APPRAISAL CURVE.

The appraisal curve of the Glenn pool is shown in figure 43. The trustworthiness of the information on which these curves are based should render estimates of future and ultimate production very trustworthy.

COMPOSITE DECLINE CURVE.

Figure 44 gives the average yearly decline of many properties in the Glenn pool. A few of the records extended over a period of 11 years, so that the accurate projection of the average curve should not be difficult. However, the limitations in using this curve should be remembered, for it is based upon the average performance of many wells on about 60 properties. Some of these wells may have been gushers that yielded large quantities of oil the first year, and others may have been small "pumpers." Other average decline curves, showing the decline of particular groups of properties, are shown in figure 3 (p. 22).

ESTIMATING CHART.

Figure 45 shows the chart, prepared from the appraisal curve, for use in making rapid estimates of the future output of producing wells. On account of the size of the wells in the Glenn pool and the rapidity with which some of them decline, the curves shown in figure 45 are not so regular as those shown in other similar charts.

[a] Smith, C. D., The Gleen oil and gas pool and vicinity, Oklahoma: U. S. Geol. Survey Bull. 541, pt. 2, 1912, pp. 34–48.

GENERALIZED DECLINE CURVE.

The generalized decline curves of wells that produce along maximum, average, and minimum curves are shown by figure 46.

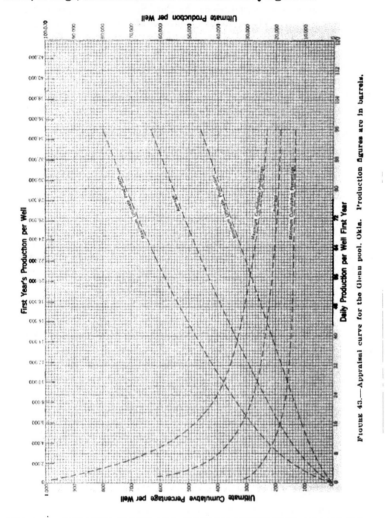

FIGURE 45.—Appraisal curve for the Glenn pool, Okla. Production figures are in barrels.

RELATION OF SPACING OF WELLS TO ULTIMATE CUMULATIVE PERCENTAGES.

Figure 47 gives the determined limits of ultimate cumulative percentages determined by plotting the average acreage per well against the ultimate cumulative percentage per well of different properties.

The use of these curves in narrowing the limits of estimates of future and ultimate production has been explained.

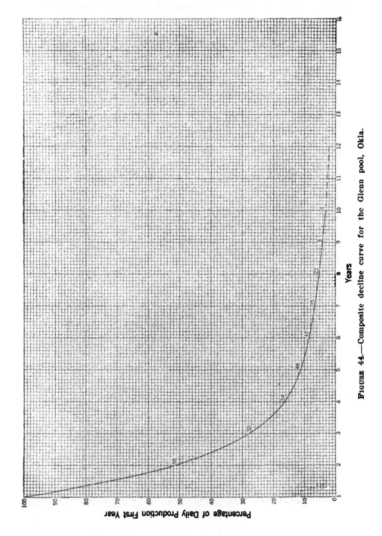

FIGURE 44.—Composite decline curve for the Glenn pool, Okla.

DATA ON TOTAL AND ULTIMATE PRODUCTION.

The total marketed production for the Glenn pool, according to statistics collected by the United States Geological Survey, is approximately 140,000,000 barrels, and the drilled area includes about

19,000 acres. Hence the average production per acre is 7,000 to 7,500 barrels. If the assumed thickness of the productive sand is

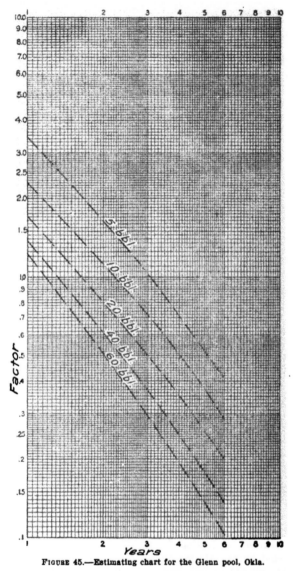

FIGURE 45.—Estimating chart for the Glenn pool, Okla.

approximately 30 feet, the average production per acre-foot is about 245 barrels. This last estimate can not be trusted, however, because of the difficulty in obtaining accurate information on the thickness of the sand that produces oil. Table 3 shows the total production to date for many of the sections in the Glenn pool, and Table 4

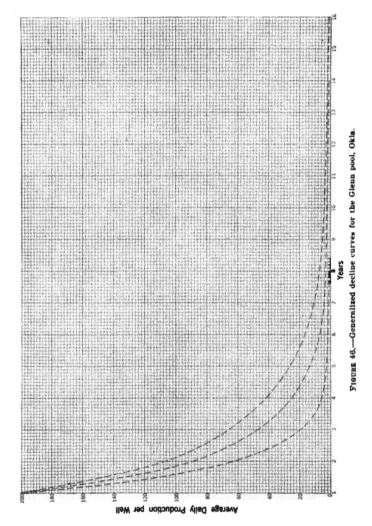

FIGURE 46.—Generalized decline curves for the Glenn pool, Okla.

following gives the production per acre, the sand producing, and the estimated production per acre-foot on several individual leases selected at random.

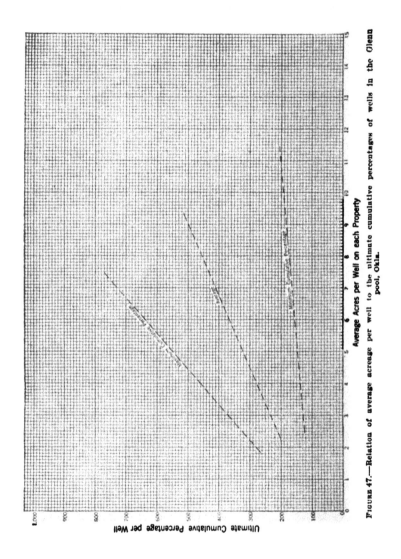

FIGURE 47.—Relation of average acreage per well to the ultimate cumulative percentages of wells in the Glenn pool, Okla.

Table 4.—*Production per acre, estimated output per acre-foot, and productive sand on several leases selected at random in the Glenn pool, Oklahoma.*

Section	Township N.	Range E.	Sand producing	Total production per acre	Total production per acre-foot
27	18	12	Glenn	2,450	90
20	17	12	....do....	9,650	270
27	18	12	....do....	2,000	75
7	17	12	....do....	1,130	32
9	17	13	Dutcher	738	50
14	18	11	Glenn	2,000	91
7	18	12	....do....	1,400	41
12	18	11	Glenn and Taneha	680	18
34	18	12	Glenn	870	24
23	18	11	....do....	2,600	130
21	18	11	....do....	2,900	100
34	18	12	....do....	4,000	110
18	18	12	....do....	5,700	95
18	18	12	....do....	1,760	30
5	17	12	....do....	3,250	72
13	18	11	Glenn and Taneha	8,600	107
1	18	11	....do....	750	13
13	18	11	....do....	2,220	22
13	18	11	Glenn	4,150	58
17	18	12	....do....	2,500	85
33-28	18	12	....do....	3,800	108
21	17	12	....do....	10,850	285
12	18	11	Glenn and Taneha	6,250	153
8-19	18	12	Glenn	5,400	96
12	18	11	Glenn and Taneha	6,080	275
12	18	11	....do....	6,900	345
7	17	12	Glenn	14,560	415
4	18	13	....do....	740	37
30	18	13	(?)	960	(?)
16	19	13	(?)	650	(?)
1	18	11	Glenn	8,900	390
5-31	17-18	12	....do....	9,950	270
7	17	12	....do....	14,500	416
7	18	12	....do....	640	10
6	18	12	....do....	1,780	75
34	18	12	....do....	2,350	66
10-28	18	12	....do....	4,950	141
21	18	12	....do....	2,400	160
12-20	18	12	....do....	3,800	62
17-20	18	12	....do....	8,650	190
20	18	12	....do....	4,000	64
20-34	18	12	....do....	5,750	115
33	18	12	....do....	2,200	59
34	18	12	....do....	3,800	105
17	18	12	....do....	2,350	78
20	18	12	....do....	4,900	77
20-21	17	12	....do....	7,050	190
1	17	12	....do....	5,200	326
7	17	12	....do....	930	58
12	17	12	Glenn and Taneha	1,580	52
3	17	12	Glenn	800	26
1	18	11	....do....	16,250	700
7	18	12	....do....	560	17
1	18	11	....do....	13,200	570
14	18	11	....do....	4,000	180
6-36	17-18	12	....do....	1,720	57
34-35	19	12	(?)	775	(?)
18	18	13	Perryman	1,550	65
7	18	13	....do....	1,700	74

THE CUSHING FIELD.

GENERAL STATEMENT.

The Cushing field lies in the western part of the Creek Nation, Okla. (Pl. I, p. 106, and fig. 50, p. 147), principally in T. 16 N., R. 7 E.; T. 17 N., R. 7 E.; T. 18 N., R. 7 E., and in the south part of T. 19 N., R. 7 E. Its productive area is about 19,000 acres, and it has produced to date more than 200,000,000 barrels, or an average of about 11,000 barrels per acre.

PRODUCTIVE SANDS.

There are three principal productive sands—the Layton, the Wheeler, and the Bartlesville. The Layton sand ranges in depth from 1,200 to 1,500 feet and is 20 to 100 feet thick, although the latter thickness is not common. The sand is porous and comparatively soft. Approximately 14 square miles of this sand produces oil, and 12 square miles of it originally carried much gas. The Wheeler sand which lies 600 to 900 feet below the Layton, ranges in thickness from 50 to 100 feet, and its lower sandy member is claimed by some to be equivalent to the "Oswego lime" (Fort Scott limestone) of northeastern Oklahoma and southeastern Kansas. Approximately 11 square miles of the Wheeler sand originally carried oil and 21 square miles produced gas exclusively. The Bartlesville sand lies 350 to 550 feet below the Wheeler and is by far the most productive oil sand in the Cushing field. This sand attains a thickness of 200 feet and is lenticular, its thickest part being in the crest of the dome of the northern part of the field. The total oil and gas producing area of the Bartlesville sand on the main Cushing anticline is about 20 square miles, of which only two square miles originally carried gas alone.

Other sands have produced oil and gas in the Cushing field, but they are not so important as the three mentioned. For instance, the Jones and the Cleveland sands lie between the Layton and the Wheeler. The Skinner sand lies just above the Bartlesville, and the Tucker sand closely underlies the Bartlesville. In some places the wells drilled to the Tucker sand have been extremely productive. Figure 48 shows the generalized columnar section of the formations penetrated by the drill in this field, and also the stratigraphic relations of the oil and gas sands.

In the Cushing field the average spacing of the wells is 6 to 10 acres per well for each sand. The average daily production per well during the first year differs with the sand to which the well is drilled. The average for the Layton sand is 34 barrels for about 100 wells; for the Wheeler sand 53 barrels for nearly 150 wells; and for the Bartlesville sand 208 barrels for about 100 wells. Wells to the Layton sand make initial productions that range from a few barrels to usually not more than 500 or 600 barrels the first 24 hours, although in one locality in the Cushing field wells to the Layton produced the first 24 hours between 500 and 1.000 barrels. The Wheeler sand has been by no means so productive as the Layton or Bartlesville sands, and ordinarily the wells to it have not produced more than 200 to 300 barrels during the first 24 hours. The Bartlesville sand on the crest of one of the domes in the northern part of the field, furnished many wells of an initial daily production between 5.000 and 10.000 barrels, but in the southern part of the field few of the wells drilled into this sand produced more than 3.000 barrels the first 24 hours. The Tucker sand, although less extensive than any others in the area, has furnished some excellent wells, the largest producing during the first 24 hours nearly 15,000 barrels.

GEOLOGIC STRUCTURE.

The general structure of the Cushing field is that of a broad north-south anticline along whose major axis are domes of varying size and importance. The major anticline, one of the largest in the oil and gas bearing formations in Oklahoma, is complicated by small subsidiary terraces along its sides. The general structure as shown

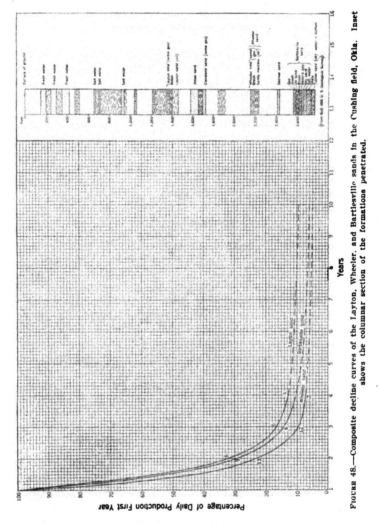

FIGURE 48.—Composite decline curves of the Layton, Wheeler, and Bartlesville sands in the Cushing field, Okla. Inset shows the columnar section of the formations penetrated.

by contours may be seen by referring to Plate I (p. 106). These contours have been generalized from Plate VIII of a report prepared by the author.[a] The contours have been generalized from the Wheeler

[a] Beal, C. H., Geologic structure of the Cushing oil and gas field. Okla., and its relations to oil, gas, and water : U. S. Geol. Survey Bull. 658, 1917, pp. 34-35.

sand because of its supposed equivalence to the Fort Scott limestone ("Oswego lime") upon which have been drawn the contours in the other parts of Oklahoma (Pl. I, p. 106).

COMPOSITE DECLINE CURVE.

Oil was discovered in the Cushing field near its center in March, 1912, in a well drilled to the Wheeler sand. For nearly two years the total output of the Cushing field came from the Wheeler and the Layton sands. The Bartlesville sand was discovered in December, 1913, and after that date drilling progressed rapidly. On many leases the wells produced from two or three sands, so that it has been difficult to obtain many records of production that give the actual output of wells producing oil from only one sand. A few such records have been collected, however, and composite decline curves constructed to show the average rate of decline of these wells. Figure 48 shows such decline curves for the Layton, Wheeler, and Bartlesville sands.

RATE OF DECLINE.

A word should be said as to the conditions influencing the rate of decline in production from these different sands. The Wheeler was principally a gas sand: as the gas usually overlies the oil it was difficult to drill into the oil sand until the pressure had fallen. The gas accordingly was allowed to waste in tremendous volumes for weeks at a time for the purpose of reducing the pressure, so that drilling could be resumed to the oil sand. As a result the wells of the Wheeler sand declined in output very rapidly, as shown in figure 48. Much the same condition was encountered in drilling wells to the Bartlesville sand, although this sand is much thicker and conditions were more complicated. The thickness of the sand, however, and the enormous gas pressure associated with the oil prevented such a rapid decline of oil production.

DATA ON TOTAL AND ULTIMATE PRODUCTION.

Table 3 (p. 118) shows the average production per acre of several different sections in the Cushing field. The following table gives the total production per acre of different sands on a few leases in the Cushing field. None of the tracts for which statistics are given supported wells more than three or four years old.

TABLE 5.—*Total production per acre of different sands to Jan. 1, 1916, on several leases in the Cushing field (Okla.).*

Lessor.	Section.	Township.	Range.	Approximate average total output per acre.
Layton sand.				*Barrels,*
Lussie Heneha a	9	16	7	300
Johnny Jacobs	15	18	7	2,600
Emma Derrisaw	32	18	7	3,100
Wheeler sand.				
Bettie Cain	18	17	7	1,400
Chas. Kernal	30	18	7	900
Aggie Wacoche	29	18	7	500
Beeley Derrisaw (80 acres)	6	17	7	1,100
Dewey Bruner	18	17	7	1,000
Beeley Derrisaw (40 acres)	6	17	7	1,200
Newman Dere	7	17	7	1,100
David Barnett	19-20	17	7	1,100
Minnie Bearhead (80 acres)	7	17	7	1,600
Beeley Derrisaw (40 acres)	6	17	7	900
Mattie Coachman	20	17	7	1,000
Polly Derrisaw	29	18	7	3,500
Minnie Bearhead (80 acres)	7	17	7	1,800
Johnson Wacoche	20	18	7	·2,600
Salo Fulsom	6	17	7	1,300
Miller Tiger b	17	17	7	1,900
William Jones	31	18	7	300
Bartlesville sand.				
Lessey Yarhala c	8	17	7	9,800
Emma Coker	16	18	7	5,500
S. Long	9	18	7	13,100
Mattie Jones	5	18	7	4,300
Lizzie Brown	9	17	7	22,100
Thos. Long	3	17	7	9,600
Moses Wiley	9	17	7	8,100
Jesse Tiger	17	17	7	2,700
Jenetta Tiger	16	17	7	17,300
Jenetta Richards	4	17	7	9,000
Katie Brown	9	17	7	9,600
Walter Starr	8, 17	18	7	5,100
Amy Simpson	8	18	7	1,000

a Production about 1 year old.   c Includes oil from 1 Tucker sand well.
b Includes oil from 1 Layton sand well.

THE PONCA CITY FIELD.

The Ponca City field is in the north central part of Oklahoma, west of the Osage Indian Reservation, and chiefly in T. 25 N., R. 2 E. According to Ohern and Garrett,[a] the first well was drilled in this field in 1905. There are several productive sands that range in depth from about 500 feet to 2,000 feet. The first year's average daily production per well, as determined from the records of several wells selected at random, is 32 barrels. The general structure of the field is that of a well-defined anticline lying on the westward dipping monocline that constitutes the general structural feature of northeastern Oklahoma.

[a] Ohern, D. W., and Garrett, R. E., The Ponca City oil and gas field: Oklahoma Geol. Survey Bull. 16, 1912, 30 pp.

COMPOSITE DECLINE CURVE.

Not enough information was available for preparing appraisal curves or estimating charts for this field. From the few records

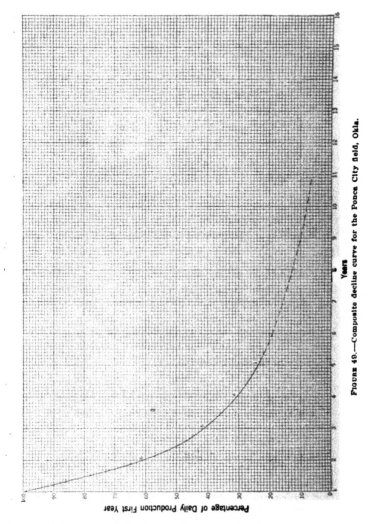

FIGURE 49.—Composite decline curve for the Ponca City field, Okla.

that could be obtained, however, the composite decline curve shown in figure 49 has been prepared.

FIELDS IN EAST CENTRAL OKLAHOMA.

Considerable information has been collected on the scattered fields in east central Oklahoma (fig. 50). These areas include what have

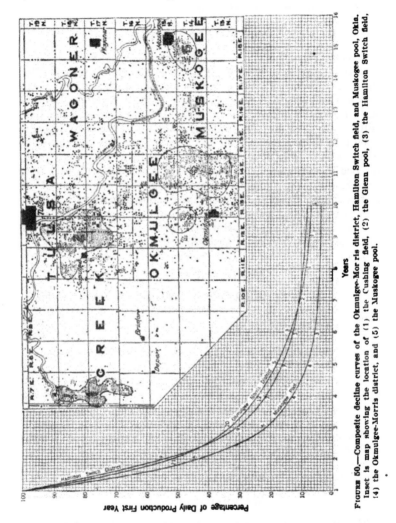

FIGURE 50.—Composite decline curves of the Okmulgee-Morris district, Hamilton Switch field, and Muskogee pool, Okla. Inset is map showing the location of (1) the Cushing field, (2) the Glenn pool, (3) the Hamilton Switch field, (4) the Okmulgee-Morris district, and (5) the Muskogee pool.

been called (1) the Okmulgee-Morris district, (2) the Hamilton Switch field, and (3) the Muskogee pool. All three are rather similar, except that the oil is not obtained wholly from the same sand.

Figure 50 shows the composite decline curves for the output of the three districts, as well as a small inset map giving the location of the fields in this part of Oklahoma.

OKMULGEE-MORRIS DISTRICT.

The Okmulgee-Morris district covers parts of what in other publications have been called the Bald Hill pool, the Morris pool, and the Booch sand area. Those properties for which production records were available are in T. 15 N., R. 14 E.; T. 15 N., R. 13 E.; T. 13 N., R. 14 E.; and T. 13 N., R. 13 E. The productive sands in the Bald Hill pool lie 1,300 to 1,700 feet deep, and that of the Morris pool is 1,500 to 2,000 feet deep. The Red Fork sand, which lies above the Glenn sand is 28 to 34 feet thick; the Glenn sand averages 16 feet, the Booch sand ranges from 28 to 34 feet; and the Morris sand from 12 to 34 feet. From 5 to 8 acres are allotted each well. The oil lies on terraces and noses on the westward dipping monocline that forms the chief structural feature of that part of Oklahoma. All the producing sands are in the Cherokee formation, the lower member of the Pennsylvanian series, of Carboniferous age, and vary in thickness and porosity. The daily production per well the first year of those wells for which the records were obtainable averaged 38 barrels.

HAMILTON SWITCH FIELD.

The Hamilton Switch field lies northwest of the Okmulgee-Morris district in T. 15 N., R. 12 E.; T. 14 N., R. 13 E.; and T. 14 N., R. 12 E. The producing sands lie 1,000 to 2,000 feet below the surface and are 19 to 50 feet thick. From 5 to 8 acres are allotted each well. Those wells for which records were obtainable averaged 56 barrels daily the first year.

MUSKOGEE POOL.

The Muskogee pool lies near the city of Muskogee. The composite decline curve shown on figure 50 was derived from the records of several properties and fairly represents the average rate at which the oil is obtained in that area.

THE HEALDTON FIELD.

The Healdton field, which lies in the southern part of Oklahoma, is on an anticlinal fold that strikes northwest, parallel to the trend of the Arbuckle Mountains. Wegemann and Heald [a] have

---

[a] Wegemann, C. H., and Heald, K. C., The Healdton oil field, Carter County, Okla.: U. S. Geol. Survey Bull. 621, 1915, pp. 13–30.

issued a preliminary report on this field, and Powers[a] has discussed in some detail the correlation of the sands and the structure of the field. The field was discovered in 1913 and at present comprises an area of approximately 7,000 acres. The sands range in depth from a few hundred feet to more than 2,000 feet, and some of the wells drilled have initial productions the first 24 hours of several thousand barrels. On account of the small size of some of the leases and the sands being in places shallow and in places unusually thick, the spacing of the wells is close, about 3 acres per well in many parts of the field. Had it not been for the thickness of the oil sand the decline of the wells in the field would have been much more rapid than it has been. On the few properties for which records were available the average daily production of the wells was 67 barrels during the first year. Figure 51 gives the composite decline curve for this field.

Several of the Healdton properties have remarkable records. A property in section 31, T. 3 S., R. 3 W., after producing only one year, made about 11,500 barrels per acre. Another property in section 4, T. 4 S., R. 3 W., at an average age of $1\frac{1}{2}$ years made 5,300 barrels per acre. A third property first drilled in April, 1914, with 4 wells producing that year, 13 wells the next year, and 53 wells the next year, yielded 29,450 barrels per acre to the end of 1916. However, as has been stated, the spacing of the wells in the Healdton field is close, and although the depths are not great, the sands are so thick as to yield a large output in a short time. This does not mean, however, that the Healdton field will not continue producing for several years; the thickness of the sands guarantees this.

FIELDS IN SOUTHEASTERN KANSAS.

Several shallow and at present rather unimportant oil fields in southeastern Kansas were discovered several years before oil was found in Oklahoma. Oil is obtained from the Cherokee member of the Pennsylvanian series, and in structure and character the fields are like the Nowata and Bartlesville districts in northeastern Oklahoma. Figure 52 gives composite decline curves for wells of three different sizes in the Neodesha field, Wilson County, Kans. In this figure the upper curve shows the average decline of five properties with wells that averaged less than one barrel daily the first year; the lower curve represents the composite decline of the wells on three properties on which the wells during the first year produced 9 to 11 barrels daily; and the middle curve gives the average decline of the wells on all the properties for which production records were available.

[a] Powers, Sydney, Age of the oil in the southern Oklahoma fields: Am. Inst. Min. Eng. Bull. 131, November, 1917, pp. 1971–1982.

The first curve is of considerable interest, as it shows that during the second year a well of this output will make an average of about 86 per cent of its first year's production. For wells that during the

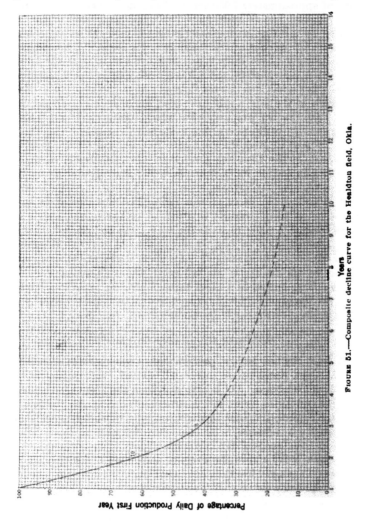

FIGURE 51.—Composite decline curve for the Healdton field, Okla.

present year produced 9 to 11 barrels daily, the second year's production is about 57 per cent as much. The average initial production of the wells on 14 properties from which the average curve was compiled amounted to only 3.9 barrels a day during the first year.

At present the most important fields in Kansas are the Augusta and the El Dorado, a few miles west of the shallow district. Several of the wells in these fields have been very productive. Table 6 gives

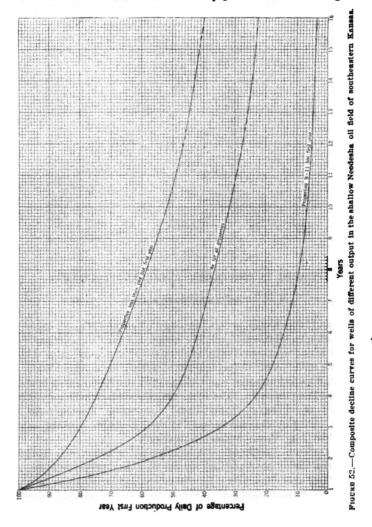

FIGURE 52.—Composite decline curves for wells of different output in the shallow Neodesha oil field of southeastern Kansas.

the oil per acre produced to March 31, 1917, by 27 properties in the Augusta field. The reader should note that some of these properties, in view of their age, have been very productive.

TABLE 6.—*Productivity and age of a few properties in the Augusta field, Kans.*

Property.	Average age.	Wells producing March, 1917.	Production to Mar. 31, 1917.	Total production per acre.
	Months.		Barrels.	Barrels
1	3	2	13,200	825
2	16	7	96,500	1,720
3	12	2	3,160	197
4	2	1	15,700	1,970
5	18	7	126,000	2,250
6	16	3	26,500	1,106
7	11	8	172,000	2,700
8	12	3	116,000	4,840
9	8	4	36,700	1,150
10	12	6	348,000	7,250
11	8	2	95,300	6,000
12	7	2	16,100	1,000
13	5	1	9,000	1,200
14	9	1	80,000	10,000
15	10	7	72,800	1,300
16	9	12	1,333,000	13,900
17	14	8	83,600	1,300
18	17	18	333,000	2,300
19	7	1	26,500	3,300
20	21	8	118,500	1,830
21	16	13	2,075,000	20,000
22	21	13	164,500	1,833
23	6	2	39,700	2,500
24	7	15	444,000	3,700
25	16	13	421,000	4,900
26	11	2	12,360	770
27	10	2	31,400	2,950
Total		163	6,309,520	
Average	11			4,800

FIELDS IN NORTHERN TEXAS AND LOUISIANA.

GENERAL STATEMENT.

Plate II shows the general geology of eastern Texas, Louisiana, and southern Oklahoma, and the situation of the oil and gas fields. The fields of northern Texas, at the time of the compilation of data for this report, were not so important as they have since become and only a little information was obtained for the Electra, Burkburnett, and the Petrolia fields. The rocks near the surface in these fields are part of the Permian "Red Beds" (shown as Carboniferous on Pl. II), but the oil sands in some of these fields are probably within the underlying Pennsylvanian formations.

In northern Louisiana, considerable information was gathered from the Caddo field and the records of a few properties were collected from the Red River, Crichton, and De Soto fields. The surface formations of the Caddo, De Soto, and Red River fields in northern Louisiana are of Tertiary age, but the oil is obtained from the underlying Cretaceous formations. These fields are located on the Sabine uplift, a great anticline that extends across northern Louisiana.

BURE

General

FIELDS OF NORTHERN TEXAS.

ELECTRA FIELD.

Lack of information on the fields in northern Texas made impossible the construction of curves other than the composite decline

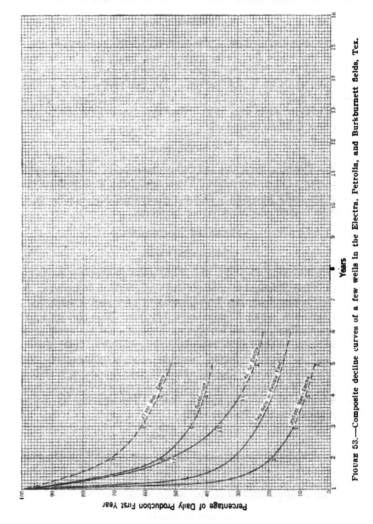

FIGURE 53.—Composite decline curves of a few wells in the Electra, Petrolia, and Burkburnett fields, Tex.

curves shown in figure 53. The average daily production per well on those properties in the Electra field for which information is available is about 75 barrels. During the first 24 hours, the ordinary well, when the field was in its prime, made 50 to 250 barrels. The geologic structure is that of a monocline. Figure 53 shows the decline of a well that made 37 barrels daily the first year, as contrasted with the decline of a well that made 350 barrels daily the first year. The average decline curve of the Electra field is also shown.

BURKBURNETT FIELD.

In the Burkburnett field the average daily production the first year of the wells for which records are available was 36 barrels. During the first 24 hours most of the wells, when the field was in its prime, made 25 to 250 barrels. The productive sand is 1,700 to 1,800 feet deep. Figure 53 shows the average decline of the wells on five properties in this field. The data were collected before the recent new drilling was done.

PETROLIA FIELD.

In the Petrolia field, the individual records of only two wells on one property were available. These wells the first year averaged 60 barrels daily per well. The curve showing the decline of these two wells is given for what it may be worth.

FIELDS OF LOUISIANA.

THE CADDO FIELD.

The Caddo field (Pl. II), one of the most productive in that part of the United States, is near Shreveport, in northern Louisiana. Most of the oil comes from the Woodbine sand of Upper Cretaceous age, which lies at a depth of 2,000 to 3,000 feet. The structure of the field is anticlinal, the oil being found usually on the crest or sides of the arch. Production records were available from about 34 properties in three different districts—the Mooringsport, Jeemes Bayou, and Monterey. During the first year the average daily production per well for the wells on all the properties was 66 barrels, with a range of 7 to 385 barrels daily. The spacing of the wells ranges from 5 to 10 acres per well.

APPRAISAL CURVE.

Figure 54 shows the appraisal curve constructed from the records of the production of 34 properties. The curves are fairly trustworthy and estimates of future and ultimate production may be made in this field with considerable assurance of reliability.

COMPOSITE DECLINE CURVE.

Figure 55 shows the composite decline curve for the Caddo field as a whole and also for two of the more important districts. As the figure shows, from the third to the eighth year the average curve for the whole field is below that of the two districts. The cause of this variation is that records from other districts were used in the calculations for the average curve for the field. Figure 55 also shows, in the inset, the decline of the first year's daily production of the wells drilled during consecutive years (see p. 51).

(see p. 51)

ESTIMATING CHART.

Figure 56 shows the chart prepared for rapidly estimating future production in the Caddo field. Being based on the appraisal curve of the field, which is considered fairly accurate, it may be used within the limitations for the use of such charts, with reasonable certainty.

GENERALIZED DECLINE CURVE.

Figure 57 has already been mentioned, but the use of this curve in estimating future production may be mentioned again. On the curves the decline in production of various properties has been plotted by the use of different symbols representing the daily production per well for each year. It is noteworthy that practically none of the declines from individual properties fall outside the limits established by these generalized curves.

DATA ON ULTIMATE PRODUCTION.

Some of the leases in the Caddo field have been very productive. The following table gives the production per acre and age of three of the more productive leases in this field:

Production and age of three leases in the Caddo field.

Property.	First produced.	Number of wells.	Total produced to 1916.	Production per acre.	Approximate depth of sand.
	Year.		*Barrels.*	*Barrels.*	*Feet.*
A	1911	8	1,407,240	22,000	2,350
B	1911	4	390,381	12,200	2,300
C	1913	11	494,820	44,800	2,250

RED RIVER, CRICHTON, AND DE SOTO FIELDS.

The Red River, Crichton, and De Soto fields lie southeast of the Caddo field on the Red River (Pl. II), and are not of great importance because of their small output. Most of the oil comes from the Woodbine sand, of Cretaceous age, which lies 2,450 to 2,550 feet deep. Between five and eight acres are allotted to each well. The average daily production per well during the first year of five proper-

ties in the Crichton pool was 85 barrels; of five properties in the Red
River field, 425 barrels; and of six properties in the De Soto field,
190 barrels.

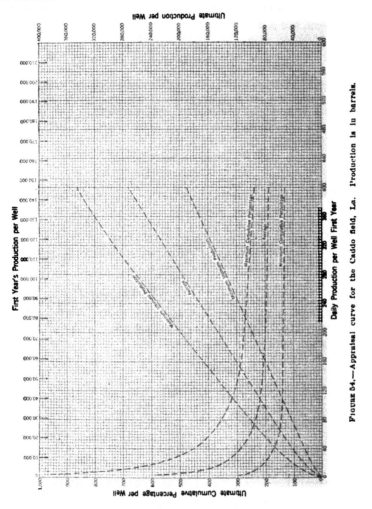

FIGURE 54.—Appraisal curve for the Caddo field, La.   Production is in barrels.

RAPID DECLINE OF PRODUCTION.

In all these fields the decline of the properties is extremely rapid.
Figure 58 shows the composite decline curves of those properties
for which production records were obtainable. As the figure shows,

wells in the De Soto field decline most slowly, for the wells on six properties during the second year produced on an average about 35 per cent of their first year's output. The most rapid decline is that

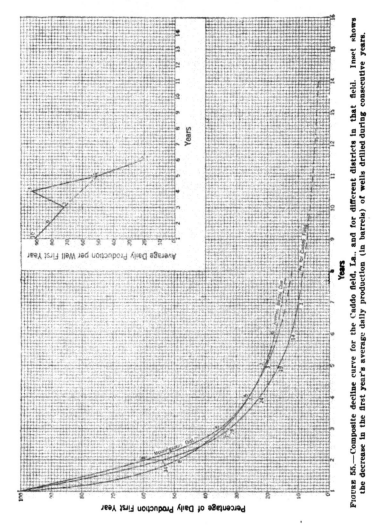

FIGURE 55.—Composite decline curve for the Caddo field, La., and for different districts in that field. Inset shows the decrease in the first year's average daily production (in barrels) of wells drilled during consecutive years.

of the Red River field, where the wells on five properties produced during the second year about 23 per cent of the first year's output. This is partly due to the difference in initial yearly output of the wells. Figure 58 shows also the average decline curve for all three fields.

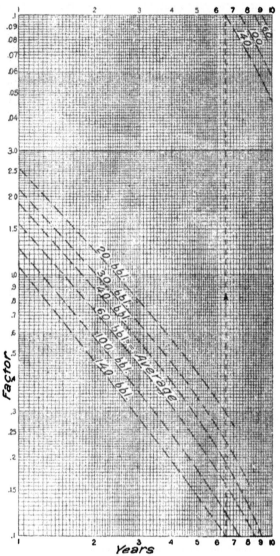

FIGURE 56.—Estimating chart for the Caddo field, La.

In all these fields the maintenance of a set production seems practically impossible on account of the extremely rapid decline in the output of old wells and in the initial output of new wells. For instance, from one lease during the second year, although seven new wells were brought in, the production per well was 18 per cent of

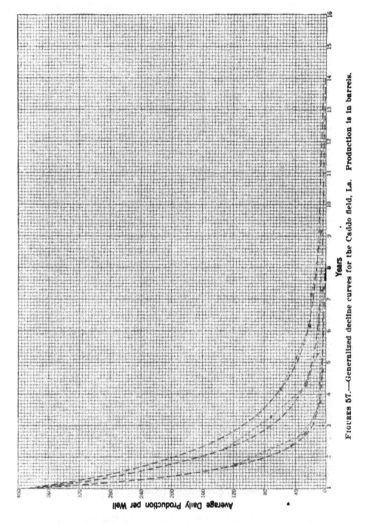

FIGURE 57.—Generalized decline curves for the Caddo field, La. Production is in barrels.

that during the first year. On another lease the average daily production per well for the second year was 32 per cent of that for the first year. During the first year there were six wells and during the second year nine wells producing. As a third example, the composite decline of the production of 12 leases in the Crichton pool,

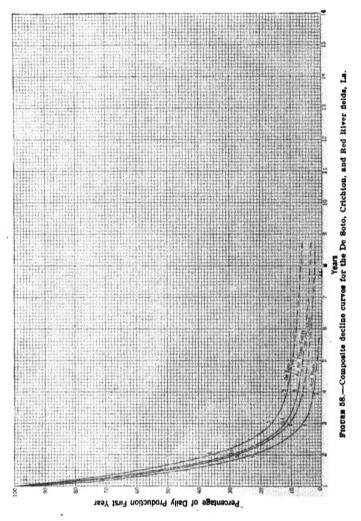

FIGURE 58.—Composite decline curves for the De Soto, Crichton, and Red River fields, La.

with many new wells producing each year, gave the figures shown in the table following:

Decline in production of 12 leases in the Crichton pool.

Year.	Wells producing.	Percentage of first year's daily production per well.
		Barrels.
1914	22	100
1915	70	50
1916	118	27
1917	130	18

Evidently, although the number of wells producing increased rapidly, the average daily output for each well fell off decidedly. In fact, the cause for the decrease in the actual production per well was the rapid decrease in initial production.

On one property in the Red River field, six wells the first year had an average daily production of 260 barrels. The next year 11 wells made an average daily output of 61 barrels, or 23.5 per cent of the average daily production for the first year. There were seven new wells drilled the third year, but their " flush " production failed to maintain the average daily production per well, which dropped to 20 barrels, or 7.9 per cent of the first year's production. The table following gives the average daily production from several properties in this field, and shows that the initial production must decline rapidly.

Decline of wells on several properties in the Red River field.

Year.	Wells producing.	Per cent of the first year's daily production per well.
1915	7	100
1916	29	22
1917	31	4

THE GULF COAST FIELD.

Plate II shows several oil fields in the area of the Quaternary formations of the Gulf coastal plain. All these fields are associated with salt domes.

CLOSE SPACING OF WELLS.

Plates III and IV show the usual practice in spacing wells in the salt-dome fields, a practice that is largely the result of subdividing the productive area into extremely small tracts. Such close spacing, however, has surprisingly little effect on the rate at which the oil is produced, although it may materially shorten the life of the wells, which often "go to water" very suddenly. From one-half acre to three or four acres is the area allotted each well in these fields.* There has been considerable discussion as to the necessity of such close drilling in salt-dome pools, but the writer believes that the practice is often justifiable because of the probable lack of communication between adjacent wells.

PRODUCTIVENESS OF FIELDS.

Some of the salt-dome fields have been extremely productive, and the figures for total oil produced per acre of drilled land show very high recoveries. For instance, Spindletop, according to Matteson.* has produced 45,000,000 barrels from 250 acres, or 180,000 barrels per acre, and other salt domes made the following outputs to January 1, 1917:

Total production to January 1, 1917, of various Gulf Coast pools.

Pool.	Barrels.
Sour Lake, Tex	45,000,000
Jennings, La	40,000,000
Saratoga, Tex	18,000,000
Batson, Tex	28,000,000
Humble, Tex	62,000,000

The average daily output per well the first year on several of these domes is as follows:

Average daily production per well the first year in various Gulf Coast pools.

Pool.	Average daily production per well, first year.	Number of properties.
	Barrels.	
Sour Lake	84	6
Spindletop	92	5
Saratoga	23	2
Humble Deep Sand a	280	5
Humble Cap Rock b	47	2

a Wells drilled to the sands that flank the Humble salt dome.
b Wells in the crest of the salt dome as distinguished from "deep-sand" wells in the flanks of the dome.

DECLINE OF FIELDS.

It is realized that a study of the decline of output in pools of this kind and under these conditions has small value, but enough information has been collected from the field, some of it on individual wells,

---

* From report of J. F. Seeman. Hearings before the Committee on Public Lands, House of Representatives, 65th Cong., 2d sess., p. 223.
b Matteson, W. G., Principles and problems of oil prospecting in the Gulf coast country: Am. Inst. Min. Eng. Bull. 134, February, 1918, pp. 430–431.

EXAMPLE OF CLOSE SPACING OF WELLS IN THE SALT-DOME POOLS OF THE GULF COAST. SHOWS A VIEW IN THE SOUR LAKE POOL. (TEXAS).

to permit the construction of the composite decline curves shown in figure 59.

Much to the author's surprise the average decline of some of these pools was by no means as rapid as anticipated, in view of the close

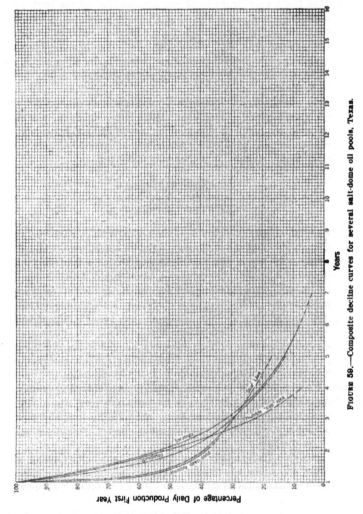

FIGURE 59.—Composite decline curves for several salt-dome oil pools, Texas.

spacing of the wells and the uncertainty as to conditions in the reservoir from which the oil comes. The lowest decline of any pool shown in figure 59 is that of the Humble Deep Sand, where the wells tap a sand similar to the ordinary oil sand of other fields. The records of "Cap Rock" properties—properties with wells drilled on the crest of the Humble dome—show that during the second year the output of the wells averages about 51 per cent of the first year's output, but during the third year drops to 23 per cent of that of the first year. These averages are given for what they are worth, as it is realized that other wells may be drilled that will show widely different curves; moreover, it is not certain that the curves shown represent averages, for the small number of properties included may cause the curves to differ much from what they would be if based on several hundred properties.

Two wells in the Sour Lake field illustrate how small wells " hold up " better than large wells. The largest well for which a record was obtained, during the first year made 196 barrels daily, the other made 30 barrels. The next year the first well made 7 per cent and the second well 61 per cent of the first year's production.

ILLINOIS FIELDS.

GENERAL STATEMENT.

Most of the oil produced in Illinois comes from three fields, the Clark County, the Crawford County, and the Lawrence County (fig. 60). These fields lie in the southeastern part of the State on the La Salle anticline, which plunges southeast. Smaller pools lie farther west in the Carlyle and Sandoval fields, and several less important pools occur in Montgomery, Macoupin, and Morgan Counties. The producing territory in the State covers about 230 square miles. and it had yielded more than 300,000.000 barrels of oil to the end of 1916. giving an average production per acre of about 2,000 barrels.

In the Clark County and Crawford County fields the productive sands lie from about 300 feet to more than 1,000 feet deep; in the Lawrence County pool the depth ranges from about 800 to 2,000 feet. In the three main fields the oil lies in sandstones in formations of Lower Pennsylvanian and Mississippian age. According to Kay[a], the Robinson sand in Crawford County averages about 25 feet thick, and the "pay" about 7 feet.

The area usually allotted each well in the Clark County and Crawford County fields is four to eight acres, and in the Lawrence County field about seven to 10 acres. In the Crawford County field the daily production per well the first year of the wells for which records were available was 11 barrels, whereas the similar average for both the Clark County and the Crawford County fields was about 10

---

[a] Kay, F. H., The oil fields of Illinois: Bull. Geol. Soc. Am., September, 1917, p. 666.

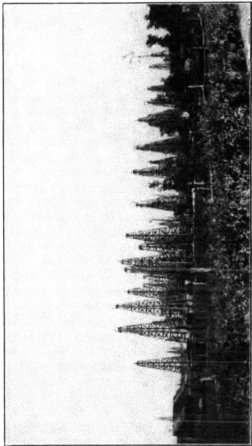

ANOTHER EXAMPLE OF CLOSELY SPACED WELLS IN THE SALT-DOME POOLS. SHOWS A GROUP OF WELLS IN THE HUMBLE POOL (TEXAS).

barrels. In the Lawrence County field the average daily production
the first year of wells on properties from which records were available

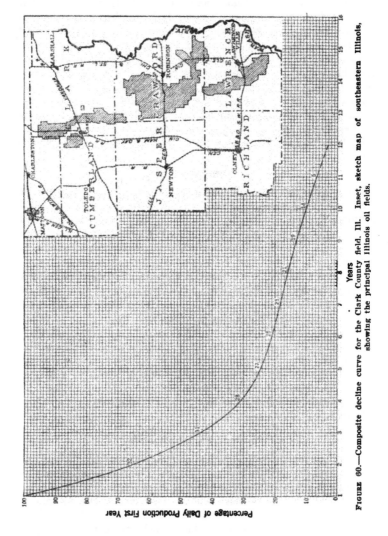

FIGURE 60.—Composite decline curve for the Clark County field, Ill. Inset, sketch map of southeastern Illinois, showing the principal Illinois oil fields.

varied from two to three barrels to about 100 barrels. The relation
of initial production the first 24 hours to the average daily production the first year in the Lawrence County field is shown in figure 15,
page 61.

CLARK COUNTY AND CRAWFORD COUNTY FIELDS.

On account of the similarity of geologic conditions and drilling practice in the Clark County and Crawford County fields, these fields are discussed together. The sands from which the wells produce are the same, although they come nearer the surface in Clark County. The production records of 60 properties in the Crawford County field and of 32 properties in the Clark County field were available for study. In the excellent reports of the Geological Survey of Illinois much information is presented regarding the underground conditions in the oil fields of the State. The data on the thickness of the sands in different fields, the depth of wells, and the detailed information on underground structure given in those reports have been of material aid in the preparation of the present bulletin.

APPRAISAL CURVE.

The appraisal curve shown in figure 5, page 33, is based on the records of properties in both the Crawford County and Clark County fields, and includes a few records from the Flat Rock pool, which lies in the southeastern part of the Crawford County field. In figure 5 the crosses represent the ultimate cumulative percentages of properties in the Clark County field and the dots those of properties in the Crawford County field. In general, the ultimate cumulative percentages of wells in the Clark County field having a given initial yearly output are a trifle higher than those of wells of the same initial yearly output in the Crawford County field. Persons who desire to make estimates of future and ultimate production of properties in the Clark County field should raise the maximum ultimate production line slightly, although the average and minimum curves for such estimates are practically the same as those for properties in the Crawford County field. Records of 89 properties were used in preparing this chart, and it is believed the curves can be used with little fear of large error.

COMPOSITE DECLINE CURVES.

For preparing composite decline curves, the records of properties in the Clark County field were separated from those of properties in the Crawford County field. Figure 60 gives the average decline of 32 properties in the Clark County field, and figure 61 that of 60 properties in the Crawford County field. As the figures show, an average well in the former field yielded during its second year about 65 per cent of its first year's output, whereas in the latter field a well during its second year yielded about 52 per cent of its first year's output. In figure 60 the curve trends rather suddenly downward after the eighth year. This was because of the reduction in the number of properties used in determining the average, and does not fairly represent the actual action of the average well in that field.

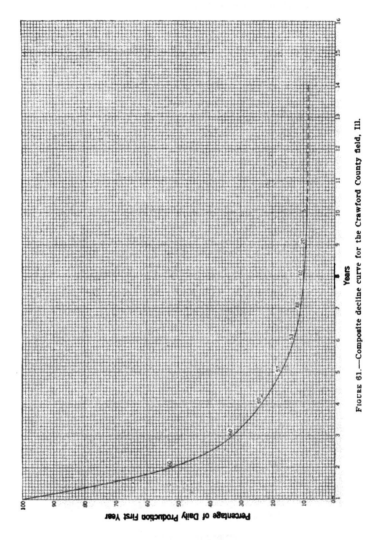

FIGURE 61.—Composite decline curve for the Crawford County field, Ill.

ESTIMATING CHART.

Figure 21, page 78, shows the chart prepared from the appraisal curve for making rapid estimates of future production of properties in the Clark County and Crawford County fields. As it is based on the appraisal curve, which is assumed to be fairly accurate, it can be used without fear of any great error.

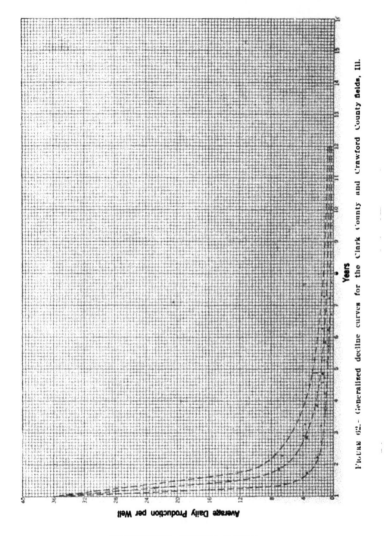

FIGURE 62.— Generalized decline curves for the Clark County and Crawford County fields, Ill.

GENERALIZED DECLINE CURVES.

Figure 62 shows the generalized decline curves prepared from the appraisal curve (fig. 5, p. 33) for this area. The actual declines of several different properties is indicated by different symbols.

CURVES SHOWING SAND THICKNESS, WELL DEPTHS, AND SPACING.

On account of the wealth of information available, the ultimate cumulative percentages of different properties in the Clark County and Crawford County fields have been plotted against (1) the approximate number of acres per well, (2) the average depths of the productive sand under each property, and (3) the average thickness of these sands. The charts thus prepared have been printed as figure 7 (p. 44), figure 9 (p. 46), and figure 8 (p. 45), respectively to show the manner in which the ultimate cumulative percentage varies with each of these three factors. In plotting figure 7 and figure 9, records of the Clark County properties were omitted because of the difference in acreage per well and average depth of sand. In figure 8, showing the relation of the ultimate cumulative percentage to the average thickness of sand, records of properties in the Clark County field were included.

DATA ON TOTAL PRODUCTION.

The following table gives statistics on the productiveness of 48 leases in the Clark County and Crawford County fields.

TABLE 7.—*Average total production per acre of 48 properties in Clark County and Crawford County, Ill.*

Property.	Location.		Number of years producing.	Average total production per acre.
	Section.	Township.		
Clark County.				*Barrels.*
1	14	Johnson	10	600
2		do	9	400
3	15	do	9	200
4	22	do	10	1,100
5	22,23	do	10	1,400
6	26	do	10	11,000
7	15	do	10	2,900
8	15	do	10	10,000
9	15	do	10	3,000
10	27	do	10	9,000
Crawford County.				
1	6	Oblong	10	3,100
2	34	Martin	10	1,700
3	3	do	10	400
4	16	5 N., 11 W	9	1,700
5	12,13	Oblong	9	1,130
6	30	do	9	4,200
7	50	do	9	700
8	30	do	9	1,300
9	25	Martin	9	2,300
10			9	1,500
11			9	1,500
12			9	1,500
13			9	1,300
14	10	Oblong	9	2,800
15			9	1,300
16	28	Martin	9	4,400
17	16	Oblong	9	2,000
18	15	Montgomery	8	1,000
19	14	do	8	800
20	15	do	8	750
21	14	7 N., 13 W	8	2,300

170 DECLINE AND ULTIMATE PRODUCTION OF OIL WELLS.

TABLE 7.—*Average total production per acre of 48 properties in Clark County and Crawford County, Ill.*—Continued.

| Property. | Location. | | Number of years producing. | Average total production per acre. |
	Section.	Township.		
Crawford County—Continued.				*Barrels.*
22	21	.....do............	8	3,200
23	5	.....do............	8	5,000
24	13	.....do............	8	1,000
25	15	5 N., 13 W........	8	1,300
26	14,24	7 N., 13 W........	8	3,000
27	13	.....do............	8	2,200
28		Oblong............	8	1,700
29	20	.....do............	8	2,800
30			8	1,500
31			8	800
32	21	6 N., 12 W........	8	2,900
33	20	.....do............	7	2,000
34	6	5 N., 12 W........	8	1,500
35	22	Honey Creek......	8	1,400
36	3	7 N., 13 W........	6	5,800
37	28	Martin............	7	1,500
38	21,35	Petty.............	5	2,400

LAWRENCE COUNTY FIELD.

In general, the properties in this field are much more productive than those in Clark and Crawford Counties. There are more productive sands, they lie deeper, and some of the sands are much thicker than any in the other two fields.

APPRAISAL CURVE.

Figure 63 shows the appraisal curve constructed from the production records of properties in Lawrence County. Not as much information was available as for the properties in Clark and Crawford Counties, so the curve is less trustworthy.

COMPOSITE DECLINE CURVE.

Figure 64 shows the composite decline curve of the properties in Lawrence County. The reader should note that the average well in this field during its second year makes about 61 per cent of its first year's production, whereas the corresponding percentages for wells in the Clark County and Crawford County fields are 65 and 52 per cent, respectively.

ESTIMATING CHART.

Figure 65 shows the chart prepared from the appraisal curve for use in making rapid estimates of future production. If the average daily yield per well during the first year's production of a property is not known, it may be safely taken as 10 barrels.

THICKNESS AND DEPTH OF SAND.

Figure 66 shows that the average thickness of sand on properties for which the records were available ranged from about 14 to 40 feet.

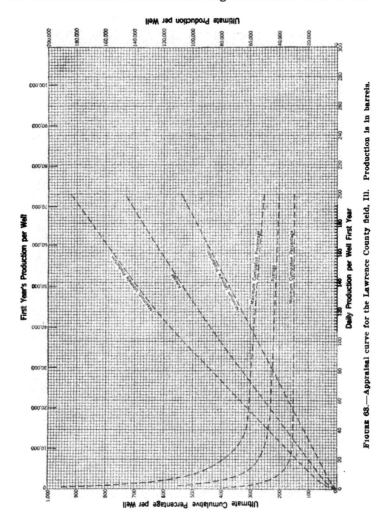

FIGURE 68.—Appraisal curve for the Lawrence County field, Ill. Production is in barrels.

The figure also shows the relation of well depths to the ultimate cumulative percentages.

Some of the properties in Lawrence County have been among the most productive in the United States. This is because of the num-

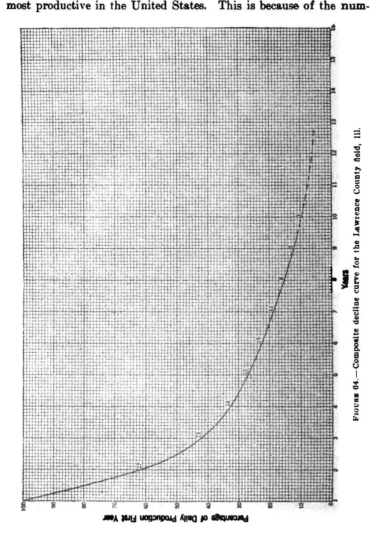

FIGURE 64.—Composite decline curve for the Lawrence County field, Ill.

ber of sands producing and the great thickness of some of them. On many properties the wells tap four sands; these range in depth from 800 to 2,000 feet. The table following has been taken directly

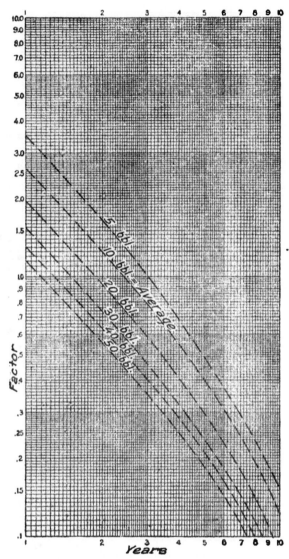

FIGURE 65.—Estimating chart for the Lawrence County field, Ill.

from an article by Kay.[a] It also includes information on the production from the Casey and Robinson sands in Clark and Crawford Counties.

[a] Kay, F. H., The oil fields of Illinois: Bull. Geol. Soc., America, September, 1917, p. 666.

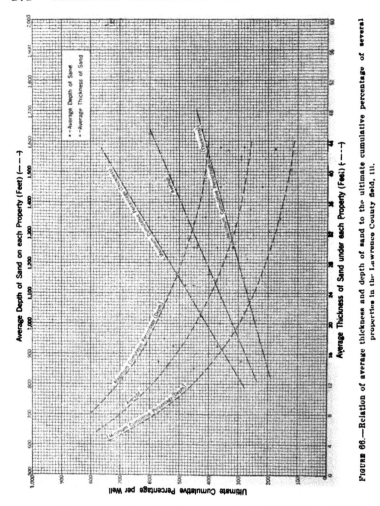

FIGURE 66.—Relation of average thickness and depth of sand to the ultimate cumulative percentage of several properties in the Lawrence County field, Ill.

TABLE 8.—*Average total production per acre for typical areas in Illinois.*

Sand.	Depth.	Period.	Production per acre.
	Feet.	*Years.*	*Barrels.*
Casey	350	10	5,309.93
Do	350	10	2,919.37
Robinson	a 900	9	719.14
Bridgeport	800–1,150	9	b 8,390.40
Buchanan	1,150–1,350	10	b 36,233.98
Kirkwood	1,350–1,650	9	b 2,546.22
McCloskey	1,750–2,000	8	b 15,672.80

a Average thickness, 7 feet.
b Many farms in Lawrence County have produced from four last sands a total of 62,000 barrels per acre, and the field is not yet exhausted.

The following table showns the average production per well of the wells on 19 properties in Lawrence County. The production per acre could not be determined because of lack of data as to the drilled acreage on each property.

TABLE 9.—*Average total production per well* [a] *on 19 properties in Lawrence County, Ill.*

Property.	Location.		Number of years produc- ing.	Average production per well.[a]
	Section.	Township.		
				Barrels.
1	8	Bridgeport	10	30,000
2	19	Petty	10	32,000
3	13	....do	8	27,000
4	29	....do	10	26,000
5	31	Bridgeport	10	20,000
6	5	Dennison	8	12,000
7	24	....do	9	8,000
8	2	....do	9	14,000
9	35	....do	9	29,000
10	22	....do	10	20,000
11	4	Lawrence	9	25,000
12	8	Bridgeport	11	24,000
13	2	2 N., 12 W	8	33,000
14	25	Petty	7	21,000
15	22	Dennison	9	18,000
16		Bridgeport	9	12,000
17	7	Petty	9	20,000
18	13	....do	5	23,000
19	5	Bridgeport	10	34,000

a Statistics on number of acres drilled on each property could not be obtained. Average acres per well in this field range from 7 to 10, so that an idea of the approximate productivity can be had by dividing production per well by the average acreage per well.

DECLINE CURVES OF THE CARLYLE AND SANDOVAL FIELDS, ILL.

Figure 67 shows decline curves of the two small and rather unimportant fields that lie west of the principal productive area in Illinois. The upper curve shows the decline of a property in the Carlyle, Clinton County, oil field; on it five wells are producing, the depth of the oil sand is 1,800 feet, and the average production per well during the first year was 2.3 barrels daily. The lower curve represents the decline curve of a property with 10 wells in the Sandoval pool, Marion County. The depth of the producing sand is approximately 1,500 feet, and the average daily production per well the first year was 15.5 barrels. These curves may or may not be representative, but they represent graphically all the available information on the decline of these two unimportant fields.

THE LIMA-INDIANA FIELD.

GENERAL STATEMENT.

One of the oldest fields in this country is the Lima-Indiana field, which lies in northwestern Ohio and northeastern Indiana. The oil comes mainly from the porous Trenton limestone, of Ordovician age, which is 1,000 to 1,600 feet deep. As the field, though once a large producer, is now practically exhausted, little time was spent in collecting

data for use in this bulletin. However, information was gathered on
the total production per acre of properties in different townships
throughout the whole field, and these figures, because of the age of

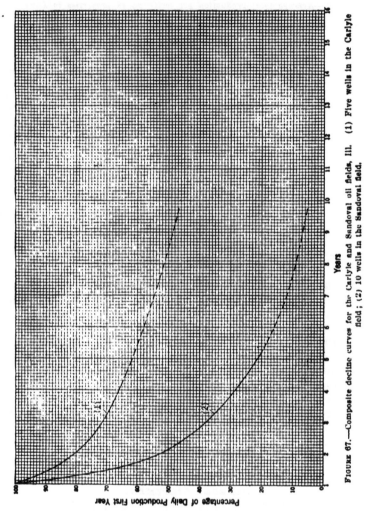

FIGURE 67.—Composite decline curves for the Carlyle and Sandoval oil fields, Ill.   (1) Five wells in the Carlyle field; (2) 10 wells in the Sandoval field.

the field and the abandoning of many wells, may be taken as practically representing ultimate production.

AVERAGE AGE AND PRODUCTIVENESS OF PROPERTIES.

Figure 68 is a map showing the greater part of the Lima-Indiana
field. On this map the approximate average age and total produc-

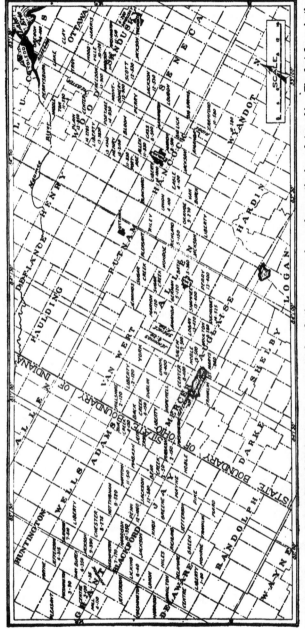

FIGURE 68.—Map of the Lima-Indiana field showing the average total production per acre of several properties in each township. The first figure in the lower right-hand corner of the township represents the approximate average age of production of the different properties in that township, and the second figure represents the total production per acre.

duction per acre on several properties in each township are shown, as indicated in the title to the figure.

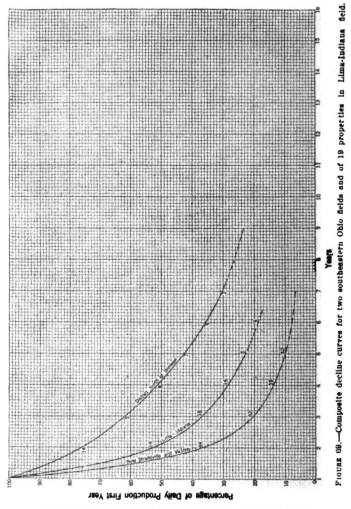

FIGURE 69.—Composite decline curves for two southeastern Ohio fields and of 19 properties in Lima-Indiana field.

COMPOSITE DECLINE CURVE OF 19 PROPERTIES.

Figure 69 shows the composite decline curve of 19 properties in the Lima-Indiana field. As these properties were drilled several years after the discovery of the field, and, in fact, after intensive drilling had been concluded, the curve represents the average decline of properties of small initial output. The properties were selected at random from Hancock and Allen Counties, Ohio, and from Blackford County, Ind., but nearly all the records were taken from the Hancock County area. On these properties the average daily production the first year was only 1½ barrels per well.

THE SOUTHEASTERN OHIO FIELDS.

Some data were collected from two different localities in southeastern Ohio. Production records, depth of sand, thickness, and other information were obtained for 88 individual wells in the New Straitsville pool, and production records were collected for several isolated and disconnected areas north of Bremen, these areas being in Hopewell and Licking Townships, Licking County; Hopewell and Licking Townships, Muskingum County; Hopewell Township, Licking County; Jackson Township, Knox County; and Pike Township, Coshocton County. All of the production records were for individual wells.

THE NEW STRAITSVILLE POOL.

In this pool the production comes from the Clinton sand, which lies 3,000 to 3,200 feet deep. In the 88 wells of which records are available the thickness of sand is 17 to 38 feet, although the actual thickness that produces the oil is probably less than 10 feet. For these wells, the average daily production the first year was 22 barrels. The spacing of the wells is about eight acres per well; some "town-lot" drilling was done, but not enough to affect the data given.

APPRAISAL CURVE.

Figure 70 shows the appraisal curve constructed by using the records of the 88 wells in the New Straitsville pool and vicinity. It should be noted that the maximum and minimum limits established are rather narrow, because the factors controlling production are approximately identical over the whole area.

ESTIMATING CHART.

Figure 71 shows the estimating chart prepared from the appraisal curve. In order to avoid the use of a large sheet of logarithmic coordinate paper the chart had to be divided into two parts—one lying below the heavy diagonal line and the other above.

GENERALIZED DECLINE CURVE.

The generalized decline curves for this field are given in figure 72.

COMPOSITE DECLINE CURVE.

Figure 69, on page 178, shows the composite decline curve of the 88 wells in the New Straitsville pool and vicinity.

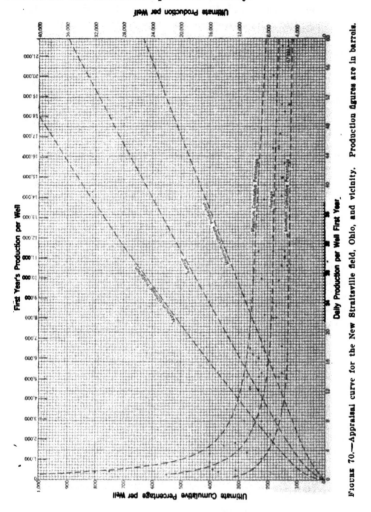

FIGURE 70.—Appraisal curve for the New Straitsville field, Ohio, and vicinity. Production figures are in barrels.

DISTRICT NORTH OF BREMAN.

In figure 69 is shown the composite curve for the properties in the isolated areas north of Breman. Although these wells are scattered they all produce from the Clinton sand, which lies 2,800 to 3,200 feet deep. This sand is 31 to 58 feet thick, averaging about 43 feet; the pay is 9 to 20 feet thick, averaging 10 to 12 feet. About 12 to

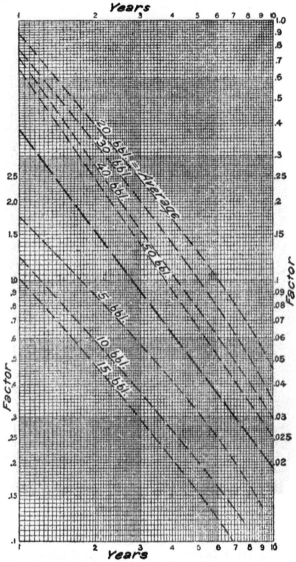

FIGURE 71.—Estimating chart for the New Straitsville field, Ohio, and vicinity.

18 acres were allotted each well, as compared with eight in the New Straitsville pool, and the average daily production per well the

first year was 12 barrels. Because of the larger acreage per well, the
curve showing the average decline of these different properties differs

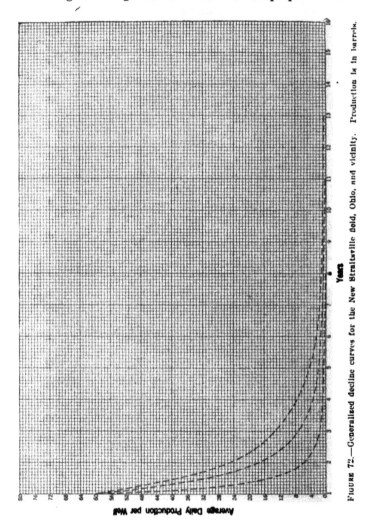

FIGURE 72.—Generalized decline curves for the New Straitsville field, Ohio, and vicinity. Production in in barrels.

widely from that of the average decline for New Straitsville and
vicinity. The other factors that influence production in the two
fields are approximately equivalent.

WEST VIRGINIA AND KENTUCKY.

GENERAL CONSIDERATIONS.

The Appalachian oil field extends southwest into West Virginia and Kentucky. In West Virginia the drilled areas cover many square miles, but the extent of the productive sands dwindles toward the southwest, so that the extension of the field in Kentucky has not been very productive. Not much information was collected in the States of West Virginia and Kentucky, the properties for which production records were gathered lying in Calhoun, Roane, Kanawha, Clay, and Lincoln Counties, W. Va., and in Lawrence and Morgan Counties, Ky. Few data were collected from the much-drilled area lying in West Virginia just across the Pennsylvania line. Depths of the producing sands differ greatly with the locality.

Fuller[a] gives some interesting statistics on the extent of the Appalachian oil and gas fields. The area of the Appalachian synclinorium is 70,000 square miles, and that of the oil field, including some potential oil land, is estimated at 2,504 square miles, or 3.6 per cent of the whole basin.

BLUE CREEK FIELD (W. VA.)

The Blue Creek field lies in Kanawha County, W. Va. The information obtained covers the Big Sandy, Wier Sand, and Elk districts, as well as the Rock Creek part of the Walton district.[b] Wells on these properties whose records were used in this study are drilled to the Squaw sand, 1,600 to 1,800 feet deep, and the Weir sand 1,700 to 2,100 feet deep. The Squaw sand, according to the records of wells, is 8 to 30 feet thick, whereas the average thickness of the Weir sand is about 13 feet. On properties producing from the Squaw sand the usual allotment is about 5 to 12 acres per well; and on those producing from the Weir sand, 8 to 12 acres per well. Some production is obtained from the Injun sand. Available records of production of wells to the Squaw sand for the properties of which records were available, is 15.2 barrels daily for the first year; for the Weir sand, 23 barrels; and for the one property producing from the Injun sand the average was 10 barrels.

Figure 73, which shows the composite decline curve of the wells on the Blue Creek field, indicates that the decline is very rapid, as the average well during the second year produces only 27 per cent of the amount it produced the first year. Curves are also shown for the decline of wells that produce from the Weir and the Squaw sands exclusively.

---

[a] Fuller, M. L., Appalachian oil field: Bull. Geol. Soc. Am., vol. 28, Sept. 30, 1917, p. 645.

[b] For the sake of uniformity the author has adopted the same classification of oil districts as that used by some of the oil companies.

THE LINCOLN COUNTY (W. VA.) AREA.

The information collected in Lincoln County covers entirely what is known by the companies as the Duvall district. In this district

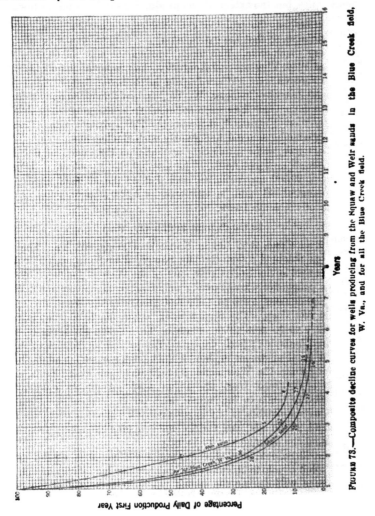

FIGURE 73.—Composite decline curves for wells producing from the Squaw and Weir sands in the Blue Creek field, W. Va., and for all the Blue Creek field.

the Berea sand is the productive formation. The sand has an average thickness of 20 to 22 feet under the 13 properties from which records were available, although the "pay" in many wells averages about 9 feet. This sand lies 2,000 to 2,600 feet deep, and each well is allotted from 6 to 10 acres. On the 13 different properties the average daily production per well the first year was 17 barrels. The composite decline curve of 12 of these properties is shown on figure 74.

THE ROANE COUNTY (W. VA.) AREA.

Two curves on figure 74 represent the average decline of the Spencer and Rock Creek districts, respectively. These lie in Roane County, W. Va., the former including the Smithfield district and the Johnson Creek part of the Walton district, and the latter including the Harper district.

In the Spencer district the records of 21 properties were available. Wells on these properties produce from the Injun sand, and the average daily production per well during the first year was 22 barrels. The average thickness of the sand is 30 to 50 feet, although the pay is probably not as thick. Under the different properties the depth of the sand ranges from 1,800 to 2,100 feet, and the properties are drilled so that there are 7 to 10 acres for each well. The composite curve shown on figure 74 indicates that the average property in this district declines slowly for the first two or three years, for during the second year the output of the average well is about 69 per cent of its output for the first year. In the Rock Creek district the records of 10 properties were studied. All these properties produced entirely from the Injun sand, which ranges from 29 to 46 feet in thickness, and lies 1,800 to 2,100 feet deep. From 9 to 10 acres are allotted each well, and the average daily production the first year for all the wells on the 10 properties was 8 barrels. The curve showing the decline of seven of these properties is shown in figure 74.

CLAY COUNTY, W. VA.

Records of only two properties in Clay County were available. On these properties the wells produce from the Injun sand, the thickness of which in three wells on one property averages 24 feet, and in nine wells on the other property averages 32 feet. The depth of the sand is about 2,000 feet, and 10 acres are allotted each well. Not enough information was available to construct a composite decline curve, but the records of the two properties studied show that the decline for these properties, at any rate, is very slow, for during the second, third, fourth, and fifth years, the average percentage of the first year's production is successively 95, 94, 88, and 79.

LAWRENCE COUNTY, KY.

Five properties in Lawrence County were studied. The oil comes from the Berea sand, which is 40 to 65 feet thick, although the pay sand is presumably not more than 10 feet thick, and the depth of the sand under the properties studied ranges from 1,650 to 1,800 feet. From 8 to 10 acres contribute to each well. The average daily

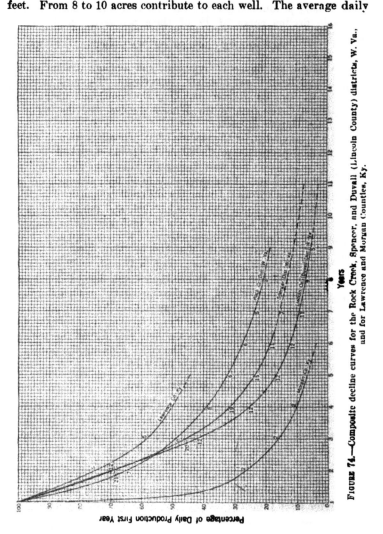

FIGURE 74.—Composite decline curves for the Rock Creek, Spencer, and Duvall (Lincoln County) districts, W. Va., and for Lawrence and Morgan Counties, Ky.

production per well the first year on five properties was only 3 barrels. The composite decline curve of these properties is shown in figure 74.

MORGAN COUNTY, KY.

Records of only two properties in Morgan County were available for study. The composite decline curve constructed is shown in figure 74. Most of the production comes from the Clinton sand, which ranges in thickness from 8 to 15 feet, and in depth from 1,800 to 2,000 feet. Each well drains oil from about eight acres, and the average daily production per well during the first year is 10 barrels.

The decline of these two properties should be compared with that of the New Straitsville field, Ohio (fig. 69, p. 178). In that field the oil comes from the Clinton sand, the spacing of the wells—eight acres each—is the same, but the initial yearly production and the depth of the sand are different. In the Morgan County field the decline is more rapid, probably because the Clinton sand is less productive than in the New Straitsville field.

DATA ON TOTAL PRODUCTION.

Data relating to the production of oil on a number of properties in West Virginia are given in the table following. The data comprise the total production per acre, the depth and thickness of sand, and the name of the chief producing sand.

TABLE 10.—*Total production per acre of several properties in different parts of West Virginia.*

Property.	District.	Years producing.	Average thickness of sand.	Approximate depth.	Principal sand producing.	Average production per acre.
Blue Creek field.			*Feet.*	*Feet.*		*Barrels.*
1	Big Sandy....	5	18	1,600	Squaw	1,300
2	....do....	5	27	1,900	....do....	2,300
3	Weir Sand....	5	30	2,100	Weir.	300
4	....do....	3		1,600-2,100	....do....	4,100
5	Elk..........	5	13	1,800	....do....	2,000
6	....do....	6	13	1,700-1,800	Squaw	1,300
7	....do....	5	19	1,800	....do....	3,600
8	....do....	6	17	1,600	....do....	3,400
9	....do....	5	17	2,000	....do....	1,000
10	....do....	6	13		....do....	2,100
11	....do....	5	13	1,900	....do....	600
12	....do....	5	13		....do....	2,100
13	....do....	6		1,800	....do....	1,400
14	....do....	5	13	1,700	....do....	2,600
15	....do....	5			....do....	2,200
16	....do....	5	17		....do....	800
17	Big Sandy....	5	8	1,900	Squaw and Weir..	5,800
Lincoln County.						
1	Dewall........	9	22	2,000	Berea.	2,400
2	....do....	9	22	2,200-2,500	....do....	2,800
3	....do....	9	21	2,200	....do....	7,000
4	....do....	9	21	2,400-2,500	....do....	2,700
5	....do....	8	23		....do....	1,900
6	....do....	8	20	2,200-2,400	....do....	1,300
7	....do....	8	22	2,400	....do....	3,600
8	....do....	8	21		....do....	1,600

TABLE 10.—*Total production per acre of several properties in different parts of West Virginia*—Continued.

Property.	District.	Years producing.	Average thickness of sand.	Approximate depth.	Principal sand producing.	Average production per acre.
Roane County.			*Feet.*	*Feet.*		*Barrels.*
1....................	Rock Creek...	8	39	2,000	Big Injin...............	1,400
2....................	.....do........	8	39	2,000	.....do...............	2,600
3....................	.....do........	7	42	1,900	.....do...............	2,100
4....................	.....do........	10	43	2,000–2,200	.....do...............	1,900
5....................	.....do........	10	43	2,000–2,100	Berea...............	1,800
6....................	Spencer........	8	35	2,000	Big Injin...............	1,400
7....................	.....do........	8	38	1,800	.....do...............	2,200
8....................	.....do........	7	13	1,900	.....do...............	1,000
9....................	.....do........	7	50	2,000	.....do...............	2,200
10....................	.....do........	6	47		.....do...............	2,100
11....................	Smithfield....	7	50	2,000	.....do...............	2,600
12....................	.....do........	6	38	2,000	.....do...............	1,200
13....................	.....do........	6	50		.....do...............	1,900
14....................	.....do........	6	36		.....do...............	1,700
15....................	.....do........	5	39		.....do...............	3,000
16....................	.....do........	5	31		.....do...............	8,700
17....................	.....do........	5	32		.....do...............	6,200

PENNSYLVANIA.

GENERAL STATEMENT.

The Pennsylvania fields were not studied in as great detail as the other fields of the United States for these reasons: (1) A great variety of conditions affects production; (2) the occurrence of oil and gas vary widely; (3) the age of many of the wells renders the collection of data extremely difficult; and (4) the fields lack importance as possible contributors of much more oil than they now contribute annually to the production of the United States. Some of the properties in Pennsylvania are more than 50 years old; fields have been discovered, pumped to exhaustion, and abandoned: even the location of many of the wells is not known. The information obtained by the author is not only meager, but is scattered widely over the productive part of the State.

AVERAGE DECLINE CURVE.

Figure 75 shows the average decline of the wells on 13 properties in the Oil City field, Venango County. In that field production, which never was large, is now slight, and the first year's production of the wells for which records were available was only 1.2 barrels daily. Most of the wells on these properties are pumped not oftener than once a week, leases with wells averaging one-tenth of a barrel a day being common, and 18 wells on one of the leases used in determining the average decline during 1916 produced only 0.07 barrel daily per well. In this district the spacing of the wells is about five acres per well. The initial production the first 24 hours of wells drilled at the present time is very small.

Pipe-line runs of one of the transportation companies that takes oil each month from about 32,000 wells in Pennsylvania show that

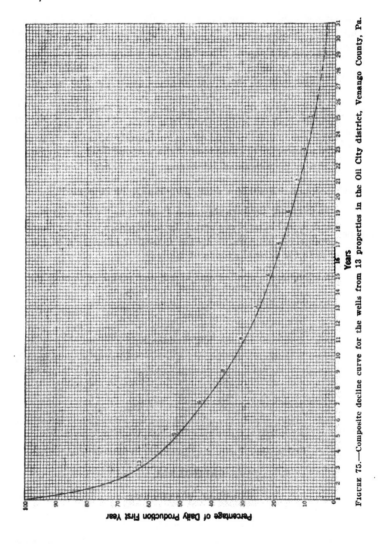

Percentage of Daily Production First Year

FIGURE 75.—Composite decline curve for the wells from 13 properties in the Oil City district, Venango County, Pa.

the average decline per well during 1916 was about 9 per cent. One company in the Oil City district allows the average well a decline each year of about 10 per cent of the preceding year's output, whereas another company operating many hundred wells of small output states that the decline per well is about 6 per cent a year.

WYOMING.

The largest field in Wyoming is the Salt Creek, which lies about 45 miles north of Casper. This field is on a large dome and has been limited by "dry holes" drilled on all sides of the dome to "edge water." The bulk of the oil is obtained from the first and second Wall Creek sands. All the outcropping formations and those penetrated in drilling to the oil-bearing beds are of Cretaceous age.

Figure 76 gives the composite decline curve, which is based on the individual production records of approximately 50 wells. The average daily production the first year of the wells utilized in the preparation of this curve was about 100 barrels.

DATA ON TOTAL AND ULTIMATE PRODUCTION.

The Salt Creek field, to January 1, 1918, had produced between 15,000 and 20,000 barrels per acre. In some places the field is exceedingly productive, as is shown by the following tabulation of the output per acre for several groups of wells:

TABLE 11.—*Total production per acre of different groups of wells in the Salt Creek oil field, Wyoming.*

Group of wells.	Location.			Years producing.	Number of wells.	Production per acre.
	Section.	Township, N.	Range, W.			
						Barrels.
1	36	40	79	6	7	63,530
2	25	40	79	6	4	31,000
3	25	40	79	4	a 1	23,000
4	36	40	79	6	b 9	5,000
5	26	40	79	4	a 1	13,000
6	26	40	79	40	a 1	5,000
7	25	40	79	6	5	20,000
8	25	40	79	4	a 1	9,000
9	27	40	79	5	3	4,000

a Each well assumed to drain 10 acres.   b Majority of wells drilled during last two years.

CALIFORNIA.

GENERAL STATEMENT.

Although the California oil fields were not investigated as thoroughly as the available information warrants, such data as were collected are given because of its possible use to others. Only the San Joaquin Valley fields were studied. Some of the composite decline curves given were prepared by the appraisal committee of the Independent Oil Producer's Agency during 1914 and 1915. As the time the author had for collecting data was short, not enough information was obtained for preparing similar curves for any of the

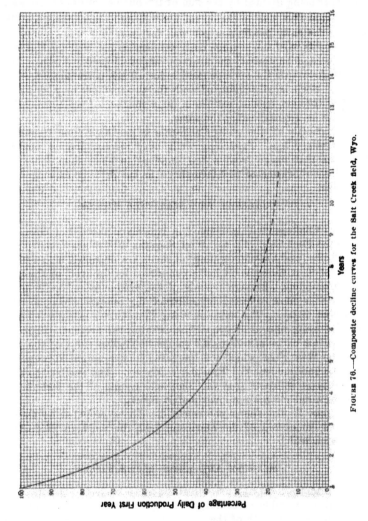

FIGURE 76.—Composite decline curves for the Salt Creek field, Wyo.

fields excepting the Midway and the Coalinga, where the curves prepared check fairly well with the older curves of the appraisal committee that recently were published by Requa.[a] That committee

---

[a] Requa, M. L., Method of valuing oil lands: Am. Inst. Min. Eng. Bull. 134, February, 1918, pp. 409–428.

in preparing composite decline curves used a method different from
that of the author, the sum of the production of all the wells for
each year being divided by the sum of the number of days each well
produced that year to obtain the average daily production per well.
Then the daily production per well for each year was expressed as
a percentage of the average daily production per well the first year.
The writer determined the average composite decline for each well,
and afterwards the average decline for all the percentages for each
year. For a small number of wells the two methods may give somewhat different results, but for a large number of wells the results
should be approximately the same.

Practically all the oil in California is produced from formations
of Tertiary age. Producing wells have reached the oil sands at
depths that range from a few hundred to more than 4,000 feet,
and the character of the formations makes drilling expensive and
difficult.

COALINGA AND MARICOPA FIELDS.

The Coalinga field is divided in two main parts—the West Side
and the East Side. On the West Side the oil has accumulated on a
monocline and the outcropping oil sands are cemented with asphalt.
On the East Side the oil has accumulated on the crest of the Coalinga anticline. Enough records of the production of wells on the
East Side were collected to construct the composite decline curve
for that district. The numbers along these curves indicate the number of wells, instead of properties, entering the average. Figure 77
shows this curve as well as the composite decline curve of the wells
on the West Side, which was prepared by the appraisal committee
of the Independent Oil Producer's Agency. The similarity of the
two curves indicates that the composite results of the factors governing production are approximately equal.

Figure 77 also shows the composite curve, prepared by the appraisal committee, of the Maricopa field, the southern part of the
Midway field.

MIDWAY AND KERN RIVER FIELD.

The Midway field may be divided into two parts or districts. In
one of these the oil has accumulated along a monocline and on an anticline, so that the depth of the sands has a rather wide range; in the
other, which lies in the Buena Vista Hills, the oil accumulated on a
large well-defined anticline. The principal part of the Buena Vista
Hills has been set aside by the Federal Government as Naval Petroleum Reserve No. 2.

COMPOSITE DECLINE CURVES FOR ALL WELLS.

As the two districts differ in depth of oil sands, quality of oil, and other conditions a composite decline curve was prepared for each. These curves are shown in figure 78.

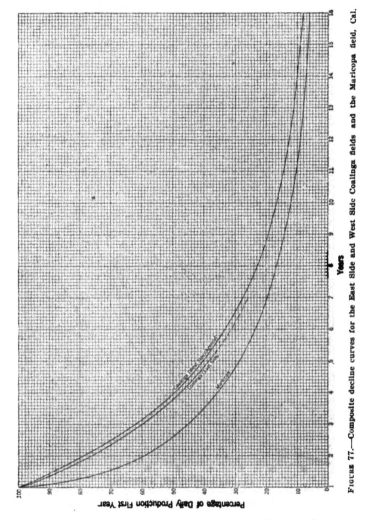

FIGURE 77.—Composite decline curves for the East Side and West Side Coalinga fields and the Maricopa field, Cal.

For the wells used in preparing the curve for the Midway field, excluding Naval Petroleum Reserve No. 2, the average daily production per well the first year was 115 barrels. The curve showing the average decline for the whole Midway field, including Naval Petroleum Reserve No. 2, was compiled from the records of more than 150 wells.

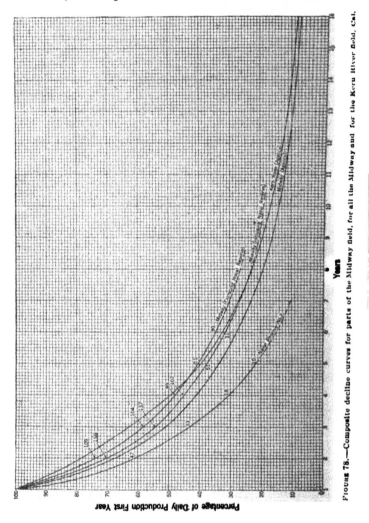

FIGURE 78.—Composite decline curves for parts of the Midway field, for all the Midway and for the Kern River field, Cal.

For these wells the average daily production per well the first year was 189 barrels. The curve prepared by the appraisal committee (fig. 78) shows a more rapid rate of decline than the curve prepared by the author. This difference may be a result of the different methods used in preparing the curves, or it may be due to the fact that the average well as determined by the appraisal committee was larger than that determined by the author.

The curve that shows the average decline of Naval Petroleum Reserve No. 2 was prepared from the individual production records of about 50 wells. As the figure shows, these wells decline much more rapidly than the wells for the Midway field with Naval Petroleum Reserve No. 2 excluded. The curve prepared by the appraisal committee for the Kern River field is also shown in figure 78.

In order to compare the rate at which these different groups of wells decline the table following has been prepared:

Production during second year, decline during second year, and ultimate cumulative percentage of wells in the Midway field.

Group.	First year's daily production.	Ratio of second year's production to first.	Ultimate cumulative percentage.
	Barrels.	*Per cent.*	*Per cent.*
Midway field, excluding Naval Petroleum Reserve No. 2...........	115	74	575
Midway field, including Naval Petroleum Reserve No. 2............	180	70	520
Naval Petroleum Reserve No. 2.....................	461	61	300
Midway field (I. O. P. Agency figures).............................	....	67.5	500

COMPOSITE DECLINE CURVES FOR WELLS OF DIFFERENT SIZES.

Conditions affecting production in the Midway field are so variable that much more information would have been necessary to prepare appraisal curves than was available; consequently another method was adopted to determine the approximate decline of wells of different sizes. The wells were divided into classes in accordance with the average daily production the first year. In Naval Petroleum Reserve No. 2, these classes were as follows: Wells that made daily the first year from zero to 100 barrels, from 101 to 200 barrels, from 201 to 300 barrels, and from 301 to 400 barrels. The average daily production per well the first year for each class was, in order, 66, 154, 262, and 334 barrels. Figure 79 gives the composite decline curve of the wells for each class. By use of these curves estimates of future production in this district can be made with greater accuracy when the first year's production of a well is known. Although the curves show the average decline of wells of different sizes within certain limits, they should not be used to estimate future production if more exact curves can be made by collecting additional records, for in the construction of some of these curves, the records of only a few wells were available.

Similar calculations were made for the Midway field with the wells in Naval Petroleum Reserve No. 2 excluded. The record of

only one well producing between 400 and 500 barrels was available.
In addition, the decline of a well making 624 barrels daily the first
year and one making 1,400 barrels daily the first year are given.

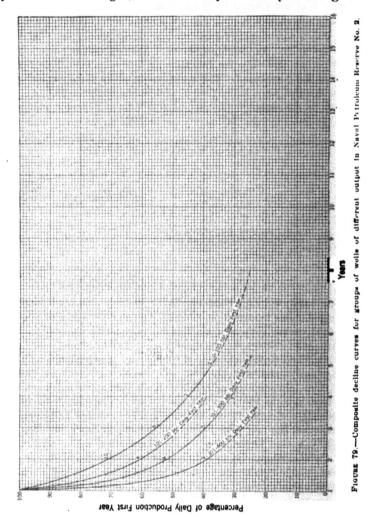

FIGURE 79.—Composite decline curves for groups of wells of different output in Naval Petroleum Reserve No. 2.

The difference in the rate of decline of these wells clearly shows the
fallacy of using the composite decline curve. It should never be
used if other more complete curves are available. These decline
curves are given in figure 80.

For wells that made between zero and 100 barrels or less daily the first year the average daily production per well the first year was 49

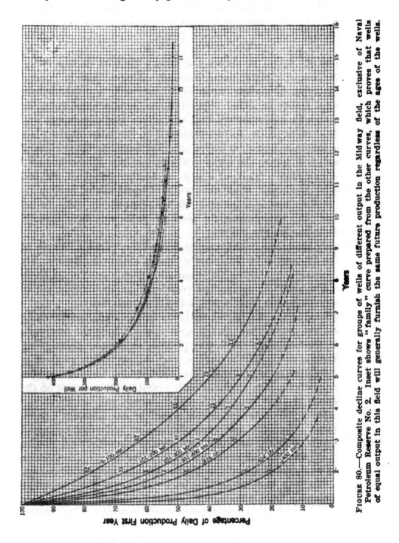

FIGURE 80.—Composite decline curves for groups of wells of different output in the Midway field, exclusive of Naval Petroleum Reserve No. 2. Inset shows "family" curve prepared from the other curves, which proves that wells of equal output in this field will generally furnish the same future production regardless of the ages of the wells.

barrels; for wells that made 101 to 200 barrels it was 148 barrels; and for wells that made between 201 and 300 barrels it was 237 barrels. Furthermore, a curve of two wells averaging 350 barrels

92436°—19——14

daily the first year is given. Because of the large number of individual well records utilized, the two uppermost curves, those showing the decline of wells that made 100 barrels or less and 101 to 200 barrels daily the first year may be used with considerable reliance.

THE " FAMILY " CURVE.

The inset on figure 80 shows a different kind of decline curve, which, for lack of a better name, has been called a " family " curve, because the types of decline curves of all the wells in a field belong to the same family or class. For instance, as a general rule, the decline curve of a medium-sized well will follow that of a larger well if the initial point of the former is placed on the latter at the place where the same production is indicated. Furthermore, the decline of a small well will follow the curve of the larger ones if " hooked on " at the proper place. This, of course, is further proof of the law advanced by Lewis and Beal[a] that wells of equal settled output, regardless of their age, as a general rule will furnish the same future production. For this reason the lives of wells will vary directly with their initial yearly output.

The heavy black line, which may be termed the " family " curve of wells in the Midway field (exclusive of Naval Petroleum Reserve No. 2), was prepared by taking the composite decline curves of those wells that averaged during the first year less than 100 barrels, 101 to 200 barrels, 201 to 300 barrels, 350, and 416 barrels daily.

The decline of the 416-barrel well was first plotted, as shown by the dashed line drawn through the open circles. Next the composite decline of the 350-barrel well was plotted, the initial point being the intersection of the 350-barrel line and the curve first plotted, as shown by the line passing through the filled circles, which represents the daily production each year of such a well. The segment of the curve between the first open circle at 416 barrels and the first filled circle at 350 barrels was considered a part of the " family " curve of that field. Then the composite decline of wells that averaged from 201 to 300 barrels daily the first year was plotted in a similar manner, as shown by the open triangles. Similarly the composite declines of wells that made 101 to 201 barrels and zero to 100 barrels daily the first year were plotted, as indicated by the filled triangles and the crosses. Thus the segments of the " family " curve were determined from point to point, the second segment being shown between the filled circles and the open triangles, the third between the open triangles and the closed triangles, and the fourth between the first closed triangle and the first cross. From that point on—a little past the sixth year—the curve follows the composite decline curve determined by the closed triangles and the crosses.

---

[a] Lewis, J. O., and Beal, C. H., Some new methods for estimating the future production of oil wells: Am. Inst. Min. Eng. Bull. 184, February, 1918.

One of the most striking characteristics of these curves is the narrow variation between the composite decline curves of wells of different output. Thus the well that made 416 barrels daily the first year produced approximately 43 barrels daily during the sixth year. During the same sixth year a new well was brought in that averaged approximately 50 barrels daily during its first year. Both wells, although they differ in age by 5 years, will produce approximately the same amounts in the future.

There can be no question as to the utility of such a curve or its superiority to a composite decline curve and possibly, if enough data are available, to an appraisal curve as the yearly production and thus the decline of a well of almost any output can be determined by reading directly. For example, if a well during the first year produced 260 barrels daily, measuring to the right a distance of one time unit gives the probable production for the second year, or 150 barrels. Similarly the production for the third, fourth, fifth, sixth, and seventh years is successively 100, 75, 55, 45, and 35 barrels. Furthermore, the future production can easily be determined by adding the future yearly production, and the life of the well may be determined by counting the years to the point of the well's exhaustion.

In addition limits of the decline of wells following this curve may be determined by plotting on the " family " curve, the actual decline of individual wells.

The " family " curve has many advantages over any other type of curve, but because of lack of time the one shown was the only one prepared. The generalized decline curves, derived from the appraisal curves for each field and shown throughout the report are based on the same principle and should be noted.

In using this or similar curves, the reader should note that the production of a well must be one year old before its future can be determined. As an aid, general curves could be prepared for determining the approximate relation of the initial production the first 24 hours, or at the end of 30 days, to the average daily production the first year. With such general curves, the probable daily production the first year can be approximated when the well is a day or a month old.

DATA ON TOTAL AND ULTIMATE PRODUCTION OF CALIFORNIA OIL FIELDS.

Accurate statistics have been kept by different companies of the total production each year in the different California fields. Statistics showing the amount produced to December 31, 1917, the proved acreage, and the total production per acre are given in the following tabulation taken from the Standard Oil Bulletin.[a] As regards the figures for area, the Bulletin says:

In determining the figures the boundary lines of the proven area are drawn 200 or 300 feet outside of the proven field. In case of outlying single wells the field is credited with about 15 acres.

The figures therefore represent the actual proven area, and give no consideration to territory that is generally regarded as proven but is not fully drilled. For instance, large areas of undrilled territory in the Buena Vista Hills, although regarded as proven, are not included in the following tabulation.

The proven acreage as shown is, therefore, low as compared with figures made by others. Other estimates have run as high as 110,000 acres, or 171.88 square miles.

TABLE 12.—*Proven acreage, total production to Dec. 31, 1917, and the total production per acre for the different California oil fields.*

Field.	Proven acreage.	Total production to Dec. 31, 1917.	Total production per acre to Dec. 31, 1917.
		Barrels.	*Barrels.*
Kern River	7,730	198,645,210	25,698
McKittrick	1,635	52,114,761	31,874
Midway-Sunset	40,204	291,822,154	7,259
Lost Hills-Belridge	4,476	28,426,055	6,351
Coalinga	14,771	196,872,731	13,328
Lompoc and Santa Maria	7,710	80,913,461	10,495
Ventura County and Newhall	4,514	19,924,745	4,414
Los Angeles and Salt Lake	2,700	52,902,331	19,593
Whittier-Fullerton	4,575	115,584,105	25,264
Summerland	[1] 230	2,180,334	9,480
Miscellaneous	[1] 200	964,727	4,824
	88,745	1,040,350,614	11,723

[1] Estimated.

The McKittrick field, the most productive in the State, has yielded approximately 32,000 barrels an acre. For the whole State the average yield per acre is very high as compared with averages for other States, such as Oklahoma.

An accurate inventory of the present undrilled locations in California has been taken by the State oil and gas supervisor, R. P. McLaughlin, and submitted to the petroleum committee of the California Council of Defense.[b] These locations were classified according to their probable productiveness the first year, and the writer has used the statistics in estimating the probable future production of the undrilled areas in the California fields.

On the assumption that the average ultimate cumulative percentage of the California wells is approximately 500, the undrilled wells in the State, provided they are drilled during the next few years,

[a] Proven territory in California: Standard Oil Bulletin, April, 1918, p. 11.

[b] Report of Committee on Petroleum, California Council of Defense, 1917, pp. 131, 140, and 141.

will ultimately produce about 1,300,000,000 barrels of oil. The probable future production of the producing wells has been estimated at approximately 1,000,000,000 barrels, so that the estimated total future production for California is about 2,300,000,000 barrels. The present proved territory, according to Table 12, is 88,745 acres, but it includes only land on which wells have been drilled. The undrilled areas comprise about 72,000 acres, making the total productive acreage 160,745 acres. As the past production has been a little more than 1,000,000,000 barrels, and the estimated total future production is 2,300,000,000 barrels, the ultimate production per acre (3,300,000,000÷160,745) is about 20,500 barrels.

THE KUROKAWA OIL FIELD, JAPAN.

Some information has been collected on individual wells belonging to the Nippon Oil Co., in the Kurokawa oil field, Province of Ugo, Japan. Figure 13 (p. 54) shows the average daily production the first week of wells drilled during succeeding months. The composite decline of several scores of wells the first three years has been determined to be, successively, 100, 49, and 29 per cent. As shown by figure 13, the ultimate cumulative percentage is approximately 300. The oil sand is reported as loose and fine, 2 to 5 feet thick, and 1,100 to 1,400 feet below the surface. The initial production of the wells (daily average the first week) ranges from 2 to more than 5,000 barrels.

COMPARISON OF THE DECLINE AND APPRAISAL CURVES OF SEVERAL FIELDS.

Interesting and fruitful subjects for future investigation are the study of the factors governing output in different fields, the way in which each factor influences the average rate at which the oil is obtained, and effect of each factor on the ultimate cumulative percentage or ultimate production. By determining these factors, the probable future of other fields and properties, where the same factors have approximately equal value, can be much more easily estimated.

For instance, let it be assumed that the individual and composite effect of all the important factors influencing the rate of production in the Bartlesville field (Okla.), are known. The thickness of the oil sand does not vary much, the wells are spaced a certain distance, and, as the depth is fairly uniform, the rock pressure is about the same. By an analysis of these data, one can determine how almost any important factor affects the ultimate production per acre and the rate at which the oil is obtained. Now if the value of each production factor in a field can be determined, the estimating of the possibilities of properties in other fields where one or more of the important production factors are similar, will be greatly facilitated. The problem is to determine the individual effects of the different factors.

To determine how various field conditions affect the decline curve and the ultimate cumulative percentages, tables have been prepared that recapitulate the conditions influencing production in the different fields studied. Table 13 gives data on the principal fields for which enough information was available to justify inclusion here. Column 2 shows the average daily production the first year of the wells used in constructing the composite curve; column 3. gives the second year's percentage of the first year's average daily production. which is an excellent indication of the probable productiveness of the field as a whole; and column 4 shows the approximate ultimate cumulative percentage of the average well in each field.

TABLE 13.—*Comparison of the ultimate and the second year's percentages of average wells in different fields.*

Field.	Average daily production first year.	Second year's percentage of first year's production.	Approximate ultimate cumulative percentage.
1	2	3	4
	Barrels.	Per cent.	Per cent.
Oklahoma:			
Bartlesville...	17	49	310
Osage (eastern)...	35	63	330
Bird Creek-Flat Rock...	30	60	300
Nowata...	19	50	270
Glenn pool...	45	51	240
Okmulgee-Morris...	38	52	320
Hamilton Switch...	56	55	320
Muskogee pool...	24	44	210
Ponca City...	32	61	320
Cushing:			
Layton sand...	34	32	215
Wheeler sand...	53	23	175
Bartlesville sand...	208	30	220
Healdton...	67	62	420
Kansas (shallow)...	4	75	750
North Texas: Electra...	75	51	330
North Louisiana:			
Caddo...	66	54	290
Red River...	475	23	....
De Soto...	190	35	....
Crichton...	85	28	....
Gulf Coast:			
Humble pool (cap rock)...	47	51	....
Humble pool (deep sand)...	289	38	....
Sour Lake pool...	84	39	....
Spindletop pool...	92	45	....
Saratoga pool...	23	53	....
Illinois			
Crawford County...	11	52	280
Clark County...	3	65	330
Lawrence County...	100	61	370
Southeastern Ohio: New Straitsville pool...	22	37	210
West Virginia:			
Blue Creek...	17	27	170
Duvall district (Lincoln County)...	17	60	280
Rock Creek...	8	65	470
Spencer...	22	70	350
Wyoming: Salt Creek...	100	69	460
California:			
Coalinga (East Side)...	180	76	480
Coalinga (West Side)...		79	574
Kern River...		78	550
Midway...	189	70	520
Maricopa...		59	445
Japan: Kurokawa...	80	49	300

UNIFORMITY OF THE SECOND YEAR PERCENTAGE AND THE ULTIMATE CUMULATIVE PERCENTAGE FOR OKLAHOMA FIELDS.

It is interesting to note the uniformity of the second year's percentage in the individual pools of a large area where the conditions that affect production are approximately the same. This uniformity is well shown by the fields of Oklahoma, if the Cushing field be excluded, because of extraordinary conditions and the Muskogee pool because of few data. For the remaining Oklahoma fields the lowest percentage for the second year is 49 per cent in the Bartlesville field, whereas the highest percentage, 63 per cent, is the Osage Indian Reservation, a few miles west. The probable cause for this difference is that the area allotted each well in the Bartlesville field is only about one-half of that allotted in the Osage Indian Reservation. Inasmuch as the rate of decline of oil wells is influenced by the spacing of the wells such a difference in spacing ought to exert a marked influence on the decline of the average wells.

In the eight Oklahoma fields under consideration the average second year's percentage is 55 per cent; the maximum limit 63 per cent; the minimum limit 49 per cent; and the maximum and minimum variation above and below the average are respectively 15 and 11 per cent. Hence the following general rule can be laid down: *Average* wells in Oklahoma fields producing under ordinary conditions will average during the second year 55 per cent of their first year's yield, and the chances are that the second year's yield will not exceed this amount more than 15 per cent nor fall below it more than 11 per cent. In other words, it is fairly safe to assume that a well which averaged 38 barrels daily the first year, which is the average for all pools, will average between 50 and 60 per cent of that much the second year.

For these normal fields in Oklahoma the ultimate percentages show the same uniformity, ranging from 240 to 330. In other words, the average well in these fields will produce ultimately not less than about 2.4 times and not more than about 3.3 times its first year's production. Other similarities of fields in other areas may be noted, especially where the conditions that affect production do not vary widely.

In the Cushing field three different sands have yielded the bulk of the oil. Production from all these sands declined much more rapidly than from sands in the other Oklahoma fields. The principal cause of this rapid decline was the great waste of gas. Furthermore, the wells were unsystematically cased and water was allowed to flood some of the sands.

The percentages given in columns 3 and 4 of Table 13 vary because of the range of the conditions affecting production in the different fields. The magnitude of the effect of each factor can not always be determined, and the rate at which the output of a well declines is a resultant of the composite influence of all the factors that in any manner govern the production of oil. However, the influence of one factor may predominate. For instance, the dominating factor in causing the decline of wells in the Bartlesville field to differ from that of wells in the Osage Indian Reservation seems to be the difference in the spacing of the wells. Unquestionably, however, the greater depth of the wells in the Osage Indian Reservation has influenced the rate of decline of the wells there considerably, because of the increased gas pressure and the consequently higher initial output. . Other factors evidently must have an opposite effect, thus counterbalancing the effect of the difference in depth.

GENERAL TENDENCIES OF FACTORS CONTROLLING OUTPUT.

The questions naturally arise: What are the general tendencies of the output of wells producing under certain conditions? Under what conditions do wells tend toward large and small ultimate productions? Answers may be found in Table 14, suggested to the author by J. O. Lewis, of the Bureau of Mines, which gives the general tendency of the ultimate cumulative percentage and the ultimate production of wells producing under certain conditions.

TABLE 14.—*General tendencies of wells producing under specified conditions.*[a]

Condition.	Rate of decline.	General tendency of ultimate cumulative percentage.	General tendency of ultimate production.
Deep wells.	Rapid.	Small.	Large.
Shallow wells.	Slow.	Large.	Small.
High rock pressure.	Rapid.	Small.	Large.
Low rock pressure.	Slow.	Large.	Small.
Large initial production.	Rapid.	Small.	Large.
Small initial production.	Slow.	Large.	Small.
Close spacing.	Rapid.	Small.	Small.[b]
Wide spacing.	Slow.	Large.	Large.[c]
Thick sand.	...do.	...do.	Large.
Thin sand.	Rapid.	Small.	Small.
Large pore space.	...do.	...do.	Large.
Small pore space.	Slow.	Large.	Small.
High-gravity oil.	Rapid.	Small.	Large.
Low-gravity oil.	Slow.	Large.	Small.
With water trouble.	Rapid.	Small.	Do.[d]
No water trouble.	Slow.	Large.	Large.
Properly operated.	...do.	...do.	Do.
Poorly operated.	Rapid.	Small.	Small.

a In this table the influence of any one factor is given on the assumption that all other factors are without influence.
b Small per well but large per acre.
c Large per well but small per acre.
d In some cases, when water floods in, the ultimate production will be greatly increased.

TENDENCIES IN THE NEW STRAITSVILLE POOL, OHIO.

An excellent example of the way in which these tendencies work can be had by comparing the second year's percentage and the ultimate cumulative percentage of wells in the New Straitsville pool,

Ohio, as shown by Table 13 (p. 202), with the conditions prevailing there. In that pool the second year's production is 37 per cent of the first year, and the ultimate cumulative percentage is 210; the average daily production of wells the first year is low; the pay sand is probably not more than 10 feet thick and the walls are approximately 3,000 feet deep and rather closely spaced. According to Table 14, the depth of these wells, the high rock pressure, the thin sand and the closeness of spacing all favor rapid decline and small ultimate cumulative percentages, and the small initial production is the only factor favoring a large ultimate cumulative percentage. Unquestionably, the composite effect of the four factors tending toward a small ultimate cumulative percentage is much stronger than the tendency of the one factor favoring a large ultimate cumulative percentage. For other fields the effects of similar factors and the tendency toward high or low ultimate cumulative percentages can be determined in much the same way.

VARIATIONS IN ESTIMATES OF ULTIMATE PRODUCTION BASED ON APPRAISAL CURVES.

Variations above and below the average in estimates of ultimate production made from the appraisal curves of the different fields are summarized in the table following, which shows the maximum and minimum error that may be expected in estimating the ultimate output.

TABLE 15.—*Probable error above and below the average of the maximum and minimum estimates of ultimate production.*

Field.	Daily production per well.a	Maximum ultimate production.	Average ultimate production.	Minimum ultimate production.	Possible error above average.	Possible error below average.
	Barrels.	Barrels.	Barrels.	Barrels.	Per cent.	Per cent.
Osage Nation, Okla.	150	140,000	103,000	78,000	36	24
Bartlesville field, Okla.	75	65,000	50,000	35,500	30	29
Bird Creek-Flat Rock, Okla.	65	71,000	55,000	39,500	29	28
Nowata field, Okla.	60	36,000	30,600	25,500	18	17
Glenn Pool, Okla.	100	106,500	81,000	54,500	31	33
Caddo field, La.	200	199,000	146,000	98,000	36	33
Crawford and Clark Counties, Ill.	24	20,300	16,000	11,700	23	27
Lawrence County, Ill.	50	54,600	40,000	25,600	37	36
New Straitsville, Ohio.	30	25,700	19,400	13,300	32	31

a Selected by determining the mean average daily production as shown in the respective appraisal curves.

In selecting the output of the well for which the ultimate production was to be estimated, the mean between the two extremes of daily production shown on the lower side of the appraisal curve was selected in order to have all estimates on the same basis. For instance, in the Osage Nation, Okla., the average daily production the first year ranged up to 300 barrels. Therefore, the ultimate production of a 150-barrel well was determined and the variation above and below the average estimate was computed. Except for the Nowata field (Okla.), the percentage of variation of the maximum and minimum limits above and below the average is fairly uniform, the maximum limits ranging from 23 to 37 per cent above the average and the minimum limits from 24 to 36 per cent below the average. The table thus serves as an index of the range of error that may be expected in appraisal curves for the different fields.

SELECTED BIBLIOGRAPHY.

THE DECLINE AND ULTIMATE PRODUCTION OF OIL PROPERTIES.

GOVERNMENT AND STATE REPORTS—ARTICLES IN TECHNICAL JOURNALS AND PROCEEDINGS OF SOCIETIES.·

ARNOLD, RALPH, The petroleum resources of the United States: Economic Geology, vol. 10, December. 1915, pp. 695–712.

BUCKLEY, E. R., Building and ornamental stones in Wisconsin: Wisconsin Geol. and Nat. Hist. Survey, Bull. 4, econ. ser. No. 2, 1898, p. 402.

DAY, DAVID T., The petroleum resources of the United States, a chapter in the conservation of mineral resources: U. S. Geol. Survey, Bull. 394, 1909, pp. 30–50.

HAGER, DORSEY, Geological factors in oil production: Min. and Sci. Press, vol. 109, December 9, 1911, pp. 738–741.

HUNTLEY, L. G., Possible causes of the decline of oil wells and suggested methods of prolonging yield: Tech. Paper 51, U. S. Bureau of Mines, 1913, 32 pp.

LEWIS, J. O., and BEAL, C. H., Some new methods for estimating future production of oil properties: Am. Inst. Min. Eng., Bull. 134, February, 1918, pp. 477–504.

LOMBARDI, M. E., The cost of maintaining production in the California oil fields: Trans. Am. Inst. Min. Eng., 1916, pp. 218–224.

McLAUGHLIN, R. P., Petroleum industry of California: California State Mining Bureau, Bull. 69, 1914. 519 pp.

PACK, R. W., The estimation of petroleum reserves: Trans. Am. Inst. Min. Eng., vol. 57, 1918, pp. 968–981.

THELEN, MAX, BLACKWELDEN, ELIOT, and FOLSOM, D. M., Report of the committee on petroleum: California State Council of Defense, July 7, 1917, 191 pp.

WASHBURNE, C. W., The estimation of oil reserves: Am. Inst. Min. Eng., Bull. 98, February, 1915, pp. 469–471.

BOOKS.

BRINTON, W. C., Graphic methods for presenting facts. 1914. 371 pp.

HAGER, DORSEY, Practical oil geology. 1916. 187 pp.

JOHNSON, ROSWELL H., and HUNTLEY, J. G., Principles of oil and gas production. 1916. 371 pp.

PEDDLE, JOHN B., The construction of graphical charts. 1910. 109 pp.

THOMPSON, A. BEEBY, Oil-field development. 1916. 684 pp.

THE VALUATION AND TAXATION OF OIL PROPERTIES.

GOVERNMENT REPORTS—ARTICLES IN TECHNICAL JOURNALS AND PROCEEDINGS OF SOCIETIES.

ASHLEY, G. H., The valuation of public coal lands: U. S. Geol. Survey Bull. 424, 1910, 75 pp.

BYLLESBY, H. M., Responsibilities of engineers in making appraisals: Trans. Am. Inst. Elect. Eng., vol. 32, pt. 2, 1911, pp. 1251–1265.

CABELL, R. E., Basis for estimating special excise tax on corporations: Oil Age, vol. 5, March 1, 1912, p. 2.

FORSTNER, W., The valuation of oil lands: Min. and Sci. Press, vol. 103, 1911. p. 578.

HAGER, DORSEY, Valuation of oil properties: Eng. and Min. Jour., vol. 101, May 27, 1916, pp. 930–932.

HAYS, V. A., Depreciation of natural gas properties: Progressive Age, vol. 30, pp. 787–788.

HENRY, P. W., Depreciation as applied to oil properties: Trans. Am. Inst. Min. Eng., vol. 51, 1916, pp. 560–570.

JOHNSON, ROSWELL, Methods of prospecting, development, and appraisement in the Mid-Continent field: Oil Investor's Jour., vol. 8, February 20, 1910, pp. 70–73.

KNIGHT, ALFRED, Discussion of depreciation: Journal of Accountancy, vol. 5, 1907, p. 189.

REQUA, M. L., Methods of valuing oil lands: Am. Inst. Min. Eng., Bull. 134, February, 1918, pp. 409–428.

BOOKS.

FLOY, HENRY, Valuation of public utility properties. 1912. 390 pp.

HOOVER, H. C., Principles of mining. 1901. 109 pp.

HOSKOLD, H. D., Engineer's valuing assistant. 1905, 185 pp.

HUMPHREYS, A. C., Business features of engineering practice. 1905. 187 pp.

JOHNSON, R. H., and HUNTLEY, L. G., Principles of oil and gas production. 1916. 371 pp.

LEAKE, P. D., Depreciation of wasting assets. 1917. 233 pp.

MACLEOD, H. D., Elements of economics. 1881–1886. 2 vols.

O'DONAHUE, T. A., The valuation of mineral property. 1910. 158 pp.

SCHOOLING, W., Inwood's tables for the purchasing of estates, etc. 1913. pp.

THOMPSON, A. BEEBY, Oil-field development. 1916. 684 pp.

WYER, S. S., Regulation, valuation, and depreciation of public utilities. 1913. 313 pp.

INDEX.

O

Bulletin 178-A

DEPARTMENT OF THE INTERIOR

FRANKLIN K. LANE, Secretary

BUREAU OF MINES

VAN. H. MANNING, Director

WAR GAS INVESTIGATIONS

ADVANCE CHAPTER FROM BULLETIN 178
WAR WORK OF THE BUREAU OF MINES

BY

VAN. H. MANNING

WASHINGTON
GOVERNMENT PRINTING OFFICE
1919

First edition. May, 1919.

CORRECTIONS.

At the end of page 39 add—

At the University of Michigan Dr. M. Gomberg did some valuable work on mustard gas and other toxic-gas syntheses.

On page 38, seventh line from bottom, add—

Dr. James Withrow did especially valuable work on prussic acid, cyanogen chloride, arsenic trichloride, and other substances.

Last three lines on page 20 and first line on page 21 should read—
the life of canisters.

Lieut. J. H. Yoe aided in devising a canister for protection against carbon monoxide. Also, a special canister for protection against hydrocyanic acid was developed by Lieuts. Yoe and Beattie.

CONTENTS.

WAR GAS INVESTIGATIONS.

GENERAL STATEMENT.

Of all the war work by the Bureau of Mines none ultimately covered such a variety of processes and equipment, employed as many men, or called for as large an outlay as that started in connection with the solution of problems relating to the use of noxious gases in warfare. Beginning with an investigation to develop the best type of gas mask, the scope of the work extended until it included researches relating to a wide range of devices, such as different types of poisonous and irritating gases and smokes, smoke screens, gas shells and gas bombs, flame throwers, trench projectors for firing gas bombs, signal lights, and incendiary bombs. The Bureau of Mines in starting this work selected the necessary personnel, procured the early equipment, and for some months, from the beginning to June 30, 1917, paid the cost of the work from its own funds.

The Bureau of Mines built up a research staff of more than 700 chemists, including many of the most prominent chemists in the country, and obtained the cooperation of many of the universities and chemical companies. When the results obtained through this research were to be applied by the War Department, the Director of the Bureau of Mines cooperated in the selection of men to have charge of manufacture or further development, recommending chemists who subsequently received commissions in the Army. The personal cooperation with the War Department was thus very close.

The bureau can fairly claim that because it started this work on gas warfare and received the hearty support of the Department of War, the Navy Department, the National Research Council, and State, educational, and private institutions, the country was in July, 1918, months ahead of where it would otherwise have been in the production of gas masks and other devices. The production of toxic gases then far exceeded the supply of shells. In addition the later work in gas-mask manufacture by the Surgeon General's Office of the Army, which resulted in the development of the best mask produced anywhere; in gas manufacture and gas proving-ground tests by the Bureau of Ordnance, and in the chemical warfare program of the Navy, including the use of smoke screens, shells and toxic gases, were the direct results of the bureau's experimental work.

The American charcoal and soda lime, developed through the research work, are the best used by any nation. The Bureau of

Mines was instrumental in developing satisfactory methods of manufacturing chlorpicrin, phosgene, mustard gas, brombenzylcyanide, cyanogen chloride, and other noxious substances on a large scale. The bureau also developed a smoke funnel for the Navy which no doubt saved many ships from submarine attacks, and devised smoke barrages, signal lights, incendiary darts, etc., for the Army. The Ordnance Department of the Army asked for incendiary bombs and were furnished with bombs better than any made previously.

Perhaps the greatest single result of the research work has been the demonstration of the value of the services that the chemist and chemical engineer can render in time of need and the recognition of chemical investigations as a necessary function of the Government.

GENERAL HISTORY OF THE RESEARCH WORK.

INCEPTION OF THE RESEARCH WORK.

, In February, 1917, when war between the United States and Germany seemed inevitable, the Bureau of Mines took up the question of what it could do to most advantage in the event of war. Since its establishment in 1910, it has maintained a staff of investigators studying poisonous and explosive gases in mines, the use of self-contained breathing apparatus for exploring mines filled with noxious gases, the treatment of men overcome by gas, and similar problems. At a conference of the director of the bureau with his division chiefs, on February 7, 1917, the matter of national preparedness was discussed, and especially the manner in which the bureau could be of most immediate assistance with its personnel and equipment. One of the things decided was to investigate gas masks and rescue apparatus for military and naval purposes.

On February 8, the director wrote C. D. Walcott, chairman of the military committee of the National Research Council, pointing out that the Bureau of Mines could immediately assist the Navy and the Army in developing for naval or military use special oxygen breathing apparatus similar to that used in mining. He also stated that the bureau could be of aid in testing types of gas masks used on the fighting lines, and had available testing galleries at the Pittsburgh experiment station and an experienced staff which included W. E. Gibbs, the inventor of the Gibbs breathing apparatus. Dr. Walcott replied on February 12 that he was bringing the matter to the attention of the military committee.

On February 7, a request for information regarding the construction of the Gibbs apparatus came from the Brooklyn Navy Yard. At a conference in New York on February 20, Asst. Naval Constructor G. W. Fulton and an assistant, of the Brooklyn Navy Yard, and G. S. Rice, chief mining engineer, and O. P. Hood, chief mechanical engineer of the Bureau of Mines being present, breathing apparatus constructed in the navy yard instrument shop was shown. Then Mr. Gibbs went to the Brooklyn yard to offer his help in designing suitable short-time apparatus, and also to give information he had received from Dr. J. S. Haldane, the English physiologist, when the latter visited the Pittsburgh station in 1916, regarding methods of absorbing CO₂ in the atmosphere of submarines. In addition to the need of developing a short-time apparatus for use on ships, there was also need for a half-hour apparatus for use in mines and for general purposes. Subsequently Mr. Gibbs spent much time working with officials of the Navy Department on problems relating to the use of breathing apparatus and the absorption of CO₂ on submarines.

On February 20, the Naval Consulting Board submitted to the bureau two questions: One related to the development of some substance which on the detonation of a high explosive in a confined space would render innocuous the gases of the explosion; the other related to putting out fires in confined spaces aboard ship. The inquiries were referred to C. E. Munroe, consulting chemist of the bureau; his recommendations were forwarded to the naval board on March 9, 1917.

On April 4 a conference was held at the War College at Washington. At this meeting there were present Gen. Kuhn, then president of the War College; Col. Aultman, of the college; Maj. Williamson, of the Army Medical Corps; Director Manning, Mr. Rice, Dr. Yandell Henderson, Mr. Gibbs, Mr. Burrell, and Mr. Fieldner, of the Bureau of Mines. The conferees discussed what could be done by the Bureau of Mines in testing gas masks and self-contained breathing apparatus for military and naval use. Director Manning stated that he had just received notice that the War and Navy Departments, through the military committee of the National Research Council, had placed the investigation of this matter in the hands of the bureau. Gen. Kuhn agreed that this arrangement was satisfactory, and detailed Maj. Williamson to represent the War College.

On April 5, at a meeting at the office of Director Manning, Maj. Williamson and Messrs. Burrell, Henderson, Fieldner, Gibbs, Cottrell, and Rice being present, Maj. Williamson gave a review of the use of noxious gases for offensive warfare, of the measures taken for protection against them, and of the work being done in England. He said that the gas masks collected by military observers had been sent to the Picatinny Arsenal, and some tests might be made there. Director Manning assigned a gas chemist to assist Maj. Williamson in going over the masks collected by the Army, and on April 6 this work began.

The director offered the supervision of the research on gases to G. A. Burrell, who had been for a number of years in charge of the chemical work done by the bureau in connection with the investigation of mine gases and natural gas. In April, 1917, Mr. Burrell was in commercial work in Pittsburgh. He accepted the offer on April 7.

COMMITTEE ON NOXIOUS GASES, NATIONAL RESEARCH COUNCIL.

The National Research Council was organized to act as an intermediary on research between the scientists and the universities of the country and the various departments of the Government and to suggest and consider research problems.

On April 6, 1917, the Director of the Bureau of Mines informed the Secretary of the Interior of the appointment of a committee on noxious gases, National Research Council. With the approval of the National Research Council and the War and Navy Departments, the following men were designated as members of the committee: Chairman, Van. H. Manning, Director of the Bureau of Mines; Col. E. B. Babbitt, Ordnance Department, United States Army; Lieut. T. S. Wilkinson, Bureau of Ordnance, United States Navy; Maj. L. P. Williamson, Medical Department, United States Navy; Dr. M. T. Bogert, professor of organic chemistry at Columbia University; and Dr. C. L. Alsberg, Chief of the Bureau of Chemistry, Department of Agriculture. On April 12 Lieut. Commander A. H. Marks, Bureau of Construction and Repairs, Navy Department, was designated a member of the committee, and on April 25, Past Asst. Surg. E. F. Dubois, Bureau of Medicine and Surgery, Navy Department.

This committee took up various problems relating to the use of gases in warfare amd made recommendations to the National Research Council, which transmitted those it approved to the organization interested. The membership of the committee changed somewhat from time to time through the transfer of Army and Navy officers; it continued its activities until August, 1918.

After the transfer of the war gas investigations to the War Department in June, 1918, the committee on noxious gases was dissolved on August 10, 1918. The membership of the committee then was as follows: Chairman, Van. H. Manning, Director Bureau of Mines; Passed Asst. Surg. E. F. Dubois, Bureau of Medicine and Surgery, United States Navy; Lieut. Col. R. A. Millikan, Signal Corps, United States Army; G. E. Hale, National Research Council; Lieut. Col. E. J. W. Ragsdale, Ordnance Department, United States Army; Col. Bradley Dewey, Chemical Warfare Service, United States Army; Lieut. Col. M. T. Bogert, Chemical Warfare Service, United States Army; Lieut. Col. E. J. Atkisson, Engineers, United States Army; Lieut. Commander T. S. Wilkinson, Bureau of Ordnance, United States Navy; C. L. Alsberg, Bureau of Chemistry.

DETERMINATION OF RESEARCH PROBLEMS.

As a first step the Bureau of Mines assembled all available reports on the use of gases in warfare that had come to the War or the Navy Department. At that time a little work had been undertaken by the Army in regard to the development of gas masks and the use of smoke screens.

Various lines of research were marked out, that of most immediate importance being the development of gas masks for the Army. This problem required much research on the materials to be used in the masks and much experimenting to develop the various parts. Maj. Williamson cooperated actively in this work, as did Lieut. Commander A. H. Marks, who was vice president of the Goodrich Rubber Co. before offering his services to the Government and was the pioneer in developing a gas mask for the Navy. Having the facilities of the Goodrich Company at his command he took an active part in the new work.

As soon as the chief lines of research had been determined, the next question was the selection of the best available chemists and engineers, many of whom were offering their services to the Government for the conduct of each new line of investigation as it came up.

By June 30, 1917, the personnel engaged in research work comprised 50 paid investigators, and the work had expanded from the devising of gas masks to the study of poison gases and chemical appliances for offensive warfare.

ADVISORY BOARD.

In November, 1917, Secretary Lane appointed the following advisory board to the Director of the Bureau of Mines on war problems: Dr. Wm. H. Nichols, Prof. E. C. Franklin, Dr. C. L. Parsons, Mr. Wm. Hoskins, Prof. H. P. Talbot, Dr. F. P. Venable, Dr. Ira Remsen, and Prof. T. W. Richards.

This board acted in an advisory capacity to the director on the gas work, both individually and as a body. The board first met on December 17, 1917, when it visited the laboratories, consulted with the research staff, and witnessed a demonstration of the work in progress. A second meeting was held May 15 and 16, 1918, when the work was again reviewed.

COOPERATION WITH WAR DEPARTMENT, NAVY DEPARTMENT, AND NATIONAL RESEARCH COUNCIL.

Throughout the course of the war-gas investigations the Bureau of Mines kept in close touch with officials of the War and Navy Department and with the Navy and Army officials of the Council of National Research; these received copies of reports telling of the progress of the work. When some fact had been determined that seemed to have a practical application, officials who would be interested were notified, and they decided whether further study should be given it.

At the American University station, the officers and enlisted men reported to the commanding officer there, who assigned them to work in the laboratories or testing units under the direction of the chiefs of the Bureau of Mines. The supervision of the work was under the director of the bureau, through the assistant in charge of gas investigations. This arrangement worked smoothly, as was testified by the manner in which the reasearch work kept in advance of the large-scale production of the gases, explosives, and devices that were approved.

CHEMICAL SERVICE OF THE NATIONAL ARMY.

One of the early results of the research work undertaken by the Bureau of Mines was the establishment of the chemical service as a unit of the National Army. This unit was established in December, 1917, as a result of conferences held at the Bureau of Mines with officers from the Navy, the Medical Department, the War College, and the General Staff of the Army, and civilian chemists. Lieut. Col. W. H. Walker commanded the American branch of the Chemical Service, reporting to Col. C. L. Potter, of the Gas Warfare Division, and Lieut. Col. R. F. Bacon was chief of the Chemical Service Section in France. Subsequently Lieut. Col. Walker was transferred to the Ordnance Department and put in charge of enormous plants for making chemicals for gas warfare that were erected at a new arsenal, known as Edgewood Arsenal, near Baltimore. His successor in charge of the Chemical Service Section was Lieut. Col. M. T. Bogert.

The establishment of this unit is a noteworthy fact in the history of applied chemistry, for never before in any war or in any country had chemistry been recognized as a separate branch of the military service.

ESTABLISHMENT OF AMERICAN UNIVERSITY EXPERIMENT STATION.

On April 30, 1917, B. F. Leighton, president of the board of trustees of the American University, made a formal tender to President Wilson of the university's buildings and grounds, who referred the offer to the Secretary of War.

On the university grounds were two large stone buildings, the College of History building and Ohio Hall. The grounds were isolated enough to permit the making of field tests and the training of men in gas warfare and yet were readily accessible.

On May 26, 1917, the committee on noxious gases of the National Research Council, Van. H. Manning, chairman, discussed the need of establishing a central laboratory in Washington to coordinate the investigations on noxious gases, and of having work begun on a large scale under the joint cooperation of the Army, the Navy, and the Bureau of Mines of the Department of the Interior. The committee resolved to recommend to the Secretary of War and the Secretary of the Navy that such a research laboratory be established at the American University, the War and Navy Departments to provide funds for equipment and personnel of the laboratory, to be expended by those departments under the direction of the committee on noxious gases. This resolution was approved by the National Research Council and transmitted through the Munitions Board to the Council of National Defense, which referred it to the Secretary of War, the Secretary of the Navy, and the Secretary of the Interior.

On recommendation of Admirals D. W. Taylor and Ralph Earle, of the Bureau of Ordnance of the Navy Department, an allotment of $50,000 for this purpose was authorized by the Secretary of the Navy on June 6, 1917. In a letter to the Secretary of the Interior, under date of June 15, the Director of the Bureau of Mines outlined the cooperative arrangement, stating that preliminary steps had already been taken for installing a central laboratory at the American University, and that the Secretary of the Navy had allotted $50,000 for this work. The arrangement outlined was approved by Secretary Lane. On June 25, 1917, Secretary Baker notified Secretary Lane that he had authorized the expenditure of $125,000 by the War Department for this purpose to be expended under the direction of the Bureau of Mines.

On July 6, 1917, Director Manning wrote the chancellor of the American University, Rev. J. W. Hamilton, accepting the offer of the use of Ohio Hall; the formal grant by the university through the president of the board of trustees was made July 21, 1917. By the terms of the grant the Government was to have the use of the building for two years from June 30, 1917, and as much longer as the war might last, and the Government was not to remove any of the permanent additions made.

The needed changes and improvements were quickly made, and the American University station of the Bureau of Mines came into being. A temporary laboratory building and other structures were completed in September, 1917. Meanwhile the War Department had begun to prepare a camp, named Camp Leach, on the grounds, for accommodating the men assigned to this work and for the training of enlisted men and officers in gas warfare.

By October the need of more room for laboratories and for offices became pressing. The chancellor on October 13 granted the Bureau of Mines permission to use part of the College of History building. Subsequently the bureau occupied nearly the whole of this building in addition to Ohio Hall.

ADDITIONAL ALLOTMENTS OF FUNDS.

To carry on the research necessary for developing the best devices and methods for having them manufactured as soon as possible on a large scale, a considerable increase in personnel and in testing facilities was needed; hence it was clear that more funds was required. When this was brought to the attention of the Navy Department, Secretary Daniels asked that $100,000, from the $100,000,000 appropriation expendable by the direction of the President for national defense, be allotted to the Bureau of Mines for conducting necessary experiments. The President's reply to Secretary Daniels, under date of September 17, follows:

17 SEPTEMBER, 1917.

THE WHITE HOUSE,
    Washington.

MY DEAR MR. SECRETARY:

I have your request, sent me under date of September 10, that $100,000 be allotted to the Navy Department from the $100,000,000 appropriation from the national security and defense, to be expended by the Bureau of Mines in the conduct of experiments in the uses of gases in war, in addition to the $50,000 already allotted under date of July 7. I am very glad to make this allotment inasmuch as I realize the critical importance of the investigations referred to.

Cordially and sincerely, yours,

(Signed)   WOODROW WILSON.

Hon. JOSEPHUS DANIELS,
    Secretary of the Navy.

The rapid expansion of the research work to keep pace with the demands made on it in connection with outfitting the Army and Navy necessitated additional expenditures that could not be foreseen. A new gas might assume extreme importance over night and an expenditure of $50,000 might be required to put it into production. A large number of chemists worked on each important problem to get out results as quickly as possible. The necessary funds were supplied by the War Department and the Navy Department.

The total sum allotted by the Army for war gas investigations during the fiscal year ended June 30, 1918 was $2,212,000; the total sum allotted by the Navy was $250,000.

TRANSFER OF RESEARCH WORK TO WAR DEPARTMENT.

By May, 1918, the question had arisen as to whether the research work on war gases should be transferred to the War Department or should continue under the Bureau of Mines. Certain Army officials believed that the work could be coordinated better by having it under military control. On the other hand it was held that there was nothing to be gained by a transfer of authority as the research work had kept well in advance of the manufacturing development in progress under the War Department and the development of methods of manufacture had been conducted so expeditiously that the supply of toxic materials exceeded the supply of shells. The Advisory Chemical Board said of the work being done by the Bureau of Mines: "The efficiency, success, fine spirit, and enthusiasm under the leadership of the Bureau of Mines is a matter upon which we wish to congratulate the bureau, as well as upon the splendid group of unselfish, self-sacrificing men who carried on this arduous and dangerous work." In regard to the organization of the work the board said: "The organization is complex and delicate but well articulated and working with an efficiency and enthusiasm which have impressed us greatly."

However, the counsel of those Army officers who believed the research work should be transferred to the War Department prevailed, and on June 25, 1918, under the authority given by what is known as the Overman Act, approved May 20, 1918 (Public, No. 152), President Wilson transferred the work being done at the American University from the Bureau of Mines to the Chemical Service of the Army. In letters relating to this transfer, the Secretary of War and the President have expressed their appreciation of the bureau's work, as follows:

JUNE 25, 1918.

MY DEAR MR. PRESIDENT:

In connection with the proposed transfer of the chemical section at American University from the Bureau of Mines to the newly constituted and consolidated Gas Service of the War Department, which you are considering, I am specially concerned to have you know how much the War Department appreciates the splendid services which have been rendered to the country and to the Army by the Department of the Interior, and especially by the Bureau of Mines, under the direction of Dr. Manning. In the early days of preparation and organization Dr. Manning's contact with scientific men throughout the country was indispensably valuable. He was able to summon from the universities and the technical laboratories of the country men of the highest quality and to inspire them with enthusiastic zeal in attacking new and difficult problems which had to be solved with the utmost speed. I do not see how the work could have been better done than he did it, and the present suggestion that the section now pass under the direction and control of the War Department grows out of the fact that the whole subject of gas warfare has assumed a fresh pressure and intensity, and the director of it must have the widest control so as to be able to use the resources at his command in the most effective way possible. The proposal does not involve the disruption of the fine groups of scientific men Dr. Manning has brought together, but merely their transfer to Gen. Sibert's direction.

Respectfully, yours,

NEWTON D. BAKER.

THE PRESIDENT.

_____

26 JUNE, 1918.

MY DEAR DR. MANNING:

I have had before me for some days the question presented by the Secretary of War involving the transfer of the chemical section established by you at the American University from the Bureau of Mines to the newly organized Division of Gas Warfare, in which the War Department is now concentrating all the various facilities for offensive and defensive gas operations. I am satisfied that a more efficient organization can be effected by having these various activities under one direction and control, and my hesitation about acting in the matter has grown only out of a reluctance to take away from the Bureau of Mines a piece of work which it thus far has so effectively performed. The Secretary of War has assured me of his own recognition of the splendid work you have been able to do, and I am taking the liberty of inclosing a letter which I have received from him, in order that you may see how fully the War Department recognizes the value of the services.

I am to-day signing the order directing the transfer. I want, however, to express to you my own appreciation of the fine and helpful piece of work which you have done, and to say that this sort of team work by the bureaus outside of the direct war-making agency is one of the cheering and gratifying evidences of the way our official forces are inspired by the presence of a great national task.

    Cordially, yours,

                                      WOODROW WILSON.

Dr. VAN. H. MANNING,

    Chief, Bureau of Mines, Department of the Interior.

EXECUTIVE ORDER.

It is hereby ordered that the experiment station at American University, Washington, D. C., which station has been established under the supervision of the Bureau of Mines, Interior Department, for the purpose of making gas investigation for the Army, under authority of appropriations made for the Ordnance and Medical Departments of the Army, together with the personnel thereof, be, and the same is hereby, placed under the control of the War Department for operation under the Director of Gas Service of the Army.

                                        WOODROW WILSON.

25 JUNE, 1918.

With this transfer there passed from under the control of the Director of the Bureau of Mines to that of the Director of the Chemical Warfare Service 1,662 employees who were engaged in the investigations. These employees were classified as follows:

Classification and number of appointees transferred.

	Military.			Civilian.	Grand total.
	Commissioned.	Noncommissioned.	Total.		
Technical	114	596	710	324	1,034
Nontechnical	9	122	131	517	648
Grand total	123	718	841	841	1,682

After the transfer of the research work from the Bureau of Mines, the War Department, on July 1, 1918, combined this work with the large-scale manufacturing gases under the Ordnance Department, Col. W. H. Walker in charge; the gas-mask manufacturing of the Medical Department, Col. Bradley Dewey in charge; the proving ground at Lakehurst, N. J., Lieut. Col. W. S. Bacon in charge; the development division, Col. F. M. Dorsey in charge; and the overseas division, Brig. Gen. Amos A. Fries in charge; and created the Chemical Warfare Service, under the direction of Maj. Gen. W. L. Sibert.

SOME OF THE MORE IMPORTANT FEATURES AND ACCOMPLISHMENTS.

THE FIRST GAS MASKS.

Only the best material is permissible in each and every part of a gas mask. Hence, the rubber companies, in cooperation with the research staff of the bureau, had to work out problems that were entirely new to the rubber industry. Every joint in a mask must be absolutely tight, and the sewing must be perfect. For assembling the various parts of the mask skilled organizations had to be developed out of inexperienced men. The final accomplishment, the best gas mask in existence, testifies to the value of the extensive research work involved in the development of each detail.

About May 1, 1917, the War Department requested the bureau to have manufactured 25,000 gas masks for shipment overseas. Then came a search for makers of cans, buckles, straps, rubber, eyepieces, knapsacks, and other materials who could do acceptable work and do it quickly.

By June 11, 1917, two factories of the American Can Co. in Brooklyn, N. Y., were at work assembling masks. To these factories came the canisters, knapsacks, rubber face pieces, metal parts, eyepieces, anti-dimming compound, instruction cards, repair kits, etc., made elsewhere.

The masks were of the box respirator type, used by the English, the wearer breathing through his mouth and all inhaled air passing through a box or canister containing absorbents. An improvement on the English type was making the charcoal canister larger. The filling absorbent was tested against chlorine, phosgene, and prussic acid, and proved to be highly efficient for absorbing these gases. However, it was not then known that chlorpicrin was to be one of the most important war gases, also that the rubberized cloth for the face piece had to be highly impermeable to gases, and that the soda-lime granules had to be so hard that rough jolting would not produce fines which would clog the canister and make resistance to breathing excessive. Before the end of June 20,086 masks were shipped overseas; another 7,000 were finished during the next two weeks but were never shipped. The masks shipped overseas were tested by the English, who found that they would not give the protection desired against chlorpicrin and other new gases being introduced by the Germans. Subsequently the masks were used in training troops.

This first lot of masks was made and delivered in record time, but owing to the desire of the Army to have manufacturing details directly under its control, the actual manufacture of masks was transferred to the Surgeon General's Office late in July, 1917.

CHARCOAL AND OTHER ABSORBENTS FOR GAS MASKS.

CHARCOAL.

As charcoal had been demonstrated by actual use in masks to be
a good gas absorbent, the Bureau of Mines undertook the preparation of a charcoal of maximum efficiency.   This search involved much
study and experiment.   Nearly every chemist in this country who
had had special experience in making charcoal was asked for information, and the problem was taken up with concerns making charcoal on a large scale.

At the laboratories of the National Carbon Co. and the National
Electric Lamp' Works in Cleveland, at the University of Chicago,
the Forest Products Laboratory at Madison, Wis., and the Bureau
of Chemistry in Washington there were experts on charcoal who
were glad to be of service.   H. D. Bachelor and M. K. Chaney, of
the National Carbon Co. laboratories, and F. M. Dorsey, of the
National Electric Lamp Works, and Dr. C. H. Hudson, of the Bureau
of Chemistry, Department of Agriculture, were largely responsible
for the excellence of the charcoal finally made.

Many different substances were carbonized, and the resulting
charcoal tried for a gas-mask absorbent.   Different kinds of wood
were tried, also nut shells—including coconut shells—lamp black,
carbon black, blood, seaweed, and ivory nuts.   The first coconut
charcoal came from Dr. H. B. Lemon, of the University of Chicago,
and by December, 1917, nut shells formed the basis of most of the
charcoal used in gas masks.

In preparing highly efficient charcoal the main idea is to clean
thoroughly the pores of the charcoal of any hydrocarbon residues.
which greatly reduce the absorptive power.   One method of cleaning
is to let the charcoal cool in air so as to absorb as much oxygen as
possible, and then reheat it to a comparatively low temperature.
Oxidizing with steam has certain advantages, the charcoal being
ground to about 8 to 14 mesh and steam treated.

SODA LIME.

Soda lime, a mixture of caustic soda and calcium hydroxide, is
used in gas masks to remove acid vapors or gases from the air.   A
great deal of study and experiment was given to the making of a
suitable product, and by December, 1917, through the efforts of the
bureau, a factory, capable of producing 800 pounds a day of this
material was actively at work.   The soda lime finally produced was
much superior to that made by the English.

SODIUM PERMANGANATE.

Sodium permanganate is used in gas masks to oxidize gases or
vapors and facilitate their absorption.   As the substance was not
produced in this country when the work on war gases began, methods

of production had to be devised and a supply of the necessary raw material provided. The methods of manufacture devised, when put in operation at a large factory in the fall of 1917, increased the production 50 per cent over methods proposed at the beginning, and gave a much better product.

ABSORBENTS FOR CARBON MONOXIDE.

At the branch laboratory established at Johns Hopkins University Dr. J. C. W. Frazer, formerly a chemist of the bureau, and his assistants, Dr. E. E. Reed, Dr. B. F. Lovelace, and Dr. W. A. Patrick, developed an absorbent in gas masks for carbon monoxide, a problem that had baffled scientists for years. The absorbent is a combination of substances, which largely through catalytic action removes carbon monoxide from air at ordinary temperatures. Twenty grams placed in an Army canister will afford protection against an atmosphere containing 1 per cent of carbon monoxide gas for one hour. This discovery, of high importance in warfare, is also of value in the industries, wherever men may be exposed to an atmosphere containing carbon monoxide. In this work Dr. Frazer was aided by Dr. Bray, of the University of California. In its later phases the work was under the general direction of A. B. Lamb, who in cooperation with C. R. Hoover, of Wesleyan University, also developed an efficient absorbent. Thus a hitherto baffling problem was solved in two ways.

SMOKE SCREENS FOR SHIPS.

One of the problems studied early in the gas investigations was that of devising mixtures for producing smoke screens to hide the movements of troops and to protect ships from attack. The smoke must be intense, heavy and lasting, and a relatively thin layer must completely obscure any object behind it. Also, the smoke cloud must be easily produced and easily controlled, and the materials required for producing the cloud must be readily available, easy to transport, and not dangerous to handle.

A large number of substances were tested by the investigative staff of the bureau. Phosphorus proved the best smoke producer, but it had the disadvantage of being somewhat difficult to control on shipboard. Other substances included titanium tetrachloride, alone and with ammonia; sulphur dioxide and ammonia; zinc dust and carbon tetrachloride; and silicon tetrachloride.

The bureau also studied the best methods of forming smoke clouds. It devised a smoke funnel, in which the chemicals mixed and reacted to form the smoke cloud, and also a smoke box to be dropped from a fleeing ship. In addition the bureau took up the

114018°—19——3

use of poisonous gases in smoke boxes, which would make trouble
for the crew of a pursuing ship that charged through a smoke screen.

Under date of July 30, 1917, Secretary Lane called the attention
of the United States Shipping Board to these investigations.   On
August 14, a successful demonstration on board a tug was given.

In mentioning the importance of this research work the Director
of the Bureau of Mines said:

I am informed that the Navy Department has placed an order for about $350,000
worth of smoke mixtures, developed by the research group, to screen merchant ships
from submarines. Further, that of 8 ships which were attacked, and which employed
the smoke screen to avoid destruction, 6 escaped, while of 100 ships which did not
employ smoke screens 75 were torpedoed.

TYPES OF WAR GASES.

The so-called war gases, many of which are liquids at ordinary
temperatures and pressures, can be divided into two general classes,
(a) lethal substances, generally those that kill by asphyxiation,
and (b) neutralizing substances.

The neutralizing materials are less poisonous but are capable of
putting men out of action for shorter or longer periods of time.   To
this class belong lachrymators (or tear gases), sternutators (or sneezing
gases), and eye, lung, and skin irritants, which inflame the eyes,
cause severe respiratory distress, and blister the skin.

Absorbent substances like charcoal, soda lime, sodium phenate,
hexamethylamine tetramine, caustic soda, zinc oxide, etc., absorb or
neutralize such gases as chlorine, phosgene, prussic acid, chlorpicrin,
mustard gas, or xylyl bromide, and when used in gas masks protect
against finely divided toxic solids such as diphenylchlorarsine; special
clothing is needed for protection against skin irritants such as mustard gas.

SELECTION AND PRODUCTION OF NEW GASES AND SMOKES.

In order to be deemed worthy of large-scale manufacture a new
gas had to possess some quality or qualities that rendered it decidedly
better for military use than its predecessors.   It had to have high
lethal value, be a powerful lachrymator or a good sternutator, or
vesicant, or be more highly penetrative.   The materials for its manufacture had to be plentiful and the process of manufacture could not
be too difficult.

When a new gas was under investigation the procedure was about
as follows: In the laboratories its physical, chemical, and physiological properties were determined, and methods of preparing it were
investigated; then a plant for small-scale manufacture was designed.
The gas was examined to determine how readily it penetrated American, English, French, and British masks.   Also, tests were made to
determine the best way of serving it to the enemy, whether in shells
or from cylinders.   In the meantime a thorough search was made

for raw materials for its manufacture. If all investigations proved satisfactory, it was turned over to the Ordnance Department—either to the development division, to build the first large unit, or to the gas manufacturing division. Chemists from the research division, who were familiar with the process of manufacture, were transferred along with the latter when it left the research division.

As soon as the research staff had completed work on some of the more important substances, larger plants were designed for manufacturing the materials in quantities at the Edgewood Arsenal.

In the first program, based on the experience of the English and French and the experiments made by the Bureau of Mines, as set forth in a letter dated July 27, 1917, from the Director of the Bureau of Mines to the Chief of Ordnance, the offense gases selected were, chlorpicrin, phosgene, xylyl bromide, and prussic (hydrocyanic) acid. As none of these substances had been manufactured in this country on a large scale, the technique of such manufacture had to be developed. The Director of the Bureau of Mines recommended manufacture by private concerns at that time, and not at Government plants because of the delay that would necessarily follow. The first large-scale work on chlorpicrin was done at the plant of the American Synthetic Color Co., at Stanford, Mass. The xylyl bromide process was worked out with the Dow Chemical Co. Prussic acid was to be furnished from sodium cyanide made by the Bucher process. A plant for making cyanide was erected later by the Bureau of Mines at Saltsville, Va.

The manufacturing program outlined was started by the research staff, under the direction of Dr. William McPherson, of Ohio University, who was transferred from the research staff to the Ordnance Department early in 1918 when that department began a definite gas-manufacturing program. His successor, W. S. Rowland, chemical engineer with the Stanley Works, of New Britain, Conn., worked out details of the process adopted for manufacturing mustard gas.

Huge plants for making mustard gas, phosgene, and chlorpicrin were built at the Edgewood Arsenal. The Oldbury Electrochemical Co. built a phosgene plant at Niagara Falls. This was under the direction of F. A. Lidbury, who also made tons of phosphorus for the American and the allied Governments.

Phosphorus throughout the war was the banner smoke producer for military purposes. Whether or not the use of gas is abolished in warfare, smokes have a definite place. They can be used in small candles to raise a smoke screen preceding the advance of troops to the attack or in shells in order to block off a part of the enemy's forces. Phosphorus has, by weight, 40 per cent more smoke-screening power than other substances the research chemists could find, but it is used wholly in shells.

As a result of all the experimental research work and the arrangements made to produce gases on a large scale, the United States was in a position at the signing of the armistice, on November 11, 1918, to manufacture poisonous gases in quantity equal to the combined production of France and England.

MUSTARD GAS.

Of all the gases used during the war none equaled mustard gas in military effectiveness. This substance, like some others used, is not a gas, but a liquid, dichloro-ethyl sulphide; it has a boiling point somewhat higher than water and volatilizes readily. It clings to the ground, to trees, or the walls of buildings, and its vapor being much heavier than air sinks into trenches, shell holes, and other depressions. The liquid readily penetrates ordinary clothing, and even leather. It causes burns that appear 4 to 12 hours after exposure and heal slowly. The vapor inflames the eyes, causing temporary blindness, and attacks the throat and bronchial tubes, causing bronchitis or broncho-pneumonia.

For some time after mustard gas was introduced by the Germans the accounts of its effectiveness were contradictory, and reliable reports of its efficacy and the service the enemy was getting from it did not reach this country until about December, 1917.

The first work on the production of the gas in this country was done under the direction of Dr. J. F. Norris in the fall of 1917 in cooperation with the Commercial Research Co. After extensive experimentation the ethylene-sulphur-monochloride reaction was adopted, the problem being solved at practically the same time in England. Thereafter work on the gas proceeded rapidly, and when the armistice was signed a plant with a capacity of 200 tons a day was being erected for its manufacture, under the direction of the War Department, at the Edgewood Arsenal.

The experimental work was conducted in various laboratories— one group worked in the research laboratories at Washington; another at the Dow Chemical Co., under the direction of Dr. A. W. Smith; and still another group at Cleveland, under F. M. Dorsey. The best features of the processes so discovered, with those used by the French and by the English, were incorporated in the Edgewood plant.

DETAILED ACCOUNT OF ORGANIZATION, PROGRESS OF WORK, RESULTS.

The more important features of the war gas work conducted under the Bureau of Mines have been outlined. On account of the scope and manifold character these investigations assumed, and the variety of problems attacked, it is not possible in this report to give more than a brief summary of the organization of the research staff and the individual results achieved by the various groups of workers.

As previously related, the rapid growth and increasing importance of the research work resulted in the establishment of a central station at American University in September, 1917, to coordinate the work and effect an efficient unified organization.

PLAN OF ORGANIZATION.

On September 1, 1917, the war-gas investigations of the bureau were organized as follows:

Head, Van. H. Manning, Director of the Bureau of Mines.

In direct charge, G. A. Burrell, assistant to the director in charge of gas investigations.

Defense problems (including research work relating to gas masks and protective devices), W. K. Lewis.

Chemical research (including analytical methods, gases for shells, smoke clouds and bombs, incendiary bombs, absorbents for gas masks, signal lights and gases for balloons), J. F. Norris.

Medical-science research (including physiological investigations of gas masks, pharmacological gassing experiments on men and on animals, pathological gross and microscopic study of gassed animals, and pathological chemistry of disorders of gassed animals), Yandell Henderson.

Pyrotechnic research (including demonstrations and large-scale investigations of gas shells, smoke clouds, signal lights, incendiary bombs at proving grounds and aboard ships, etc.), George Richter.

Chemical manufacture (covering the manufacture of chemicals, on a semicommercial scale, for demonstration purposes), William McPherson.

Mechanical research (including mechanical details of gas-mask design), H. H. Clark.

Investigations of submarine gases (including the removal of hydrogen, carbon dioxide, and engine gases from the atmosphere within submarines), W. E. Gibbs.

Gas-mask research (including tests of canisters and masks and the efficiency of absorbents), A. C. Fieldner.

Of the investigations relating to the chemical examination of balloon gases and the building of plants for their manufacture, those dealing with helium were subsequently placed in charge of F. G. Cottrell, chief metallurgist of the Bureau of Mines, as related in another chapter of this bulletin.

The branch laboratories and the investigators in charge were as follows:

John Hopkins University, J. W. C. Frazer and E. E. Reid.

Princeton University, G. A. Hulett and F. Neher.

Nela Park, Cleveland, Ohio, H. D. Bachelor and N. K. Cheney.
Harvard University, E. P. Kohler and G. P. Baxter.
Bryn Mawr College, R. F. Brunel.
Ohio State University, C. E. Boord.
Massachusetts Institute of Technology, S. P. Naulliken.
Yale University, F. P. Underhill.

With the expansion of the research work new groups of chemists were organized and the personnel rapidly increased. On June 1, 1918, the organization of the war-gas investigations was as follows:

Head, Van. H. Manning, Director Bureau of Mines.

In direct charge, G. A. Burrell, assistant to the director in charge gas investigations.

Defense problems, W. K. Lewis.

Offense problems, E. P. Kohler.

Physiological and pharmacological problems, Yandell Henderson.

Editorial work and miscellaneous research, W. D. Bancroft.

Chemical research (defense), A. B. Lamb; (offense), J. F. Norris.

Manufacturing development (offense), W. S. Rowland; (defense), W. K. Lewis.

Pyrotechnic research, G. A. Richter.

Mechanical research (Army gas masks), B. B. Fogler; (Navy and miscellaneous), H. H. Clark.

Therapeutic research, F. P. Underhill.

Pharmacological research (offense), E. K. Marshall; (defense), A. S. Lovenhart.

Toxicity tests, M. C. Winternitz.

Dispersoid research, R. C. Tolman.

GAS-MASK RESEARCH.

When investigations relating to gas masks began in March, 1917, A. C. Fieldner, chemist in charge of the chemical laboratories at the Pittsburgh station of the bureau, was placed in charge of methods for testing the efficiency of various absorbents, and of testing canisters, and finished masks. Bradley Dewey, research chemist of the American Tin Plate Co., at Pittsburgh, Pa., entered the service of the bureau on May 7, 1917, to take charge of the development of gas-mask manufacture. On August 1, 1917, he was commissioned a major, in the Surgeon General's Office. Dr. W. K. Lewis, assistant professor of chemical engineering at the Massachusetts Institute of Technology, who joined the research staff in May, 1917, was placed in charge of the development of absorbents. These two men also took a large part in obtaining the cooperation of chemists at many places in studying special problems connected with the production of a satisfactory gas mask and with other work related to the use of gases in warfare.

TESTING OF ABSORBENTS, CANISTERS, FABRICS, AND FINISHED MASKS.

The work of the gas-mask research division, under Mr. Fieldner, was as follows:

Samples of soda lime, charcoal, and other materials were sent to the Pittsburgh station for test by the new standard methods being devised there.

By April 1, 1917, work had begun on developing analytical methods for testing one new gas against another.

When the Army requested the bureau on May 1 to have manufactured 25,000 masks, so many canisters came to the Pittsburgh station for test that temporary laboratories had to be built. Up to this time attention had been chiefly directed to making tests against chlorine, as this gas was used in the first gas attacks made by the Germans. The chlorine testing machine was devised by Mr. Fieldner and by A. W. Gauger, assistant chemist of the bureau, with the aid of the regular staff of the gas laboratory.

When, by June 1, information came that phosgene, chlorpicrin, and hydrocyanic acid were being used, methods were devised for testing absorbents and canisters against these gases.

On July 1, 1917, the testing of canisters and absorbents was organized into two units; the absorbent testing was placed in charge of G. G. Oberfell, and the canister testing in charge of A. W. Gauger, and later of M. C. Teague.

In the summer of 1917 a section was created to determine the penetrability of fabrics for masks and protective clothing by various gases, with G. St. J. Perrott in charge. Mr. Perrott developed the standard methods for testing fabrics that were used later by the Chemical Warfare Service. By June, 1918, approximately 1,000 fabrics had been tested.

In September, 1917, all the gas-warfare work at Pittsburgh was transferred to the American University station in Washington, and a man-test unit was formed with Capt. J. N. Lawrence in charge, assisted by W. J. Harper. A large number of masks were tested against practically all the war gases then known.

As soon as the methods of testing absorbents, canisters, and mask fabrics were completed and standardized, an extensive investigation was begun, early in November, 1917, to determine the best proportion, fineness, position, etc., of the absorbents in a canister. These tests were made on men and on machines.

As each new gas appeared, methods of analysis and of determining the protection various canisters afforded against it were devised. This work was under the direction of C. V. Smith and, later, of Dr. R. E. Nelson.

One-half per cent of all the canisters manufactured by the Gas Defense Service of the Surgeon General's Office was tested on men and on machines, in order to maintain a high standard of manufacture. Lieut. R. N. Pease had direct charge of the machine testing under Capt. Teague. Early in the spring of 1918, a new man house was completed, having a capacity of more than 100 canister tests a day. About 15,000 canisters of all types were tested on men, including the standard Army types, Navy drum type, and various foreign canisters. A change in the filling of the Navy drum improved its efficiency 100 per cent and made it protective against stannic chloride and similar smokes.

On March 1, 1918, a canister-filling section was formed under charge of Lieut. H. J. Beattie, and equipment was installed.

In mechanical testing of canisters, various machines were tried out until a satisfactory type was devised. Later the volume and rate of breathing of men at rest and at various degrees of work were determined, so that results with a machine would duplicate the results of tests on men under any conditions. The first mechanical tests of this type were made about April 1, 1918, against phosgene. Later machines were built for testing against chlorine, hydrocyanic acid, and chlorpicrin, facilities being provided for testing about 150 canisters a day.

Many filtering materials for removing smoke were tried. Lieut. Perrott first suggested and tested the bag type filter and the use of paper as a smoke filterer. The fabric-testing section conducted a long series of weathering tests of gas-mask fabrics at Washington, D. C., and at Pensacola, Fla. Important problems solved by January 1, 1918, included the development of a satisfactory mustard-gas detector, of an impregnating medium for protective clothing, of a method for renovating cloth contaminated with mustard gas, and the devising of so-called "solid mustard" for accustoming troops to its odor.

At a request from Gen. Pershing, in January, 1918, for a cheap method of impregnating dugout blankets, an eminently satisfactory screen was devised by impregnating an ordinary Army horse blanket with a mixture of mineral and vegetable oils. Lieut. Perrott designed and shipped some impregnating machinery.

In January, 1918, a miscellaneous section was formed at Pittsburgh under Lieut. S. H. Katz, formerly assistant chemist of the Bureau of Mines. Among the problems investigated were the effect of temperature on the absorption of gases by charcoal, detection of gases in the atmosphere of submarines, effect of field conditions on the life of canisters, and protection afforded by canisters against industrial gases. This section cooperated in devising a canister for protection against carbon monoxide, also a special canister to give long life against hydrocyanic acid. A method of determining the course of gas through a canister, known as the "wave-front" method, developed by Lieut. Beattie, was of high value in detecting defects in canister design and filling. In May, 1918, detailed tests were made of the protection afforded by standard canisters against various gas mixtures.

MECHANICAL RESEARCH.

Research on the mechanical problems of gas masks began under W. E. Gibbs, inventor of the Gibbs breathing apparatus, in April, 1917; later the division of mechanical research was organized under H. H. Clark, electrical engineer of the bureau. This division had charge of the machine shop at the American University station and the design of gas masks. In May, 1918, the work was divided, B. B. Fogler taking charge of research relating to the Army masks and H. H. Clark taking charge of the mechanical research work for the Navy and miscellaneous research work. In June, 1918, a new machine shop, measuring 120 by 50 feet and accommodating 50 machinists, was nearly completed at the American University.

At first the principal mechanical problems related to the improvement of different parts of the Army mask for soldiers. Next the Tissot type, designed by the French for artillery service, was studied, and a mask developed to a point where the Gas Defense Service put it into production after making certain modifications. This mask has no mouthpiece or nose clip, and the entering air impinges on the eye pieces and reduces fogging. In collaboration with Dr. J. A. E. Eyster a similar type of mask was developed for the Aviation Service.

In collaboration with the United States Rubber Co. a naval mask that carries the canister on the head was developed which affords protection against both war gases and carbon monoxide. The original Navy mask was modified and standardized in the same manner as the Army respirator.

A horse mask and a trench fan similar to those used by the British were designed and specifications submitted to the Gas Defense Service.

In collaboration with the chemical research division, canisters for absorbing carbon monoxide were developed. One of these was standardized for use with the mask adopted by the Navy. Much experimental work was done in the developing of smoke filters for gas masks and the use of baffles and screens for mask canisters. The baffling increases the capacity of the canister, or permits a considerable decrease in size without reduction of capacity. A small canister giving at least 30 minutes' protection against chlorine was developed for use in submarines; also a special canister for the use of workmen in poison-gas plants.

Special clothing was developed for the use of workmen in poison-gas plants. One type of suit was provided with a helmet and a hose pipe to supply fresh air. Another type of suit had a mask, an oil-cloth hood, and one of the special canisters.

Other work included the devising of a mechanism for firing smoke boxes from ships, and miscellaneous tests of mask parts, such as eye-pieces, nose clips, valves, etc.

PHYSIOLOGICAL RESEARCH.

On April 4, 1917, Dr. Yandell Henderson, of Yale University, who as consulting physiologist of the Bureau of Mines had investigated the improvement of rescue breathing apparatus and resuscitation from gas poisoning, was put in charge of the physiological, or medical science, research of the war gas investigations. The work under his direction grew and expanded so rapidly that by June, 1918, it included ten sections employing several hundred scientific men. The chiefs of these sections formed a board of which Dr. Henderson was chairman.

ADVICE ON GAS MASKS.

Dr. Henderson acted as technical advisor regarding various features of gas masks, being associated with W. E. Gibbs, and helped in arranging for the manufacture of the first 20,000 masks for the American Expeditionary Forces. During the summer of 1917 Dr. Henderson gave scientific supervision to tests of masks on men in the gas chamber and later, with Mr. Gibbs, studied the possible development of a cloth mask of the French type as a secondary mask for American troops. In October, 1917, he had general direction of an investigation conducted by L. F. Rettger, at Yale University, of the most effective ways of sterilizing masks.

TOXICITY AND GASSING TESTS.

Pending the opening of the central laboratory at the American University, Dr. Henderson organized, in New Haven, in May, 1917, tests with animals to determine the toxicity of gases, and the development of better methods of treating men who had been gassed. Drs. F. P. Underhill, H. G. Barbour, M. C. Winternitz, E. K. Marshall, H. W. Haggard, D. W. Wilson, Samuel Goldschmidt, H. F. Pierce, and others were appointed consulting physiologists or junior physiologists of the Bureau of Mines, and in June, 1917, began work in the Yale buildings and in a temporary laboratory at Yale Field. Methods were developed for the accurate quantitative gassing of animals, a new departure, essential for determining the degree of gassing and the methods of treatment, that the possible cure might have a scientific basis. The essential apparatus was the gassing chamber; the initial features of this were designed by Dr. Henderson, but the apparatus was rapidly improved and developed by Prof. Underhill and Drs. Barbour, Marshall, and Haggard. The results achieved were superior to any obtained abroad up to that time.

REORGANIZATION OF THE MEDICAL SCIENCE WORK.

In October, 1917, the pharmacological and toxicological work was moved to the American University at Washington, where it was organized as the division of pharmacological research. This division was in charge of Dr. Marshall, under the supervision of Dr. Henderson. In November, 1917, the other investigations under Dr. Henderson were organized as follows: Dr. F. P. Underhill was placed in charge of therapeutic investigations, at New Haven; Dr. H. C. Bradley, in charge sanitary supervision at gas factories; Dr. Reid Hunt, in charge toxicological research; Lieut. W. L. Bacon, in charge gas-shell tests. As the work increased it was subdivided, Dr. M. C. Winternitz taking charge of pathological research and Dr. Underhill retaining therapeutic research.

SUPERVISION OF SANITATION CONDITIONS AT POISON-GAS FACTORIES.

In August, 1917, when plans were being made for the large scale manufacture of poison gas, Dr. Henderson recognized the need of careful supervision of sanitation conditions and of measures to prevent poisoning among employees. Accordingly, Dr. H. C. Bradley, of the University of Wisconsin, was put in charge of factory supervision. Dr. Bradley immediately began inspection of chlorine and other plants, and selected and trained inspectors. Thus poisoning among the workmen in the plants was largely reduced. Dr. Bradley also began investigations of the treatment of chronic gas poisoning, being aided by Profs. J. A. E. Eyster and A. S. Lovenhart, and Dr. W. J. Meek, of the University of Wisconsin. These investigations resulted in developing a large experimental laboratory at the university, although Dr. Bradley and Prof. Lovenhart were later moved to Washington.

FIELD TESTS OF GASES.

In August, 1917, when the Ordnance Department stated that large-scale field tests of gases would be necessary and physiological advice on the results obtained would be required, W. S. Bacon was placed in immediate charge of such work by the Bureau of Mines, under the general supervision of Dr. Henderson. Mr. Bacon investigated the various proving grounds to determine their suitability for the tests and developed apparatus for testing the toxicity of gases under field conditions. He was eventually commissioned major in the Army, and developed the large proving ground at Lakehurst, N. J.

INVESTIGATIONS ON SHOCK, ASPHYXIA, HEMORRHAGE, AND CARBON MONOXIDE POISONING.

In England and America scientific men were investigating the problems of shock in wounded soldiers. As the solution seemed to depend largely on a better knowledge of the changes in the blood during the different forms of asphyxia, Dr. Henderson had a section of eight men, under Dr. H. W. Haggard at New Haven, investigate these problems. The results obtained show the nature of the changes that take place and afford new conceptions of their character and proper mode of treatment.

AVIATION INVESTIGATIONS.

In September, 1917, Dr. Henderson was appointed chairman of the medical research board of the Aviation Service. The purpose of this board was to develop methods for testing the ability of aviators to withstand altitude and to devise an apparatus to supply oxygen to aviators ascending to great heights. An allotment of $100,000 was transferred from the Air Service to the Bureau of Mines for these purposes.

OXYGEN APPARATUS FOR AVIATORS.

W. E. Gibbs, mechanical engineer of the Bureau of Mines, was assigned to develop the oxygen apparatus. Mr. Gibbs, with the aid of Dr. Henderson, developed an apparatus that can be manufactured on a large scale and is markedly superior to the British type. This apparatus was ready for acceptance when the armistice was signed.

TESTS FOR AVIATORS.

In developing tests for aviators, Dr. Henderson began work at the American University experiment station, and put Prof. E. C. Schneider in immediate charge of the physiological research, and Prof. Knight Dunlap in charge of the psychological research. Dr. Henderson devised for these tests the so-called rebreathing apparatus; this has a tank containing air, which the subject inhales through a mouthpiece and tubing, the exhaled air passing back through a cartridge of alkali which removes the carbon dioxide. As the subject breathes, the oxygen content of the air in the tank becomes progressively lower and his physiologic condition becomes equivalent to that of altitude. Some results of the tests are discussed in seven papers, published in the Journal of the American Medical Association by Dr. Henderson and his colleagues.[a] These tests

---

[a] Organization and objects of the Medical Research Board, Air Service U. S. Army, by Yandell Henderson, and E. G. Seibert; Physiologic observations and methods, by E. C. Schneider; Cardiovascular observations, by J. L. Whitney; Psychologic observations and methods, by Knight Dunlap; The effect of altitude on ocular functions, by W. H. Wilmer and Conrad Berens, jr.; Influence of altitude on the hearing and the motion-sensing apparatus of the ear, by E. R. Lewis; Effects of low oxygen pressure on the personality of the aviator, by Stewart Paton. 1918. Jour. Am. Med. Assoc., vol. 71, Oct. 26, pp. 1382-1400.

developed the fact that 14 to 15 per cent of the candidates for the Air Service who passed the ordinary medical examination were really unfit to fly at the altitude necessary for active service, but a small group, between 5 and 10 per cent of the total number, can be picked who are capable of ascending to the greatest altitudes without collapse.

In January, 1918, a laboratory was established at the flying field at Mineola, Long Island, for investigative research and for training medical officers to fit them to take charge of similar laboratories at the other flying fields. When the armistice was signed nearly 20 subsidiary laboratories had been established or planned, and a party headed by Col. W. H. Wilmer had been sent to France to establish similar laboratories at the American aerodromes there.

THERAPEUTIC RESEARCH.

Research to determine the changes that take place in the body as the result of exposure to lethal gases, and thus to devise methods of treatment for gas poisoning was organized under Dr. Henderson at Yale University in May, 1917, with Dr. F. P. Underhill in charge. After the pharmacological and toxicological research work was moved to the American University station, therapeutic work was continued under Dr. Underhill, in the laboratories at Yale. Ultimately some 50 men were busy there making examinations of gassed animals, and studying the effects and the treatment of gas poisoning. The work involved the application of chemical methods to physiological processes. The first gas considered was chlorine. By January, 1918, a promising method of treatment for chlorine poisoning was devised. Dr. D. W. Wilson and Dr. Samuel Goldschmidt received commissions in the Army and went to England, where they did further work on phosgene poisoning, finally winning the British authorities over to recognize the fundamental correctness of the method of treatment developed by the work at New Haven. An intensive study of phosgene was made at New Haven, and by July, 1918, a method of treatment had been evolved, which was later indorsed by the Medical Department of the Army. Later it was shown that chlorine, phosgene, and chlorpicrin act fundamentally alike.

Another investigation was undertaken to determine whether inhalation of oxygen would reduce the mortality among persons who had breathed poison gas; another demonstrated that many persons who survive gassing subsequently contract pneumonia or other pulmonary diseases through self-infection with bacteria.

PATHOLOGICAL RESEARCH.

For studying the organs and tissues of animals killed in gassing tests, a pathological section under the general supervision of Dr. Henderson and in charge of Dr. M. C. Winternitz was established in the summer of 1917 with laboratories at the American University station and at Yale University.

The pathological division was established at the American University station under Dr. Winternitz in May, 1918. Prior to that date pathological tests at the station had been made by Dr. G. M. McKenzie.

Among the problems attacked by this division was the effect of mustard gas on the mucous membrane of the eyes and respiratory organs and on the skin. No specific remedy for mustard-gas burns was found, although it was demonstrated that burns could be prevented by washing and scrubbing the skin with kerosene a few minutes after exposure. Mustard gas was shown to be highly toxic.

Tests to determine the minimum effective concentration showed that 1 part of mustard gas in 12,500,000 parts of air would make a soldier a casualty in a few hours, chiefly through inflammation of the eyes.

PHARMACOLOGICAL RESEARCH.

The pharmacological division was established at American University on November 27, 1917, with Dr. E. K. Marshall, jr., in charge.

In May, 1918, the need of determining the toxic value of new gases, the best methods of treating gassed persons, and determining the susceptibility of different individuals led to the enlargement of the toxicologic work. Dr. Marshall was commissioned a captain in the Medical Reserve Corps of the Army and put in charge of the pharmacological research division, which took up all pharmacological problems relating to defense, and Dr. A. S. Lovenhart, of the University of Wisconsin, took charge of the pharmacological and toxicological division which studied offense problems.

On July 1, 1918, the pharmacological research division became the pharmacological research section of the Chemical Warfare Service.

The work of the pharmacological research division comprised the following investigations:

1. Devising toxicity tests by exposing mice, guinea pigs, rats, rabbits, cats, and dogs to poison gases, and by studying the pharmacological and toxicological effects.

2. Testing tear gases and sneezing gases on men.

3. Determining the skin-irritant effect of gases.

4. Testing fabrics for their permeability to mustard gas.

5. Devising analytical methods for the control of the concentration of the gases used in tests.

6. Determining the sensitivity to gas of the skin of different individuals.

Toxicity tests with animals covered a long list of poisonous liquids and vapors. Many of the tests with dogs were for varying lengths of time to determine the relation between the lethal effect and the concentration and length of exposure. Tests were also made with animals in a bomb pit at the American University station and at the Indianhead Proving Ground to determine the effective area of shells charged with the different toxic substances.

A large number of fabrics were tested for permeability to mustard gas, both dogs and men being used in the tests, and results of the tests were transmitted to the Army and Navy.

Many tests with men and animals were made to determine the effectiveness of tear gases and the minimum concentration at which these gases could be detected by their odor or their irritating effects. One result of these tests was to show that man was more than one thousand times as susceptible to tear gas (xylyl bromide) as the horse, and more than ten times as sensitive as the dog.

TOXICOLOGICAL RESEARCH.

Toxicity tests on dogs and experiments to devise a rational method of treating gassed animals were begun by Dr. Henderson at Yale University in May, 1917. These investigations were afterward supervised by Dr. H. G. Barbour and later by Dr. F. P. Underhill.

Preliminary tests were made chiefly with mice. Most of the early work was on chlorine; later, tests were made with chlorpicrin, superpalite, and cyanogen bromide. The analytical control of the toxicity tests was done by L. L. Satler. The lethal values determined for various substances in the early experiments were confirmed by later work in this country and abroad.

In September, 1917, the apparatus accumulated was transferred to Dr. Underhill's laboratory at Yale. Dr. Marshall, as chief of the pharmacological section, went to the American University and took with him Dr. A. C. Kolls, Messrs. L. L. Satler, H. A. Kuhn, and Harry Pettingill. At American University the section was rapidly enlarged.

In November, 1917, F. W. Sherwood and T. W. Snyder, under the direction of Dr. Marshall, took up the determination of the least detectable concentration of poison gases. The section also tested the skin irritant effect of gases, especially mustard gas and the homologues of mustard gas, then that of the chlorine compounds, and then that of the arsenicals, Dr. R. P. Gilbert being assigned to these tests.

At the station as many as 700 dogs were kept at one time. Besides dogs and mice, rats, rabbits, and guinea pigs were used.

In February, 1918, a series of tests in cooperation with the pyrotechnic division of certain toxic shell fillers were made at the American University grounds and at the Indianhead testing ground. Pharmacological work on mustard gas showing its absorption by the lungs and the systemic action of the gas was done by Vernon Lynch.

In May, 1918, Dr. A. S. Lovenhart, of the University of Wisconsin, was placed in charge of pharmacological problems relating to offense, Capt. Marshall retaining those relating to defense. Messrs. A. C. Kolls, L. L. Satler, H. A. Kuhn, and others were transferred to Dr. Lovenhart's division.

GASES IN SUBMARINES.

In February, 1917, W. E. Gibbs, engineer of the Bureau of Mines and inventor of the Gibbs breathing apparatus, was authorized by the director of the bureau to offer the Navy Department the results of experiments with soda lime as an absorber of carbon dioxide, because of its possible use in submarines. A few days later Mr. Gibbs conferred with Naval Constructor Wright and Lieut. G. C. Fuller at the Brooklyn Navy Yard.

Details having been arranged, Mr. Gibbs had made early in March, at the navy yard, about 500 pounds of soda lime, after the formula given him in September, 1916, by Dr. J. S. Haldane of England. At about the same time an apparatus was built at the Navy yard for making use of this soda lime in submarines. In this apparatus a blower circulated the air of the submarine through trays of soda lime contained in a large box.

After experiments at the navy yard, the first long submergence with this apparatus began at New London on June 16, when submarine G-1 remained submerged for 39 hours. A few days later the same boat remained submerged for 30 hours. A short time after the same boat remained submerged for 48 hours, the air within the boat being as pure at the end of the test as at the beginning. The test was concluded at the end of 48 hours, not because the capacity of the soda lime had been exhausted, but because the supply of food gave out.

After these tests, Lieut. Commander Yates Sterling, then in command of the submarine base at New London, officially reported that these were record submergences for the United States Navy. The longest previous submergence had been 12 hours.

In these tests at the navy yard the soda lime was in tin cartridges containing 50 pounds each. After opening the top and bottom of a cartridge it could be placed directly over the outlet of a small electrically driven blower. The exhausted cartridges were thrown overboard. These cartridges were easily stowed and eliminated the handling of soda lime in bulk.

By September 15, when Mr. Gibbs gave up further work on this problem in order to undertake problems of supplying oxygen to aviators, he was informed that 72 submarines had been equipped with this system. Later, under Dr. Eugene F. Dubois of the Navy, the soda lime was improved in porosity, but the final and best form was that developed at the American University station of the Bureau of Mines.

As elimination of hydrogen from the atmosphere of submarines was necessary to the safety of the crews, Mr. Gibbs had J. F. Duggar, jr., a gas chemist of the Bureau of Mines, attack this problem, which was solved with much ingenuity.

In the 48-hour submergence,. Mr. Duggar's apparatus, used for the first time in its improved form, kept the hydrogen down to a negligible quantity. The naval committee in charge of the submergence did not, however, officially report on this performance, because the trial was regarded as a private test and part of the research.

Later, after Mr. Duggar had resigned from the bureau, Mr. Gibbs turned over the data to Dr. E. F. Dubois, who perfected the device.

Early in the spring of 1917 Mr. Gibbs helped in the development of a half-hour breathing apparatus, which was being worked out by John Lind, of the hull division of the Brooklyn Navy Yard.

In September, 1917, Mr. Gibbs was detailed by the bureau to work in cooperation with Dr. Yandell Henderson on the problem of supplying oxygen to aviators when at great altitudes. Shortly thereafter Mr. Gibbs was made a member of the board of medical research on the Aviation Section of the Signal Corps, and until the signing of the armistice was engaged in devising a suitable mask for aviators.

PYROTECHNIC RESEARCH.

The pyrotechnic division was organized in June, 1917. G. A. Richter, formerly chemical engineer of the Berlin Mills Co., of Berlin Mills, N. H., was in charge of the division from the start.

GAS-SHELL SECTION.

Among the tasks undertaken was the determining of the stability of various gases and toxic solids when used in shells. This section also devised a training bomb for use by the gas-defense officers in cantonments, and studied lachrymatory bombs, smoke bombs, and so-called "noise" bombs. Also, several gas-producing chemical combinations were submitted to the Bureau of Ordnance of the Navy for use in armor-piercing shell.

SMOKE-SCREEN SECTION.

The smoke-screen section prepared specifications for a Navy smoke funnel which was accepted by the Bureau of Ordnance. Several hundred of these funnels, made by that bureau, were being supplied to the merchant marine in May, 1918. A smoke-box float was designed and accepted by the Navy; large-scale production had begun in May, 1918. The section perfected a smoke signal for airplane bombs that was accepted by the Ordnance Department, also a similar device for use in the Navy dummy airplane bomb. A portable smoke apparatus using silicon tetrachloride was designed for the Army, a smoke mixture of phosphorus and TNT, for use in shells and in candles, a smoke bomb of the Liven's type, a smoke grenade, and a "noiseless" nozzle for gas attacks. A method of using oleum (fuming sulphuric acid) to produce a smoke screen was accepted by the Navy.

MUNITIONS AND ORDNANCE.

A new form of Stokes mortar was designed that gave greater accuracy and longer range. A long series of tests with Liven's projectors and projectiles were made to reduce the weight of the unit and to increase the range. Results were submitted to the Ordnance Department.

HAND-GRENADE SECTION.

The hand-grenade section was organized at the American University station on October 1, 1917, to continue work on grenades, but as the grenades already adopted by the Army were standardized, the section took up the testing of explosives used in pyrotechny and in gas warfare, and later, testing of high explosives. The sensitivity in use of various picrates was studied in order to assure proper safety measures in storing and handling being taken at the plant making chlorpicrin, for the Ordnance Department; the recommendations were adopted at the plant. Fragmentation tests were made of hand grenades charged with TNT, amatol, and victorite, the last a chlorate explosive; the results were forwarded to the Army and the Navy. Preliminary studies were made of the properties and usefulness of parazol and other chlornitro products as explosives. Studies of chlorate and perchlorate explosives led to the finding of satisfactory chlorate and perchlorate powders for use in hand grenades. A new explosive called anilite, which employs liquid NO_2 as an oxidizing agent, and is used by the French, was studied.

After July 1, 1918, the section was designated the explosives section of the Chemical Warfare Service.

INCENDIARY SECTION.

The incendiary section was organized as a separate unit in October, 1917, under Lieut. A. B. Ray. This section as stated on page 32, originated under the chemical research division (offense), being transferred to the pyrotechnic division in April, 1918.

This section perfected a scatter type of bomb for the Army and an intensive type of bomb for the Army, devised an incendiary dart to be used from air planes, an incendiary projectile for use with the Liven's gun, and prepared recommendations for the Signal Corps as to the best type of explosives to be used with an airplane destroyer.

After July 1, 1918, the section was reorganized as the inorganic section of the Chemical Warfare Service.

FLAMING-LIQUID SECTION.

Two portable flaming-liquid guns were designed and perfected. Tests to determine fuel mixtures for use in connection with the flaming-liquid guns were made and recommendations submitted to the Ordnance Department of the Army.

SIGNAL-LIGHT SECTION.

The signal section conducted a thorough investigation of green flares, devised several colored smokes for use in connection with land signals and air signals, studied methods of improving the white flare, designed an illumination float to be dropped from an airplane, and constructed a special device to be used by submarines in signaling to surface craft.

LABORATORY SECTION.

Linings, packings, and cements for gas shell were carefully studied. A shell that uses a special type of lead lining was recommended to the Ordnance)Department of the Army. Further work toward the development of enamel linings and glass linings was under way in June, 1918. Other work included the development of a systematic method for analyzing pyrotechnic materials, and a thorough laboratory study of the usefulness of paper and pulp containers for powder. The recommendations submitted were adopted by the Ordnance Department, which had the tubes made in large quantities by June, 1918.

MISCELLANEOUS SECTION.

This section designed a smoke shell for the Navy, tested a special shrapnel shell with special loading for the Chemical Service of the Army, and made tests of silicon tetrachloride, stannic tetrachloride, and titanium tetrachloride for smoke grenades, to determine which smoke is the most obnoxious.

CHEMICAL RESEARCH (OFFENSE).

The chemical research division, when first established in June, 1917, was in charge of J. F. Norris, with W. K. Lewis as assistant. In October, 1917, the work was divided; A. B. Lamb was placed in charge of problems relating to defense, and J. F. Norris of those relating to offense, also certain problems were turned over to the divisions of pyrotechnic research and manufacturing development. Early in June, 1918, when Dr. Norris received a commission as lieutenant colonel in the Army and went to England, L. W. Jones was put in charge of the offense chemical research work. The problems studied by the division up to May, 1918, were as follows: Preparing new toxic substances; developing the manufacture of

toxic materials; development of smoke mixtures and incendi
mixtures; research work on inorganic compounds; and analytic
research.

PREPARATION OF NEW TOXIC MATERIALS.

By June 1, 1918, more than 250 toxic substances had been prepared
and submitted to the division of pharmacological research for toxicity tests. Many proved promising, but because of the inadvisa-
bility of abandoning materials already known to be excellent, comparatively few were recommended. In addition, a number of toxic
solids were prepared and delivered to the dispersoid division.

DEVELOPMENT WORK FOR THE MANUFACTURE OF TOXIC MATERIALS.

The section in charge of the development of the manufacture of
new toxic materials, later designated organic research section No. 1,
came into being in September, 1917, with J. B. Conant in charge.

As almost none of the toxic materials used in modern warfare were
being manufactured in the United States, development of manufacturing methods received chief attention up to the end of June, 1918.
Investigations, on a laboratory scale, of manufacturing methods for
seven toxic substances, including acrolein, martonite, and mustard
gas, had been completed in May, 1918.

Practically nothing was known of the manufacture of mustard gas,
and many laboratory experiments had to be made to determine
details of preparation, purification, and manufacture. By June 1,
1918, most of the experimental work on preparation had been completed and semilarge-scale manufacture had started. As a result
of this research, manufacturing was planned by the ethylenechlorhy-
drin synthesis, which was to be used for the Ordnance Department of
the Army at a plant in Flushing, Long Island, and by the sulphur
tmonochloride-ethylene synthesis which was used later in the plant a
Edgewood Arsenal. Most of the work on mustard gas was under
the direction of J. B. Conant. Among those who worked with him
were E. B. Harlstrom, G. O. Richardson, and others. W. E. Lewis
developed a new gas more effective than mustard gas.

Methods of making important arsenic derivatives were developed.
Diphenyl chlorarsine was made in large quantities in the laboratories
for the general use of the American University experiment station,
as the material could not be obtained elsewhere.

PROBLEMS IN INORGANIC CHEMISTRY.

An incendiary section was organized in October, 1917, under
A. B. Ray. This section did early work on flaming mixtures and on
incendiary bombs of the scatter type and the intensive type. Incendiary materials were developed for bombs, shells, and projectiles,
and for hand darts and grenades. Each of these devices required

special incendiary materials, which were adopted by the Ordnance Department of the Army.

The section was reorganized under Lieut. Ray as the section of inorganic chemical research in April, 1918, when it turned over its remaining incendiary problems to the pyrotechnic division.

This section, in cooperation with the pyrotechnic division, prepared specifications for smoke mixtures which were adopted by the Navy for its smoke screens. The section also studied the production of arsine, of calcium and magnesium arsenide, of hydrofluoric acid and fluorides, and of nitrogen tetroxide, the last for use in the new explosive, anilite, devised by the French. The section also studied such toxic materials as cyanides, the selenides and tellurides of arsine, and cyanogen sulphide.

Colored rockets for day signaling were developed and recommended to the Ordnance Bureau of the Navy. The Ordnance Department of the Army adopted the specifications for red and yellow smokes. The smokes prepared were far superior to any then in use in intensity of color and in volume. In February, 1918, the work on colored smokes was turned over to the pyrotechnic division.

Through the work of Lieut. Ray and John Gore the mixtures for use in smoke boxes for the Army were greatly improved.

A suitable resistant enamel for lining shells charged with corrosive substances was devised and turned over to the pyrotechnic division to incorporate in shell design.

ANALYTICAL RESEARCH.

The analytical research section devised new methods of analyzing various materials as they came into use in gas warfare. A matter that received much study was the analysis of gases from the explosion of a gas shell in order to determine whether the toxic material used was decomposed by the explosion. In May, 1918, the section was in charge of Albert Finck.

CHEMICAL RESEARCH (DEFENSE).

When the division of defense chemical research was organized in January, 1918, under A. B. Lamb, it took over problems relating to gas masks, such as the absorbent qualities of the charcoal and the chemicals used, the dimming of eyepieces, the penetrability by war gases of gas-mask fabrics, the development of special absorbents or filters for such irritating or toxic substances as diphenylchlorarsine, carbon monoxide, ammonia, hydrofluoric acid, and cyanogen chloride, and also problems bearing on the detection and neutralization of gases and the production of smoke screens.

The work on the absorbents used in the canisters of gas masks included studies of the effectiveness of charcoal under different conditions, the effect of moisture, aging, etc., by G. A. Hulett at Princeton University and N. K. Chaney at Cleveland, Ohio, and D. W. Wilson. One result of the investigation was the development of an impregnated charcoal superior to any in use abroad. It was subsequently manufactured at a plant in Astoria, N. Y. Investigations of the improvement of the efficiency of soda-lime-permanganate by R. E. Wilson, T. W. B. Welsh, and C. P. McNeil comprised studies of the effect of the size of the grains of this absorbent, the use of cement in obtaining grains of proper hardness, and various problems bearing on the production and use of sodium permanganate and of calcium permanganate.

A carbon monoxide absorbent composed of various metallic oxides was developed by J. C. W. Fraser, J. S. Chamberlain, and others. The absorption of various gases by the canister material was studied by R. E. Wilson, E. W. Fuller, and others.

Ammonia absorbents for use in a Navy mask were investigated by Leo Finkelstein; arsine absorbents were studied by C. C. Scallione and others; paper and other filters for removing diphenylchlorarsine were investigated by R. E. Wilson, F. S. Pratt, and others.

Protective ointments for use against mustard gas and methods of neutralizing mustard gas on the ground were investigated by R. E. Wilson and E. W. Fuller.

In the study of the dimming of the eyepieces of gas masks and the devising of efficient antidimmer compounds much work was done by P. W. Carleton and others. The antidimming compounds were prepared by different manufacturers, or made in the laboratories at the American University. The work on smoke screens included investigations of different types of smoke boxes and the study of smoke-box mixtures by R. E. Wilson, R. H. Ryerson, and others.

MANUFACTURING DEVELOPMENT (OFFENSE).

Research work relating to the manufacture of poison gas was begun by the division of chemical manufacture under Dr. William McPherson, formerly dean of the chemical department at Ohio University, in August, 1917. In December, 1917, Dr. McPherson was transferred to large-scale manufacture, and the division of small-scale chemical manufacture was organized under W. S. Rowland. Later, in January, 1918, the work was again reorganized, Dr. Rowland taking charge of that for defense. The work of the division was aided by the cooperation of the chemical research (offense) division, at first under J. F. Norris and later under L. W. Jones. The results of the chief problems that fell to the offense division up to the end of June, 1918, may be divided into four parts as follows: (1) Determining the most suitable method of production on a large scale; (2) production of gases for testing purposes; (3) production and purchase of chemicals not readily obtainable; (4) storage and shipment of toxic chemicals produced at the American University Station and branch laboratories.

Provisional apparatus, specifications and operating directions for making mustard gas were prepared. Work on other methods was done at Hastings-on-Hudson; N. Y., Cleveland, Ohio, and Midland, Mich. Among the men who worked on mustard gas were L. G. Wesson and E. D. Streeter. Work on manufacturing development of mustard gas was discontinued at the American University in May, 1918.

A small-scale plant for making hydrocyanic acid was put into successful operation.

A group of men in charge of R. H. Uhlinger built a small-scale plant to make superpalite.

In cooperation with the division of chemical research, a small-scale plant was put in operation to make cyanogen chloride. Dr. J. R. Withrow and Dr. E. J. Witzemann worked on this problem.

A small-scale plant for making brombenzyl cyanide was tried out and a large-scale plant designed. This work was done by E. M. Hayden.

In May, 1918, a ton of such mixtures was made up and shipped. More than 500 pounds of magnesium arsenide was made for field tests by June, 1918.

Among the other chemicals made in small lots by this division up to June, 1918, were strontium chlorate, strontium permanganate, nitrogen peroxide, aluminum arsenide, ammonium cyanide, arsenic trichloride, and arsenic trifluoride.

MANUFACTURING DEVELOPMENT (DEFENSE).

When the division of small-scale chemical manufacture was reorganized in January, 1918, W. K. Lewis took charge of the work relating to defense. The results of the principal problems attacked by this division are given below:

A horse mask for the Army was developed that was radically different from the British mask and more efficient.

Work on the production of hydrogen for balloons involved the supervision, construction, and testing·of a hydrogen plant for the Army at Langley Field; the control of the raw materials; and the installation of a laboratory equipment.

The division developed testing methods for mustard gas, including tests of absorbent materials such as charcoal, and of protective fabrics. Suitable clothing, including gloves, were developed for protection against mustard gas, and the results turned over to the Gas Defense Service of the Army. A large number of materials were tested for the Gas Defense Service, the work being done in cooperation with the Gas Defense Service and the mechanical development division. The testing methods developed were adopted by the Gas Defense Service.

Members of the division assisted the Gas Defense Service in solving difficulties in the production of absorbent charcoal at Astoria, N. Y., and in the manufacture of "batchite" at Springfield, Mass.

In cooperation with the mechanical development division, an exhaustive study was made of filter materials, such as paper and felt, for gas masks.

Much analytical work was done in the attempt to determine the effect of baffling on the absorption of toxic gas by the canister filling materials.

A new method of investigating toluol, benzol, and xylol in the presence of each other was studied, and fairly quantitative methods for the hydrolysis of the sulphonic acids were developed. These methods promised to be of great value.

In May, 1918, a subdivision was organized to provide satisfactory gas masks for industrial use. Several industries manufacturing poisonous gases for war were supplied with these masks.

DISPERSOID DIVISION.

The dispersoid division was established April 1, 1918, to study the small-scale production of smokes or mists and the best way of protecting men against them, R. C. Tolman being in charge.

Facilities for analytical work with toxic and nontoxic smoke materials, and a dark room for testing the apparatus for optical analyses were provided through the Catholic University, Washington, D. C.

Among the apparatus developed was a Tyndall meter for determining the rate of dissipation of smokes under varying conditions. Tests were started to obtain systematic data on the toxicity and rate of dissipation of smokes, and rate of penetration through filters and mask canisters. One of the problems taken up was the development of smoke candles.

Besides the work on smokes, the dispersoid division continued some of the work on primers that was started at Urbana, Ill., using the Tolman hangfire measurer. Routine tests of airplane ammunition were made with this machine for the Ordnance Department.

EDITORIAL WORK AND MISCELLANEOUS RESEARCH.

In October, 1917, Dr. Wilder D. Bancroft, professor of physical chemistry at Cornell University, joined the gas research staff of the Bureau of Mines. At first he helped keep the various branch laboratories in touch with the Washington laboratory, and later organized a staff which edited and scrutinized all reports leaving the American University station. Dr. Bancroft also had charge of a group of chemists that worked on various problems.

WORK AT THE CARNEGIE INSTITUTE OF TECHNOLOGY.

Work in the war-gas research laboratory at the Carnegie Institute of Technology at Pittsburgh, Pa., began about June 20, 1917, under the general direction of Dr. W. K. Lewis, and the immediate supervision of Dr. R. E. Wilson, formerly assistant professor of chemistry at the Massachusetts Institute of Technology. The work was transferred to Washington about September 15, 1917.

Four main problems engaged the attention of this laboratory: The development of soda lime, the production of smoke screens, the removal of CO from the air in submarines, and the development of incendiary bombs.

In the work on soda lime, a study was made of the effect of different processes of manufacture on the composition and physical properties of the product, as related to its absorptive capacity for various war gases, and stability in use. As a result it was possible, on September 1, to improve greatly the method of manufacture being used by the Army. This investigation also served as a groundwork for further research, which eventually resulted in the development of the best soda lime used by any nation. This work was chiefly done by Dr. P. W. Carleton and Dr. C. P. McNeil.

In the work on smoke screens, substances for producing smokes for screening ships from submarines were studied. Silicon tetrachloride and ammonia were recommended, also a method of production on a large scale. The apparatus devised was developed further at the American University station and placed on practically all ships about December, 1917. The general principles developed in smokes proved of great value later in connection with the work on toxic gases. Paul G. Woodward worked on this problem.

A method of absorbing CO from the air of submarines by ozone in the presence of a catalytic agent was found to be effective, but the size of the necessary ozone generating unit prohibited its use. Dr. F. H. Smyth and Dr. A. B. Ray were engaged upon this problem.

An incendiary bomb for airplanes was designed that gave good results; it was considerably improved later by the incendiary section at the American University station. Dr. A. B. Ray did most of the work.

The laboratory also contributed to the solution of several minor problems, including the development of methods of testing absorbents against superpalite and phosgene, the improvement of antidimming compounds for gas-mask eyepieces, the testing of eyepieces, preliminary work on the impregnation of charcoal, development of signal smokes, etc. This work was done largely by W. D. Wolfe and E. W. Fuller.

Not the least important work of this early laboratory was the development of a trained personnel. Messrs. Smyth, Ray, Carleton, McNeil, Woodward, and Wilson were later all placed in charge of investigations at Washington or elsewhere.

WORK AT CATHOLIC UNIVERSITY AT WASHINGTON.

One of the largest of the branch laboratories engaged in the study of war gases was that of the Catholic University at Washington. The work done there included the study of various problems in organic chemistry under the direction of Dr. W. K. Lewis; investigations of smokes and mists (dispersoids) under Dr. R. C. Tolman; and the study of various problems of physical chemistry under Dr. W. D. Bancroft, including catalytic investigations, such as the catalytic action of charcoal, the production of fluorine and fluorine compounds, the conversion of phosgene into superpalite, and the oxidation of alcohol to acetic acid. At the time of the transfer of the research work to the War Department, June 27, 1918, there were about 75 persons, civilians and enlisted men, busy at the Catholic University.

Dr. J. J. Griffin, the head of the chemical department of the university, greatly facilitated the investigations by his work and cooperation.

WORK AT OTHER UNIVERSITIES.

At Clark University Dr. C. A. Kraus worked on the dimming of eyepieces of gas masks; on metal Dewar flasks for providing liquid oxygen for aviators and submarines; on the stability of gases; on a heat interchanger for use on submarines; and on boosters for gas shells.

At Yale University Dr. T. B. Johnston had charge of a staff of organic chemists that worked chiefly on the halogen ethers, on hydrogen selenide, and on certain selenocyanides.

At Bryn Mawr Dr. R. F. Brunel studied the preparation of diazomethane and chlor and brom ketones. He was assisted by three volunteer chemists.

At Ohio State University Dr. C. A. Boord worked on mustard gas, phenylchlorarsine, diphenylchlorarsine, and on the selenium and tellurium derivatives of mustard gas, which for a time threatened to displace mustard gas.

At Harvard University Dr. G. P. Baxter did much work on the physical constants of the war gases, especially their vapor pressures, thus supplying data that were essential for the proper use of these materials.

Dr. E. E. Reid, of Johns Hopkins University, who became a consulting chemist of the Bureau of Mines in May, 1917, helped greatly in interesting the chemists of the country in the work on noxious gases, and in having them submit organic preparations. He had charge of a group of investigators working on problems of organic chemistry at the university.

Dr. Reid Hunt, at Harvard University, with a staff of pharmacologists, investigated toxicological problems.

Dr. G. A. Hulett, head of the department of physical chemistry at Princeton University was sent abroad in April, 1917, by the Bureau of Mines as a member of a commission appointed by the National Research Council to gather data on technical war problems. He gave especial attention to the use of noxious gases in warfare, and brought back information that proved of immense value. After his return he organized a research group at Princeton, that studied the physical properties of charcoal. These studies brought out new facts regarding the absorption of gas by charcoal. Dr. Hulett also developed a novel method for the removal of iron oxide from glass sand by the use of phosgene. This reaction has other possible industrial applications.

Bulletin 178-B

DEPARTMENT OF THE INTERIOR
FRANKLIN K. LANE, Secretary
BUREAU OF MINES
VAN. H. MANNING, Director

WAR MINERALS NITROGEN FIXATION AND SODIUM CYANIDE

—————

Advance Chapter from Bulletin 178
War Work of the Bureau of Mines

BY

VAN. H. MANNING

WASHINGTON
GOVERNMENT PRINTING OFFICE
1919

CONTENTS.

WAR MINERALS, NITROGEN FIXATION, AND SODIUM CYANIDE.

WAR MINERALS INVESTIGATIONS.

GENERAL STATEMENT.

The term war minerals has been applied to those ores and minerals that were largely imported before the war. Among the more important of these are manganese, essential for making high-grade steel for munitions and industrial use; graphite, for making crucibles; tin, for plating utensils and for bearing-metal; mercury, used as fulminate to explode shells; potash, for making fertilizer and explosives; tungsten and molybdenum, for high-speed tool steel; antimony, for hardening bullet lead; chromite, for tool steel, for tanning leather, and as a refractory lining in furnaces; magnesite, for refractory linings; mica, as insulating material; platinum, for the manufacture of sulphuric acid and for electrical apparatus.

When the United States entered the war it was clear that every ship would be needed and that the number available for importing minerals would be small. Hence a quick and thorough survey of domestic resources was necessary.

Throughout the war the scope and volume of the war-minerals work increased until it covered practically every mineral that was known to be or was liable to be in short supply.

BEGINNING OF THE WORK.

Before this country entered the war, the Bureau of Mines on its own initiative or in cooperation with other Government organizations, had already begun investigations of the more important minerals or mineral products. Among them were nitric acid, sulphuric acid, and potash. In 1916, as related elsewhere (p. 55), the chief chemist of the bureau inspected nitrogen fixation plants in Europe, as the accredited representative of the War Department, and after his return devised a highly efficient method of manufacturing nitric acid, which was used in the development of a Government plant. When the blockade cut off the importation of potash from Germany, the Bureau of Mines in cooperation with the United States Geological Survey, Bureau of Soils, and other Government agencies began an investigation of sources of potash and methods of manufacture. In cooperation with the War Department, a study of the sulphuric acid situation was begun about the same time that of nitric acid was taken up.

After the declaration of war the bureau took a leading part in investigations aimed to prevent deficiencies in the supplies of needed minerals. The director of the bureau served as a member of the military committee of the National Research Council and of the various committees on mineral products under the Council of National Defense. Representatives of the bureau served on and took an active part in the work of the War Minerals Committee, established in July, 1917, for studying the minerals situation, and of the Joint Information Board on Minerals and Derivatives. The bureau has had charge of the extensive investigations of war minerals conducted under the $150,000 emergency fund appropriated in March, 1918.

The Bureau of Mines has endeavored to aid producers of war minerals by taking up questions of priority, transportation, fuel, and equipment, construction of highways and roads, labor supply, and the study of the best mining methods and concentrating practice for the development of these new materials. The results of the bureau's investigations have been condensed into summarized reports for the War Industries Board, the United States Shipping Board, and other Government establishments.

WAR MINERALS COMMITTEE.

The War Minerals Committee, established in July, 1917, was composed of William Young Westervelt, representing the American Institute of Mining Engineers and the Mining and Metallurgical Society of America; W. O. Hotchkiss, representing the Association of State Geologists; David White, of the United States Geological Survey; and A. G. White, of the Bureau of Mines; later H. S. Mudd acted as assistant secretary. This committee aided in establishing cooperation between the bureau and the various agencies interested in the development of minerals; it also assisted in outlining investigations and in obtaining the best available men for special work.

JOINT INFORMATION BOARD.

The Joint Information Board on Minerals and Derivatives was organized to act as a clearing house for information to all of the Government departments interested in mineral problems. H. S. Mudd acted as the representative of the Bureau of Mines on this board.

SPECIAL APPROPRIATION.

In the latter part of March, 1918, Congress in the urgent deficiency bill appropriated $150,000, to be expended under the Bureau of Mines, to extend and continue investigations relating to minerals of military importance. The general purpose of this work is indicated by the heading under which the appropriation was made:

War materials investigation.—For inquiries and scientific and technologic investigations concerning the mining, preparation, treatment, and utilization of ores and other mineral substances which are particularly needed for carrying on the war, in connection with military and manufacturing purposes, and which have heretofore been largely imported, with a view to developing domestic sources of supply and substitutes for such ores and mineral products as are particularly needed, and conserving resources through the prevention of waste in the mining, quarrying, metallurgical, and other mineral industries.

This appropriation permitted organization and work on a much more extensive scale, although a large amount of war minerals work had already been undertaken by the Bureau of Mines.

MINERALS ADMINISTRATION ACT.

The so-called war minerals bill was submitted to Congress in December, 1917. Originally limited to a few special minerals, the bill was later broadened to include all minerals, and finally was again restricted to a group consisting principally of minor and rare metals and minerals. The bill was passed by the House on April 30, and in modified form by the Senate on September 11. It was approved by the President on October 5, 1918.

By Executive order of November 11, 1918, the administration of the act was delegated to the Secretary of the Interior. On that date the armistice ended hostilities. As the act primarily provided for insuring production for war purposes, it was not a reconstruction measure and further legislation was required for that purpose.

An amendment to the minerals control act, authorizing the Secretary of the Interior to examine claims and pay off financial losses of the producers of certain war minerals, where it could be shown that such production had taken place as a result of Government action, was attached to the Dent military bill, which was signed by the President on March 3, 1919. The measure authorizes investigation of net losses incurred by producers of manganese, chrome, pyrite, and tungsten. The maximum amount to be expended under the provision is limited to $8,500,000. A commission of three members, to be known as the War Minerals Relief Commission, will review the claims and make the awards. The Secretary of the Interior has authorized the Director of the Bureau of Mines to conduct the field engineering and accounting investigations, also the office routine and administrative work for the commission. The members of the commission are J. F. Shafroth, M. D. Foster, and P. N. Moore.

PERSONNEL OF WAR MINERALS INVESTIGATIONS.

At the close of the war the personnel of the war minerals investigations included some 90 mining engineers, metallurgists, and chem-

ists. Of these about two-thirds were engaged in the staff and field
work of the longer investigations, the rest were consulting engineers
working on special problems. The war minerals investigations were
in charge of J. E. Spurr as chief executive. At first they were temporarily under the direction of D. A. Lyon, supervisor of the bureau's
mining experiment stations.

The general staff consisted of H. H. Porter, in charge of priority
matters; F. W. Paine, in charge of shipping problems; H. S. Mudd.
representative on the Joint Information Board, and matters relating to highways and roads; A. G. White, matters of organization
and planning, and excess-profits taxes; R. R. Hornor, mining
methods; J. H. Mackenzie, mining costs, and ore markets; H. C.
Morris, matters relating to work of the Capital Issues Committee:
J. E. Orchard, political and commercial control of minerals; Oliver
Bowles, informational, and nonmetallic minerals; C. T. Robertson.
files and editing of manuscripts. In addition to this general staff
were the engineers in charge of the researches on the different
minerals.

The work of the war minerals investigations has been closely associated with the work done at the mining experiment stations of the
bureau under the direction of the division chiefs of the bureau and
the supervisor of the experiment stations. The Minneapolis, Minn..
station worked entirely on problems related to manganese ores: the
Columbus, Ohio, station took up special investigations of domestic
clays, graphite, and refractories; the Urbana, Ill., station worked at
the recovery and use of pyrite from coal mines; the Golden. Colo..
station studied the recovery of potash, tungsten, and other rare
metals; the Tucson, Ariz., station made field investigations of manganese ores; the Salt Lake City, Utah, station investigated graphite
milling; the Berkeley, Calif., station gave attention to the concentration and recovery of quicksilver and sulphur; the Seattle, Wash..
station studied concentration of chrome and tin.

MANGANESE.

Manganese was in some respects the most important of the war
minerals. The United States requirements for 1918 were estimated
to be 798,000 tons of high-grade ore. Because of the need in the
steel industry, little reduction of consumption was feasible; hence
the problem confronting the Bureau of Mines was to endeavor to
stimulate domestic production so as to release bottoms that would
otherwise be used in importing some 650,000 tons of manganese ore.

Because of the importance of the investigations a large number
of engineers were assigned to them. Some engineers investigated all
the known manganese deposits in this country as well as deposits
in Canada and in Cuba, other engineers assisted mine operators in

utilizing the best mining and concentrating methods and aided the small producers, others endeavored to develop improved methods of milling and concentrating, others devoted their attention to metallurgical practice at blast furnaces smelting manganese alloys.

As a result of the demand for manganese and the work done by the Government the domestic production of manganese in 1918 was about 294,497 tons, as compared with a production in 1913, before the war, of 4,048 tons.

Practically all the work of the Minneapolis station in 1918 was on manganese, especially the concentration and beneficiation of low-grade ores. Edmund Newton, superintendent of the station, was charged with the preliminary organization of all the manganese investigations. H. C. Morris organized the field investigation in cooperation with Mr. Newton. In July, 1918, the work on manganese as related to the war minerals investigations was placed in charge of C. M. Weld.

FIELD EXAMINATION OF MINING AND MILLING METHODS.

W. C. Phalen made a preliminary investigation of the manganese mines of the country. In the fall of 1917, through cooperation with the War Minerals Committee and the American Institute of Mining Engineers, P. N. Moore, A. H. Rogers, R. H. Richards, and F. Lynwood Garrison, mining engineers, spent several weeks in studying mining and washing methods in Virginia. W. R. Crane investigated the Georgia deposits in cooperation with the State Geological Survey; E. R. Eaton studied the improvement of mining and beneficiation methods in the South; and E. G. Spilsbury, a consulting engineer of the bureau, examined one of the new Virginia districts. C. E. Van Barneveld, superintendent of the Tucson station, investigated many of the western deposits. G. D. Louderbach examined the California deposits in cooperation with the United States Geological Survey and with State organizations. G. H. Clevenger studied the recovery of manganese from the ores of the Cripple Creek district. T. M. Bains and G. E. Ingersoll examined the manganiferous deposits of the Cuyuna and other ranges in Minnesota.

In April and May, 1918, Albert Burch, for the Bureau of Mines, and E. T. Burchard, for the Geological Survey, investigated the manganese resources of Cuba.

BLAST-FURNACE PRACTICE.

Early in 1918 H. D. Hibbard and J. E. Johnson, consulting engineers, and Edmund Newton, superintendent of the Minneapolis station, investigated metallurgical practice at steel plants, using low-manganese alloys produced largely from domestic ores.

USE OF MANGANESE ALLOYS IN OPEN-HEARTH STEEL PRACTICE.

A study of the use of manganese alloys in open-hearth steel practice was undertaken to determine the most suitable means of utilizing domestic manganese and the extent to which low-grade alloys could be substituted for high-grade alloys without impairing the steel produced either as to quantity or quality. The principal furnaces were visited and the reactions taking place in the furnace studied, and samples of metal and slag taken. A report on the results of the work was prepared for publication. This work was conducted by S. L. Hoyt, F. B. Foley, and R. L. Dowdell.

PRODUCTION OF FERRO-MANGANESE AND SPIEGELEISEN.

In order to collect and prepare data on the production of ferro-manganese and of spiegeleisen in blast furnaces, 18 furnaces, including all furnaces in blast on manganese alloys, were visited. Furnace records, charge sheets, and analyses over long periods were compared, and various factors, such as blast temperature, composition of alloy and of slag, etc., were studied during representative periods of furnace operation. A report was prepared on the results of this work, which was conducted by P. H. Royster.

ELECTRIC SMELTING OF SILICO-MANGANESE.

As silico-manganese had been used to some extent by steel makers and as an increase in its use would have helped to relieve the manganese situation, laboratory experiments on the electric smelting of silico-manganese were made. This work was done by H. W. Gillett, alloy chemist of the bureau, who prepared a report on the experiments and the results obtained.

ORE-DRESSING TESTS.

Samples of ore from different States were tested at the Minneapolis station in the effort to devise improved methods of dressing and to assist operators in preparing their ores for market. These tests were conducted by J. W. Norton and H. H. Wade.

TESTS OF JONES AND BOURCOUD PROCESSES.

Two new processes for utilizing low-grade and complex ores were examined at the Minneapolis station. The Jones process for the direct reduction of manganiferous iron ores was shown to be metallurgically feasible. A similar series of experiments on the Bourcoud process was in progress on March 1, 1919. These tests were made under the direction of Peter Christianson, consulting engineer.

When all the manganese work is completed the various reports on manganese and manganese alloys will be assembled and published as a bulletin of the bureau.

RECONSTRUCTION WORK.

After the armistice was signed the demand for domestic ores fell. The Bureau of Mines at once instructed its field men to discourage further expansion of the manganese mining, and to advise producers as to the most efficient methods, so that they might have a better chance of meeting the competition of imported ores.

COLLECTION OF STATISTICS ON CONSUMPTION.

From the results of information obtained in the field and through monthly reports from consumers of manganese ore and of ferro-alloys, the available supplies of ore were ascertained. A careful study of the requirements of furnaces was also made and valuable supplies were conserved. This statistical work was in charge of W. C. Phalen and W. R. Crane.

CHROMITE AND CHROMIUM.

Estimates made soon after the bureau's war investigations began placed the country's requirements for 1918 at 130,000 tons of chromite. In 1917 the imports from Rhodesia and New Caledonia amounted to some 43,000 tons. As there seemed little probability that production from domestic sources could be increased enough, effort was made to stimulate production in other countries of America as well as in the United States. The domestic output in 1918, however, was about 64,590 long tons of chromite (50 per cent Cr_2O_3), or 14,590 tons more than was anticipated. Thus the bureau's investigations demonstrated that the chrome resources of the United States would yield at war prices a large production for some time. The chromite investigations were under the general charge of J. H. Mackenzie and J. E. McGuire.

Efforts were also made to devise substitutes for chromite used in refractories and to demonstrate the possibility of decreasing the amount of chromium salts used for tanning leather and for other purposes. A special study of substitutes for chrome brick was started at the Columbus experiment station under R. T. Stull, superintendent. H. D. Hibbard has submitted a report on the use of chromite as a refractory.

As the chromium-bearing iron deposits in Cuba were reported to promise a source of chrome, Albert Burch of the bureau, accompanied by E. F. Burchard of the United States Geological Survey, visited them and subsequently prepared a report on the probable extent of the principal deposits and the feasibility of developing them.

Deposits in Newfoundland, in North Carolina, in Pennsylvania, and in the Western States were also investigated. These investigations were made by Albert Burch, assisted by field engineers J. E. McGuire, E. G. Hill, and F. H. Probert.

Concentrating tests of low-grade chrome ore were made at the bureau's experiment stations at Berkeley, Cal., and at Seattle, Wash. A survey of the ferrochrome situation was made by R. M. Keeney.

TIN.

The United States uses annually about 70,000 tons of tin, but produces only 100 tons, chiefly from Alaska. As a large part of the tin consumed was used as tin plate for food containers and these were of vital importance to the Army and Navy, every effort was made to insure an adequate supply.

Domestic deposits of tin ore were investigated and partly as a result of such investigations a Virginia deposit was commandeered by the War Department and turned over to a Boston company to develop. R. R. Hornor examined the tin deposits in the Black Hills of South Dakota and the samples he collected were subjected to milling tests at the Seattle experiment station. Thomas Leggett investigated the tin area of North and South Carolina.

The use of substitutes for tin and the reclamation of used tin were investigated at the Seattle station.

Encouragement of the smelting in this country of Bolivian tin ore so as to reduce the need for importing Straits tin also seemed desirable. A commission headed by Charles Janin, of the Bureau of Mines, and including Howland Bancroft, of the bureau, investigated the Bolivian tin situation. H. Foster Bain, assistant director of the bureau, was in charge of the work on tin.

PYRITE.

Pyrite is used in making sulphuric acid, which is indispensable in the manufacture of explosives. After this country entered the war practically every known pyrite deposit of importance in the United States was investigated by representatives of the Bureau of Mines. These investigations have demonstrated that there are large pyrite deposits which can be developed to meet any shortage that may arise. Valuable assistance was rendered to new mines in the matter of priorities, labor supply, and improved methods of mining.

The recovery of copper from pyrite cinder was studied by D. E. Fogg and tests were made to determine whether the cinder could be used for making low-phosphorous iron.

A complete survey was made of the producing pyrite mines in Colorado and the Eastern States. C. E. Julihn investigated deposits in the Southern States; R. R. Hornor did field work in Virginia, New York, Wisconsin, Missouri, and Colorado. In cooperation with the

State geological surveys of Pennsylvania, West Virginia, Ohio, Kentucky, Tennessee, Indiana, Illinois, Michigan, Missouri, and Iowa; E. A. Holbrook investigated the possible recovery of pyrite at coal mines. The results of work at the Urbana experiment station indicate that much pyrite may be obtained from this source. The pyrrhotite deposits at Pulaski, Va., have been thoroughly examined. A method for concentrating pyrrhotite by electrostatic concentration was developed at the Golden station.

The work on pyrite was under the supervision of H. A. Buehler.

SULPHUR.

In October, 1917, the possible output of the two chief commercial deposits of native sulphur were examined by J. Parke Channing, J. W. Malcolmson, and A. B. W. Hodges, consulting engineers of the Bureau of Mines, W. O. Hotchkiss, representing the War Minerals Committee, and P. S. Smith, representing the United States Geological Survey. This investigation was made to determine whether the sulphur output could meet the war requirements.

In 1918 C. O. Lindberg, of the Bureau of Mines, made a thorough examination of western sulphur deposits, with a view to determining the possibilities of their development commercially or to meet an emergency. Concentration tests of samples from the deposits were made by J. M. Hyde at the Berkeley mining experiment station.

SULPHURIC ACID.

A. E. Wells, superintendent of the Salt Lake City mining experiment station, was put in charge of the sulphuric acid investigations; he cooperated with the War Industries Board and with the Chemical Alliance. A statement was prepared for the Railroad Administration and for the Shipping Board on the transportation of brimstone to Atlantic ports and to interior districts.

A complete field survey was made of plants producing concentrated acid for munitions purposes; a report was prepared on the possibility of using acid from fumes of western smelters; the possibility of increasing acid production at the zinc and copper smelters in the Eastern States was studied.

Reports were submitted to other Government establishments in connection with the construction of new acid plants and the allocation of raw materials. Largely because of the need of acid in making explosives the demand for acid during the war increased approximately two and one-half times.

The attention of the War Industries Board was called to the fact that if necessity should arise the supply of acid to the steel pickling industry could be curtailed 50 per cent through the substitution of niter cake, without producing hardships to that industry.

D. E. Fogg assisted in the work on sulphuric acid.

GRAPHITE.

Before the war most of the flake graphite used in the manufacture of crucibles for making brass and crucible steel came from Ceylon and Madagascar. In 1917 the domestic production of flake graphite of crucible grade was only about 3,400 tons, whereas the foreign imports were about 27,000 tons. It was felt that domestic production could easily be stimulated by creating a market for domestic flake and that the market could be created if the bureau's engineers demonstrated that satisfactory crucibles could be made with a larger percentage of domestic flake than had commonly been used.

FIELD INVESTIGATION OF COMMERCIAL PRACTICE.

A field investigation was made of commercial methods of sampling and analysis with a view to establishing more efficient relations between graphite producers and consumers, by recommending uniform specifications. The leading domestic producing areas were visited and a detailed report entitled "Preparation of Crucible Graphite" was published. On August 10, 1918, the War Industries Board requested that all crucible makers use 20 per cent domestic flake in their crucible mixtures for the balance of 1918 with an increase to 25 per cent in 1919. This action, which was taken on the recommendation of the Bureau of Mines, resulted in the establishment of a market for domestic flake graphite. This investigation was conducted by G. D. Dub.

EXPERIMENTS ON FINISHING GRAPHITE.

The Salt Lake City, Utah, experiment station of the bureau studied the processes of preparing commercial graphite. An improved method of finishing graphite, by which milling losses are greatly decreased, was devised. Under existing practice an unduly large proportion of the flakes was crushed to a size too small for crucible grade. Experiments were made with different devices and with different methods of finishing graphite, and the possibilities of each demonstrated. This work was under the direction of F. G. Moses, acting superintendent of the station.

EXPERIMENTS ON CRUCIBLE MANUFACTURE.

At the Columbus, Ohio, experiment station an extensive investigation of the possibilities of using larger proportions of domestic flake in the crucible mixtures was made and numerous tests were conducted. This study was given satisfactory results. This work was under the direction of E. A. Holbrook, then superintendent of the station.

COLLECTION OF GENERAL DATA.

At the Washington office, data were collected on methods of analyzing graphite and graphitic ores to determine the best practice, and on mill capacity of all plants in operation or under construction, their actual production, labor conditions, efficiency, and grades of graphite manufactured. H. S. Mudd conducted this work.

POTASH.

Because of the importance of potash in agriculture and explosives. a review of the potash situation was made by A. W. Stockett. Although the normal annual consumption of potash in this country is probably 250,000 tons, with a domestic production in 1917 of only 32,000 tons, Mr. Stockett showed that the deficiency could be met by utilizing the Searles Lake deposit in southeastern California, the greensands deposits of New Jersey and adjoining States, and by erecting potash-recovery plants at cement works and at blast furnaces.

. The bureau prepared for the War Department a report on the potash situation, and advised the Capital Issues Committee on applications for the construction of potash-recovery plants.

The Bureau of Mines, through its experiment stations, has done much experimental work on potash and has cooperated with manufacturers. The Berkeley experiment station cooperated with the California State Council of Defense in studying methods of stimulating production from saline deposits. An engineer of the bureau examined the salt lakes of Nebraska and the plants there. The Salt Lake City station conducted some experiments in the recovery of potash from leucite and other potash-bearing rocks.

LIMESTONE.

Soon after this country entered the war, owners of limestone quarries faced a serious shortage of labor caused by the drafting of men and the shifting of workers. There was danger that any considerable decrease in the output of fluxing limestone might ultimately result in a decreased production of pig iron. To meet this situation the Bureau of Mines issued and distributed to the limestone quarries a paper on "Labor Saving in Limestone Quarrying," prepared by Oliver Bowles, quarry technologist.

WHITE ARSENIC.

By the midsummer of 1917, it became evident that on account of the practically complete stoppage of arsenic imports, there would be considerable difficulty in obtaining from domestic sources enough white arsenic for insecticides, glass manufacture, sheep and cattle dip, and other nonwar purposes; and the prospective shortage would be augmented by the probable manufacture of certain toxic gases for the War Department.

A survey of the situation was made in September, 1917, by A. E. Wells, who visited all of the arsenic plants in the United States and studied the possibilities of increasing production. An accumulation of dust in an old flue system of a smeltery at Great Falls, Mont., was disclosed, which yielded an unexpected supply of arsenic, large enough with the regular production of the other arsenic plants to supply the demand for nonwar purposes. In October, 1918, the War Department put in a heavy requirement, and steps were taken immediately to meet the demand. As it was believed that the war minerals act provided means for the Government to encourage directly an immediately increased production, Mr. Wells and H. S. Mudd examined several prospective sources of arsenic in the eastern States, and producers in the West were encouraged to plan expenditures to increase their output. However, at the signing of the armistice the War Department's requirements were cancelled, and all the new projects for increasing arsenic production were discontinued at once before extra expenditures were incurred, except a new plant at Anaconda, Mont. That plant is being built primarily for other purposes, the arsenic being merely a by-product.

PLATINUM.

Hennen Jennings, consulting engineer, and C. L. Parsons, chief chemist, of the Bureau of Mines, cooperated with the War Industries Board in the steps taken to reserve an adequate supply of platinum for war requirements. By act of Congress the Bureau of Mines was given authority to control the production and sale of platinum during the war. This work was conducted through the license and inspection system built up for administering the explosives regulation act.

TUNGSTEN.

Because of the value of tungsten in the manufacture of high-speed tool steels, needed in munition plants, the Bureau of Mines sought by advice and encouragement to aid the development of domestic deposits and thus increase imports and release foreign production to the allies. Early in 1918 the allied governments proposed to allocate the world's production for the year, estimated at about 24,000 tons, the United States to receive half, and the other half to go to England, France, and Italy. However, the United States requirements for 1918 were estimated at about 16,000 tons, and if the war had continued, nearly 18,000 tons for the year 1919.

At first the work on tungsten was in charge of J. H. Mackenzie, assisted by J. E. Maguire; J. S. Means made a field examination of the principal tungsten districts in the United States. Later, H. C. Morris took charge of the work on the rare metals.

A series of special tungsten steels were prepared for tests by the Ordnance Department by H. W. Gillett, alloy chemist of the bureau, as related in another chapter of this bulletin. The tungsten for these experiments was prepared at the Golden station. Also at this station the conditions under which tungstic acid obtained in the treatment of tungsten ores may be efficiently reduced to metallic tungsten have been studied and the results published.

REPORTS ON POLITICAL AND COMMERCIAL CONTROL OF MINERAL RESOURCES.

In May, 1918, the Bureau of Mines, the United States Geological Survey, and other Government bureaus were invited to contribute to a series of reports on the political and commercial control of the mineral resources of the world initiated by Dr. J. E. Spurr, then a member of the Committee on Mineral Imports of the United States Shipping Board. J. E. Orchard, assistant mine economist of the Bureau of Mines, was placed in charge of the editing and publication of the reports, which are confidential and are being supplied to Government officials only. At present they are being used by the economic advisers of the American delegation at the peace conference. Steps are being taken by other departments of the Government to make similar studies of the other raw materials.

FIXATION OF NITROGEN AND OXIDATION OF AMMONIA.

The work of the Bureau of Mines in connection with the development of processes and plants for insuring an adequate supply of nitric acid and nitrates in this country ranks among its important achievements. This work, conducted in cooperation with the War Department, was the direct outgrowth of the necessity of rendering the Nation independent as regards a supply of nitric acid, essential in the manufacture of high explosives, and of the nitrogen compounds used in agriculture and other industries. Such a need was recognized in the national defense act of June 3, 1916, which (sec. 124) authorized the President to have made an investigation to determine the best, cheapest, and most available means of producing nitrates and nitrogenous materials used in munitions and in the manufacture of fertilizers and other products, and to have erected such plants as might be deemed necessary.

About two months previous to the passage of this act, the Secretary of the Interior had, April 7, 1916, offered the Secretary of War the aid of the Interior Department in any capacity that would be useful for national preparedness, and called attention to the fact that the Bureau of Mines could aid in the study of methods and materials necessary for the large-scale manufacture of nitrogen products.

During April and May, 1916, C. L. Parsons, chief chemist and chief of the division of mineral technology of the Bureau of Mines, had several informal conferences with Brig. Gen. William Crozier, Chief of Ordnance, in regard to nitrate supply, and at Gen. Crozier's request Dr. Parsons made a tentative report on June 6, 1916, on the outlook for the fixation of atmospheric nitrogen by existing methods.

On June 9 the Secretary of War inquired whether the Bureau of Mines was in position to undertake such researches as would demonstrate the practicability of oxidizing ammonia to nitric acid, and asked for an estimate of the funds that would be needed to carry on this work, in view of the appropriations that had been made to the War Department under section 124 of the act approved June 3, 1916. The Secretary of the Interior replied on July 13 that the Bureau of Mines would be glad to undertake this work and outlined a tentative plan for procedure.

On August 1 the Secretary of War proposed that qualified representatives be sent abroad to study the methods of manufacturing nitric acid, otherwise than from sodium nitrate, followed in the various

54

countries of Europe. On August 3 the Secretary of the Interior informed the Secretary of War that arrangements were in progress, after consultation with the Chief of Ordnance, for a cooperative agreement between the Bureau of Mines and the Semet Solvay Co. for demonstrating on a plant scale the possibility of oxidizing ammonia produced from by-product coke ovens, and that preliminary investigations had already begun. Secretary Lane suggested that the War Department set aside funds to assist in this work and also to pay the expenses abroad of the investigators proposed by the Secretary of War in his letter of August 1. Secretary Lane also said that he had designated Dr. C. L. Parsons, chief chemist of the Bureau of Mines, to make this report as the representative of the Interior Department, and suggested that Mr. Eysten Berg, an engineer long familiar with the nitrogen fixation processes used in Norway, might be associated with him in making the investigation. On August 14 the Secretary of War accepted the suggested arrangement and allotted $10,000 and outlined a method of procedure for the expenditure of this sum.

A contract with the Semet Solvay Co. was signed August 10, 1916, whereby the Semet Solvay Co., in cooperation with the Bureau of Mines and with the approval of Gen. Crozier, undertook to erect in Syracuse a plant, from plans furnished by the Bureau of Mines, for the chief purpose of demonstrating whether ammonia could be successfully oxidized on a commercial scale to nitric acid, and especially whether ammonia produced by the destructive distillation of coal was as suitable for the purpose as cyanamid.

Early in 1916 the American Cyanamid Co. had carried on experiments on the oxidation of ammonia at its plant at Niagara Falls, Ontario, and during the summer of 1916 had erected at Warner's, N. J., a small plant having a capacity of approximately 1 ton of nitric acid a day, for producing the acid from cyanamid ammonia by the catalytic action of electrically heated platinum. During the summer of 1916 Dr. Parsons was allowed to visit this plant, with the distinct understanding that although certain details of the process would not be revealed to him, nevertheless he was at liberty to use anything that he himself saw during his visit.

Dr. G. B. Taylor and J. H. Capps, of the Bureau of Mines, began laboratory studies of oxidation methods in the Pittsburgh experiment station of the bureau and J. D. Davis began similar studies with chemists of the Semet Solvay Co. at Syracuse, N. Y. The Bureau of Soils, Department of Agriculture, also sent a representative to the laboratories of the Semet Solvay Co., but he remained only a short time. Mr. Davis had associated with him in his early experimental work Bryan Handy, of the Semet Solvay Co., who received advice

from Dr. L. C. Jones, chief chemist of the company, and G. N. Ter-
ziev, who had previously worked on the oxidation of ammonia by the
use of nonmetallic catalytic agents. C. D. Davis, of the Pittsburgh
station, was assigned to assist J. D. Davis in this work.

Before he left for Europe early in October, 1916, Dr. Parsons, assisted by others, prepared general plans for the erection of a small
plant at Split Rock near Syracuse. The plant was to be much like the
plant at Warner's, N. J., platinum gauze being used as a catalyzer.
When Dr. Parsons returned from Europe in December, 1916, this
plant was nearing completion. The chemists mentioned above made
many experiments with the hope of utilizing nonmetallic catalyzers.
Many such catalyzers were found that worked successfully for a short
period, but nothing to compare with platinum was ever discovered.
Hence this line of experiment was abandoned, only apparatus in which
platinum was the basic agent for the conversion of ammonia to nitric
acid was tried.

Dr. Parsons was transferred to the War Department for the last
three months of the year 1916. As chemical engineer to the Ordnance
Department he visited, with Mr. Eysten Berg, the plants in France
and Italy, and later alone visited the plants in England, Norway, and
Sweden that were working on the fixation of atmospheric nitrogen.
He returned to the United States on December 24, 1916; made his
preliminary report to the Ordnance Department on January 17, 1917;
and gave his final conclusions on April 30, 1917.

Meanwhile, a committee of the National Academy, appointed at the
request of the Secretary of War, had reported its findings to the Ord-
nance Department. The members of this committee were Dr. A. A.
Noyes, chairman, Dr. L. H. Baekeland, Dr. Gano Dunn, Dr. Charles
Herty, Dr. W. K. Lewis, Mr. M. I. Pupin, Prof. T. W. Richards, Mr.
Elihu Thompson, and Prof. W. R. Whitney. In order to harmonize
the reports made by the committee and by Dr. Parsons, the Secretary
of War appointed a committee on nitrate supply consisting of Brig.
Gen. William Crozier, Admiral Ralph Earle, Brig. Gen. William M.
Black, Frederick Brown (Bureau of Soils), Dr. L. H. Baekeland, Dr.
Gano Dunn, Dr. Charles Herty, Dr. W. F. Hillebrand (Bureau of
Standards), Dr. A. A. Noyes, Dr. W. R. Whitney, and Dr. C. L. Parsons (Bureau of Mines). This committee met on May 11, 1917, and
made its final report to the Secretary of War. It adopted practically
all of the recommendations made in the reports of the chief chemist
of the Bureau of Mines. These reports, with the exception of the
report of the committee of the National Academy, were printed in the
Journal of Industrial and Engineering Chemistry, September, 1917.
and in a pamphlet issued by the Ordnance Department, entitled "A
Statement of Action Taken and Contemplated Looking to the Fixation of Nitrogen."

The work of constructing and operating the plants for the fixation of nitrogen was turned over to a special division of the Army known as the nitrate division, Ordnance Department, under Col. J. W. Joyes. Work on the oxidation of ammonia continued for some time at Syracuse, and the nitrate division also established a small experimental laboratory of its own at Sheffield, Ala., for studying the oxidation of ammonia.

Early in 1917 G. A. Perley, of New Hampshire College, was assigned to the ammonia oxidation work at Syracuse as a member of the Bureau of Mines. Shortly after he was commissioned a first lieutenant in the Ordnance and later promoted to captain. He proved an able associate of J. D. Davis and the other chemists there in developing the ammonia oxidation process.

It was early determined that the German apparatus, as modified by W. L. Landis, of the American Cyanamid Co., would readily oxidize ammonia from by-product coke plants with high efficiency, but so much electricity was needed to maintain the heat of the single sheet of platinum gauze used that a method using no external heat was plainly desirable, especially as there was a large amount of waste heat from the reaction. The first modification of the apparatus was to use three or four layers of platinum gauze in the hope of reducing the proportion of heat radiated. This hope was realized and apparatus of essentially the same shape as that previously used, but having multiple gauze, worked fairly well and gave fairly high efficiencies.

Dr. Parsons had already given the experimenters data on apparatus used in England and France. Additional information on the multiple-gauze apparatus was now sent to the European representatives of the Semet Solvay Co. and became the basis of experimental plants that were constructed in England except the large plant erected at Dagenham Docks, which used the Ostwald process developed at Vilvorde, Belgium, before the war.

Although the experiments with the multiple-gauze apparatus gave fairly good results, they were not entirely satisfactory, as the temperature of the gauze was not maintained at the point most desirable for high efficiencies. After considerable thought, an apparatus was developed and patented by Dr. Jones, of the Semet Solvay Co., and Dr. Parsons. In this apparatus the layers of platinum gauze were arranged within a cylinder made of refractory material. The layers of gauze radiated to each other and also raised the walls of the refractory cylinder to a red heat so that it in turn helped to maintain a high temperature. With pure ammonia this apparatus proved a complete success; one single apparatus was in continual use day and night for six months at the plant of the Semet Solvay Co. and gave an average efficiency of more than 90 per cent. This form of apparatus was adopted for Chemical Plant No. 1 at Sheffield, Ala.

Another important development of this work was the determining of the impurities that are liable to be present in ammonia and must be removed to insure oxidation with high efficiency. These impurities are chiefly iron in any form, oil or similar organic material carried mechanically, and phosphine. The last has a very poisonous action, but it is present only in cyanamid ammonia and can be readily removed.

DEVELOPMENT OF PLANT BUILT FOR PRODUCING SODIUM CYANIDE BY THE BUCHER PROCESS.

Reports from France received by the Bureau of Mines indicated strongly that hydrocyanic acid was desirable for gas warfare. The French were preparing to use it in shells and were trying to purchase cyanide in America.

On September 7, 1917, the Secretary of the Interior wrote to the Secretary of War calling attention to the Bucher process of the Nitrogen Products Co., and offering the services of the Bureau of Mines in constructing and operating a plant for making sodium cyanide. This letter was acknowledged by the War Department on September 24, with the statement that the matter was being investigated. On November 26 a letter from the Assistant Secretary of War definitely accepted the offer of the Bureau of Mines to undertake the work. On December 5 the Director of the Bureau of Mines asked the Chief of Ordnance to allot the necessary funds, and said that Dr. C. L. Parsons, chief chemist of the Bureau of Mines, had been designated to take charge of the engineering, building, construction, and operation of the plant.

In the meantime Capt. E. J. W. Ragsdale, of the Ordnance Department, had been making preliminary arrangements, through conferences with the officials of the Nitrogen Products Co. C. W. Marsh was appointed consulting engineer. Dr. Norman E. Holt, formerly of the Nitrogen Products Co., and thoroughly familiar with the experimental development of its process, was commissioned a captain in the Ordnance Department. R. M. Ross was appointed assistant engineer, and was commissioned a first lieutenant in the Ordnance Department some months later. Plans for the plant were prepared as rapidly as possible after the funds were made available. The early plans were for a 15-ton plant. As later advices from France, transmitted to Capt. Ragsdale on December 14, indicated that the probable use of cyanide would be less than previously anticipated, the plans were changed to a 5-ton basis.

On December 27, 1917, the Chief of Ordnance set aside $750,000 to cover the cost of a 5-ton plant. This fund was retained by the Ordnance Department, but was subject to vouchers duly approved by the Bureau of Mines.

On January 8, 1918, a communication received from the Director of the Research Division of the Gas Warfare Service stated that a committee consisting of Dr. J. F. Norris, Dr. John Johnston, Col. William McPherson, and Col. William Walker, which had considered the need for cyanide, believed that a 5-ton plant was not large enough and recommended the building of a 10-ton plant. Because of this recommendation the Ordnance Department was asked for more funds on an estimate that the 10-ton project would cost approximately $2,000,000, as cantonments and other requisites for an enlisted personnel of 350 men and 25 officers would have to be provided at the site chosen.

CONSTRUCTION.

The preliminary engineering work proceeded rapidly and was nearly completed by the time contracts were signed, February 28. In the meantime, at the recommendation of the Council of National Defense, it had been deemed wise to ask the Cantonment Division of the Army to assign officers who would take charge of actual construction work, and to designate a contractor. Of the funds placed to the credit of the Bureau of Mines and the Ordnance Department $500,000 were accordingly transferred to the construction division of the Quartermaster General's Office. Maj. R. A. Widdicombe was chosen as construction quartermaster and Maj. J. R. Werth as supervising constructing quartermaster. Frazer, Brace & Co. were designated contractors.

Work at the site begun March 1 and construction proceeded rapidly, the only delays being those from inability to obtain structural steel in the months of May and June. The construction work carefully followed the plans of the engineering staff employed by the bureau and was subject to the direction and approval of Dr. C. L. Parsons, in charge of the work.

In order that the plant might be ready at the date indicated, a corps of officers and men were trained in a small experimental plant of the Nitrogen Products Co. at Saltville, Va. Capt. C. O. Brown, chemical engineer of the Ordnance Department, was chosen commanding officer in charge of plant operations. He was assisted by Capt. N. E. Holt, who proved invaluable both in planning the plant construction and in seeing that the process worked smoothly. Conferences were held from time to time with the engineers and officers of the Ordnance Department as the work progressed.

Outside of the cantonment buildings, roads, filtering plant, electric installations, sewers, fire protection, and other subsidiary details. the main plant itself had three essential divisions housed in three main buildings, known as the mechanical, retort, and lixiviator buildings.

The mechanical building, for the preparation of the raw material and its formation into special briquets, was designed largely by Capt. Holt and C. W. Marsh, assisted by Lieut. R. M. Ross. The retort building was designed chiefly by Mr. Marsh, assisted by Capt. Holt, Lieut. Ross, and others. The furnaces in the retort building were designed by special furnace engineers under Mr. Alfred Ernst with the assistance of Messrs. E. A. W. Jefferies and J. W. Loomis. The plans for the producer-gas equipment, flues, etc., for these furnaces were furnished by the Morgan Construction Co. The planning and engineering of the lixiviator building was done by Capt. Holt, assisted by J. W. Emig of the York Manufacturing Co., and other engineers of that company which supplied the apparatus.

The original intent was to obtain the nitrogen from the waste gases of the carbonating towers of the plant of the Mathieson Alkali Co., at Saltville, and the original contract provided for the nitrogen being furnished by that company. It early became evident, however, that the cost of a plant for producing this nitrogen would be excessive and that there was no definite assurance that nitrogen could be procured regularly. Accordingly, the plans for the production of pure nitrogen were radically changed and arrangements were made with the Air Reduction Co. of New York for erecting three Claude towers. These towers were built and installed.

The land within the plant survey comprised some 35 acres rented from the Mathieson Alkali Co. with the privilege of purchase. The plant required some 40 buildings of which 25 were for the men and officers and 15 for the plant proper. When the construction division finally turned the plant over as ready for operation on November 4, 1918, Maj. C. O. Brown was in local command, reporting to Dr. Parsons. Under him were 17 officers besides the enlisted men.

PLANT PUT IN OPERATION.

The few units that had been previously turned over to the operating department functioned well. The plant was running successfully. though not on full scale, when the armistice was signed on November 11. Orders were immediately issued by the Chief of Ordnance to slow down work, and thereafter for a few weeks one bank of furnaces was run to obtain data. A few changes in burner design and furnace construction were necessary, but these were minor and easily made.

There is no question whatever that the plant will produce sodium cyanide to the full capacity for which it was planned. On November 11 the plant was 190 men short in its personnel for full operation.

The total funds allotted for the plant amounted to $2,800,000 and approximately $2,500,000 was spent in construction and early operation.

PLANT TURNED OVER TO ORDNANCE DEPARTMENT.

On November 26, the Secretary of the Interior informed the Secretary of War that the plant had been tried out and was working properly, and that he was ready to turn it over to the Ordnance Department for operation. The Secretary of War then assigned the future control of the plant to the nitrate division and on December 21, 1918, the control of the plant and its personnel passed from the Bureau of Mines to the Ordnance Department.

In considering the success achieved one should remember that the Bucher process was new and had never been tried on an extensive manufacturing scale, although the experimental plant erected at Saltville gave much data for the construction of the larger plant. The plant as constructed will produce cyanide at a price much lower than it could have been purchased in the open market.

Special credit for the engineering is due to Capt. N. E. Holt, Mr. C. W. Marsh, and Lieut. R. M. Ross, and Maj. C. O. Brown. Credit for the plant construction is due Maj. R. A. Widdicombe of the construction quartermaster's office and his assistants, and Maj. J. R. Werth, supervising constructing quartermaster. Many important details of the early plant operations were worked out by other officers assigned to the work in Saltville. Every officer carried out faithfully and well the work assigned to him.

Dr. Parsons requests that special mention be made of the excellent work of Lieut. B. V. Reeves in charge of the retort building, Lieut. W. M. Bowman in charge of the lixiviator building; and Lieut. F. A. Vestal in charge of the mechanical preparation building. Capt. E. J. Mullen, detachment commander, a chemical engineer of experience who was especially detailed for the procurement and selection of the personnel, handled plant operations with ability. Maj. C. O. Brown, from the time he was appointed commanding officer and assigned to the Bureau of Mines, displayed intelligence and good judgment in all his work. The ability he showed in handling the plant and its personnel merits the highest commendation.

O

Bulletin 178C

DEPARTMENT OF THE INTERIOR
FRANKLIN K. LANE, Secretary

BUREAU OF MINES
VAN. H. MANNING, Director

PETROLEUM INVESTIGATIONS

AND

PRODUCTION OF HELIUM

ADVANCE CHAPTER FROM BULLETIN 178
WAR WORK OF THE BUREAU OF MINES

BY

VAN. H. MANNING

WASHINGTON
GOVERNMENT PRINTING OFFICE

The Bureau of Mines, in carrying out one of the provisions of its organic act—to disseminate information concerning investigations made—prints a limited free edition of each of its publications.

When this edition is exhausted, copies may be obtained at cost price only through the Superintendent of Documents, Government Printing Office, Washington, D. C.

The Superintendent of Documents *is not an official of the Bureau of Mines.* His is an entirely separate office and he should be addressed:

SUPERINTENDENT OF DOCUMENTS,
Government Printing Office,
Washington, D. C.

The general law under which publications are distributed prohibits the giving of more than one copy of a publication to one person. The price of this publication is —— cents.

First edition. June, 1919.

CONTENTS.

iii

PETROLEUM INVESTIGATIONS AND PRODUCTION OF HELIUM.

PETROLEUM INVESTIGATIONS.

The work of the petroleum division of the Bureau of Mines during the period of the war was under the supervision of Chester Naramore, chief petroleum technologist. In its petroleum investigations the bureau cooperated freely with other Government bureaus and acted as a source of information on petroleum matters.

COOPERATIVE WORK ON INTERALLIED PETROLEUM COMMISSION.

On July 15, 1918, Mr. Naramore and W. E. Perdew, chemical engineer of the bureau, left for London to act as representatives of the United States Fuel Administration and the United States Shipping Board on the Inter-allied Petroleum Commission. Numerous conferences were held with representatives from France, England, and Italy on specifications and requirements of the various allied Governments for petroleum products, and for this purpose meetings were held in London, Paris, and Rome.

EXAMINATION OF PETROLEUM FACILITIES IN FRANCE.

The storage facilities and the transportation system in France were studied, and recommendations based on the information gained were made to the director of the oil division of the Fuel Administration and to the chairman of the Shipping Board.

Refinery plants in France were inspected, as well as filling stations and shipping facilities from ports in France and Italy to the various fronts and the petroleum base of the American Expeditionary Force at Romorantin, France. At the request of Ambassador Sharp, a report was prepared on the advisability of developing the petroleum resources in France.

AIRPLANE FUELS.

Conferences were held with the various officials of the American Expeditionary Force regarding airplane fuels and lubricants. A large amount of data was collected on the tests conducted by the French and British Governments on different grades of gasoline and

63

on special blends for airplane fuel. Various French laboratories were inspected to ascertain the methods used in testing petroleum products for airplane engines.

STUDY OF OIL-SHALE INDUSTRY IN SCOTLAND.

The oil-shale industry in Scotland was studied with a view to determining the practicability of applying the methods used there to the treatment of oil shales in the United States.

SPECIFICATIONS FOR MOTOR FUELS.

The petroleum division assisted several branches of the United States Government in drawing up specifications for the purchase of various grades of motor fuel. Specifications were written for the Panama Canal and for both motor and airplane gasoline, which were adopted by the Panama Canal Commission.

Specifications for three grades of airplane gasoline for the use of the United States Government were adopted at a conference held May 30, 1918, at the office of Mark L. Requa, director of the oil division of the Fuel Administration. At this conference Mr. Naramore and Mr. Perdew represented the Bureau of Mines. Largely through the efforts of the petroleum division, the Atlantic Refining Co. agreed to make five drums of fighting gasoline, the first gasoline of that grade to be made in this country, for tests at the Bureau of Standards. Through its experience in making this sample, the Atlantic company was enabled to manufacture fighting gasoline on a large scale at a later date. This company is the only one in the United States that has attempted to make gasoline that would meet the specifications for the fighting grade.

In August, 1918, an Executive order authorized the establishment of a committee on the standardization of specifications for petroleum products. The duties of this committee were to standardize specifications for all petroleum products used by Government departments and to make suggestions on specifications for the allied governments. Shortly after the organization of this committee, a technical subcommittee was appointed to study existing specifications and to make recommendations to the general committee. The Bureau of Mines was represented on the general committee by C. H. Beal, petroleum technologist, and on the subcommittee by H. H. Hill, assistant chemist. Later Mr. Beal was transferred to San Francisco and Mr. Hill represented the bureau on both committees. Also Mr. Hill acted in an advisory capacity to the Inter-Allied Conference at its meetings, held in this country, on specifications.

The committee on specifications adopted specifications for three grades of aviation gasoline, one grade of motor gasoline, three grades of fuel oil, kerosene, long-time burning oil, signal oil, and "mineral seal oil," and is considering specifications for lubricants, particularly those used by the Railroad Administration.

The investigations relating to specifications for petroleum products were conducted by Messrs. Naramore, Perdew, Hill, and Beal.

INVESTIGATIONS OF AIRPLANE-MOTOR FUELS.

About August 2, 1917, the bureau arranged to cooperate with the Aviation Section of the Signal Corps in studying fuels for airplane engines. Information collected through visits to a number of aviation fields, by conferences with organizations testing airplane motors, and by correspondence indicated that no reliable data were available regarding the proper fuel for aviation motors. This work was done by E. W. Dean, petroleum chemist.

TESTS WITH DIFFERENT FUELS AND ENGINES.

The bureau arranged with the engineers at Langley field to make some actual flying tests to ascertain the advantages of certain grades of gasoline. Flying tests with Curtis training planes, indicated that plain "motor" gasoline gave as good results as a "high test" Pennsylvania product. The Italian fliers at Langley reported, however, that their engines tended to overheat when "motor" gasoline was used. The tests were made by C. Netzen, chemical engineer, and E. W. Dean.

After it was seen that the problem was what increase in efficiency would be possible through improved design if the fuel were better, an arrangement was made with the officials of the McCook aviation field and with the reseach organization of the Dayton-Wright Airplane Co. to cooperate in thorough tests of different engines and gasolines. This work included tests with a small air-cooled high-compression engine, tests with a single-cylinder high-compression Liberty engine, the preparation of special fuels (including cyclohexane), flying tests at the Dayton-Wright and Wilbur Wright fields with De Haviland airplanes equipped with Liberty motors, and tests in the altitude chamber of the Bureau of Standards.

Many interesting and important facts were learned. The general conclusion was that a mixture of 70 per cent cyclohexane and 30 per cent benzol (so-called "hecter") seemed the most desirable for use in fighting airplanes. This fuel permitted the use of engines having compression ratios as high as 7.5 to 1, which applied to the Liberty motor would develop perhaps 10 per cent more power.

The cessation of hostilities prevented the development of commercial cyclohexane plants. Flying tests at the Wilbur Wright field, completed after the signing of the armistice, showed that the use of

a high-compression motor with cyclohexane as fuel gave a standard
De Havilaad No. 4 airplane a full thousand feet extra of "ceiling"
(maximum flying altitude).

The investigation of aviation gasoline was conducted chiefly by
Clarence Netzen and J. P. Smootz, assistant petroleum technologists.

COOPERATIVE TESTS WITH BUREAU OF STANDARDS.

It was realized soon after this country entered the war that ordinary motor gasoline was not suitable for use in airplanes, particularly
at high altitudes. The Bureau of Standards constructed an altitude
chamber for testing airplane engines under conditions similar to those
encountered at various altitudes from sea level to 30,000 feet. Different
types of fuels were tested to determine which were most satisfactory
at high altitudes. The Bureau of Mines cooperated by collecting
samples of various grades of gasoline, special fuel mixtures, and
close-cut gasoline. In April and May, 1918, arrangements were made
with the Atlantic Refining Co. and the Union Oil Co. of California
to make special close-cut gasoline with a boiling range of 10° C. if
possible. A special fuel was made from absolute alcohol and gasoline,
the absolute alcohol and a blend of gasoline and benzol being prepared
in the Washington laboratory of the Bureau of Mines. All the fuels.
about 100 samples, tested by the Bureau of Standards in airplane engines were submitted to distillation tests in that laboratory. W. E.
Perdew collected the samples; C. P. Bowie and M. J. Gavin, assistant
refinery engineers, assisted in obtaining gasoline cuts from the Union
Oil Co. of California.

ADVICE ON GASOLINE PROBLEMS.

Considerable time was spent in advising the oil division of the
Fuel Administration on problems particularly relating to tests and
manufacture of gasoline. At the request of Mr. Requa, director of
the oil division, a report was prepared on the potential production of
airplane gasoline in the United States. This report, made by H. H.
Hill with the assistance of E. W. Dean and J. P. Smootz, showed
that the maximum estimated requirement could be met if the use
of gasoline for motor cars was restricted.

SUPPLEMENTARY INVESTIGATIONS IN CONNECTION WITH AIRPLANE FUELS.

In some airplane engine tests at the McCook field aluminum pis-
tons were seriously corroded. Investigation showed that the corrosion was not caused by impurities in the fuel and could be controlled by lubrication. The corrosion tests were made by Messrs.
Netzen and Smootz.

Two samples of German airplane gasoline, analyzed at the Pittsburgh laboratory of the bureau by H. H. Hill, proved to be petroleum gasoline, carefully refined.

Samples of gasoline from Borneo and from Sumatra were found to be suited for aviation fuel. The analyses were made by Mr. Hill and by N. A. C. Smith, assistant petroleum chemist.

INSPECTION OF OVERSEAS SHIPMENTS OF GASOLINE.

In the latter part of July, 1918, a report was received in this country that some gasoline shipped in May, 1918, had given trouble in airplane engines in France, and that the loss of a number of planes had been attributed to its use. In a letter dated August 1, 1918, Col. U. G. Lyons, of the Fuel and Forage Division of the War Department, requested the Bureau of Mines to furnish inspectors for testing all shipments of gasoline.

Details were arranged at a conference between Col. Lyons and H. H. Hill in New York, August 26, 1918. C. R. Bopp was assigned to the task and spent practically his entire time from August 25 to December 1, 1918, inspecting shipments of gasoline for service overseas, from the Atlantic Refining Co., Philadelphia; Gulf Refining Co., Philadelphia; and the Standard Oil Co. of New Jersey; and the Tidewater Oil Co., Bayonne, N. J. Mr. Bopp inspected shipments representing 8,000,000 gallons of motor gasoline, 2,600,000 gallons of export aviation gasoline, and 725,000 gallons of fighting gasoline. W. G. Hiatt inspected at Gulf ports 55,000 barrels of motor gasoline for the American Expeditionary Forces.

EFFICIENCY IN RECOVERY OF GASOLINE FROM NATURAL GAS.

The attention of oil men was called to the feasibility of treating natural gas of any gasoline content, however small, at any pressure by the absorption process. As tests of the residual gases from compression plants showed incomplete recovery of gasoline vapors, the bureau pointed out that the gasoline in the residual gas can be saved at small cost by an auxiliary absorption unit.

Through the bureau's efforts companies recovering gasoline from natural gas by compression have come to realize the waste of gasoline in the gases discharged from water-cooled coils, and have put in compression or absorption units to save this gasoline. The average loss in the gas being discharged from the plants examined was about 0.3 gallon per 1,000 cubic feet. In one plant the adoption of recommendations by bureau engineers effected a saving equivalent to $120 a day.

The investigations of absorption and compression plants were conducted by W. P. Dykema, assistant petroleum engineer, aided by R. O. Neal, assistant chemical engineer.

COLLECTION AND DISSEMINATION OF DATA ON THE REFINERY INDUSTRY.

In the fall of 1917 maps showing the oil fields of the United States, storage centers, refinery capacity, and consumption statistics were prepared for the Fuel Administration by H. F. Mason, petroleum economist.

During the summer of 1917 a census of the refineries of the United States was made in order to determine the status of the refining industry, and to furnish the War and Navy Departments with reliable information as to the consumption of crude oil and the output of gasoline, kerosene, and other products. Practically every refinery was visited. The figures thus obtained were grouped according to districts in a report compiled at the Washington office. Messrs. Bowie, Tough, Beal, Wadsworth, Netzen, Hill, Jacobs, and Wiggins were engaged in this work.

After the United States Fuel Administration was established, the petroleum division of the Bureau of Mines, by request, continued to compile monthly statistics of the refining industry. The work was in charge of Mr. Mason.

On account of the value of petroleum coke in the electrode business and in the production of caustic soda and chlorine, a census of the manufacturers and the production of that commodity was made by Mr. Mason in December, 1917.

On account of the possible shortage of ammonia during the summer of 1918, estimates were prepared by Mr. Mason, in May, 1918, for the Fuel Administration, as to the amount of ammonia the oil refineries would need for the year. Mr. Mason made a similar investigation of the sulphuric acid used in the oil-refining industry for the first seven months of 1918. Also, he prepared other special reports for the Fuel Administration.

In September, 1917, a report on the supply of fuel oil available for naval purposes was prepared for the Shipping Board by J. H. Wiggins. A report was prepared in cooperation with members of the United States Geological Survey for the Council of National Defense, giving a resumé of the condition of the petroleum industry during April and May, 1917, especially as regards the supply of and the demand for crude petroleum.

A map and report showing the gasoline storage facilities in the United States was prepared for the Signal Corps in August, 1918, by Mr. Wiggins. He also prepared a report for the War Department in February, 1918, giving comparative data on costs, labor, etc., for transporting gasoline by pipe line and tank car in France.

In March, 1918, a map showing the oil fields of the world and their production was prepared for the United States Fuel Administration by A. W. Ambrose.

In January, 1918, C. H. Beal and A. W. Ambrose, in cooperation with members of the Geological Survey, prepared for the Shipping Board a report dealing with the need for Mexican oil in this country.

ESTIMATING FUTURE PRODUCTION OF OIL AND GAS, AND APPLICATION TO WAR REVENUE TAXATION.

Three months were spent in cooperative work with the Bureau of Internal Revenue on methods for estimating depletion charges in determining excess profits and income taxes for oil and gas properties. Because of the difficulties involved, no satisfactory method had been worked out and no final settlement of taxes on oil and gas properties had been made for three years. The method adopted by the Bureau of Internal Revenue was based on the methods devised by C. H. Beal and J. O. Lewis, petroleum engineers of the Bureau of Mines. Mr. Lewis had charge of the compilation of curves and data for the Mid-Continent oil fields, and Mr. Beal for the fields of California. A. R. Elliott assisted in compiling curves for the report. The principles adopted should benefit the petroleum industry and enable the Treasury Department to compute depletion allowances by an easy and equitable method.

During 1918, reports were prepared from time to time for the Capital Issues Committee on the issuance of stock by oil companies. For many of the properties a detailed field investigation and a thorough search of all available literature were required. Messrs. Dykema, Beal, and Ambrose conducted this work.

ESTIMATE OF COST OF GAS FOR NAVAL ORDNANCE PLANT.

The Navy Department requested the United States Geological Survey to prepare a report on the amount of gas available for the armor-plate plant to be constructed at Charleston, W. Va., and asked the Bureau of Mines to estimate the cost of producing the gas and of piping. The request to the Bureau of Mines was made about November 10, 1917, by Rear Admiral Ralph Earle, chief of the Bureau of Ordnance of the Navy.

A detailed estimate was made of the cost of drilling, of laying pipe, and of installing a compression plant for gas from the properties described by the Geological Survey, the cost of the gas being based on the survey's estimate of the volume. The report was made December 18, 1918, by Messrs. Wiggins and Dykema.

In September and October, 1918, at the request of the Secretary of the Navy, through Rear Admiral Earle, a report was made by R. V. Mills, petroleum technologist, and A. R. Elliott, assistant engineer, on the gas property from which it was proposed to furnish gas to the naval ordnance plant near Charleston.

VALUATION OF OIL PROPERTIES IN THE NAVAL PETROLEUM RESERVES.

In a letter dated January 18, 1918, Hon. Scott Ferris, chairman of the House Public Lands Committee, requested Secretary Lane to supply certain information on the petroleum industry. One of the most important and difficult questions asked was the value of the oil properties in the naval reserves. The information was compiled during February, 1918. The appraisal of the lands was made by Messrs. Beal and Ambrose.

EFFICIENCY IN DEVELOPMENT OF CALIFORNIA OIL FIELDS.

At the request of the Fuel Administration an investigation was made to determine the possibility of maintaining and of increasing the production of oil in California by drilling promiscuously, as in the past, and by drilling at selected locations, without encroaching upon the naval petroleum reserves. At that time the outlook was that the production of oil in California would have to increase to meet the war demand, and the only practicable way of doing this was to drill in selected localities where the maximum amount of oil could be obtained with the least labor and cost.

The work was done during April and May, 1918, by Messrs. Beal and Ambrose, with the assistance of R. E. Collom, chief deputy State oil and gas inspector of the California State Mining Bureau.

CENSUS OF OIL-WELL CASING.

The War Trade Board requested the Bureau of Mines to furnish information as to whether exports of oil-well casing should be restricted. A census of the amount of casing in reserve in the different oil fields of the country was begun about December 20, 1918, and lasted until February 1, 1919. Oil-pipe companies gave information as to past and probable future demand, past and estimated future production, and amount on hand of all kinds of casing. A careful study of the foreign oil fields was included as it was necessary to provide pipe for those fields that were supplying petroleum to the allied nations. Mr. Ambrose was in charge of this work.

ECONOMY IN USE OF OIL FUELS.

Various problems in connection with oil-fuel economies were investigated for the western division of the Fuel Administration. A number of fuel-saving devices were examined, such as burners, carburetor attachments, gages for small oil tanks in buildings, and gas radiators. Twenty-five boiler plants about San Francisco Bay and 226 power plants in California using fuel oil were inspected and recommendations made that effected a saving of 5 to 25 per cent at each plant. Demonstrations were given at San Francisco of the use of powdered coal as a substitute for oil fuel. This work was done by C. P. Bowie of the bureau, who acted as technical adviser of the western division, and by J. M. Wadsworth, petroleum engineer of the bureau, who served as State administrative engineer for the California Fuel Administration.

The Bureau of Mines, in cooperation with the National Automobile Chamber of Commerce and the Council of National Defense distributed copies of a poster showing how gasoline is wasted and giving instructions for preventing waste.

TESTS OF FUEL OIL MIXED WITH POWDERED COAL.

From May 1 to June 1, 1918, cooperative tests of colloidal fuel oil were conducted for the Submarine Defense Association of America. The fuel was composed of Navy fuel oil so mixed with powdered coal as to retain the coal in suspension. The tests were made on the U. S. S. *Gem* by Messrs. Wadsworth and Perdew.

PROSPECTING FOR OIL IN THE BRITISH ISLES.

In cooperation with the British Government, through its agents, Pearson & Son, the Bureau of Mines selected drillers and drilling material for prospecting for oil in the British Isles. The cooperative work was conducted chiefly by C. H. Beal, A. W. Ambrose, V. L. Conaghan, and M. A. La Velle, Messrs. Conaghan and La Velle, expert drillers of the Bureau of Mines, leaving the service of the bureau to take charge of the drilling work.

UTILIZATION OF OIL-SHALES.

In July, 1917, W: E. Perdew, in a trip through the oil-shale fields of Colorado and Utah, collected information on the companies organized, the proposed methods of retorting, and details of some special types of retorts. Samples of shale from various districts in this country were tested for their oil content. Mr. Perdew also compiled a résumé of the oil-shale industry. The work on oil shale was continued by M. J. Gavin and W. D. Bonner. Under supervision of D. T. Day, consulting chemist of the bureau, a retorting plant is in process of construction at Elko, Nev. The funds for erecting the plant have been provided by the Southern Pacific Co.

PROTECTION OF OIL AND GAS IN MID-CONTINENT FIELDS.

Work to prevent waste of oil and gas became of vital importance in war time. Much effort was spent on introducing methods for protecting oil and gas sands by mudding and cementing, and in assisting operators in difficult drilling jobs. Particularly valuable results were obtained in the Cushing field, Oklahoma, where the production of oil was increased several thousands of barrels daily by excluding water from the wells with cement on the recommendations of A. A. Hammer and B. H. Scott. In the Butler County field, Kansas, where water threatened to cause much damage, the bureau's engineers rendered timely assistance. Also, an educational campaign was conducted to arouse producers and call their attention to methods for preventing oil and gas sands from being flooded. The use of cement to shut off bottom water materially increased the production of many oil wells in Kansas and Oklahoma. This work was conducted by A. A. Hammer, B. H. Scott, H. R. Reuch, J. O. Lewis, F. B. Tough, and Thomas Curtin.

A similar investigation of the water problems in the Illinois fields was conducted in cooperation with the State geological survey during six months, April to September, 1918. In a well repaired according to the recommendations of the bureau engineers, the flow of water decreased to less than one-third and the settled production of oil more than doubled. As a common valuation of settled production in Illinois is $2,000 a barrel of daily output, the value of this property was increased by about $20,000.

The work was done by F. B. Tough, assisted by Frank Madden. of the Bureau of Mines, and by M. L. Nebel, assisted by S. W. Williston, of the State geological survey.

EXAMINATION OF THE SALT CREEK FIELD, WYOMING.

In September, 1918, the Secretary of the Interior requested the Bureau of Mines to examine the Salt Creek oil field of Wyoming. F. B. Tough and B. H. Scott, of the bureau, spent a month in that field and prepared a preliminary report, which showed that present methods of drilling and producing are wasteful. They suggested means of checking the waste and recommended that steps be taken to safeguard the equity that the Government has in the field.

MISCELLANEOUS LABORATORY TESTS AND ANALYSES.

· Fuel oil used by the War Department at the Watertown Arsenal. Watervliet Arsenal, Fort Sill, and Panama is purchased on specifications. Samples of the consignments received are analyzed at the bureau's petroleum laboratory at Pittsburgh. During 1917 and 1918, samples representing purchases of more than $1,000,000 were analyzed.

At the request of the Ordnance Department of the Army, the Bureau of Mines inspected the wash oil used at certain toluene recovery plants. About 15 samples were examined, the work being done principally by N. L. Shoop, D. C. Dunn, and N. A. C. Smith.

While in London, Mr. Naramore was requested to obtain information concerning the available supply of American fuel oil that would meet British Admiralty specifications, which are more exacting than those of the American Navy. Tests conducted at the Pittsburgh petroleum laboratory showed that practically all distillates and some residuum fuel oils would meet the British viscosity requirement. The laboratory investigation was undertaken by J. P. Smootz and N. A. C. Smith.

Some experiments were conducted by H. H. Hill at the Pittsburgh petroleum laboratory for the American University station to ascer-· tain the best oil for saturating Army blankets to be used as screens to exclude poison gases from dugouts.

During the summer of 1917, 25 samples of emulsified lubricating oil from the Washington Navy Yard were analyzed and tested for viscosity and water content by J. P. Smootz, N. L. Shoop, and N. A. C. Smith.

Various preparations alleged to increase the efficiency of gasoline were tested and found to be worthless. This work was done chiefly by Clarence Netzen, H. H. Hill, and C. R. Bopp.

Various products that were claimed to be satisfactory substitutes for motor gasoline were analyzed and engine tests of some of them were made. The analyses were made by E. W. Dean, H. H. Hill, and C. R. Bopp.

Among devices tested by the bureau which had been offered the Government for the duration of the war were the Swan process for reclaiming waste crank-case oils, the Armstrong-Godward vaporizer, and a so-called motor perfector, the last two being designed to increase the efficiency of gasoline.

Mr. Perdew investigated the Swan process for the bureau and made a favorable report. The bureau recommended the adoption of the process to the Council of National Defense, and later to the French Scientific Commission during its visit to Washington. The process was put into use by the French Government.

The Armstrong-Godward device was designed to vaporize the gasoline-air mixture delivered by a metering carbureter. After mechanical developments had been in progress about a year the device was pronounced ready for final tests at the Bureau of Standards. These tests had not been completed in November, 1918. The investigation was conducted by C. Naramore, E. W. Dean, W. E. Perdew, and E. F. Hewitt, consulting engineer.

Tests made by W. A. Jacobs in August, 1917, at Waukesha, Wis., of the "motor perfector," a device for adding moisture to the gasoline-air mixture in a carbureter, indicated that it was no better than others already on the market.

CRACKING OF OILS.

Experiments in cracking oil from the Humble field, Texas, showed that the distillates were suitable for the manufacture of gasoline by cracking processes. The experiments were made in May and June, 1918, by W. A. Jacobs and D. C. Dunn.

In May, 1918, through an arrangement with the Gasoline Corporation, which controls the Greenstreet cracking process, six engineers of the petroleum division witnessed a series of test runs with the process at a plant in East St. Louis, Ill.

RIFLE CORROSION PROBLEMS.

At the suggestion of Gen. Ainsworth, United States Army, retired, the petroleum division of the bureau studied the causes of corrosion in Army rifles after being cleaned. A number of preparations designed to prevent corrosion with a single cleaning were collected and analyzed. Results of firing tests, made in cooperation with the Pittsburgh division of the Ordnance Office, have been negative, as no signs of corrosion have appeared. The rifles were stored in a dry room after cleaning, and failure to develop aftercorrosion may have been due to this cause. The investigation is being continued. The tests were made by H. H. Hill, W. A. Jacobs, and N. A. C. Smith.

TESTS OF THE DUNN TRENCH HEATER.

Tests for the War Department of the Dunn oil heater showed that the heater had some advantages over the regulation Army stove for heating tents. C. R. Bopp supervised the tests, which were made at the War College Barracks, Washington, D. C., June 26 and 27, 1918.

PRODUCTION OF HELIUM FOR USE IN AIRSHIPS.

One of the great scientific and technical triumphs resulting from the application of exact knowledge and inventive genius to problems of military importance has been the large-scale production of helium, making this hitherto exceedingly rare gas available for use in balloons and dirigibles. The following account of the bureau's work in connection with helium was prepared by Dr. Andrew Stewart.

DISCOVERY OF HELIUM.

During the solar eclipse of August 18, 1868, P. J. C. Janssen noted in the solar chromosphere a bright yellow line near to but not identical with the D_1 and D_2 lines of sodium. Frankland and Lockyer called the line D_3 and, as it was referable to no known terrestrial substance, ascribed it to a hypothetical element which they called *helium*, from the Greek word *helios*, the sun.

In 1889, Dr. W. F. Hillebrand, of the United States Geological Survey, reported in the American Journal of Science that in analytical work on uraninite, which he began in 1888, he had obtained up to $2\frac{1}{2}$. per cent by weight of a peculiar gas by treating the mineral with nonoxidizing inorganic acids. This gas had the characteristics of nitrogen, yet was different from any nitrogen known, and was thought by Hillebrand, on purely negative grounds, to be some allotropic form of that element. Lack of proper equipment, however, precluded any further researches on the subject. As a matter of fact, a certain percentage of the gas so found by Hillebrand was nitrogen.

The attention of Sir William Ramsay, the distinguished British chemist, was later drawn to Hillebrand's work, and he repeated the latter's experiments. He separated from cleveite, a variety of uraninite, a gas identical with that of Hillebrand, in the spectrum of which was found, with Sir William Crookes' aid, the D_3 line of Janssen, proving that in the gas there was present a new element, to which the name "helium," borrowed from Frankland and Lockyer, was given. On sparking this gas with oxygen to eliminate nitrogen, removing the excess oxygen with phosphorus, introducing the residual gas into an exhausted spectrum tube and passing an electric current through it, a bright yellow glow was obtained. Examined with a spectroscope, this glow gave rise to a series of lines in the red, green, and blue, but most prominent was the bright yellow line of

75

Janssen. In 1895 the characteristic line of this gas was discovered in the spectrum of the atmosphere by Kayser, and the element was isolated from the air by Ramsay and Travers in the same year.

OCCURRENCE AND PROPERTIES.

Helium is widely distributed in nature, but generally in minute quantities. The amount of it in the earth's atmosphere is exceedingly small, being present in the proportion of only 1 volume to 250,000 volumes of air. It has been found, as mentioned above, imprisoned within radioactive mineral substances, in the gases evolved from some thermal springs, in those emanating from volcanoes, and also in the natural gas of several gas fields in the United States (notably in Kansas, Oklahoma, Ohio, and also Texas). In some of the natural gas of this country helium is present in amounts as high as 2½ per cent. Of late, emissions of helium in comparatively large quantity have been reported from the boric acid soffione of Larderello, Tuscany, Italy. It issues from the earth at Stassfurt, Germany, and at Karlsbad. The natural gas in certain parts of Europe also contains helium, but in much smaller quantity than in our own.

Helium is one of a series of very rare, inert gases—other members being neon, argon, krypton and xenon—all of which are present in very small quantities in the atmosphere. Thus far every attempt to make these gases enter into chemical combination has failed. Helium is, next to hydrogen, the lightest of known substances. The gas although twice as heavy as hydrogen, has in balloons a buoyancy or ascensional power of 92.6 per cent, as compared with the latter. The reason for this is that the buoyancy of a gas is measured not directly by its weight, but by the difference between its weight and that of the air displaced. Both hydrogen and helium are so light as compared with air that the difference in their lifting power is relatively insignificant. For example, 1,000 cubic feet of pure hydrogen will lift a weight of 75.14 pounds whereas the same amount of pure helium will lift 69.58 pounds, and a mixture, containing 85 per cent helium and 15 per cent hydrogen, which is the gas contemplated for all-around aircraft use, will have 93.4 per cent of the lifting power of hydrogen and will lift 70.18 pounds.

In its physical behavior, helium is the nearest approximation to the ideal perfect gas. It is monatomic and liquefies at even lower temperatures than hydrogen. By its evaporation in vacuum, Onnes, of Leyden, produced it in liquid form, first cooling it in solid hydrogen and then expanding it through a small nozzle. Liquefied helium is a colorless, mobile liquid, having a density of 0.122; hence it is the lightest liquid known. It boils at 4.5° absolute, and has a critical temperature of about 5° absolute, with a critical pressure of 2.75 atmospheres. By the rapid evaporation of liquid helium, a temperature below 2.5° absolute has been reached, but solid helium has not been produced. The low dielectric strength of helium permits electric discharges to pass through it with much greater ease than through most gases, and its conductivity for heat is very high.

Helium has been proved to be the end product of the emanations of radioactive substances, but the origin of its presence in natural gas has not been established. The helium content of some natural gas is relatively so large and the radiation from radioactive material is so small that derivation from such material would seem entirely out of the question. In the whole world not more than three ounces of radium have yet been isolated; but it must be remembered that radioactive material is very widely distributed in nature and that helium generation has been going on through countless ages of past time. A great many factors enter into consideration in this question, however, and it is difficult to set up an unassailable hypothesis regarding it.

Helium is of prime importance in aeronautics because of its great buoyancy, because of its rate of diffusion and consequent wastage through fabrics being only half that of hydrogen, and, above all, because of its chemical inertness. The effectiveness of hydrogen-filled dirigibles in war, owing to the high inflammability of hydrogen, was reduced to the vanishing point when means of combating them with incendiary propectiles were developed. Even under peace conditions, the great hydrogen-filled envelope of the dirigible constitutes a serious hazard because of possible ignition from atmospheric electricity, or from flames originating in the power plant of the craft. Helium, however, being absolutely inert, can not be ignited or exploded, and even mixtures containing certain amounts of hydrogen with helium can, as has been indicated above, be used with perfect safety in lighter-than-air craft.

INITIATION OF THE RESEARCH WORK.

Up to April, 1918, helium had been obtained only in extremely small amounts, as a curiosity in scientific laboratories, so that the total quantity separated in the whole world probably did not exceed 100 cubic feet, the cost of production being about $1,700 to $2,000 a cubic foot. Hence, although the possibility of helium as a lifting gas had been realized, its practical use seemed out of the question.

In the development of a process by which helium is now being produced in large quantities, the Bureau of Mines was the pioneer and has taken the leading part. In March, 1916, the Director of the

Bureau of Mines called the attention of Dr. F. G. Cottrell, chief metallurgist of the bureau, to a new process for air separation that embodied some novel and striking features, the bureau being interested then in the cheap production of oxygen for use in blast furnaces. This process had been evolved by Fred E. Norton, a graduate of the Massachusetts Institute of Technology and an engineer of wide experience and international reputation; Mr. Norton had pooled his patents with E. A. W. Jefferies and the consolidation became known as the Jefferies-Norton processes.

On February 28, 1915, Sir William Ramsay wrote to Dr. R. B. Moore, of the Bureau of Mines, who had worked with him in investigating the rare gases of the atmosphere. Sir William said, among other things: "I have investigated blowers—that is, coal-damp rush of gas—for helium for our Government. There does not appear to be anything in the English blowers, but I am getting samples from Canada and the States. The idea is to use helium for airships."

As the United States was not yet in the war and strict neutrality was being maintained by this country, these remarks were passed over by Dr. Moore, but he remembered that in 1907 Dr. H. P. Cady, of the University of Kansas, had found more than 1 per cent of helium in some natural gas from Kansas, and he realized what it would mean if such helium could be made available for balloons and dirigibles. Dr. Cady and D. F. McFarland were the first investigators to discover helium in natural gas; in 1907 they published a paper on the subject in the Journal of the American Chemical Society.[a]

In April, 1917, Dr. Moore attended a meeting of the American Chemical Society in Kansas City, Mo., where a paper on rare gases was read by Dr. C. W. Seibel, of the University of Kansas, who had worked under Dr. Cady. In the discussion that followed Dr. Moore pointed out the possibility of helium for war-balloon use if its separation from natural gas could be accomplished at a reasonable cost. He asked Dr. C. L. Parsons, chief chemist of the Bureau of Mines, who was present at the meeting, to take up the matter with bureau officials on his return to Washington, and this Dr. Parsons did.

In April, 1917, the Director of the Bureau of Mines placed G. A. Burrell (later Col. Burrell) in charge of the bureau's war-gas work. The Bureau of Mines had initiated this work, chiefly on gas masks, smoke screens, and toxic gases, in cooperation with the Army and the Navy. Mr. Burrell, who had been in charge of gas investigations for the Bureau of Mines from 1908 until 1916, had analyzed natural gas from different fields and knew of its helium content. He had thought of helium as a possible filler for balloons and dirigibles and

[a] Cady, H. P., and McFarland, D. F., the occurrence of helium in natural gas and the composition of natural gas: Jour. Am. Chem. Soc., vol. 29, 1907, pp. 1523 et seq.

had mentioned the matter to F. A. Lidbury, of the Oldbury Chemical Co., and from him learned of Sir William Ramsay's interest in the subject and of the presence of helium-bearing gas in Canada. Prof. Satterly and Prof. Patterson, of the University of Toronto, began experiments looking to the separation of helium from Canadian natural gas January 1, 1916. They used a Claude-system unit, at Hamilton, Ontario, but had much trouble in eliminating the heavier hydrocarbons of the natural gas.

Mr. Burrell was aware of the possibilities of the gases of the Petrolia field, in Texas, in regard to helium yield, and had samples sent to Dr. Cady for analysis. These samples contained about 1 per cent of the element. Thereupon, Mr. Burrell took up the question of utilizing this gas for helium recovery, with Dr. Cottrell and Dr. Moore, and the matter was discussed at length.

On May 12, 1917, Mr. Burrell wrote to Maj. C. DeF. Chandler (now Col. Chandler, of the Air Service), asking for an expression of opinion as to whether helium possessed sufficient advantages for use in balloons to warrant undertaking its production; and Messrs. Burrell and Moore, at the latter's suggestion, conferred with Maj. Chandler to ascertain whether the Army would consider furnishing the necessary funds for the project. G. O. Carter, of the Bureau of Steam Engineering of the Navy Department, learned from Maj. Chandler of this matter and became interested in it on behalf of the Navy Department.

At this time several conferences were held to discuss helium production from every angle, and opinions and views, all favoring it, were obtained from Dr. Cady, T. B. Ford, then in charge of the low-temperature laboratory of the Bureau of Standards, and O. P. Hood, chief mechanical engineer of the Bureau of Mines.

Dr. Cottrell advised Mr. Burrell to get in touch with Mr. Norton, for the possible employment of his process to separate helium from natural gas; accordingly, Mr. Norton was asked to come to Washington, where he arrived on June 4, 1917. Asked for a minimum estimate on the cost of a small experimental plant for the purpose, Mr. Norton estimated that a plant with a capacity of about 5,000 cubic feet of helium a day, to try out his process, might be constructed for about $28,000. With this data in hand, the Director of the Bureau of Mines did his utmost to push the investigation. On July 19, 1917, he addressed a letter to the Chief Signal Officer of the War Department, requesting an allotment of $28,000 to try out the Norton process, and on July 20, 1917, Secretary Lane wrote to the Secretary of War on the same subject.

Dr. Moore had suggested to Dr. Cottrell that the Linde Air Products Co. and the Air Reduction Co., established firms in the business of gas liquefaction and separation, be brought into the helium undertaking, and in this suggestion Dr. Cottrell had heartily concurred. So these firms and one other (the Lacy process) were asked to cooperate with the Government, and developments finally culminated in a recommendation on July 26, 1917, by the joint Army and Navy Airship Board, of an allotment of $100,000, in equal shares from Navy and Army appropriations, to make helium available as a substitute for hydrogen for balloons and dirigibles. It was further recommended that arrangements be made with the Bureau of Mines to carry out the undertaking. On July 31, 1917, the Aircraft Production Board recommended that this be done.

About this time Capt. R. B. Owens, of the Signal Corps, went to England on a special mission. He had become very enthusiastic as regards helium, especially the possibilities of the Norton process, and took with him a letter from Mr. Burrell, which he presented to the British Admiralty. This led to the dispatch to this country of Commander C. D. C. Bridge and Lieut. Commander S. R. Locock, as a commission on behalf of the Admiralty to canvass the helium situation in this country and to investigate the Norton process. Through Vice Admiral Richard H. Peirse the Admiralty had written to Dr. R. A. Millikan (later Lieut. Col. Millikan), of the National Research Council, on July 25, 1917, that, on the strength of researches instituted by the Board of Inventions and Research, it had decided to fit up an experimental plant to investigate whether the use of helium for airships was practicable. The Admiralty was anxious to obtain from the United States 100,000,000 cubic feet of the gas at once and to contract for a further supply of 1,000,000 cubic feet a week, being particularly interested in the Norton process on account of the possibility of its greatly reducing the cost of producing helium, which the Admiralty feared would be prohibitive by the older, established processes. It was estimated that the best that could be expected from these older processes would be about $80 a thousand cubic feet.

DEVELOPMENT OF PLANTS FOR HELIUM PRODUCTION.

After the $100,000, mentioned previously, was allotted for helium work, the director of the Bureau of Mines placed Mr. Burrell in charge of the investigation. Mr. Norton started the design of a plant along the lines of his process, he and Mr. P. McD. Biddison, chief engineer of the Ohio Fuel & Supply Co., being appointed consulting engineers of the bureau. With others they made a field survey to determine the best site for operations. Dr. Cady was appointed consulting chemist on the staff of the Bureau of Mines; he and his assistants at the University of Kansas, particularly Dr. C. W. Seibel, did a large amount of analytical and research experimental work on helium, the results of which influenced the designs of the various helium plants that were constructed later. Dr. Cady also made valuable researches into the limits of inflammability of helium and hydrogen mixtures, and carried out a number of experiments on the permeability of balloon fabrics by helium.

The work in connection with the helium undertaking began at this time to assume such proportions and to broaden so much that Mr. Burrell felt that he would be unable to carry it on properly, his time being fully occupied with other pressing duties. Therefore, Prof. W. H. Walker, of the Massachusetts Institute of Technology, was asked to assume charge. This he did for a few weeks, and possibly it was at his suggestion that, for purposes of secrecy as a war measure, the name " argon " was substituted for helium and the three experimental plants finally constructed became known as " argon " plants. After Prof. Walker gave up this charge to become chief of the Chemical Service Section of the War Department, with the rank of lieutenant colonel, the helium work was carried on mainly by Dr. F. G. Cottrell in Washington and Mr. Norton and Mr. Biddison in the field, all under Dr. Manning's personal supervision, with Mr. Carter acting for the Navy and the Army, and the Linde Air Products Co. and the Air Reduction Co. cooperating so far as their processes were concerned. Contracts were closed with these two companies in November, 1917, and a contract was made with the Lone Star Gas Co., controlling the Petrolia field, for a supply of gas. Construction of a Linde plant and a Claude (Air Reduction Co.) plant was begun at Fort Worth, Tex., almost immediately after the contracts were executed. These plants, known respectively as "Argon Plants 1 and 2," were completed in March and in May, 1918. On May 11, 1918, Profs. Satterly and Patterson visited Fort Worth to inspect the plants and expressed themselves as highly pleased with what they saw. On February 2, 1918, Dr. Manning assigned F. C. Czarnecki to Fort Worth as superintendent of the Linde and Claude plants; also, about the same time Mr. J. R. George, jr., was assigned to assist Dr. Cottrell in Washington in conducting the helium work, with special reference to the Norton process. On January 1, 1919, Dr. Andrew Stewart succeeded Mr. George. In the spring of 1919 Dr. L. H. Duschak was for some time associated with the helium project.

In June, 1918, the director of the Bureau of Mines placed Dr. R. B. Moore in general charge of all three helium plants, as his personal representative. After visiting New York for consultation with the presidents of the Linde and the Air Reduction Cos. and to arrange for even closer cooperation of those corporations with the Government, Dr. Moore went to Worcester, where he conferred with Mr. Norton, and then went directly to Fort Worth. He has been actively connected with the helium work up to date, June, 1919.

The coming to this country of the British commission, in the fall of 1917, stimulated the interest of the Army and the Navy in helium. At a conference called by the director of the Bureau of Mines in his office, those present, besides Dr. Manning, were Commander Bridge. Lieut. Commander Locock, of the British commission; Prof. John Satterly, of the University of Toronto; Lieut. Commander Arthur H. Marks, Bureau of Construction and Repair, United States Navy; Capt. P. Pleiss, of the Signal Corps, United States Army; G. O. Carter, Bureau of Steam Engineering, United States Navy; Dr. W. H. Walker, Dr. C. L. Parsons, G. A. Burrell, P. McD. Biddison, F. E. Norton, Dr. Marston T. Bogert, National Research Council; W. W. Birge, president of the Air Reduction Co.; and Dr. B. S. Lacy. The helium project was discussed exhaustively and, as a result, recommendation was made that an appropriation of $500,000 be made available from Army and Navy funds, in equal shares, for four separate experimental plants, under the systems of Linde, Claude, Norton, and Lacy, respectively; that is to say, all systems recognized as practicable were to be given a trial. Later, the Lacy system was dropped. The recommendation was transmitted to the proper authorities of the Army and the Navy.

The $500,000 allotment was appropriated pursuant to a resolution of the Aircraft Production Board of October 17, 1917, but, on the recommendation of the Navy Department, the Norton process was excluded from the benefits of any share thereof. This unexpected action was embarassing both to the Bureau of Mines and to the British commission, but the Navy Department adhered to it and induced the War Department to take the same stand. Not until. at the Navy Department's insistence, the Norton process had been submitted to the searching examination of the National Research Council and that body had, by the unanimous action of a special committee, reported favorably on it January 14, 1918, was a further sum of $100,000 made available for the Norton process, on January 18, 1918. The special committee of the National Research Council was composed of Dr. Harvey N. Davis, of Harvard University; Dr. Edgar Buckingham and Dr. C. W. Waidner, of the Bureau of Standards; Dr. W. S. Landis, of the Air Nitrates Corporation; and S. L. G. Knox, consulting mechanical engineer, and later scientific attaché of the United States Embassy at Rome.

The Norton process is the latest practicable development in liquefying and separating gases. The Linde process (Plant No. 1) depends upon the so-called Joule-Thomson effect. obtained by the sudden expansion of a highly compressed gas through a small orifice, or nozzle, and the consequent cooling of the gas, the process being elaborated into a self-intensive or cumulative cycle of heat interchange by causing the cooled gas, on escaping, to circulate around the tube leading the initial gas into the apparatus.

George Claude, of Paris, conceived the idea of a liquefaction cycle with an expansion engine interpolated. Although the Joule-Thomson effect is used in the Claude cycle (Plant No. 2), its value is reduced to a minimum because the compression of the gas in this system is lowered. The maximum cooling effect is produced by the expansion engine because the compressed gas, on expanding in the engine cylinder, is made to do work and thus its temperature is lowered.

In the Norton process (Plant No. 3) three expansion engines are used, liquid is throttled, and the heat interchanger and fractionating still are of new design. In the Linde system an enormous expenditure of power is demanded to compress the gas in order to obtain the maximum effect of throttling, and this energy is then wasted. The Claude system requires much less compression power, but in this system the energy stored in the compressed gas is also dissipated. In the Norton system the requirement for gas compression is reduced to a minimum by the interpolation of the multiple-expansion engines, and what is needed is conserved and reapplied through the energy developed by these engines. Thus the maximum cooling effect is obtained at a minimum cost.

Construction of the Bureau of Mines (Norton process) plant, later known as Argon Plant No. 3, on the enlarged basis of a maximum production of 30,000 cubic feet of helium a day, was begun on April 3, 1918, and was completed October 1, 1918. The cost was $148,398.29, the estimate for the plant enlarged to the proportions given having been $150,000. During the early construction work at this plant Mr. Biddison's services were of especial value; succeeding him, April 20, 1918, George A. Orrok was appointed consulting engineer of the Bureau of Mines and has since been identified in this capacity with Plant 3.

The first site picked for the helium plants was at Fort Worth, Tex., selected by Messrs. Norton and Biddison, but later the Otto, Kans., region was deemed more advisable, as the natural gas there is exceptionally rich in helium; the supply, however, proved inadequate. The Bureau of Mines plant (Norton process) was then located at Petrolia, Tex., adjacent, for obvious reasons, to the properties of the Lone Star Co., but the plants of the Linde Co. and the Air Reduction Co., because of their requiring more water and more power, were built at Fort Worth, about 100 miles to the southeast, and were supplied with Petrolia gas through the Lone Star Co.'s pipe line.

The approximate cost of Plant No. 1 (Linde) was $245,000; that of Plant No. 2 (Air Reduction) was $135,000. The capacity of the first plant for producing 90 per cent helium from 0.4 per cent to 1 per cent helium-bearing natural gas, was 5,000 cubic feet a day, and that of the second was 3,000 cubic feet a day.

Plant No. 1 was the first to start, March 6, 1918. Its production of helium began April 8, 1918, when 27 per cent gas was obtained. Progressively better results were achieved until a purity of about 70 per cent on straight runs was reached. By reprocessing, the purity of this gas was raised to 92¼ per cent. Plant No. 2 began to operate May 1, 1918, and on May 13 finally produced gas of 62 per cent to 70 per cent purity, which was reprocessed by Plant No. 1 to 92¼ per cent. In all, about 200,000 cubic feet of 92¼ per cent helium has been produced by Plants 1 and 2. This gas has been stored in steel cylinders at a pressure of 2,000 pounds to the square inch, each cylinder containing about 200 cubic feet expanded to atmospheric pressure. Of these cylinders 750 were ordered sent to France for aeronautic purposes and were on the dock at New Orleans awaiting shipment when the armistice was signed. They were returned to Fort Worth.

In order to coordinate properly all the different agencies concerned in the helium project, and to take steps for controlling effectively the exploration and conservation of helium, a committee consisting of one representative from each of the three governmental departments chiefly concerned was appointed by resolution of the Aircraft Board on August 23, 1918. G. O. Carter, chairman, represented the Navy; Dr. Harvey N. Davis, the Army; and George A. Orrok, the Department of the Interior. Up to and including October 23, 1918, there had been recommended and allotted from Army and Navy appropriations, in equal share, funds aggregating $1,090,000 for Plants 1, 2, and 3 and expenses incident to their maintenance and operation. The expenditure of this $1,090,000 has not only added vastly to human knowledge, but has supplied a commodity worth, at prewar prices, approximately from $250,000,000 to $400,000,000; and has prepared a war weapon for the United States of incalculable potency.

Construction of all three helium plants was carried out by the Constructing Quartermaster Corps of the Army, cooperating with the Bureau of Mines, and the finances of the helium undertaking have been handled by the finance division of the Bureau of Aircraft Production and the accounting section of the Bureau of Mines.

In August, 1918, the War and Navy Departments, realizing the paramount importance of insuring a supply of helium for military purposes, determined to build a large plant that would produce 30,000 cubic feet of the gas a day. This plant, to be constructed along the lines of the Linde process, is to be situated at Fort Worth, Tex., and arrangements have been made for a lease of the Petrolia gas pool and for a Government pipe line to convey gas from Petrolia to Fort Worth.

On December 8, 1918, the Aircraft Board submitted a report to the Secretary of War and the Secretary of the Navy, in which the change in the military situation because of the signing of the armistice was noted, the helium situation was reviewed, and recommendations for the future were made. Four different programs for further work on helium were submitted, that designated. as "Plan C" being preferred. This was as follows:

Operation of Plant No. 3. for three months	$86,000
Construction of production Plant No. 1	1,700,000
Operation of production Plant No. 1 for 8 months, producing 7,200,000 cubic feet helium	750,000
Pipe-line construction	1,800,000
Petrolia field lease	1,500,000
Total expenditures	5,786,000
Salvage	500,000
Net cost	5,286,000
Helium production, cubic feet	7,200,000

On December 17, 1918, the Director of the Bureau of Mines, in a letter to the Secretary of the Navy and the Secretary of War, submitted a report by the engineers of the bureau. This report outlined a plan similar to "Plan C," but provided for further operation of all three experimental plants, also for a new Claude unit, and included a fund for conservation purposes. The bureau's engineers were convinced that the cost of producing helium had not been reduced to a minimum and that undoubtedly in the long run money would be saved by perfecting producing methods through further experimental work. In any event, it was felt that something should be done toward conserving the Nation's helium supply.

On December 27, 1918, the Secretary of the Navy wrote to the Director of the Bureau of Mines disapproving further experimental work by the Government, other than the small amount countenanced for Plant No. 3, under "Plan C," and advising him of the adoption of "Plan C." On January 9, 1919, Maj. Gen. Jervey, of the General Staff of the Army, wrote advising that the War Department concurred in the action of the Secretary of the Navy and that the accomplishment of "Plan C" would be entrusted to the Navy Department.

Pursuant to this plan, the director of the Bureau of Mines ordered Plants 1 and 2 shut down on January 23, 1919, but asked that Plant 2 be permitted to continue work, at its own expense, for a limited time in order to try out an improvement in its apparatus. On this basis, the Navy and War Departments consented to the further operation of Plant No. 2, the Navy agreeing to furnish gas of a higher helium content from its new pipe line (contemplated under " Plan C ") when this would be completed, about May, 1919.

The Bureau of Yards and Docks of the Navy Department took physical possession of Plant No. 2 and of the Government property of Plant No. 1, and assumed all expenses incurred by the Government incident to these plants as of April 1, 1919. Plant 1 has been dismantled and the Linde Air Products Co. is to repurchase such of its apparatus and equipment as it agreed to do under its contract with the Bureau of Mines. All helium produced at these plants has been placed in the custody of the Navy Department at its special request.

Although only $36,000 was provided for further operation of Plant No. 3, under " Plan C," a visit of inspection to this plant by Commanders H. N. Jenson and H. T. Dyer, Lieut. Commander Smith and G. O. Carter, on behalf of the Navy Department, about March 15, 1919, so convinced the Navy officials of the exceptional merits of the Norton process and the possibilities of the plant in respect to producing high-purity helium at a minimum cost that they became strongly in favor of affording it every opportunity to prove itself.

After overcoming many obstacles that could not be foreseen, Plant No. 3, on April 2, 1919, began to produce helium, and by 5.30 a. m. April 3, a purity of 19.8 per cent was reached. Helium production continued until that afternoon, when it became necessary to shut down for some minor repairs. On April 17, 21 per cent helium was made, and it is confidently expected that helium of the highest purity will soon be produced by this plant on a large scale.

On April 15, 1919, the Director of the Bureau of Mines called a conference in his office to discuss matters incident to the further operation of Plant 3 and kindred matters. There were present, besides the Director, Commander Jenson, representing the Navy; Col. C. DeF. Chandler, Col. A. L. Fuller, and Dr. H. N. Davis, representing the Army; and George A. Orrok, Fred E. Norton, L. H. Duschak, W. A. Ambrose, and Andrew Stewart, representatives of the Bureau of Mines. Fred E. Norton, who had come from Texas for the purpose, gave a detailed report on the operation of Plant No. 3 and described the need for alterations in and additions to its apparatus to insure the greatest possible efficiency. It was the sense of the meeting that liberal allotments of funds should be made for the continued experimental operation of the plant, with regard to its ultimately being run as a production unit. As a result of this conference, the Director of the Bureau of Mines recommended an allotment of $100,000 to make the alterations and acquire the equipment referred to. This fund has been made available, one-half by the Army and one-half by the Navy.

COOPERATING AGENCIES.

During the course of the work of the Bureau of Mines on processes of helium extraction, G. S. Rogers, of the United States Geological Survey, undertook a reconnoissance of the natural-gas fields in the United States with regard to their possible helium supply, so far as that might be judged by the analysis of samples of gas from existing wells and the study of geological conditions. In this work Dr. Cady and Dr. C. W. Seibel rendered valuable assistance in analytical investigations.

On April 3, 1918, the petroleum division of the Bureau of Mines began work on the possible discovery of natural gas yielding helium, and from April 3, 1918, to January 1, 1919 conducted intermittently administrative work relating to the conservation of such natural gas. Examinations as to the possibility of obtaining helium in Montana, North Dakota, Washington, California, and southern Canada were made by C. H. Beal, J. O. Lewis, and A. W. Ambrose.

Following a conference between Dr. Cottrell and members of the staff of the Bureau of Standards, the Director of the Bureau of Mines, on December 14, 1917, requested the Bureau of Standards to undertake the determination of certain physical properties of methane, especially with regard to its latent and specific heats and specific volumes over a wide range of pressures and temperatures. The purpose of this inquiry was to facilitate the experimental operation of the helium plants, in which liquid methane plays an important part.

Bulletin 178–D

DEPARTMENT OF THE INTERIOR

FRANKLIN K. LANE, Secretary

BUREAU OF MINES

VAN. H. MANNING, Director

EXPLOSIVES AND MISCELLANEOUS INVESTIGATIONS

EXPLOSIVES RESEARCH, REGULATION OF EXPLOSIVES
AND PLATINUM, MARINE-BOILER TESTS, UNDERGROUND
SOUND RANGING, TRAINING IN FIRST AID AND RESCUE
WORK, CENSUS OF MINING ENGINEERS AND CHEMISTS,
PREPARATION OF ALLOY STEELS, LIGHTING
AVIATION FIELDS, COOPERATION WITH
CAPITAL ISSUES COMMITTEE

––––––

Advance chapter from Bulletin 178
War Work of the Bureau of Mines

WASHINGTON
GOVERNMENT PRINTING OFFICE
1919

CONTENTS.

EXPLOSIVES AND MISCELLANEOUS INVESTIGATIONS.

EXPLOSIVES RESEARCH.

As the physical laboratories of the explosives section of the Pittsburgh experiment station were equipped for testing the physical properties of explosives, both the Army and Navy Departments for many years had requested the Bureau of Mines to make such tests.

Chemical control of the physical tests through analyses, heat tests, analyses of gaseous products of combustion, and sand tests was exercised through the explosives chemical laboratory.

Often an investigation was cooperative between two laboratories. The explosives chemical laboratory also aided the work of the Explosives Regulation both by chemical tests and investigation of accidents in explosives plants.

COOPERATIVE WORK FOR ARMY AND NAVY DEPARTMENTS.

The explosives chemical laboratory devised, at the request of the Bureau of Ordnance of the Navy, a more economical method of making hexanitrodiphenylamine. This problem was assigned to the Bureau of Mines by Dr. C. E. Munroe, chairman of the committee on explosives of the National Research Council, acting as the department of science and research of the Council of National Defense.

Before the United States entered the war the following reports of investigations were submitted to the Army or the Navy:

1. Report of sensitiveness and relative efficiency of TNT and ammonium picrate, dated February 6, 1913, undertaken at the request of the Chief of Ordnance.

2. Report of tests and examination of two samples of TNT and one of TNA for the Ordnance Department of the Army, dated August 5, 1915, was undertaken at the request of commanding officer, Frankford Arsenal.

3. Report of tests for sensitiveness to detonation of mercury fulminate of 21 samples of TNT crystallized from various solvents, for the Bureau of Ordnance, Navy Department, dated June 17, 1916.

4. Report of tests of two samples of mercury fulminate for Frankford Arsenal, dated December 2, 1916, undertaken at the request of the commanding officer, Frankford Arsenal.

After the United States entered the war the following reports of investigations were submitted:

5. Report of pendulum friction and large impact tests on two samples of " sodatol " and two samples of " amatol." The final report was dated July 20, 1918; and the preliminary report, May 15, 1918. This investigation was undertaken at the request of the Ordnance

Department, War Department, to determine the sensitiveness to shock and friction of sodatol and amatol at high temperatures.

6. Fifteen reports of tests of explosives for the United States Engineers, to determine the relative strength and efficiency of eight samples of explosives for use in military mining and demolition operations.

The data obtained from these tests will be used for preparing a manual on demolition work and explosives for officers and noncommissioned officers of the United States Engineers, on which the captain assigned to aid in the tests is now working.

7. Seventeen reports of tests on explosives for the Ordnance Department.

This investigation was undertaken at the request of Engineering Explosives, Ordnance Department of the Army, to determine the physical characteristics, relative efficiency, and strength under varying conditions of 13 explosives used as bursting charges and boosters. In addition, special minor investigations were carried out.

The results of the completed work will be used as the " Standard physical properties of these explosives " by the War Department.

8. Miscellaneous tests. The Bureau of Mines physical testing laboratory was called upon to make special tests for the American University experiment station. These included products-of-combustion tests of "tetryl" and Bichel pressure gage tests of "carolnite."

9. Almost immediately after this country entered the war, the explosives section of the Bureau of Mines was requested by both the Army and Navy Departments to furnish all information available and reports of tests on numerous explosives. In addition, drawings and advice on the construction of testing apparatus was furnished to different stations of the Army and Navy.

INVESTIGATIONS OF EXPLOSIONS.

In connection with the work of explosives regulation members of the explosives section and the explosives chemical laboratory inspected explosives plants and investigated accidents in explosives plants and attended conferences.

The accidents investigated and reported upon were as follows:

Accidents at explosive plants.

Plant.	Location.	Date of accident.
		1917.
Grasselli Powder Co.	New Castle, Pa.	June 12
O. R. McAbee Powder & Oil Co.	Tunnelton, Pa.	Nov. 7
American Zinc Co.	Mascot, Tenn.	Nov. 22
Aetna Chemical Co.	Heidelberg, Pa.	Dec. 5
Aetna Explosives Co.	Carnegie, Pa.	Do.
Bethlehem Steel Co.	New Castle, Del.	Dec. 12
Newark Rubber Co.	Newark, N. J.	Dec. 22

Accidents at explosive plants—Continued.

Plant.	Location.	Date of accident.
		1918.
Atlas Powder Corporation	Patterson, Okla	Jan. 10
Hercules Powder Co	Kenvil, N. J	Jan. 22
Do	Valley Falls, N. Y	Jan. 31
E. I. du Pont de Nemours & Co	Butte, Mont	Do.
Do	Welpen, Minn	Feb. 20
Do	Repauno, N. J	Feb. 21
Do	Wayner N. J	Feb. 25
Do	Hopewell, Va	Mar. 2
Hercules Powder Co	Gillespie, N. J	Mar. 18
Trojan Powder Co	Iron Bridge, Pa	Mar. 26
Jarvis Warehouse	Jersey City, N. J	Do.
International Explosives Co	Swanton, Vt	Mar. 28
Atlas Powder Co	Scottdale, Pa	Mar. 29
Aetna Explosives Co	Poacher Plant	Do.
Hercules Powder Co	Gillespie, N. J	Apr. 5
National Chemical Mfg. Co	Hunkers, Pa	Apr. 16
E. I. du Pont de Nemours & Co	Carneys Point, N. J	Apr. 19
Hercules Powder Co	Carthage, Mo	Apr. 22
Do	Hercules, Calif	Apr. 25
E. I. du Pont de Nemours & Co	Deepwater, N. J	Apr. 26
Hercules Powder Co	Bacchus, Utah	May 6
Do	Kenvil, N. J	May 8
Do	Youngstown, Ohio	May 9
Do	Marlow, Tenn	May 11
E. I. du Pont de Nemours & Co	Jermyn, Pa	May 17
Aetna Chemical Co	Oakdale, Pa	May 18
Aetna Explosives Co	Emporium, Pa	May 31
Hercules Powder Co	Kenvil, N. J	June 2
Do	....do	June 10
Aetna Explosives Co	Ishpeming, Mich	June 12
Hercules Powder Co	Ferndale, Pa	June 14
Do	Gillespie, N. J	June 19
Grasselli Powder Co	New Castle, Pa	June 25
Aetna Explosives Co	Mount Union, Pa	July 2
Semet Solvay Co	Syracuse, N. Y	Do.
Hercules Powder Co	Kenvil, N. J	Do.
Newtons California Manufacturing Co	San Francisco, Calif	July 3
Hercules Powder Co	Kenvil, N. J	July 5
E. I. du Pont de Nemours & Co	Carneys Point, N. J	July 10
Aetna Explosives Co	Ishpeming, Mich	July 19
Do	Goes, Ohio	July 20
Bethlehem Steel Co	Redington, Pa	Do.
Standard Powder Co	Pittsburgh, Pa	July 22
Atlas Powder Co	Reynolds, Pa	July 26
Aetna Explosives Co	Aetna, Ind	July 30
E. I. du Pont de Nemours & Co	Carneys Point, N. J	Aug. 1
Do	Barksdale, Wis	Aug. 2
Grasselli Powder Co	New Castle, Pa	Aug. 5
Hercules Powder Co	Kenvil, N. J	Aug. 7
Illinois Powder Manufacturing Co	Easton, Ill	Aug. 8
American Standard Metal Products Co	Paulsbury, N. J	Do.
G. R. McAbee Powder & Oil Co	Tunnelton, Pa	Aug. 10
Hercules Powder Co	Kenvil, N. J	Aug. 14
Trojan Powder Co	Seiple, Pa	Do.
Stauffer Chemical Co	Chauncey, N. Y	Aug. 19
Standard Powder Co	Pittsburgh, Pa	Aug. 10
Eddystone Munitions Co	Eddystone, Pa	Sept. 11
E. I. du Pont de Nemours & Co	Barksdale, Wis	Sept. 15
Aetna Chemical Co	Carnegie, Pa	Sept. 17
E. I. du Pont de Nemours & Co	Wilmington, Del	Sept. 22
Trojan Powder Co	Allentown, Pa	Oct. 2
Aetna Chemical Co	Carnegie, Pa	Oct. 9
E. I. du Pont de Nemours & Co	Wilpen, Minn	Oct. 12
T. A. Gillespie Loading Co	Morgan, N. J	Oct. 14
Western Powder Manufacturing Co	Edwards, Ill	Nov. 12
Grasselli Powder Co	Quaker Falls, Pa	Nov. 25
E. I. du Pont de Nemours & Co	Oliver Mills, Pa	Dec. 7
Do	Moosie, Pa	Dec. 19
		1919.
Hercules Powder Co	Hercules, Calif	Jan. 7
E. I. du Pont de Nemours & Co	Deepwater Point, N. J	Jan. 10
Grasselli Powder Co	New Castle, Pa	Feb. 3
Egyptian Powder Co	Pollard, Ill	Feb. 10
General Explosives Co	Joplin, Mo	Feb. 19
Winchester Repeating Arms Co	New Haven, Conn	Mar. 1
Western Powder Co	Edwards, Ill	Mar. 2
Grasselli Powder Co	Wayside, Pa	Mar. 5
E. I. du Pont de Nemours & Co	Gibbstown, N. J	Mar. 11
Aetna Explosives Co	North Birmingham, Ala	Do.
Ordnance Department, United States Army	Aberdeen Proving Grounds, Md	Mar. 28

The following plants were inspected: Grasselli Powder Co., New Castle, Pa.; Aetna Explosives Co., Mount Union, Pa.

The explosives section was also requested by the Department of Justice to investigate explosions, give advice on destruction of confiscated explosives, and examine incendiary and explosive bombs.

EXPLOSIVES REGULATION.

The preamble to the explosives-regulation act, approved October, 1917 (40 Stat., 385), sets forth briefly the purpose of the act as follows:

Be it enacted by the Senate and House of Representatives of the United States of America in Congress assembled, That when the United States is at war it shall be unlawful to manufacture, distribute, store, use, or possess powder, explosives, blasting supplies, or ingredients thereof, in such manner as to be detrimental to the public safety, except as in this act provided.

This act was passed after seven months of preparatory work, during which time officials of the large powder companies, officers of the Ordnance Department of the United States Army, and various experts on explosives and on legal matters were consulted. The administration of the act was delegated to the Director of the Bureau of Mines, as this bureau, through its work on mine explosives, was best equipped for the purpose.

One of the primary objects of the act was to prevent explosives from getting into the hands of persons hostile to the United States Government. An essential provision was the requirement of licenses, no person being permitted to manufacture or possess explosives or their ingredients without holding a license, in order to protect these articles from theft and also from loss or destruction.

It was evident that the enforcement of such a law throughout the United States, Alaska, and insular possessions would entail an extensive and thorough organization, as under the law not only were all common explosives included, but also ingredients that are household necessities, and would involve the issuing, after investigation, of more than 1,000,000 licenses. To this end an organization was established as follows:

The Director of the Bureau of Mines, under section 18 of the act, was authorized to make rules and regulations for carrying its provisions into effect, subject to the approval of the Secretary of the Interior. A new division of the bureau, known as the Explosives Regulation, was organized by the director, and F. S. Peabody was placed in charge, as assistant to the director in charge of explosives. The chief branches of this office were: (1) Administrative and questions of policy; (2) investigation of applications and the issuing of manufacturer's, exporter's, and importer's licenses, which have numbered approximately 918; (3) examination and appointment of field employees; (4) investigation and prosecution of violations of the act; (5) proper storage of explosives or ingredients.

In each State and in the Territory of Alaska, an explosive inspector was appointed by the President, with the approval of the Senate. Each field inspector acted under the supervision of the Washington office.

Under each explosives inspector was an advisory committee, which was made up, when practicable, of representatives of those interests most affected, such as the Council of National Defense, the Department of Justice, Department of Agriculture, a fire insurance company, a casualty company, a large dealer in and user of explosives, or a manufacturer of explosives. The members of the advisory committee were designated assistant inspectors, and rendered the explosive inspector such assistance as lay within their power. They were appointed by the Director of the Bureau of Mines, with the approval of the Secretary of the Interior.

The Bureau of Explosives, New York City, acted in close cooperation with the explosives division of the Bureau of Mines; it handled all cases relative to explosives or ingredients while in transit in public carriers, and its inspectors throughout the country assisted the Federal explosives inspector in every way possible.

For licensing users of explosives and ingredients a field force of over 15,000 men was selected and appointed by the Director of the Bureau of Mines after investigation was made as to loyalty and as to their having unlimited power to administer oath and their not being connected or interested in any way with the manufacture or sale of explosives or ingredients.

Less than 1 per cent of the entire force under the explosives act received salaries for this work, and the work was so systematized that the active force required in the Washington office numbered not more than 50 persons.

In the enforcement of the act it was found that in many mining districts the miners not only had free access to explosives, but made a common practice of carrying explosives home. It was not uncommon to find 500 to 800 pounds of explosives in a single miners' lodging house, and statistics show that many accidents have happened from miners filling their paper cartridges by the light of ordinary oil lamps. The Explosives Regulation made vigorous efforts to stop this practice. In Kansas and Alabama, for instance, about 80 per cent of the practice was largely abolished through the cooperation of the mine operators in issuing or selling the explosives to the men at the mine, thereby reducing the necessity of carrying the explosives to lodging houses.

In examining magazines it was found that in thousands of instances explosives were stored so carelessly that theft would have been easy, or were stored in such close proximity to dwellings, highways, or railroads as to constitute a serious menace. Quantities up to 10,000 pounds in a single magazine were found in the heart of a town, where an explosion would have resulted in great loss of life and property. Several attempts to dynamite railroad bridges and other structures, with the object of hampering the transportation or equipment of troops, were reported to the bureau by the War Department. Also on several occasions the War Department appealed to the Bureau of Mines to remove the menace caused by the proximity to mobilization camps of improperly constructed, unguarded magazines. Among these may be mentioned three powder storehouses at Possession Point, State of Washington, considered to be insufficiently and improperly guarded; magazines of two powder companies situated in the vicinity of St. Paul, Minn.; magazines owned by three powder companies in the vicinity of Camp Zachary Taylor, Ky.

The number of magazines listed exceeds 8,000, particulars of additional ones being received daily in owner's reports, in the form of an affidavit, showing detailed construction, situation, and contents. These are checked up and recommendations are made whereby the magazines will not only be reasonably proof against theft but also safely situated with respect to life, limb, and property.

Acting upon instructions, special investigations were made in the field by the inspectors or their assistants of refusals or revocations of licenses, improper storage facilities, violations of the act, and outrages against life and property by the use of bombs and other explosives. There have been over 16,000 special investigations made and over 100 convictions obtained for violation of the act.

The personnel of the division has always received the active cooperation of the officials of the Army, Navy, and other Federal departments, and of State authorities. In making investigations the Explosives Regulation is authorized by law to request the assistance of Federal or State authorities.

The policy in general as regards technical offenses was to seek fines rather than terms of imprisonment; however, when the offense seemed grave enough, prosecution was vigorously instigated. Sentences pronounced have ranged all the way up to $5,000 and 11 months' imprisonment, the maximum sentence possible being one year's imprisonment and $5,000 fine.

In a recent case in Detroit the defendant said that he could not buy powder, so had purchased cartridges at different stores, had cut them open and used the powder contained therein for making a bomb. The jury rendered a verdict of guilty and the judge gave an option of $500 fine and three months' imprisonment or one year's imprisonment without the fine. The inability to obtain powder, except through purchasing cartridges, was a result of the explosives act.

Through the field force, numerous cases of disloyalty have been reported which, as they do not come under the jurisdiction of the Department of the Interior, were turned over to the Department of Justice, with the result that many enemies were sentenced either to prison or to the internment camps. As further evidence of the benefits which accrued under the act, it should be mentioned that various States are now planning effective legislation to replace the act when it becomes inoperative.

The personnel of the explosives division has remained practically the same, except that F. S. Peabody, acting as assistant to the director in charge of explosives, resigned on July 17, 1918, and was succeeded by Clarence Hall on July 18, 1918, with the title of chief explosives engineer.

On November 14, 1918, some of the commodities listed as requiring license were exempted as no longer coming under the law except under certain special circumstances. This modification was made because of the signing of the armistice and applied to ingredients of explosives and fireworks.

REGULATION OF PLATINUM AND ALLIED METALS.

Platinum being essential to the production of explosives, it became imperative, when the United States entered the war, to ascertain the demand for and the supply of the metal in this country. In the early part of 1917, the chemical division of the War Industries Board, through an order from the Secretary of the Treasury, took the first steps toward conserving the supply, withholding all platinum that passed into the United States Mint. It was found that the different departments of the Government were unable to furnish exact estimates of their requirements, and as the demand for the metal grew, such estimates as were submitted had to be constantly increased. As an example, orders for 20,000 ounces were placed in June, 1917, in Russia, but by the time the metal was delivered in San Francisco the Government's requirements exceeded that amount.

By requisitioning the supplies in the hands of 14 of the leading firms in this country a certain amount was obtained, but owing to the inadequacy of the supply it became necessary to turn to the jewelers and appeal to the people at large throughout the country. Colombia was able to supply only about 2,500 ounces a month, and the collapse of the Russian Government aggravated the situation. A loan of 10,000 ounces from England helped somewhat. By this time the estimated requirements for 1918 amounted to 36,400 ounces, but, as previously stated, these estimates kept increasing as the country took a larger part in the prosecution of the war. Because of the larger demand, Government regulation of the use and sale of platinum became imperative. Immediately the question arose as to which industries should be curtailed in their use of the metal, and it was decided that nonessential or "luxurious" uses could be eliminated, and that for various purposes, as far as possible, substitutes should be used to release platinum.

Both the War Department and the War Industries Board appreciated the necessity of licensing the users of platinum so that proper control over the metal could be obtained. It was decided, after due consideration, that the explosives regulation division of the Bureau of Mines would be the best organization to handle the matter, because of the licensing system it had established and the distribution of its 15,000 agents, working under a law that was controlling other commodities as a war measure. Accordingly, Congress, by the act of October 6, 1917 (40 Stat., 385), amended by the act of July 1, 1918 (Public, 181), authorized the Director of the Bureau of Mines, under rules and regulations approved by the Secretary of the Interior, to limit, during the period of the war, the sale, possession, and use of platinum, palladium, and iridium and compounds thereof.

On account of the signing of the armistice the restrictions this act placed on the manufacture, possession, purchase, sale, and use of platinum, palladium, iridium, and compounds were removed by the Director of the Bureau of Mines, with the approval of the Secretary of the Interior, on November 14, 1918.

TESTS OF MARINE BOILERS FOR THE UNITED STATES SHIPPING BOARD.

In February, 1918, F. W. Dean, of the Emergency Fleet Corporation, arranged with O. P. Hood, chief mechanical engineer, that the Bureau of Mines take temperature measurements at some tests of water-tube boilers being designed and built for the emergency fleet. Henry Kreisinger, fuel engineer of the bureau, arranged details and conducted the tests.

In April, 1918, one of the boilers was ready for test at the Erie City Iron Works, Erie, Pa. Mr. Kreisinger went to Erie about April 10, met Mr. Conti and Mr. Dean, of the Emergency Fleet Corporation, and discussed with them the plans of the work. It was agreed that the work the bureau should undertake was to make all temperature measurements and analyze the furnace gases. The Emergency Fleet Corporation did not have a skilled fireman, so that an experienced fireman from the Pittsburgh experiment station of the bureau was used.

The first test was on April 18, and in all 22 tests were made at Erie with George's Creek coal. During each test temperatures were measured at several places inside the boiler setting, a total of several hundred readings being taken. These temperature measurements indicated that the performance of the boiler could be improved by changes in the baffling. These changes were made and the efficiency of the boiler was considerably increased.

During the first test the flue gases contained a considerable percentage of combustible matter. It was agreed that the bureau should study in more detail the process of combustion in the furnace. In this study a large number of samples of the furnace gases were taken and the analyses of these samples were carefully investigated. The investigation showed that it was difficult, if not impossible, to get air enough over the fuel bed to insure complete combustion. Hence a larger number of holes were made in the firing doors and in the furnace front to admit secondary air. These openings improved the combustion, but the furnace was small and not enough air could be admitted in this way, nor could the air be mixed thoroughly with the combustible gases rising from the fuel bed. Therefore it was decided to admit additional air at the rear of the furnace by using a Wager bridgewall, which is made of iron bars, somewhat like a grate, and admits air through the openings between the bars. Thus the combustible gases rising from the fuel bed were between two streams of air, one coming from the front of the furnace and one through the bridgewall. In consequence better mixing and improved combustion was obtained.

In the early tests the flue gases often contained so much combustible matter that they ignited and burned with a flame at the base of the stack. After admitting air through the bridgewall and the front of the furnace all flaming at the base of the stack ceased, showing that combustion within the furnace was more nearly complete. By these improvements in the furnace and the changes in the baffling, the over-all boiler efficiency was raised from about 60 per cent to 72 per cent. This high efficiency of 72 per cent was obtained when the boiler evaporated 5½ to 6 pounds of water per square foot of heating surface, a rate considerably higher than is the practice in the average stationary plant. Thus the fuel section of the Bureau of Mines helped to develop a highly efficient boiler. The tests of this furnace substantiated the results obtained by the bureau in laboratory investigations, namely, that enough air can not be supplied through the fuel bed to obtain complete combustion, and additional air must be introduced over the fuel bed.

The bureau developed a method for studying the flow of gases through the boiler by measuring the temperatures across the path of gases. It is believed work of this character had never before been undertaken. The engineers of the Emergency Fleet Corporation appreciated the value of the work done by the fuel section and gave the bureau due credit.

In August another boiler, at the Murray Iron Works, Burlington, Iowa, was ready for test. This boiler, designed by John F. Bell, of New York, for the Power Specialty Co., differed somewhat from the boiler tested at Erie, Pa., being fitted with a Foster superheater. The Emergency Fleet Corporation had ordered a large number of these boilers for steel ships and was interested in their performance as compared with that of the Erie boilers. It desired that the boiler be tested first with oil, using Dahl and Coen burners, and that then a hand-fired furnace be substituted and further tests made with Georges Creek coal as fuel. Consequently, the Emergency Fleet Corporation asked the bureau to do in these boiler tests work similar to that done at Erie, Pa., and, particularly, to furnish the same skilled firemen to do the firing in the test with coal.

In all, 32 tests were made with oil and 12 tests were made with coal. Members of the fuel section of the bureau made three trips to Iowa. There were three men on the first trip. On the second trip the Emergency Fleet Corporation asked that Mr. Kreisinger take charge of the whole test during the daytime and one of its men take charge at night. This necessitated the bureau supplying an extra man. On the third trip, during which the coal tests were made, the bureau was asked to furnish a fireman, so that it supplied five men on this trip. The coal tests were supervised by Mr. Kreisinger as the representative of the Emergency Fleet Corporation's interest in the tests.

The chief result of the work done by the bureau in these tests was the demonstration that a larger combustion space for burning oil gave better combustion and higher efficiency. In the test with coal the results showed that introducing secondary air through the firing doors, the furnace front, and a Wager bridge wall improved the combustion and increased the efficiency of the boiler. As the coal was fired by the same firemen and the tests were supervised by the same engineer, the results are directly comparable with those obtained at Erie.

As the expense connected with making the tests at Burlington was far in excess of the allotment of bureau funds available for that purpose, the Emergency Fleet Corporation was asked to bear part of the expense of the work. This the corporation agreed to do, and the agreement was used as a basis for further cooperation. The corporation intends to use the methods of the Bureau of Mines in testing all designs of boilers for its ships, the same crew and firemen to conduct the tests.

Subsequently the bureau tested another boiler at Erie, Pa., and in January, 1919, was planning to test two more—a large Scotch marine boiler at the Sun Shipbuilding Co. yards at Chester, Pa., and a boiler at the works of the Heine Boiler Co. at Phoenixville, Pa. All three boilers were to be tested with hand-fired furnaces and also with oil, using Dahl and Coen burners. In addition, the boiler at Erie, Pa., will be tested with a spreading stoker and with a type E underfeed stoker furnished by the Combustion Engineering Co.

Devices to supply additional air over the fuel bed at high velocity, so as to improve the mixing and thus the combustion in the furnace and increase the efficiency of the boiler, will also be tested. The devices will be tested with the object of using, in the new ships, those that prove practical and economical.

The United States Shipping Board is setting a new standard in marine boiler-plant engineering. It is developing highly efficient boilers and supplying trained men to operate the power plants of the ships.

By cooperating in the work on boilers the Bureau of Mines is enabled to apply on a practical scale some of the principles developed in its fuel laboratory at Pittsburgh, and to determine how far the deductions drawn from laboratory experiments are applicable to boilers of marine type. The data collected during the tests will be of value to all persons interested in the design of boiler furnaces and boilers, as work of this kind is believed to be new.

SOUND DETECTION AND SOUND RANGING.

An investigation of decided military importance was that of sound detection and sound-ranging underground. By means of delicate listening devices—geophones and microphones—it is possible to tell by the sound waves transmitted through earth and rock the distance and the direction of points at which underground sapping or mining operations are being conducted. The attention of the mining division of the Bureau of Mines was called to the problem in 1917, through drawing specifications for self-contained breathing apparatus for use in underground work and in subsequently getting the Edison Co. to build Gibbs breathing apparatus for the Army. Capt. H. D. Trounce, of the Army, who had two years experience with the Royal Engineers before he was transferred to the American forces, was interested in the matter of detecting the sounds made by mining machines and through him the Bureau of Mines ascertained that Maj. W. D. Young, of the Engineers, had been assigned a problem relating to sound detection. Accordingly, the Director of the Bureau of Mines took up the offering of assistance to Maj. Gen. Black, Chief of Engineers, in the matter of sound detection and the use of the experimental mine for making the tests.

On November 30, 1917, the director wrote to Maj. Gen. Black introducing G. S. Rice, chief of the mining division, "whom I have intrusted with an investigation, suggested by you, into means of detecting sounds through the ground." In the letter the director called attention to Mr. Rice's experience in the detection of shock waves in air at the experimental mine of the bureau, 12 miles from Pittsburgh, where he had the assistance of W. L. Egy, physicist, and other men of scientific training. The director authorized Mr. Rice to obtain the services of such other engineers and scientists as might be available.

At that time investigations were being carried on in conjunction with the Bureau of Standards, and the original French geophones and other listening devices and instruments were under test. A map of the experimental mine was submitted to Lieut. E. B. Stevenson, afterwards captain, with the thought that this mine might be particularly useful in studying sound waves through earth and rock. At a conference between Gen. Black and Mr. Rice, regarding the general features of sound waves, Gen. Abbott was called in, who described sound measurements made at the time of the Hellgate blast in 1877. Mr. Rice also conferred with Gen. Winslow, who had charge of the division of work in which sound detection and ranging had been placed.

Early in December Lieut. Stevenson conferred with Mr. Rice in regard to what might be done through cooperation and the facilities available at the experimental mine. • Lieut. Stevenson acknowledged the advantage of standardizing in point of sound production all the different kinds of noises from machines and other sources, made and transmitted through the strata in mining.

On December 15, Gen. Black asked that the services of G. S. Rice and W. L. Egy be placed at his disposal. In reply, Director Manning said that Messrs. Rice and Egy could not be spared from other important duties, but he would be glad to have them contribute what work they could, especially in connection with the use of the experimental mine for testing.

The project of undertaking the investigations at the experimental mine was taken up definitely with Mr. Egy on December 28. On January 4, 1918, Lieut. Stevenson inquired as to the character of underground sounds, the kind of detection so far employed, and the laws of sound transmission.

On January 11 and 17, 1918, Mr. Rice, in conference with Dr. W. J. Humphries, of the United States Weather Bureau, discussed the subject of sound or shock waves through ground. On the latter date the Bureau of Standards furnished plans of geophones and earpieces that had been developed at the Bureau of Standards by Lieut. Stevenson.

Later in the month Mr. Egy visited Prof. D. C. Miller, at Cleveland, Ohio, different physicists at Urbana, Ill., and Prof. Foley, of the University of Indiana, to discuss various questions of the physics of sound production. A report on the information gathered was submitted to Maj. Young, who had been placed in charge, for the Army, of all investigations relating to sound detection.

About January 20 a conference was held in Washington, at which Maj. Young, Capt. Wood, Mr. Egy, and Mr. Rice were present. As a result of this conference it was agreed that the bureau should take up (1) the standardization of sound waves by character and by the blows or forces employed under ground; and (2) should study the mechanical development of recording instruments for sound waves transmitted through earth or rock.

Under the first of these investigations, the problems proposed were determining the distance at which the blow of a pick, boring with a hand auger, cutting chips with a chisel by pressure, and talking, could be heard, one of the purposes of the tests being to procure information to guide in selecting the best types of boring and tunneling machines for military work. The investigations involved the detection of sound waves passing through air, unconsolidated material (soil and clay), and rock.

The work undertaken included reviewing the literature of shock and sound waves and method of detecting them; constructing geophones of the French type; training the personnel at the experimental mine in their use; placing listening stations in the experimental mine; and determining the distances at which different sounds could be heard through coal, shale, clay, and earth.

Work at the experimental mine began on January 22, 1918, under Mr. Egy. The first tests were with geophones to determine the distances at which different sounds could be heard in different directions through coal. Subsequent tests in February, March, and April showed that sounds were transmitted more readily through coal than through shale, and more readily through shale than through clay. Other tests determined how far standard sounds could be detected through the top soil, shale, and sandstone overlying the coal. Two types of French microphones were tested at the experimental mine in June, the tests including the distance at which the explosion of small charges of dynamite could be detected.

In September Alan Leighton, assistant physicist, went to Camp Humphreys, Md., and cooperated with officers of the Engineers Training School in tests with geophones and French microphones to determine the distances at which sounds could be heard through clay.

Eight reports were made on the various tests, and copies of the reports were sent to the office of the Chief of Engineers of the Army.

Mathematical analysis of the data obtained in the tests showed that the relationship between the energy of a blow or of an explosion of a given charge of dynamite can be expressed by an empyrical formula that had been developed years before by Brig. Gen. H. L. Abbott.

Subsequently tests at the experimental mine to determine the applicability of geophones for use in detecting sounds made by men entombed by mine disasters were planned.

A seismograph devised by Prof. H. V. Carpenter, of the University of Washington, for detecting the shock of explosives was tested at the experimental mine.

TRAINING IN FIRST-AID AND RESCUE WORK.

Perhaps the most important work done by mine safety stations and rescue cars of the Bureau of Mines was the training of soldiers in the use of oxygen breathing apparatus and in first-aid and rescue methods. Rescue car No. 8, with its crew, spent several weeks at Camp Meade, Md., training men for mining and sapping regiments in the use and care of breathing apparatus and in first aid, 184 being trained in mine rescue and 158 in first aid. At Edgewood Arsenal, where poison gases were made on a huge scale, the crew of rescue car 6 trained 66 men in the use of oxygen breathing apparatus and 65 in methods of resuscitation from gas. At Houghton, Mich., 809 soldiers of the Student Army Training Corps, Michigan College of Mines, were trained in the use of apparatus and 136 in first aid. Incidentally, training was given some of the Army officers detailed to the Pittsburgh station. The total number of officers and soldiers trained was 1,055. These men formed a neuclus from which instructors could be selected for training other groups.

Tests were made for the War Department of 200 sets of Gibbs oxygen breathing apparatus, at Orange, N. Y., also comparative tests of the Draeger (Atmos) and Fleuss half-hour apparatus were made at the Pittsburgh station for the War Department. These tests were under the supervision of D. J. Parker, mine safety engineer.

The mine-safety section at the Pittsburgh station assisted in testing soda-lime regenerators, gas masks, and gas detectors for the war-gas investigations. At Camp Dix assistance was rendered in some tests of gases from the detonation of high explosives.

In addition, thousands of miners throughout the country were trained in mine rescue and first-aid methods. Probably one-half of them were ultimately called to the colors. During the war period 10,665 men were trained in first aid and 4,075 in mine rescue methods. It is estimated that 6,509 of these men subsequently entered the Army or the Navy.

The crews of the cars and stations assisted at various mine disasters. By their experience in mine recovery work they were frequently able to give aid or advice in exploring workings, extinguishing fires, or restoring ventilation that resulted in saving life and in hastening the resumption of production.

CENSUS OF MINING ENGINEERS, METALLURGISTS, AND CHEMISTS.

As a measure toward national preparedness for war, the Bureau of Mines, at the request of the Council of National Defense, under date of February 24, 1917, undertook a census of mining engineers and chemists. The need for such a census was obvious. In the conduct of war to-day chemists and engineers play a far greater part than ever before. The products of the mines, furnaces, factories, and chemical plants are consumed so rapidly that the highest possible efficiency is required to keep pace with the enormous demand. In the organization of a great army many classes of specialists are needed. Men with a knowledge of sanitation were essential at the various training camps; men with an intimate knowledge of pyrotechnics could find a place in the manufacture and use of signal devices; chemists and engineers were needed at munitions plants. Coal and iron are absolute necessities for the manufacture of arms and munitions. Many naval vessels were dependent on petroleum as a fuel, and aircraft and motor trucks were useless without gasoline, or spirits obtained with greater difficulty.

Moreover, it was believed that mining engineers, under military control, could be of great assistance in directing sapping operations; in digging trenches, dugouts, and tunnels behind the lines; and in reinforcing, ventilating, and draining such excavations; in rehabilitating wrecked coal and iron mines taken from the enemy, and in increasing the output of minerals for military uses in other districts. It was on the theory of being able to place the right man in the right place at the right time that the census was conducted.

Work on this census began at once, mailing lists and questionnaires being compiled in cooperation with the American Institute of Mining Engineers and the American Chemical Society. A circular letter under date of April 6 was addressed to approximately 5,000 members of the American Institute of Mining Engineers and 9,000 members of the American Chemical Society. This letter requested that each recipient fill out and return a card accompanying the letter and make himself a committee of one to see that every chemist and engineer of his acquaintance likewise filled out a card.

As a result, the Bureau of Mines received replies from 7,500 men engaged in mining and from 15,000 men engaged in the various chemical industries. These 22,500 names were classified according to the character of the work in which each writer claimed proficiency. If any industrial plant engaged on Government contracts desired the services of special men, the bureau, on request, endeavored to furnish such names.

In classifying the engineers and chemists no attention was paid to any occupation in which the individual had less than a year's experience, except where it was known that there were only a few experts in a certain branch. As many men had an experience of more than one year under several items, for some of these as many as five or six cards were made out. The total number of index cards for chemists was approximately 28,000 or an average of two for each person, and that for mining engineers was 30,000, or four for each engineer. The list of chemists and engineers having had experience abroad included only the names of those who had spent more than one year in a foreign land.

The census returns were followed by postal inquiries when the selective draft had been completed and the men knew their military status. This questionnaire asked each man in the Government service to give his new address, the name of the branch with which he was connected, and his official title. Then, if the War Department called for the names of men in certain lines of mining or chemistry, the bureau could select from the postal-card returns the names of men already in the military service who could be transferred where their work would be of greatest benefit.

On account of the importance of petroleum and its products, the Bureau of Mines prepared a classified list of men engaged in the industry. A circular letter was addressed to the principal operating companies requesting the names of their managers, drillers, pump men, transportation engineers, chemists, and refiners. The companies promptly furnished the required lists, and a registration card was sent each man named to fill out and return to the bureau. As a result a classified list of men in the petroleum industry containing 2,000 names was obtained.

Although the bureau's schedule was sent primarily to mining engineers and chemists, the questions relating to metallurgy were identical and were so arranged that anyone with experience in metallurgy could find the proper place to indicate the line along which he had worked. The number of metallurgists, by principal groups, were classified as from the returns of the mining engineers and chemists.

In order to furnish certain information—the names of men available for research work—to the chemistry committee of the National Research Council, a special letter and questionnaire were added to the regular chemist's form. When these forms were received at the office of the bureau they were detached and forwarded to Dr. Marston Taylor Bogert, chairman of the committee.

The taking of this census and the compilation of the information obtained by the questionnaires were in charge of A. H. Fay, engineer.

INVESTIGATIONS OF ALLOY STEELS.

COOPERATIVE WORK WITH THE ORDNANCE DEPARTMENT (WATERTOWN ARSENAL).

During February, 1917, the question of the value of uranium in steel for the liners of cannon became more pressing. According to rumors from abroad the Germans were using uranium steel in some of their special guns. After some preliminary discussion a conference was called at the Watertown Arsenal March 24, 1917, between Dr. H. W. Gillett, alloy chemist of the bureau, and officials appointed by the arsenal. On March 27 the Director of the Bureau of Mines received from the commanding officer of the arsenal a letter stating the determination of the influence of uranium on gun erosion was a matter that warranted thorough investigation. Accordingly work was begun at Ithaca by Dr. Gillett under the direction of C. L. Parsons, chief chemist of the bureau.

On April 12, 1917, the commanding officer of the Watertown Arsenal requested the Director of the Bureau of Mines to extend the investigation to molybdenum and tungsten steels. In the early fall of 1917, J. G. Thompson and Louis S. Deitz, jr., were inducted into the service, assigned to the Bureau of Mines, and later commissioned as second lieutenants to assist in this work. On July 17, 1918, Private E. O. Denzler was inducted into the Chemical Warfare Service and assigned to the alloy steel work at Ithaca.

As a result of the work done at Ithaca a large number of uranium and tungsten steels were prepared and submitted to the Watertown Arsenal. Because of delay in testing, samples were reshipped to the Bureau of Standards for investigation of their physical and chemical properties. Tests were not completed when the armistice was signed.

COOPERATIVE WORK WITH THE NAVY DEPARTMENT AND THE BUREAU OF STANDARDS.

A question as to the light armor used in airplanes and tanks having arisen, it was deemed advisable to arrange a cooperative investigation with the Army, Navy, Bureau of Standards, and War Industries Board, and a conference was called by Commander Holmes of the Navy, to meet at the Bureau of Mines on August 9, 1918. After preliminary discussion the conferees adjourned until August 12 and met again on the 13th.

As a result a cooperative investigation of ferrozirconium and of zirconium steel as light armor was arranged with the Navy and the Bureau of Standards. This investigation was in charge of the Bureau of Mines, C. L. Parsons, chief chemist, acting as chairman of the joint board. The Ordnance Department of the Army at the same time began investigations of its own in cooperation with the

Ford Motor Co., of Detroit, Mich. The results achieved to February 1 did not show definitely that zirconium has any beneficial effect when added to steel, but it seems certain that the investigations will result in the manufacture of bullet-resisting armor much more effective than any heretofore used.

Ferrozirconium and some 60 steels alloyed with zirconium have been produced by the Bureau of Mines and were under test at the Bureau of Standards on February 1, 1919.

LIGHTING DEVICE FOR AVIATION FIELDS.

In December, 1917, Secretary Lane had his attention called to a portable lighting device for assisting aviators in night flying. This device consisted essentially of a motor-driven generating unit and suitable electric lights or projectors, both white and colored, mounted in such manner as to facilitate signaling to an aviator, illuminating the ground, and indicating in what direction a landing could be best be made, the whole device being mounted on a small motor vehicle. In January, 1918, the inventor, Otto H. Mohr, of Chicago, Ill., gave a demonstration of the device at the American University station of the Bureau of Mines, Dr. Durand, Chief of the Board of Aeronautics, and Director Manning being present. Dr. Durand strongly advised that the device be offered to the War and Navy Departments. On January 25 Secretary Lane called the device to the attention of Secretary Baker and to Secretary Daniels, suggesting they might wish to have it tried out under actual conditions. The Signal Corps of the Army finally took up the matter and the device received a preliminary test on April 6 at Langley Field, at the direction of Lieut. Col. R. A. Millikan, of the Signal Corps and the National Research Council. As a result of these tests a few changes were made in the device and a second trial made at Langley Field on April 25, with Lieut. J. T. Tait, representing the Science and Research Division of the Signal Corps, acting as observer. As a result of this second test the Science and Research Division recommended that a completely equipped unit be constructed on a light truck body and that further tests be made.

In making the tests and in building this unit, the Bureau of Mines, in which Mr. Mohr was a consulting electrical illuminating engineer, cooperated with the National Research Council and the Signal Corps, Lieut. Tait working under the direction of Mr. Mohr in completing the first unit, and the Signal Corps providing the truck and the generating equipment.

This unit was thoroughly tested at Ellington field, Tex., on August 9 to 20. As a result of these tests another unit was constructed under the direction of Lieut. Tait. In its final form, which differed from the original chiefly in providing a more brilliant illumination over a larger area of ground, the device was approved by the Signal Corps and two units were built for each of the Government aviation fields.

COOPERATION WITH CAPITAL ISSUES COMMITTEE.

In accordance with its general policy of requesting those departments of the Government most thoroughly conversant with various subjects to cooperate with it, the Capital Issues Committee of the Treasury Department in May, 1918, asked the aid of the Bureau of Mines in establishing or checking statements of fact regarding mining companies applying for permission to issue new securities. In order to obtain the necessary information from the mining companies applying, the Capital Issues Committee requested the bureau to assist in drawing up a questionnaire or report form. In June the Bureau of Mines appointed a committee of its members to advise the Capital Issues Committee on mining and related applications. This committee consisted at different times of J. H. Mackenzie, H. H. Porter, H. S. Mudd, A. W. Stockett, and H. C. Morris. The committee carefully reviewed the evidence submitted through the Capital Issues Committee, and when necessary called for additional information, and sometimes had examinations and reports made by its own field or consulting staff.

Final decisions and matters of policy rested entirely with the Capital Issues Committee. The Bureau of Mines assisted in checking statements of fact and acted in an advisory capacity.

O

Lightning Source UK Ltd.
Milton Keynes UK
UKHW020936051118
331793UK00011B/1035/P